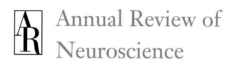

Annual Review of
Neuroscience

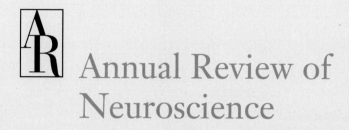

Annual Review of Neuroscience

Volume 33, 2010

Steven E. Hyman, *Editor*
Harvard University

Thomas M. Jessell, *Associate Editor*
Columbia University

Carla J. Shatz, *Associate Editor*
Stanford University

Charles F. Stevens, *Associate Editor*
Salk Institute for Biological Studies

www.annualreviews.org • science@annualreviews.org • 650-493-4400

Annual Reviews
4139 El Camino Way • P.O. Box 10139 • Palo Alto, California 94303-0139

Annual Reviews
Palo Alto, California, USA

International Standard Serial Number: 0147-006X
International Standard Book Number: 978-0-8243-2433-9

TYPESET BY APTARA
PRINTED AND BOUND BY SHERIDAN BOOKS, INC., CHELSEA, MICHIGAN

Annual Review of
Neuroscience

Volume 33, 2010

Contents

Indexes

Errata

An online log of corrections to *Annual Review of Neuroscience* articles may be found at
http://neuro.annualreviews.org/

Related Articles

Attention, Intention, and Priority in the Parietal Lobe

James W. Bisley[1] and Michael E. Goldberg[2]

[1]Department of Neurobiology and Jules Stein Eye Institute, David Geffen School of Medicine, and Department of Psychology and the Brain Research Institute, University of California, Los Angeles, California 90095; email: jbisley@mednet.ucla.edu

[2]Mahoney Center for Brain and Behavior and Departments of Neurology and Psychiatry, Columbia University College of Physicians and Surgeons and the New York State Psychiatric Institute, New York, NY 10032; email: meg2008@columbia.edu

Annu. Rev. Neurosci. 2010. 33:1–21

First published online as a Review in Advance on February 19, 2010

The *Annual Review of Neuroscience* is online at neuro.annualreviews.org

This article's doi:
10.1146/annurev-neuro-060909-152823

Key Words

lateral intraparietal area, LIP, saccade, visual search, salience, vision

Abstract

For many years there has been a debate about the role of the parietal lobe in the generation of behavior. Does it generate movement plans (intention) or choose objects in the environment for further processing? To answer this, we focus on the lateral intraparietal area (LIP), an area that has been shown to play independent roles in target selection for saccades and the generation of visual attention. Based on results from a variety of tasks, we propose that LIP acts as a priority map in which objects are represented by activity proportional to their behavioral priority. We present evidence to show that the priority map combines bottom-up inputs like a rapid visual response with an array of top-down signals like a saccade plan. The spatial location representing the peak of the map is used by the oculomotor system to target saccades and by the visual system to guide visual attention.

Contents

INTRODUCTION

Much of modern thinking about attention stems from William James' classic description, "Everyone knows what attention is... the taking possession by the mind, in clear and vivid form, of one out of what seem several simultaneously possible objects or trains of thought. Focalization, concentration, of consciousness are of its essence. It implies withdrawal from some things in order to deal effectively with others" (James 1890). James went on to describe two different varieties of attention: "Passive, reflex, nonvoluntary, effortless; Active and voluntary," a classification that modern jargon has named "bottom-up" and "top-down", respectively. James also pointed out the linkage between attention and orienting: "When we look or listen we accommodate our eyes and ears involuntarily, and we turn our head and body as well." The parietal lobe of the human and nonhuman primate brain has long been associated with attention as well as sophisticated motor planning on the basis of clinical and physiological evidence (Critchley 1953): Patients with parietal lesions exhibit extinction and neglect, deficits of attention; and apraxia, difficulty with planning sophisticated movements.

The human parietal lobe (**Figure 1a**) was traditionally labeled as an association cortex, appearing to play a role in the processing of sensory information, including perception, decision making, numerical cognition, integration, speech comprehension, and spatial awareness (Critchley 1953). We focus on spatial processing, which combines a number of these processes. While a combination of lesion, predominantly from strokes, functional imaging and electrical stimulation studies have broadened our knowledge of how the parietal lobe functions, the majority of the details have been gained by examining the responses of neurons in monkey parietal cortex.

Within the monkey intraparietal sulcus (**Figure 1b**) are a number of areas that integrate information from multiple senses and appear to be important for guiding behavior within specific workspaces. For example, medial intraparietal area (MIP) responses are related to reaching (Johnson et al. 1996), anterior intraparietal area (AIP) responses are related to grasping (Sakata et al. 1995), and lateral intraparietal area (LIP) responses are related to eye movements (Gnadt & Andersen 1988). However, these overt actions are not the only way that information within these maps is used. To illustrate this we focus on LIP. Originally thought of as a saccade planning area (Gnadt & Andersen 1988), a debate arose as to whether the responses in LIP are related to visual attention (Colby & Goldberg 1999) or to the intention to make saccades (Andersen & Buneo 2002). We propose that LIP acts as a priority map in which stimuli in the world are represented according to their behavioral priority, incorporating both visual and cognitive inputs. We suggest that the map is then read out by the visual system to guide attention and by the oculomotor system to guide eye movements.

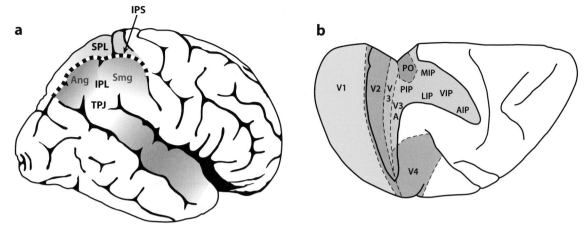

Figure 1

Posterior parietal cortex of human (*left*) and macaque monkey (*right*). (*a*) The human posterior parietal cortex (PPC) is divided by the intraparietal sulcus (IPS) into the superior parietal lobe (SPL) and the inferior parietal lobe (IPL). The IPL consists of the angular gyrus (Ang) and supramarginal gyrus (Smg) and borders the superior temporal gyrus (*purple*) at a region often referred to as the temporoparietal junction (TPJ). (*b*) The lunate and intraparietal sulci are opened up to show the locations of several extrastriate areas in addition to the visually responsive areas within the intraparietal sulcus. These include the parieto-occipital area (PO), the posterior intraparietal area (PIP), the medial intraparietal area (MIP), the lateral intraparietal area (LIP), the ventral intraparietal area (VIP), and the anterior intraparietal area (AIP). Adapted from Husain & Nachev (2007) and Colby et al. (1988).

In this chapter, we begin by describing the model upon which we base our hypothesis. We show data that led to this hypothesis and data that suggest how the map is composed and updated. We also discuss how this model of LIP is consistent with the plethora of results that have been collected from LIP under a range of conditions. In doing so, we hope to convince the reader that activity in LIP does not represent visual attention or motor intention per se; rather, it creates a map that the brain uses to guide both of these processes.

THE PRIORITY MAP

Our hypothesis is that neurons in LIP act as a priority map. This hypothesis is strongly based on the concept of the saliency maps of Koch, Itti, and colleagues (Itti & Koch 2000, Koch & Ullman 1985, Walther & Koch 2006). A saliency map is a theoretical model that guides attention primarily based on bottom-up inputs. The concept is that the original image is filtered through several preattentive feature detection filters, such as orientation, intensity, or color. This creates a map for each feature in which the activity represents the strength of the salience of that feature across the image. These feature maps are then combined to create a map of general salience. Attention is then allocated to the peak of the map, using a winner-take-all method. We have chosen to use the term priority map rather than saliency map because the term salience connotes bottom-up influences, as it does in the Koch and Itti models. Using the term priority map (Bisley et al. 2009, Fecteau & Munoz 2006, Ipata et al. 2009, Serences & Yantis 2006) implies that both bottom-up and top-down influences play a major role in the selection of objects for eye movements and attention.

In this view, LIP represents a map of the visual world in which locations or objects with high behavioral priority are represented by greater activity in LIP, and in which locations or objects with low behavioral priority are represented by lower activity. We propose that on a moment-by-moment basis, attention is allocated based on the topography of activity across the map and that, when appropriate, saccades are made to the peak of the map. In the next five sections we illustrate that LIP has all of the characteristics of a priority map: It highlights bottom-up signals; it is strongly biased by top-down inputs; and the activity appears to guide both visual attention and eye movements.

VISUAL INPUT AND SALIENCE

We usually relegate stimuli that are not of interest to the background of a scene and, as such, we would expect them to not stand out on a priority map. For example, when we enter a room we often do not notice stable, uninteresting objects in the environment unless they are brought to our attention. Gottlieb and colleagues (Gottlieb et al. 1998, Kusunoki et al. 2000) showed that this observation is true for neuronal responses in LIP. In this study, an array of eight objects remained on the screen for large periods of time and the animals were rewarded for fixating a small point, rather than to anything related to the stable array of stimuli per se. When the monkeys made a saccade to the fixation point, they brought one or more of the stable, task-irrelevant stimuli into the receptive field of the neuron. A typical LIP neuron responded briskly when a stimulus flashed in its receptive field (**Figure 2b**); the abrupt onset of a visual stimulus is a salient event. When a saccade brought a stable stimulus, which had not been made salient, into the receptive field the neuron responded much less (**Figure 2c**). However, when the monkey made a saccade that brought a recently flashed stimulus into the receptive field, the cell responded as briskly as it did to the abrupt onset of the stimulus in its receptive field (**Figure 2d**). Thus, the so-called visual response of parietal neurons (Robinson et al. 1978, Yin & Mountcastle 1977) reports not the arrival of photons in the receptive field, but rather the salience of the objects emitting those photons. The vast majority of LIP neurons recorded behaved in this way, suggesting that when a stimulus is part of the background, it is represented by low activity in LIP.

Stimuli that are inherently salient are represented in LIP by greater responses. This was best demonstrated by Balan & Gottlieb (2006) who had a monkey perform a search task in which one of a number of placeholders morphed into the search target. Although the array was stable, before the search target appeared they introduced a perturbation to a member of the array (Balan & Gottlieb 2006). Five forms of perturbation all induced an increased response: an increase in luminance, a decrease in luminance, an equiluminant change in color, the onset of a frame, and a back-and-forth radial movement (**Figure 3**). Of note is the fact that moving or flashing (in terms of a change in luminance) perturbations induced greater responses in LIP, consistent with our general experience of having attention captured by transients in the visual world (Yantis & Jonides 1984).

THE PRIORITY MAP INCORPORATES TOP-DOWN FEEDBACK

One of the important features of a priority map is the incorporation of top-down feedback. It has been suggested that earlier visual areas, such as V1 (Li 2002) and V4 (Mazer & Gallant 2003), may act as salience maps, but to be a true priority map, top-down information needs to have a strong influence on the bottom-up visual information. In LIP, top-down information has been shown both to enhance and to suppress activity.

Experimentally, enhanced activity in LIP is often driven by task demands. Thus, when a monkey makes a saccade bringing a task-irrelevant stable object into the receptive field of a neuron the response is usually weak (**Figure 2c**), but if the monkey knows that the object will be the target of the next saccade, this top-down consideration results in a robust post-saccadic response (**Figure 4**) (Gottlieb et al. 1998, Kusunoki et al. 2000). Similarly, the response of LIP neurons to perturbations is modulated by whether the search target will appear at or away from the perturbed site (**Figure 3**, right panel). Variations of top-down biases have been shown in a number of visual search tasks

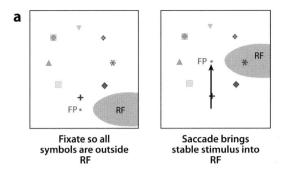

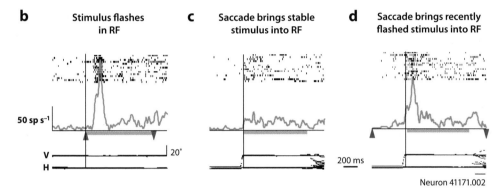

Fixate so all
symbols are outside
RF

Saccade brings
stable stimulus into
RF

b Stimulus flashes in RF

c Saccade brings stable stimulus into RF

d Saccade brings recently flashed stimulus into RF

50 sp s⁻¹

V

H

20°

200 ms

Neuron 41171.002

Figure 2

Effect of a recent flash on the response of an LIP neuron in the stable array task. (*a*) In this task the eight objects remained on the screen for the duration of the experiment. The monkey initiated the trial by fixating a small spot (FP) positioned such that none of the stable objects appeared in the receptive field (RF). After a short delay the fixation point jumped to the center of the array, bringing one of the objects into the receptive field. (*b*) The response to a single stimulus flashing in the receptive field during a fixation task. No other stimuli were present on the screen. Activity is aligned on the stimulus onset. (*c*) The response to the same stimulus, as part of the stable array, moving into the receptive field by a saccade. Activity is aligned by the end of the saccade. (*d*) A saccade brings a recently flashed stimulus into the receptive field. The stimulus appears approximately 500 ms before the saccade, and the data are aligned by the end of the saccade. The gray bars beneath the spike density functions show when, during the trial, the stimulus was in the receptive field of the neuron. The up arrows represent the onset of a flashed stimulus, and the down arrows represent its disappearance. Activity is aligned on the saccade end. Adapted from Kusunoki et al. (2000).

in which a particular target must be differentiated from an array of distractors (Buschman & Miller 2007, Ipata et al. 2006a, Mirpour et al. 2009, Oristaglio et al. 2006, Thomas & Pare 2007). In each case, the activity in LIP ended up being stronger for the target of the search than for any distractor.

A large component of this top-down influence may be related to the relationship between the stimulus in the receptive field and the likelihood that it will provide the animal with a reward. In the example in **Figure 4**, the top-down enhancement could be due to the highlighting of a stimulus as a target, but it could also be due to the increase in behavioral relevance driven by the relationship between the item and the reward now linked with it. A number of studies have examined the responses of LIP neurons in terms of their reward-related properties, and it is clear that part of the top-down influence is related not just to the absolute value of a reward, but to its relative value

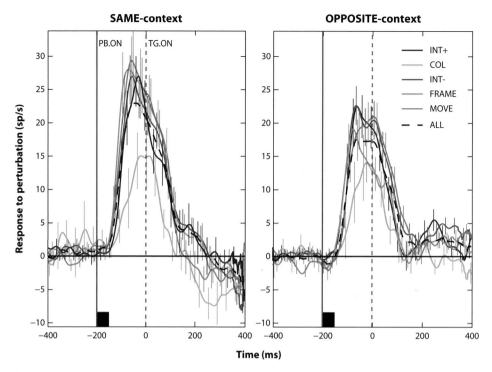

SAME-context   **OPPOSITE-context**

Figure 3

Responses of a population of LIP neurons to salient events (perturbations) that occurred 200 ms before stimuli were revealed in a covert visual search task. Data show the responses to perturbations as the difference between responses on trials with and without the perturbation occurring on trials in which the animal knew that the target would appear on the SAME or OPPOSITE location as the perturbation. The perturbations were: an increase in luminance (INT+); a change in color (COL); a decrease in luminance (INT-); the appearance of frame surrounding one pattern (FRAME); and a back-and-forth radial movement (MOVE). Perturbation onset (PB.ON) is indicated by the solid vertical line; and search target onset (TG.ON) is indicated by the dashed vertical line. Reproduced from Balan & Gottlieb (2006) with permission.

(Bendiksby & Platt 2006, Dorris & Glimcher 2004, Platt & Glimcher 1999, Seo et al. 2009, Sugrue et al. 2004). Initially, Platt & Glimcher (1999) showed that LIP neurons responded more to a stimulus that would provide a bigger reward than to a stimulus that would provide a smaller reward, but more recent studies have suggested that the activity actually represents "expected value" (Sugrue et al. 2004) or "subjective desirability" (Dorris & Glimcher 2004). In other words, the neural responses to a stimulus are related, in a nuanced way, to the expected outcome of responding to that stimulus. Note that the reward does not have to be in the form

---

Figure 4

Response of an LIP neuron in a stable array task requiring a saccade to a cued object. While the monkey was fixating outside of the array, a cue was flashed. The fixation point (FP) then jumped into the center of the array bringing an object into the receptive field. The animal then waited until the fixation point was extinguished and made a saccade to the cued object. (*a*) The cued object was within the receptive field after the first saccade; (*b*) the same object, this time not the target of the saccade, was brought into the receptive field by the first saccade. Each trio of raster plots shows the response of the neuron in the same trials synchronized on the cue (*left*), first saccade beginning (*middle*) and second saccade beginning (*right*). Adapted from Kusunoki et al. (2000) with permission.

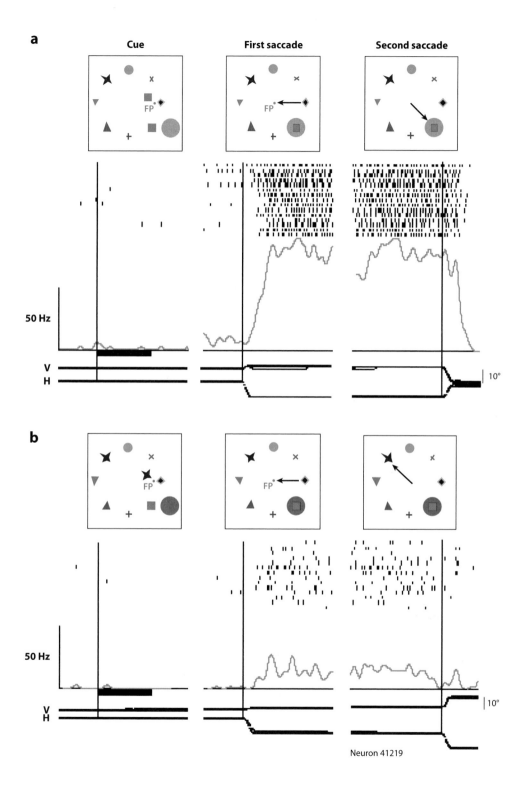

of juice; similar modulation is seen when part of the reward is the chance to view social information not normally readily available, such as to stare at a more dominant male's face (Klein et al. 2008). For this review, the key point is that because most experiments on monkeys rely on rewards related to eye movements to collect data, it is almost certain that a large part of the top-down enhancement seen related to "behavioral importance" is related directly to the reward related to that stimulus expected by the animals. However, Maunsell (2004) pointed out that reward evokes attention, and no study has yet addressed the issue of whether a stimulus that predicts an aversive result is as effective a stimulus for LIP as a stimulus that predicts a reward.

Top-down suppression in LIP is less common. The response to a popout distractor in a visual search task is suppressed when an overtrained animal is able to successfully ignore it (Ipata et al. 2006b). By definition, a highly salient stimulus should be represented by high activity on a priority map, so one would expect that a popout stimulus should have elevated activity and attract attention (Balan & Gottlieb 2006). In this case, monkeys were trained in a free-viewing search task, in which they had to release a bar with either the left or right hand to indicate the orientation of a target. Apart from an initial resetting of the eye to the center of the screen at the beginning of each trial, there were no restrictions on the animals' eye movements. They were not punished if they looked at a distractor and did not have to look at the target to get a reward (Ipata et al. 2009; Ipata et al. 2006a, 2006b). On some trials all of the objects in the array were black; on some trials one stimulus popped out by virtue of being green and brighter than the other array members. The monkeys' performance in this task resembled that of humans: There was a clear set-size effect on reaction time when all members of the array were black, but little if any when the target popped out. Reaction time was even longer when a distractor popped out (Bisley et al. 2009). During the recording experiments, after the psychophysics had been established, the popout stimulus was never the target

of the search. Although looking at the popout did not change the probability of the monkey getting a reward (there were no rules governing eye movements), it did delay the getting of the reward, as it increased the number of saccades the monkeys made before fixating the target. As in normal primate behavior, the animals almost always looked at the target before releasing the bar and, presumably to optimize performance, they made a minimal number of eye movements, looking straight to the target soon after the array appeared on about half the trials (Bisley et al. 2009). As such, over a period of many tens of thousands of trials, the monkeys began to ignore the popout stimulus, as measured by a reduction in first eye movements made to it. When the neural responses were recorded from LIP, the responses to the popout were usually less than to a distractor (**Figures 5a,b**). However, when the mean responses from individual sessions were plotted as a function of how likely the animal was to make its first saccade to the popout within the session, a relationship was seen between the difference in response to the popout and distractor and the salience (i.e., automatic grabbing of attention) of the popout in that session (**Figure 5c**). On days in which the animal could not help but look at the popout first on more than 5% of trials, the mean response to the popout was the same as or greater than the response to a distractor. However, on days in which the animal almost completely ignored the popout (i.e., when the proportion of first saccades went to the popout on less than 2% of trials), the response to the popout was always lower than to a distractor. These data indicate that the normal high response to a salient feature can be suppressed by training, resulting in a reduced attentional capture by the popout (Ipata et al. 2006b).

Top-down suppression of normally enhanced responses in LIP also provides a way of keeping track of which stimuli have been examined in goal-directed search (Mirpour et al. 2009). As described so far, the priority map sets the scene for guiding visual search; targets are highlighted and distractors are suppressed. But

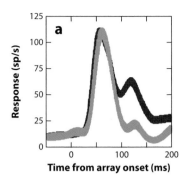 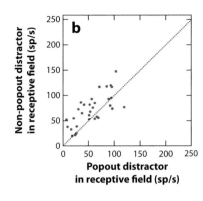 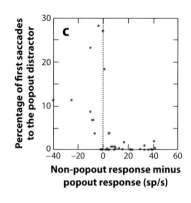

Figure 5

LIP responses to a task-irrelevant popout distractor. (*a*) Responses of a single neuron to the appearance of an array object in the receptive field are plotted against time from target onset. Gray trace: response to a non-popout distractor in the receptive field when the monkey made a saccade to the target elsewhere. Green trace: response to the popout distractor in the receptive field when the monkey made a saccade to the target elsewhere. (*b*) The response of each cell from a 50-ms epoch, starting 40 ms after the latency, to the non-popout distractor is plotted against the response of the same cell to the popout distractor. (*c*) Cell by cell correlation of response difference with saccade suppression. The percentage of trials in which the first saccade went to the popout distractor for each cell is plotted against the difference in the number of spikes between the responses (80–130 ms) to the non-popout and popout distractors for the cell recorded in that session. Reproduced from Ipata et al. (2006b) with permission.

searching for an item is a dynamic process; stimuli are examined and, if the target of the search is not found, the eyes move on. If eye movements are guided by the activity on the priority map, then something must be done to make sure that the eyes will not come back to the previously checked stimulus. In designing their saliency map models, Koch, Itti, and colleagues took a behavioral phenomenon called inhibition of return as their guide (Itti & Koch 2000, Koch & Ullman 1985). Inhibition of return is the slowing down of reaction times in response to a stimulus that is placed at a previously attended location (Klein 2000). Described as a mechanism that biases the system to attend to novel locations or items, Koch & Ullman (1985) implemented this into their model by suppressing the activity representing a stimulus that was being attended. Thus, once attention shifted away from the stimulus, it was no longer strongly represented on the map and would not draw attention back immediately. Such inhibition has been seen in LIP, where responses representing potential targets are reduced after the potential targets have been examined and ruled out as being reward-loaded (Mirpour et al. 2009). This activity can explain why patients with parietal

lesions often revisit items they have already examined (Mannan et al. 2005). As in most examples of response modulation in LIP, this can be explained by reward likelihood; a potential target has a particular reward probability. However, once it has been examined and does not give the reward, then its reward probability drops significantly. In this case, the suppression is the result of top-down information based on the guidance of visual search, but it can also be seen as a form of short-term spatial memory; reducing the activity of seen stimuli allows the animal to remember which potential targets he has seen. The responses to the remaining potential targets do not increase as the reward probability increases, which suggests that LIP activity is not a simple, constantly updated representation of reward probability (Mirpour et al. 2009).

CONSTRUCTION OF THE PRIORITY MAP

Inputs to LIP come from the visual system, the oculomotor system, and a host of other cortical and subcortical areas. Thus far, we have talked about bottom-up and top-down inputs to LIP based on its physiological responses to given stimuli. However, these results are supported by neuroanatomical studies that have shown LIP receiving inputs from both the traditional dorsal and ventral streams of visual processing, including areas V2, V3, V3a, MT, MST, V4, and IT (Baizer et al. 1991, Blatt et al. 1990, Distler et al. 1993, Lewis & Van Essen 2000). It also receives input from a wide range of other cortical and subcortical areas, such as the frontal eye field, anterior cingulate cortex, the claustrum, and a host of thalamic nuclei (Baizer et al. 1993, Blatt et al. 1990).

The various inputs to LIP, as defined physiologically, appear to provide distinctly different components to the activity of neurons in LIP (Ipata et al. 2009). When the monkeys perform the free-viewing search task described above, three different signals are present (**Figure 6a**) in LIP. For the first 70 or so ms after the appearance of the array the responses to a target or a distractor in the receptive field are identical. This is an undifferentiated, bottom-up visual response, which carries no information about the nature of the object in the receptive field or the direction of the impending saccade (Ipata et al. 2009, Thomas & Pare 2007). Soon thereafter, however, the responses diverge according to whether or not the monkey will actually make a saccade to the receptive field or elsewhere (Ipata et al. 2006a, Thomas & Pare 2007). The difference in response, which is dependent upon the direction of the impending saccade, is a

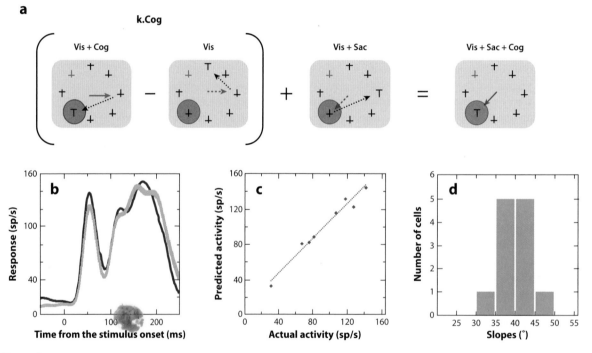

Figure 6

Summation in LIP. (*a*) A cartoon illustration of the test of summation. The cognitive signal (Cog) was in trials in which the target was in the receptive field. k is the constant that accounts for the different strength of the cognitive signal (see text for more details). The visual signal (Vis) was in all trials in which a stimulus appeared in the receptive field. The saccadic signal (Sac) was in trials in which the monkey made a saccade toward the receptive field. (*b*) Single cell responses from trials in which the saccade was made to the target inside the receptive field (*orange trace*) are compared to the calculated signal obtained by summing the three components obtained in different trial types (*blue trace*). (*c*) The mean activity in 20-ms time intervals from 80 ms to 240 ms after array onset measured in saccade-to-target-in-the-receptive-field trials (abscissa) against the activity in the same intervals calculated from the other three trial types (ordinate) for the same single neuron. Least-squares correlation line is shown with a dashed gray line. (*d*) Bar plot of the distribution of the slopes, in degrees, from the regression analyses for each cell. Adapted from Ipata et al. (2009) with permission.

top-down "saccadic" signal. However, the response is also modulated by the nature of the object in the receptive field even when the monkey is not going to make a saccade to that object on the next or even the second saccade of the trial. The difference between the response to the target and a distractor is a top-down catch-all described as the "cognitive" signal. The cognitive signal is found also on trials in which the monkey makes a saccade to the receptive field.

Of most interest to this review, Ipata and colleagues showed that if the three signals are broken down from three different trial conditions, they can be summed up to give an almost perfect representation of the fourth trial condition, which incorporates all three signals (Ipata et al. 2009). The logic behind this deconstruction followed by summation is illustrated in **Figure 6a**. First, the responses containing both a visual and cognitive signal were collected from trials in which the target appeared in the receptive field, but to which the first saccade was not made. Thus, this signal is devoid of the saccadic signal. The authors then subtracted the activity from trials in which only a visual signal was present (i.e., trials in which a distractor that was not the goal of the first saccade was in the receptive field) from the activity from the visual and cognitive trial type. This left the cognitive signal alone. The authors found that a multiplicative factor was needed to amplify the cognitive signal when the saccadic signal was absent. They calculated this factor using mean data from the appropriate animal, but excluding the neuron being examined (Ipata et al. 2009). After multiplying the cognitive signal by this factor, it was added to the activity recorded from trials in which a saccade was made directly to a distractor in the receptive field. This activity represents the visual and saccadic signals. Thus, the resultant signal contained all three components. The authors then compared this response to the response from trials in which all three signals were present, when the first saccade was made towards the target that was in the receptive field. **Figure 6b** shows the summed and actual responses from a single neuron. To quantify the similarity, the authors compared the responses over 20-ms blocks from 80 to 240 ms after array onset and plotted a line of best fit. They found that the mean slopes of the lines of best fit from the population of neurons were not significantly different from 45 deg (**Figures 6c,d**), suggesting that the summation analysis holds up over single neurons as well as over the population.

THE RELATIONSHIP BETWEEN LIP AND COVERT ATTENTION

Covert attention is the generation of attention without making a saccade to the attended object. If, as we have claimed, a priority map is used to guide covert attention, then there should be a relationship between the activity in LIP and measures of covert attention. Numerous early studies had claimed to show such a relationship (Bushnell et al. 1981, Lynch et al. 1977, Robinson et al. 1978), yet covert attention was never actually measured. For example, recordings from posterior parietal cortex, of which LIP was a part at the time, showed an enhanced response when a stimulus had to be monitored than when it was task irrelevant (Bushnell et al. 1981). This was later explicitly shown in LIP (Colby et al. 1996). However, these studies relied on the assumption that if an animal has to respond to a change in stimulus intensity, then it will attend to that location, and if it does not have to respond to the stimulus, then it will not attend to that location. On first glance, this assumption appears valid; however, it is possible that attention could be grabbed by the task-irrelevant peripheral stimulus. Furthermore, it is also possible that the monkeys did not pin their attention on the peripheral stimulus in the attention task, as a change in luminance itself grabs attention (Yantis & Jonides 1984) and would allow the animal to perform the task without maintaining peripheral attention at the stimulus location.

To link the activity in LIP and covert attention, Bisley & Goldberg trained monkeys on a dual task in which they identified the locus of attention by measuring contrast sensitivities (Bisley & Goldberg 2003, 2006). In this task

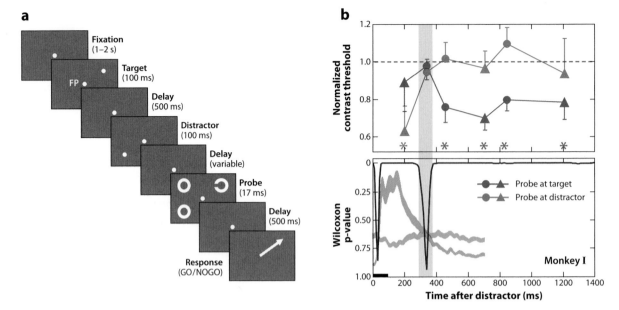

Figure 7

The task and data collected to compare the locus of attention with activity in LIP. (*a*) The task was based on a memory-guided saccade task, with a task-irrelevant distractor. In this task, before the fixation point (FP) was extinguished four rings appeared. One of the rings had a gap on either the left or right (the probe). The monkey had to identify the side of ring that the gap was on and indicate it either by making the planned memory-guided saccade when the fixation point was extinguished (GO) or by canceling the saccade and maintaining fixation until the end of the trial (NOGO). (*b*) Pooled behavioral and physiological data from a single animal. The thin traces in the top panel show the animal's behavioral performance plotted as normalized threshold. Points that are significantly beneath the black dashed line indicate an attentional advantage (*). The thick traces in the lower panel show the mean spike density functions from a population of 23 neurons (the width of the trace shows the SEM). Blue traces show data from trials in which the probe was placed at the target site, and the distractor had flashed elsewhere. Red traces show data from trials in which the probe was placed at the distractor site and the target had flashed elsewhere. The thin gray trace shows the result of a running statistical test showing when the thick red and blue traces were indistinguishable (*gray block*). The black bar shows the onset and duration of the distractor. From Bisley & Goldberg (2003) as modified in Bisley & Goldberg (2006) with permission.

(**Figure 7a**), monkeys were trained to plan a memory-guided eye movement to a particular location in space (the saccade goal). While waiting for the instruction to make the saccade, an array of 4 rings flashed for about 17 ms. One of the rings (the probe) had a gap on one side and the monkey had to determine whether the gap was on the left or right. The animal indicated its choice by either canceling the saccade plan (if the gap was on one side) or not canceling it (if the gap was on the other side). By varying the contrast of the four rings, the authors could create contrast functions at each location in which the probe could appear. By definition, a location with enhanced contrast sensitivity is an attended location. On some trials, a task-irrelevant distractor flashed at a location opposite the saccade goal, and this was included to briefly capture attention.

In performing the task, the animals first pinned their attention at the saccade goal. On trials in which the distractor was presented, an attentional benefit in contrast threshold was seen at the distractor location, but not the saccade goal, 200 ms following the distractor onset (**Figure 7b**, upper panel). This was inverted 500 ms later; contrast thresholds were better at the saccade goal than at the distractor location. Thus, the monkey's attention started out at the saccade goal, was briefly captured by the distractor, and then returned to the saccade goal.

The activity and the time course of the activity in LIP were related to the locus of attention. Typically, LIP neurons respond with a visual burst when the target for a memory-guided saccade appears; they then retain an elevated firing rate ("persistent" or "delay" activity) until the saccade is made, at which time they often show a small burst in activity (Barash et al. 1991). The neurons in this task behaved no differently, despite the presentation of a distractor or the array of rings (**Figure 7b**, lower panel).

The two traces in the lower panel of **Figure 7b** really represent the population responses at two locations on the priority map: the response of the neurons at the location where the target appeared (blue) and the response of the neurons at the location where the distractor flashed (red). The data can be thought of as a sustained peak of activity at the saccade goal along with a transient burst at the distractor location. If these data are compared to the behavioral data, it becomes clear that when the response at the distractor site is higher than the delay activity at the saccade goal, then there was an attentional benefit at the distraction location (the first red triangle in upper panel of **Figure 7b** is lower than the blue triangle). However, when the response at the saccade goal is higher than the response to the distractor (at 455, 700, 840, and 1200 ms), an attentional benefit was seen at the saccade goal (the blue points are less than the red point in the upper panel of **Figure 7b**).

To confirm the relationship between the activity in LIP and the attentional benefit, the authors looked at the behavioral performance at three times during the task based on the LIP responses in two animals. At 340 ms, the activity at the distractor site began to drop below the delay response at the saccade goal in this animal. This is indicated by the increase in p-value on the inverted scale in the lower plot in **Figure 7b**. At 455 ms, the activity at the distractor site began to drop below the delay response at the saccade goal in the second animal, but was already low in this animal. At 840 ms, the activity was clearly greater at the saccade goal in both animals. The prediction was that if attention is allocated to the site of greatest activity, then at the time when the two traces are crossing, there should be no difference in performance at the two locations. This was what the authors found; at 340 ms there was no attentional advantage in the behavioral data at either the saccade goal or the distractor location. However, in the second monkey, attention was still at the distractor location at 340 ms, consistent with higher activity at the distractor site in that animal. Conversely, at the time the activity was equal in the second monkey (455 ms), no attentional benefit was seen in either location in that animal, but the distractor activity had already dropped in this animal and an attentional benefit was already seen at the saccade goal. The time at which attention shifted from the distractor back to the saccade goal was present at the level of the single neuron (Bisley & Goldberg 2006, Ganguli et al. 2008), suggesting that this behavioral trait is hardwired into the network properties of the neurons. In any case, this double dissociation suggests that the activity in LIP is related to the allocation of covert attention in a way consistent with a priority map: Attention is allocated to the peak of activity in LIP.

Previous studies of attention in the parietal cortex in general (Bushnell et al. 1981, Robinson et al. 1978) and LIP in specific (Colby et al. 1996) emphasized the enhanced response evoked by the attended object, a tradition that goes back to the early days of attention studies in behaving monkeys (Goldberg & Wurtz 1972, Moran & Desimone 1985). The assumption was that the enhanced response to the object in the receptive field correlated with the monkey's attention to it. In the contrast sensitivity experiment, however, the determinant of attention was not the activity evoked by the probe (Bisley & Goldberg 2003, 2006). On successful trials the response to the GO probe in the receptive field was not different from the response to a ring (null probe) in the receptive field when the successfully perceived probe lay elsewhere (blue and red traces, **Figure 8**). Instead, the determinant of attention was the peak of the priority map when the probe appeared (time 0 in **Figure 8**), which had set up a locus

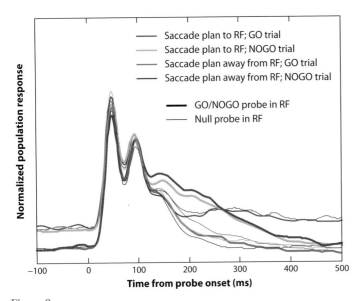

Figure 8

The normalized population responses from one monkey to the GO, NOGO and null probes in the contrast sensitivity task (**Figure 7a**). The color coding indicates the direction of saccade plan and whether the trial was a GO or NOGO trial, and the thickness of the line indicates whether a GO, NOGO, or null probe was in the receptive field. Adapted from Bisley & Goldberg (2006) with permission.

of attention that was operating when the probe activity reached the visual area performing the discrimination. The only case in which a probe evoked a greater response was when the monkey was planning a saccade and a NOGO probe appeared within the receptive field. In this case the response to the probe remained high much longer than when the probe confirmed the saccade plan (purple and orange traces, **Figure 8**). This finding is consistent with the intuitive idea that a stimulus signaling a change in motor output has a higher behavioral priority than one that confirms the status quo and is consistent with results from LIP in a reach-or-saccade task (Snyder et al. 1998).

The effects of reversible lesions in LIP are also consistent with LIP's importance in the generation of attention. Transient inactivation of LIP using muscimol injected into multiple sites in LIP caused monkeys to have longer reaction times to locate a target in search tasks when the target was in the contralateral visual field, but had no effect on reaction times

in the ipsilateral visual field (Wardak et al. 2004).

THE RELATIONSHIP BETWEEN LIP AND THE GENERATION OF SACCADES

Saccades and attention are ordinarily tightly linked. Most of the time humans and nonhuman primates make saccades to their object of attention. LIP was first identified as a unique area within posterior parietal cortex by virtue of its connections to the oculomotor system (Andersen et al. 1985), suggesting that it played a role in eye movement control (Gnadt & Andersen 1988). In fact, for many years it was suggested that LIP played a role in motor intention (Mountcastle et al. 1975), which was defined as explicitly planning the next saccade (Mazzoni et al. 1996), or planning the next saccade that may or may not be made (Andersen & Buneo 2002). This hypothesis was supported by data that showed elevated responses to targets of saccades under a number of conditions. However, unlike the frontal eye field (Bruce et al. 1985, Robinson & Fuchs 1969) and the intermediate layers of the superior colliculus (Robinson 1972, Schiller & Stryker 1972) only high current microstimulation of LIP could produce saccades (Constantin et al. 2007, Shibutani et al. 1984, Thier & Andersen 1998), so although it was implied that LIP activity was related to saccades, no study had shown a direct correlation between eye movements and the activity in LIP.

Further evidence for the intention theory of LIP arose from experiments in which monkeys had simultaneously to plan a reaching movement and a saccade to different targets (Snyder et al. 1997). Delay period activity in LIP was greater when the monkey was planning a saccade to the target than when the monkey planned a reach to it. A reanalysis of the data showed that for the first 300 ms there was no difference between saccade and reach locations (Quian Quiroga et al. 2006). Although the authors used these data to claim that LIP was more related to saccade intention than to attention, they did not actually measure the locus of the monkey's attention, but rather asserted that the monkey's visual attention lay at both the reach and saccade targets throughout the delay period. In fact, at least in humans, attention as measured by perceptual threshold stays at the saccade goal for the entire delay period, but leaves the reach goal after 300 ms (Deubel & Schneider 2003). It is not unreasonable to assume that in the dual task recordings attention left the reach goal at the same time at which the activity in LIP declined. Therefore the decline in activity in LIP during a delayed reach is quite consistent with a priority map interpretation of activity in LIP.

More convincing evidence for the role of LIP in the generation of saccades came from the free-viewing search task described above (**Figures 5** and **6**). Because the monkeys were not punished for making saccades to a distractor, they made quite short latency saccades (151 and 146 ms for the two monkeys, respectively) (Ipata et al. 2006a). More surprisingly, activity in LIP correlated with the saccadic reaction time (**Figure 9**). In an easier search task, in which the monkey only had to find a popout stimulus in an array of eight objects but had to report the results of the search by making a saccade to the popout, saccadic latency was much longer (192 ms), and frontal eye field visuomovement neurons did not predict saccadic reaction time (Thompson et al. 1996), presumably because punishment for making the wrong saccade caused the animals to recheck their saccade plan after the FEF decision to make the saccade.

The antisaccade task, in which a subject has to make a saccade away from a target, has been used to separate stimulus responses from movement responses (Hallett & Adams 1980). The great bulk of neurons in LIP respond to the stimulus away from which the monkey must move its eyes, some of which subsequently respond to the saccade when its spatial location is in their receptive field (Gottlieb & Goldberg 1999), even when the same neurons do not have a traditional presaccadic response,

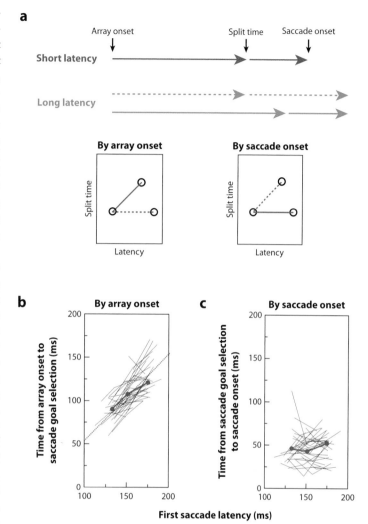

Figure 9

Relationship between LIP activity and first saccadic latency. (*a*) An example of a short latency trial is compared to two possible long latency trials showing the two extreme possibilities in how the extra time is added to latency. In the upper (*dashed*) example, the time from array onset until the split time is identical, so all the variability in latency time comes in to the process later than LIP. This would suggest LIP is not involved in saccadic selection. In the lower (*solid*) example, the extra latency time comes before the split time in LIP. This would suggest that LIP is driving the saccade. The two sets of hypothesized results (*dashed* and *solid*) are plotted in the small panels comparing the split time calculated by array onset and split time calculated by saccade onset. (*b*) The time from array onset to split time is plotted against the mean first saccadic latency for each group for each cell. The dotted line shows an example slope of 1. (*c*) The time from the split time to saccade onset is plotted against the mean first saccadic latency for each group for each cell. The flat lines suggest that a saccade is generated a set time after a peak is identified in LIP. For (*b*) and (*c*), the black lines connect points from the same cell and the solid red lines connect the population means. Adapted from Ipata et al. (2006a) with permission.

as tested using a memory-guided delayed saccade task (Zhang & Barash 2000).

BEYOND ATTENTION AND SACCADES: OTHER FINDINGS IN LIP

Activity in LIP also reflects the dynamics of a decision process. In tasks in which monkeys must evaluate evidence to determine the direction of an impending saccade, activity in LIP gradually increases in the neurons representing the chosen goal and decreases at the rejected goal (Gold & Shadlen 2007). The more difficult the decision, the slower the increase of neuronal activity. Whether the activity is a direct measure of the decision process or a representation about how sure that the animal will get a reward for making an eye movement to that stimulus (which is a direct measure of the output of the decision process), the activity in LIP appears to represent the temporal dynamics of the process (Gold & Shadlen 2007). Thus, activity in LIP has been shown to illustrate the process in estimating time (Janssen & Shadlen 2005, Leon & Shadlen 2003, Maimon & Assad 2006), the direction of motion of a noisy stimulus (Churchland et al. 2008, Roitman & Shadlen 2002, Shadlen & Newsome 2001), confidence (Kiani & Shadlen 2009), and probabilistic reasoning (Yang & Shadlen 2007).

Not all neurons in LIP may be operating as a priority map; a subset with both spatial and nonspatial encoding properties may play additional roles but do not disrupt the function of the area as a priority map. In a number of studies, activity in LIP neurons has been found to contain nonspatial information, although in almost all cases this is superimposed on spatial responses. In one exception, weak, but significant, categorization activity is found in neurons with receptive fields far removed from the stimulus location (Freedman & Assad 2009). In another exception, activity across LIP weakly differentiates between instructions to make eye or arm movements before a target is presented (Dickinson et al. 2003). However, in both cases the responses are substantially less than the same responses when a relevant stimulus is presented in the receptive field, so it is not clear whether these weak global responses are being processed in LIP or are a remnant of one of its inputs.

Within the spatial confines of an LIP neuron's receptive field, information pertaining to stimulus shape (Sereno & Maunsell 1998), direction of motion (Fanini & Assad 2009), categorization (Freedman & Assad 2006), color (Toth & Assad 2002), handedness (Oristaglio et al. 2006), and even numerosity (Roitman et al. 2007) has been found in LIP. Generally, these nonspatial responses have been found in a subset of neurons and a recent study has shown a distinct subset of LIP neurons that are spatially tuned and do not show any obvious nonspatial activity (Ogawa & Komatsu 2009). Thus, whether or not neurons that also have nonspatial activity contribute to the priority map role of LIP, there is a clear spatial-only response across LIP that is strong enough to act as a priority map. Indeed, inactivation of LIP biases spatial processing, but does not effect the handedness of response (Balan & Gottlieb 2009) nor the discrimination of shape (Wardak et al. 2004). In addition, LIP inactivation has no effect on the processing of this nonspatial information when the monkey signals an answer by making a saccade to a point outside the receptive fields of the inactivated neurons, suggesting that the relevant activity seen in LIP is indeed just the output of a decision process occurring elsewhere (Balan & Gottlieb 2009).

ATTENTION, MOTOR INTENTION, DECISION MAKING, AND THE PRIORITY MAP

The experiments described above led to a welter of frankly confusing results about the role of LIP in the generation of behavior in general and of visual attention and saccadic eye movements. For example, LIP responds more strongly to a stimulus that tells the monkey to cancel a saccade plan than it does to a stimulus that confirms the plan. Conversely, under conditions of visual free search, LIP predicts not only the goal but also the latency of visually guided saccades, yet at the same time it describes the nature of the object in the receptive field independently of the direction of the current or next planned saccade. One way to understand these apparently contradictory responses is by a mode switch: Sometimes LIP describes a saccade; sometimes it drives attention; sometimes it accumulates evidence. If this were the case, the brain would somehow have to know how to turn the mode switch, when to use the LIP signal for attention, for saccade planning, and for decision making.

The alternative hypothesis is that LIP provides a priority map that describes the behavioral importance of objects in the visual field. This priority map is built up of a number of disparate top-down and bottom-up signals describing, among others, the abrupt onset of objects in the visual field, saccade planning, value, the nature of the pattern in the visual field, its category, and how close it is to being the chosen object in a decision process. The spikes that contribute to the map lose their identity: Saccade planning spikes are counted no differently than are visual spikes. The oculomotor system drives saccades to the peak of the priority map when saccades are appropriate and the visual system pins attention to the same peak. This makes psychological sense: Attention usually lies at the goal of planned memory-guided saccade unless it is pulled away by the abrupt onset of a salient, task-irrelevant stimulus; and saccades can rapidly be made to the abrupt onset of a visual stimulus (Boch et al. 1984). This hypothesis is consistent with the results from a wide range of studies on LIP and represents the integration of information that acts in far space, the space explored by eye movements. We propose that other maps in the parietal lobe may play similar functions in other workspaces, such as near space (VIP), immediate extrapersonal space (MIP), and hand space (AIP).

Furthermore, because most LIP neurons display perisaccadic remapping (Duhamel et al. 1992, Kusunoki & Goldberg 2003), the priority map is updated around the time of each saccade. This means that the combined top-down and bottom-up information is always present at the correct retinal location. Thus at any given time, both the oculomotor and visual systems have a spatially appropriate priority map that they can use to guide both visual attention and saccades.

DISCLOSURE STATEMENT

The authors are not aware of any affiliations, memberships, funding, financial holdings, or any other conflicts of interests that might be perceived as affecting the objectivity of this review.

ACKNOWLEDGMENTS

J.W.B. is supported by a Klingenstein Fellowship Award in the Neurosciences, an Alfred P. Sloan Foundation Research Fellowship, a McKnight Scholar Award, and the National Eye Institute (R01 EY019273). M.E.G. is supported by the Zegar, Keck and Dana Foundations and the National Eye Institute (R01 EY014978, R01 EY017039).

LITERATURE CITED

Andersen RA, Asanuma C, Cowan WM. 1985. Callosal and prefrontal associational projecting cell populations in area 7a of the macaque monkey: a study using retrogradely transported fluorescent dyes. *J. Comp. Neurol.* 232:443–55

Andersen RA, Buneo CA. 2002. Intentional maps in posterior parietal cortex. *Annu. Rev. Neurosci.* 25:189–220

Baizer JS, Desimone R, Ungerleider LG. 1993. Comparison of subcortical connections of inferior temporal and posterior parietal cortex in monkeys. *Visual Neurosci.* 10:59–72

Baizer JS, Ungerleider LG, Desimone R. 1991. Organization of visual inputs to the inferior temporal and posterior parietal cortex in macaques. *J. Neurosci.* 11:168–90

Balan PF, Gottlieb J. 2006. Integration of exogenous input into a dynamic salience map revealed by perturbing attention. *J. Neurosci.* 26:9239–49

Balan PF, Gottlieb J. 2009. Functional significance of nonspatial information in monkey lateral intraparietal area. *J. Neurosci.* 29:8166–76

Barash S, Bracewell RM, Fogassi L, Gnadt JW, Andersen RA. 1991. Saccade-related activity in the lateral intraparietal area. I. Temporal properties; comparison with area 7a. *J. Neurophysiol.* 66:1095–108

Bendiksby MS, Platt ML. 2006. Neural correlates of reward and attention in macaque area LIP. *Neuropsychologia* 44:2411–20

Bisley JW, Goldberg ME. 2003. Neuronal activity in the lateral intraparietal area and spatial attention. *Science* 299:81–86

Bisley JW, Goldberg ME. 2006. Neural correlates of attention and distractibility in the lateral intraparietal area. *J. Neurophysiol.* 95:1696–717

Bisley JW, Ipata AE, Krishna BS, Gee AL, Goldberg ME. 2009. The lateral intraparietal area: a priority map in posterior parietal cortex. In *Cortical Mechanisms of Vision*, ed. M Jenkin, L Harris, pp. 9–34. Cambridge: Cambridge Univ. Press

Blatt GJ, Andersen RA, Stoner GR. 1990. Visual receptive field organization and cortico-cortical connections of the lateral intraparietal area (area LIP) in the macaque. *J. Comp. Neurol.* 299:421–45

Boch R, Fischer B, Ramsperger E. 1984. Express-saccades of the monkey: reaction times versus intensity, size, duration and eccentricity of their targets. *Exp. Brain Res.* 55:223–31

Bruce CJ, Goldberg ME, Bushnell MC, Stanton GB. 1985. Primate frontal eye fields. II. Physiological and anatomical correlates of electrically evoked eye movements. *J. Neurophysiol.* 54:714–34

Buschman TJ, Miller EK. 2007. Top-down versus bottom-up control of attention in the prefrontal and posterior parietal cortices. *Science* 315:1860–62

Bushnell MC, Goldberg ME, Robinson DL. 1981. Behavioral enhancement of visual responses in monkey cerebral cortex. I. Modulation in posterior parietal cortex related to selective visual attention. *J. Neurophysiol.* 46:755–72

Churchland AK, Kiani R, Shadlen MN. 2008. Decision-making with multiple alternatives. *Nat. Neurosci.* 11:693–702

Colby CL, Duhamel JR, Goldberg ME. 1996. Visual, presaccadic, and cognitive activation of single neurons in monkey lateral intraparietal area. *J. Neurophysiol.* 76:2841–52

Colby CL, Gattass R, Olson CR, Gross CG. 1988. Topographical organization of cortical afferents to extrastriate visual area PO in the macaque: a dual tracer study. *J. Comp. Neurol.* 269:392–413

Colby CL, Goldberg ME. 1999. Space and attention in parietal cortex. *Annu. Rev. Neurosci.* 22:319–49

Constantin AG, Wang H, Martinez-Trujillo JC, Crawford JD. 2007. Frames of reference for gaze saccades evoked during stimulation of lateral intraparietal cortex. *J. Neurophysiol.* 98:696–709

Critchley M. 1953. *The Parietal Lobes.* London: Edward Arnold

Deubel H, Schneider WX. 2003. Delayed saccades, but not delayed manual aiming movements, require visual attention shifts. *Ann. NY Acad. Sci.* 1004:289–96

Dickinson AR, Calton JL, Snyder LH. 2003. Nonspatial saccade-specific activation in area LIP of monkey parietal cortex. *J. Neurophysiol.* 90:2460–64

Distler C, Boussaoud D, Desimone R, Ungerleider LG. 1993. Cortical connections of inferior temporal area TEO in macaque monkeys. *J. Comp. Neurol.* 334:125–50

Dorris MC, Glimcher PW. 2004. Activity in posterior parietal cortex is correlated with the relative subjective desirability of action. *Neuron* 44:365–78

Duhamel JR, Colby CL, Goldberg ME. 1992. The updating of the representation of visual space in parietal cortex by intended eye movements. *Science* 255:90–92

Fanini A, Assad JA. 2009. Direction selectivity of neurons in the macaque lateral intraparietal area. *J. Neurophysiol.* 101:289–305

Fecteau JH, Munoz DP. 2006. Salience, relevance, and firing: a priority map for target selection. *Trends Cogn. Sci.* 10:382–90

Freedman DJ, Assad JA. 2006. Experience-dependent representation of visual categories in parietal cortex. *Nature* 443:85–88

Freedman DJ, Assad JA. 2009. Distinct encoding of spatial and nonspatial visual information in parietal cortex. *J. Neurosci.* 29:5671–80

Ganguli S, Bisley JW, Roitman JD, Shadlen MN, Goldberg ME, Miller KD. 2008. One-dimensional dynamics of attention and decision making in LIP. *Neuron* 58:15–25

Gnadt JW, Andersen RA. 1988. Memory related motor planning activity in posterior parietal cortex of macaque. *Exp. Brain Res.* 70:216–20

Gold JI, Shadlen MN. 2007. The neural basis of decision making. *Annu. Rev. Neurosci.* 30:535–74

Goldberg ME, Wurtz RH. 1972. Activity of superior colliculus in behaving monkey. II. Effect of attention on neuronal responses. *J. Neurophysiol.* 35:560–74

Gottlieb J, Goldberg ME. 1999. Activity of neurons in the lateral intraparietal area of the monkey during an antisaccade task. *Nat. Neurosci.* 2:906–12

Gottlieb JP, Kusunoki M, Goldberg ME. 1998. The representation of visual salience in monkey parietal cortex. *Nature* 391:481–84

Hallett PE, Adams BD. 1980. The predictability of saccadic latency in a novel voluntary oculomotor task. *Vision Res.* 20:329–39

Husain M, Nachev P. 2007. Space and the parietal cortex. *Trends Cogn. Sci.* 11:30–36

Ipata AE, Gee AL, Bisley JW, Goldberg ME. 2009. Neurons in the lateral intraparietal area create a priority map by the combination of disparate signals. *Exp. Brain Res.* 192:479–88

Ipata AE, Gee AL, Goldberg ME, Bisley JW. 2006a. Activity in the lateral intraparietal area predicts the goal and latency of saccades in a free-viewing visual search task. *J. Neurosci.* 26:3656–61

Ipata AE, Gee AL, Gottlieb J, Bisley JW, Goldberg ME. 2006b. LIP responses to a popout stimulus are reduced if it is overtly ignored. *Nat. Neurosci.* 9:1071–76

Itti L, Koch C. 2000. A saliency-based search mechanism for overt and covert shifts of visual attention. *Vision Res.* 40:1489–506

James W. 1890. *The Principles of Psychology*. New York: Holt

Janssen P, Shadlen MN. 2005. A representation of the hazard rate of elapsed time in macaque area LIP. *Nat. Neurosci.* 8:234–41

Johnson PB, Ferraina S, Bianchi L, Caminiti R. 1996. Cortical networks for visual reaching: physiological and anatomical organization of frontal and parietal lobe arm regions. *Cereb. Cortex* 6:102–19

Kiani R, Shadlen MN. 2009. Representation of confidence associated with a decision by neurons in the parietal cortex. *Science* 324:759–64

Klein JT, Deaner RO, Platt ML. 2008. Neural correlates of social target value in macaque parietal cortex. *Curr. Biol.* 18:419–24

Klein RM. 2000. Inhibition of return. *Trends Cogn. Sci.* 4:138–47

Koch C, Ullman S. 1985. Shifts in selective visual attention: towards the underlying neural circuitry. *Hum. Neurobiol.* 4:219–27

Kusunoki M, Goldberg ME. 2003. The time course of perisaccadic receptive field shifts in the lateral intraparietal area of the monkey. *J. Neurophysiol.* 89:1519–27

Kusunoki M, Gottlieb J, Goldberg ME. 2000. The lateral intraparietal area as a salience map: the representation of abrupt onset, stimulus motion, and task relevance. *Vision Res.* 40:1459–68

Leon MI, Shadlen MN. 2003. Representation of time by neurons in the posterior parietal cortex of the macaque. *Neuron* 38:317–27

Lewis JW, Van Essen DC. 2000. Corticocortical connections of visual, sensorimotor, and multimodal processing areas in the parietal lobe of the macaque monkey. *J. Comp. Neurol.* 428:112–37

Li Z. 2002. A saliency map in primary visual cortex. *Trends Cogn. Sci.* 6:9–16

Lynch JC, Mountcastle VB, Talbot WH, Yin TC. 1977. Parietal lobe mechanisms for directed visual attention. *J. Neurophysiol.* 40:362–89

Maimon G, Assad JA. 2006. A cognitive signal for the proactive timing of action in macaque LIP. *Nat. Neurosci.* 9:948–55

Mannan SK, Mort DJ, Hodgson TL, Driver J, Kennard C, Husain M. 2005. Revisiting previously searched locations in visual neglect: role of right parietal and frontal lesions in misjudging old locations as new. *J. Cogn. Neurosci.* 17:340–54

Maunsell JH. 2004. Neuronal representations of cognitive state: reward or attention? *Trends Cogn. Sci.* 8:261–65

Mazer JA, Gallant JL. 2003. Goal-related activity in V4 during free viewing visual search. Evidence for a ventral stream visual salience map. *Neuron* 40:1241–50

Mazzoni P, Bracewell RM, Barash S, Andersen RA. 1996. Motor intention activity in the macaque's lateral intraparietal area. I. Dissociation of motor plan from sensory memory. *J. Neurophysiol.* 76:1439–56

Mirpour K, Arcizet F, Ong WG, Bisley JW. 2009. Been there, seen that: a neural mechanism for performing efficient visual search. *J. Neurophysiol.* 102:3481–91

Moran J, Desimone R. 1985. Selective attention gates visual processing in the extrastriate cortex. *Science* 229:782–84

Mountcastle VB, Lynch JC, Georgopoulos A, Sakata H, Acuna C. 1975. Posterior parietal association cortex of the monkey: command functions for operations within extrapersonal space. *J. Neurophysiol.* 38:871–908

Ogawa T, Komatsu H. 2009. Condition-dependent and condition-independent target selection in the macaque posterior parietal cortex. *J. Neurophysiol.* 101:721–36

Oristaglio J, Schneider DM, Balan PF, Gottlieb J. 2006. Integration of visuospatial and effector information during symbolically cued limb movements in monkey lateral intraparietal area. *J. Neurosci.* 26:8310–19

Platt ML, Glimcher PW. 1999. Neural correlates of decision variables in parietal cortex. *Nature* 400:233–38

Quian Quiroga R, Snyder LH, Batista AP, Cui H, Andersen RA. 2006. Movement intention is better predicted than attention in the posterior parietal cortex. *J. Neurosci.* 26:3615–20

Robinson DA. 1972. Eye movements evoked by collicular stimulation in the alert monkey. *Vision Res.* 12:1795–808

Robinson DA, Fuchs AF. 1969. Eye movements evoked by stimulation of frontal eye fields. *J. Neurophysiol.* 32:637–48

Robinson DL, Goldberg ME, Stanton GB. 1978. Parietal association cortex in the primate: sensory mechanisms and behavioral modulations. *J. Neurophysiol.* 41:910–32

Roitman JD, Brannon EM, Platt ML. 2007. Monotonic coding of numerosity in macaque lateral intraparietal area. *PLoS Biol.* 5:e208

Roitman JD, Shadlen MN. 2002. Response of neurons in the lateral intraparietal area during a combined visual discrimination reaction time task. *J. Neurosci.* 22:9475–89

Sakata H, Taira M, Murata A, Mine S. 1995. Neural mechanisms of visual guidance of hand action in the parietal cortex of the monkey. *Cereb. Cortex* 5:429–38

Schiller PH, Stryker M. 1972. Single-unit recording and stimulation in superior colliculus of the alert rhesus monkey. *J. Neurophysiol.* 35:915–24

Seo H, Barraclough DJ, Lee D. 2009. Lateral intraparietal cortex and reinforcement learning during a mixed-strategy game. *J. Neurosci.* 29:7278–89

Serences JT, Yantis S. 2006. Selective visual attention and perceptual coherence. *Trends Cogn. Sci.* 10:38–45

Sereno AB, Maunsell JH. 1998. Shape selectivity in primate lateral intraparietal cortex. *Nature* 395:500–3

Shadlen MN, Newsome WT. 2001. Neural basis of a perceptual decision in the parietal cortex (area LIP) of the rhesus monkey. *J. Neurophysiol.* 86:1916–36

Shibutani H, Sakata H, Hyvarinen J. 1984. Saccade and blinking evoked by microstimulation of the posterior parietal association cortex of the monkey. *Exp. Brain Res.* 55:1–8

Snyder LH, Batista AP, Andersen RA. 1997. Coding of intention in the posterior parietal cortex. *Nature* 386:167–70

Snyder LH, Batista AP, Andersen RA. 1998. Change in motor plan, without a change in the spatial locus of attention, modulates activity in posterior parietal cortex. *J. Neurophysiol.* 79:2814–19

Sugrue LP, Corrado GS, Newsome WT. 2004. Matching behavior and the representation of value in the parietal cortex. *Science* 304:1782–87

Thier P, Andersen RA. 1998. Electrical microstimulation distinguishes distinct saccade-related areas in the posterior parietal cortex. *J. Neurophysiol.* 80:1713–35

Thomas NW, Pare M. 2007. Temporal processing of saccade targets in parietal cortex area LIP during visual search. *J. Neurophysiol.* 97:942–47

Thompson KG, Hanes DP, Bichot NP, Schall JD. 1996. Perceptual and motor processing stages identified in the activity of macaque frontal eye field neurons during visual search. *J. Neurophysiol.* 76:4040–55

Toth LJ, Assad JA. 2002. Dynamic coding of behaviourally relevant stimuli in parietal cortex. *Nature* 415:165–68

Walther D, Koch C. 2006. Modeling attention to salient proto-objects. *Neural. Netw.* 19:1395–407

Wardak C, Olivier E, Duhamel JR. 2004. A deficit in covert attention after parietal cortex inactivation in the monkey. *Neuron* 42:501–8

Yang T, Shadlen MN. 2007. Probabilistic reasoning by neurons. *Nature* 447:1075–80

Yantis S, Jonides J. 1984. Abrupt visual onsets and selective attention: evidence from visual search. *J. Exp. Psychol. Hum. Percept. Perform.* 10:601–21

Yin TCT, Mountcastle VB. 1977. Visual input to the visuomotor mechanisms of the monkey's parietal lobe. *Science* 197:1381–83

Zhang M, Barash S. 2000. Neuronal switching of sensorimotor transformations for antisaccades. *Nature* 408:971–75

The Subplate and Early Cortical Circuits

Patrick O. Kanold[1] and Heiko J. Luhmann[2]

[1]Department of Biology, University of Maryland, College Park, Maryland 20742;
email: pkanold@umd.edu

[2]Institute of Physiology and Pathophysiology, University Medical Center of the Johannes
Gutenberg University, D-55128 Mainz, Germany; email: luhmann@uni-mainz.de

Annu. Rev. Neurosci. 2010. 33:23–48

First published online as a Review in Advance on
March 4, 2010

The *Annual Review of Neuroscience* is online at
neuro.annualreviews.org

This article's doi:
10.1146/annurev-neuro-060909-153244

Key Words

cerebral cortex, development, plasticity, cortical column, neuronal
network, connectivity, cell death, neuropeptides, MAP2,
neurotransmitters

Abstract

The developing mammalian cerebral cortex contains a distinct class of
cells, subplate neurons (SPns), that play an important role during early
development. SPns are the first neurons to be generated in the cerebral
cortex, they reside in the cortical white matter, and they are the first
to mature physiologically. SPns receive thalamic and neuromodulatory
inputs and project into the developing cortical plate, mostly to layer
4. Thus SPns form one of the first functional cortical circuits and are
required to relay early oscillatory activity into the developing cortical
plate. Pathophysiological impairment or removal of SPns profoundly
affects functional cortical development. SPn removal in visual cortex
prevents the maturation of thalamocortical synapses, the maturation of
inhibition in layer 4, the development of orientation selective responses
and the formation of ocular dominance columns. SPn removal also alters
ocular dominance plasticity during the critical period. Therefore, SPns
are a key regulator of cortical development and plasticity. SPns are vulnerable to injury during prenatal stages and might provide a crucial link
between brain injury in development and later cognitive malfunction.

Contents

INTRODUCTION

The subplate represents a transient layer in the developing cerebral cortex, which is located directly under the cortical plate and which consists of a heterogeneous neuronal population according to morphology and neurotransmitter identity (Kostovic & Rakic 1980, 1990; Luskin & Shatz 1985). The subplate plays an important role in the pathfinding of corticopetal and corticofugal axonal projections (Ghosh et al. 1990, McConnell et al. 1989), in the development of the cortical columnar architecture (Ghosh & Shatz 1992a, 1994; Kanold 2004; Kanold et al. 2003), in developmental plasticity, and in the maturation of cortical inhibition (Kanold 2009, Kanold & Shatz 2006). SPns (subplate neurons) possess a number of structural and functional properties, which put them in an ideal position to be critically involved in all these developmental processes, i.e., they show relatively mature electrophysiological properties and they are well interconnected in the developing cortical network (Friauf et al. 1990, Hanganu et al. 2002).

In 1994, Allendoerfer and Shatz published a review in *Annual Review of Neuroscience* (Allendoerfer & Shatz 1994) summarizing the role of the subplate in the development of connections between the thalamus and neocortex. Over the past 15 years, we have learned a lot more about the function and diverse roles of SPns in neocortical development. The present review aims to provide a summary of our current knowledge on the development, connectivity, function, and plasticity of SPns in the cerebral cortex.

DEVELOPMENTAL ORIGINS OF SUBPLATE NEURONS

Cortical neurons are generated in the ventricular zone (VZ). The first postmitotic neurons are the preplate cells (Bystron et al. 2008) (**Table 1**). The subsequent rounds of cell divisions give rise to neurons forming cortical layers 2–6. Via radial migration, these neurons split

Table 1 Subplate neurons across species

Species	Mouse	Rat	Cat	Primate	Human
Gestation	19.5	21	65	167	40GW
Birth	E11–13 (visual, somato, auditory) (Del Rio et al. 2000, Price et al. 1997, Wood et al. 1992, Zeng et al. 2009)	E12–15 (Al-Ghoul & Miller 1989, Bayer & Altman 1990)	E24–E30 (visual) (Allendoerfer et al. 1990, Luskin & Shatz 1985)	E38–E43 (somato) (Kostovic & Rakic 1980) E43–E45 (Visual) (Kostovic & Rakic 1980)	GW5–6 (Bayer et al. 1993, Kostovic & Rakic 1990)
Waiting	E14–P0 (Del Rio et al. 2000, Deng & Elberger 2003)	E16–17, none (Catalano et al. 1991, Erzurumlu & Jhaveri 1992, Kageyama & Robertson 1993)	E36–E50 (Ghosh & Shatz 1992b)	E78–E124 (Kostovic & Rakic 1984)	GW20–26 (Hevner 2000, Kostovic et al. 2002, Kostovic & Judas 2002, Kostovic & Rakic 1984)
Death (0–80%)	E18–P21 (McQuillen et al. 2002, Price et al. 1997, Torres-Reveron & Friedlander 2007, Wood et al. 1992)	E20–P30 (Al-Ghoul & Miller 1989, Ferrer et al. 1990, Robertson et al. 2000)	P0–P28 (Chun & Shatz 1989a)	E104–P7 (visual) (Kostovic & Rakic 1990) E120–P7 (Somato) (Kostovic & Rakic 1990)	GW34–41 (Kostovic & Rakic 1990, Samuelsen et al. 2003) 2 years (PFC) (Delalle et al. 1997)
Birth.%GP	56%–66%	57%	36%–46%	22%–27%	13%–15%
Waiting%GP	71%–100%	76%–80%	55%–76	45%–85	50%–65%
Death%GP	200%	200%	150%	110%	100%–240%

the preplate into two regions, the marginal zone (containing Cajal-Retzius cells) and the subplate zone (**Figure 1a**). The relative size of the subplate in comparison to the overlying cortical plate varies among species (Aboitiz & Montiel 2007). The subplate is largest in monkeys and humans (Molnár et al. 2006), suggesting that SPns are not a vestige of earlier neuronal structures but rather a key structure enabling radial organization and higher intercortical connectivity (Aboitiz 1999, Aboitiz et al. 2005). SPns have been identified in placental mammals such as rodents, cats, ferrets, primates, and humans (Molnár et al. 2006). The existence of a subplate in marsupials is controversial (Harman et al. 1995, Marotte & Sheng 2000, Reep 2000, Reynolds et al. 1985). Since all experiments discussed in this review have been performed on placental animals, we focus on placental mammals.

MORPHOLOGICAL AND ELECTROPHYSIOLOGICAL PROPERTIES OF SUBPLATE NEURONS

In all mammalian species studied so far, SPns are characterized by their relatively mature structural and functional properties. In newborn rodents and fetal cats, subplate neurons show an extensive axonal and dendritic arborization pattern (**Figure 1b**) (Friauf et al. 1990; Hanganu et al. 2001, 2002). SPns are capable of integrating synaptic inputs over their large dendritic tree, which in rodents can span in horizontal and vertical directions over a few hundred micrometers (**Figure 1c–e**). In human prenatal cortex, the dendrites of SPns may extend up to 1 mm and significantly exceed the size of the basal dendrites of pyramidal neurons (Mrzljak et al. 1988, 1992). Short dendritic spines can be observed on the dendrites of a subpopulation of SPns (Mrzljak et al. 1988), indicating an excitatory function. SPns are characterized not only by their large diversity in the expression pattern of molecular markers (**Table 2**) (Hoerder-Suabedissen et al. 2009, Osheroff & Hatten 2009), but also by their variability in morphological appearance (**Figure 1c–e**). On the basis of their

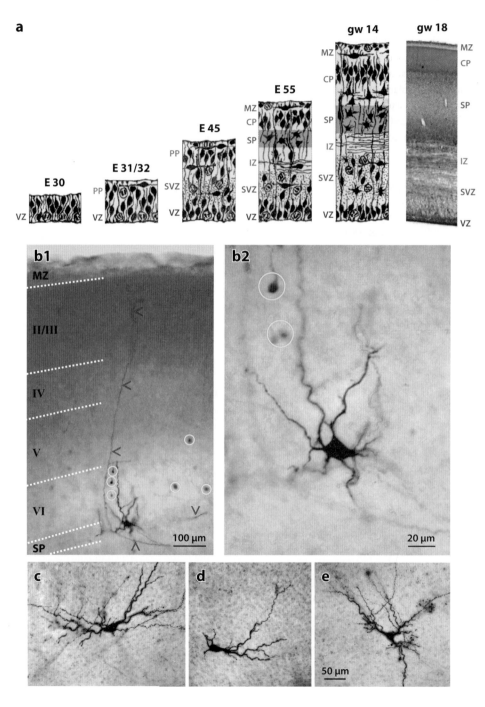

somato-dendritic morphology (i.e., the form of the soma and the orientation of the dendritic tree), at least five to six different neuronal types of SPns can be distinguished in rodent (Hanganu et al. 2002) and human cerebral cortex (Mrzljak et al. 1988), i.e., bitufted and monotufted horizontal, multipolar, inverted pyramidal, polymorphous, and fusiform SPns.

SPns show not only a dense axonal arborization within the subplate and axonal projections to the cortical plate and marginal zone/layer I (**Figure 1*b1***) (Clancy & Cauller 1999, Finney et al. 1998, Friauf et al. 1990) but also long-range axons to the thalamus (De Carlos & O'Leary 1992; Kim et al. 1991; McConnell et al. 1989, 1994) and to more distant neocortical regions (Higo et al. 2007, Tomioka et al. 2005). Some of these long-distance projections arise from GABAergic subplate cells (Luhmann et al. 2009). A subset of SPns persisting in adult rats, called subgriseal neurons by Clancy & Cauller (1999), have cortico-cortical projections of more than 4 mm. SPns are connected not only extensively via chemical synapses, but also locally via electrical synapses. The spatial extent of this gap junction–coupled syncytium can be visualized after intracellular filling of a single subplate neuron with a dye that passes through gap junctions (**Figure 1*b***). In newborn rats, one SPn is on average electrically coupled to about 9 other neurons in the subplate or cortical plate (Dupont et al. 2006). The average distance of the coupled neurons is ~100 μm in the medio-lateral direction and ~125 μm in the dorso-ventral direction, thereby forming a columnar network of about 100 μm in diameter. The average coupling conductance between two neighboring subplate neurons is in the range of 1.2 nS (Dupont et al. 2006).

In addition to this high level of morphological differentiation, SPns also have rather mature functional properties, as judged by their ability to fire repetitive overshooting action potentials (**Figure 2*a***) and by the presence of chemical synaptic inputs with fast kinetics. Intracellular or whole-cell patch-clamp recordings in neocortical slices from mice (Hirsch & Luhmann 2008), rats (Hanganu et al. 2001, 2002; Luhmann et al. 2000), cats (Friauf et al. 1990), and humans (Moore et al. 2009) demonstrated relatively mature passive and active membrane properties. When compared to other neurons at the same developmental stage, SPns reveal the largest amplitudes and fastest kinetics in voltage-dependent sodium and calcium currents (Luhmann et al. 2000).

THALAMIC INNERVATION, CORTICAL MICROCIRCUITRY, AND NEUROMODULATION OF SUBPLATE NEURONS

SPns receive prominent synaptic inputs from various presynaptic sources (**Figure 2*b***). Electron microscopical studies in different species have documented the presence of symmetrical and asymmetrical synapses on subplate cells (Chun & Shatz 1988b; Herrmann et al. 1994;

Figure 1

Development of the subplate and morphological properties of SPn. (*a*) Prenatal development of the human cerebral cortex from embryonic day (E) 30 to gestational week (gw) 18. Photograph to the right shows coronal section of gw 18 human cortex stained with cresyl violet. CP, cortical plate; IZ, intermediate zone; MZ, marginal zone; PP, preplate; SP, subplate; SVZ, subventricular zone; VZ, ventricular zone. Drawing from Pasko Rakic, reproduced and modified with permission from Bystron et al. (2008). Photograph of gw 18 human cortex reproduced with permission from Kostovic et al. (2002). (*b–e*) Morphology of biocytin-stained subplate neurons in newborn rat cerebral cortex. (*b1, b2*) Subplate neuron in a coronal section of a P3 rat. Note axonal collaterals ascending into upper layers and projecting horizontally within subplate (marked by blue <). Several cells are dye coupled and are marked by yellow circles. (*c*) Postnatal day (P) 3 horizontal bitufted subplate neuron with horizontal dendrites. (*d*) P2 horizontal monotufted subplate cell. (*e*) P2 inverted pyramidal neuron with triangular soma and dendrite oriented towards white matter. Scale bar in *e* corresponds to *c* to *e* and pial surface is located at the top. Panel *b* is reproduced and modified with permission from Luhmann et al. (2003); panels *c–e* are reproduced with permission from Hanganu et al. (2002).

Table 2 Expression of markers on subplate neurons

Marker	Subtype	Species, cortical area, age	Reference for mRNA expression	Reference for immunocytochemistry
Ca^{2+} binding proteins	Calbindin	Mouse, >E13		Del Río et al. 2000
	Calbindin	Rat, >E18		Liu & Graybiel 1992
	Calbindin	Ferret, visual cortex		Antonini & Shatz 1990
	Calbindin	Human, >g.w. 20		Ulfig 2002
	Calretinin	Mouse, >E13		Del Río et al. 2000, Hevner et al. 2003
	Calretinin	Human, >g.w. 20		Ulfig 2002
	Parvalbumin	Rat, E16-P10		Csillik et al. 2002
	Parvalbumin	Ferret, visual cortex, >P28		Finney et al. 1998
	Hippocalcin	Mouse, >E14.5	Osheroff & Hatten 2009	Osheroff & Hatten 2009
Extracellular matrix-associated proteins	Connective tissue growth factor	Mouse, visual and somatosensory cortex, >E18 (in situ), >P8 (immuno)	Hoerder-Suabedissen et al. 2009	Hoerder-Suabedissen et al. 2009, Molyneaux et al. 2007
	Connective tissue growth factor	Rat, >E16	Heuer et al. 2003	
	Chondroitin sulfate proteoglycan	Mouse, >E16		Bicknese et al. 1994
	Chondroitin sulfate proteoglycan (neurocan)	Rat, >E16		Fukuda et al. 1997, Miller et al. 1995
	Fibronectin	Cat, >E50	Chun & Shatz 1988a	Chun & Shatz 1988a
Growth factors	p75NTR	Monkey, visual cortex, >E56		Meinecke & Rakic 1993
	p75NTR	Mouse, >E14	McQuillen et al. 2002	
	p75NTR	Rat, >E17	DeFreitas et al. 2001, Koh & Higgins 1991	DeFreitas et al. 2001, Koh & Higgins 1991
	NGF receptor	Human, >g.w. 16		Kordower & Mufson 1992
	NGF receptor	Cat, >E43		Allendoerfer et al. 1990
	NGF receptor	Ferret, >P2		Allendoerfer et al. 1990
Transcription factors, guidance molecules	Nuclear receptor-related 1/Nr4a2	Mouse, visual and somatosensory cortex, >E18 (in situ), >E20 (immuno)	Hoerder-Suabedissen et al. 2009	Arimatsu et al. 2003, Hoerder-Suabedissen et al. 2009, Molyneaux et al. 2007
	SOX5	Mouse, >E14	Kwan et al. 2008	Kwan et al. 2008
	Dlx	Mouse, >E13		Hevner et al. 2003
	Ephrin-A5	Rat, >E17	Mackarehtschian et al. 1999	
	Ephrin-A4, -A7	Mouse, >E15	Yun et al. 2003	

(Continued)

Table 2   (*Continued*)

Marker	Subtype	Species, cortical area, age	Reference for mRNA expression	Reference for immunocytochemistry
Nitric oxide synthase		Ferret, visual cortex, >P28		Finney et al. 1998
		Rat E16-P10		Csillik et al. 2002
		Rat, visual cortex, >P4		Clancy et al. 2001
		Human, >g.w. 15		Judas et al. 1999
	nNOS	Cat		Higo et al. 2007
Chemokine, cytokine	Cxcr4	Mouse, >E14	Tissir et al. 2004	
	TNF-alpha and IL-1beta	Sheep, >E40		Dziegielewska et al. 2000
Steroid hormones	Beta-estradiol (estrogen)	Mouse, >E15	Osheroff & Hatten 2009	
	Progesterone receptor	Rat, >E18	López & Wagner 2009	López & Wagner 2009, Wagner 2008
Others	Monooxygenase Dbh-like 1	Mouse, visual and somatosensory cortex, >E18 (in situ), >P8 (immuno)	Hoerder-Suabedissen et al. 2009	Hoerder-Suabedissen et al. 2009
	Subplate −1	Cat, visual cortex		Dunn et al. 1995, Wahle et al. 1994
	Subplate −1	Rat, mouse, >E18	Fairen et al. 1992	Fairen et al. 1992
	Paired-immunoglobulin–like receptor B (PirB)	Mouse	Syken et al. 2006	
	Complexin 3	Mouse, visual and somatosensory cortex, >E18 (in situ), >P8 (immuno)	Hoerder-Suabedissen et al. 2009	Hoerder-Suabedissen et al. 2009
	Cadherin-related neuronal receptor (CNR)/protocadherin (Pcdh)	Mouse	Morishita et al. 2004	Morishita et al. 2004
	G protein-gated inwardly rectifying K-channels (GIRK)	Mouse	Wickman et al. 2000	
	Phosphodiesterase 1C	Mouse, >E13.5	Osheroff & Hatten 2009	Osheroff & Hatten 2009

Kostovic & Rakic 1980, 1990), indicating that SPns receive GABAergic as well as glutamatergic synaptic inputs. As initially suggested by Kostovic & Rakic, glutamatergic inputs onto SPns arise from the thalamus and other neocortical areas, whereas GABAergic synaptic inputs originate from GABAergic interneurons located in the subplate (Kostovic & Rakic 1980). GABAergic and glutamatergic receptors and markers can be demonstrated in various species at the earliest developmental stages (**Table 3**). Functionally, spontaneous synaptic inputs with fast kinetics mediated by AMPA, NMDA, and $GABA_A$ receptors have been recorded in SPns

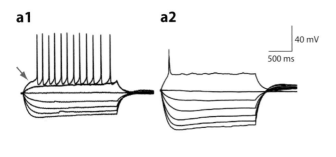

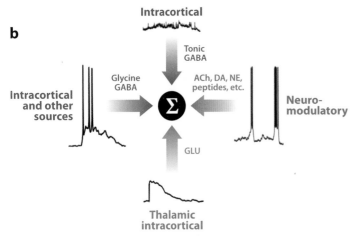

Figure 2

Firing pattern and synaptic inputs of SP. (*a*) Whole-cell patch-clamp recordings from a subplate neuron (*a*1) and in comparison from an immature cortical plate pyramidal cell (*a*2) in newborn rat neocortical slice. Current–voltage relationship and firing pattern illustrates the relative mature electrophysiological properties of the subplate cell with large and repetitive action potentials. Note presence of an A-current in the SPn (*arrow*). Reproduced and modified with permission from Luhmann et al. (2003). (*b*) Subplate neurons integrate afferent inputs from various presynaptic sources. A glutamatergic input innervates the subplate from the thalamus and cerebral cortex. A cortical phasic and tonic GABAergic input depolarizes SPn. The origin of the glycinergic input is unknown, but glycinergic receptors may be tonically activated by taurine. Various neuromodulatory inputs (ACh, DA, 5-HT, NE, peptides, etc.) transiently innervate the subplate and have a profound influence on cortical network activity.

in newborn rats (Hanganu et al. 2001). SPns receive a glutamatergic input from the thalamus mediated via ionotropic glutamate receptors (Hanganu et al. 2002, Herrmann et al. 1994, Higashi et al. 2002, Hirsch & Luhmann 2008). Thalamic axons arrive in the subplate around the time that layer 4 cells are born and wait in the subplate before growing into layer 4 (**Figure 3***a*). The duration of the waiting period varies considerably between species and is longer in species with longer gestation times (**Table 1**). In marsupials, however, thalamic axons seem to directly innervate layer 4 neurons without waiting in the subplate (Molnár et al. 1998, Pearce & Marotte 2003).

Due to the changing nature of subplate circuits, the pattern of thalamocortical activation of cortex varies over development (**Figure 3***a*), as shown by studies in brain slices of cat visual (Friauf & Shatz 1991) and rodent somatosensory cortex (Higashi et al. 2002, Molnár et al. 2003). Electrical white matter stimulation in cat visual cortex at birth results in short latency responses in the subplate and long latency responses in layer 4. This latency difference likely indicates disynaptic responses, suggesting that SPns strongly excite layer 4 neurons (Friauf & Shatz 1991). At later ages, short latency responses to white matter stimulation start to emerge in layer 4. This indicates that now thalamic activity directly activates layer 4 neurons, consistent with mature thalamocortical circuits. Similar results were obtained from imaging experiments and current source density analyses in slices from rodent somatosensory cortex (Higashi et al. 2002, Molnár et al. 2003). In rodents thalamic stimulation activates SPNs by embryonic day 16 (E16) while cortical plate activation is seen at E21. However, in these studies, disynaptic cortical activation was absent. The delay in the emergence of cortical responses in both species reflects the "waiting period" and time needed for synapses to mature. The difference in timing (prenatal versus postnatal) between these studies might reflect an early maturation of the somatosensory relative to the visual system or species differences. The absence of disynaptic responses in the imaging studies could be due to different stimulation sites (thalamus versus white matter) recruiting fewer thalamocortical fibers. However, these data together show that thalamocortical transmission undergoes a functional reorganization from activating subplate neurons to activating layer 4 neurons.

Another glutamatergic input onto SPn arises from the cortical plate and from glutamatergic

Table 3   Expression of transmitter receptors and subtypes on subplate neurons

Transmitter	Receptor subtype	Species, cortical area, age	Reference for mRNA expression	Reference for immunocytochemistry	Reference for electrophysiology
GABA		Mouse, >E13		Del Río et al. 2000	
		Rat, >E16		Lauder et al. 1986, Robertson et al. 2000	
		Cat, >E50		Chun & Shatz 1989b	
		Human, gestation week >7		Zecevic & Milosevic 1997	
	Glutamic acid decarboxylase (GAD)	Rat, >E18		Arias et al. 2002	
	GAD-67	Ferret, visual cortex, >P28		Finney et al. 1998	
	GABA-A	Rat, somatosensory cortex, >P0			Hanganu et al. 2001, 2002
Glycine		Rat, somatosensory cortex, >P0			Kilb et al. 2008
Glutamate	VGLUT1, VGLUT2	Mouse, >E13	Ina et al. 2007		
	AMPA (GluR 2/3)	Rat, >E18		Arias et al. 2002	
	AMPA (GluR2/3)	Sheep, >E60		Furuta & Martin 1999	
	Glutamate	Ferret, visual cortex, >P28		Finney et al. 1998	
	AMPA, kainate	Mouse, somatosensory cortex >P0			Hirsch & Luhmann 2008
	AMPA, kainate	Rat, somatosensory cortex >P0			Hanganu et al. 2001, 2002
	Kainate (GluR6/7)	Sheep, >E60		Furuta & Martin 1999	
	NMDA, kynurenine aminotransferase (KAT)-I	Rat E16-P7		Csillik et al. 2002	
	NMDA	Rat, visual and somatosensory cortex, >P0			Hirsch & Luhmann 2008; Hanganu et al. 2001, 2002; Torres-Reveron & Friedlander 2007
	NR2A	Rat P1-P7		Csillik et al. 2002	
	NR2A, NR2B, NR2D	Mouse, somatosensory cortex >P0	Hirsch & Luhmann 2008		Hirsch & Luhmann 2008

(*Continued*)

Table 3   (*Continued*)

Transmitter	Receptor subtype	Species, cortical area, age	Reference for mRNA expression	Reference for immunocytochemistry	Reference for electrophysiology
Acetylcholine, nicotinic	Alpha4	Human, frontal cortex, >17 weeks of gestation	Schröder et al. 2001	Schröder et al. 2001	
	Alpha4, beta2	Rat, somatosensory cortex >P0			Hanganu & Luhmann 2004
	Alpha5	Rat >E18	Winzer-Serhan & Leslie 2005		
	Alpha7	RAT, >P1		Csillik et al. 2002	
Acetylcholine, muscarinic	M1–m5	Rat, somatosensory cortex >P0	Hanganu et al. 2009		Dupont et al. 2006, Hanganu et al. 2009
Dopamine	DOPA decarboxylase	Mouse, visual and somatosensory cortex, >E18 (in situ), >P8 (immuno)	Hoerder-Suabedissen et al. 2009	Hoerder-Suabedissen et al. 2009	
Neuro-peptides	NPY	Mouse, >E16		Del Río et al. 2000	
	NPY	Rat, >E18, P7-P10		Arias et al. 2002, Csillik et al. 2002, Robertson et al. 2000	
	NPY	Ferret, visual cortex, >P28		Antonini & Shatz 1990, Finney et al. 1998	
	NPY	Cat, >E50		Chun & Shatz 1989b	
	NPY	Monkey, visual cortex, >E75		Mehra & Hendrickson 1993	
	NPY	Human, >14 weeks of gestation		Delalle et al. 1997	
	CCK	Mouse, >E16		Del Río et al. 2000	
	CCK	Cat, >E60		Chun & Shatz 1989b	
	Somatostatin	Rat		Robertson et al. 2000	
	Somatostatin	Ferret, visual cortex, >P28		Antonini & Shatz 1990, Finney et al. 1998	
	Somatostatin	Cat, >E50		Chun & Shatz 1989b	
	Somatostatin	Human, frontal cortex, >22 weeks of gestation		Kostovic et al. 1991	
	Substance P	Mouse, >P0		Del Rio et al. 1991	
	Substance P	Monkey, visual cortex, >E90		Mehra & Hendrickson 1993	
	Hypocretin-orexin (Hcrtr2-OX2)	Rat, different cortical areas, >P15			Bayer et al. 2004

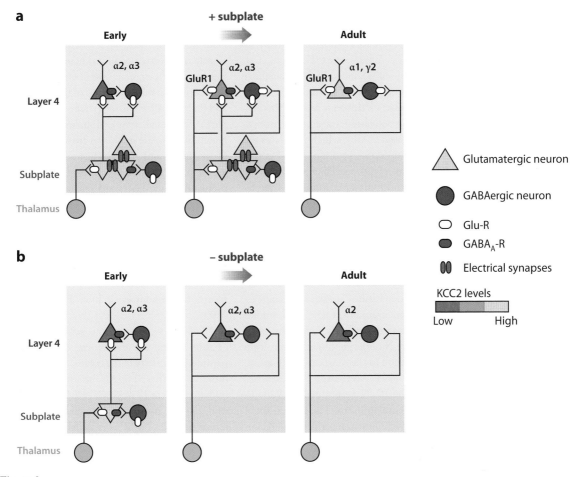

Figure 3

Subplate neurons affect thalamocortical circuit development and inhibitory maturation (*a*) Developmental changes in thalamocortical projections and intracortical inhibition over development. Early (*left*): Thalamus projects to subplate, which in turn projects to layer 4. Potential targets for subplate neurons are both GABAergic and excitatory layer 4 neurons. Subplate neurons and cortical neurons are coupled via electrical synapses. At these ages GABA is depolarizing in layer 4 due to low KCC2 levels (see cell shading). Subplate neurons have higher KCC2 levels due to their advanced maturity. Over development (*middle*) thalamic axons grow into layer 4 and contact layer 4 cells. KCC2 levels in layer 4 increase. Adult (*right*): The thalamocortical synapse has matured. KCC2 levels are high, thus GABA is hyperpolarizing. At early ages GABAergic receptors are composed of α2 and α3 containing subunits, whereas mature receptors contain the α1 and γ2 subunit. (*b*) Summary of circuit changes after early subplate ablation. The α2 and α3 receptor mRNA are expressed at high levels in layer 4, whereas KCC2, α1, γ2, and Glur1 mRNA levels remain low. Since Glur1 mRNA levels are low, the cortex is decoupled from its inputs.

SPns (Hanganu et al. 2002, Hirsch & Luhmann 2008). Both inputs are mediated via ionotropic glutamate receptors, but a significant proportion of the intrasubplate input is mediated via NMDA receptors, which can be activated at more hyperpolarized membrane potentials (Hanganu et al. 2002). These intrasubplate synaptic connections show a pronounced paired-pulse facilitation and temporal summation and at postnatal day 0 (birth) (P0) contain a large amount of the NR2D subunit (Hirsch & Luhmann 2008).

A GABAergic input may originate from local as well as from remote GABAergic subplate neurons, which in the mouse cerebral cortex can project over distances of up to 2 mm

(Higo et al. 2007, Tomioka et al. 2005). Activation of GABA$_A$ receptors as well as glycine or taurine receptors elicits a depolarizing response in newborn rat cerebral cortex (Hanganu et al. 2002, Kilb et al. 2008) due to high intracellular chloride concentrations (Yamada et al. 2004). Besides this phasic GABAergic input, SPns also receive a tonic activation via ambient nonsynaptically released GABA, which facilitates the generation of up states in the neonatal cortex (Hanganu et al. 2009). GABAergic SPns may contribute to this tonic GABA release, thereby modulating proliferation and migration of neuronal progenitors (Maric et al. 2001). Nonsynaptically released taurine may have a similar role by tonic activation of glycine receptors (Flint et al. 1998).

SPns receive a diversity of neuromodulatory inputs from various presynaptic sources. Anatomical and immunocytochemical studies have demonstrated a selective innervation of the subplate by cholinergic fibers arising from the basal forebrain in the newborn rat (Calarco & Robertson 1995, Mechawar & Descarries 2001) and in the 18–22 gestational week in human cortex (Kostovic 1986). The expression of nicotinic acetylcholine transcripts and receptors on SPn have been documented in the rat (Csillik et al. 2002, Winzer-Serhan & Leslie 2005) and human (Schröder et al. 2001) cortex. Patch-clamp recordings from SPns in neonatal rats showed a strong nicotinic excitation mediated by alpha4beta2 receptors (Hanganu & Luhmann 2004). Nicotine, at concentrations similar to the amount that reaches the developing human brain through maternal smoking, induced in SPns a prominent desensitization of nicotinic acetylcholine receptors (Hanganu & Luhmann 2004), suggesting that exposure to nicotine during prenatal stages may disturb developmental processes that are influenced by acetylcholine. SPns in neonatal rat cortex also have M1 to M5 muscarinic receptors, as shown by single-cell PCR studies (Hanganu et al. 2009). Activation of muscarinic M1 receptors causes a membrane depolarization and repetitive ~20 Hz burst discharges in SPns (**Figure 2b**) (Hanganu et al. 2009). Due to their intense coupling via chemical and electrical synapses to other SPns and to cortical plate neurons, these burst discharges are efficiently transmitted to a local neuronal network (see next section). The first fibers approaching the subplate before the arrival of the cholinergic and thalamic afferents seem to be monoaminergic (Mrzljak et al. 1988). Monoaminergic fibers reach the subplate in human cortex at 12 weeks of gestation and immunohistochemical studies in rodents have demonstrated that these monoaminergic fibers are serotonergic, noradrenergic, and dopaminergic (Kalsbeek et al. 1988, Molliver 1982). The function of these monoaminergic inputs onto SPns is currently unknown, but activation of metabotropic receptors may cause subplate-driven network oscillations similar to those shown for muscarinic receptors (Dupont et al. 2006, Hanganu et al. 2009).

SPns also express various peptide receptors (**Table 3**) and are the source of the earliest peptidergic activity in the cortex. Somatostatin-immunoreactive SPns can be identified in the human cortex at 22 weeks of gestation (Kostovic et al. 1991). The exact functional role of the different peptides on SPns is poorly understood. In juvenile rat cortex, application of cholecystokinin (CCK) to layer 6b neurons (subplate) causes a strong excitation via CCK(B) receptors (Chung et al. 2009). It has been further demonstrated that hypocretin-orexin neurons in the lateral hypothalamus innervate layer 6b and that activation of Hcrtr2-OX2 receptors causes a closure of a potassium conductance, thereby promoting widespread activation of layer 6b/SPns (Bayer et al. 2004).

Subplate Projections

SPns have diverse axononal output patterns. Subplate axons project into the developing cortical plate (Friauf et al. 1990; Friauf & Shatz 1991; Hanganu et al. 2001, 2002; Hanganu & Luhmann 2004; Luhmann et al. 2000; Piñon et al. 2009) and also pioneer the corticogeniculate projection (De Carlos & O'Leary 1992, McConnell et al. 1989, Molnár & Cordery

1999). In higher mammals, but not in rodents, SPns also project through the corpus collosum (Antonini & Shatz 1990, deAzevedo et al. 1997, Del Rio et al. 2000). However, it is unknown if these three different projection patterns are subserved by different classes of SPns. Of the three projection targets, the feed-forward cortical projection is the best-studied projection. Subplate projections to the cortical plate are radially oriented, show some collateral axon branches, and predominantly target layer 4 (Dupont et al. 2006, Friauf et al. 1990, Friauf & Shatz 1991, Piñon et al. 2009). Most SPns projecting to the cortical plate are glutamatergic (Finney et al. 1998), and recent physiological experiments show that selective subplate stimulation evokes excitatory synaptic currents in layer 4 (Zhao et al. 2009).

THE SUBPLATE: AN ACTIVE HUB STATION

Numerous studies have demonstrated with tracing methods that the subplate receives a transient input from the specific thalamic nuclei and that the subplate serves as a waiting station for the ingrowing thalamocortical axons (see above). Friauf et al. were the first to demonstrate a functional synaptic input from the thalamus onto SPns (Friauf et al. 1990). In subsequent studies, different groups confirmed these results for rats and mice (Hanganu et al. 2002, Higashi et al. 2002, Molnár et al. 2003, Zhao et al. 2009). Furthermore, electrophysiological studies demonstrated functional intracortical GABAergic inputs (Hanganu et al. 2001, 2002, 2009), an intracortical and thalamocortical glutamatergic input (Hanganu et al. 2002, Hirsch & Luhmann 2008), and a cholinergic input mediated via muscarinic receptors (Dupont et al. 2006, Hanganu et al. 2009). In addition, functional nicotinic alpha4beta2 receptors (Hanganu & Luhmann 2004) and glycinergic (Kilb et al. 2008) receptors have been demonstrated on SPns in newborn rodent cortex. All these functional data demonstrate that the subplate may have a more important function than just serving as a

rather passive waiting station of the ingrowing thalamocortical afferents. Voigt and colleagues (Voigt et al. 2001) have demonstrated in dissociated neuronal cell cultures from embryonic rat cerebral cortex that a distinct population of large GABAergic neurons is a key element in the generation of synchronous oscillatory network activity. The authors have suggested that SPns function as an integrating element that synchronizes neuronal activity by collecting incoming extrinsic and intrinsic signals and distributing them effectively throughout the developing cortical plate. A minimal number of two large GABAergic SPns per square millimeter were required for the occurrence of synchronous activity. The pivotal role of SPns in generating synchronous oscillatory network activity has been confirmed by multichannel recordings from acute neocortical slices of newborn rodents. Electrical stimulation of the subplate in 800–1000 μm thick slices with a sufficiently preserved neuronal network elicits synchronized oscillatory activity (Sun & Luhmann 2007). Carbachol-induced synchronized network oscillations with similar properties can be elicited in intact cortices of the newborn rat only when the subplate is intact (Dupont et al. 2006). These in vitro observations have been recently confirmed by in vivo experiments demonstrating a clear participation of the subplate in generating locally synchronized oscillatory network activity (Yang et al. 2009).

Together these data show that SPns play a very active role in cortical processing. The subplate functions not only as a passive relay or waiting zone, but rather as an active hub station of the developing cortical network! This is due to their unique anatomical properties (extensive dendritic arborization, widespread axonal projections), their relative mature functional state (firing pattern, etc.), their gap junction mediated electrical coupling to other SPns and cortical plate neurons, their substantial glutamatergic or GABAergic synaptic inputs from thalamic, intrasubplate and cortical plate sources, and their strong synaptic inputs from neuromodulatory systems (e.g., the selective innervation of the SP by the cholinergic basal forebrain). Thus SPns possess key attributes and are in a key position to affect cortical development.

ROLE OF THE SUBPLATE IN REGULATING MATURATION OF CORTICAL INHIBITION (AND EXCITATION)

The maturation of cortical circuits involves the functional maturation of neurons, the maturation of their capability to release neurotransmitters, and the increased expression of excitatory and inhibitory neurotransmitter receptors.

Role of Subplate Neurons in Maturation of Thalamocortical Synapses

Selective lesioning of SPns can be achieved by excitotoxic injections (Ghosh et al. 1990, Ghosh & Shatz 1992a, Kanold et al. 2003, Kanold & Shatz 2006, Lein et al. 1999) or by exploiting the selective expression of p75 in subplate neurons (**Table 2**) and using injections of p75-immunotoxin (Kanold et al. 2003, Kanold & Shatz 2006). Such selective subplate lesions have been used to elucidate the role of SPns in cortical circuit maturation. Ablations in cat during the first postnatal week when thalamocortical excitation is immature revealed a profound effect of SPns on thalamocortical maturation when animals were examined ∼3–4 weeks later (Kanold et al. 2003).

Ablation of SPns in visual cortex prevents the developmental increase in expression of glutamate receptor subunits (GluR1) mRNA specifically in layer 4 (Kanold et al. 2003) (**Figure 3b**). The low mRNA levels are paralleled functionally by weak thalamocortical synapses (Kanold et al. 2003). However, despite the low thalamocortical synaptic strength, an increase in spontaneous synaptic events and spiking activity is seen (Kanold et al. 2003). Thus, visually driven thalamic activity is unable to strongly drive cortical neurons, and therefore the visual cortex becomes functionally decoupled from the visual thalamus (LGN). SPns might strengthen thalamocortical synapses by interacting with synaptic plasticity rules. Simulations using a computational model have shown that strong subplate input to layer 4 can entrain correlations between thalamic activity and layer 4 activity that lead to strengthening of thalamocortical synapses via Hebbian plasticity rules (Kanold & Shatz 2006).

Role of Subplate Neurons in Maturation of Cortical Inhibition

Maturation of inhibition involves the maturation of inhibitory neurons to express GABA synthesizing enzymes (glutamate-decarboxylase, GAD) and postsynaptic expression of a mature complement of GABA receptors. In addition, fast GABAergic inhibition via GABA$_A$ receptors involves the influx of chloride (Cl$^-$) ions. Thus the intracellular Cl$^-$ concentration and thereby E$_{Cl}$ determine the functional effect of GABAergic inhibition. KCC2 removes Cl$^-$ from the cytosol and thus can control Cl$^-$ levels (and E$_{Cl}$) (Blaesse et al. 2009). Low E$_{Cl}$ renders GABA hyperpolarizing, whereas high E$_{Cl}$ renders GABA depolarizing, which can act excitatory or inhibitory (shunting), depending on the size of the depolarization (Achilles et al. 2007, Blaesse et al. 2009). KCC2 levels increase over development and render GABA$_A$ receptors hyperpolarizing (Blaesse et al. 2009) (**Figure 3a**). KCC2 expression can be regulated by neuronal activity (Fiumelli et al. 2005, Ganguly et al. 2001, Kriegstein & Owens 2001, Ludwig et al. 2003), BDNF (Aguado et al. 2003; Rivera et al. 2002, 2004), or injury (Cramer et al. 2008, Rivera et al. 2004, Shimizu-Okabe et al. 2007).

Subplate ablations in cat during the first postnatal week when intracortical inhibition is immature revealed a profound effect of SPns on inhibitory maturation when animals were examined ∼3–4 weeks later (Kanold & Shatz 2006). Ablation of SPns prevents the developmental increase in expression of KCC2 and "mature" GABA$_A$ receptor subunits such as the alpha1 and gamma2 subunit (Kanold & Shatz 2006) (**Figure 3b**). This immature expression pattern is paralleled functionally by a sustained presence of depolarizing responses to GABAergic stimulation (Kanold & Shatz 2006). SPns might regulate the expression levels of KCC2 and GABA receptors by providing depolarization to layer 4 that is able to increase the expression of these genes. This view is supported by in vivo experiments in which glutamatergic signaling was blocked during development. In these experiments, KCC2 mRNA levels also failed to increase, suggesting that a glutamatergic input is required to induce KCC2 and GABA$_A$ alpha1 mRNA expression (Kanold & Shatz 2006). During development, there are three sources of glutamatergic excitation to cortical neurons (see above): thalamic inputs, intracortical inputs, and subplate inputs. Since both thalamic and intracortical inputs are present after subplate ablation, but fail to cause increased expression of KCC2 and GABA$_A$ alpha1 (Kanold & Shatz 2006), these data suggest that specific glutamatergic input from the subplate is needed for their increased expression.

The lower expression levels of GluR1, KCC2, GABA receptors, and increased spontaneous activity levels are paralleled by increased expression of BDNF mRNA (Lein et al. 1999). Since BDNF mRNA levels can be regulated by neural activity (Castren et al. 1998, Lein et al. 2000), overall activity levels in layer 4 after ablation might be higher than normal, despite reduced thalamic inputs. Alternatively, the regulation of BNDF might be altered after subplate ablation. Subplate ablation also results in increased levels of GAD (Lein et al. 1999) suggesting that interneurons are present and active. The increased cortical GAD levels after ablation might indicate that GABAergic neurons are hyperactive. In unmanipulated cortex, increased activity can lead to increased inhibitory tone and decreased excitatory tone possibly via BDNF signaling (Turrigiano 2007). However, BDNF levels can also lead to a reduction in KCC2 levels (Molinaro et al. 2009; Rivera et al. 2002, 2004). Thus, increased BDNF levels after subplate ablation might prevent inhibitory maturation. The observed increased inhibitory activity together with high BDNF levels after ablation might be indicative of dysfunctional homeostatic regulation of cortical activity (Turrigiano 2007) due to immature inhibitory maturation. Because levels of KCC2 remain low after subplate ablation, increased GABAergic activity due to possibly hyperactive GABAergic neurons (Lein et al. 1999) can further increase cortical activity levels, possibly contributing to seizure activity following SPn ablation (Lein et al. 1999).

ROLE OF THE SUBPLATE IN SCULPTURING NEOCORTICAL ARCHITECTURE (COLUMNS)

One hallmark of neocortical organization especially primary sensory cortices are functional columns that group neurons with similar stimulus selectivity (Mountcastle 1997). On a small scale, columns are formed by grouping several microcolumns (radial units) and on a larger scale, columns are organized into cortical maps. SPns are involved in setting up this architecture at multiple levels and at multiple developmental time points.

Role of the Subplate in Area Identity and Radial Unit Formation

SPns aid in the guidance of thalamic axons into layer 4. Removal of SPns before thalamic axons enter layer 4 redirects these axons to an area where SPns are present (Ghosh et al. 1990). In addition to providing guidance cues, SPns might contain direct cues directing the positioning of cortical maps. FGF8 gradients are involved in positioning sensory maps in the rostro-caudal axis (Fukuchi-Shimogori & Grove 2001). Misexpressing FGF8 in the cortical plate causes thalamocortical axons to enter the cortical plate and then turn posterior to innervate layer 4 (Shimogori & Grove 2005). However, if FGF8 was also misexpressed in subplate then thalamocortical axons travel further posterior within the subplate and innervate the cortical plate radially, suggesting that SPns contained positional information (Shimogori & Grove 2005). Such position information can be conveyed by graded expression of guidance molecules in the subplate, such as p75 (McQuillen et al. 2002) and ephrinA5 (Mackarehtschian et al. 1999, Yun et al. 2003) (**Table 2**).

In addition to radial glia cells (Rakic 1988), SPns may represent an additional cell type contributing to the radial organization of the cerebral cortex (Mountcastle 1997). SPns are coupled via electrical synapses to cells in the cortical plate (Dupont et al. 2006) and could aid in establishing cortical microcolumns by defining coupled radial units (Mountcastle 1997). This is consistent with evolutionary hypotheses about the role of SPns in allowing the radial organization of the mammalian cerebral cortex (Aboitiz 1999, Aboitiz et al. 2005).

Role of Subplate Neurons in Establishing the Functional Cortical Architecture

Neurons in the visual cortex respond selectively to lines of a particular orientation and in higher mammals these neurons are grouped in orientation columns and orientation maps (Hubel & Wiesel 1977). In binocular animals, thalamic afferents segregate into ocular dominance columns (ODCs) in layer 4 (Hubel & Wiesel 1977) (**Figure 4a**). The development of ODCs and orientation tuning has been a model system to investigate mechanisms of development. Subplate ablation after thalamocortical axons have innervated layer 4, but before ODCs and orientation maps have formed, prevents the formation of ODCs and functional orientation maps, even though both thalamic fibers and layer 4 neurons are present (Ghosh & Shatz 1992a, Kanold et al. 2003) (**Figure 4b**). Single unit recordings show that subplate ablation prevents the acquisition of normal visual responses (Kanold et al. 2003). While a large fraction of neurons is unresponsive to visual stimuli after ablation, consistent with weak thalamocortical synapses, the remaining neurons show weak orientation tuning (Kanold et al. 2003). Since orientation-tuned responses and ODCs require the refinement of thalamocortical projections, these results suggest that the refinement of LGN projections to layer 4 did not occur.

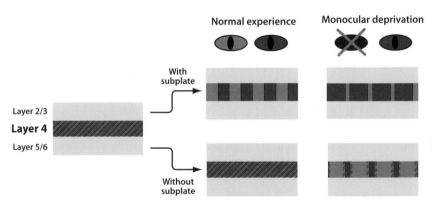

Figure 4

Subplate neurons are required for formation and normal plasticity of ocular dominance columns. A schematic of the development of ocular dominance columns (ODCs) under two conditions with and without subplate neurons: normal visual experience and monocular deprivation. Initially in development, thalamocortical projections representing the two eyes (*red* and *blue*) overlap. With normal experience, equally spaced ODCs emerge in V1 (*top left*). If one eye is closed during the critical period with subplate neurons present, then open eye projections expand and closed eye projections contract (*top right*). Without subplate neurons thalamocortical projections do not segregate, and no ODCs are observed (*bottom left*). If one eye is closed in the absence of subplate neurons, then deprived eye projections are retained and open eye projections are removed (*bottom right*).

Neuronal activity and normal sensory experience are required for ODCs and orientation maps to emerge (Crair et al. 1998, Hensch 2004, Reiter et al. 1986, Stryker & Harris 1986). Since ablation of subplate prevents the maturation of thalamocortical synapses, the visual cortex is decoupled from its inputs and thus deprived of visual inputs (Kanold et al. 2003). This deprivation prevents the emergence of the functional architecture of the visual cortex. Alternatively, computational modeling studies have suggested that the functional architecture of the cortex might develop in the subplate and be transferred into the developing cortical plate (Grossberg & Seitz 2003). If such a scenario were true, then subplate ablation would remove the organizational template in the subplate.

Thus, SPns contain molecular cues that direct thalamic axons to the right cortical area and also enable cortical neurons to respond to early spontaneous and later sensory evoked neuronal activity in order to develop the functional architecture of the cortex.

ROLE OF THE SUBPLATE IN DEVELOPMENTAL PLASTICITY

The lack of functional cortical organization following subplate ablation does not imply that sensory experience has no influence on cortical organization. Sensory imbalances such as monocular deprivation (MD) during early life, in particular during the critical period, are able to alter the functional organization of the cortex (Hensch 2004). Following MD, there is an expansion of the cortical territory innervated by thalamic projections representing the open eye, whereas there is a loss of projections representing the closed eye. This results in a OD shift toward the open eye (**Figure 4a**). The maturation of inhibition has been shown to be a crucial regulator in allowing OD shift to occur in the critical period (Hensch 2004). If inhibitory circuits are weakened, then no OD shift is observed (Hensch 2004). However, even though inhibitory circuits remain immature after subplate removal, sensory imbalances can change ODCs in a "paradoxical" manner (Kanold et al.

2003). In contrast to normal OD plasticity, after ablation concurrent with MD, thalamocortical projections representing the deprived eye are retained, whereas thalamocortical projections representing the open eye are removed (Kanold et al. 2003) (**Figure 4b**). Thus, mechanisms underlying OD plasticity still operate in the absence of SPns. A paradoxical shift of OD toward the less active eye is also observed in experiments where the cortex has been pharmacologically silenced (Hata et al. 1999, Hata & Stryker 1994).

Simulations using a computational model of ODC development (Kanold & Shatz 2006) based on circuits shown in **Figure 3a** suggest that decorrelation of thalamic and cortical activity after subplate removal can lead to such paradoxical shifts. One assumption of the model is that a spike-time–dependent learning rule (STDP) exists in layer 4 (Kanold & Shatz 2006). Cortical STDPs show a longer time window for synaptic depression (LTD) than for synaptic strengthening (LTP) (Abbott & Nelson 2000). Thus, if synaptic inputs are uncorrelated with cellular firing, a net weakening occurs and more active synapses become weakened than less active synapses. Thus open eye projections would be weakened more than closed eye projections and paradoxical plasticity results (Kanold & Shatz 2006). These simulations point to a key role of SPns in promoting the correlation of cortical activity with thalamic activity that enables the strengthening and refinement of thalamocortical connections by Hebbian learning rules such as STDP (Kanold & Shatz 2006). SPns can promote such correlations by providing excitatory inputs to layer 4 (Finney et al. 1998, Zhao et al. 2009) and also by controlling the balance of excitation and inhibition within layer 4 (Kanold & Shatz 2006).

Therefore, by controlling the maturation of excitatory and inhibitory circuits, SPns enable cortical circuits to reorganize correctly following sensory manipulations. The disappearance of SPns over development might restrict this ability and might restrict the observed circuit plasticity to a limited critical period.

CONSEQUENCES OF EARLY HYPOXIA, ISCHEMIA, ETC., ON SUBPLATE (DYS-)FUNCTION

Animal studies and clinical evidence indicate that the subplate is critically involved in various brain developmental disorders including cerebral palsy, periventricular leukomalacia (PVL), autism, schizophrenia, and epilepsy. An enhanced vulnerability of subplate neurons to early hypoxia-ischemia resulting in PVL has been documented in neonatal rats (Csillik et al. 2002, McQuillen et al. 2003). In vitro electrophysiological recordings in neocortical slices from newborn rats have demonstrated a pronounced functional impairment of SPns following a combined oxygen and glucose deprivation (Albrecht et al. 2005). In humans, the peak of subplate development coincides with the gestational age of highest vulnerability to perinatal brain injury in the premature infant (McQuillen & Ferriero 2005). It has been postulated that the second trimester represents the "window of vulnerability" for selective subplate injury and that defects in prefrontal cortical regions are related to schizophrenia (Bunney et al. 1997). An immunohistochemical analysis on neonatal telencephalon samples obtained postmortem from infants with white matter lesions and born at 25–32 weeks of gestation has shown a significant loss of GABAergic SPns, indicating that this subpopulation of SPns may be more vulnerable to perinatal systemic insults (Robinson et al. 2006). The mechanisms of SPn selective susceptibility are unknown, but their unique molecular, structural and functional properties may well explain this vulnerability. SPns express glutamate receptors at the earliest stages (**Table 3**), receive functional glutamatergic synaptic inputs (**Figure 2b**), and possess NMDA receptors that can be activated at resting membrane potentials (Hirsch & Luhmann 2008).

Disturbances in the programmed cell death of SPns may also cause long-term neurological deficits. It has been proposed that cortical dysplasia associated with pharmaco-resistant epilepsy could be the consequence of postnatal retention of some SPns (Cepeda et al. 2007). An increased density of interstitial cells in the white matter have been found in the frontal and temporal cortex of schizophrenic patients (Kirkpatrick et al. 1999) and this has been attributed to alterations in the pattern of programmed cell death (Akbarian et al. 1996). If these surviving SPns maintain their extensive local and long-range synaptic connections, they may disturb cortical processing (Bunney & Bunney 2000) or may function as pacemaker regions for the generation of epileptic activity (Luhmann et al. 2003) via their feed forward excitatory projections (**Figure 3a**) (Zhao et al. 2009).

SUMMARY AND PERSPECTIVES

Accumulating evidence points to a key role of SPns in neocortical development. The number of SPns increases with increasing brain complexity. Not only is the subplate larger in primates than in rodents, it also persists for a much longer developmental period, the longest being in humans. To date, investigations of SPns have mostly focused on their role in thalamocortical processing. However, SPns also project back to the thalamus and to the opposite hemisphere, but the function of the corticothalamic and callosal projections has not been investigated to date.

Unfortunately, despite their demonstrated importance, SPns are woefully understudied. For example, the role of the various neuromodulatory systems innervating the subplate at earliest stages is mostly unknown. The lack of information about these neurons might derive from the fact that they are only present in very young animals and that they are located deep in the brain and hence not easily accessible. In rodents the subplate is only very thin, making an analysis or manipulation of the subplate difficult. In addition, manipulations of subplate function have to be precisely targeted to avoid affecting other cortical neurons that migrate through the subplate. Thus, better selective markers to specifically target subplate neurons are needed. Genetic profiling of SPns is the first step to unequivocally identify these neurons and to categorize subpopulations of SPns. The search for SPn-specific genetic markers has recently begun (Hoerder-Suabedissen et al. 2009, Osheroff & Hatten 2009) and most likely will open the door for new experimental strategies to better understand the role of the subplate in the developing and mature cerebral cortex. It is even now possible to immunopurify subplate neurons for in vitro cellular, molecular, and physiological studies of synaptogenesis and to study mechanisms of subplate neuron death (McKeller & Shatz 2009, DeFreitas et al. 2001). Such studies might also contribute to the urgently needed development of genetic techniques to silence or activate subplate neurons to investigate their role in cortical development.

One key question that needs exploration is why SPns exist in the first place and why and how do they die. Ablation data show that SPns are required for the functional maturation and plasticity of thalamocortical connections. However, in many other areas of the brain (such as the thalamus) connections mature and refine without the aid of a transient cell population. Thus the function of the subplate might be related to unique properties of the neocortex, such as its radial organization and increased lateral connectivity. It might be that subplate neurons are needed to generate and control activity patterns to set up areas that show a large-scale systematic organization (such as orientation maps).

Another unsolved question is why a larger fraction of SPns seem to survive in rodents versus carnivores and primates. Whereas in the mature rodent neocortex layer 6b neurons play an important function in cortical processing, surviving SPns in the adult human cortex seem to be involved in pathophysiological disturbances such as epilepsy and schizophrenia. It may be most relevant clinically to identify the genetic disorders and the environmental risk factors that cause SPn dysfunction during early developmental stages.

In summary, subplate neurons are closely intertwined with the developing cortical circuit and play key roles at multiple stages of development to ensure normal emergence of the complex circuitry of the cerebral cortex.

DISCLOSURE STATEMENT

The authors are not aware of any affiliations, memberships, funding, or financial holdings that might be perceived as affecting the objectivity of this review.

ACKNOWLEDGMENTS

P.O.K. is supported by NIDCD R01DC009607, NIDCD R21DC009454, and the International Cerebral Palsy Research Association. H.J.L. received support from the DFG, the EC (LSH-CT-2006-037315, EPICURE), and the Schram Stiftung.

LITERATURE CITED

Abbott LF, Nelson SB. 2000. Synaptic plasticity: taming the beast. *Nat. Neurosci.* 3(Suppl.):1178–83

Aboitiz F. 1999. Evolution of isocortical organization. A tentative scenario including roles of reelin, p35/cdk5 and the subplate zone. *Cereb. Cortex* 9:655–61

Aboitiz F, Montiel J. 2007. Origin and evolution of the vertebrate telencephalon, with special reference to the mammalian neocortex. *Adv. Anat. Embryol. Cell Biol.* 193:1–112

Aboitiz F, Montiel J, Garcia RR. 2005. Ancestry of the mammalian preplate and its derivatives: evolutionary relicts or embryonic adaptations? *Rev. Neurosci.* 16:359–76

Achilles K, Okabe A, Ikeda M, Shimizu-Okabe C, Yamada J, et al. 2007. Kinetic properties of Cl uptake mediated by Na^+-dependent K^+-2Cl cotransport in immature rat neocortical neurons. *J. Neurosci.* 27(32):8616–27

Aguado F, Carmona MA, Pozas E, Aguilo A, Martinez-Guijarro FJ, et al. 2003. BDNF regulates spontaneous correlated activity at early developmental stages by increasing synaptogenesis and expression of the K$^+$/Cl$^-$ cotransporter KCC2. *Development* 130:1267–80

Akbarian S, Kim JJ, Potkin SG, Hetrick WP, Bunney WE Jr, Jones EG. 1996. Maldistribution of interstitial neurons in prefrontal white matter of the brains of schizophrenic patients. *Arch. Gen. Psychiatry* 53(5):425–36

Albrecht J, Hanganu IL, Heck N, Luhmann HJ. 2005. In vitro ischemia induced dysfunction in the somatosensory cortex of the newborn rat. *Eur. J. Neurosci.* 22:2295–305

Al-Ghoul WM, Miller MW. 1989. Transient expression of Alz-50 immunoreactivity in developing rat neocortex: a marker for naturally occurring neuronal death? *Brain Res.* 481:361–67

Allendoerfer KL, Shatz CJ. 1994. The subplate, a transient neocortical structure: its role in the development of connections between thalamus and cortex. *Annu. Rev. Neurosci.* 17:185–218

Allendoerfer KL, Shelton DL, Shooter EM, Shatz CJ. 1990. Nerve growth factor receptor immunoreactivity is transiently associated with the subplate neurons of the mammalian cerebral cortex. *Proc. Natl. Acad. Sci. USA* 87(1):187–90

Antonini A, Shatz CJ. 1990. Relation between putative transmitter phenotypes and connectivity of subplate neurons during cerebral cortical development. *Eur. J. Neurosci.* 2:744–61

Arias MS, Baratta J, Yu J, Robertson RT. 2002. Absence of selectivity in the loss of neurons from the developing cortical subplate of the rat. *Dev. Brain Res.* 139(2):331–35

Arimatsu Y, Ishida M, Kaneko T, Ichinose S, Omori A. 2003. Organization and development of corticocortical associative neurons expressing the orphan nuclear receptor Nurr1. *J. Comp. Neurol.* 466(2):180–96

Bayer L, Serafin M, Eggermann E, Saint-Mleux B, Machard D, et al. 2004. Exclusive postsynaptic action of hypocretin-orexin on sublayer 6b cortical neurons. *J. Neurosci.* 24(30):6760–64

Bayer SA, Altman J. 1990. Development of layer I and the subplate in the rat neocortex. *Exp. Neurol.* 107:48–62

Bayer SA, Altman J, Russo RJ, Zhang X. 1993. Timetables of neurogenesis in the human brain based on experimentally determined patterns in the rat. *Neurotoxicology* 14:83–144

Bicknese AR, Sheppard AM, O'Leary DD, Pearlman AL. 1994. Thalamocortical axons extend along a chondroitin sulfate proteoglycan-enriched pathway coincident with the neocortical subplate and distinct from the efferent path. *J. Neurosci.* 14(6):3500–10

Blaesse P, Airaksinen MS, Rivera C, Kaila K. 2009. Cation-chloride cotransporters and neuronal function. *Neuron* 61:820–38

Bunney BG, Potkin SG, Bunney WE. 1997. Neuropathological studies of brain tissue in schizophrenia. *J. Psychiatr. Res.* 31(2):159–73

Bunney WE, Bunney BG. 2000. Evidence for a compromised dorsolateral prefrontal cortical parallel circuit in schizophrenia. *Brain Res. Rev.* 31(2–3):138–46

Bystron I, Blakemore C, Rakic P. 2008. Development of the human cerebral cortex: Boulder Committee revisited. *Nat. Rev. Neurosci.* 9(2):110–22

Calarco CA, Robertson RT. 1995. Development of basal forebrain projections to visual cortex: DiI studies in rat. *J. Comp. Neurol.* 354:608–26

Castren E, Berninger B, Leingartner A, Lindholm D. 1998. Regulation of brain-derived neurotrophic factor mRNA levels in hippocampus by neuronal activity. *Prog. Brain Res.* 117:57–64

Catalano SM, Robertson RT, Killackey HP. 1991. Early ingrowth of thalamocortical afferents to the neocortex of the prenatal rat. *Proc. Natl. Acad. Sci. USA* 88:2999–3003

Cepeda C, Andre VM, Wu N, Yamazaki I, Uzgil B, et al. 2007. Immature neurons and GABA networks may contribute to epileptogenesis in pediatric cortical dysplasia. *Epilepsia* 48(Suppl. 5):79–85

Chun JJ, Shatz CJ. 1988a. A fibronectin-like molecule is present in the developing cat cerebral cortex and is correlated with subplate neurons. *J. Cell Biol.* 106:857–72

Chun JJ, Shatz CJ. 1988b. Redistribution of synaptic vesicle antigens is correlated with the disappearance of a transient synaptic zone in the developing cerebral cortex. *Neuron* 1(4):297–310

Chun JJ, Shatz CJ. 1989a. Interstitial cells of the adult neocortical white matter are the remnant of the early generated subplate neuron population. *J. Comp. Neurol.* 282:555–69

Chun JJ, Shatz CJ. 1989b. The earliest-generated neurons of the cat cerebral cortex: characterization by MAP2 and neurotransmitter immunohistochemistry during fetal life. *J. Neurosci.* 9(5):1648–67

Chung L, Moore SD, Cox CL. 2009. Cholecystokinin action on layer 6b neurons in somatosensory cortex. *Brain Res.* 1282:10–19

Clancy B, Cauller LJ. 1999. Widespread projections from subgriseal neurons (layer VII) to layer I in adult rat cortex. *J. Comp. Neurol.* 407(2):275–86

Clancy B, Silva M, Friedlander MJ. 2001. Structure and projections of white matter neurons in the postnatal rat visual cortex. *J. Comp. Neurol.* 434(2):233–52

Crair MC, Gillespie DC, Stryker MP. 1998. The role of visual experience in the development of columns in cat visual cortex. *Science* 279:566–70

Cramer SW, Baggott C, Cain J, Tilghman J, Allcock B, et al. 2008. The role of cation-dependent chloride transporters in neuropathic pain following spinal cord injury. *Mol. Pain* 4:36

Csillik AE, Okuno E, Csillik B, Knyihar E, Vecsei L. 2002. Expression of kynurenine aminotransferase in the subplate of the rat and its possible role in the regulation of programmed cell death. *Cereb. Cortex* 12(11):1193–201

deAzevedo LC, Hedin-Pereira C, Lent R. 1997. Callosal neurons in the cingulate cortical plate and subplate of human fetuses. *J. Comp. Neurol.* 386:60–70

De Carlos JA, O'Leary DDM. 1992. Growth and targeting of subplate axons and establishment of major cortical pathways. *J. Neurosci.* 12:1194–211

DeFreitas MF, McQuillen PS, Shatz CJ. 2001. A novel p75NTR signaling pathway promotes survival, not death, of immunopurified neocortical subplate neurons. *J. Neurosci.* 21(14):5121–29

Delalle I, Evers P, Kostovic I, Uylings HB. 1997. Laminar distribution of neuropeptide Y-immunoreactive neurons in human prefrontal cortex during development. *J. Comp. Neurol.* 379(4):515–22

Del Río JA, Martínez A, Auladell C, Soriano E. 2000. Developmental history of the subplate and developing white matter in the murine neocortex. Neuronal organization and relationship with the main afferent systems at embryonic perinatal stages. *Cereb. Cortex* 10:784–801

Del Rio JA, Soriano E, Ferrer I. 1991. A transitory population of substance P-like immunoreactive neurones in the developing cerebral cortex of the mouse. *Dev. Brain Res.* 64:205–11

Deng J, Elberger AJ. 2003. Corticothalamic and thalamocortical pathfinding in the mouse: dependence on intermediate targets and guidance axis. *Anat. Embryol.* 207:177–92

Dunn JA, Kirsch JD, Naegele JR. 1995. Transient immunoglobulin-like molecules are present in the subplate zone and cerebral cortex during postnatal development. *Cereb. Cortex* 5:494–505

Dupont E, Hanganu IL, Kilb W, Hirsch S, Luhmann HJ. 2006. Rapid developmental switch in the mechanisms driving early cortical columnar networks. *Nature* 439:79–83

Dziegielewska KM, Møller JE, Potter AM, Ek J, Lane MA, Saunders NR. 2000. Acute-phase cytokines IL-1beta and TNF-alpha in brain development. *Cell Tissue Res.* 299(3):335–45

Erzurumlu RS, Jhaveri S. 1992. Emergence of connectivity in the embryonic rat parietal cortex. *Cereb. Cortex* 2:336–52

Fairén A, Smith-Fernández A, Martí E, DeDiego I, de la Rosa EJ. 1992. A transient immunoglobulin-like reactivity in the developing cerebral cortex of rodents. *NeuroReport* 3(10):881–84

Ferrer I, Bernet E, Soriano E, del Rio T, Fonseca M. 1990. Naturally occurring cell death in the cerebral cortex of the rat and removal of dead cells by transitory phagocytes. *Neuroscience* 39:451–58

Finney EM, Stone JR, Shatz CJ. 1998. Major glutamatergic projection from subplate into visual cortex during development. *J. Comp. Neurol.* 398(1):105–18

Fiumelli H, Cancedda L, Poo MM. 2005. Modulation of GABAergic transmission by activity via postsynaptic Ca^{2+}-dependent regulation of KCC2 function. *Neuron* 48:773–86

Flint AC, Liu XL, Kriegstein AR. 1998. Nonsynaptic glycine receptor activation during early neocortical development. *Neuron* 20(1):43–53

Friauf E, McConnell SK, Shatz CJ. 1990. Functional synaptic circuits in the subplate during fetal and early postnatal development of cat visual cortex. *J. Neurosci.* 10:2601–13

Friauf E, Shatz CJ. 1991. Changing patterns of synaptic input to subplate and cortical plate during development of visual cortex. *J. Neurophysiol.* 66:2059–71

Fukuchi-Shimogori T, Grove EA. 2001. Neocortex patterning by the secreted signaling molecule FGF8. *Science* 294:1071–74

Fukuda T, Kawano H, Ohyama K, Li HP, Takeda Y, et al. 1997. Immunohistochemical localization of neurocan and L1 in the formation of thalamocortical pathway of developing rats. *J. Comp. Neurol.* 382:141–52

Furuta A, Martin LJ. 1999. Laminar segregation of the cortical plate during corticogenesis is accompanied by changes in glutamate receptor expression. *J. Neurobiol.* 39(1):67–80

Ganguly K, Schinder AF, Wong ST, Poo M. 2001. GABA itself promotes the developmental switch of neuronal GABAergic responses from excitation to inhibition. *Cell* 105:521–32

Ghosh A, Antonini A, McConnell SK, Shatz CJ. 1990. Requirements of subplate neurons in the formation of thalamocortical connections. *Nature* 347:179–81

Ghosh A, Shatz CJ. 1992a. Involvement of subplate neurons in the formation of ocular dominance columns. *Science* 255:1441–43

Ghosh A, Shatz CJ. 1992b. Pathfinding and target selection by developing geniculocortical axons. *J. Neurosci.* 12:39–55

Ghosh A, Shatz CJ. 1994. Segregation of geniculocortical afferents during the critical period: a role for subplate neurons. *J. Neurosci.* 14:3862–80

Grossberg S, Seitz A. 2003. Laminar development of receptive fields, maps and columns in visual cortex: the coordinating role of the subplate. *Cereb. Cortex* 13:852–63

Hanganu IL, Kilb W, Luhmann HJ. 2001. Spontaneous synaptic activity of subplate neurons in neonatal rat somatosensory cortex. *Cereb. Cortex* 11:400–10

Hanganu IL, Kilb W, Luhmann HJ. 2002. Functional synaptic projections onto subplate neurons in neonatal rat somatosensory cortex. *J. Neurosci.* 22:7165–76

Hanganu IL, Luhmann HJ. 2004. Functional nicotinic acetylcholine receptors on subplate neurons in neonatal rat somatosensory cortex. *J. Neurophysiol.* 92(1):189–98

Hanganu IL, Okabe A, Lessmann V, Luhmann HJ. 2009. Cellular mechanisms of subplate-driven and cholinergic input-dependent network activity in the neonatal rat somatosensory cortex. *Cereb. Cortex* 19:89–105

Harman AM, Eastough NJ, Beazley LD. 1995. Development of the visual cortex in a wallaby—phylogenetic implications. *Brain Behav. Evol.* 45:138–52

Hata Y, Stryker MP. 1994. Control of thalamocortical afferent rearrangement by postsynaptic activity in developing visual cortex. *Science* 265:1732–35

Hata Y, Tsumoto T, Stryker MP. 1999. Selective pruning of more active afferents when cat visual cortex is pharmacologically inhibited. *Neuron* 22:375–81

Hensch TK. 2004. Critical period regulation. *Annu. Rev. Neurosci.* 27:549–79

Herrmann K, Antonini A, Shatz CJ. 1994. Ultrastructural evidence for synaptic interactions between thalamocortical axons and subplate neurons. *Eur. J. Neurosci.* 6:1729–42

Heuer H, Christ S, Friedrichsen S, Brauer D, Winckler M, et al. 2003. Connective tissue growth factor: a novel marker of layer VII neurons in the rat cerebral cortex. *Neuroscience* 119(1):43–52

Hevner RF. 2000. Development of connections in the human visual system during fetal mid-gestation: a DiI-tracing study. *J. Neuropathol. Exp. Neurol.* 59:385–92

Hevner RF, Neogi T, Englund C, Daza RA, Fink A. 2003. Cajal-Retzius cells in the mouse: transcription factors, neurotransmitters, and birthdays suggest a pallial origin. *Dev. Brain Res.* 141(1–2):39–53

Higashi S, Molnár Z, Kurotani T, Toyama K. 2002. Prenatal development of neural excitation in rat thalamocortical projections studied by optical recording. *Neuroscience* 115(4):1231–46

Higo S, Udaka N, Tamamaki N. 2007. Long-range GABAergic projection neurons in the cat neocortex. *J. Comp. Neurol.* 503(3):421–31

Hirsch S, Luhmann HJ. 2008. Pathway-specificity in N-methyl-d-aspartate receptor-mediated synaptic inputs onto subplate neurons. *Neuroscience* 153:1092–102

Hoerder-Suabedissen A, Wang WZ, Lee S, Davies KE, Goffinet A, et al. 2009. Novel markers reveal subpopulations of subplate neurons in the murine cerebral cortex. *Cereb. Cortex* 19:1738–50

Hubel DH, Wiesel TN. 1977. Ferrier lecture. Functional architecture of macaque monkey visual cortex. *Proc. R. Soc. London Ser. B* 198:1–59

Ina A, Sugiyama M, Konno J, Yoshida S, Ohmomo H, et al. 2007. Cajal-Retzius cells and subplate neurons differentially express vesicular glutamate transporters 1 and 2 during development of mouse cortex. *Eur. J. Neurosci.* 26(3):615–23

Judas M, Sestan N, Kostovic I. 1999. Nitrinergic neurons in the developing and adult human telencephalon: transient and permanent patterns of expression in comparison to other mammals. *Microsc. Res. Tech.* 45(6):401–19

Kageyama GH, Robertson RT. 1993. Development of geniculocortical projections to visual cortex in rat: evidence early ingrowth and synaptogenesis. *J. Comp. Neurol.* 335:123–48

Kalsbeek A, Voorn P, Buijs RM, Pool CW, Uylings HB. 1988. Development of the dopaminergic innervation in the prefrontal cortex of the rat. *J. Comp. Neurol.* 269(1):58–72

Kanold PO. 2009. Subplate neurons: crucial regulators of cortical development and plasticity. *Front. Neuroanat.* 3:16

Kanold PO. 2004. Transient microcircuits formed by subplate neurons and their role in functional development of thalamocortical connections. *NeuroReport* 15:2149–53

Kanold PO, Kara P, Reid RC, Shatz CJ. 2003. Role of subplate neurons in functional maturation of visual cortical columns. *Science* 301:521–25

Kanold PO, Shatz CJ. 2006. Subplate neurons regulate maturation of cortical inhibition and outcome of ocular dominance plasticity. *Neuron* 51:627–38

Kilb W, Hanganu IL, Okabe A, Sava BA, Shimizu-Okabe C, et al. 2008. Glycine receptors mediate excitation of subplate neurons in neonatal rat cerebral cortex. *J. Neurophysiol.* 100(2):698–707

Kim GJ, Shatz CJ, McConnell SK. 1991. Morphology of pioneer and follower growth cones in the developing cerebral cortex. *J. Neurobiol.* 22(6):629–42

Kirkpatrick B, Conley RC, Kakoyannis A, Reep RL, Roberts RC. 1999. Interstitial cells of the white matter in the inferior parietal cortex in schizophrenia: an unbiased cell-counting study. *Synapse* 34(2):95–102

Koh S, Higgins GA. 1991. Differential regulation of the low-affinity nerve growth factor receptor during postnatal development of the rat brain. *J. Comp. Neurol.* 313(3):494–508

Kordower JH, Mufson EJ. 1992. Nerve growth factor receptor-immunoreactive neurons within the developing human cortex. *J. Comp. Neurol.* 323(1):25–41

Kostovic I. 1986. Prenatal development of nucleus basalis complex and related fiber systems in man: a histochemical study. *Neuroscience* 17(4):1047–77

Kostovic I, Judas M. 2002. Correlation between the sequential ingrowth of afferents and transient patterns of cortical lamination in preterm infants. *Anat. Rec.* 267:1–6

Kostovic I, Judas M, Rados M, Hrabac P. 2002. Laminar organization of the human fetal cerebrum revealed by histochemical markers and magnetic resonance imaging. *Cereb. Cortex* 12(5):536–44

Kostovic I, Rakic P. 1980. Cytology and time of origin of interstitial neurons in the white matter in infant and adult human and monkey telencephalon. *J. Neurocytol.* 9:219–42

Kostovic I, Rakic P. 1984. Development of prestriate visual projections in the monkey and human fetal cerebrum revealed by transient cholinesterase staining. *J. Neurosci.* 4:25–42

Kostovic I, Rakic P. 1990. Developmental history of the transient subplate zone in the visual and somatosensory cortex of the macaque monkey and human brain. *J. Comp. Neurol.* 297:441–70

Kostovic I, Stefulj-Fucic A, Mrzljak L, Jukic S, Delalle I. 1991. Prenatal and perinatal development of the somatostatin-immunoreactive neurons in the human prefrontal cortex. *Neurosci. Lett.* 124:153–56

Kriegstein AR, Owens DF. 2001. GABA may act as a self-limiting trophic factor at developing synapses. *Sci. STKE* 2001:PE1

Kwan KY, Lam MM, Krsnik Z, Kawasawa YI, Lefebvre V, Sestan N. 2008. SOX5 postmitotically regulates migration, postmigratory differentiation, and projections of subplate and deep-layer neocortical neurons. *Proc. Natl. Acad. Sci. USA* 105(41):16021–26

Lauder JM, Han VK, Henderson P, Verdoorn T, Towle AC. 1986. Prenatal ontogeny of the GABAergic system in the rat brain: an immunocytochemical study. *Neuroscience* 19:465–93

Lein ES, Finney EM, McQuillen PS, Shatz CJ. 1999. Subplate neuron ablation alters neurotrophin expression and ocular dominance column formation. *Proc. Natl. Acad. Sci. USA* 96:13491–95

Lein ES, Hohn A, Shatz CJ. 2000. Dynamic regulation of BDNF and NT-3 expression during visual system development. *J. Comp. Neurol.* 420:1–18

Liu F-C, Graybiel AM. 1992. Transient calbindin-D_{28K}-positive systems in the telencephalon: ganglionic eminence, developing striatum and cerebral cortex. *J. Neurosci.* 12:674–90

López V, Wagner CK. 2009. Progestin receptor is transiently expressed perinatally in neurons of the rat isocortex. *J. Comp. Neurol.* 512(1):124–39

Ludwig A, Li H, Saarma M, Kaila K, Rivera C. 2003. Developmental up-regulation of KCC2 in the absence of GABAergic and glutamatergic transmission. *Eur. J. Neurosci.* 18:3199–206

Luhmann HJ, Hanganu IL, Kilb W. 2003. Cellular physiology of the neonatal rat cerebral cortex. *Brain Res. Bull.* 60(4):345–53

Luhmann HJ, Kilb W, Hanganu-Opatz IL. 2009. Subplate cells: amplifiers of neuronal activity in the developing cerebral cortex. *Front. Neuroanat.* 3:19

Luhmann HJ, Reiprich RA, Hanganu I, Kilb W. 2000. Cellular physiology of the neonatal rat cerebral cortex: intrinsic membrane properties, sodium and calcium currents. *J. Neurosci. Res.* 62:574–84

Luskin MB, Shatz CJ. 1985. Studies of the earliest generated cells of the cat's visual cortex: cogeneration of subplate and marginal zones. *J. Neurosci.* 5:1062–75

Mackarehtschian K, Lau CK, Caras I, McConnell SK. 1999. Regional differences in the developing cerebral cortex revealed by ephrin-A5 expression. *Cereb. Cortex* 9(6):601–10

Maric D, Liu QY, Maric I, Chaudry S, Chang YH, et al. 2001. GABA expression dominates neuronal lineage progression in the embryonic rat neocortex and facilitates neurite outgrowth via GABA$_A$ autoreceptor/Cl$^-$ channels. *J. Neurosci.* 21(7):2343–60

Marotte LR, Sheng X. 2000. Neurogenesis and identification of developing layers in the visual cortex of the wallaby (*Macropus eugenii*). *J. Comp. Neurol.* 416:131–42

McConnell SK, Ghosh A, Shatz CJ. 1989. Subplate neurons pioneer the first axon pathway from the cerebral cortex. *Science* 245:978–82

McConnell SK, Ghosh A, Shatz CJ. 1994. Subplate pioneers and the formation of descending connections from cerebral cortex. *J. Neurosci.* 14:1892–907

McKellar CE, Shatz CJ. 2009. Synaptogenesis in purified subplate neurons. *Cereb. Cortex.* 19(8):1723–37

McQuillen PS, DeFreitas MF, Zada G, Shatz CJ. 2002. A novel role for p75NTR in subplate growth cone complexity and visual thalamocortical innervation. *J. Neurosci.* 22(9):3580–93

McQuillen PS, Ferriero DM. 2005. Perinatal subplate neuron injury: implications for cortical development and plasticity. *Brain Pathol.* 15:250–60

McQuillen PS, Sheldon RA, Shatz CJ, Ferriero DM. 2003. Selective vulnerability of subplate neurons after early neonatal hypoxia-ischemia. *J. Neurosci.* 23(8):3308–15

Mechawar N, Descarries L. 2001. The cholinergic innervation develops early and rapidly in the rat cerebral cortex: a quantitative immunocytochemical study. *Neuroscience* 108(4):555–67

Mehra RD, Hendrickson AE. 1993. A comparison of the development of neuropeptide and MAP2 immunocytochemical labeling in the macaque visual cortex during pre- and postnatal development. *J. Neurobiol.* 24:101–24

Meinecke DL, Rakic P. 1993. Low-affinity p75 nerve growth factor receptor expression in the embryonic monkey telencephalon: timing and localization in diverse cellular elements. *Neuroscience* 54(1):105–16

Miller B, Sheppard AM, Bicknese AR, Pearlman AL. 1995. Chondroitin sulfate proteoglycans in the developing cerebral cortex: the distribution of neurocan distinguishes forming afferent and efferent axonal pathways. *J. Comp. Neurol.* 355(4):615–28

Molinaro G, Battaglia G, Riozzi B, Di Menna L, Rampello L, et al. 2009. Memantine treatment reduces the expression of the K$^+$/Cl$^-$ cotransporter KCC2 in the hippocampus and cerebral cortex, and attenuates behavioural responses mediated by GABA(A) receptor activation in mice. *Brain Res.* 1265:75–79

Molliver ME. 1982. Role of monoamines in the development of the neocortex. *Neurosci. Res. Program Bull.* 20(4):492–507

Molnár Z, Cordery P. 1999. Connections between cells of the internal capsule, thalamus, and cerebral cortex in embryonic rat. *J. Comp. Neurol.* 413:1–25

Molnár Z, Knott GW, Blakemore C, Saunders NR. 1998. Development of thalamocortical projections in the South American gray short-tailed opossum (*Monodelphis domestica*). *J. Comp. Neurol.* 398:491–514

Molnár Z, Kurotani T, Higashi S, Yamamoto N, Toyama K. 2003. Development of functional thalamocortical synapses studied with current source-density analysis in whole forebrain slices in the rat. *Brain Res. Bull.* 60(4):355–71

Molnár Z, Metin C, Stoykova A, Tarabykin V, Price DJ, et al. 2006. Comparative aspects of cerebral cortical development. *Eur. J. Neurosci.* 23(4):921–34

Molyneaux BJ, Arlotta P, Menezes JR, Macklis JD. 2007. Neuronal subtype specification in the cerebral cortex. *Nat. Rev. Neurosci.* 8(6):427–37

Moore AR, Filipovic R, Mo Z, Rasband MN, Zecevic N, Antic SD. 2009. Electrical excitability of early neurons in the human cerebral cortex during the second trimester of gestation. *Cereb. Cortex* 19:1795–805

Morishita H, Murata Y, Esumi S, Hamada S, Yagi T. 2004. CNR/Pcdhalpha family in subplate neurons, and developing cortical connectivity. *NeuroReport* 15(17):2595–99

Mountcastle VB. 1997. The columnar organization of the neocortex. *Brain* 120(Pt. 4):701–22

Mrzljak L, Uylings HB, Kostovic I, Van Eden CG. 1988. Prenatal development of neurons in the human prefrontal cortex. I. A qualitative Golgi study. *J. Comp. Neurol.* 271:355–86

Mrzljak L, Uylings HB, Kostovic I, Van Eden CG. 1992. Prenatal development of neurons in the human prefrontal cortex. II. A quantitative Golgi study. *J. Comp. Neurol.* 316:485–96

Osheroff H, Hatten ME. 2009. Gene expression profiling of preplate neurons destined for the subplate: genes involved in transcription, axon extension, neurotransmitter regulation, steroid hormone signaling, and neuronal survival. *Cereb. Cortex* 19(Suppl. 1):i126–34

Pearce AR, Marotte LR. 2003. The first thalamocortical synapses are made in the cortical plate in the developing visual cortex of the wallaby (*Macropus eugenii*). *J. Comp. Neurol.* 461:205–16

Piñon MC, Jethwa A, Jacobs E, Campagnoni A, Molnár Z. 2009. Dynamic integration of subplate neurons into the cortical barrel field circuitry during postnatal development in the Golli-tau-eGFP (GTE) mouse. *J. Physiol.* 587:1903–15

Price DJ, Aslam S, Tasker L, Gillies K. 1997. Fates of the earliest generated cells in the developing murine neocortex. *J. Comp. Neurol.* 377:414–22

Rakic P. 1988. Specification of cerebral cortical areas. *Science* 241(4862):170–76

Reep RL. 2000. Cortical layer VII and persistent subplate cells in mammalian brains. *Brain Behav. Evol.* 56:212–34

Reiter HO, Waitzman DM, Stryker MP. 1986. Cortical activity blockade prevents ocular dominance plasticity in the kitten visual cortex. *Exp. Brain Res.* 65:182–88

Reynolds ML, Cavanagh ME, Dziegielewska KM, Hinds LA, Saunders NR, Tyndale-Biscoe CH. 1985. Postnatal development of the telencephalon of the tammar wallaby (*Macropus eugenii*). An accessible model of neocortical differentiation. *Anat. Embryol.* 173:81–94

Rivera C, Li H, Thomas-Crusells J, Lahtinen H, Viitanen T, et al. 2002. BDNF-induced TrkB activation down-regulates the K^+-Cl^- cotransporter KCC2 and impairs neuronal Cl^- extrusion. *J. Cell Biol.* 159:747–52

Rivera C, Voipio J, Thomas-Crusells J, Li H, Emri Z, et al. 2004. Mechanism of activity-dependent down-regulation of the neuron-specific K-Cl cotransporter KCC2. *J. Neurosci.* 24:4683–91

Robertson RT, Annis CM, Baratta J, Haraldson S, Ingeman J, et al. 2000. Do subplate neurons comprise a transient population of cells in developing neocortex of rats? *J. Comp. Neurol.* 426(4):632–50

Robinson S, Li Q, Dèchant A, Cohen ML. 2006. Neonatal loss of gamma-aminobutyric acid pathway expression after human perinatal brain injury. *J. Neurosurg.* 104(6 Suppl.):396–408

Samuelsen GB, Larsen KB, Bogdanovic N, Laursen H, Graem N, et al. 2003. The changing number of cells in the human fetal forebrain and its subdivisions: a stereological analysis. *Cereb. Cortex* 13:115–22

Schröder H, Schütz U, Burghaus L, Lindstrom J, Kuryatov A, et al. 2001. Expression of the alpha4 isoform of the nicotinic acetylcholine receptor in the fetal human cerebral cortex. *Dev. Brain Res.* 132:33–45

Shimizu-Okabe C, Okabe A, Kilb W, Sato K, Luhmann HJ, Fukuda A. 2007. Changes in the expression of cation-Cl^- cotransporters, NKCC1 and KCC2, during cortical malformation induced by neonatal freeze-lesion. *Neurosci. Res.* 59:288–95

Shimogori T, Grove EA. 2005. Fibroblast growth factor 8 regulates neocortical guidance of area-specific thalamic innervation. *J. Neurosci.* 25:6550–60

Stryker MP, Harris WA. 1986. Binocular impulse blockade prevents the formation of ocular dominance columns in cat visual cortex. *J. Neurosci.* 6:2117–33

Sun JJ, Luhmann HJ. 2007. Spatio-temporal dynamics of oscillatory network activity in the neonatal mouse cerebral cortex. *Eur. J. Neurosci.* 26(7):1995–2004

Syken J, Grandpre T, Kanold PO, Shatz CJ. 2006. PirB restricts ocular-dominance plasticity in visual cortex. *Science* 313:1795–800

Tissir F, Wang CE, Goffinet AM. 2004. Expression of the chemokine receptor Cxcr4 mRNA during mouse brain development. *Dev. Brain Res.* 149(1):63–71

Tomioka R, Okamoto K, Furuta T, Fujiyama F, Iwasato T, et al. 2005. Demonstration of long-range GABAergic connections distributed throughout the mouse neocortex. *Eur. J. Neurosci.* 21(6):1587–600

Torres-Reveron J, Friedlander MJ. 2007. Properties of persistent postnatal cortical subplate neurons. *J. Neurosci.* 27(37):9962–74

Turrigiano G. 2007. Homeostatic signaling: the positive side of negative feedback. *Curr. Opin. Neurobiol.* 17:318–24

Ulfig N. 2002. Calcium-binding proteins in the human developing brain. *Adv. Anat. Embryol. Cell Biol.* 165:1–92

Voigt T, Opitz T, De Lima AD. 2001. Synchronous oscillatory activity in immature cortical network is driven by GABAergic preplate neurons. *J. Neurosci.* 21(22):8895–905

Wagner CK. 2008. Progesterone receptors and neural development: a gap between bench and bedside? *Endocrinology* 149(6):2743–49

Wahle P, Lübke J, Naegele JR. 1994. Inverted pyramidal neurons and interneurons in cat cortical subplate zone are labeled by monoclonal antibody SP1. *Eur. J. Neurosci.* 6:1167–78

Wickman K, Karschin C, Karschin A, Picciotto MR, Clapham DE. 2000. Brain localization and behavioral impact of the G-protein-gated K$^+$ channel subunit GIRK4. *J. Neurosci.* 20(15):5608–15

Winzer-Serhan UH, Leslie FM. 2005. Expression of alpha5 nicotinic acetylcholine receptor subunit mRNA during hippocampal and cortical development. *J. Comp. Neurol.* 481(1):19–30

Wood JG, Martin S, Price DJ. 1992. Evidence that the earliest generated cells of the murine cerebral cortex form a transient population in the subplate and marginal zone. *Brain Res. Dev. Brain Res.* 66:137–40

Yamada J, Okabe A, Toyoda H, Kilb W, Luhmann HJ, Fukuda A. 2004. Cl$^-$ uptake promoting depolarizing GABA actions in immature rat neocortical neurones is mediated by NKCC1. *J. Physiol.* 557(Pt. 3):829–41

Yang JW, Hanganu-Opatz IL, Sun JJ, Luhmann HJ. 2009. Three patterns of oscillatory activity differentially synchronize developing neocortical networks in vivo. *J. Neurosci.* 29:9011–25

Yun ME, Johnson RR, Antic A, Donoghue MJ. 2003. EphA family gene expression in the developing mouse neocortex: regional patterns reveal intrinsic programs and extrinsic influence. *J. Comp. Neurol.* 456(3):203–16

Zecevic N, Milosevic A. 1997. Initial development of gamma-aminobutyric acid immunoreactivity in the human cerebral cortex. *J. Comp. Neurol.* 380:495–506

Zeng SJ, Lin YT, Tian CP, Song KJ, Zhang XW, Zuo MX. 2009. Evolutionary significance of delayed neurogenesis in the core versus shell auditory areas of *Mus musculus*. *J. Comp. Neurol.* 515:600–13

Zhao C, Kao JP, Kanold PO. 2009. Functional excitatory microcircuits in neonatal cortex connect thalamus and layer 4. *J. Neurosci.* 29(49):15479–88

Fly Motion Vision

Alexander Borst, Juergen Haag, and Dierk F. Reiff

Department of Systems and Computational Neurobiology, Max-Planck-Institute of Neurobiology, Martinsried, Germany; email: borst@neuro.mpg.de

Annu. Rev. Neurosci. 2010. 33:49–70

First published online as a Review in Advance on March 12, 2010

The *Annual Review of Neuroscience* is online at neuro.annualreviews.org

This article's doi:
10.1146/annurev-neuro-060909-153155

0147-006X/10/0721-0049$20.00

Key Words

optic flow, ego-motion, motion detection, receptive field, lobula plate, tangential cell

Abstract

Fly motion vision and resultant compensatory optomotor responses are a classic example for neural computation. Here we review our current understanding of processing of optic flow as generated by an animal's self-motion. Optic flow processing is accomplished in a series of steps: First, the time-varying photoreceptor signals are fed into a two-dimensional array of Reichardt-type elementary motion detectors (EMDs). EMDs compute, in parallel, local motion vectors at each sampling point in space. Second, the output signals of many EMDs are spatially integrated on the dendrites of large-field tangential cells in the lobula plate. In the third step, tangential cells form extensive interactions with each other, giving rise to their large and complex receptive fields. Thus, tangential cells can act as matched filters tuned to optic flow during particular flight maneuvers. They finally distribute their information onto postsynaptic descending neurons, which either instruct the motor centers of the thoracic ganglion for flight and locomotion control or act themselves as motor neurons that control neck muscles for head movements.

Contents

INTRODUCTION

In an animal's daily life, visual motion is abundant. Motion occurs, for example, when another animal, be it predator or prey, is moving in the observer's environment. Its motion usually increases the saliency of the moving animal, attracting the attention of the observer to the patch of the image where motion occurred. Therefore, many species developed a particular pattern of locomotion during which the episodes of self motion are as short as possible, with the animal being frozen in between, as if the animal knows how its motion takes off the magic hood provided otherwise by its camouflaging body pattern. Besides attracting visual attention, motion cues can segregate objects from background and indicate what is moving: The man in the dark with light bulbs on the joints of his arms and legs is a noninterpretable collection of points as long as he is at rest, but he becomes a man as soon as he starts walking (Johansson 1973). Although these two examples include a passive observer, motion cues also occur when the observer itself is actively moving. Then, the behavior of the observer largely determines the sensory input and causes the whole image to move across the observer's retinae. The resulting distribution

of motion vectors is called optic flow (Gibson 1950). Optic flow depends on two things: first, on the observer's type of movement in three-dimensional (3D) space, and second, on the 3D structure of the environment in which the observer is moving (Koenderink & van Dorn 1987). However, the simultaneous extraction of exact information about the ego-motion and the structure of the environment from the optic flow represents an ill-posed problem because both parameters mutually depend on each other. Nevertheless, organisms seem to make meaningful assumptions on both aspects. They interpret a particular optic flow in terms of ego-motion as well as in terms of the environment's 3D structure. An expanding flow field with the pole of expansion in front can signal forward motion of the animal with an impending collision (Braitenberg & Taddei Ferretti 1966; Borst & Bahde 1988a,b; Rind & Simmons 1992; Hatsopoulos et al. 1995). Similarly, a high-velocity patch embedded in a low-velocity surround may indicate a nearby object in front of a more distant background (Reichardt & Poggio 1979).

Neural mechanisms underlying the analysis of optic flow may have evolved particularly well in animals that move fast and have poor spatial vision, leaving them with motion vision as their primary source of visual information. These considerations make flies favorable subjects in which to study optic flow processing. Moreover, the fly's nervous system contains only a few hundred thousand neurons as opposed to billions and more found in the vertebrate central nervous system. This simplicity makes circuit analysis in the fly's nervous system, compared with a vertebrate system, a more manageable task, at least to some extent. Finally, the combination of physiological recording and genetic manipulation of neuronal function has now been established in the fruit fly *Drosophila melanogaster*. Thus, activity recording as well as behavioral studies can be combined with interfering with the fly's nervous system using sophisticated genetic tools.

OPTOMOTOR RESPONSE AND THE ELEMENTARY MOTION DETECTOR

The optomotor response represents the behavioral paradigm that has most influenced the study of insect motion vision. When a fly is tethered in the center of a striped drum (**Figure 1a**) and the drum is rotating clockwise, the insect tries to turn clockwise, too. When the drum is moving in the opposite direction, the insect turns counterclockwise (**Figure 1b**). Thus, the optomotor response consists of a following reaction, syndirectional with the motion of the surround, that builds up slowly over several seconds. Measuring it in a tethered animal that cannot move its head offers two advantages: First, it isolates the visual response component from the proprioceptive, vestibular one; second, it allows investigators to allocate the visual motion stimulus on the animal's retina with ultimate precision. This approach was pioneered by Hassenstein & Reichardt (1956), who analyzed the turning tendency of the beetle *Chlorophanus viridis* walking on a spherical Y-maze built from straws (**Figure 1c**). Their experiments finally led them to propose a specific model of elementary motion detection that accounts for all their observations in a quantitative way. This algorithmic model for elementary motion detection consists of two subunits, which are mirror-symmetrical to each other (**Figure 1d**) (Reichardt 1961, 1987; Borst & Egelhaaf 1989; Borst 2007). Each subunit reads the luminance values measured in two adjacent ommatidia and multiplies them after one has been processed (i.e., delayed) by a low-pass filter. The output values of both subunits finally become subtracted. In contrast with a simple speedometer, whose output linearly increases with image speed, the model predicts a speed optimum at which the response is maximal. This optimum speed is set by the time constant of the low-pass filter. Beyond the optimum speed, the response decreases again. Furthermore, the optimum speed is a linear function of the pattern wavelength. Thus, optimum speed divided by pattern wavelength remains constant. The dimension of this ratio is a temporal frequency.

Thus, the Reichardt detector responds maximally to a certain number of spatial periods passing by a single photoreceptor, not to a certain image speed. In more general terms, the model predicts a highly counterintuitive dependence of the motion-detection process on pattern properties, such as its spatial wavelength as well as its contrast. Following these seminal studies of Hassenstein and Reichardt, sophisticated devices, such as the torque meter (**Figure 1e**)(Goetz 1964), the wing beat analyzer (Goetz 1987), or a patterned Styrofoam ball, the movement of which was automatically detected (Buchner 1976), were introduced to measure the insect's turning tendency in flight or during walking. Using these kinds of setups, a number of experiments were performed showing the Reichardt detector to underlie motion vision in houseflies (*Musca domestica*; Fermi & Reichardt 1963, Eckert 1973) and fruit flies (*Drosophila melanogaster*; Goetz 1964, 1965). The optimum temporal frequency of both species was determined at ~1 Hz, which allows investigators to infer the time constant of the filters involved. To assess the sampling base of the elementary motion detector, sine-gratings of different wavelengths were used: For pattern wavelengths smaller than twice the sampling base, an inversion of the response (spatial aliasing) is expected (Goetz 1964). Determining the wavelength at which the response becomes negative revealed a sampling base of ~2 degrees in *Musca* (Eckert 1973) and 4.6 degrees for *Drosophila* (Goetz 1964, 1965; Buchner 1976). This conclusion fits exactly the interommatidial angle of each of the two species. Thus, nearest-neighbor interactions within the retina form the input to the Reichardt detector in the fly.

Although the optomotor response of tethered insects proved to be seminal for the discovery of the elementary motion-detection process, its role in free flight is more complex and more difficult to address. Under free-flight conditions, experimental parameters are less well defined, and multiple sensory inputs are integrated. In addition to the visual input, the

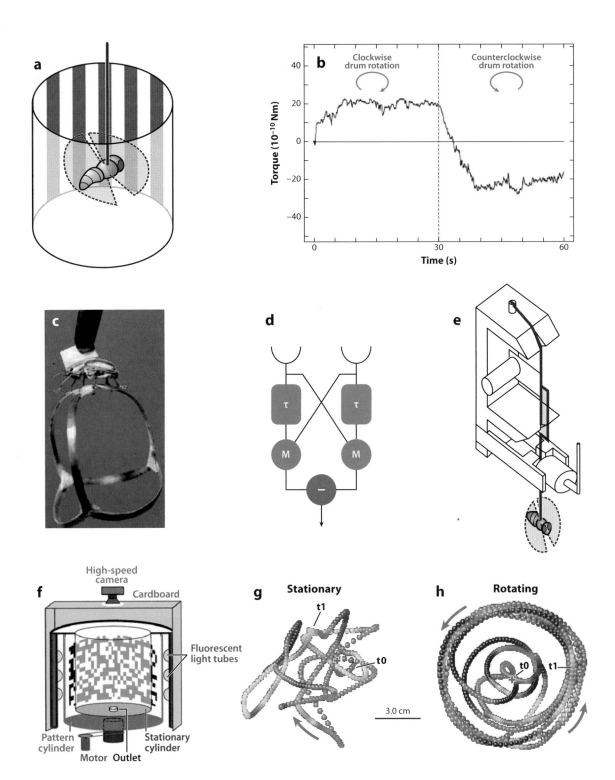

animal is informed about its ego-motion by numerous proprioreceptors, first of all its halteres (Mayer et al. 1988, Nalbach & Hengstenberg 1994, Chan et al. 1998, Sherman & Dickinson 2002). Nevertheless, high-speed video analysis of fruit flies flying inside a transparent cylinder (**Figure 1f**) showed that free-flight behavior is dramatically influenced by rotation of a surrounding textured drum (Mronz & Lehmann 2008). When the drum is stationary, flies display their typical saccade-like flight structure with rather straight episodes interspersed by rapid changes in their flight direction (**Figure 1g**). In contrast, when the drum is rotating, the flight path becomes much more curved, syndirectional to the drum's rotation, with straight flight episodes and saccades almost absent (**Figure 1h**).

NEURAL ARCHITECTURE OF THE OPTIC LOBES

The fly's nervous system consists of two ganglia: the head and the thoracic ganglion (for an overview, see Strausfeld 1976). Because the head is covered with various sensory organs, most conspicuously the eyes, large parts of the head ganglion are devoted to the processing of information coming from these sensory organs. In the present context, the visual ganglia (**Figure 2**) are of specific interest. They consist of four different layers called the lamina, the medulla, the lobula, and the lobula plate. All these layers exhibit the same columnar structure as the retina. The principle underlying this building plan is retinotopy, i.e. neighboring image points are processed by neighboring facets in the eye and by neurons within neighboring columns in each of the layers of the visual ganglia. There exist two large chiasms between the

optic ganglia, reversing the image along the antero-posterior axis twice: The first, known as the outer chiasm, occurs between the lamina and the medulla, and the second one, the inner chiasm, occurs between the medulla and the lobula complex. At the lobula plate level, a set of large motion-sensitive neurons can be found, known as lobula plate tangential cells. These tangential cells are key players with respect to optic flow processing and visual course control.

The neural processing of motion vision starts in the eye. Each eye is composed of facets or ommatidia, which are each equipped with a set of eight photoreceptors, with six outer photoreceptors, R1-6, surrounding the two central ones, R7 and R8. The photoreceptors carry their densely packed photopigment in rhabdomeres. Hardie (1986) described five populations of photopigments with different spectral properties in the fly retina. Their expression is strictly regulated during development (Wernet et al. 2006). Pigment Rh1 found in all R1-6 cells throughout the retina has two absorption peaks, one in the ultraviolet and the other one in the green. Four other pigments with single absorption peaks are expressed in R7 or R8, forming a stochastic matrix for color vision. When activated by light, all fly photoreceptors become depolarized through the opening of trp and trp-like channels (for review, see Hardie & Raghu 2001, Wang & Montell 2007). This depolarization is a fast process; thus, fly photoreceptors can follow sinusoidal luminance modulations up to 200–300 Hz depending on the fly species. The different photoreceptors in one ommatidium have different optical axes, but certain groups of photoreceptors within neighboring ommatidia have parallel optical axes. By connecting these groups of photoreceptors to

---

Figure 1

Optomotor behavior and elementary motion detection. (*a*) A fly tethered to a torque meter is surrounded by a striped drum. (*b*) When the drum is rotating clockwise, the fly exerts a clockwise turn; when the drum is rotating counterclockwise, the fly tries to turn counterclockwise, too (from Heisenberg & Wolf 1984). This reaction is called an optomotor response. (*c*) A beetle walking on a spherical Y-maze (from Hassenstein 1991). (*d*) The Reichardt detector model for elementary motion detection. (*e*) A torque meter as devised by Goetz (1964). (*f–h*) Flight arena with tracks from individual flies, with a stationary panorama (*g*) and while the panorama is rotating at a constant speed (*h*) (from Mronz & Lehmann 2008).

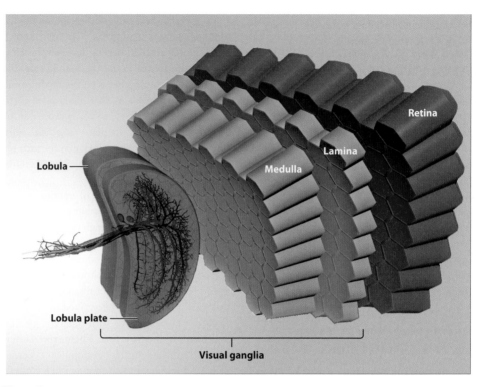

Figure 2

Schematic of the fly optic lobe. In the lobula plate, the group of vertical system (VS) cells is shown as three-dimensional reconstructions from 2-photon image stacks of single dye-filled cells (from Cuntz et al. 2007).

the same postsynaptic target, a so-called optic cartridge (Vigier 1908, Trujillo-Cenoz & Melamed 1966, Braitenberg 1967), the sensitivity of the system is increased without sacrificing acuity (Kirschfeld 1967). This principle is called neural superposition. Whereas the axons of photoreceptors R1–6 stop in the lamina, where they connect to large monopolar cells and amacrine cells, the axons of photoreceptors R7,8 run through the lamina without forming synapses and terminate in specific layers of the medulla.

The lamina contains, in addition to wide-field amacrine cells, eight different cell types per column, which connect it to the medulla: five lamina monopolar cells, L1–5, two centrifugal cells called C2 and C3, and the T1 cell. Ultrastructural studies on the connectivity within the lamina by serial sectioning transmission electron microscopy and subsequent

3D reconstruction (Meinertzhagen & O'Neill 1991) revealed that only L1-3 and the amacrine cell receive direct input from photoreceptor R1–6 terminals via tetradic synapses. At these tetrads photoreceptor synaptic transmission involves the release of histamine by R1-6 (Hardie 1989). Histamine gated chloride channels defective in the *ort* mutation (O'Tousa et al. 1989) and encoded by the gene *hclA* (Gengs et al. 2002) are expressed on the postsynaptic target cells of R1–6 and mediate signal-inverting synaptic communication: a strong, transient hyperpolarization upon illumination onset of the eye, which is followed by a sustained component that disappears with increasing light intensity (Jaervilehto & Zettler 1971, Straka & Ammermueller 1991, Zheng et al. 2009). When the light is switched off, a rebound depolarization is observed. The response of large monopolar cells L1 and L2 readily adapts to the mean luminance over several orders of magnitude while leaving its contrast sensitivity almost unchanged (Laughlin & Hardie 1978, Laughlin et al. 1987). Within each cartridge, L4 receives its exclusive input from L2 (Braitenberg 1970, Strausfeld 1970) and connects to two neighboring posterior cartridges by synapsing again onto L2 (Braitenberg 1970).

Starting with the work of Cajal & Sanchez (1915), the columnar cell types of the medulla, lobula, and lobula plate have all been identified and described on the basis of Golgi impregnations in the housefly (Strausfeld 1976) as well as in *Drosophila* (Fischbach & Dittrich 1989). In addition to the terminals of R7/8 and the terminals of all the lamina neurons (except the lamina intrinsic amacrine cells), each medulla column houses more than 60 different cells per column. All incoming terminals ramify in different layers of the medulla (Takemura et al. 2008). This layout suggests a splitting of photoreceptor signals into several parallel pathways that might supply different functions such as the detection of form, polarization patterns, ultraviolet, color, and motion processing. However, because of the small diameter of columnar neurons' processes, until now, only a few electrophysiological recordings from identified neurons have described the visual response properties of some of them (DeVoe 1980; Gilbert et al. 1991; Douglass & Strausfeld 1995, 1996).

In the lobula plate, large neurons run perpendicular to the columns covering many hundreds or thousands of them with their dendrites. These are the lobula plate tangential cells, investigated in great detail first by Hausen and Hengstenberg (Hausen 1984, Hengstenberg et al. 1982). A total of 60 different cells are found in the blow fly *Calliphora vicina* all of which are motion sensitive. Some of these cells have also been described for *Drosophila* (Fischbach & Dittrich 1989, Scott et al. 2002, Raghu et al. 2007, Joesch et al. 2008, Maimon et al. 2010, Schnell et al. 2010). Interestingly, the same cells in different individuals turn out to be highly stereotyped with respect to the area covered by their dendrites within the lobula plate, but not with respect to the branching pat-

tern (Cuntz et al. 2008). Using ablation experiments (Heisenberg et al. 1978, Geiger & Nässel 1981, Hausen & Wehrhahn 1983), investigators concluded that lobula plate tangential cells are involved in the fly's optomotor response.

REICHARDT-TYPE MOTION COMPUTATION IN THE OPTIC LOBES

The most significant response characteristic of the lobula plate tangential cells is their directional selectivity (**Figure 3** and see sidebar, Simulation Details): If a grating moves in one direction (the cell's preferred direction), the cell depolarizes or fires a train of action potentials. When the grating moves in the opposite direction (the cell's null direction), the cell hyperpolarizes or ceases to fire. In contrast, the photoreceptor signal is nondirectional, i.e. a single photoreceptor displays the same response regardless of whether the grating moves in one or the opposite direction. Thus somehow a nondirectional response at the photoreceptor level is transformed into a directional signal at the lobula plate tangential cell level. The Reichardt detector describes this transformation in amazing detail.

As mentioned above, one of the hallmarks of the Reichardt detector is its temporal frequency optimum: the larger the pattern wavelength, the higher the optimum speed (**Figure 3b**, left panel). As already verified in the optomotor response, this property is also found in the visual responses of lobula plate tangential cells of the blow flies (**Figure 3b**, middle panel) (Haag et al. 2004) and of fruit flies (**Figure 3b**, right panel) (Joesch et al. 2008). For both species, this optimum is found at ~1 Hz. In contrast with the optomotor response, which is inherently slow, recordings in blow fly tangential cells also allowed for a comparison between the response transients of cellular responses and the ones of the Reichardt detector. When the velocity of a grating is stepped from zero to a constant value, the Reichardt detector exhibits a transient ringing at the temporal frequency of the pattern

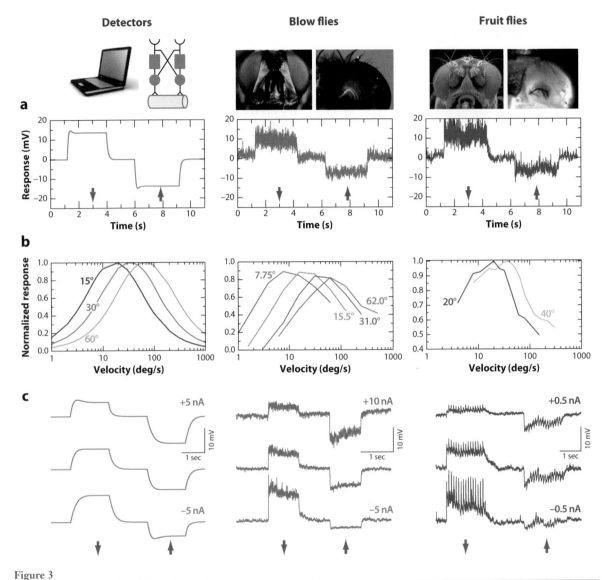

Figure 3

Comparison of the responses of an array of Reichardt detectors (*left column*) and of lobula plate tangential cells in blow flies (*middle column*) as well as in fruit flies (*right column*). (*a*) Visual response to preferred direction (downward, *indicated by the arrow*) and null direction (upward) motion. (*b*) Steady-state responses to sine-gratings with different spatial wavelengths drifting at constant velocities. The optimum is shifted toward larger velocities with increasing wavelength of the pattern in such a way that the optimum is always at the same temporal frequency (ratio of velocity and pattern wavelength). This optimum temporal frequency is roughly 1 Hz in both blow flies and *Drosophila*. (*c*) Evidence for a push-pull configuration of local motion input to lobula plate tangential cells. Visual motion along the preferred and null directions was presented during injection of depolarizing and hyperpolarizing current in the recorded cell. When the cell is artificially depolarized, the preferred-direction (PD) response becomes smaller; when the cell is hyperpolarized, the null-direction response becomes smaller. Experimental data are from Haag et al. (2004), Joesch et al. (2008), and J. Haag & A. Borst, unpublished observations.

motion before settling at the steady state (Borst & Bahde 1986). Such a transient ringing could indeed be observed in lobula plate tangential cells of the blow fly (Egelhaaf & Borst 1989). Furthermore, the amplitude of the ringing showed a characteristic time constant by which it decays to the steady state, which depended on the contrast of the moving pattern as well as on the pattern contrast before the onset of motion (Reisenman et al. 2003, Joesch et al. 2008). These findings led to an elaborated Reichardt detector with high-pass filters inserted in the cross-arms, the time constant of which rapidly adapts (Borst et al. 2003). Given these modifications, the Reichardt model can account for both the steady state and all transient response features of the lobula plate tangential cells in a detailed way. This statement applies not only to stimulus situations using velocity steps, but also to Gaussian white-noise velocity profiles. When using such stimuli with different standard deviations, the response exhibited a velocity gain control; i.e., the response-velocity function was found to be steeper the smaller the velocity fluctuations (Brenner et al. 2000). Astonishingly and completely counterintuitively, the Reichardt detector replicates this velocity-gain control even when all its filter time constants are fixed (Borst et al. 2005, Safran et al. 2007). Last, even though tangential cells spatially integrate the output signals of local motion detectors and, thus, should represent their summated output, the signals of individual motion detectors can also be observed experimentally, either when spatial integration is prevented by presentation of grating motion through a slit or by local calcium measurements in fine dendritic branches. Both these techniques revealed local signals that have all the characteristics of local motion detectors of the Reichardt type (Egelhaaf et al. 1989, Single & Borst 1998, Haag et al. 2004).

Given the evidence that has been accumulated for Reichardt-like motion computation in the optic lobes of different fly species, the question naturally arises about its neural implementation: Which neurons form the input to the Reichardt detector? Which neurons constitute the Reichardt detector? What are the biophysical mechanisms underlying mathematical operations such as low-pass and high-pass filtering and multiplication? As for the question about the input, it is fairly undisputed that motion vision is fed primarily by signals from photoreceptors R1–6, but not from R7 and 8. This statement is supported by the observation that the optomotor response in *Drosophila* is abolished by genetic elimination of R1–6, but unaffected when R7 is missing (Heisenberg & Buchner 1977). Furthermore, the optomotor response turned out to be color-blind under certain experimental conditions: When presenting a grating of alternating color, there is a brightness ratio, the so-called point of equi-luminance, at which the optomotor response is

SIMULATION DETAILS

The visual pattern consisted of a sine grating with 100% contrast covering a visual angle of 60 degrees. It was moved at a precision of 0.001 degrees/ms. The detector array comprised 32 detectors with a sampling base of 1.875 deg. Signals from each photoreceptor were low-pass filtered (1st order filter, time constant = 100 ms) and multiplied with the high-pass filtered (1st order filter, time constant = 200 ms) signal from the adjacent receptor. A DC value of 100 nS was added to the summed output of these multiplications, and the resulting signal was clipped when negative. This signal provided the excitatory conductance to a passive one-compartment model neuron. A mirror-symmetrical operation was used to provide the inhibitory conductance to the model neuron. The membrane potential V_m was calculated as

$$V_m = \frac{E_{exc} \cdot g_{exc} + E_{inh} \cdot g_{inh} + I_{inj}}{g_{exc} + g_{inh} + g_{leak}},$$

with $E_{exc} = 40\,\text{mV}$, $E_{inh} = -40\,\text{mV}$, and $g_{leak} = 100\,\text{nS}$. In panels *a* and *c*, the pattern had a spatial wavelength of 60 degrees and was moved at 15 degrees/s. In panel *b*, the patterns had a wavelength of 15, 30, and 60 degrees, respectively, moving at velocities between 1 and 1000 degrees/s. In panel *c*, a current of +5 nA and −5 nA was injected permanently, resulting in an offset of the membrane potential of + and − 16.6 mV, respectively, corresponding to an input resistance of 3.33 MΩ while the pattern was at rest. During motion, this input resistance dropped by ∼5%–10%, depending on the specifics of the stimulus conditions.

zero (Yamaguchi et al. 2008). With respect to the other questions raised above, experimental evidence is rare, leaving room for many speculations. At present, it is not clear which columnar neurons provide synaptic input to the lobula plate tangential cells. Most evidence speaks in favor of the bushy T cells, T4 and T5, as potential input candidates, one of which (T5) has been reported to respond to moving gratings in a directionally selective way, the other one (T4) to be only weakly directionally selective (Douglass & Strausfeld 1995, 1996). So far, a single study has shown unequivocally a chemical synapse between a horizontal system (HS)-cell dendrite and a columnar T4 cell (Strausfeld & Lee 1991). Additional circumstantial evidence in favor of T4 and T5 cells includes the observation that these cell types exist in four different subtypes per column, each of which ramifies in a different stratum of the lobula plate. Anatomical investigations have revealed that horizontally and vertically sensitive lobula plate tangential cells extend their dendrites to four different strata of the lobula plate, according to their preferred direction. These four strata have also been labeled in the *Drosophila* brain by using the 2-deoxyglucose (2-DG) method (Buchner et al. 1984, Bausenwein & Fischbach 1992) simultaneously with the most proximal layer of the medulla, exactly where T4 cells ramify, and the posterior layer of the lobula, where T5 cells extend their branches. The direction of motion that activates a specific stratum, as labeled using the 2-DG method, matches the preferred direction of those lobula plate tangential cells that extend their dendrite in this stratum.

Although it is still unclear which neurons constitute the Reichardt detector, good evidence indicates that motion-sensitive neurons with opposite preferred directions provide excitatory and inhibitory input to the dendrites of lobula plate tangential cells. In terms of the Reichardt model, these inputs correspond to the mirror-symmetrical detector subunits. A conductance-based model of an isopotential compartment that receives input from two arrays of such subunits predicts the following

(**Figure 3c**, left panel): Depolarizing the postsynaptic compartment by a tonic injection of positive current decreases the preferred-direction response amplitude while increasing the null-direction response amplitude. Hyperpolarizing the postsynaptic compartment by a tonic injection of negative current increases the preferred-direction response amplitude while decreasing the null-direction response amplitude. The reason for this effect is simply the reduction of the respective driving force by manipulating the postsynaptic membrane potential. This exact effect can be observed in tangential cells of blow flies (**Figure 3c**, middle panel) (J. Haag and A. Borst, unpublished observations) and fruit flies (**Figure 3c**, right panel) (Joesch et al. 2008). These results support the subtraction stage in the Reichardt detector to be realized on the tangential cells' dendrites (see also Borst & Egelhaaf 1990). Given this push-pull input organization, a moving pattern of increasing size is expected to stimulate increasingly more local motion detectors and, thus, to decrease the input resistance of the integrating cell. Thus the response as a function of pattern size will saturate while still being sensitive to image velocity. This so-called gain control is indeed observed in blow fly tangential cells under various conditions (Haag et al. 1992, Borst et al. 1995, Single et al. 1997).

The chemical identity of the transmitter systems involved in this push-pull input organization was clarified by in vitro studies of blow fly lobula plate tangential cells and revealed excitatory nicotinic acetylcholine receptors (nAChRs) as well as inhibitory γ-aminobutyric acid receptors (GABARs) on these cells (Brotz & Borst 1996). Blocking the inhibitory input in vivo by injecting the GABAR-antagonist Picrotoxinin leads to an enhanced preferred-direction response, whereas the null-direction response is reversed (Egelhaaf et al. 1990, Single et al. 1997). Because blocking the inhibitory input should isolate the excitatory input, investigators thought it indicated a weak direction selectivity of each of the two subunits: The enhanced preferred-direction response revealed an inhibitory activity during preferred-direction motion, and the positive null-direction response uncovered an excitatory activity during null-direction response. Accordingly, the full direction selectivity as observed under control conditions in the tangential cells is the result of subtracting two inputs with opposite preferred directions realized by the push-pull input organization.

Drosophila offers the possibility to visualize the intracellular distribution of certain transmitter receptors with high resolution. This visualization was done first by proving the expression of a particular receptor on a given cell by antibody staining. Then, a labeled version of the same receptor subtype could be expressed in the same cell in an otherwise unlabeled brain (**Figure 4**). Using a Gal4-driver line that led to expression in lobula plate tangential cells and two types of labeled reporter genes, excitatory and inhibitory transmitter receptors were found to be located on the fine dendritic branches of HS and VS cells (Raghu et al. 2007, 2009). One such reporter gene encodes the GABA receptor subunit Rdl (resistance against Dieldrin, Dieldrin being a potent insecticide; Ffrench-Constant et al. 1990) fused to a small hemagglutinin (HA) tag (Sanchez-Soriano et al. 2005). This way the receptor subunit can be visualized by antibody staining against the HA tag. The other transgene encodes the alpha

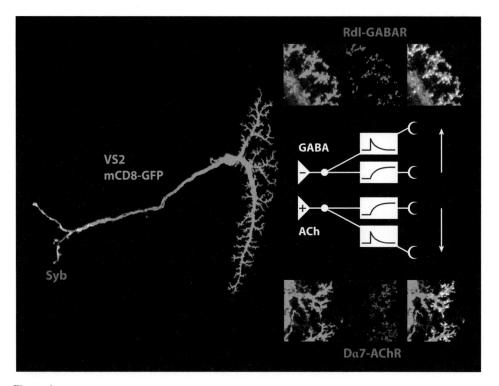

Figure 4

Immunohistochemistry reveals synaptic polarity in *Drosophila* VS cells. Staining of an individual VS2 cell was obtained by a mosaic analysis with repressible cell marker (MARCM) technique (Lee & Luo 1999). The anatomy of the cell is visualized using an mCD8-GFP transgene (*green*). Presynaptic terminals are labeled by a DsRed fluorophore tagged to synaptobrevin (magenta). Driving expression of receptor subunits of the acetylcholine receptor Dα7 [*bottom three insets*: anatomy (*left*), Dα7 fluorescence (*center*), overlay (*right*)], or the GABA receptor Rdl [*top three insets*: anatomy (*left*), Rdl fluorescence (*center*), overlay (*right*)] visualizes the location of excitatory and inhibitorys input onto the VS2 cell on the small higher-order branchlets of its dendrite (compiled from Raghu et al. 2007, 2009). These inputs may correspond to directionally selective subunits of the Reichardt type, schematically represented to the right.

subunit 7 (Dα7) of the *Drosophila* nicotinic acetylcholine receptor tagged with green fluorescent protein (GFP). Taken together with the available pharmacology of lobula plate tangential cells (Brotz & Borst 1996) and immunohistochemical data in blow flies (Brotz et al. 2001), these data strongly suggest that retinotopically organized local motion detectors with opposite-direction selectivity provide excitatory and inhibitory input onto the dendrites of tangential cells, endowing them with direction selectivity (**Figure 4**).

GLOBAL OPTIC-FLOW ANALYSIS IN THE LOBULA PLATE AND BEYOND

Local motion detection constitutes the first step in optic flow analysis by providing the nervous system with a vector field as represented by the output signals from the retinotopic array of Reichardt-type motion detectors. This optic flow information is now processed within the lobula plate by the so-called tangential cells. All these cells have large dendrites by which they spatially integrate over various subpopulations of local motion detectors. According to their overall preferred direction, they are grouped into horizontal (H) and vertical (V) cells, respectively (for details, see Hausen 1984, Borst & Haag 2002). Cells of the horizontal system have their dendrites located in the anterior layer of the lobula plate. Well-studied representatives of this group are the three HS cells (Hausen 1982a,b), the two CH cells (Eckert & Dvorak 1983, Egelhaaf et al. 1993, Gauck et al. 1997), H1, and and H2 (Hausen 1984). The vertical system is composed of 10 VS cells (Hengstenberg 1982, Hengstenberg et al. 1982) in large fly species and presumably only 6 VS cells in *Drosophila* (Scott et al. 2002). VS cells orient their dendrites along the dorso-ventral axis in the posterior layer of the lobula plate (**Figure 2**). VS cells are numbered sequentially according to the location of their dendrite from most lateral (VS1) to proximal (VS10).

Most tangential cells (HS and VS cells) respond to visual motion in a graded way: In response to motion along their preferred direction, they depolarize, and this depolarization is superimposed by action potentials of irregular amplitude (Hengstenberg 1977, Haag & Borst 1996). In response to null-direction motion, they hyperpolarize. However, some tangential cells such as H1, H2, H3, H4, or V1 produce regular action potentials. These spiking neurons extend their axon across the midline of the brain to contact neurons of the contralateral lobula plate. Passive and active membrane properties of HS, CH, and VS cells were investigated by current- and voltage-clamp experiments and optical recording of calcium concentration, accompanied by detailed biophysical modeling (Egelhaaf & Borst 1995; Borst & Haag 1996; Haag et al. 1997, 1999; Borst & Single 2000; Haag & Borst 2000; Oertner et al. 2001; Single & Borst 2002). In addition, the contribution of these active membrane properties to the encoding of motion information as well as the impact of photon noise on the response reliability could also be clarified (Haag & Borst 1997, 1998; Borst & Haag 2001; Borst 2003; Shi & Borst 2006).

According to the retinotopic layout of the lobula plate, the location of a cell's dendrite within the lobula plate is a good predictor of its receptive field center. Thus, the three HS cells, which cover the lobula plate in the northern (HSN), equatorial (HSE), and southern (HSS) parts, have their receptive field centers in the dorsal, middle, and ventral parts of the fly's visual field. Even within the dendrite of a single cell, the retinotopic arrangement of the lobula plate becomes evident when local motion stimuli are presented at different positions within the receptive field while visualizing dendritic activity via calcium imaging (Borst & Egelhaaf 1992, Borst & Single 2000).

However, when investigating the receptive fields of lobula plate tangential cells in detail, Krapp and Hengstenberg (Krapp & Hengstenberg 1996, Krapp et al. 1998) discovered that the receptive fields extend over a much larger area along the azimuth than expected from their dendritic field within the lobula plate. Furthermore, they found that the receptive

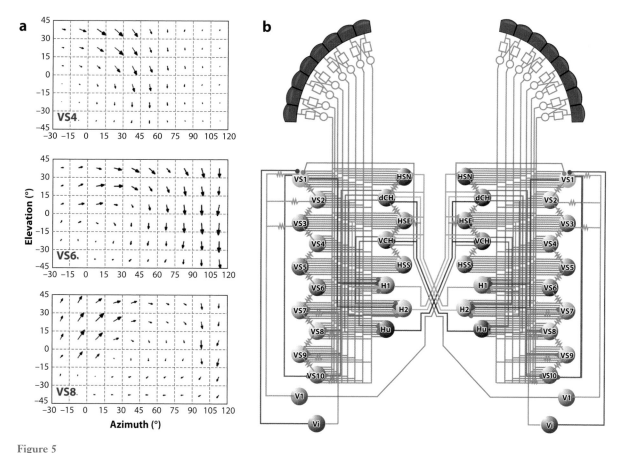

Figure 5

(*a*) Receptive fields of three VS cells (from Wertz et al. 2009). (*b*) Network circuitry of the different tangential cells of the blow fly lobula plate. In addition to receiving retinotopic input from arrays of local motion detectors, cells are strongly interconnected either within one hemisphere or between the two hemispheres. Excitatory and inhibitory chemical synapses are symbolized by triangles and circles, respectively. Resistor symbols represent electrical synapses.

fields are composed of areas with different preferred directions. This property is shown in **Figure 5a** for three different VS cells. The receptive fields of VS cells exhibit maximum sensitivity to downward motion that corresponds with their location within the lobula plate. In addition, they are sensitive to horizontal motion in the dorsal part of the visual field as well as to upward motion at a position that is ~180 degrees displaced along the azimuth. In sum, the receptive fields have the appearance of curled vector fields, such as an optic flow occurring when the animal rotates around a particular body axis. Because each cell had a

different receptive field, this finding gave rise to the notion that the tangential cells could act as matched filters, responding maximally during certain flight maneuvers (Franz & Krapp 2000). This hypothesis was indeed confirmed experimentally (Karmeier et al. 2005).

Although this observation puts the lobula plate tangential cells on center stage for visual course control, the question remains of how these receptive fields come about. If acting in isolation and strictly in parallel, the receptive fields of all these cells should be much narrower. In addition, their elementary motion detector input is expected to have a mostly uniform preferred direction, given that most of the cells ramify within only one layer of the lobula plate. A solution to this problem was provided by a series of experiments during which the signals of two tangential cells were recorded simultaneously. In these experiments, current was injected in one of the cells while the response to the current injection was recorded in the respective other cell (Haag & Borst 2004). These and other experiments revealed an intriguing network within the lobula plate (**Figure 5b**); most of the tangential cells were connected to each other, within each hemisphere as well as between the two hemispheres (Hausen 1984; Horstmann et al. 2000; Haag & Borst 2001, 2002, 2003, 2004, 2005, 2007, 2008; Kurtz et al. 2001; Farrow et al. 2006; Kalb et al. 2006). Many of these connections are based on electrical instead of chemical synapses. This connectivity was hypothesized to account for the large and complex receptive fields: While one part of the receptive field would be brought into the cell via its dendrite, additional information should arrive at the cell indirectly via its neighbors. Therefore, ablating certain cells within the lobula plate should affect the receptive fields of the remaining cells. Performing such experiments via single-cell photoablation in blow flies indeed revealed defective receptive fields in the remaining cells (Farrow et al. 2003, 2005), as correctly predicted by detailed computer simulations of the lobula plate network (Cuntz et al. 2003, 2007). Furthermore, these computer simulations predicted that, based on the electrical compartmentalization of VS cells and the specific contact site between neighboring VS cells at the axon terminal, different receptive fields should be observable in the dendrite and in the axon terminal (Cuntz et al. 2007): Using calcium imaging to visualize such signals even in the thinnest branches, Elyada et al. (2009) confirmed this prediction experimentally. Thus much experimental evidence indicates that the receptive fields of the lobula plate tangential cells come about by dendritic integration of local, motion-sensitive input elements in addition to the interconnectivity among the tangential cells themselves.

Much optic flow analysis is already performed at the lobula plate level. In the next step toward flight control, lobula plate tangential cells synapse onto descending neurons that either connect to the motor centers in the thoracic ganglion or directly innervate the neck muscles for head motion control (Strausfeld & Bassemir 1985, Strausfeld & Seyan 1985, Milde et al. 1987, Strausfeld et al. 1987, Gronenberg et al. 1995, Huston & Krapp 2008). As two representatives of such neurons, DNOVS1 and DNOVS2 (descending neurons of the ocellar and vertical system) have been recently examined in great detail (Haag et al. 2007; Wertz et al. 2008, 2009). Using current injection during dual intracellular recording from DNOVS cells and various VS cells, their connectivity to VS cells was established. It appeared that the two DNOVS cells are tuned to two different axes of rotation similar to the tuning of their input VS cells (Wertz et al. 2009). Also, the tuning width of DNOVS cells turned out to be similar to those of their input VS cells. However, during rotation of naturalistic images, the responses of DNOVS cells are rather smooth, whereas the signals of VS cells strongly fluctuate over time (Wertz et al. 2009). This effect can be attributed to the axo-axonal gap junctions between the VS cell terminals, which perform a linear interpolation of the output signals (Cuntz et al. 2007, Weber et al. 2008, Elyada et al. 2009) and which become fully visible in the membrane potential of the postsynaptic cells. Therefore, it is not the selectivity for particular optic flows that increases when going from the lobula plate to descending neurons, but rather the robustness of the responses against the particular layout of the visual environment. Of course, the small number of descending neurons that have been studied in such detail does not allow for any generalization at the moment, and indeed, extracellular recordings from the fly cervical connective, which contains, depending on the species, between 3600 and 8000 axons of ascending and descending neurons (Coggshall et al. 1973), revealed a large number of rather diverse and often highly nonlinear response types (Borst 1991).

FUTURE ISSUES

Approximately a half-century after the Reichardt detector correctly described the process of elementary motion vision in insects, it is still unclear which cells are responsible for the computations as defined in this model. However, this situation may change in the future. Promise comes from recently developed genetic techniques in *Drosophila* (for an introduction, see Borst 2009). Here, combining cell-specific expression lines (enhancer trap or Gal4-lines; Brand & Perrimon 1993) with genetically encoded indicators of neural activity (Miyawaki et al. 1997) or blockers of synaptic transmission (van der Bliek & Meyerowitz 1991) provides the tools to identify those columnar elements involved in motion processing. In a series of experiments, our lab has tested a large set of genetically encoded calcium indicators under identical conditions at the neuromuscular junction of *Drosophila* larvae (Guerrero et al. 2005, Reiff et al. 2005, Mank et al. 2006, Hendel et al. 2008). One of the indicators, TN-XXL (Mank et al. 2008), proved to be best suited for in vivo imaging in the visual system of adult flies with respect to signal-to-noise ratio, calcium sensitivity range, and kinetics. Using this indicator, we started to record the activity of columnar neurons in the optic lobes of *Drosophila* in response to visual motion stimuli. In a different approach, selected sets of columnar neurons can be removed from the circuit while recording from the lobula plate tangential cells during motion stimulation: Any alteration of the wild-type motion response (Joesch et al. 2008) will indicate the participation of the respective neurons in the motion-detection circuitry. As appealing as this approach may look initially, its biggest caveat concerns the often-variable expression level and lack of selectivity of the different driver lines available. A negative result (wild-type-like motion response) during blockade of cell X can mean that cell X does not participate in the circuit or that the expression level of the toxin was not high enough to suppress synaptic transmission. A positive result can mean that cell X does indeed participate in the circuit or that another cell Y, which somehow went unnoticed, is part of the expression pattern and is the real player. Such effects may explain why three different studies, performed to determine which lamina cells represent the input channels to motion vision, came to rather divergent conclusions (Rister et al. 2007, Katsov & Clandinin 2008, Zhu et al. 2009). Controlled and standardized expression levels in well-defined and small sets of neurons are highly desirable, such as those resulting from enhancer fragment lines currently being produced (Pfeiffer et al. 2008). Given these new developments, we are confident that both techniques outlined above represent the way to answer the long-standing question about the cellular nature of elementary motion detection.

Another important question concerns the performance of the system under natural conditions. How does fly motion vision cope with natural images? How is it adapted to the animal's specific flight style, with its rapid saccadic turns interleaved by fairly straight flight episodes, and what are the specific roles of the different tangential cells under these conditions? Investigating the performance of Reichardt detectors when confronted with natural image sequences, O'Carroll and colleagues found the ambiguity of the Reichardt detector with respect to velocity estimation to vanish under these conditions owing to the predictable spatial frequency content of natural scenes (Dror et al. 2001). When recording the responses of HS cells in hoverflies, they indeed found these neurons to encode the velocity of natural images independently of the particular image used, despite large differences in contrast between the images (Straw et al. 2008). To assess the information encoded in the tangential cell's signals, Egelhaaf, van Hateren, and colleagues used flight trajectories of blow flies recorded by a coil system (Schilstra & van Hateren 1999) and reconstructed the exact retinal motion sequences experienced by the fly during flight. Playing back these stimuli to a tethered fly while recording intracellularly from tangential cells, HS cells encoded information about the spatial structure of the environment during straight flight segments between saccadic turns (Karmeier et al. 2006). As a complementary approach, *Drosophila* offers the possibility to ablate individual neurons or subpopulations of the set of tangential cells genetically using targeted expression of, e.g., translational blockers such as Ricin A (Moffat et al. 1992). By testing these flies in free flight or walking paradigms, differences in behavior should be attributable to the specific loss of genetically determined functional classes of neurons. Thus, a combined genetic, physiological, and behavioral approach should shed further light on the cellular processing of optic flow and the role of different neurons in behavior.

DISCLOSURE STATEMENT

The authors are not aware of any affiliations, memberships, funding, or financial holdings that might be perceived as affecting the objectivity of this review.

LITERATURE CITED

Bausenwein B, Fischbach KF. 1992. Activity labeling patterns in the medulla of *Drosophila melanogaster* caused by motion stimuli. *Cell Tissue Res.* 270:25–35

Borst A. 1991. Fly visual interneurons responsive to image expansion. *Zool. Jb. Physiol.* 95:305–13

Borst A. 2003. Noise, not stimulus entropy, determines neural information rate. *J. Comput. Neurosci.* 14:23–31

Borst A. 2007. Correlation versus gradient type motion detectors—the pros and cons. *Phil. Trans. R. Soc. B* 362:369–74

Borst A. 2009. *Drosophila*'s view on insect vision. *Curr. Biol.* 19:36–47

Borst A, Bahde S. 1986. What kind of movement detector is triggering the landing response of the housefly? *Biol. Cybern.* 55:59–69

Borst A, Bahde S. 1988a. Spatio-temporal integration of motion: a simple strategy for safe landing in flies. *Naturwiss* 75:265–67

Borst A, Bahde S. 1988b. Visual information processing in the fly's landing system. *J. Comp. Physiol. A* 163:167–73

Borst A, Egelhaaf M. 1989. Principles of visual motion detection. *Trends Neurosci.* 12:297–306

Borst A, Egelhaaf M. 1990. Direction selectivity of fly motion-sensitive neurons is computed in a two-stage process. *Proc. Natl. Acad. Sci. USA* 87:9363–67

Borst A, Egelhaaf M. 1992. In vivo imaging of calcium accumulation in fly interneurons as elicited by visual motion stimulation. *Proc. Natl. Acad. Sci. USA* 89:4139–43

Borst A, Egelhaaf M, Haag J. 1995. Mechanisms of dendritic integration underlying gain control in fly motion-sensitive interneurons. *J. Comput. Neurosci.* 2:5–18

Borst A, Flanagin V, Sompolinsky H. 2005. Adaptation without parameter change: dynamic gain control in motion detection. *Proc. Natl. Acad. Sci. USA* 102:6172–76

Borst A, Haag J. 1996. The intrinsic electrophysiological characteristics of fly lobula plate tangential cells. I. Passive membrane properties. *J. Comput. Neurosci.* 3:313–36

Borst A, Haag J. 2001. Effect of mean firing on neural information rate. *J. Comput. Neurosci.* 10:213–21

Borst A, Haag J. 2002. Neural networks in the cockpit of the fly. *J. Comp. Physiol.* 188:419–37

Borst A, Reisenman C, Haag J. 2003. Adaptation of response transients in fly motion vision. II: Model studies. *Vision Res.* 43:1309–22

Borst A, Single S. 2000. Local current spread in electrically compact neurons of the fly. *Neurosci. Lett.* 285:123–26

Braitenberg V. 1967. Patterns of projection in visual system of fly. 1. Retina-Lamina projections. *Exp. Brain. Res.* 3:271–98

Braitenberg V. 1970. Order and orientation of elements in the visual system of the fly. *Kybernetik* 7:235–42

Braitenberg V, Taddei Ferretti C. 1966. Landing reaction of Musca domestica induced by visual stimuli. *Naturwiss* 6:155

Brand AH, Perrimon N. 1993. Targeted gene expression as a means of altering cell fates and generating dominant phenotypes. *Development* 118:401–15

Brenner N, Bialek W, de Ruyter van Steveninck R. 2000. Adaptive rescaling maximizes information transmission. *Neuron* 26:695–702

Brotz T, Borst A. 1996. Cholinergic and GABAergic receptors on fly tangential cells and their role in visual motion detection. *J. Neurophysiol.* 76:1786–99

Brotz T, Gundelfinger E, Borst A. 2001. Cholinergic and GABAergic pathways in fly motion vision. *Bio. Med. Cent. Neurosci.* 2:1

Buchner E. 1976. Elementary movement detectors in an insect visual system. *Biol. Cybern.* 24:86–101

Buchner E, Buchner S, Bülthoff I. 1984. Deoxyglucose mapping of nervous activity induced in *Drosophila* brain by visual movement. *J. Comp. Physiol. A* 155:471–83

Cajal SR, Sanchez D. 1915. *Contribucion al Conocimiento de los Centros Nerviosos de los Insectos.* Madrid: Imprenta de Hijos de Nicholas Moja

Chan WP, Prete F, Dickinson MH. 1998. Visual input to the efferent control system of a fly's "gyroscope." *Science* 280:289–92

Coggshall JC, Boschek CB, Buchner SM. 1973. Preliminary investigations on a pair of giant fibers in the central nervous system of dipteran flies. *Z. Naturforsch.* 28c:783–84

Cuntz H, Foerstner F, Haag J, Borst A. 2008. The morphological identity of insect dendrites. *PLoS Comp. Biol.* 4: doi:10.1371/journal.pcbi.1000251

Cuntz H, Haag J, Borst A. 2003. Neural image processing by dendritic networks. *Proc. Natl. Acad. Sci. USA* 100:11082–85

Cuntz H, Haag J, Foerstner F, Segev I, Borst A. 2007. Robust coding of flow-field parameters by axo-axonal gap junctions between fly visual interneurons. *Proc. Natl. Acad. Sci. USA* 104:10229–33

DeVoe RD. 1980. Movement sensitivities of cells in the fly's medulla. *J. Comp. Physiol. A* 138:93–119

Douglass JK, Strausfeld NJ. 1995. Visual motion detection circuits in flies: peripheral motion computation by identified small-field retinotopic neurons. *J. Neurosci.* 15:5596–611

Douglass JK, Strausfeld NJ. 1996. Visual motion–detection circuits in flies: parallel direction- and nondirection-sensitive pathways between the medulla and lobula plate. *J. Neurosci.* 16:4551–62

Dror RO, O'Carroll DC, Laughlin SB. 2001. Accuracy of velocity estimation by Reichardt correlators. *J. Opt. Soc. Am. A* 18:241–52

Eckert H. 1973. Optomotorische untersuchungen am visuellen system der stubenfliege musca domestica L. *Kybernetik* 14:1–23

Eckert H, Dvorak DR. 1983. The centrifugal horizontal cells in the lobula plate of the blowfly *Phaenicia sericata*. *J. Insect Physiol.* 29:547–60

Egelhaaf M, Borst A. 1989. Transient and steady-state response properties of movement detectors. *J. Opt. Soc. Am. A* 6:116–27

Egelhaaf M, Borst A. 1995. Calcium accumulation in visual interneurons of the fly: stimulus dependence and relationship to membrane potential. *J. Neurophysiol.* 73:2540–52

Egelhaaf M, Borst A, Pilz B. 1990. The role of GABA in detecting visual motion. *Brain Res.* 509:156–60

Egelhaaf M, Borst A, Reichardt W. 1989. Computational structure of a biological motion detection system as revealed by local detector analysis in the fly's nervous system. *J. Opt. Soc. Am. A* 6:1070–87

Egelhaaf M, Borst A, Warzecha AK, Wildemann A, Flecks S. 1993. Neural circuit tuning fly visual neurons to motion of small objects. II. Input organization of inhibitory circuit elements revealed by electrophysiological and optical recording techniques. *J. Neurophysiol.* 69:340–51

Elyada Y, Haag J, Borst A. 2009. Different receptive fields in axons and dendrites underlie robust coding in motion-sensitive neurons. *Nat. Neurosci.* 12:327–33

Farrow K, Borst A, Haag J. 2005. Sharing receptive fields with your neighbors: tuning the vertical system cells to wide field motion. *J. Neurosci.* 25:3985–93

Farrow K, Haag J, Borst A. 2003. Input organization of multifunctional motion sensitive neurons in the blowfly. *J. Neurosci.* 23:9805–11

Farrow K, Haag J, Borst A. 2006. Nonlinear, binocular interactions underlying flow field selectivity of a motion-sensitive neuron. *Nat. Neurosci.* 9:1312–20

Fermi G, Reichardt W. 1963. Optomotorische reaktionen der fliege *Musca domestica*. *Kybernetik* 2:15–28

Ffrench-Constant RH, Roush RT, Mortlock D, Dively GP. 1990. Isolation of dieldrin resistance from field populations of *Drosophila melanogaster* (Diptera: Drosophilidae). *J. Econ. Entomol.* 83:1733–37

Fischbach KF, Dittrich APM. 1989. The optic lobe of *Drosophila melanogaster*. I. A Golgi analysis of wild-type structure. *Cell Tissue Res.* 258:441–75

Franz MO, Krapp HG. 2000. Wide-field, motion-sensitive neurons and matched filters for optic flow fields. *Biol. Cybern.* 83:185–97

Gauck V, Egelhaaf M, Borst A. 1997. Synapse distribution on VCH, an inhibitory, motion-sensitive interneuron in the fly visual system. *J. Comp. Neurol.* 381:489–99

Geiger G, Nässel DR. 1981. Visual orientation behavior of flies after selective laser beam ablation of interneurons. *Nature* 293:398–99

Gengs CX, Leung HT, Skingsley DR, Iovchev MI, Yin Z, et al. 2002. The target of *Drosophila* photoreceptor synaptic transmission is a histamine-gated chloride channel encoded of ort (hclA). *J. Biol. Chem.* 277:42113–20

Gibson JJ. 1950. *Perception of the Visual World*. Boston: Houghton Mifflin

Gilbert C, Penisten DK, DeVoe RD. 1991. Discrimination of visual motion from flicker by identified neurons in the medulla of the fleshfly *Sarcophaga bullata*. *J. Comp. Physiol. A* 168:653–73

Goetz KG. 1964. Optomotorische untersuchung des visuellen systems einiger augenmutanten der fruchtfliege Drosophila. *Kybernetik* 2:77–92

Goetz KG. 1965. Die optischen übertragungseigenschaften der komplexaugen von *Drosophila*. *Kybernetik* 2:215–21

Goetz KG. 1987. Course-control, metabolism and wing interference during ultralong tethered flight in *Drosophila melanogaster*. *J. Exp. Biol.* 128:35–46

Gronenberg W, Milde JJ, Strausfeld NJ. 1995. Oculomotor control in Calliphorid flies—organization of descending neurons to neck motor-neurons responding to visual-stimuli. *J. Comp. Neurol.* 361:267–84

Guerrero G, Reiff DF, Agarwal G, Ball RW, Borst A, et al. 2005. Heterogeneity in synaptic transmission along a *Drosophila* larval motor axon. *Nat. Neurosci.* 8:1188–96

Haag J, Borst A. 1996. Amplification of high-frequency synaptic inputs by active dendritic membrane processes. *Nature* 379:639–41

Haag J, Borst A. 1997. Encoding of visual motion information and reliability in spiking and graded potential neurons. *J. Neurosci.* 17:4809–19

Haag J, Borst A. 1998. Active membrane characteristics and signal encoding in graded potential neurons. *J. Neurosci.* 18:7972–86

Haag J, Borst A. 2000. Spatial distribution and characteristics of voltage-gated calcium currents within visual interneurons. *J. Neurophysiol.* 83:1039–51

Haag J, Borst A. 2001. Recurrent network interactions underlying flow-field selectivity of visual interneurons. *J. Neurosci.* 21:5685–92

Haag J, Borst A. 2002. Dendro-dendritic interactions between motion-sensitive large-field neurons in the fly. *J. Neurosci.* 22:3227–33

Haag J, Borst A. 2003. Orientation tuning of motion-sensitive neurons shaped by vertical-horizontal network interactions. *J. Comp. Physiol. A* 189:363–70

Haag J, Borst A. 2004. Neural mechanism underlying complex receptive field properties of motion-sensitive interneurons. *Nat. Neurosci.* 7:628–34

Haag J, Borst A. 2005. Dye-coupling visualizes networks of large-field motion-sensitive neurons in the fly. *J. Comp. Physiol. A* 191:445–54

Haag J, Borst A. 2007. Reciprocal inhibitory connections within a neural network for rotational optic-flow processing. *Front. Neurosci.* 1:111–21

Haag J, Borst A. 2008. Electrical coupling of lobula plate tangential cells to a heterolateral motion-sensitive neuron in the fly. *J. Neurosci.* 28:14435–42

Haag J, Denk W, Borst A. 2004. Fly motion vision is based on Reichardt detectors regardless of the signal-to-noise ratio. *Proc. Natl. Acad. Sci. USA* 101:16333–38

Haag J, Egelhaaf M, Borst A. 1992. Dendritic integration of motion information in visual interneurons of the blowfly. *Neurosci. Lett.* 140:173–76

Haag J, Theunissen F, Borst A. 1997. The intrinsic electrophysiological characteristics of fly lobula plate tangential cells. II. Active membrane properties. *J. Comput. Neurosci.* 4:349–69

Haag J, Vermeulen A, Borst A. 1999. The intrinsic electrophysiological characteristics of fly lobula plate tangential cells. III. Visual response properties. *J. Comput. Neurosci.* 7:213–34

Haag J, Wertz A, Borst A. 2007. Integration of lobula plate output signals by DNOVS1, an identified premotor descending neuron. *J. Neurosci.* 27:1992–2000

Hardie RC. 1986. The photoreceptor array of the dipteran retina. *Trends Neurosci.* 9:419–23

Hardie RC. 1989. A histamine-activated chloride channel involved in neurotransmission at a photoreceptor synapse. *Nature* 339:704–6

Hardie RC, Raghu P. 2001. Visual transduction in *Drosophila*. *Nature* 413:186–93

Hassenstein B. 1991. *Freiburger Universitaetsblaetter*. Freiburg, Germ.: Rombach

Hassenstein B, Reichardt W. 1956. Systemtheoretische analyze der zeit-, reihenfolgen- und vorzeichenauswertung bei der bewegungsperzeption des rüsselkäfers chlorophanus. *Z. Naturforsch.* 11b:513–24

Hatsopoulos N, Gabbiani F, Laurent G. 1995. Elementary computation of object approach by a wide-field neuron. *Science* 270:1000–3

Hausen K. 1982a. Motion sensitive interneurons in the optomotor system of the fly. I. The horizontal cells: structure and signals. *Biol. Cybern.* 45:143–56

Hausen K. 1982b. Motion sensitive interneurons in the optomotor system of the fly. II. The horizontal cells: receptive field organization and response characteristics. *Biol. Cybern.* 46:67–79

Hausen K. 1984. The lobula-complex of the fly: structure, function and significance in visual behavior. In *Photoreception and Vision in Invertebrates*, ed. MA Ali, pp. 523–59. New York/London: Plenum

Hausen K, Wehrhahn C. 1983. Microsurgical lesion of horizontal cells changes optomotor yaw response in the blowfly *Calliphora erythocephala*. *Proc. R. Soc. London Ser. B* 219:211–16

Heisenberg M, Buchner E. 1977. The role of retinula cell types in visual behavior of *Drosophila melanogaster*. *J. Comp. Physiol. A* 117:127–62

Heisenberg M, Wolf R. 1984. *Vision in Drosophila*. Berlin, Heidelberg: Springer

Heisenberg M, Wonneberger R, Wolf R. 1978. Optomotor-blind (H31): a *Drosophila* mutant of the lobula plate giant neurons. *J. Comp. Physiol. A* 124:287–96

Hendel T, Mank M, Schnell B, Griesbeck O, Borst A, Reiff DF. 2008. Fluorescence changes of genetic calcium indicators and OGB-1 correlated with neural activity and calcium in vivo and in vitro. *J. Neurosci.* 28:7399–411

Hengstenberg R. 1977. Spike response of "nonspiking" visual interneurone. *Nature* 270:338–40

Hengstenberg R. 1982. Common visual response properties of giant vertical cells in the lobula plate of the blowfly Calliphora. *J. Comp. Physiol. A* 149:179–93

Hengstenberg R, Hausen K, Hengstenberg B. 1982. The number and structure of giant vertical cells (VS) in the lobula plate of the blowfly *Calliphora erytrocephala*. *J. Comp. Physiol. A* 149:163–77

Horstmann W, Egelhaaf M, Warzecha AK. 2000. Synaptic interactions increase optic flow specificity. *Eur. J. Neurosci.* 12:2157–65

Huston SJ., Krapp HG. 2008. Visuomotor transformation in the fly gaze stabilization system. *PLoS Biol.* 6:e173

Jaervilehto M, Zettler F. 1971. Localized intracellular potentials from pre- and postsynaptic components in the external plexiform layer of an insect retina. *Z. Vergl. Physiol.* 75:422–40

Joesch M, Plett J, Borst A, Reiff DF. 2008. Response properties of motion-sensitive visual interneurons in the lobula plate of *Drosophila melanogaster*. *Curr. Biol.* 18:368–74

Johansson G. 1973. Visual perception of biological motion and a model of its analysis. *Percept. Psychophys.* 14:201–11

Kalb J, Egelhaaf M, Kurtz R. 2006. Robust integration of motion information in the fly visual system revealed by single cell photoablation. *J. Neurosci.* 26:7898–906

Karmeier K, Krapp HG, Egelhaaf M. 2005. Population coding of self-motion: applying Bayesian analysis to a population of visual interneurons in the fly. *J. Neurophysiol.* 94:2182–94

Karmeier K, van Hateren JH, Kern R, Egelhaaf M. 2006. Encoding of naturalistic optic flow by a population of blowfly motion-sensitive neurons. *J. Neurophysiol.* 96:1602–14

Katsov AY, Clandinin TR. 2008. Motion processing streams in *Drosophila* are behaviorally specialized. *Neuron* 59:322–35

Kirschfeld K. 1967. Die projektion der optischen umwelt auf das raster der rhabdomere im komplexauge von MUSCA. *Exp. Brain Res.* 3:248–70

Koenderink JJ, van Doorn AJ. 1987. Facts on optic flow. *Biol. Cybern.* 56:247–54

Krapp HG, Hengstenberg B, Hengstenberg R. 1998. Dendritic structure and receptive-field organization of optic flow processing interneurons in the fly. *J. Neurophysiol.* 79:1902–17

Krapp HG, Hengstenberg R. 1996. Estimation of self-motion by optic flow processing in single visual interneurons. *Nature* 384:463–66

Kurtz R, Warzecha AK, Egelhaaf M. 2001. Transfer of visual motion information via graded synapses operates linearly in the natural activity range. *J. Neurosci.* 21:6957–66

Laughlin SB, Hardie RC. 1978. Common strategies for light adaptation in the peripheral visual systems of fly and dragonfly. *J. Comp. Physiol. A* 128:319–40

Laughlin SB, Howard J, Blakeslee B. 1987. Synaptic limitations to contrast coding in the retina of the blowfly Calliphora. *Proc. R. Soc. London B* 231:437–67

Lee T, Luo L. 1999. Mosaic analysis with a repressible cell marker for studies of gene function in neuronal morphogenesis. *Neuron* 22:451–61

Maimon G, Straw AD, Dickinson MH. 2010. Active flight increases the gain of visual motion processing in *Drosophila*. *Nat. Neurosci.* 13:393–99

Mank M, Ferrão Santos A, Direnberger S, Mrsic-Flogel TD, Hofer SB, et al. 2008. A genetically encoded calcium indicator for chronic in vivo two photon imaging. *Nat. Methods* 5:805–11

Mank M, Reiff DF, Heim N, Friedrich MW, Borst A, et al. 2006. A FRET-based calcium biosensor with fast signal kinetics and high fluorescence change. *Biophys. J.* 90:1790–96

Mayer M, Vogtmann K, Bausenwein B, Wolf R, Heisenberg M. 1988. Flight control during free yaw turns' in *Drosophila melanogaster*. *J. Comp. Physiol. A* 163:389–99

Meinertzhagen IA, O'Neil SD. 1991. Synaptic organization of columnar elements in the lamina of the wild type in *Drosophila melanogaster*. *J. Comp. Neurol.* 305:232–63

Milde JJ, Seyan HS, Strausfeld NJ. 1987. The neck motor system of the fly *Calliphora erythrocephala*. 2. Sensory organization. *J. Comp. Physiol. A* 160:225–38

Miyawaki A, Llopis J, Heim R, McCaffery JM, Adams JA, et al. 1997. Fluorescent indicators for Ca^{2+} based on green fluorescent proteins and calmodulin. *Nature* 388:882–87

Moffat KG, Gould JH, Smith HK, O'Kane CJ. 1992. Inducible cell ablation in Drosophila by cold-sensitive Ricin A chain. *Development* 114:681–87

Mronz M, Lehmann FO. 2008. The free-flight response of *Drosophila* to motion of the visual environment. *J. Exp. Biol.* 211:2026–45

Nalbach G, Hengstenberg R. 1994. The halteres of the blowfly Calliphora. II. Three-dimensional organization of compensatory reactions to real and simulated rotations. *J. Comp. Physiol. A* 175:695–708

Oertner TG, Brotz T, Borst A. 2001. Mechanisms of dendritic calcium signaling in fly neurons. *J. Neurophysiol.* 85:439–47

O'Tousa JE, Leonard DS, Pack WL. 1989. Morphological defects in oraJK84 photoreceptors caused by mutation in R1-6 opsin gene of *Drosophila*. *J. Neurogenetics* 6:14–52

Pfeiffer BD, Jenett A, Hammonds AS, Ngo TB, Misra S, et al. 2008. Tools for neuroanatomy and neurogenetics in *Drosophila*. *Proc. Natl. Acad. Sci. USA* 105:9715–20

Raghu SV, Joesch M, Borst A, Reiff DF. 2007. Synaptic organization of lobula plate tangential cells in *Drosophila*: GABA-receptors and chemical release sites. *J. Comp. Neurol.* 502:598–610

Raghu SV, Joesch M, Sigrist S, Borst A, Reiff DF. 2009. Synaptic organization of lobula plate tangential cells in *Drosophila*: Dα7 cholinergic receptors. *J. Neurogenetics* 23:200–9

Reichardt W. 1961. Autocorrelation, a principle for the evaluation of sensory information by the central nervous system. In *Sensory Communication*, ed. WA Rosenblith, pp. 303–17. New York/London: MIT Press/Wiley

Reichardt W. 1987. Evaluation of optical motion information by movement detectors. *J. Comp. Physiol. A* 161:533–47

Reichardt W, Poggio T. 1979. Figure-ground discrimination by relative movement in the visual system of the fly. Part I: Experimental results. *Biol. Cybern.* 35:81–100

Reiff DF, Ihring A, Guerrero G, Isacoff EY, Joesch M, et al. 2005. In vivo performance of genetically encoded indicators of neural activity in flies. *J. Neurosci.* 25:4766–78

Reisenman C, Haag J, Borst A. 2003. Adaptation of response transients in fly motion vision. I: Experiments. *Vision Res.* 43:1291–307

Rind FC, Simmons PJ. 1992. Orthopteran DCMD neuron—a reevaluation of responses to moving objects. 1. Selective responses to approaching objects. *J. Neurophysiol.* 5:1654–66

Rister J, Pauls D, Schnell B, Ting CY, Lee CH, et al. 2007. Dissection of the peripheral motion channel in the visual system of *Drosophila melanogaster*. *Neuron* 56:155–70

Safran M, Flanagin V, Borst A, Sompolinsky H. 2007. Adaptation and information transmission in fly motion detection. *J. Neurophysiol.* 98:3309–20

Sanchez-Soriano N, Bottenberg W, Fiala A, Haessler U, Kerassoviti A, et al. 2005. Are dendrites in *Drosophila* homologous to vertebrate dendrites? *Dev. Biol.* 288:126–38

Schilstra C, van Hateren JH. 1999. Blowfly flight and optic flow. II. Head movement during flight. *J. Exp. Biol.* 202:1491–500

Schnell B, Joesch M, Foerstner F, Raghu SV, Otsuna H, et al. 2010. Processing of horizontal optic flow in three visual interneurons of the *Drosophila* brain. *J. Neurophysiol.* 103:1646–57

Scott EK, Raabe T, Luo L. 2002. Structure of the vertical and horizontal system neurons of the lobula plate in *Drosophila*. *J. Comp. Neurol.* 454:470–81

Sherman A, Dickinson MH. 2002. A comparison of visual and haltere-mediated equilibrium reflexes in the fruit fly *Drosophila melanogaster*. *J. Exp. Biol.* 206:295–302

Shi L, Borst A. 2006. Propagation of photon noise and information transfer in motion vision. *J. Comput. Neurosci.* 20:167–78

Single S, Borst A. 1998. Dendritic integration and its role in computing image velocity. *Science* 281:1848–50

Single S, Borst A. 2002. Different mechanisms of calcium entry within different dendritic compartments. *J. Neurophysiol.* 87:1616–24

Single S, Haag J, Borst A. 1997. Dendritic computation of direction selectivity and gain control in visual interneurons. *J. Neurosci.* 17:6023–30

Straka H, Ammermüller J. 1991. Temporal resolving power of blowfly visual system: effects of decamethonium and hyperpolarization on responses of laminar monopolar neurons. *J. Comp. Physiol. A* 168:129–39

Strausfeld NJ. 1970. Golgi studies on insects. 2. Optic lobes of diptera. *Phil. Trans. R. Soc. London B* 258:135–223

Strausfeld NJ. 1976. *Atlas of an Insect Brain*. Berlin/Heidelberg: Springer

Strausfeld NJ, Bassemir UK. 1985. Lobula plate and ocellar interneurons converge onto a cluster of descending neurons leading to neck and leg motor neuropil in *Calliphora-Erythrocephala*. *Cell Tissue Res.* 240:617–40

Strausfeld NJ, Lee JK. 1991. Neuronal basis for parallel visual processing in the fly. *Vis. Neurosci.* 7:13–33

Strausfeld NJ, Seyan HS. 1985. Convergence of visual, haltere and prosternal inputs at neck motor neurons of *Calliphora erythrocephala*. *Cell Tissue Res.* 240:601–15

Strausfeld NJ, Seyan HS, Milde JJ. 1987. The neck motor system of the fly *Calliphora erythrocephala*. 1. Muscles and motor neurons. *J. Comp. Physiol. A* 160:205–24

Straw AD, Rainsford T, O'Carroll DC. 2008. Contrast sensitivity of insect motion detectors to natural images. *J. Vision* 8:1–9

Takemura SY, Lu Z, Meinertzhagen IA. 2008. Synaptic circuits of the *Drosophila* optic lobe: the input terminals to the medulla. *J. Comp. Neurol.* 509:493–513

Trujillo-Cenoz O, Melamed J. 1966. Compound eye of dipterans: anatomical basis for integration—an electron microscopic study. *J. Ultrastruct. Res.* 16:395–98

van der Bliek AM, Meyerowitz EM. 1991. Dynamin-like protein encoded by the *Drosophila* shibire gene associated with vesicular traffic. *Nature* 351:411–14

Vigier P. 1908. Mecanisme de la synthese des impressions lumineuses receuellies par les yeux compose des Diptereres. *C. R. Soc. Biol. Paris* 64:1221–23

Wang T, Montell C. 2007. Phototransduction and retinal degeneration in *Drosophila*. *Pflugers Arch.* 454:821–47

Weber F, Eichner H, Cuntz H, Borst A. 2008. Eigenanalysis of a neural network for optic flow processing. *New J. Phys.* 10:1–21

Wernet MF, Mazzoni EO, Celik A, Duncan DM, Duncan I, Desplan C. 2006. Stochastic spikeless expression creates the retinal mosaic for color vision. *Nature* 440:174–80

Wertz A, Borst A, Haag J. 2008. Nonlinear integration of binocular optic flow by DNOVS2, a descending neuron of the fly. *J. Neurosci.* 28:3131–40

Wertz A, Gaub B, Plett J, Haag J, Borst A. 2009. Robust coding of ego-motion in descending neurons of the fly. *J. Neurosci.* 29:14993–5000

Wertz A, Haag J, Borst A. 2009. Local and global motion preferences in descending neurons of the fly. *J. Comp. Physiol. A* 195:1107–20

Yamaguchi S, Wolf R, Desplan C, Heisenberg M. 2008. Motion vision is independent of color in *Drosophila*. *Proc. Natl. Acad. Sci. USA* 105:4910–15

Zheng L, Nikolaev A, Wardill TJ, O'Kane CJ, de Polavieja GG, Juusola M. 2009. Network adaptation improves temporal representation of naturalistic stimuli in *Drosophila* eye: I. Dynamics. *PLoS One* 4(1):e4307

Zhu Y, Nem A, Zipursky SL, Frye MA. 2009. Peripheral visual circuits functionally segregate motion and phototaxis behaviors in the fly. *Curr. Biol.* 19:1–7

Molecular Pathways of Frontotemporal Lobar Degeneration

Kristel Sleegers,[1,2,3] Marc Cruts,[1,2,3] and Christine Van Broeckhoven[1,2,3]

[1]Neurodegenerative Brain Diseases Group, Department of Molecular Genetics, VIB, [2]Laboratory of Neurogenetics, Institute Born-Bunge, and [3]University of Antwerp, Universiteitsplein 1, B-2610, Antwerpen, Belgium; email: kristel.sleegers@molgen.vib-ua.be, marc.cruts@molgen.vib-ua.be, christine.vanbroeckhoven@molgen.vib-ua.be

Annu. Rev. Neurosci. 2010. 33:71–88

First published online as a Review in Advance on March 22, 2010

The *Annual Review of Neuroscience* is online at neuro.annualreviews.org

This article's doi: 10.1146/annurev-neuro-060909-153144

Key Words

progranulin, TDP-43, neurodegeneration, etiological heterogeneity

Abstract

Frontotemporal lobar degeneration (FTLD) is a neurodegenerative condition that predominantly affects behavior, social awareness, and language. It is characterized by extensive heterogeneity at the clinical, pathological, and genetic levels. Recognition of these levels of heterogeneity is important for proper disease management. The identification of progranulin and TDP-43 as key proteins in a significant proportion of FTLD patients has provided the impetus for a wealth of studies probing their role in neurodegeneration. This review highlights the most recent developments and future directions in this field and puts them in perspective of the novel insights into the neurodegenerative process, which have been gained from related disorders, e.g., the role of FUS in amyotrophic lateral sclerosis.

Contents

INTRODUCTION

Frontotemporal dementia (FTD), semantic dementia (SD), and progressive nonfluent aphasia (PNFA) are gravely disabling and irreversible clinical conditions that are characterized by progressive neuronal loss in the frontal and/or temporal cortices, collectively referred to as frontotemporal lobar degeneration (FTLD). Different patterns of neurodegeneration can be distinguished, with predominant prefrontal and anterotemporal atrophy in FTD, atrophy of the middle and inferotemporal cortex in SD, and asymmetric atrophy of the left frontal and temporal cortices in PNFA (Neary et al. 2005). These topological differences are reflected in the clinical presentation. FTD, often referred to as behavioral variant FTLD, presents clinically with changes in behavior and social conduct, including loss of social awareness, poor impulse control, and stereotypic, ritualized behavior. Patients with SD develop a loss of semantic (long-established) knowledge of faces, emotions, objects, and language, resulting in impaired word comprehension with otherwise fluent and grammatically faultless speech. In contrast, patients with PNFA progressively lose fluency of speech with intact word comprehension. In later stages of FTLD, the distinct clinical phenotypes usually start to show overlap, possibly reflecting the progressive involvement of other brain regions. Moreover, motor neuron disease (MND) and parkinsonism complicate the disease in up to 15% of patients, and overlap with symptoms of Alzheimer disease (AD), corticobasal syndrome (CBS), and progressive supranuclear palsy (PSP) is not uncommon (Boeve 2007). Onset age of FTLD ranges on average from 45 to 65 years (Neary et al. 2005), often affecting people who are mid-career and still raising a family. Because of insidious onset, clinical heterogeneity, and the current lack of rapid and conclusive diagnostic tests, erroneous referrals tend to prolong the diagnostic trajectory, with an average diagnostic delay of three to four years (Hodges et al. 2003). At present, no cure exists. In addition to clinical heterogeneity, FTLD is markedly heterogeneous at the pathological as well as genetic level. As long as this etiological heterogeneity is not fully characterized and recognized, it will seriously impede the development of diagnostic markers and of drugs and/or treatments. Breakthroughs in recent years, however, have significantly advanced our understanding of the complexity and heterogeneity of FTLD, opening up avenues for further research, several of which are highlighted in this review.

ETIOLOGICAL HETEROGENEITY

The intraneuronal accumulation of filamentous, hyperphosphorylated microtubule associated protein tau is the most thoroughly investigated molecular signature of FTLD (FTLD-tau), driven by A. Pick's (1892) report of characteristic lesions in the brain of an FTD patient, hence the eponym Pick's disease (Pick 1892), and the affirmative observation more than 100 years later that mutations in the gene encoding microtubule associated protein tau (*MAPT*) cause FTLD-tau (Hutton et al. 1998). Tau is

FTD: frontotemporal dementia

SD: semantic dementia

PNFA: progressive nonfluent aphasia

MND: motor neuron disease

FTLD-tau: frontotemporal lobar degeneration with tau pathology

abundantly expressed in the central nervous system, where it interacts with microtubules to stabilize the microtubule network and regulates axonal transport, mediated by three or four microtubule binding repeat regions (3R or 4R) in tau. Most *MAPT* mutations identified to date are missense, deletion, or silent mutations or intronic mutations affecting splice sites, resulting in decreased ability to bind microtubules, increased tendency to form filaments, or altered ratio of tau 4R and 3R isoforms by alternative splicing of exon 10, which encodes the second of the four microtubule binding repeat regions (see the FTD mutation database, **http://www.molgen.ua.ac.be/FTDmutations**; for review, see Rademakers et al. 2004). In addition to FTLD-tau, tauopathy is a pathological hallmark of other neurodegenerative diseases as well, including AD, CBS, PSP, and argyrophilic grain disease (AGD). Although several questions remain to be resolved about the ways in which aberrant tau leads to neuronal death (Wolfe 2009), it indisputably plays a role in neurodegeneration.

In most FTLD patients who come to autopsy, however, tau pathology cannot be detected. In most of these FTLD brains, neuronal cytoplasmic and intranuclear inclusions are present that are immunoreactive to ubiquitin (Bergmann et al. 1996, Johnson et al. 2005, Josephs et al. 2004). Adding to this ambiguity was the observation that several multiplex tau-negative FTLD families showed conclusive genetic linkage to the chromosomal region 17q21 harboring *MAPT*, but they carried no mutations in *MAPT* (e.g., Rademakers et al. 2002, van der Zee et al. 2006). The fog began to clear with the 2006 discovery of mutations in progranulin (*GRN*), located 1.7 Mb centromeric of *MAPT*, explaining disease in these tau-negative, ubiquitin-positive FTLD families linked to 17q21 (Baker et al. 2006, Cruts et al. 2006). Since then, the cause of FTLD has rapidly been resolved in many patients; with 68 *GRN* mutations in 210 families currently known worldwide (Gijselinck et al. 2008a), *GRN* mutations are as important a cause of FTLD as *MAPT* mutations (**http://www.molgen.ua.ac.be/FTDmutations**) (**Table 1**).

The story quickened further with the near-simultaneous discovery of pathological 43 kDa TAR DNA-binding protein (TDP-43) as the principal component of the ubiquitin-immunoreactive inclusions in tau-negative FTLD-brains (FTLD-TDP) (Arai et al. 2006, Neumann et al. 2006). Taken together, tau and TDP-43 pathology still does not fully explain the occurrence of FTLD. Some FTLD brains show inclusions that are immunoreactive only to proteins of the ubiquitin-proteasome system (FTLD-UPS), and some have no inclusions (FTLD-ni) (Mackenzie et al. 2009). And other genes besides *MAPT* and *GRN* cause FTLD (**http://www.molgen.ua.ac.be/FTDmutations**): Valosin-containing protein gene (*VCP*) mutations cause FTLD-TDP in the frame of a broader syndrome including Paget's disease of the bone and inclusion body myopathy (IBMPFD; Watts et al. 2004), rare mutations in the charged multivesicular body protein 2B gene (*CHMP2B*) are found in FTLD-UPS (Holm et al. 2007, Skibinski et al. 2005, van der Zee et al. 2008), and numerous families with both FTLD-TDP and MND show linkage to a region on chromosome 9, which suggests that at least one other gene for FTLD exists (Gijselinck et al. 2010, Le Ber et al. 2009, Morita et al. 2006, Valdmanis et al. 2007, Vance et al. 2006) (**Table 1**). Nevertheless, with the discovery of *GRN* and TDP-43, the knowledge of FTLD is rapidly crystallizing, which has already led to a highly sensitive and specific biomarker for FTLD-TDP caused by *GRN* mutations (Ghidoni et al. 2008, Finch et al. 2009, Sleegers et al. 2009) as a first step toward personalized health care for FTLD patients.

TDP-43

TDP-43 is a DNA-, RNA-, and protein-binding nucleoprotein implicated in the regulation of numerous processes, including transcription, splicing, cell cycle regulation, apoptosis, microRNA biogenesis, mRNA transport to and local translation at the synapse,

TDP-43: 43 kDa transactivating responsive sequence DNA-binding protein

FTLD-TDP: frontotemporal lobar degeneration with TDP-43 pathology

FTLD-UPS: frontotemporal lobar degeneration with tau- and TDP-43-negative inclusions, which are immunoreactive to proteins of the ubiquitin-proteasome system

Table 1 Genetic heterogeneity of FTLD[a]

Locus	Gene	Mutation frequency	Mutation type	Phenotype	Pathology[b]	References
3p11	CHMP2B	<1%	missense	FTLD	FTLD-UPS	Skibinski et al. 2005
9p13	VCP	<1%	missense, nonsense	IBMPFD	FTLD-TDP	Watts et al. 2004
9p	?	?	?	FTLD+MND	FTLD-TDP	Morita et al. 2006[c]
17q21	MAPT	5–10%	missense, deletion, silent, splice-site	FTLD	FTLD-tau	Hutton et al. 1998
17q21	GRN	5–10%	deletion, nonsense, frameshift, met1, missense, splice-site	FTLD	FTLD-TDP	Baker et al. 2006, Cruts et al. 2006

[a]Abbreviations: CHMP2B, charged multivesicular body protein 2B; VCP, valosin-containing protein; MAPT, microtubule associated protein tau; GRN, progranulin; FTLD, frontotemporal lobar degeneration; IBMPFD, inclusion body myopathy associated with Paget's disease of bone and frontotemporal dementia; MND, motor neuron disease; UPS, ubiquitine-proteasome system; TDP, TDP-43.
[b]Nomenclature according to Mackenzie et al. 2009.
[c]First family published.

and scaffolding for nuclear bodies (for review, see Buratti & Baralle 2008). Structurally, TDP-43 contains two RNA-recognition motifs and a carboxyl (C)-terminal glycine-rich region involved in protein-binding (Wang et al. 2004). Its exact role in neurodegeneration is still a topic of intense investigation. Inclusion-bearing cells (neurons, but also glia) often show a change in subcellular distribution of TDP-43, with loss of nuclear TDP-43 and cytoplasmic sequestration (Arai et al. 2006, Neumann et al. 2006). These inclusions contain both full-length TDP-43 and N-terminally truncated fragments, which are hyperphosphorylated and ubiquitinated. These posttranslational modifications are late events in the pathological process (Dormann et al. 2009, Neumann 2009), and N-terminal truncation is not required for aggregation (Dormann et al. 2009). Like tauopathy, TDP-43 proteinopathy is not unique to FTLD but is also a pathological hallmark of other neurodegenerative disorders, e.g., amyotrophic lateral sclerosis (ALS) with or without dementia (Arai et al. 2006, Neumann et al. 2006) and Perry syndrome, an autosomal dominant form of parkinsonism (Wider et al. 2009). Furthermore, investigators have observed secondary TDP-43 pathology in various other neurodegenerative disorders, including AD, hippocampal sclerosis,

α-synucleinopathies [Parkinson disease (PD), dementia with Lewy bodies (DLB)] and Huntington disease (for review, see Geser et al. 2009). Inclusions in cortical brain regions are rich in C-terminal fragments, whereas inclusions in spinal cord contain more full-length TDP-43 (Igaz et al. 2008, Neumann et al. 2009), suggesting different pathomechanisms in at least some TDP-43 proteinopathies, but phosphorylation at serine residues 409 and 410 is a shared feature of all known TDP-43 proteinopathies (Neumann et al. 2009). Nevertheless, the inclusions contain TDP-43, leaving the possibility that TDP-43 accumulation is not a primary event in the pathogenesis, but rather a by-product of neurodegeneration sequestered by other aggregated components. Strongest evidence that aberrant TDP-43 is sufficient to trigger neurodegeneration, however, comes from the detection of autosomal dominant mutations in the gene encoding TDP-43 (TARDBP) in familial ALS (Kabashi et al. 2008, Sreedharan et al. 2008, Van Deerlin et al. 2008). To date, most mutations identified reside in the C-terminal glycine-rich region of the gene (http://www. molgen.ua.ac.be/FTDmutations), which is necessary for binding of heterogeneous nuclear ribonucleoproteins (hnRNP) in exon skipping and splicing activity (Buratti et al. 2005).

Proteinopathy: pathology characterized by the abnormal aggregation of proteins, e.g., TDP-43 or tau; a frequent neuropathological observation in neurodegenerative diseases

ALS: amyotrophic lateral sclerosis

Several mutations are predicted to affect phosphorylation by introducing new phosphorylation sites or increasing phosphorylation at adjacent residues, suggesting a role for abnormal phosphorylation in neurodegeneration (Neumann et al. 2009), although phosphorylation was found to be a relatively late event in the conversion of soluble to insoluble TDP-43 (Dormann et al. 2009). The observation of pathological TDP-43 inclusions in the cytoplasm suggests that a loss of normal function in the nucleus could result in neurodegeneration, e.g., through a dysfunctional shuttling of TDP-43 to and from the nucleus and the cytoplasm (Winton et al. 2008a). *Drosophila melanogaster* depleted of TDP-43 shows atrophic presynaptic terminals at the neuromuscular junction, impaired locomotion, and reduced life span, which can be rescued by expressing human TDP-43; this observation is indicative of a pathogenic loss-of-function mechanism (Feiguin et al. 2009). The pattern of nuclear clearing of TDP-43 is in line with a loss-of-function mechanism, but it is not obvious how the intranuclear inclusions of TDP-43, which are frequently observed in *GRN* and *VCP* mutation carriers (Mackenzie et al. 2006, Sampathu et al. 2006), for example, fit into this picture. Conversely, a toxic gain of function is also conceivable. In a yeast model, several *TARDBP* mutations accelerated aggregation of TDP-43 and were more toxic than wild-type TDP-43 expressed at equally high levels, suggesting a toxic gain of function (Johnson et al. 2009). Other investigators (Igaz et al. 2009, Zhang et al. 2009) have also observed cytotoxicity of C-terminal fragments. Overexpression of TDP-43 in rat substantia nigra through an adeno-associated virus vector brought about a neurodegeneration pattern that resembles human TDP-43 proteinopathy as well as behavioral motor dysfunction (Tatom et al. 2009), an observation that was further strengthened by TDP-43 transgenic mouse lines (whether overexpressing mutant or wild-type human TDP-43) that displayed a neurodegenerative phenotype (Wegorzewska et al. 2009, Wils et al.

2010), with homozygous and hemizygous wild-type TDP-43 transgenic mice showing dose-dependent neurodegeneration (Wils et al. 2010). On the other hand, both overexpression of mutant human TDP-43 and knockdown of tardbp in *Danio rerio* embryos caused a similar motor phenotype, which suggested that both gain- and loss-of-function of TDP-43 can induce neurodegeneration (Kabashi et al. 2010). Further proof could come from the observation of genetic variants affecting gene dosage, but so far, no copy-number variants have been detected in ALS or FTLD (Gijselinck et al. 2009). Of note, although *TARDBP* missense mutations appeared to be unique for ALS, missense mutations have been reported in patients with a clinical diagnosis of FTLD with or without MND (Benajiba et al. 2009, Winton et al. 2008b), and Kovacs et al. (2009) recently described a novel missense mutation in a patient with pathologically confirmed FTLD-TDP with concomitant supranuclear palsy and choreatic movements, but without signs of MND. Although the evidence supporting a pathogenic nature of some of these variants is not yet conclusive (e.g., lack of family data for segregation analysis or functional assay), these data suggest that genetic screening for *TARDBP* mutations should be considered in the broader spectrum of neurodegeneration.

PROGRANULIN

Progranulin (PGRN) is a ubiquitously expressed secreted precursor protein that contains tandem repeats of a unique (10- or) 12-cysteine (Cys) motif, which are proteolytically cleaved to form seven granulin (grn) peptides (grn A-G) (He & Bateman 2003). Both full-length PGRN and the grn peptides are implicated in a wide variety of biological functions, starting at embryonic development, including cell cycle regulation, wound repair, tumor growth, and inflammation, sometimes with opposite effects (e.g., Daniel et al. 2003, He et al. 2003, He & Bateman 2003, Plowman et al. 1992). The balance between PGRN and grn peptides

Loss-of-function mutation: any mutation that gives rise to a gene product with a complete or partial loss of its function

PGRN (or *GRN*): progranulin (protein/*gene*)

Nonsense mediated mRNA decay: a cellular surveillance mechanism that degrades aberrant mRNAs containing premature translation termination codons

Null allele mutation: a mutation in an allele that prevents transcription of the gene and/or translation into a functional protein

Haploinsufficiency: condition in which loss of one functional allele of a gene is sufficient to cause an abnormal phenotype

is regulated by secretory leukocyte protease inhibitor (SLPI), which binds to PGRN to prevent proteolysis by elastase (Zhu et al. 2002). PGRN is abundantly expressed in mitotically active tissues or upon injury in tissues that are less mitotically active (Daniel et al. 2000). Elevated PGRN is a frequent observation in tumors (He & Bateman 2003). PGRN is also expressed in the central nervous system, already at the embryonic stage, and significant expression has been observed in the pyramidal and granule cells of the hippocampus, the Purkinje cells, and cortical neurons (Daniel et al. 2000, Daniel et al. 2003), but its exact role in the postmitotic neurons of the adult nervous system is less well-characterized.

All pathogenic *GRN* mutations identified in FTLD so far are dominant loss-of-function mutations, including whole gene deletions (Gijselinck et al. 2008b) and nonsense and frameshift mutations leading to premature termination codons with subsequent nonsense mediated mRNA decay of the mutant transcript, as well as mutations in the signal peptide leading to protein mislocalization and degradation and mutations in the initiation codon preventing translation (i.e., *GRN* null allele mutations) (for review, see Cruts & Van Broeckhoven 2008, Gijselinck et al. 2008a). These dominant null allele mutations in *GRN* invariably result in PGRN haploinsufficiency (i.e., loss of 50% PGRN), implying that PGRN is critical for survival of neurons in the adult brain. Several observations further support this hypothesis. Investigators have documented significantly increased levels of PGRN in diseased tissue in ALS (Malaspina et al. 2001, Irwin, Lippa & Rosso 2009), Creutzfeldt-Jakob disease (Baker & Manuelidis 2003), and AD (Baker et al. 2006; Pereson et al. 2009). Furthermore, full-length PGRN enhanced axonal outgrowth, and grn E promoted neuronal survival of cultured neurons (Van Damme et al. 2008). Last, *GRN* mutation carriers can show a wide range of clinical presentations, including AD, PD, and CBS (e.g., Benussi et al. 2009, Brouwers et al. 2007, Le Ber et al. 2008, Rademakers et al. 2007), which could be interpreted to reflect a loss of neuroprotection against other looming neurodegenerative injuries.

GRN null allele mutations have not yet been observed in patients with ALS, despite the presence of a similar TDP-43 proteinopathy, which suggests that motor neurons are less sensitive to a 50% reduction in PGRN. Partial loss of GRN is clearly not detrimental for all functions of PGRN because patients carrying a GRN null allele mutation have no other apparent abnormalities, e.g., in development, growth, or wound repair. There could be many reasons why GRN is not always harmful, including functional redundancy in these important biological processes, regulation of expression of the functional allele of the gene, differences in signal transduction (Zanocco-Marani et al. 1999), differences in receptors (which are currently unknown), and/or the presence of strong modifying factors. Homozygous Grn knockout mice show only a mild male-type behavioral phenotype (Kayasuga et al. 2007), underscoring that PGRN is not indispensable for many of the biological processes in which it is involved. Of note, even though some patients develop first symptoms of FTLD in their early forties, some mutation carriers live well beyond 70 years without any cognitive complaints, even within one family (e.g., Brouwers et al. 2007), which strongly indicates modifying factors. Further exploration of these aspects will likely be instrumental in identifying targets for therapy.

CAN THE UNDERLYING ETIOLOGY BE ESTABLISHED IN VIVO?

In postmortem tissue, distinct pathological profiles of FTLD-TDP can be discerned that appear to correlate well with the clinical and genetic subtypes (Mackenzie et al. 2006, Sampathu et al. 2006). Small neurites, cytoplasmic inclusions, and intranuclear inclusions are frequent in *GRN* mutation carriers and in FTD or PNFA patients. Intranuclear inclusions are less frequent in families linked to chromosome 9 and patients with FTD-MND or SD, and the latter often have a predominance of long neurites. Patients with IBMPFD with a *VCP* mutation typically have more intranuclear than cytoplasmic inclusions (Mackenzie et al. 2006, Sampathu et al. 2006). Conversely, the ability to distinguish between the different etiological pathways during life is less straightforward, even though this knowledge is crucial to make the correct treatment decisions when drugs targeting the pathological proteins are developed. *VCP* mutations cause the rare disorder IBMPFD (Watts et al. 2004), but this correlation is not perfect; even in carriers of an identical *VCP* mutation, investigators have observed marked heterogeneity in clinical expression, such as *VCP* p.Arg159His, which presented with FTD only in one family, with FTD and/or Paget's disease of the bone in another family, and with predominant inclusion body myopathy or Paget's disease in a third family (van der Zee et al. 2009). Nevertheless, presence of Paget's disease or inclusion body myopathy in a patient with FTD strongly suggests an underlying *VCP* mutation. Apart from FTLD-TDP caused by rare mutations in *VCP*, however, no set of clinical symptoms is specific for one genetic or pathological subtype of FTLD (Boeve & Hutton 2008). Roughly half of patients with FTD will have FTLD-tau, and the other half will have FTLD-TDP (Forman et al. 2006, Johnson et al. 2005). The presence of MND is thought to be more frequent in FTLD-TDP, but not in those patients carrying a *GRN* mutation (e.g., Pickering-Brown et al. 2008). Several symptoms are present at an unusually high frequency in *GRN* mutation carriers, including early ideomotor apraxia, early memory impairment, and hallucinations (Rademakers et al. 2007, Le Ber et al. 2008), the latter two sometimes leading to an initial clinical diagnosis of AD or DLB. Furthermore, onset age tends to be earlier in *MAPT* mutation carriers than in *GRN* mutation carriers (Cruts & Van Broeckhoven 2008). However, none of these features allows perfect discrimination, and they might also be dependent on the presence of modifying factors in families. In addition, *GRN* mutation carriers may have an atypical clinical presentation, such as CBS or

AD [15% in French series (Le Ber et al. 2008)], and go unnoticed when no relatives are affected. At neuroimaging of *GRN* mutation carriers, asymmetrical atrophy seems to be a frequent feature (e.g., Beck et al. 2008, Le Ber et al. 2008, van der Zee et al. 2007). For positron emission tomography, development of radioisotopes and tracers specific for tau or TDP-43 will likely be an important step forward to allow in vivo imaging of the proteinopathies comparable to imaging of amyloid deposition in AD using Pittsburgh compound B (Klunk et al. 2004). The inconsistent results of biomarker studies in FTLD focusing on levels of tau in cerebrospinal fluid (CSF) (Green et al. 1999, Grossman et al. 2005) may have been caused by unrecognized etiological heterogeneity in the study population, with an expected 50% of tau-negative patients (Bian & Grossman 2007). The recent advances, however, have stimulated the exploration of disease-specific biomarkers. Foulds et al. (2008) reported TDP-43 protein plasma levels to be elevated in approximately half of the patients with clinical FTD, compatible with what is expected based on neuropathological studies. This biomarker assay will most likely facilitate drug trials targeting TDP-43 pathology by allowing a more careful trial design. Plasma TDP-43 levels were also elevated in 20% of AD patients and 8% of cognitively healthy individuals (Foulds et al. 2008). Similarly, TDP-43 levels in CSF were elevated in patients with FTLD and ALS but showed considerable overlap with control individuals (Kasai et al. 2009, Steinacker et al. 2008). In contrast, reduced PGRN levels in serum and plasma proved to be up to 100% sensitive and specific for underlying *GRN* mutations (Finch et al. 2009, Ghidoni et al. 2008, Sleegers et al. 2009) already in unaffected mutation carriers. Because this biomarker accurately reflects the underlying disease entity and can already predict disease in a prodromal stage, it holds great promise for the future of FTLD disease management. Furthermore, it may have an application in drug development aiming to restore PGRN levels because elevated levels are associated with various cancers. A panel of rapid, noninvasive tests covering all distinct etiological entities should ultimately facilitate the diagnostic process.

CAN GRN MUTATIONS OTHER THAN NULL ALLELE MUTATIONS CAUSE DISEASE?

Large-scale sequencing projects of *GRN* have uncovered other genetic variants that occur only, or more frequently, in patients, but which are not manifestly null allele mutations (**http://www.molgen.ua.ac.be/ FTDmutations**). The pathogenic potential of some of these genetic variants is still hotly debated. Patient-specific missense mutations have been detected, for example, in the evolutionarily highly conserved grn domains (Brouwers et al. 2008, van der Zee et al. 2007). Mutations that introduce or replace a Cys-residue are especially predicted to be pathogenic because Cys-residues are required to stabilize the fold of four stacked beta-hairpins characteristic for the grn domains through disulfide bridges between these residues (Hrabal et al. 1996, Tolkatchev et al. 2000). These mutations could result in misfolding of the protein, which may lose its normal activity (Tolkatchev et al. 2008) or be destined for degradation in the endoplasmic reticulum (ER), resulting in reduced amounts of functional protein. In support of this hypothesis, pathogenic Cys-residue mutations in several other proteins containing disulfide bond-rich epidermal-growth-factor (EGF)-like domains [fibrillin-1 (*FBN1*), uromodulin (*UMOD*)] (Robinson et al. 2002, Scolari et al. 2004) indeed induce retention of the mutant transcript in the ER because of abnormal protein folding (Bernascone et al. 2006, Whiteman & Handford 2003). Alternatively, because gain or loss of a Cys-residue results in an odd number of Cys-residues, the unpaired Cys-residue in turn could have an increased propensity to form multimers through aberrant disulfide bond formation, as was recently reported for pathogenic mutations in *NOTCH3*, which caused cerebral autosomal dominant arteriopathy with subcortical

infarcts and leukoencephalopathy (CADASIL) (Opherk et al. 2009). Pathogenic mutations in *NOTCH3* almost all lead to an odd number of Cys-residues in one of its EGF-like domains (Federico et al. 2005, Joutel et al. 1997). In a yeast-two hybrid assay, a grn dimeric repeat consisting of grn B and A was shown to bind human immunodeficiency virus (HIV) Tat proteins, and mutation of grn Cys-residues weakened this binding ability (Trinh et al. 1999), further highlighting the pathogenic potential of Cys-residue mutations.

Other *GRN* missense mutations have affected highly conserved Pro-residues occurring in a loop of the protein backbone of the grn domain. Molecular modeling predicted amino acid substitutions at this position to affect folding and stability of the grn domain (Brouwers et al. 2008, van der Zee et al. 2007). Arguing against a putative pathogenic nature of *GRN* missense mutations is the fact that some, albeit different, missense mutations also occur in healthy control individuals. FTLD-TDP has not been pathologically confirmed, and segregation data are sparse. One mutation, p.Arg432Cys, was observed in three (genetically distantly) related patients with clinical FTLD (van der Zee et al. 2007); p.Cys521Tyr was observed in four affected relatives from a PNFA family but also in several unaffected siblings, although cognitive tests of carriers predicted a mild cognitive impairment (Cruchaga et al. 2009). However, in cultured cells, two predicted pathogenic mutations, p.Pro248Leu and p.Arg432Cys, showed less efficient transport through the secretory pathway, which led to a significant (70% and 45%) reduction in PGRN secretion (Shankaran et al. 2008). This partial loss of functional protein suggests that these rare mutant alleles act as low penetrant risk factors, as opposed to the autosomal dominant nature of null allele mutations that create complete haploinsufficiency. This hypothesis is very well compatible with the presence of some of these mutations in control individuals and with the lack of segregation data. Of note, the frequency of these missense mutations in both patients and healthy individuals is usually rare, with minor allele frequencies less than 1%. In addition, missense mutations are also detected in other neurodegenerative disorders, such as AD (Brouwers et al. 2008), PD (Nuytemans et al. 2008), and ALS (Schymick et al. 2007, Sleegers et al. 2008), which again suggests that these rare mutant alleles might increase susceptibility for neurodegeneration by failing neurotrophic properties. In line with this speculation, in vivo data on circulating protein levels in plasma or serum show a partial reduction of PGRN in some predicted pathogenic missense mutation carriers [e.g., unrelated carriers of p.Cys139Arg had partially reduced PGRN levels in two independent studies (Finch et al. 2009, Sleegers et al. 2009)], with levels intermediate between noncarriers and null mutation carriers, regardless of their clinical phenotype. Because of their low frequency and possibly low penetrance, definite proof of the pathogenicity of some of these mutations relies on continued functional experiments, which could, among others, address the question of whether loss of one grn peptide is sufficient to affect disease risk, and if so, which grn peptides are relevant for neuroprotection, whether acting alone or in multimers. Of note, the ability to adopt a grn fold is necessary but not sufficient for biological activity of the grn peptides (Tolkatchev et al. 2008), which suggests that not every grn domain will be equally sensitive to amino acid substitutions. Common *GRN* genetic variants have also been implicated in disease risk, affecting onset age and survival in patients with ALS (Sleegers et al. 2008) among others, and a common 3' UTR variant affected risk of FTLD by influencing a microRNA binding site (Rademakers et al. 2008); however, other investigators have been unable to replicate these findings (Rollinson et al. 2010). Because of their potential therapeutic implications, however, these findings warrant follow-up.

PGRN AND TDP-43

One of the puzzles that remain to be resolved is whether and how loss of PGRN is linked with the pathological accumulation of TDP-43. Mutant *TARDBP* is clearly sufficient to induce TDP-43 pathology in ALS patients, but mutations in *TARDBP* are at best a rare occurrence in FTLD (Benajiba et al. 2009, Gijselinck et al. 2009, Kovacs et al. 2009, Winton et al. 2008b), and conversely, despite similarities at the pathological level between FTLD-TDP and ALS, *GRN* null allele mutations have not been found in ALS (Schymick et al. 2007, Sleegers et al. 2008). *VCP* mutations, which are associated with predominant neuronal intranuclear TDP-43 inclusions (Mackenzie et al. 2006, Sampathu et al. 2006), altered localization of TDP-43 between nucleus and cytosol, and a minority of TDP-43-positive inclusions was also VCP-immunoreactive (Gitcho et al. 2009); for *GRN* mutations, however, no such evidence exists at present. Knockdown of neuronal *GRN* expression using small interfering RNA in cell culture models led to abnormal caspase 3-mediated cleavage of TDP-43 (Zhang et al. 2007), but this finding has been challenged by others (Shankaran et al. 2008). Brains of 7–8-month-old PGRN knockout mice showed no evidence for accumulation of C-terminal fragments of TDP-43 or of phosphorylated TDP-43 (Dormann et al. 2009). In an axotomy model in mice, TDP-43 showed increased expression and relocalization to the cytosol following acute neuronal injury, while a neuronal decrease in PGRN was observed (Moisse et al. 2009). The increase in TDP-43 upon injury indicates a physiological role in neuronal repair. The concomitant decrease in neuronal PGRN is less well interpretable but could reflect secretion of PGRN, given the neurotrophic properties of secreted PGRN (Moisse et al. 2009). However, Matzilevich et al. (2002) also reported a delayed response of PGRN expression to acute neuronal injury, which suggests that it has a more prominent role in long-term neuronal survival (Ahmed et al. 2007). In surrounding microglia, expression of PGRN was upregulated in response to axotomy, which is compatible with a role in inflammation and repair (Ahmed et al. 2007). Nevertheless, this occurrence does not imply a direct functional link between TDP-43 and PGRN. Given the possible role of grn peptides in HIV Tat protein binding (Trinh et al. 1999), and the ability of TDP-43 to bind to the HIV TAR DNA element to which Tat proteins bind to activate gene expression (Ou et al. 1995), Ahmed et al. (2007) have speculated that PGRN and TDP-43 may be able to interact directly; this awaits further exploration, however.

TDP-43 AND FUS SUGGEST A ROLE FOR ABERRANT RNA PROCESSING IN NEURODEGENERATION

Much may be learned from searching for unifying themes in related neurodegenerative disorders. For FTLD-TDP, ALS seems to be particularly relevant. Like FTLD, ALS is a mechanistically heterogeneous disorder, but one of the major ALS pathologies is TDP-43 proteinopathy (Arai et al. 2006, Neumann et al. 2006). Substantial overlap also exists at the clinical level. FTD is complicated by MND in a significant proportion of patients, and up to 50% of ALS patients develop symptoms of frontotemporal dysfunction (Lomen-Hoerth et al. 2003). Moreover, multiplex families in which FTLD-TDP and ALS cosegregate have been linked to a region on chromosome 9p, which implies that mutations in one gene could induce both clinical phenotypes (Le Ber et al. 2009, Morita et al. 2006, Valdmanis et al. 2007, Vance et al. 2006; Gijselinck et al. 2010). These data support the notion that FTLD-TDP and ALS are closely related conditions in a continuum of neurodegeneration and that scientific breakthroughs in one area may have a ripple effect extending to other disorders along the continuum. Dominant familial ALS linked to chromosome 16 was recently shown to result from pathogenic mutations in *FUS* (fusion in sarcoma) (Kwiatkowski et al. 2009, Vance et al. 2009). FUS is an RNA/DNA processing protein like TDP-43, and its implication in ALS bears remarkable resemblance to TDP-43 (Sleegers & Van Broeckhoven 2009). Like TDP-43, FUS is associated with hnRNP, and

its functions include DNA repair, transcription, RNA splicing, transport of mRNA to and rapid local RNA translation at the synapse, and microRNA processing (for review, see Lagier-Tourenne & Cleveland 2009). Most mutations identified in *FUS* to date reside in the C-terminus (Kwiatkowski et al. 2009, Vance et al. 2009), resembling the C-terminal clustering of pathogenic mutations in *TARDBP*. In spinal cord and brain tissue of *FUS* mutation carriers, cytoplasmic inclusions containing mutant FUS protein could be observed, and in vitro experiments confirmed relocation of mutant FUS to the cytoplasm (Kwiatkowski et al. 2009, Vance et al. 2009). *FUS* mutation carriers did not show TDP-43 pathology, which implies the involvement of different pathways that lead to neurodegeneration. Mutations could lead to a loss of normal function or a gain of toxic function of FUS in the nucleus, resulting in (potentially general) impaired RNA processing. Instead of RNA metabolism per se, a specific RNA target of FUS could be crucial for neuron viability; identifying such RNA targets may provide a link between FUS and TDP-43 but also among other ALS-associated RNA processing proteins such as ELP3 (Simpson et al. 2008). Mislocalization of mutant FUS to the cytoplasm could create a loss of its function in the nucleus, but cytoplasmic aggregates of FUS could also be toxic to the cell or lead to sequestration of other proteins in the cytoplasm. Last, mutant FUS or TDP-43 could perturb normal mRNA transport to the dendrites and/or impair rapid RNA translation at the postsynaptic site (e.g., in response to stimulation by neurotrophic factors). Future research will likely shed more light on the mechanisms by which aberrant FUS as well as TDP-43 affect neuron viability (**Figure 1**). The identification of aberrant TDP-43 and FUS in motor neuron degeneration provides further support for the hypothesis that impaired RNA processing plays a crucial role in neurodegeneration (Gallo et al. 2005). Examples of impaired RNA processing seem especially numerous in MND (for review, see Simpson et al. 2008). Overexpression of TDP-43 in vitro enhances the transcript inclusion of exon 7 of survival of motor neuron 2 (*SMN2*) (Bose et al. 2008), encoding an RNA-binding protein implicated in the neurodegenerative disorder spinal muscular atrophy (Lefebvre et al. 1995). In FTLD, RNA processing is already indirectly implicated through alternative splicing of *MAPT* exon 10. Thus it is also of interest that synthetic multimers of grn repeats have been implicated in RNA processing by modulating transcription elongation through an interaction with cyclin T1 (Hoque et al. 2003). Broader involvement of RNA processing in FTLD is likely because it is involved in numerous biological processes specific for proper neuronal functioning, including synaptic plasticity (Steward & Schuman 2001). This, and the overlap between ALS and FTLD-TDP, raises the question of whether FUS could also be involved in FTLD (Sleegers & Van Broeckhoven 2009). Although the research is in its early days, Mackenzie (2009) recently reported that some FTLD brains that were previously classified as FTLD-UPS because of the presence of tau- as well as TDP-43-negative inclusions harbor FUS-positive inclusions; likewise, we have recently identified an FTLD patient carrying a *FUS* missense mutation (p.Met254Val) affecting an evolutionarily conserved residue in the glycine-rich region of *FUS*, which is predicted to affect protein function (Van Langenhove et al. 2010).

CONCLUDING REMARKS

The identification of mutations in *MAPT*, *GRN*, *VCP*, and *CHMP2B* in FTLD patients and the observation of pathological intraneuronal accumulation of tau, TDP-43, and FUS protein in brains of FTLD patients demonstrate the diversity of molecular pathways involved in this condition. Rare variants in *TARDBP* and *FUS* in FTLD patients may further contribute to the genetic heterogeneity of FTLD, and at least one other genetic cause for FTLD (on 9p) remains to be uncovered. These recent observations suggest that the field of FTLD research might be on the verge of a new wave of discoveries. Although the search

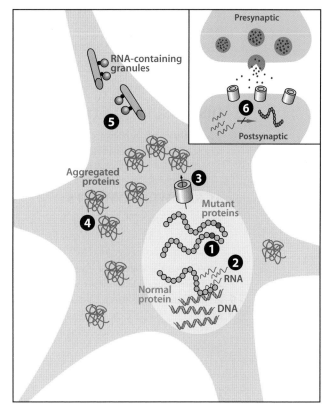

Figure 1

Schematic representation of possible mechanisms by which mutant TARDBP or FUS might affect neuronal viability. (1) Mutant TDP-43 or FUS could lead to a loss of normal function or a gain of toxic function in the nucleus, resulting in impaired RNA processing; (2) a specific RNA target (possibly shared by FUS and TDP-43) could be crucial for neuron viability; (3) mislocalization of mutant FUS or TDP-43 to the cytoplasm could create a loss of its function in the nucleus; (4) cytoplasmic aggregates of FUS or TDP-43 (full-length or C-terminal fragments, ubiquitinated and phosphorylated) could be toxic to the cell; (5) mutant FUS or TDP-43 could perturb normal mRNA transport or (6) impair local RNA translation at the synapse in response to synaptic stimulation.

for mechanistic links between these pathways may provide novel insights into the neurodegenerative process, recognition of the etiologial diversity will be crucial to advance patient care. It allows studies focusing on more homogeneous etiological subclasses, which can facilitate development of diagnostic procedures and drugs, but it may also increase our understanding of disease mechanisms. The first genome-wide association study on FTLD, e.g., including only patients with FTLD-TDP, recently uncovered common genetic variation in the gene *TMEM106B* as a putative risk factor for FTLD-TDP (Van Deerlin et al. 2010). Further characterization of this transmembrane protein of unknown function may provide one more piece of the puzzle of FTLD. Hopefully, in the not-too-distant future the various etiological entities underlying FTLD will be fully characterized.

DISCLOSURE STATEMENT

The authors are not aware of any affiliations, memberships, funding, or financial holdings that might be perceived as affecting the objectivity of this review.

ACKNOWLEDGMENTS

Research in the authors' research group was funded in part by the Special Research Fund of the University of Antwerp, the Fund for Scientific Research Flanders (FWO-V), the Interuniversity Attraction Poles program (IAP) P6/43 of the Belgian Science Policy Office, the Medical Foundation Queen Elisabeth, the Foundation for Alzheimer Research (SAO/FRMA), the Association for Frontotemporal Dementias (AFTD), and a Methusalem Excellence Grant of the Flemish Government, Flanders, Belgium, and a Zenith award from the Alzheimer's Association USA to C.V.B., a Marie-Thérèse de Lava Fund of the King Baudouin Foundation Belgium to M.C., and a Santkin award from the National Alzheimer League Belgium to M.C. and K.S. K.S. is a postdoctoral fellow of the FWO-V.

LITERATURE CITED

Ahmed Z, Mackenzie IR, Hutton ML, Dickson DW. 2007. Progranulin in frontotemporal lobar degeneration and neuroinflammation. *J. Neuroinflamm.* 4:7

Arai T, Hasegawa M, Akiyama H, Ikeda K, Nonaka T, et al. 2006. TDP-43 is a component of ubiquitin-positive tau-negative inclusions in frontotemporal lobar degeneration and amyotrophic lateral sclerosis. *Biochem. Biophys. Res. Commun.* 351(3):602–11

Baker CA, Manuelidis L. 2003. Unique inflammatory RNA profiles of microglia in Creutzfeldt-Jakob disease. *Proc. Natl. Acad. Sci. USA* 100:675–79

Baker M, Mackenzie IR, Pickering-Brown SM, Gass J, Rademakers R, et al. 2006. Mutations in progranulin cause tau-negative frontotemporal dementia linked to chromosome 17. *Nature* 442(7105):916–19

Beck J, Rohrer JD, Campbell T, Isaacs A, Morrison KE, et al. 2008. A distinct clinical, neuropsychological and radiological phenotype is associated with progranulin gene mutations in a large UK series. *Brain* 131(Pt. 3):706–20

Benajiba L, Le Ber I, Camuzat A, Lacoste M, Thomas-Anterion C, et al. 2009. TARDBP mutations in motoneuron disease with frontotemporal lobar degeneration. *Ann. Neurol.* 65(4):470–73

Benussi L, Ghidoni R, Pegoiani E, Moretti DV, Zanetti O, Binetti G. 2009. Progranulin Leu271LeufsX10 is one of the most common FTLD and CBS associated mutations worldwide. *Neurobiol. Dis.* 33(3):379–85

Bergmann M, Kuchelmeister K, Schmid KW, Kretzschmar HA, Schröder R. 1996. Different variants of frontotemporal dementia: a neuropathological and immunohistochemical study. *Acta Neuropathol.* 92(2):170–79

Bernascone I, Vavassori S, Di Pentima A, Santambrogio S, Lamorte G, et al. 2006. Defective intracellular trafficking of uromodulin mutant isoforms. *Traffic* 7:1567–79

Bian H, Grossman M. 2007. Frontotemporal lobar degeneration: recent progress in antemortem diagnosis. *Acta Neuropathol.* 114(1):23–29

Boeve BF. 2007. Links between frontotemporal lobar degeneration, corticobasal degeneration, progressive supranuclear palsy, and amyotrophic lateral sclerosis. *Alzheimer Dis. Assoc. Disord.* 21:S31–38

Boeve BF, Hutton M. 2008. Refining frontotemporal dementia with Parkinsonism linked to chromosome 17: introducing FTDP-17 (MAPT) and FTDP-17 (PGRN). *Arch. Neurol.* 65(4):460–64

Bose JK, Wang IF, Hung L, Tarn WY, Shen CK. 2008. TDP-43 overexpression enhances exon 7 inclusion during the survival of motor neuron pre-mRNA splicing. *J. Biol. Chem.* 283(43):28852–59

Brouwers N, Nuytemans K, Van Der Zee J, Gijselinck I, Engelborghs S, et al. 2007. Alzheimer and parkinson diagnoses in progranulin null mutation carriers in an extended founder family. *Arch. Neurol.* 64(10):1436–46

Brouwers N, Sleegers K, Engelborghs S, Maurer-Stroh S, Gijselinck I, et al. 2008. Genetic variability in progranulin contributes to risk for clinically diagnosed Alzheimer disease. *Neurology* 71(9):656–64

Buratti E, Baralle FE. 2008. Multiple roles of TDP-43 in gene expression, splicing regulation, and human disease. *Front. Biosci.* 13:867–78

Buratti E, Brindisi A, Giombi M, Tisminetzky S, Ayala YM, Baralle FE. 2005. TDP-43 binds heterogeneous nuclear ribonucleoprotein A/B through its C-terminal tail: an important region for the inhibition of cystic fibrosis transmembrane conductance regulator exon 9 splicing. *J. Biol. Chem.* 280(45):37572–84

Cruchaga C, Fernández-Seara MA, Seijo-Martínez M, Samaranch L, Lorenzo E, et al. 2009. Cortical atrophy and language network reorganization associated with a novel progranulin mutation. *Cereb. Cortex* 19(8):1751–60

Cruts M, Gijselinck I, Van Der Zee J, Engelborghs S, Wils H, et al. 2006. Null mutations in progranulin cause ubiquitin-positive frontotemporal dementia linked to chromosome 17q21. *Nature* 442(7105):920–24

Cruts M, Van Broeckhoven C. 2008. Loss of progranulin function in frontotemporal lobar degeneration. *Trends Genet.* 24(4):186–94

Daniel R, Daniels E, He Z, Bateman A. 2003. Progranulin (acrogranin/PC cell-derived growth factor/granulin-epithelin precursor) is expressed in the placenta, epidermis, microvasculature, and brain during murine development. *Dev. Dyn.* 227(4):593–99

Daniel R, He Z, Carmichael KP, Halper J, Bateman A. 2000. Cellular localization of gene expression for progranulin. *J. Histochem. Cytochem.* 48(7):999–1009

Dormann D, Capell A, Carlson AM, Shankaran SS, Rodde R, et al. 2009. Proteolytic processing of TAR DNA binding protein-43 by caspases produces C-terminal fragments with disease defining properties independent of progranulin. *J. Neurochem.* 110(3):1082–94

Federico A, Bianchi S, Dotti MT. 2005. The spectrum of mutations for CADASIL diagnosis. *Neurol. Sci.* 26(2):117–24

Feiguin F, Godena VK, Romano G, D'Ambrogio A, Klima R, Baralle FE. 2009. Depletion of TDP-43 affects Drosophila motoneurons terminal synapsis and locomotive behavior. *FEBS Lett.* 583(10):1586–92

Finch N, Baker M, Crook R, Swanson K, Kuntz K, et al. 2009. Plasma progranulin levels predict progranulin mutation status in frontotemporal dementia patients and asymptomatic family members. *Brain* 132(Pt. 3):583–91

Forman MS, Farmer J, Johnson JK, Clark CM, Arnold SE, et al. 2006. Frontotemporal dementia: clinicopathological correlations. *Ann. Neurol.* 59(6):952–62

Foulds P, McAuley E, Gibbons L, Davidson Y, Pickering-Brown SM, et al. 2008. TDP-43 protein in plasma may index TDP-43 brain pathology in Alzheimer's disease and frontotemporal lobar degeneration. *Acta Neuropathol.* 116(2):141–46

Gallo JM, Jin P, Thornton CA, Lin H, Robertson J, et al. 2005. The role of RNA and RNA processing in neurodegeneration. *J. Neurosci.* 25(45):10372–75

Geser F, Martinez-Lage M, Kwong LK, Lee VM, Trojanowski JQ. 2009. Amyotrophic lateral sclerosis, frontotemporal dementia and beyond: the TDP-43 diseases. *J. Neurol.* 256(8):1205–14

Ghidoni R, Benussi L, Glionna M, Franzoni M, Binetti G. 2008. Low plasma progranulin levels predict progranulin mutations in frontotemporal lobar degeneration. *Neurology* 71(16):1235–39

Gijselinck I, Engelborghs S, Maes G, Cuijt I, Peeters K, et al. 2010. A genome wide linkage study in a multiplex FTLD-ALS family identifies two loci at chromosomes 9 and 14. *Arch. Neurol.* In press

Gijselinck I, Sleegers K, Engelborghs S, Robberecht W, Martin JJ, et al. 2009. Neuronal inclusion protein TDP-43 has no primary genetic role in FTD and ALS. *Neurobiol. Aging* 30(8):1329–31

Gijselinck I, Van Broeckhoven C, Cruts M. 2008a. Granulin mutations associated with frontotemporal lobar degeneration and related disorders: an update. *Hum. Mut.* 29(12):1373–86

Gijselinck I, Van Der Zee J, Engelborghs S, Goossens D, Peeters K, et al. 2008b. Progranulin locus deletion in frontotemporal dementia. *Hum. Mutat.* 29(1):53–58

Gitcho MA, Strider J, Carter D, Taylor-Reinwald L, Forman MS, et al. 2009. VCP mutations causing frontotemporal lobar degeneration disrupt localization of TDP-43 and induce cell death. *J. Biol. Chem.* 284(18):12384–98

Green AJ, Harvey RJ, Thompson EJ, Rossor MN. 1999. Increased tau in the cerebrospinal fluid of patients with frontotemporal dementia and Alzheimer's disease. *Neurosci. Lett.* 259(2):133–35

Grossman M, Farmer J, Leight S, Work M, Moore P, et al. 2005. Cerebrospinal fluid profile in frontotemporal dementia and Alzheimer's disease. *Ann. Neurol.* 57(5):721–29

He Z, Bateman A. 2003. Progranulin (granulin-epithelin precursor, PC-cell-derived growth factor, acrogranin) mediates tissue repair and tumorigenesis. *J. Mol. Med.* 81(10):600–12

He Z, Ong CH, Halper J, Bateman A. 2003. Progranulin is a mediator of the wound response. *Nat. Med.* 9(2):225–29

Hodges JR, Davies R, Xuereb J, Kril J, Halliday G. 2003. Survival in frontotemporal dementia. *Neurology* 61:349–54

Holm IE, Englund E, Mackenzie IR, Johannsen P, Isaacs AM. 2007. A reassessment of the neuropathology of frontotemporal dementia linked to chromosome 3. *J. Neuropathol. Exp. Neurol.* 66:884–91

Hoque M, Young TM, Lee CG, Serrero G, Mathews MB, Pe'ery T. 2003. The growth factor granulin interacts with cyclin T1 and modulates P-TEFb-dependent transcription. *Mol. Cell. Biol.* 23(5):1688–702

Hrabal R, Chen Z, James S, Bennett HP, Ni F. 1996. The hairpin stack fold, a novel protein architecture for a new family of protein growth factors. *Nat. Struct. Biol.* 3(9):747–52

Hutton M, Lendon CL, Rizzu P, Baker M, Froelich S, et al. 1998. Association of missense and 5′-splice-site mutations in tau with the inherited dementia FTDP-17. *Nature* 393(6686):702–5

Igaz LM, Kwong LK, Chen-Plotkin A, Winton MJ, Unger TL, et al. 2009. Expression of TDP-43 C-terminal fragments in vitro recapitulates pathological features of TDP-43 proteinopathies. *J. Biol. Chem.* 284(13):8516–24

Igaz LM, Kwong LK, Xu Y, Truax AC, Uryu K, et al. 2008. Enrichment of C-terminal fragments in TAR DNA-binding protein-43 cytoplasmic inclusions in brain but not in spinal cord of frontotemporal lobar degeneration and amyotrophic lateral sclerosis. *Am. J. Pathol.* 173(1):182–94

Irwin D, Lippa CF, Rosso A. 2009. Progranulin (PGRN) expression in ALS: an immunohistochemical study. *J. Neurol. Sci.* 276(1–2):9–13

Johnson BS, Snead D, Lee JJ, McCaffery JM, Shorter J, Gitler AD. 2009. TDP-43 is intrinsically aggregation-prone and ALS-linked mutations accelerate aggregation and increase toxicity. *J. Biol. Chem.* 284(30):20,329–39

Johnson JK, Diehl J, Mendez MF, Neuhaus J, Shapira JS, et al. 2005. Frontotemporal lobar degeneration: demographic characteristics of 353 patients. *Arch. Neurol.* 62(6):925–30

Josephs KA, Holton JL, Rossor MN, Godbolt AK, Ozawa T, et al. 2004. Frontotemporal lobar degeneration and ubiquitin immunohistochemistry. *Neuropathol. Appl. Neurobiol.* 30(4):369–73

Joutel A, Vahedi K, Corpechot C, Troesch A, Chabriat H, et al. 1997. Strong clustering and stereotyped nature of Notch3 mutations in CADASIL patients. *Lancet* 350(9090):1511–15

Kabashi E, Lin L, Tradewell ML, Dion PA, Bercier V, et al. 2010. Gain and loss of function of ALS-related mutations of TARDBP (TDP-43) cause motor deficits in vivo. *Hum. Mol. Genet.* 19(4):671–83

Kabashi E, Valdmanis PN, Dion P, Spiegelman D, McConkey BJ, et al. 2008. TARDBP mutations in individuals with sporadic and familial amyotrophic lateral sclerosis. *Nat. Genet.* 40(5):572–74

Kasai T, Tokuda T, Ishigami N, Sasayama H, Foulds P, et al. 2009. Increased TDP-43 protein in cerebrospinal fluid of patients with amyotrophic lateral sclerosis. *Acta Neuropathol.* 117(1):55–62

Kayasuga Y, Chiba S, Suzuki M, Kikusui T, Matsuwaki T, et al. 2007. Alteration of behavioural phenotype in mice by targeted disruption of the progranulin gene. *Behav. Brain Res.* 185(2):110–18

Klunk WE, Engler H, Nordberg A, Wang Y, Blomqvist G, et al. 2004. Imaging brain amyloid in Alzheimer's disease with Pittsburgh Compound-B. *Ann. Neurol.* 55(3):306–19

Kovacs GG, Murrell JR, Horvath S, Haraszti L, Majtenyi K, et al. 2009. TARDBP variation associated with frontotemporal dementia, supranuclear gaze palsy, and chorea. *Mov. Disord.* 24(12):1843–47

Kwiatkowski TJ Jr, Bosco DA, Leclerc AL, Tamrazian E, Vanderburg CR, et al. 2009. Mutations in the FUS/TLS gene on chromosome 16 cause familial amyotrophic lateral sclerosis. *Science* 323(5918):1205–8

Lagier-Tourenne C, Cleveland DW. 2009. Rethinking ALS: the FUS about TDP-43. *Cell* 136(6):1001–4

Le Ber I, Camuzat A, Berger E, Hannequin D, Laquerrière A, et al. 2009. Chromosome 9p-linked families with frontotemporal dementia associated with motor neuron disease. *Neurology* 72(19):1669–76

Le Ber I, Camuzat A, Hannequin D, Pasquier F, Guedj E, et al. 2008. Phenotype variability in progranulin mutation carriers: a clinical, neuropsychological, imaging and genetic study. *Brain* 131(Pt. 3):732–46

Lefebvre S, Bürglen L, Reboullet S, Clermont O, Burlet P, et al. 1995. Identification and characterization of a spinal muscular atrophy-determining gene. *Cell* 80(1):155–65

Lomen-Hoerth C, Murphy J, Langmore S, Kramer JH, Olney RK, Miller B. 2003. Are amyotrophic lateral sclerosis patients cognitively normal? *Neurology* 60(7):1094–97

Mackenzie IR. 2009. *TDP-43, FUS and new neuropathological insights.* Presented at Int. Res. Workshop on Frontotemporal Dementia in ALS, 3rd, London, Ontario, Can.

Mackenzie IR, Baborie A, Pickering-Brown S, Du Plessis D, Jaros E, et al. 2006. Heterogeneity of ubiquitin pathology in frontotemporal lobar degeneration: classification and relation to clinical phenotype. *Acta. Neuropathol.* 112(5):539–49

Mackenzie IR, Neumann M, Bigio EH, Cairns NJ, Alafuzoff I, et al. 2009. Nomenclature for neuropathologic subtypes of frontotemporal lobar degeneration: consensus recommendations. *Acta Neuropathol.* 117(1):15–18

Malaspina A, Kaushik N, de Belleroche J. 2001. Differential expression of 14 genes in amyotrophic lateral sclerosis spinal cord detected using gridded cDNA arrays. *J. Neurochem.* 77(1):132–45

Matzilevich DA, Rall JM, Moore AN, Grill RJ, Dash PK. 2002. High-density microarray analysis of hippocampal gene expression following experimental brain injury. *J. Neurosci. Res.* 67(5):646–63

Moisse K, Volkening K, Leystra-Lantz C, Welch I, Hill T, Strong MJ. 2009. Divergent patterns of cytosolic TDP-43 and neuronal progranulin expression following axotomy: implications for TDP-43 in the physiological response to neuronal injury. *Brain. Res.* 1249:202–11

Morita M, Al-Chalabi A, Andersen PM, Hosler B, Sapp P, et al. 2006. A locus on chromosome 9p confers susceptibility to ALS and frontotemporal dementia. *Neurology* 66(6):839–44

Neary D, Snowden J, Mann D. 2005. Frontotemporal dementia. *Lancet Neurol.* 4(11):771–80

Neumann M. 2009. Molecular neuropathology of TDP-43 proteinopathies. *Int. J. Mol. Sci.* 10(1):232–46

Neumann M, Kwong LK, Lee EB, Kremmer E, Flatley A, et al. 2009. Phosphorylation of S409/410 of TDP-43 is a consistent feature in all sporadic and familial forms of TDP-43 proteinopathies. *Acta Neuropathol.* 117(2):137–49

Neumann M, Sampathu DM, Kwong LK, Truax AC, Micsenyi MC, et al. 2006. Ubiquitinated TDP-43 in frontotemporal lobar degeneration and amyotrophic lateral sclerosis. *Science* 314(5796):130–33

Nuytemans K, Pals P, Sleegers K, Engelborghs S, Corsmit E, et al. 2008. Progranulin variability has no major role in Parkinson disease genetic etiology. *Neurology* 71(15):1147–51

Opherk C, Duering M, Peters N, Karpinska A, Rosner S, et al. 2009. CADASIL mutations enhance spontaneous multimerization of NOTCH3. *Hum. Mol. Genet.* 18(15):2761–67

Ou SH, Wu F, Harrich D, García-Martínez LF, Gaynor RB. 1995. Cloning and characterization of a novel cellular protein, TDP-43, that binds to human immunodeficiency virus type 1 TAR DNA sequence motifs. *J. Virol.* 69(6):3584–96

Pereson S, Wils H, Kleinberger G, McGowan E, Vandewoestyne M, et al. 2009. Progranulin expression correlates with dense-core amyloid plaque burden in Alzheimer disease mouse models. *J. Pathol.* 219(2):173–81

Pick A. 1892. Uber die beziehungen der senilen hirnatrophie zur aphasie. *Prag. Med. Wochenschr.* 17:165–67

Pickering-Brown SM, Rollinson S, Du Plessis D, Morrison KE, Varma A, et al. 2008. Frequency and clinical characteristics of progranulin mutation carriers in the Manchester frontotemporal lobar degeneration cohort: comparison with patients with MAPT and no known mutations. *Brain* 131(Pt. 3):721–31

Plowman GD, Green JM, Neubauer MG, Buckley SD, McDonald VL, et al. 1992. The epithelin precursor encodes two proteins with opposing activities on epithelial cell growth. *J. Biol. Chem.* 267(18):13073–78

Rademakers R, Baker M, Gass J, Adamson J, Huey ED, et al. 2007. Phenotypic variability associated with progranulin haploinsufficiency in patients with the common 1477C → T (Arg493X) mutation: an international initiative. *Lancet Neurol.* 6(10):857–68

Rademakers R, Cruts M, Dermaut B, Sleegers K, Rosso SM, et al. 2002. Tau negative frontal lobe dementia at 17q21: significant finemapping of the candidate region to a 4.8 cM interval. *Mol. Psychiatry* 7(10):1064–74

Rademakers R, Cruts M, Van Broeckhoven C. 2004. The role of tau (MAPT) in frontotemporal dementia and related tauopathies. *Hum. Mutat.* 24(4):277–95

Rademakers R, Eriksen JL, Baker M, Robinson T, Ahmed Z, et al. 2008. Common variation in the miR-659 binding-site of GRN is a major risk factor for TDP43-positive frontotemporal dementia. *Hum. Mol. Genet.* 17(23):3631–42

Robinson PN, Booms P, Katzke S, Ladewig M, Neumann L, et al. 2002. Mutations of FBN1 and genotype-phenotype correlations in Marfan syndrome and related fibrillinopathies. *Hum. Mutat.* 20:153–61

Rollinson S, Rohrer JD, Van Der Zee J, Sleegers K, Mead S, et al. 2010. No association of PGRN 3′UTR rs5848 in frontotemporal lobar degeneration. *Neurobiol. Aging.* In press

Sampathu DM, Neumann M, Kwong LK, Chou TT, Micsenyi M, et al. 2006. Pathological heterogeneity of frontotemporal lobar degeneration with ubiquitin-positive inclusions delineated by ubiquitin immunohistochemistry and novel monoclonal antibodies. *Am. J. Pathol.* 169(4):1343–52

Schymick JC, Yang Y, Andersen PM, Vonsattel JP, Greenway M, et al. 2007. Progranulin mutations and amyotrophic lateral sclerosis or amyotrophic lateral sclerosis-frontotemporal dementia phenotypes. *J. Neurol. Neurosurg. Psychiatry* 78(7):754–56

Scolari F, Caridi G, Rampoldi L, Tardanico R, Izzi C, et al. 2004. Uromodulin storage diseases: clinical aspects and mechanisms. *Am. J. Kidney Dis.* 44:987–99

Shankaran SS, Capell A, Hruscha AT, Fellerer K, Neumann M, et al. 2008. Missense mutations in the progranulin gene linked to frontotemporal lobar degeneration with ubiquitin-immunoreactive inclusions reduce progranulin production and secretion. *J. Biol. Cell.* 283(3):1744–53

Simpson CL, Lemmens R, Miskiewicz K, Broom WJ, Hansen VK, et al. 2008. Variants of the elongator protein 3 (ELP3) gene are associated with motor neuron degeneration. *Hum. Mol. Genet.* 18(3):472–81

Skibinski G, Parkinson NJ, Brown JM, Chakrabarti L, Lloyd SL, et al. 2005. Mutations in the endosomal ESCRTIII-complex subunit CHMP2B in frontotemporal dementia. *Nat. Genet.* 37:806–8

Sleegers K, Brouwers N, Maurer-Stroh S, van Es MA, Van Damme P, et al. 2008. Progranulin genetic variability contributes to amyotrophic lateral sclerosis. *Neurology* 71(4):253–59

Sleegers K, Brouwers N, Van Damme P, Engelborghs S, Gijselinck I, et al. 2009. Serum biomarker for progranulin-associated frontotemporal lobar degeneration. *Ann. Neurol.* 65(5):603–9

Sleegers K, Van Broeckhoven C. 2009. Motor-neuron disease: rogue gene in the family. *Nature* 458(7237):415–17

Sreedharan J, Blair IP, Tripathi VB, Hu X, Vance C, et al. 2008. TDP-43 mutations in familial and sporadic amyotrophic lateral sclerosis. *Science* 319(5870):1668–72

Steinacker P, Hendrich C, Sperfeld AD, Jesse S, von Arnim CA, et al. 2008. TDP-43 in cerebrospinal fluid of patients with frontotemporal lobar degeneration and amyotrophic lateral sclerosis. *Arch. Neurol.* 65(11):1481–87

Steward O, Schuman EM. 2001. Protein synthesis at synaptic sites on dendrites. *Annu. Rev. Neurosci.* 24:299–325

Tatom JB, Wang DB, Dayton RD, Skalli O, Hutton ML, et al. 2009. Mimicking aspects of frontotemporal lobar degeneration and Lou Gehrig's disease in rats via TDP-43 overexpression. *Mol. Ther.* 17(4):607–13

Tolkatchev D, Malik S, Vinogradova A, Wang P, Chen Z, et al. 2008. Structure dissection of human progranulin identifies well-folded granulin/epithelin modules with unique functional activities. *Protein Sci.* 17(4):711–24

Tolkatchev D, Ng A, Vranken W, Ni F. 2000. Design and solution structure of a well-folded stack of two beta-hairpins based on the amino-terminal fragment of human granulin A. *Biochemistry* 39(11):2878–86

Trinh DP, Brown KM, Jeang KT. 1999. Epithelin/granulin growth factors: extracellular cofactors for HIV-1 and HIV-2 Tat proteins. *Biochem. Biophys. Res. Commun.* 256(2):299–306

Valdmanis PN, Dupre N, Bouchard JP, Camu W, Salachas F, et al. 2007. Three families with amyotrophic lateral sclerosis and frontotemporal dementia with evidence of linkage to chromosome 9p. *Arch. Neurol.* 64(2):240–45

Vance C, Al-Chalabi A, Ruddy D, Smith BN, Hu X, et al. 2006. Familial amyotrophic lateral sclerosis with frontotemporal dementia is linked to a locus on chromosome 9p13.2–21.3. *Brain* 129(Pt. 4):868–76

Vance C, Rogelj B, Hortobágyi T, De Vos KJ, Nishimura AL, et al. 2009. Mutations in FUS, an RNA processing protein, cause familial amyotrophic lateral sclerosis type 6. *Science* 323(5918):1208–11

Van Damme P, Van Hoecke A, Lambrechts D, Vanacker P, Bogaert E, et al. 2008. Progranulin functions as a neurotrophic factor to regulate neurite outgrowth and enhance neuronal survival. *J. Cell. Biol.* 181(1):37–41

Van Deerlin VM, Leverenz JB, Bekris LM, Bird TD, Yuan W, et al. 2008. TARDBP mutations in amyotrophic lateral sclerosis with TDP-43 neuropathology: a genetic and histopathological analysis. *Lancet Neurol.* 7(5):409–16

Van Deerlin VM, Sleiman PMA, Martinez-Lage M, Chen-Plotkin A, Wang L-S, et al. 2010. Common variants at 7p21 are associated with frontotemporal lobar degeneration with TDP-43 inclusions. *Nat. Genet.* 42:234–39

Van Der Zee J, Le Ber I, Maurer-Stroh S, Engelborghs S, Gijselinck I, et al. 2007. Mutations other than null mutations producing a pathogenic loss of progranulin in frontotemporal dementia. *Hum. Mut.* 2(4):416

Van Der Zee J, Pirici D, Van Langenhove T, Engelborghs S, Vandenberghe R, et al. 2009. Clinical heterogeneity in 3 unrelated families linked to VCP p.Arg159His. *Neurology* 73(8):626–32

Van Der Zee J, Rademakers R, Engelborghs S, Gijselinck I, Bogaerts V, et al. 2006. A Belgian ancestral haplotype harbours a highly prevalent mutation for 17q21-linked tau-negative FTLD. *Brain* 129(Pt. 4):841–52

Van Der Zee J, Urwin H, Engelborghs S, Bruyland M, Vandenberghe R, et al. 2008. CHMP2B C-truncating mutations in frontotemporal lobar degeneration are associated with an aberrant endosomal phenotype in vitro. *Hum. Mol. Genet.* 17(2):313–22

Van Langenhove T, van der Zee J, Sleegers K, Engelborghs S, Vandenberghe R, et al. 2010. Genetic contribution of FUS to frontotemporal lobar degeneration. *Neurology* 74(5):366–71

Wang HY, Wang IF, Bose J, Shen CK. 2004. Structural diversity and functional implications of the eukaryotic TDP gene family. *Genomics* 83(1):130–39

Watts GD, Wymer J, Kovach MJ, Mehta SG, Mumm S, et al. 2004. Inclusion body myopathy associated with Paget disease of bone and frontotemporal dementia is caused by mutant valosin-containing protein. *Nat. Genet.* 36:377–81

Wegorzewska I, Bell S, Cairns NJ, Miller TM, Baloh RH. 2009. TDP-43 mutant transgenic mice develop features of ALS and frontotemporal lobar degeneration. *Proc. Natl. Acad. Sci. USA* 106(44):18809–14

Whiteman P, Handford PA. 2003. Defective secretion of recombinant fragments of fibrillin-1: implications of protein misfolding for the pathogenesis of Marfan syndrome and related disorders. *Hum. Mol. Genet.* 12:727–37

Wider C, Dickson DW, Stoessl AJ, Tsuboi Y, Chapon F, et al. 2009. Pallidonigral TDP-43 pathology in Perry syndrome. *Parkinsonism Relat. Disord.* 15(4):281–86

Wils H, Kleinberger G, Janssens J, Pereson S, Joris G, et al. 2010. TDP-43 transgenic mice develop spastic paralysis and neuronal inclusions characteristic of ALS and frontotemporal lobar degeneration. *Proc. Natl. Acad. Sci. USA* 107:3858–63

Winton MJ, Igaz LM, Wong MM, Kwong LK, Trojanowski JQ, Lee VM. 2008a. Disturbance of nuclear and cytoplasmic TAR DNA-binding protein (TDP-43) induces disease-like redistribution, sequestration, and aggregate formation. *J. Biol. Chem.* 283(19):13302–9

Winton MJ, Van Deerlin VM, Kwong LK, Yuan W, Wood EM, et al. 2008b. A90V TDP-43 variant results in the aberrant localization of TDP-43 in vitro. *FEBS Lett.* 582(15):2252–56

Wolfe MS. 2009. Tau mutations in neurodegenerative diseases. *J. Biol. Chem.* 284(10):6021–25

Zanocco-Marani T, Bateman A, Romano G, Valentinis B, He ZH, Baserga R. 1999. Biological activities and signaling pathways of the granulin/epithelin precursor. *Cancer Res.* 59(20):5331–40

Zhang YJ, Xu YF, Cook C, Gendron TF, Roettges P, et al. 2009. Aberrant cleavage of TDP-43 enhances aggregation and cellular toxicity. *Proc. Natl. Acad. Sci. USA* 106(18):7607–12

Zhang YJ, Xu YF, Dickey CA, Buratti E, Baralle F, et al. 2007. Progranulin mediates caspase-dependent cleavage of TAR DNA binding protein-43. *J. Neurosci.* 27(39):10530–34

Zhu J, Nathan C, Jin W, Sim D, Ashcroft GS, et al. 2002. Conversion of proepithelin to epithelins: roles of SLPI and elastase in host defense and wound repair. *Cell* 111(6):867–78

Error Correction, Sensory Prediction, and Adaptation in Motor Control

Reza Shadmehr,[1] Maurice A. Smith,[2]
and John W. Krakauer[3]

[1] Department of Biomedical Engineering, Johns Hopkins School of Medicine, Baltimore, Maryland 21205; email: shadmehr@jhu.edu

[2] School of Engineering and Applied Science, Harvard University, Cambridge, Massachusetts 02138

[3] The Neurological Institute, Columbia University College of Physicians and Surgeons, New York, NY 10032

Annu. Rev. Neurosci. 2010. 33:89–108

First published online as a Review in Advance on March 22, 2010

The *Annual Review of Neuroscience* is online at neuro.annualreviews.org

This article's doi:
10.1146/annurev-neuro-060909-153135

Key Words

forward models, reaching, saccades, motor adaptation, error feedback, sensorimotor integration

Abstract

Motor control is the study of how organisms make accurate goal-directed movements. Here we consider two problems that the motor system must solve in order to achieve such control. The first problem is that sensory feedback is noisy and delayed, which can make movements inaccurate and unstable. The second problem is that the relationship between a motor command and the movement it produces is variable, as the body and the environment can both change. A solution is to build adaptive internal models of the body and the world. The predictions of these internal models, called forward models because they transform motor commands into sensory consequences, can be used to both produce a lifetime of calibrated movements, and to improve the ability of the sensory system to estimate the state of the body and the world around it. Forward models are only useful if they produce unbiased predictions. Evidence shows that forward models remain calibrated through motor adaptation: learning driven by sensory prediction errors.

Contents

INTRODUCTION

We have the ability to control our movements during a vast array of behaviors ranging from making simple limb movements to dribbling a basketball, throwing a baseball, or juggling. We marvel at the accomplishments of athletes, yet, from a theoretical standpoint, even the ability to make the simplest eye and arm movements accurately is quite extraordinary. Consider that the motors that actuate robots reliably produce the same force for a given input. Yet, our muscles quickly fatigue, altering their responses from one movement to the next. The sensors that record motion of a robot do so with far more precision than one finds in the response of our proprioceptive neurons. The transmission lines that connect a robot's motors and sensors to the controller move information at the speed of light, and the controller can process sensory information to issue commands in microseconds. In contrast, our transmission lines (axons) move information slower than the speed of sound, and neural computations often require tens of milliseconds. Therefore, our ability to produce a lifetime of accurate movements lies not in the fact that we are born with an invariant set of actuators, precise set of sensors, or fast transmission lines, but rather in that we are born with a nervous system that adapts to these limitations and continuously compensates for them. If left uncompensated, these inherent limitations could give rise to systematic errors in our movements. How the brain is able to predict and correct systematic errors to produce a lifetime of accurate movements is the subject of this review.

CORRECTING MOVEMENT ERRORS WITHOUT SENSORY FEEDBACK

A typical saccade takes less than 80 ms to complete and moves the eyes at more than 400 deg/s. Such movements are too brief for visual or proprioceptive feedback to influence control of the eyes during the saccade (Keller & Robinson 1971, Guthrie et al. 1983). However, a fundamental problem is that the motor commands that initiate the movement are highly variable and this variability is related to the context in which the eye movement is made. For example, people make saccades with higher velocities in anticipation of seeing a more interesting visual stimulus (e.g., image of a face) (Xu-Wilson et al. 2009b). If a target is presented and they are instructed to look in the opposite direction, saccade velocities are much lower than if they are asked to look to the target (Smit et al. 1987). Repeating a visual target (Straube et al. 1997, Chen-Harris et al. 2008, Golla et al. 2008) or reducing the reward associated with that stimulus (Takikawa et al. 2002) also reduces saccade velocities. On the other hand, increasing the reward associated with the target (Takikawa et al. 2002), making the target the goal of both the eye and the arm movements (van Donkelaar 1997, Snyder et al. 2002), or unexpectedly changing the characteristics of the target (Xu-Wilson et al. 2009a) all result in increased saccade velocities without altering saccade amplitude. Thus, the motor commands that accelerate the eyes toward a given target are affected by the expected reward, attention, or cognitive state of the subject. These factors all induce variability in the motor commands that start the movement. Despite this variability, the brain accurately guides the eyes to the target without sensory feedback.

Another problem is that we sometimes blink when we make saccades. As the eyelid comes down on the eyes, it acts as a source of mechanical perturbation that pushes the eye, altering its trajectory. Remarkably, the motor commands that guide the eyes during a blink-affected saccade appear to take into account this self-induced source of perturbation: The eyes tend to arrive on target (Rottach et al. 1998). How does the brain take into account these potential sources of variability without the use of sensory feedback?

Some three decades ago David Robinson proposed that endpoint accuracy of saccades is possible because the brain incorporates an internal feedback process through the cerebellum that monitors the motor commands and corrects them online (Robinson 1975). Corollary discharge encodes a copy of the motor command (so-called efference copy), and this efference copy could be processed to predict the consequences of actions before sensory feedback is available. Predicting the consequences of a motor command is called a forward model. In the current computational view of motor control [e.g., Shadmehr & Krakauer (2008)], the cerebellum may be a forward model (Pasalar et al. 2006) that uses an efferent copy to predict consequences of motor commands and to correct the movement as it is being generated (**Figure 1a**). For example, if the internal feedback is intact, variability in the commands that initiate the saccade might be compensated via cerebellar dependent commands that arrive later during the same saccade.

A strong prediction of this idea is that if an experiment could impose variability in the commands that accelerate the eyes, a subject who is missing this internal feedback process (e.g., cerebellar patients) would exhibit an inability to compensate for that variability, resulting in dysmetric saccades. A simple way to induce changes in the motor commands that accelerate the eyes is to present the visual targets in a repeating pattern: In healthy subjects, this repetition results in a decline in saccade velocities without affecting saccade amplitudes (Fuchs & Binder 1983, Straube et al. 1997). The origin of this decline is not well understood. However, the decline is not due to muscle fatigue but likely of neural origin (Xu-Wilson et al. 2009a). One possibility is that the repetition of the stimulus acts to devalue the target of the movement, and a reduced value associated with the target results in reduced saccade velocities (Takikawa et al. 2002). Golla et al. (2008) and Xu-Wilson et al. (2009a) used this repetition method to introduce changes in the motor commands that initiated saccades. They then examined the ability of patients with cerebellar damage to respond to these changes. Both groups of investigators found that repetition of the stimulus produced strong reductions in saccade velocities in the healthy and cerebellar subjects (**Figure 1b**). That is, if saccade velocities are viewed as a proxy for the value that the brain assigns the target (Takikawa et al. 2002, Xu-Wilson et al. 2009b), then in both healthy and cerebellar subjects the repetition of the target resulted in its devaluation. However, whereas in the healthy subjects saccade amplitudes remained accurate, in the cerebellar patients the saccades fell short of the target as the velocities declined (**Figure 1b**). It appeared that in healthy people, the changes in the motor commands that initiated the saccade were generally compensated via motor commands that arrived later in the same saccade. However, the compensation was missing in cerebellar subjects.

In summary, there is variability in the motor commands that initiate even the simplest movements like saccades. If left uncompensated, this variability would result in dysmetric movements. The brain appears to monitor the motor commands and compensate for the variability, a process consistent with that of an internal model that predicts the sensory consequences of motor commands.

CORRECTING MOVEMENT ERRORS WITH DELAYED SENSORY FEEDBACK

Unlike saccades, most movements are long enough in duration that sensory feedback

plays an essential role in their control. All goal-directed arm movements fall into this category. However, the problem is that the delays inherent in sensory feedback can destabilize movements, confining online error correction to peripheral mechanisms characterized by fast spinal reflexes and intrinsic biomechanical properties of the muscles. Such a control strategy, however, allows for only a very narrow class of error feedback behaviors. Despite these long delays, supraspinal, long-latency responses to perturbations are often much larger in amplitude than their short-latency counterparts (Strick 1978), and these responses are present even in relatively quick arm movements (Saunders & Knill 2003, Saunders & Knill 2004). Together these observations suggest that cortically modulated, in addition to spinal, error correction mechanisms play a role in even short, rapid arm movements. This ability to perform cortically driven online error feedback control is remarkable in light of the long sensorimotor delays we experience.

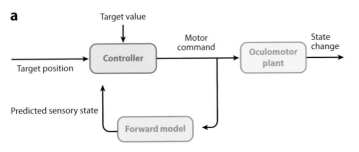

a

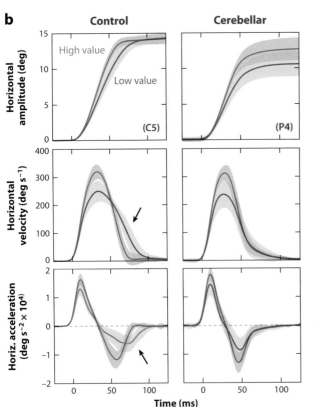

b

Figure 1

Control of movements without sensory feedback is difficult because there is variability in motor commands. The brain appears to maintain accuracy by using a forward model that predicts the sensory consequences of motor commands. (*a*) A highly stylized view of the process of generating a saccade. The motor commands depend not just on the position of the saccade target, but also on the internal value associated with that movement. For example, saccades tend to be slower toward stimuli that have a lower value. (*b*) Examples of saccades from a healthy subject and a patient with degeneration of the cerebellum. With repeated presentation of a visual target, the saccade target is devalued, and the motor commands that initiate the saccade become smaller, generating slower velocities and accelerations in both the healthy subject and the cerebellar patient (*blue lines*). In response to this variability, the healthy brain produces motor commands late in the saccade to maintain accuracy, bringing the eyes to the target. In contrast, the patient with cerebellar damage cannot correct for the reductions in the motor commands that initiated the saccade. As a result, the saccades of the cerebellar patient fall short of the target. Part *b* is reproduced from Xu-Wilson et al. (2009a).

Why do sensorimotor loop delays present a problem for real-time feedback control? Consider a simple example. Most of us have experienced the trauma of being scalded in an unfamiliar shower. This situation nicely illustrates how delayed feedback can lead to unstable control. In an unfamiliar bathroom, an unsuspecting/naïve individual may start his shower at too low a temperature and then want to increase the heat. Since the delay between adjusting the temperature control knob and feeling its effects is notoriously long, initial adjustment of the knob has no immediate effect on the water temperature, which may spur the victim-to-be to continue turning up the heat. By the time the water temperature starts to respond, the heat adjustment may already be at scalding levels. Then, when the water becomes too hot, the victim turns down the temperature control. However, the temperature continues to rise. The victim responds by turning down the temperature control even farther, and when the temperature finally responds, he is soon freezing. If he were to continue with the same pattern behavior, the victim would continue to experience large, unstable fluctuations in water temperate resulting in repeated freeze/burn cycles. Most such victims soon decide to stop adjusting the temperature control continually—abandoning rapid, real-time feedback adjustments for a prolonged "wait and see" approach after each small temperature adjustment. This avoids large temperature fluctuations at the expense of an extended time for achieving a comfortable water temperature. Note that the whole traumatic scenario could have been avoided if the victim exercised a good understanding of how the water temperature would react to each adjustment of the knob, as would be the case in his own shower. Such an understanding entails the ability to predict the sensory consequences of motor actions, i.e., a forward model.

If the motor system has the means to predict the sensory state of the motor apparatus, instabilities arising from delays in measuring that state can be effectively eliminated (Miall & Wolpert 1996, Bhushan & Shadmehr 1999). Long-latency online feedback control, therefore, might rely on a forward model of dynamics, which would enable the motor system to predict state variables such as position and velocity based on a history of motor commands (Ariff et al. 2002, Mehta & Schaal 2002, Flanagan et al. 2003). If this predicted state is an unbiased estimate of the actual state, then feedback control driven by these predicted states could essentially take place in real time and avoid the instability normally associated with feedback delays.

Feedback responses driven by internal models can also be more effective in compensating for a perturbation than responses driven solely by sensory feedback alone. For example, when the arm is moved because of an unexpected perturbation, the short-latency spinal reflexes respond solely to the muscle stretch, but longer latency reflexes produce a response that also takes into account the consequences of the net torques on the joints (Lacquaniti & Soechting 1984, Soechting & Lacquaniti 1988, Kurtzer et al. 2008). This is important because a perturbation that stretches one muscle often cannot be compensated by activating that muscle alone. That is, because of intrinsic mechanical coupling between the physical dynamics of connected joints such as the elbow and the shoulder (so-called interaction torques), responding to a stretch of a shoulder muscle by activating that muscle would produce a motion of not only the shoulder, but also of the elbow. Therefore, responses that take into account these interaction torques can counteract the consequences of external perturbations more effectively than responses specific to the muscles that were stretched by the perturbation (Kurtzer et al. 2008). These studies suggest that long-latency reflexes take into account the physical dynamics of the limb in producing a compensatory response to a perturbation.

One way to test the role of forward models during ongoing arm movements is to have people generate motor commands based on their estimate of current limb position. For example, if the hand is moving and the brain can predict the current location of the hand and its velocity, then the motor commands should reflect this predicted state. If unexpected perturbations to this state are encountered, then the predicted state will not match the actual state. However, after the sensorimotor loop delay, accurate predictions should be restored if the forward model can integrate information about motor commands with the observed sensory feedback. Wagner & Smith (2008) tested this idea by exposing subjects to novel velocity-dependent dynamics to train a relationship between hand velocity and lateral force. On some trials, the reaching movement was perturbed. In these cases, lateral force profiles, which reflected real-time internal estimates of hand velocity, were initially disrupted but became extremely accurate 150 ms after perturbation offset, even though the hand velocity during this period was clearly different from what had been planned. This observation demonstrates that a forward model is providing accurate state esti-

mates that are then fed to the previously adapted controller to provide corrections to the unexpected perturbation without a time lag.

If the brain had to rely solely on sensory feedback, then the motor commands should reflect a time-delayed estimate of hand position rather than a real-time prediction. Miall et al. (2007) used this idea to test the hypothesis that the cerebellum is involved in predicting the position of the hand during a movement. They had people move their hands laterally until they heard a tone, at which point visual feedback was removed and they reached in a forward direction toward a target location (**Figure 2a**). At the time of the tone, the brain needs to generate motor commands that bring the hand from its current position to the target. These commands depend on the estimated state of the hand. If this state estimate is primarily due to a delayed sensory feedback, then the estimate of

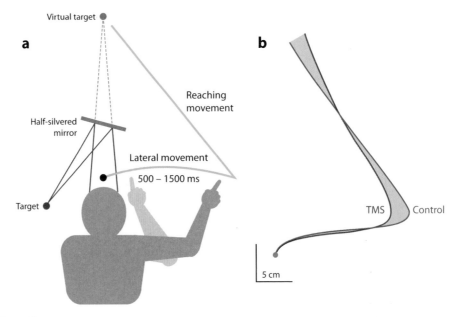

Figure 2

Control of movements with sensory feedback is difficult because feedback is time delayed. The brain appears to maintain accuracy by using a forward model that predicts the sensory consequences of motor commands. (*a*) Volunteers were asked to move their hand laterally until they heard a tone, at which point they would reach toward a target. The cerebellum was disrupted via a TMS pulse soon after the tone. (*b*) Reach trajectories from a movement in which TMS was applied to the cerebellum, and a movement in which no TMS was applied. Application of TMS produces a movement in which the estimate of the state of the arm appears to be delayed with respect to its actual state, resulting in missing the target to the right. From Miall et al. (2007).

hand position will be closer to the shoulder than in reality, in which case the motor commands will miss the target to the right. Alternatively, if this state estimate is primarily due to an unbiased prediction, then the estimate of hand position will be near its actual position, in which case the motor commands will bring the hand to the target. When people heard the tone, they brought their hand very near the target (control group, **Figure 2b**). However, if the cerebellum could be disrupted by a series of transcranial magnetic stimulation (TMS) pulses right after the tone, the motor commands that would be generated would miss the target to the right. That is, disruption of the cerebellum should make the state estimate "stale," reflecting not the current position of the hand but a position some time in the past. This is indeed what was observed; the state estimate was off by 130 ms. These results are consistent with the idea that the cerebellum predicts the state of the limb from the history of motor commands, allowing one to act on this estimate of state rather than relying solely on a delayed sensory feedback, and suggest that cerebellar output might provide a "motion update" signal that can be combined with delayed sensory feedback elsewhere in the brain in order to generate real-time state estimates for motor control.

THE EFFECT OF PREDICTING SENSORY CONSEQUENCES OF MOTOR COMMANDS ON PERCEPTION

A clear advantage of making sensory predictions is that the brain does not have to wait for the sensory measurements before it can act. However, there is a more fundamental advantage to making predictions, and this has to do with perception.

Our ability to estimate the state of our body and the external world appears to be a combination of two streams of information: one in which our brain predicts what should happen, and one in which our sensory system reports what did happen. The advantage of this is that if our predictions are unbiased, then our perception

(and the decisions that are made based on that perception) will be better than if we had to rely on sensory measurements alone. In a sense, our perception will be more accurate (e.g., display less variance) if we combine what we predicted with what our sensory system measured.

Although this may seem like a fairly new idea, it was first proposed around 1011 by Ibn al-Haytham, an Iraqi scientist (also known as Alhazen), in his *Book of Optics* in which he explained the "Moon illusion", the perception that the Moon is bigger when it is near the horizon than high in the sky, in terms of expectations about distance (although we perceive it this way, when measured by a simple camera the size of the Moon is actually a bit smaller near the horizon, as it is farther away by about the radius of the Earth). In 1781, Immanuel Kant in his theory of idealism claimed that our perceptions are not the result of a physiological process in which, for example, the eyes faithfully transmit visual information to the brain, but rather, our perceptions are a result of a psychological process in which our brain combines what it already believes with the sensory information. He wrote, "The understanding can intuit nothing, the senses can think nothing. Only through their union can knowledge arise" (Kant 1781).

If we follow this line of reasoning and return to our example of predicting the sensory consequences of the motor commands that move the eyes in a saccade, we might guess that during the postsaccadic period, the brain should have a better estimate of eye position than if it simply relied on proprioceptive information (from eye muscles) alone. In theory, the combination of the two streams of information (prediction and sensory feedback) should allow it to estimate eye position better. As a consequence, subsequent movements that depend on this position estimate should be more accurate.

For example, consider a task in which subjects are asked to make two saccades in succession in the dark so that the endpoint of the first saccade is the starting point of the second saccade. The motor commands that generate the second saccade should take into account the consequences of the motor commands for the first saccade. That is, planning of the second saccade can benefit from the efference copy of the first saccade. Indeed, a neural pathway for the transmission of this efference copy information appears to be from the superior colliculus, through the mediodorsal nucleus in the thalamus, to the frontal eye fields (Sommer & Wurtz 2002, 2006). When two eye movements are made in rapid succession, the motor command for the second saccade takes the first into account, but when this pathway is disrupted, this ability is markedly reduced (Sommer & Wurtz 2002).

Usually when we make saccades, visual information is present at the end of each saccade. This visual information is simply a remapped version of the visual field before the saccade. In effect, the eye is a camera that moves during a saccade, altering the retinal image. Therefore, the change in the visual information is a predictable event, and is part of the sensory consequences of the motor commands that moved the eye. Accordingly, neurons in posterior parietal cortex have been shown to remap their receptive fields based on such predictions before saccades occur (Duhamel et al. 1992). An interesting prediction is that the brain should have a better estimate of what it sees after a saccade than if that same visual information was only provided passively. Indeed, this was recently confirmed in an experiment in which people were asked to reach to targets that they could see but not predict versus targets that they could both predict their location [because of self-generated eye movements, i.e., remapping; see Vaziri et al. (2006)]. Being able to predict the sensory consequences of a saccade allowed subjects to have a more accurate estimate of the location of the target.

Therefore, by predicting the sensory consequences of motor commands, the brain can overcome not only delay in sensory feedback, but perhaps more importantly, it can actually sense the world better than is possible from sensory feedback alone. The latter comes about when our brain combines what it has predicted with what it has measured—two sources of information, when used properly, are better than one.

If our brain could not accurately predict sensory consequences of our motor commands, then we would not be able to sense the world around us in a normal way. An example of this is patient R.W., a 35-year-old man who was described by Haarmeier et al. (1997). R.W. suffered a stroke in a region covering parts of the parietal and occipital cortex, affecting a part of cortex that receives vestibular input, a location in which cells are sensitive to visual motion. R.W. complained of vertigo only when his eyes tracked visual objects and not when his eyes were closed. He explained that when he was watching his son run across a field (a condition in which his eyes moved smoothly to follow his son), he would see the boy running, but he would also perceive the rest of the visual scene (e.g., the trees) smoothly moving in the opposite direction. Haarmeier et al. (1997) demonstrated that when R.W. moved his eyes, his brain was unable to predict the sensory consequences of the oculomotor commands. As his eyes moved to follow his son, the trees moved in the opposite direction on his retina. The healthy brain predicts that moving the eyes will have the sensory consequence of shifting the image of the stationary world on the retina. We do not perceive this shifting image as real motion of the world because we predict it to be a consequence of motion of our eyes. In R.W., the vertigo was a symptom of the brain's inability to predict such sensory consequences.

You do not need to have a brain lesion to get a feel for what R.W. sees when he moves his eyes. Take a camera and aim it at a runner and try to move (i.e., pan) so that the image of the runner stays at the center of the picture. As you are moving the camera, take a picture. That picture will show a sharply focused runner but a blurry background that appears to be moving in the opposite direction. However, when you are looking at the runner with your naked eyes, the background appears perfectly still. The reason is that your brain predicts the background image shift that should take place on the retina as you move your eyes, and combines it with the actual shift. By combining the observed and predicted images, the parts that agree must have been stationary, and parts that disagree must have moved.

COMBINING PREDICTIONS WITH SENSORY OBSERVATIONS TO PRODUCE A MOTOR RESPONSE

To combine two streams of information, one needs to apply a weighting to each stream. In principle, the weight should be higher for the more reliable information source. For example, during a reaching movement, one can make predictions about the state of the arm from the history of motor commands, but as the movement proceeds, one faces the problem of how to combine this prediction with the sensory feedback. Some four decades ago, Rudolph Kalman proposed a principled way that this kind of problem should be solved: Combine the two sources of information in a way that minimizes the variance of the resulting estimate (Kalman 1960). If we view the prediction as the prior belief, and the sensory feedback as the current evidence, then Kalman's algorithm is equivalent to a Bayesian process of integration in which the weight associated with each piece of information depends on the uncertainty of each quantity. Therefore, the idea that emerges is that the brain should not only make predictions about sensory consequences of motor commands, but should also incorporate a measure of uncertainty about that prediction.

Körding & Wolpert (2004) tested the idea that as a movement took place, the brain combined its predictions about sensory feedback with actual sensory feedback using weights that depended on the uncertainty of each kind of information. They first trained subjects to reach to a goal location by providing them feedback via a cursor on a screen (the hand was never visible). As the finger moved from the start position, the cursor disappeared. Halfway to the target, the cursor reappeared briefly. However, its position was, on average, 1 cm to the right of the actual finger position, but on any

given trial the actual displacement was chosen from a Gaussian distribution. Because the location of the cursor was probabilistic, the variance of the Gaussian distribution described the confidence with which people could predict the sensory consequences of their motor commands. To control the confidence that the brain should have regarding sensory measurements, they added noise to the display of the cursor: The cursor was displayed as a cloud of dots. On some trials, the cursor was shown clearly so the uncertainty regarding its position was low. In other trials, the uncertainty was high as the cursor was hidden in a cloud of noise. The idea was that on a given trial, when a subject observes the cursor position midway to the target, she should issue a feedback correction to her movement based on two sources of information: the observation on that trial, and prior prediction regarding where the cursor would have been expected to be. The weighting of each source of error information should be inversely related to the variance of each distribution. Indeed, Körding & Wolpert's (2004) experimental data were consistent with this theoretical framework. Bayesian integration also explains feedback responses to force perturbations (Körding et al. 2004), feedback responses to perturbations to the properties of the target (Izawa & Shadmehr 2008), and perceptual estimates of position and velocity in the presence of noise (Ernst & Banks 2002, Weiss et al. 2002, Stocker & Simoncelli 2006, Sato et al. 2007).

In summary, the data suggest that as the brain programs motor commands, it also predicts the sensory consequences. Once the sensory system reports its measurements, the brain combines what it had predicted with the measurements to form a "belief" that represents its estimate of the state of the world. This belief is then used to issue feedback responses to current actions and to adapt our internal models of the world that will guide future actions. Thus, our actions are not simply based on our current sensory observations. Rather, our actions are often based on a statistically optimal integration of sensory observations with our predictions.

MOTOR ADAPTATION: LEARNING FROM SENSORY PREDICTION ERRORS

Combining predictions with observations is useful only if the predictions are generally accurate. If in trial after trial there are persistent differences between predictions and observations, that is, the brain's predictions are consistently biased, then there is something wrong in these predictions. The problem of forming unbiased predictions of sensory observations is a fundamental problem of learning. When you have formed an accurate representation, i.e., a forward model of how motor commands affect the motion of your arm or your eyes, you can apply motor commands to this internal model and (on average) accurately predict the motion that will result. However, during development, bones grow and muscle mass increases, changing the relationship between motor commands and motion of the limb. Disease can affect the strength of muscles that act on the eyes. In addition to such gradual variations, the arm's dynamics change over a shorter time scale when we grasp objects and perform manipulation. It follows that in order to maintain a desired level of performance, our brain needs to be "robust" to these changes. This robustness may be achieved through an updating, or adaptation, of an internal model that predicts the sensory consequences of motor commands.

In the case of arm movements, there are two well studied versions of the adaptation paradigm. In one version, called visuomotor adaptation, the investigator introduces a perturbation that distorts the visual consequences of the motor commands but leaves the proprioceptive consequences unchanged. This is typically done by wearing prism goggles, or having people move a cursor on the screen in which the relationship between cursor position and hand position is manipulated (Krakauer et al. 1999, 2000). In another more recent version of the adaptation paradigm, called force-field adaptation, the investigator introduces a physical perturbation that alters both the visual and proprioceptive consequences of motor commands.

This is typically done by having the volunteer hold the handle of an actuated manipulandum (a robotic arm) that can produce force on the hand that varies with hand motion (Shadmehr & Mussa-Ivaldi 1994). This type of adaptation can also be studied by having people hold a passive manipulandum for which the weight can be adjusted (Krakauer et al. 1999), reach in a rotating room [the rotation imposes Coriolis forces on the hand (Lackner & Dizio 1994)], or even in microgravity in which the usual forces are removed (Lackner & Dizio 1996).

In the case of eye movements, there is also a well-studied version of the adaptation experiment. In the experiment, a target is shown and as soon as the eyes begin moving toward it, the target is extinguished and a new target is displayed (McLaughlin 1967). As a result, the saccade completes with an endpoint error, i.e., the motor command produces the expected proprioceptive feedback from the eye muscles but an unexpected visual feedback from the retina. The current data suggest that in both the arm adaptation and saccade adaptation experiments, learning depends on the sensory prediction errors.

The oldest record of a visuomotor adaptation experiment is an 1867 report by Hermann von Helmholtz. In that work, he asked subjects to point with their finger at targets while wearing prism lenses that displaced the visual field laterally. When the displacement was to the left, subjects initially had errors (an overshoot) in that direction, but after some practice, they learned to compensate for the visual displacement. Helmholtz observed that as soon as the prisms were removed, subjects made erroneous movements to the right of the target. This is known as an after-effect of adaptation.

Nearly a century later, Held & Freedman (1963) repeated Helmholtz's experiment with a new twist. They compared the performance of subjects when they actively moved their arm while viewing their finger through prism glasses, versus when they viewed their fingers but their arms were passively moved for them. In both cases, the subjects viewed finger motion through a prism that induced a displacement in the visual feedback. After this viewing, subjects were tested in a pointing task. Held and colleagues (Held & Gottlieb 1958, Held & Freedman 1963) found that in the test session, subjects only showed after-effects if they had actively moved their hands while viewing them. In their words, "Although the passive-movement condition provided the eye with the same optical information that the active-movement condition did, the crucial connection between motor output and visual feedback was lacking." In our terminology, sensory prediction error was missing in the passive condition, as the subjects did not actively generate a movement, and therefore could not predict the sensory consequences.

A more recent example of visuomotor adaptation provides strong evidence for the crucial role of sensory prediction errors. Mazzoni & Krakauer (2006) had people move

their wrist so that the position of the index finger was coupled with the position of a cursor on a screen. There were always eight targets on display, spanning 360°. On a given trial, one of the targets would light up and the subject would move the cursor in an out-and-back trajectory, hitting the target and then returning to the center. After a baseline familiarization period (40 trials), the experimenters imposed a 45° counter-clockwise rotation on the relationship between the cursor and finger position (early adaptation, **Figure 3a**). Let us label this perturbation with r. Now, a motor command u that moved the hand in direction θ did not produce a cursor motion in the same direction, but in direction $\theta + r$. If we label the predicted sensory consequences $\hat{y} = \theta$ and the observed consequences $y = \theta + r$, then there is a sensory prediction error $y - \hat{y}$. The objective is to use this prediction error to update an estimate for $\hat{r}$. With that estimate, for a target at direction θ^*,

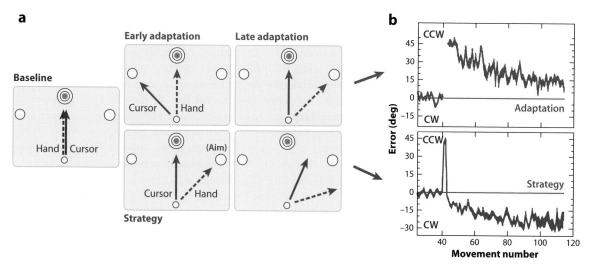

Figure 3

An example of learning from sensory prediction errors during visuomotor adaptation. (*a*) Subjects were asked to make an out-and-back motion with their hand so a cursor was moved to one of eight targets. In the baseline condition, hand motion and cursor motion were congruent. In the adaptation condition, a 45° rotation was imposed on the motion of the cursor and the hand. In the adaptation group (top two plots), the subjects gradually learned to move their hand in a way that compensated for the rotation. In the strategy group (bottom two plots), after two movements subjects were told about the perturbation and asked to simply aim to the neighboring target. (*b*) Endpoint errors in the adaptation and strategy groups. The strategy group immediately compensated for the endpoint errors, but paradoxically, the errors gradually grew. The rate of change of errors in the strategy and adaptation groups was the same. The rapid initial improvement is due to learning in the explicit memory system, whereas the gradual learning that follows is due to an implicit system. From Mazzoni & Krakauer (2006).

we can generate a motor command $u = \theta^* - \hat{r}$ to bring the cursor to the target. Indeed, after about 80 trials, in response to target at θ^* people would move their hands to $\theta^* - 40$ so the cursor would land within $5°$ of the target (as shown in the "adaptation" subplot of **Figure 3b**).

Now, Mazzoni & Krakauer (2006) took another group of naïve subjects and after they had experienced a couple of rotation trials, they simply told them: "Look, you made two movements that had large errors because we imposed a rotation that pushed you $45°$ counterclockwise. You can counter the error by aiming for the neighboring clockwise target." That is, simply issue the motor command $u = \theta^* - 45$ and as a consequence, the cursor will move at direction θ and land at the target. Indeed, the subjects followed this strategy: On the very next trial, all the error dropped to zero (strategy group, **Figure 3b**). However, something very interesting happened: As the trials continued, the errors gradually grew! What's more, the rate of change in the errors in this "strategy" group was exactly the same as the rate of change in the regular adaptation paradigm.

To explain this, Mazzoni & Krakauer (2006) hypothesized that on trial 43, when the subjects in the strategy group were producing the motor commands that brought the cursor to the target, there was still a discrepancy between the predicted and observed sensory consequences of motor commands $y - \hat{y}$. This is because whereas they had been told explicitly of the perturbation, implicitly their estimate was still around zero, $\hat{r} \approx 0$. The "learning curve" (over which the performance errors grew in the strategy group) tells us how quickly the implicit estimates became accurate. This occurred gradually over the course of about 50 trials, which closely matches the time course of normal adaptation (compare panels a and b in **Figure 3**), suggesting that the ability to predict the consequences of our actions and the ability to control our actions improve at the same rate. This finding is inconsistent with the hypothesis that accurate prediction (i.e., learning of a forward model) precedes the ability to control one's actions (Flanagan et al. 2003), and suggests instead that accurate prediction may be the limiting factor for accurate control.

However, there is another way to view the process of motor adaptation. As we perform a movement that experiences an error, we sometimes have the opportunity to correct it (for example, via reflexive pathways during that movement, or via corrective movements that occur later). These motor corrections may act as a teaching signal for the brain (Miles & Lisberger 1981, Kawato 1996). For example, when arm movements are perturbed with unexpected forces, reflex pathways respond to partially compensate for the sensory prediction errors (Thoroughman & Shadmehr 1999). When a visual target is moved during a saccade, there is a second saccade that brings the eyes to the new target position. One way to think of motor learning is to imagine that the motor commands that corrected the movement might be added with a slight time advance to the motor commands that initiate the next movement. The error feedback learning theory of the cerebellum relies on this motor correction (Kawato 1996).

To address this question, experiments have attempted to dissociate the effect of error-driven motor corrections from error signals themselves. In saccade adaptation experiments this has been done by presenting endpoint errors (sensory prediction errors) under conditions that reduce motor corrections (Wallman & Fuchs 1998, Noto & Robinson 2001). In reach adaptation paradigms this has been done by making a rapid movement so that there is a reduced possibility of correcting that movement (Tseng et al. 2007). All three experiments demonstrate that adaptation of the motor commands is driven directly by error signals rather than error corrections.

THE MULTIPLE TIMESCALES OF ADAPTATION

Recent work has shown that the interactions between adaptive processes that learn and forget on different time scales underlie several key features of motor adaptation (Smith et al. 2006). The idea that the process of memory formation proceeds on multiple time scales is, of course, not new as multiple time scales have long been observed in learning curves and forgetting curves (Rubin & Wenzel 1996, Scheidt et al. 2000, Wixted 2004). However, the ability to understand the natural interactions between different adaptive processes opens a window onto understanding several seemingly complex learning phenomena.

One such phenomenon is spontaneous recovery. Spontaneous recovery of memory refers to the natural reemergence of a learned response after that response has been extinguished through extinction training. This phenomenon has been observed in a wide range of paradigms (Rescorla 2004, Kojima et al. 2004, Stollhoff et al. 2005). Our recent work proposed a mechanism that explains spontaneous recovery as an interaction between the decay of two adaptive processes (**Figure 4**). If two processes participate in learning from error, one with a faster time course than the other, then both

processes need not be extinguished for the overall learning to disappear. Instead, learning can be extinguished if these processes come to cancel each other's effects during the extinction training. If this is the case, then the extinction training would leave two opposing memory traces that compete with each other and decay at different rates. Since the extinction training is generally quite rapid, the faster process would be expected to oppose the initial learning, and its decay would reveal the incomplete extinction of the slower process—corresponding to a spontaneous recovery of the overall learning. Decay of this slower process would eventually extinguish the spontaneous recovery.

In recent studies, the pattern of spontaneous recovery was shown to correspond to the intrinsic decay rates of two adaptive processes in exactly this way for motor adaptation in both arm (Smith et al. 2006) and eye movements (Ethier et al. 2008). In the first study, the adaptation to a velocity-dependent force-field was

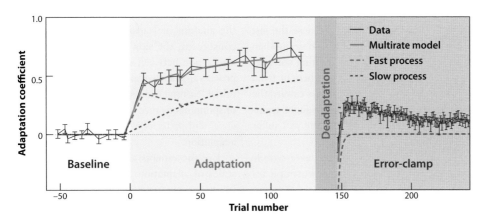

Figure 4

Spontaneous recovery in motor adaptation is explained by a two-process multi-rate model. After a baseline period (no shading), participants were trained to adapt to a velocity-dependent force-field (*beige shading*), de-adapted by exposure to the opposite force-field (*brown shading*), and tested for recovery under conditions in which lateral errors were held to zero. Adaptation levels during the recovery period were measured on error-clamp trials during which lateral errors were clamped at zero to prevent error-driven learning and so that the feed-forward patterns of lateral forces could be directly measured without contamination from error-driven feedback corrections or inertial interactions. These error-clamp trials were also occasionally interspersed into the baseline and training periods to measure the initial learning curve. The multirate model reproduces the double exponential pattern in the learning curve, and not only the amount, but the shape of the spontaneous recovery profile. According to the model, this shape emerges because of the superposition of a rapid up-going decay of the fast learning process and a prolonged down-going decay of the slow learning process. From Smith et al. (2006).

extinguished by exposure to the opposite force-field. When motor errors were subsequently clamped to zero, the adaptive response to the original force-field spontaneously reemerged with a fast exponential time course and then died away with a slower exponential time course as would be predicted by this model. In the second study, the same pattern of spontaneous recovery was replicated for saccade adaptation.

The basic idea that extinction results when the fast component of learning cancels the effects of slower components provides a possible explanation for two other key features of learning: savings and anterograde interference. If the original response is retrained after extinction, this relearning is noticeably faster than the initial learning despite the same initial performance level and the same training procedure—a phenomenon known as savings or facilitation. However, if a response opposite to the initial response is trained after extinction, this opposite learning is slower than the initial learning—a phenomenon known as anterograde interference or forward interference. Because the fast process responds more quickly to training than the slow process, subsequent training compatible with the initial learning will benefit from the bias of the slow learning process toward the initial learning. In contrast, subsequent training that opposes the initial learning will suffer from this bias.

However, neither savings nor anterograde interference can be fully explained by simple interactions between adaptive processes. A recent study showed that savings persists even when extinction is carried out gradually under conditions in which the individual adaptive processes should be extinguished (Zarahn et al. 2008). The mechanism underlying this component of savings is not clear, but it may result from a general improvement in the rate of learning or from a memory of the initial adaptation that is not expressed until it is recalled. In an earlier study, catastrophic anterograde interference occurred after delays of several days, suggesting the operation of contextual effects rather than persistence of adaptive processes (Krakauer et al.

2005). Another study found that when training was followed by "reverse-training" until extinction, the spontaneous recovery was so strong as to suggest that during reverse-training, there was effectively no unlearning, but rather only an instantiation of a new and competing fast adaptive memory (Criscimagna-Hemminger & Shadmehr 2008).

Whereas interactions between the fast and slow processes account for initial learning, spontaneous recovery, and some forms of interference, it appears that only one of these processes provides a gateway to long-term memory formation. Twenty-four-hour retention levels assessed after various amounts of initial training (i.e., at different points of the learning curve) do not directly reflect the adaptation level achieved during the initial training period, but instead reflect the predicted level of the slow component of adaptation as shown in **Figure 5** (Joiner & Smith 2008). This suggests that maximizing the long-term benefit of a training session does not necessarily come from maximizing the overall level of learning, but rather from maximizing the amount of learning achieved by a single constituent, the slow process. Thus, a simple interaction between a fast and a slow adaptive process appears to explain not only the shapes of initial learning curves, but also the phenomena of spontaneous recovery, anterograde interference, and patterns of 24-h retention in motor adaptation.

Eye movements also display multiple time scales of adaptation. For example, eye velocity in optokinetic nystagmus (involuntary eye movements in response to continuous movement of the visual field) is characterized by two components: a rapid rise followed by a slower increase to steady state (Cohen et al. 1977). The rapid rise in eye velocity has been shown to be specifically affected by particular neural lesions (Zee et al. 1987), suggesting that these time scales may have distinct neuroanatomical bases. This idea is supported by data indicating that during saccade adaptation in patients with cerebellar cortical damage, there is a profound loss in the fast time scales of adaptation but less impairment in the slower adaptive processes

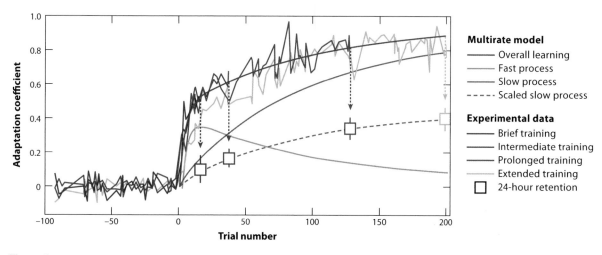

Figure 5

Long-term retention in motor adaptation is explained by the level of a slow component of motor adaptation. Four groups of participants were tested for 24-h retention after different amounts of training with a velocity-dependent force-field. The learning curves during the training period itself are plotted with lines, and the 24-h retention levels are plotted as squares (mean ± SEM). The pattern of 24-h retention has essentially the same shape as the slow learning process (r = 0.99), suggesting that this process serves as a gateway to long-term retention. From Joiner & Smith (2008).

(Xu-Wilson et al. 2009a). In reaching movements, transcranial magnetic stimulation of the lateral posterior parietal cortex (Della-Maggiore et al. 2004) or the motor cortex (Richardson et al. 2006, Hadipour-Niktarash et al. 2007) appears to specifically affect the slower adaptive processes. Interestingly, stimulation of posterior parietal cortex appears to have a stronger effect than stimulation of motor cortex. Imaging studies also suggest that there may be distinct neural networks associated with the fast and slow adaptive processes (Krakauer et al. 2004, Tunik et al. 2007).

LIMITATIONS

The framework that we focused on in this review is one in which the brain learns to accurately predict the sensory consequences of motor commands. Three important and pertinent problems in motor learning have not been discussed here: how to discern common structure in the dynamics of the tasks in which one is engaged (e.g., uncovering structure), how to produce motor commands that maximize some measure of performance (e.g., maximizing reward while minimizing effort, as in optimal control), and how the brain decides when to form new memories and when to modify existing ones.

The first problem is one of discovering the structure of the system that one is controlling. For example, inertial objects share a common structure in which acceleration and velocity are linearly separable quantities in the function that relates motion to forces. An intelligent system that interacts with a large number of inertial objects would benefit from discovering this structure, as it would allow it to learn control by simply adjusting a small number of parameters, rather than searching in the high-dimensional space of all possible motor commands (Hwang et al. 2006, Braun et al. 2009). It is possible that in the short time scale of training in a laboratory, subjects use internal models with structures that are similar to that of their body and merely tweak the parameters of these models to adapt to the imposed perturbations (Körding et al. 2007). Another way to say this is that the mechanisms involved in correcting movements in response to self-generated perturbations (natural variability in the motor commands that move our body parts) are the same as those co-opted when people adapt to externally imposed perturbations. However, long-term learning likely involves building a structural model specific to the task at hand, a process that would allow specialization and expertise that goes beyond parameter estimation with a generic model (Reis et al. 2009).

The second problem is that even if one has a perfect model that predicts the sensory consequences of motor commands, one still has to find the motor commands that maximize performance. This is the general problem of optimal control, i.e., given a relationship between motor commands and sensory consequences, find a feedback control law that achieves the greatest amount of reward (or some other performance measure) at the least amount of effort (Todorov & Jordan 2002). Evidence for this process comes from experiments that find that during adaptation to a perturbation, motor commands do not return performance to a baseline trajectory, but rather a trajectory that maximizes performance while minimizing a measure of effort (Uno et al. 1989, Emken et al. 2007, Izawa et al. 2008).

The third problem is that if adaptive changes to the mechanisms for predicting the consequences of our actions and for controlling our actions occur continuously as described here, it is difficult to imagine that each movement leads to a new memory that can be recalled. On the other hand, we know that multiple discrete memories can be formed and consolidated. Understanding the nature of the contexts that lead to the formation and recall of motor memories is clearly important (Gandolfo et al. 1996, Krakauer et al. 2006, Krakauer 2009, Cothros et al. 2009). State, as a physicist would define it—position and velocity of motion—has been shown to be an important context for the formation and recall of motor memories, but multiple memories can be associated with the same motion states. What other contexts are important, and how do some contexts come to be more important than others?

SUMMARY AND CONCLUSIONS

The motor system needs to be able to adjust for the both the presence of noise and delay in sensory feedback, and for changes in the body and the world that alter the relationship between motor commands and their sensory consequences. The common solution to these two problems is a forward model. Forward models, possibly located in the cerebellum, receive a copy of the outgoing motor command and generate a prediction of the expected sensory consequence at very short latency. This output in sensory coordinates can be used to make fast trajectory corrections with the pre-existing controller before true sensory feedback is available, and can be integrated with true sensory feedback to optimize state estimates and enhance perception. Forward models are useful only if they are accurate and evidence suggests that accuracy is maintained through adaptive processes driven by sensory prediction errors.

DISCLOSURE STATEMENT

The authors are not aware of any affiliations, membership, funding, or financial holdings that might be perceived as affecting the objectivity of this review.

ACKNOWLEDGMENTS

This work was supported by funds from NIH grants R01NS057814 and R01NS037422 to R.S., R01NS052804 to J.W.K., and a McKnight Scholar Award and a Sloan Research Fellowship to M.A.S.

LITERATURE CITED

Ariff GD, Donchin O, Nanayakkara T, Shadmehr R. 2002. A real-time state predictor in motor control: study of saccadic eye movements during unseen reaching movements. *J. Neurosci.* 22:7721–29

Bhushan N, Shadmehr R. 1999. Computational architecture of human adaptive control during learning of reaching movements in force fields. *Biol. Cybern.* 81:39–60

Braun DA, Aertsen A, Wolpert DM, Mehring C. 2009. Motor task variation induces structural learning. *Curr. Biol.* 19:352–57

Chen-Harris H, Joiner WM, Ethier V, Zee DS, Shadmehr R. 2008. Adaptive control of saccades via internal feedback. *J. Neurosci.* 28:2804–13

Cohen B, Matsuo V, Raphan T. 1977. Quantitative analysis of the velocity characteristics of optokinetic nystagmus and optokinetic after-nystagmus. *J. Physiol.* 270:321–44

Cothros N, Wong J, Gribble PL. 2009. Visual cues signaling object grasp reduce interference in motor learning. *J. Neurophysiol.* 102:2112–20

Criscimagna-Hemminger SE, Shadmehr R. 2008. Consolidation patterns of human motor memory. *J. Neurosci.* 28:9610–18

Della-Maggiore V, Malfait N, Ostry DJ, Paus T. 2004. Stimulation of the posterior parietal cortex interferes with arm trajectory adjustments during the learning of new dynamics. *J. Neurosci.* 24:9971–76

Duhamel JR, Colby CL, Goldberg ME. 1992. The updating of the representation of visual space in parietal cortex by intended eye movements. *Science* 3:90–92

Emken JL, Benitez R, Sideris A, Bobrow JE, Reinkensmeyer DJ. 2007. Motor adaptation as a greedy optimization of error and effort. *J. Neurophysiol.* 97:3997–4006

Ernst MO, Banks MS. 2002. Humans integrate visual and haptic information in a statistically optimal fashion. *Nature* 415:429–33

Ethier V, Zee DS, Shadmehr R. 2008. Spontaneous recovery of motor memory during saccade adaptation. *J. Neurophysiol.* 99:2577–83

Flanagan JR, Vetter P, Johansson RS, Wolpert DM. 2003. Prediction precedes control in motor learning. *Curr. Biol.* 13:146–50

Fuchs AF, Binder MD. 1983. Fatigue resistance of human extraocular muscles. *J. Neurophysiol.* 49:28–34

Gandolfo F, Mussa-Ivaldi FA, Bizzi E. 1996. Motor learning by field approximation. *Proc. Natl. Acad. Sci. USA* 93:3843–46

Golla H, Tziridis K, Haarmeier T, Catz N, Barash S, Thier P. 2008. Reduced saccadic resilience and impaired saccadic adaptation due to cerebellar disease. *Eur. J. Neurosci.* 27:132–44

Guthrie BL, Porter JD, Sparks DL. 1983. Corollary discharge provides accurate eye position information to the oculomotor system. *Science* 221:1193–95

Haarmeier T, Thier P, Repnow M, Petersen D. 1997. False perception of motion in a patient who cannot compensate for eye movements. *Nature* 389:849–52

Hadipour-Niktarash A, Lee CK, Desmond JE, Shadmehr R. 2007. Impairment of retention but not acquisition of a visuomotor skill through time-dependent disruption of primary motor cortex. *J. Neurosci.* 27:13413–19

Held R, Freedman SJ. 1963. Plasticity in human sensorimotor control. *Science* 142:455–62

Held R, Gottlieb N. 1958. Technique for studying adaptation to disarranged hand-eye coordination. *Percept. Motor Skills* 8:83–86

Hwang EJ, Smith MA, Shadmehr R. 2006. Adaptation and generalization in acceleration-dependent force fields. *Exp. Brain Res.* 169:496–506

Izawa J, Rane T, Donchin O, Shadmehr R. 2008. Motor adaptation as a process of reoptimization. *J. Neurosci.* 28:2883–91

Izawa J, Shadmehr R 2008. Online processing of uncertain information in visuomotor control. *J. Neurosci.* 28:11360–68

Joiner WM, Smith MA. 2008. Long-term retention explained by a model of short-term learning in the adaptive control of reaching. *J. Neurophysiol.* 100:2948–55

Kalman RE. 1960. A new approach to linear filtering and prediction problems. *Trans. ASME J. Basic Eng.* 82(Ser. D):35–45

Kant I. 1781. *Critique of Pure Reason*. Trans. NK Smith, 1965, p. 93. New York: St. Martin's Press (From German)

Kawato M. 1996. Learning internal models of the motor apparatus. In *The Acquisition of Motor Behavior in Vertebrates*, ed. JR Bloedel, TJ Ebner, SP Wise, pp. 409–30. Cambridge, MA: MIT Press

Keller EL, Robinson DA. 1971. Absence of a stretch reflex in extraocular muscles of the monkey. *J. Neurophysiol.* 34:908–19

Kojima Y, Iwamoto Y, Yoshida K. 2004. Memory of learning facilitates saccadic adaptation in the monkey. *J. Neurosci.* 24:7531–39

Körding KP, Ku SP, Wolpert DM. 2004. Bayesian integration in force estimation. *J. Neurophysiol.* 92:3161–65

Körding KP, Tenenbaum JB, Shadmehr R. 2007. The dynamics of memory as a consequence of optimal adaptation to a changing body. *Nat. Neurosci.* 10:779–86

Körding KP, Wolpert DM. 2004. Bayesian integration in sensorimotor learning. *Nature* 427:244–47

Krakauer JW. 2009. Motor learning and consolidation: the case of visuomotor rotation. *Adv. Exp. Med. Biol.* 629:405–21

Krakauer JW, Ghez C, Ghilardi MF. 2005. Adaptation to visuomotor transformations: consolidation, interference, and forgetting. *J. Neurosci.* 25:473–78

Krakauer JW, Ghilardi MF, Ghez C. 1999. Independent learning of internal models for kinematic and dynamic control of reaching. *Nat. Neurosci.* 2:1026–31

Krakauer JW, Ghilardi MF, Mentis M, Barnes A, Veytsman M, Eidelberg D, Ghez C. 2004. Differential cortical and subcortical activations in learning rotations and gains for reaching: a PET study. *J. Neurophysiol.* 91:924–33

Krakauer JW, Mazzoni P, Ghazizadeh A, Ravindran R, Shadmehr R. 2006. Generalization of motor learning depends on the history of prior action. *PLoS Biol.* 4:e316

Krakauer JW, Pine ZM, Ghilardi MF, Ghez C. 2000. Learning of visuomotor transformations for vectorial planning of reaching trajectories. *J. Neurosci.* 20:8916–24

Kurtzer IL, Pruszynski JA, Scott SH. 2008. Long-latency reflexes of the human arm reflect an internal model of limb dynamics. *Curr. Biol.* 18:449–53

Lackner JR, Dizio P. 1994. Rapid adaptation to coriolis force perturbations of arm trajectory. *J. Neurophysiol.* 72:299–313

Lackner JR, DiZio P. 1996. Motor function in microgravity: movement in weightlessness. *Curr. Opin. Neurobiol.* 6:744–50

Lacquaniti F, Soechting JF. 1984. Behavior of the stretch reflex in a multi-jointed limb. *Brain Res.* 311:161–66

Mazzoni P, Krakauer JW. 2006. An implicit plan overrides an explicit strategy during visuomotor adaptation. *J. Neurosci.* 26:3642–45

McLaughlin S. 1967. Parametric adjustment in saccadic eye movements. *Percept. Psychophys.* 2:359–62

Mehta B, Schaal S. 2002. Forward models in visuomotor control. *J. Neurophysiol.* 88:942–53

Miall RC, Christensen LOD, Owen C, Stanley J. 2007. Disruption of state estimation in the human lateral cerebellum. *PLoS Biol.* 5:e316

Miall RC, Wolpert DM. 1996. Forward models for physiological motor control. *Neural Networks* 9:1265–79

Miles FA, Lisberger SG. 1981. The "error" signals subserving adaptive gain control in the primate vestibulo-ocular reflex. *Ann. NY Acad. Sci.* 374:513–25

Noto CT, Robinson FR. 2001. Visual error is the stimulus for saccade gain adaptation. *Brain Res. Cogn. Brain Res.* 12:301–5

Pasalar S, Roitman AV, Durfee WK, Ebner TJ. 2006. Force field effects on cerebellar Purkinje cell discharge with implications for internal models. *Nat. Neurosci.* 9:1404–11

Reis J, Schambra HM, Cohen LG, Buch ER, Fritsch B, et al. 2009. Noninvasive cortical stimulation enhances motor skill acquisition over multiple days through an effect on consolidation. *Proc. Natl. Acad. Sci. USA* 106:1590–95

Rescorla RA. 2004. Spontaneous recovery. *Learn. Mem.* 11:501–9

Richardson AG, Overduin SA, Valero-Cabre A, Padoa-Schioppa C, Pascual-Leone A, et al. 2006. Disruption of primary motor cortex before learning impairs memory of movement dynamics. *J. Neurosci.* 26:12466–70

Robinson DA. 1975. Oculomotor control signals. In *Basic Mechanisms of Ocular Motility and Their Clinical Implications*, ed. P Bachyrita, G Lennerstrand, pp. 337–74. Oxford, UK: Pergamon

Rottach KG, Das VE, Wohlgemuth W, Zivotofsky AZ, Leigh RJ. 1998. Properties of horizontal saccades accompanied by blinks. *J. Neurophysiol.* 79:2895–902

Rubin DC, Wenzel AE. 1996. One hundred years of forgetting: a quantitative description of retention. *Psychol. Rev.* 103:734–60

Sato Y, Toyoizumi T, Aihara K. 2007. Bayesian inference explains perception of unity and ventriloquism aftereffect: identification of common sources of audiovisual stimuli. *Neural Comput.* 19:3335–55

Saunders JA, Knill DC. 2003. Humans use continuous visual feedback from the hand to control fast reaching movements. *Exp. Brain Res.* 152:341–52

Saunders JA, Knill DC. 2004. Visual feedback control of hand movements. *J. Neurosci.* 24:3223–34

Scheidt RA, Reinkensmeyer DJ, Conditt MA, Rymer WZ, Mussa-Ivaldi FA. 2000. Persistence of motor adaptation during constrained, multi-joint, arm movements. *J. Neurophysiol.* 84:853–62

Shadmehr R, Krakauer JW. 2008. A computational neuroanatomy for motor control. *Exp. Brain Res.* 185:359–81

Shadmehr R, Mussa-Ivaldi FA. 1994. Adaptive representation of dynamics during learning of a motor task. *J. Neurosci.* 14:3208–24

Smit AC, Van Gisbergen JA, Cools AR. 1987. A parametric analysis of human saccades in different experimental paradigms. *Vision Res.* 27:1745–62

Smith MA, Ghazizadeh A, Shadmehr R. 2006. Interacting adaptive processes with different timescales underlie short-term motor learning. *PLoS Biol.* 4:e179

Snyder LH, Calton JL, Dickinson AR, Lawrence BM. 2002. Eye-hand coordination: saccades are faster when accompanied by a coordinated arm movement. *J. Neurophysiol.* 87:2279–86

Soechting JF, Lacquaniti F. 1988. Quantitative evaluation of the electromyographic responses to multidirectional load perturbations of the human arm. *J. Neurophysiol.* 59:1296–313

Sommer MA, Wurtz RH. 2002. A pathway in primate brain for internal monitoring of movements. *Science* 296:1480–82

Sommer MA, Wurtz RH. 2006. Influence of the thalamus on spatial visual processing in frontal cortex. *Nature* 444:374–77

Stocker AA, Simoncelli EP. 2006. Noise characteristics and prior expectations in human visual speed perception. *Nat. Neurosci.* 9:578–85

Stollhoff N, Menzel R, Eisenhardt D. 2005. Spontaneous recovery from extinction depends on the reconsolidation of the acquisition memory in an appetitive learning paradigm in the honeybee (*Apis mellifera*). *J. Neurosci.* 25:4485–92

Straube A, Fuchs AF, Usher S, Robinson FR. 1997. Characteristics of saccadic gain adaptation in rhesus macaques. *J. Neurophysiol.* 77:874–95

Strick PL. 1978. Cerebellar involvement in volitional muscle responses to load change. In *Cerebral Motor Control in Man: Long Loop Mechanisms*, ed. JE Desmedt, pp. 85–93. Basel: Karger

Takikawa Y, Kawagoe R, Itoh H, Nakahara H, Hikosaka O. 2002. Modulation of saccadic eye movements by predicted reward outcome. *Exp. Brain Res.* 142:284–91

Thoroughman KA, Shadmehr R. 1999. Electromyographic correlates of learning an internal model of reaching movements. *J. Neurosci.* 19:8573–88

Todorov E, Jordan MI. 2002. Optimal feedback control as a theory of motor coordination. *Nat. Neurosci.* 5:1226–35

Tseng YW, Diedrichsen J, Krakauer JW, Shadmehr R, Bastian AJ. 2007. Sensory prediction errors drive cerebellum-dependent adaptation of reaching. *J. Neurophysiol.* 98:54–62

Tunik E, Schmitt PJ, Grafton ST. 2007. BOLD coherence reveals segregated functional neural interactions when adapting to distinct torque perturbations. *J. Neurophysiol.* 97:2107–20

Uno Y, Kawato M, Suzuki R. 1989. Formation and control of optimal trajectory in human multijoint arm movement. Minimum torque-change model. *Biol. Cybern.* 61:89–101

van Donkelaar P. 1997. Eye-hand interactions during goal-directed pointing movements. *Neuroreport* 8:2139–42

Vaziri S, Diedrichsen J, Shadmehr R. 2006. Why does the brain predict sensory consequences of oculomotor commands? Optimal integration of the predicted and the actual sensory feedback. *J. Neurosci.* 26:4188–97

Wagner MJ, Smith MA. 2008. Shared internal models for feedforward and feedback control. *J. Neurosci.* 28:10663–73

Wallman J, Fuchs AF. 1998. Saccadic gain modification: Visual error drives motor adaptation. *J. Neurophysiol.* 80:2405–16

Weiss Y, Simoncelli EP, Adelson EH. 2002. Motion illusions as optimal percepts. *Nat. Neurosci.* 5(6):598–604

Wixted JT. 2004. On common ground: Jost's (1897) law of forgetting and Ribot's (1881) law of retrograde amnesia. *Psychol. Rev.* 111:864–79

Xu-Wilson M, Chen-Harris H, Zee DS, Shadmehr R. 2009a. Cerebellar contributions to adaptive control of saccades in humans. *J. Neurosci.* 29:12930–39

Xu-Wilson M, Zee DS, Shadmehr R. 2009b. The intrinsic value of visual information affects saccade velocities. *Exp. Brain Res.* 196:475–81

Zarahn E, Weston GD, Liang J, Mazzoni P, Krakauer JW. 2008. Explaining savings for visuomotor adaptation: Linear time-invariant state-space models are not sufficient. *J. Neurophysiol.* 100:2537–48

Zee DS, Tusa RJ, Herdman SJ, Butler PH, Gucer G. 1987. Effects of occipital lobectomy upon eye movements in primate. *J. Neurophysiol.* 58:883–907

How Does Neuroscience Affect Our Conception of Volition?

Adina L. Roskies

Department of Philosophy, Dartmouth College, Hanover, New Hampshire 03755;
email: adina.roskies@dartmouth.edu

Annu. Rev. Neurosci. 2010. 33:109–30

The *Annual Review of Neuroscience* is online at
neuro.annualreviews.org

This article's doi:
10.1146/annurev-neuro-060909-153151

0147-006X/10/0721-0109$20.00

Key Words

free will, decision, intention, control, agency, determinism

Abstract

Although there is no clear concept of volition or the will, we do have
intuitive ideas that characterize the will, agency, and voluntary behavior. Here I review results from a number of strands of neuroscientific
research that bear upon our intuitive notions of the will. These neuroscientific results provide some insight into the neural circuits mediating
behaviors that we identify as related to will and volition. Although some
researchers contend that neuroscience will undermine our views about
free will, to date no results have succeeded in fundamentally disrupting
our commonsensical beliefs. Still, the picture emerging from neuroscience does raise new questions, and ultimately may put pressure on
some intuitive notions about what is necessary for free will.

Contents

INTRODUCTION

Long a topic of debate for theologians and philosophers, the will, or the faculty of volition, was an object of scientific study for psychologists of the nineteenth century. Volition fell off the scientific radar screen with the behaviorist revolutions in philosophy and psychology, for the will was deemed too thoroughgoing a mentalistic concept to be amenable to empirical approaches. However, with the more recent abandonment of dualistic perspectives and the development of novel techniques for investigating the brain, the topic of volition is enjoying something of a renaissance. Discussions of freedom of the will in philosophy are also blossoming, and much of the impetus for volition's resurrection lies in the new challenges and opportunities posed by the cognitive and neural sciences (see sidebar, The Philosophy of Free Will).

PHILOSOPHICAL AND EXPERIMENTAL APPROACHES TO VOLITION

There is no uncontroversial, univocal concept of volition to be found in philosophy or in the sciences (Audi 1993; Brass & Haggard 2008; Zhu 2004a,b). Generally speaking, volition is a construct used to refer to the ground for endogenous action, autonomy, or choice. Intuitions vary on the specifics: Some contrast voluntary actions with actions that are reflexive or specified by the environment; others claim that the will is primarily involved in making decisions that then determine action; still others combine these views in holding that choices are but mental acts. Volition is sometimes used to refer to the endogenous mental act of deciding or forming an intention (Searle 1983, Zhu 2004a); at other times it is used to refer to the decision or intention itself (Adams & Mele 1992). Some suggest that volition is independent of the successful execution of the willed act itself, and that a central aspect of volition lies in the trying (Adams & Mele 1992).

The heterogeneity of the preceding list provides one clue to the difficulties in reviewing the impact of neuroscience on our conception of volition, for the absence of a clear concept of volition complicates the task of investigating it experimentally. In addition, in order for neuroscientific research to bear upon the conception of volition, volition has to be operationalized in some way. In light of these considerations, I organize my discussion around five different experimental threads which, separately or in combination, seem to capture much of what is signified by the intuitive, but less than clear, concept of the will. These threads are (*a*) initiation of action, (*b*) intention, (*c*) decision, (*d*) inhibition and control, and (*e*) phenomenology of agency. These five themes map loosely onto Haggard & Brass' (2008) "What, when, whether" model of intentional action.

There are relatively identifiable bodies of research associated with each of the above themes, although many of them blend into each other. Moreover, each of these maps to some elements of the commonsensical conceptions of volition. For example, if one takes voluntary action to contrast with stimulus-generated action, examining the neural events that distinguish self-initiated movements from similar movements that are responses to external stimuli ought to provide insight into the proximal mechanisms underlying endogenously generated action.

Volition: the faculty that makes possible voluntary action or choice; the will

Intention: mental states representing plans for future action. There may be many kinds of intention

Phenomenology: what an experience is like from the perspective of the subject

However, if one conceives of volition as not essentially tied to motor movements, but rather to abstract plans for future action, the proximal mechanisms that lead to simple movements may be of less interest than the longer-term plans or intentions one has or forms. Some research on intentions attempts to address these higher-level aspects of motor planning. Philosophical discussions of volition often focus upon the ability to choose to act in one way or another. Although historically this emphasis on choice may be a vestige of an implicit dualism between the mentalism of choice and the physicalism of action, the processes underlying decision as a means for forming intentions seem to be a central aspect of volition even if one rejects dualism, as most contemporary philosophers and scientists have.

A different approach to volition focuses less on the prospective influence of the will on future action than on the occurrent ability of the subject to inhibit or control action. Although one might not intuitively think of control as central to the will, recognizing the importance of control for attributions of responsibility in morality and the law may help clarify the relevance of control to volition. Moreover, some lines of evidence from psychology and neuroscience suggest that actions are often initiated unconsciously (e.g., Libet 1985), and so if free will is to exist, it will take the form of control or veto power over unconsciously initiated actions. Finally, regardless of whether one believes that we can act or choose freely, we normally do perceive certain actions as self-caused and others as not. Most people concur that there is phenomenology that accompanies voluntary action, and neuroscience has begun to illuminate the physiological basis of the feeling of agency.

Work on these five strands thus may influence or illuminate our conception of volition. A few caveats are in order before we begin. First, in light of the lack of agreement about the concept of volition itself, and its close relation to discussions of the will and of agency, I feel free to use the terms volition, will, and agency interchangeably, without, I hope, additionally

THE PHILOSOPHY OF FREE WILL

Philosophers have typically framed the problem of free will in terms of determinism and indeterminism. Compatibilists try to provide accounts of freedom that are compatible with determinism; incompatibilists deny that we can be free if determinism is true. Incompatibilists come in two flavors: Hard determinists place their bet on the side of determinism's truth, and deny that we are free; Libertarians maintain that we are free in virtue of indeterministic events. The challenge for the compatibilist is to show how we can be free or morally responsible if we could not act otherwise than we do. The Libertarian's challenge is to make his picture scientifically plausible, while showing how indeterministic events can have the right connection to choice and action to confer agency and responsibility.

Traditionally, freedom is thought to be intimately tied to moral responsibility. Now, however, some philosophers try to dissociate moral responsibility from freedom; others suggest that mechanistic explanation of the causes of behavior, not determinism, poses the greatest threat to freedom and responsibility. These new angles on old philosophical questions, as well as scientific inroads into understanding the neural bases of behavior, make it an exciting time to contemplate the philosophy of free will.

muddying the waters. That is not to say that there are not substantive distinctions to be made between them. Second, because of the extensive literature in each of the areas discussed, I focus more on discussions of how neuroscience has, can, or may affect our conception of volition, rather than on an exhaustive review of the literature in each of these lines of research. Finally, I recognize that there is a considerable literature on the neural basis of perceiving and attributing agency to others (Ciaramidaro et al. 2007, Cunnington et al. 2006, de Lange et al. 2008, Fogassi et al. 2005, Hamilton & Grafton 2006, Ramnani & Miall 2004, Rizzolatti & Sinigaglia 2007). On some hypotheses, the neural systems that support perception of agency in others are the same as those operative in perception of self-agency (Ciaramidaro et al. 2007, Cunnington et al. 2006, Fogassi et al. 2005, Iacoboni et al. 2005, Lamm et al. 2007, Rizzolatti & Sinigaglia

Dualism: mental and physical belong to two fundamentally different ontological categories, and cannot be reduced to each other

2007). Although this literature may indirectly bear upon our understanding of volition, I do not discuss it here. Finally, this paper does not attempt to review the philosophical work regarding volition. I begin, however, with a brief digression on the relation between volition and freedom.

VOLITION AND FREEDOM

It is difficult, if not impossible, to disentangle our notion of volition from questions about human freedom. The construct of volition largely exists in order to explain the possibility, nature, or feeling of autonomous agency. Before discussing the neuroscientific literature relating to volition, I want to say a few words about neuroscientific approaches that do not hold much promise for adjudicating the problem of free will.

In philosophy, discussions of freedom have traditionally hinged in part upon the question of how freedom relates to determinism, and in part on whether determinism is true. Incompatibilists believe that determinism precludes freedom, whereas compatibilists believe that determinism is compatible with, or even necessary for, freedom. Although the question of whether the universe (or the brain) is deterministic is a matter of empirical fact, it is not a fact that can be established by neuroscience. Some neuroscientists seem to think that neuroscientific work is able to illuminate the truth or falsity of determinism, by identifying the neural manifestation of indeterminism in randomness, noise, or stochastic behavior of neural systems. This, I believe, is mistaken, for at least two reasons. Although neuroscience may provide data that appear to reflect randomness or stochastic behavior in the nervous system, the epistemic limits of neuroscientific investigation are such that the evidence we gather from neuroscientific techniques is an insufficient basis from which to make that determination (Roskies 2006). Moreover, merely establishing randomness in the nervous system would be insufficient to account for human freedom: Randomness would have to be shown to play the right

role in processes of volition for it to ground freedom.

Regarding the first point, in order for neuroscience to bear upon the truth or falsity of determinism, it has to have something to measure, and it does this by operationalizing determinism as predictability. However, predictability is at best a poor cousin to determinism, and one that can betray its familial roots. Although a deterministic system is in principle predictable, in practice predictability is not a guide to determinism. What appears to be stochastic behavior at one level could be the result of deterministic processes at a lower level. For example, Mainen & Sejnowski (1995) found that spike timing that appeared to be stochastic with the injection of DC current was in fact extremely reliable when neurons were injected with the same variable voltage patterns that characterized their normal inputs. This finding might suggest that neuronal firing operates deterministically (but see Dorval 2006). However, the Mainen & Sejnowski data indicated that although spike timing was remarkably reliable, it was not perfect. These discrepancies could be attributable to stochastic behavior, but also to entirely deterministic factors such as unmonitored inputs, or to other features of the neuron that varied between trials. Similar ambiguities arise when one looks at a finer grain. For example, the stochastic properties of neurotransmitter release may be due to fundamentally probabilistic processes, or to the purely deterministic operation of a system that is structurally variable over time at the subcellular level, such as the spatial distribution of vesicles in the presynaptic terminal (Franks et al. 2003). Thus, in order to make judgments about determinism from neuroscientific data, we would need to know far more about the microphysical makeup of neurons than our neurophysiological techniques tell us, as well as to have complete information about the global state of the system impinging upon the neurons from which we are recording.

While certain features of neural events may be signatures of random behavior (e.g., Poisson distributions), such features can also be generated by deterministic mechanisms (Glimcher 2005). For example, the "random" number generators in computers are merely deterministic algorithms, and many chaotic processes are deterministic. Although some neuroscientists seem to be convinced that neuronal behavior is indeterministic, the verdict is still out. References to noise in neural systems invoke the spirit of indeterminism, but one person's noise is another's signal. In the absence of a clear and complete understanding of the way the brain codes and reads out relevant information, we cannot simply label unexplained activity as noise and from there infer that brain processes are indeterministic in the sense required by the incompatibilist. That said, there may be ways to view the role of variability in neural activity as relevant to decision or action that do not hinge upon the question of determinism.

Neuroscience can affect views on free will by elucidating mechanisms underlying behavior (where mechanism is silent on the determinism question). Merely elucidating mechanism may affect the layperson's views on freedom (Monterosso et al. 2005, Nahmias et al. 2007), but on the assumption that dualism is false, mechanistic or causal explanation alone is insufficient for answering the question of freedom. The real interest lies in whether neuroscientific accounts show volition to have or lack characteristics that comport with our intuitive notions of the requirements for freedom of the will. When discussing the relevance that the neuroscientific data have for our belief in freedom of the will, I focus upon whether particular empirical characteristics of the will put pressure on ordinary notions of freedom.

NEUROSCIENTIFIC APPROACHES TO VOLITION

Volition as Initiation

The will is thought to be critical in endogenously generated or self-initiated actions, as opposed to exogenously triggered actions, like reflexes or simple stimulus-response associations.

This view may be criticized on a number of fronts, among them that no such dichotomy in action types makes sense. Nevertheless, a number of neuroscientific studies have compared brain activity during self-initiated and externally cued actions, and have found differences in the functional architecture subserving actions that are externally cued and those that are not (Haggard 2008).

Imaging studies of endogenous generation of simple motor actions compared to rest consistently show activation of primary motor cortex, SMA (supplementary motor area) and preSMA, regions in the anterior cingulate, basal ganglia, and DLPFC (dorsolateral prefrontal cortex). Cued responses seem to involve a network including parietal and lateral premotor cortices that mediate sensory guidance of action.

Prior to the availability of imaging techniques, EEG recordings at the vertex revealed a slow negative electrical potential that precedes motor activity by 1–2 seconds (**Figure 1**). This "readiness potential" (RP) was initially hypothesized to arise in the SMA (Deecke & Kornhuber 1978, Jahanshahi et al. 1995). Further studies have suggested that the RP reflects more than one component process (Haggard & Eimer 1999, Libet et al. 1982, Shibasaki & Hallett 2006), and the source of the early components of the RP has been localized to preSMA (Shibasaki & Hallett 2006). The magnitude of the RP is greater in self-paced than in cued movements, and studies indicate that the late and peak phases of this electrical signal are associated with spontaneous or self-initiated motor acts, whereas the earliest components may be more involved in cognitive processes related to preparation or motivation (Jahanshahi et al. 1995, Libet et al. 1982, Trevena & Miller 2002).

Despite abundant evidence implicating medial frontal cortex in self-initiation of movements, determination of the source and function of the differences in brain activity during self-initiated and cued actions is a matter about which there is less consensus. A PET study controlling for predictability of movement timing suggests that the signals associated with

SMA: supplementary motor area

DLPFC: dorsolateral prefrontal cortex

Readiness potential (RP) or the Bereitschafts potential: an electrical negativity recorded at the midline with EEG that precedes voluntary movement

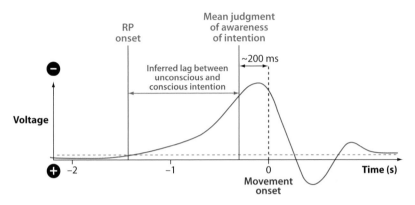

Figure 1

Schematic results of Libet's findings. On average, neural signals in the motor areas of the brain preceding finger movement (RP) begin more than 1 s before movement onset, whereas awareness of intending the movement, by contrast, is reported to be only ~200 ms before movement onset. Reprinted from Haggard (2005). [Reprinted from Haggard (2005) with permission from Elsevier]

self-paced actions arise in the rostral SMA, anterior cingulate, and DLPFC (Jenkins et al. 2000). Using fMRI, Deiber et al. (1999) mapped results onto the medial frontal anatomical areas defined by Picard & Strick (1996) and found that activation in preSMA and rCZ (rostral cingulate zone) were greater in self-initiated than in externally cued movements. Cunnington et al. (2002) found a difference in the timing, but not level of activation in preSMA with self-paced movements, and also report activation in rCZ. Lau et al. (2004b) try to disentangle attention to selection of action from initiation, and find that preSMA, but not DLPFC, is preferentially activated during initiation. However, in this study the greater activity in this region is correlated with time on task, and thus may not reflect specificity for initiation. Mueller et al. (2007) argue that once other variables are controlled for, preSMA fails to show differences between self-initiated and cued movement tasks, and they associate self-initiated movements with activity in rCZ and not preSMA.

Lesions of the preSMA in the monkey inhibit self-initiation of action, but not cued action (Thaler et al. 1995). In humans, direct electrical stimulation of regions in the SMA (including preSMA) produces an urge to move; stronger stimulation results in action (Fried et al. 1991). In addition, rTMS (repetitive

transcranial magnetic stimulation) of preSMA disrupts initiation of uncued motor sequences (Kennerley et al. 2004) during task switching. These studies provide further evidence of the involvement of these areas in action initiation. These regions also may be involved in automatic inhibition of competing responses: On the basis of unconscious priming studies with lesion patients, Sumner et al. (2007) report that SMA (but not preSMA) mediates automatic inhibition of motor plans in performance of alternative voluntary actions. Lesions in these regions prevent that inhibition (Sumner et al. 2007), and appear in some syndromes characterized by involuntary motor behavior (Haggard 2008).

In summary, the areas most consistently associated with action initiation are the rCZ and preSMA, but interpretation of their function is still controversial. A variety of factors make it difficult to reconcile results of many studies. Some paradigms require the subject to decide upon the timing of an instructed action, whereas others require choice between action alternatives. Reported results may reflect task confounds such as complexity of stimulus-action associations and conflict-induced activity rather than something particularly related to volition (Nachev et al. 2008). Indeed, there is some evidence that

rCZ: rostral cingulate zone

rTMS: repetitive transcranial magnetic stimulation

preSMA and rCZ can be subdivided into subregions preferentially involved in response conflict and initiation (Nachev et al. 2005, Picard & Strick 2001). Future work will better resolve the regions involved in self-initiated and externally cued activity, and the circuits mediating such processing. These results may provide clearer targets for future experiments. However, until more is known about the computations involved, the precise identification of regions involved in self-initiation does little to influence our conception of volition.

Volition as Intention

Intentions are representational states that bridge the gap between deliberation and action. Arguably, intentions can be conscious or unconscious. Moreover, there may be different types of intention involved in different levels of planning for action (Pacherie 2006). If we assume that intentions are the proximal cause of all voluntary movement, then studies of initiation of action and of intention may well concern the same phenomena (we might call these proximal intentions or motor-intentions, or as some call them, volitions). However, we also commonly refer to intentions in a broader, more abstract sense, as standing states that constitute conscious or purposeful plans for future action, that exist prior to and independently of action execution. In moral and legal contexts, when we ask whether a person acted intentionally, we often employ this more general notion of intention.

In general, willed action involves the intention to act, and many presume that freely willed actions must be caused by our conscious intentions. The efficacy of our conscious intentions was challenged by the studies of Benjamin Libet, who examined the relative timing of awareness of the intention to move and the neural signals reflecting the initiation of action. Libet reported that the time of onset of the readiness potential (RP) occurs approximately 350 ms or more prior to the awareness of an urge or intention to move (Libet 1985; Libet et al. 1982, 1983b,c) (**Figure 1**). Libet and others have viewed this discrepancy as evidence that actions are not consciously initiated (see Banks 2002, Libet 1985, Libet et al. 1983a). Many have taken these results as a challenge to free will, on the supposition that conscious intention must drive, and thus precede, initiation of action for that action to be freely willed. Although Libet's basic neurophysiological findings about RP timing have withstood scrutiny (Haggard & Eimer 1999, Matsuhashi & Hallett 2008, Trevena & Miller 2002), his interpretations have been widely criticized. For example, Libet's data do not enable us to determine whether the RP is always followed by a movement, and thus whether it really reflects movement initiation, as opposed to a general preparatory signal or a signal related to intention (Mele 2006, Roskies 2010).Haggard & Eimer use temporal correlation to explore the possibility that the anticipatory brain processes identified by Libet and others underlie the awareness of intention. Their results suggest that a different signal, the lateralized readiness potential (LRP), is a better candidate than the RP for a brain process related to motor intention (Haggard & Eimer 1999). Trevena & Miller agree that the LRP is more closely tied to movement initiation than is the RP, and their data suggest that awareness of intention may precede the LRP. Others argue that Libet's experimental design fails to accurately measure the onset of conscious intention to move (Bittner 1996, Lau et al. 2006, Roskies 2010, Young 2006), and may measure a different state instead (Banks 2002). Other research suggests that the Libet paradigm may bias judgments of the time of conscious awareness (Lau et al. 2006, 2007), so that inferences about relative timing may not be reliable. (For further commentary on Libet, see Banks 2002, Banks & Pockett 2007, Mele 2009, Pacherie 2006, Sinnott-Armstrong & Nadel 2010.) In sum, Libet's studies do little to undermine the general notion of human freedom, even if they do suggest that in certain kinds of repetitive motion tasks, individual motor movements may not be consciously initiated.

LRP: lateralized readiness potential

In a recent event-related study probing the timing of motor intentions, Haynes and colleagues used pattern classification techniques on fMRI data from regions of frontopolar and parietal cortex to predict a motor decision. Surprisingly, information that aided prediction was available 7–10 seconds before the decision was consciously made, although prediction success prior to awareness was only slightly better than chance (~60%) (Soon et al. 2008). This study demonstrates that prior brain states, presumably unconscious, can influence or bias decision-making. While neural precursors to decision and action and physical influences on behavior are to be expected from cognitive systems that are physically embodied, it is startling that any brain information could provide much guidance to future arbitrary decisions so long before they are made. The weak predictive success of this study does not undermine our notion of volition or freedom, but it nonetheless raises important challenges to ordinary views about choice.

Little neuroscientific work has focused explicitly on abstract human intentions, in part because it is so difficult to figure out how to measure them objectively. In one study, Lau et al. (2004a) instructed subjects to press a button at will, while attending to the timing of either their intention to move, or the movement itself. Attention to intention led to increased BOLD (blood oxygenation level dependent) signal in pre-SMA, DLPFC, and IPS (interparietal sulcus) relative to attention to movement. Relying on a large body of imaging results indicating that attention to specific aspects of a cognitive task increases blood flow to regions involved in processing those aspects (Corbetta et al. 1990, O'Craven et al. 1997), they interpreted their results to indicate that motor intention is represented in the pre-SMA. These results are consistent with the view that proximal intentions leading to self-initiated motor activity are represented in the pre-SMA, but also with the view that conscious intentions are represented there as well.

In addition to pre-SMA, Lau's study highlighted frontal and parietal regions often implicated in intentional action (**Figure 2**). Hesse et al. (2006) identify a frontoparietal network in motor planning, including left supramarginal gyrus, IPS, and frontal regions. The left anterior IPS has also been associated with goal representation, crucial in motor planning (Hamilton & Grafton 2006). These results are consistent with the view that posterior parietal regions represent motor intentions (Andersen & Buneo 2003, Cui & Andersen 2007, Quian Quiroga et al. 2006, Thoenissen et al. 2002). Sirigu et al. (2004) report that damage to parietal cortex disrupts awareness of intention to act, although voluntary action is undisturbed. The role of PPC (posterior parietal cortex) in the experience of intention is further discussed in a later section.

Often we think of intentions as more abstract plans less closely related to motor activity. Many studies indicate that dorsal prefrontal cortex (DPFC) is active in tasks involving willed action. Medial parts of DPFC may be involved in thinking about one's own intentions (den Ouden et al. 2005), whereas DLPFC may be involved in generating cognitive as well as motor responses (Frith et al. 1991, Hyder et al. 1997, Jenkins et al. 2000, Lau et al. 2004a). However, it is difficult to determine whether the activity observed corresponds to selection, control, or attention to action. Lau et al. (2004b) attempt to control for working memory and attention in a free response task in order to determine what areas are involved in selection of action. DLPFC was not more active in the free choice than in the externally specified selection condition, suggesting it had more to do with attention to selection than with choice. In contrast, preSMA was more active in free choice than in other conditions. This provides further evidence that preSMA is involved in free selection of action. Moreover, attention to selection involves DLPFC. Since attention may be required for awareness of intention, DLPFC activity may reflect conscious intention.

Thus far, the regions discussed reveal little about the content of intentions. Using pattern-analysis on fMRI data from regions of prefrontal and parietal cortex, Haynes and

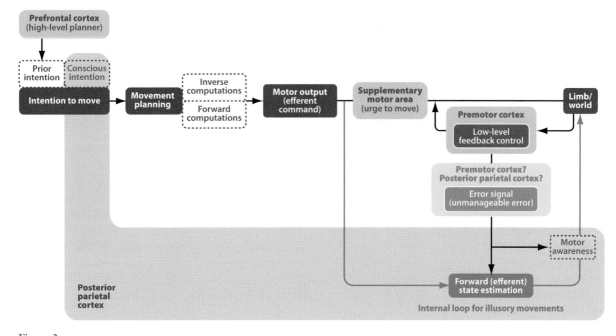

Figure 2

Partial schematic of information flow for voluntary action and awareness of agency, involving prefrontal, posterior parietal, and premotor cortices. Adapted from Desmurget & Sirigu (2009).

colleagues were able predict with up to 70% accuracy a subject's conscious but covert intention to add or subtract numbers (Haynes et al. 2007). Information related to specific intentions is thus present in these regions (including medial, lateral, and frontopolar prefrontal regions) while the subject holds his intended action in mind. Regions that are predictive appear to be distinct from the ones generally implicated in representation of intention or endogenous actions, raising the possibility that information related to intention is differentially represented depending on task.

To date, neuroscience has shown that mechanisms underlying endogenous initiation and selection of action have some features that deviate from commonsensical conceptions of volition, largely with regard to the relative timing of neural events and awareness. Although in certain contexts neural mechanisms of selection and motor intention may be unconsciously activated, once one takes into account the variety of levels at which intentions operate (Mele 2009, Pacherie 2006, Roskies 2010), none of

the current data undermines the basic notions of volition or free will. Reports of the death of human freedom have been greatly exaggerated.

Volition as Decision-Making

In one prevalent view, the paradigmatic exercise of the will lies in our ability to choose what course of action to take, rather than to initiate or represent future action. Many philosophers have located freedom of the will in the ability to choose freely which intentions to form. Decision often precedes intention and initiation.

Researchers in primate neurophysiology are constructing a rich picture of the dynamics of perceptual decision-making, using single-cell recording and population modeling. Because of its cohesiveness and breadth I concentrate on a body of work from the laboratories of William Newsome, Michael Shadlen, and colleagues, who have elucidated in detail the neural basis of decision-making under uncertainty using a visual motion paradigm. This work has been extensively reviewed elsewhere (Glimcher

2001, 2003; Gold & Shadlen 2007); I briefly summarize the main findings here.

These studies share a common paradigm: Rhesus macaques view random-dot motion displays. The monkeys' task is to fixate on the stimulus, to determine the direction of net motion, and to indicate the direction of coherent motion by making a saccade in the direction of net motion to one of two targets placed to the right and left of the fixation point. The task is made more or less difficult by changing the percentage of dots with component motion vectors in a particular direction, thus altering the coherence (or strength) of the net motion. By recording from cells in different brain areas during task performance, neuronal contributions to decision can be elucidated.

Cells in visual areas MT (middle temporal) and MST (medial superior temporal) are tuned to motion in particular directions. Recording from cells in these areas whose receptive fields are coincident with the location of the visual stimulus indicates that their neural activity reflects the momentary strength of the motion signal in the cell's preferred direction of motion (Britten et al. 1992, Celebrini & Newsome 1994, Newsome et al. 1989). Neural activity in area LIP (lateral interparietal area) shows a different profile. Neurons in LIP represent both visual and motor information (Shadlen & Newsome 1996). LIP cells appear to integrate signals from extrastriate motion areas over time (Huk & Shadlen 2005); they are also active in the planning and execution of eye movements (Andersen & Buneo 2002).

In the random-dot motion task, for example, a stimulus with a strong coherent motion signal to the right will lead to a ramping up of activity in LIP neurons whose response field encompasses the corresponding saccade target (Shadlen & Newsome 2001). The rate of increase in firing is proportional to motion strength, and when the activity in the LIP neurons reaches a certain absolute level, the monkey makes a saccade to the target, and firing ceases (Roitman & Shadlen 2002). Thus, LIP neurons seem to accumulate evidence of motion strength from sensory areas, until a

certain threshold is reached, and a decision is made (Huk & Shadlen 2005). This interpretation is strengthened by the finding that if the monkey is trained to withhold its response until cued, LIP neurons with response fields in the planned response direction maintain elevated firing rates during the delay period, and only cease their activity after the saccade. Thus, unlike neurons in MT and MST, these neurons are not purely stimulus driven, and their continued firing in the absence of the stimulus is taken to reflect the maintenance of the monkey's "decision" about the motion direction of the stimulus, until the completion of the task (Shadlen & Newsome 2001). Activity in these neurons predicts the monkey's response, in both correct and incorrect trials (Shadlen & Newsome 2001). Other experiments provide evidence for the causal involvement of LIP neurons in the decision-making process. For example, microstimulation of LIP neurons with response fields corresponding to a saccade target biases the monkey's choice and affect the timing of its responses (Hanks et al. 2006).

These neural processes have been mathematically modeled and incorporated in race-to-threshold or accumulate-to-bound mathematical models that capture well the behavioral patterns exhibited by the monkeys (Gold & Shadlen 2007, Mazurek et al. 2003). Psychophysical studies in humans indicate that monkeys and humans perform this task similarly, suggesting that analogous neuronal processes are involved in our decisions about these motion stimuli (Palmer et al. 2005).

One may, however, question whether the task of making a decision based on motion stimuli has much do to with the sorts of decisions we typically care about, especially when thinking of decision-making as a manifestation of volition. After all, one might argue (*a*) that the sorts of decisions to which moral responsibility applies, and for which the notion of voluntariness is important, involve much more complex considerations of value, consequences, reasons, feelings, and so on, than this simple perceptual system does; and (*b*) that the stimulus-driven nature of this task is precisely what we do not mean by

Saccade: an eye movement in which the eyes jump from one point in the visual field to another

LIP: lateral interparietal area

volition, which is by definition endogenous. This paradigm seems too impoverished to serve as a model for voluntary choice in the light of these considerations.

There are ways of conceiving of these studies in a more general way, making it easier to imagine how this model could serve as the core of a model of human decision-making. In these monkey studies, populations of neurons with particular response properties represent the choices the monkey can make in the task, and their relative firing rates appear to represent the weight given to them during the process leading to decision. We might conceive of these neuronal populations as representing distinct hypotheses or alternatives, such as "motion to the right/left," or alternatively "move eyes to the right/left target." If this is accurate, these neurons are part of representations with propositional content corresponding to the decision alternatives (Gold & Shadlen 2007). It is not difficult to imagine that different neural populations can represent other propositions, and although we currently lack a general framework for conceiving of how propositional content is represented in the nervous system, we know that it can be because we do represent it. Once we can conceive of the neural representation of abstract propositions, it is but a small step to think of them as representing reasons or considerations for action, and their relative firing rates as reflecting the weight given to reasons for decision or action. When we think of free actions, or actions for which we are morally responsible, those actions typically are—or are based on—our decisions in response to reasons (Fischer & Ravizza 1998).

What is more, a number of studies have extended this paradigm in novel ways, suggesting how the general paradigm can incorporate the richer, more nuanced, and abstract considerations that bear on human decision-making. For example, the firing rates of neurons in LIP that are associated with decisions in the visual motion task are also influenced by the expected value of the outcome and its probability, and these play a role in the decision-calculus (Platt & Glimcher 1999, Yang & Shadlen 2007). So

outcomes (decisions) associated with higher reward are more heavily weighted, and the time course of the rise to threshold occurs more rapidly to outcomes with higher payoff or those the animal has come to expect as more likely to occur. The firing of these neurons seems to encode the many aspects of decision-making recognized by classical decision theory and embodied in the concept of subjective value (Dorris & Glimcher 2004, Glimcher 2001, Platt & Glimcher 1999). Similar computations occur when the number of options is increased, suggesting that this sort of model can be generalized to decisions with multiple outcomes (Churchland et al. 2008). This system thus provides a basic framework for conceptualizing the main elements central to human decision-making of the most subtle and nuanced sorts.

In most trials, the random-dot motion task is perceptually driven: The nature of the stimulus itself specifies the correct choice. However, the decisions are usually made in conditions of uncertainty, and in some trials the stimulus does not provide determinative evidence for the decision. Monkeys are occasionally presented with random dot motion displays that have 0% coherent motion. Although there is a visual stimulus, the information in the stimulus is ambiguous and unrelated to a "correct" or rewarded choice. Still, in these trials monkeys choose rightward and leftward directions seemingly randomly, even in response to identical movies of 0% motion. The monkey's choices thus are not driven entirely by the external stimulus, but rather by factors internal to the monkey himself. And although the activity levels of the populations representing the alternative choices are nearly evenly matched, slight correlations are found between small fluctuations in activity in LIP in one direction or another, and the monkey's ultimate response (Britten et al. 1996, Shadlen et al. 1996). This suggests that the responses are indeed driven by competition between these neuronal populations.

Some might take the existence of the correlation between neural firing levels and choice even in these 0% motion cases to be evidence for determinism, whereas others could view the

Libertarian freedom:
a view that we have
free will in virtue of
the falsity of
determinism

stimulus-independent fluctuations as evidence for the existence and efficacy of random noise in decision-making. I think neither position is warranted, for reasons specified earlier. One person's noise is another person's signal, and without being able to record from all the neural inputs to a system, one cannot determine whether such activity is truly due to stochastic variability of neuronal firing, or is activity due to inputs from other parts of a dynamically evolving system, from local ongoing activity, or is from nonstimulus-related environmental inputs. Without being able to rule out these alternatives, we cannot ascertain whether these fluctuations are due to indeterministic processes or not, and whether the inputs should be viewed as noise or just unidentified signal. For these reasons, given our current knowledge, it seems insufficient to point to them as a basis for libertarian freedom or for the absence thereof.

Although the work on the neural basis of decision-making does not help adjudicate between traditional questions of freedom, if taken to be a general model for the neural basis of decision-making, it is illuminating. This work provides a relatively comprehensive model of a decision process in that it incorporates all the basic elements we would intuitively expect—representation of options, value, evidence, a dynamical characterization of the evolution of the system over time with changing inputs, and even confidence (Kiani & Shadlen 2009). It is only the first pass at a characterization, and there are relevant differences with human decision-making. For example, this system is tightly circumscribed by the task the animal has been trained to do, and the neural bases for decision and motor preparation are intimately related (Gold & Shadlen 2000). If the same stimulus is used but the response indicating the decision is not oculomotor, evidence suggests that other neuronal populations, not in LIP, will represent the decision of direction of motion (Cui & Andersen 2007, Gold & Shadlen 2007). In contrast, some human decision-making may operate at a more abstract level—certainly humans can make decisions in the absence of responses that necessitate concrete motor representations. Whether monkeys can also make abstract decisions remains an open question. Moreover, the picture we currently have is still only a partial and piecemeal view of what the brain is doing during any decision process. Many other brain areas also contribute to decision-making. For example, neuronal activity in DLPFC was also predictive of the monkey's decision in the random-dot motion task (Kim & Shadlen 1999), and responses were sensitive to expected reward value (Leon & Shadlen 1999). This region of monkey cortex is reciprocally connected with the parietal regions discussed above, and temporal coordination of these regions could be important in decision-making (Pesaran et al. 2008). Other areas involved in reward processing are also undoubtedly involved (see O'Doherty 2001, Schultz et al. 2000).

How does the work on decision relate to work on intention? In the random dot motion paradigm discussed above, it is tempting to identify the neural activity in LIP with intention: that activity seems to be causally linked to the production of a response, and when the monkey is required to delay its response, activity in LIP persists in the absence of the stimulus, exactly what one would expect of an intention that bridges the temporal gap between deliberation and action. However, as noted, activity in LIP is modality specific, reflecting a particular motor intention, one that involves eye movements, and not an amodal response. It is possible that most intentions, even many intentions of human animals, are realized in modality-specific motor programs. However, it is also possible that there are amodal means of representing intentions for future action for which there is no clear motor response, such as the intention to finish college, to search for a job, etc. There is some evidence in humans linking DLPFC to decisions independent of response modality (Heekeren et al. 2008). Language may make possible such representations in humans. Depending upon how linguistic abilities arise from neural computation, monkey neurophysiology may or may not provide insight into the nature of linguistically encoded intention.

Despite some shortcomings as a model of human decision-making, the monkey work on decision encourages us to think about volition mechanistically. Some philosophers argue that it is not determinism, but the recognition that mechanism underlies our decisions, that is the most potent challenge to freedom (Nahmias et al. 2007). Although there is some evidence to support this notion, there is much we do not understand about the threat of mechanism, and the relation of mechanism to reductionism. If mechanism is inimical to freedom, it may well be that our growing understanding of mechanisms underlying decision-making will undermine our conception of the will as free, but it is equally conceivable that our views about freedom will adapt to embrace the insights this research provides into the processes underlying our ability to choose among options when the correct choice is not externally dictated.

Volition as Executive Control

The control aspect of volition is the notion that higher-order cortical regions can influence the execution of action by lower regions. This may take several forms. For example, one conception is that volition involves the conscious selection of action (Bunge 2004, Donohue et al. 2008, Fleming et al. 2009, Hyder et al. 1997, Lau et al. 2004b, Matsumoto et al. 2003, Rowe et al. 2008, Rushworth 2008). Another is that monitoring can affect the form an action takes as it is executed (Barch et al. 2000, Kerns et al. 2004, Ridderinkhof et al. 2004, Schall & Boucher 2007, Schall et al. 2002). It is but a step further to think of control as including a capacity to inhibit an intended or planned action (Aron et al. 2007, Brass & Haggard 2007, Brown et al. 2008, Kühn et al. 2009b). The capacity to control one's behavior by inhibiting inappropriate actions is one that some parts of the law recognize as important for legal culpability.

Frontal cortex is generally implicated in executive control, but frontal cortex is a large and heterogeneous area, and much remains to be determined about the functional role of frontal subregions. Some regions of frontal cortex appear to be of particular importance to executive control. Numerous studies implicate interactions between PFC and regions of parietal cortex in attentional control and task switching (Badre 2008; Bode & Haynes 2009; Chiu & Yantis 2009; Dosenbach et al. 2007, 2008; Praamstra et al. 2005; Rossi et al. 2009; Serences & Yantis 2007). Other regions of cortex, such as some parietal regions, seem to play a role in guiding action that is under way (Dosenbach et al. 2007, 2008).

Several regions in frontal cortex appear time and time again in studies on volition. DLPFC is activated in many tasks involving choice or decision-making (Cunnington et al. 2006, Heekeren et al. 2006, Jahanshahi et al. 1995, Kim & Shadlen 1999, Lau et al. 2004a). DLPFC has been implicated in abstract and concrete decisions, as it is activated in choices between actions and in rule selection (Assad et al. 1998; Bunge 2004; Bunge et al. 2003, 2005; Donohue et al. 2008; Rowe et al. 2008). As noted earlier, there are competing hypotheses about the role of DLPFC in tasks involving choice and selection of action, including response selection, conscious deliberation, and conflict resolution. Although some work suggests that DLPFC activity is reflective of attention to selection of action (and thus, presumably, conscious control) (Lau et al. 2004b), other studies indicate that DLPFC activation is not always to be associated with conscious pathways (Lau & Passingham 2007). DLPFC has also been implicated in more abstract forms of control in humans. For example, Knoch & Fehr's (2007) rTMS studies indicate that the capacity to resist temptation depends on right DLPFC.

Discerning the networks subserving voluntary inhibitory control of action appears to be more straightforward. Libet, who argued on the basis of his experimental evidence that conscious intention is not causally efficacious in producing action, consoled himself with the view that the lag between the RP and action could possibly allow for inhibition of unconsciously generated actions, thus preserving the spirit of free will with "free won't" (Libet et al. 1983b) (**Figure 1**). However, he left this as pure conjecture. More recent studies have begun to shed light upon the neural mechanisms of inhibition of intended actions (although they lack the dualistic flavor Libet may have expected for "free won't" to really be free). For example, Brass & Haggard (2007) recently performed fMRI experiments in which they report increased activity in frontomedial cortical areas in Libet-like tasks in which subjects are required to intend to respond, and then to choose randomly to inhibit that response. They conjecture that these frontomedial areas are involved in voluntarily inhibiting self-generated action. Similar regions are involved in decisions to inhibit prepotent responses (Kühn et al. 2009b). Connectivity analyses suggest that medial frontal inhibition influences preSMA in a top-down fashion (Kühn et al. 2009b). Other evidence suggests that inhibition occurs at lower levels in the motor hierarchy as well, for example in local cortical networks in primary motor areas (Coxon et al. 2006).

Whereas dorsal medial frontal regions appear to be involved directly in inhibitory processes, the same regions that mediate voluntary decisions to act appear to be involved in voluntary decisions to refrain from action. Evidence from both ERP and fMRI studies demonstrate that the neural signatures of intentionally not acting, or deciding not to act after forming an intention to act, look very much like those of decisions to act (Kühn & Brass 2009b, Kühn et al. 2009a). For example, areas in anterior cingulate cortex and dorsal preSMA are active in both freely chosen button presses and free decisions not to press a button. The similar neural basis between decisions to act and to refrain from action lends credence to the commonsensical notion that both actions and omissions are acts of the will for which we can be held responsible.

Volition as a Feeling

The experience of willing is an aspect of a multifaceted volitional capacity. Some think that experience is all there is to explain because it is an experience of an otherwise illusory will (Wegner 2002). There are at least two phenomenological aspects of agency: the awareness of an intention or urge to act that we identify as prior to the action, and the post hoc feeling that an action taken was one's own.

With respect to the first, recent results reveal that the experience of voluntary intention depends upon parietal cortex (**Figure 2**). Electrical stimulation in this area elicited motor intentions, and stronger stimulation sometimes led to the erroneous belief that movement had occurred (Desmurget et al. 1999). In contrast, stimulation of premotor cortex led to movements without awareness of movement (Desmurget et al. 2009). Although this suggests that awareness of agency relies primarily on parietal rather than premotor areas, Fried et al. reported that stimulation in SMA also evoked desires to move. Intentions triggered by stimulation in SMA, in contrast to those triggered by parietal stimulation, had the phenomenology of compulsions more than of voluntary intentions (Fried et al. 1991). Although Desmurget et al. did not find that prefrontal stimulation elicited felt intention, the sites in BA6 that they stimulated tended to be more lateral than the regions stimulated by Fried et al. In addition, lesions in the inferior parietal lobe alter the awareness of timing of motor intention. Instead of becoming aware of intentions prior to movement, these lesion patients reported awareness only immediately prior to the time of movement (Sirigu et al. 2004). In contrast, their ability to report movement timing accurately was not impaired.

Considerable progress is also being made in identifying the neural signals involved in production of the feeling of agency or ownership of action (**Figure 2**). The feeling of agency seems to depend upon both proprioceptive and perceptual feedback from the effects of the action (Kühn & Brass 2009a, Moore & Haggard 2008, Moore et al. 2009, Pacherie 2008, Tsakiris et al. 2005). A number of studies indicate that plans for action are often accompanied by efferent signals that allow the system to form expectations for further sensory feedback, which, if not violated, contribute to the feeling of agency (Linser & Goschke 2007, Sirigu et al. 2004). Grafton and colleagues found activation in right angular gyrus (inferior parietal cortex) in cases of discrepancy between anticipated and actual movement outcome, and in awareness of authorship (Farrer et al. 2008). Signals from parietal cortex when predictions of a forward model match sensory or proprioceptive information may be important in creating the sense of agency. Moreover, some aspects of awareness of agency seem constructed retrospectively. A recent study shows that people's judgments about the time of formation of intention to move can be altered by time-shifting sensory feedback, leading to the suggestion that awareness of intention is inferred at least in part from responses, rather than directly perceived (Banks & Isham 2009). Expectation can also play a role (Voss et al. 2008). These studies lend credence to criticisms that the Libet measurement paradigm may affect the reported time of awareness of intention (Lau et al. 2006, 2007). In addition, perceived onset of action relative to effects is modulated by whether the actor perceives the action as volitional (Engbert et al. 2008, Haggard 2008).

As noted, frontal regions may also contribute to awareness of intention. Fried's stimulation study showed that stimulation of regions of SMA (probably pre-SMA) can lead to awareness of intention (Fried et al. 1991). TMS over SMA after action execution also affects the reported time of awareness of intention (Lau et al. 2007), further evidence that awareness of intention is in part reconstruction.

These results are consistent with a model in which parietal cortex generates motor intentions and a predictive signal or forward model for behavior during voluntary action (**Figure 2**). The motor plans are relayed to frontal regions for execution, and activation of these regions may be crucial for aspects of awareness of intention and timing. At the same time, parietal regions compare the internal predictions with sensory feedback [although a recent promising model suggests that the comparator resides in premotor cortex (Desmurget & Sirigu 2009)]. Feedback signals

alone are insufficient for a sense of authorship (Tsakiris et al. 2005). When signals match, we may remain unaware of our motor intentions (Sirigu et al. 2004), yet perceive the actions as our own. We may only be made aware of our motor intentions when discrepancies between the forward model and information from perception are detected. Thus, both an efferent internal model and feedback from the environment is important in the perception of agency and self-authorship (Moore et al. 2009).

Under normal circumstances, we experience our voluntary actions as voluntary. Under abnormal circumstances, people may wrongly attribute, or fail to attribute, agency to themselves (Wegner 2002, Wegner & Wheatley 1999). That feelings of agency have led some to suggest that it is merely an illusion that the will is causally efficacious (Hallett 2007, Wegner 2002). However, although experience of agency is not always veridical, we should not conclude that, in general, feelings of agency do not reflect actual agency, that the will is not causally efficacious, or that free will is nothing more than a feeling. The mere fact that the experience of volition has neural underpinnings is also not a basis for denying freedom of the will. Understanding better the interactions between circuits mediating the experience of agency and those involved in initiation of movement, formation of intention, etc., may explain how these various aspects of volition are related and can be dissociated, both with particular forms of brain damage, or with given certain arrangements of external events.

CONCLUDING THOUGHTS

On the whole, neuroscience has not much affected our conception of volition. It has maintained in large part notions of intention, choice, and the experience of agency. Where neuroscience has affected our conception, it has typically challenged traditional views of the relationship between consciousness and action. For example, more aspects of behavior than previously imagined are governed by unconscious processes. However, since we have little traction on the neural basis of consciousness, none of those challenges, to my mind, has succeeded in undermining traditional views. However, neuroscience promises to show volition not to be a unitary faculty, but rather a collection of largely separable processes that together make possible flexible, intelligent action. It may affect our notion of volition in the future by elucidating the neural systems and computations underlying these different aspects of volition. Further elucidation of brain networks may provide a better way of taxonomizing the elements of volition (Brass & Haggard 2008; Pacherie 2006, 2008). Although I believe that neuroscience will not bear upon the question of freedom via a frontal assault on the determinism question, increasing our understanding of the neural bases of these processes might cause us to think of volition more mechanistically than we currently do, and that may ultimately put pressure on our ordinary notions of what is required for freedom. For now, however, the most significant contribution neuroscience has made has been in allowing us to formulate novel questions about the nature of voluntary behavior, and in providing new ways of addressing them.

SUMMARY POINTS

1. What we think of as volition may not be a unitary faculty.

2. Neuroscience will not settle the question of determinism.

3. A network of frontal and parietal regions is involved in initiating, selecting, and controlling voluntary actions.

4. PreSMA and rCZ are implicated in endogenously generated movement.

5. The neural bases of many aspects of decision-making are well understood and can be mathematically modeled.

6. Choices to act and to refrain from acting seem to involve similar brain circuitry.

7. The feeling of agency is mediated in part by parietal cortex; it depends upon both predictive signals and postdictive feedback.

8. Thus far, neither the timing of conscious intention, mechanism, nor illusions of agency undermine the existence or efficacy of the will.

FUTURE ISSUES

1. What is the challenge that mechanism poses for accounts of volition and freedom?

2. How do different circuits mediating choice, planning, action initiation, control, and feelings of agency interact with each other?

3. How do those circuits involve, underlie, and interact with representations of self?

4. What neural computations underlie the signals identified in voluntary action in the preSMA and rCZ, and what roles do they play in action initiation?

5. What processes set the threshold and/or baseline activity in decision-making?

6. How do internal loops make possible action that is not stimulus-bound?

7. How do frontal areas control, regulate, and modify neural activity in other brain areas?

8. What are the sources of variability in the nervous system? What role does noise play in choice and action?

DISCLOSURE STATEMENT

The author is not aware of any affiliations, memberships, funding, or financial holdings that might be perceived as affecting the objectivity of this review.

ACKNOWLEDGMENTS

This work was supported in part by an NEH collaborative research grant to the Johns Hopkins Berman Institute of Bioethics, and by the MacArthur Project in Law and Neuroscience. I would like to thank Eran Klein, Nancy McConnell, Al Mele, Shaun Nichols, Michael Shadlen, and Walter Sinnott-Armstrong for comments on an earlier draft.

LITERATURE CITED

Adams F, Mele A. 1992. The intention/volition debate. *Can. J. Philos.* 22:323–38

Andersen RA, Buneo CA. 2002. Intentional maps in posterior parietal cortex. *Annu. Rev. Neurosci.* 25:189–220

Andersen RA, Buneo CA. 2003. Sensorimotor integration in posterior parietal cortex. *Adv. Neurol.* 93:159–77

Aron AR, Behrens TE, Smith S, Frank MJ, Poldrack RA. 2007. Triangulating a cognitive control network using diffusion-weighted magnetic resonance imaging (MRI) and functional MRI. *J. Neurosci.* 27:3743–52

Assad WF, Rainer G, Miller EK. 1998. Neural activity in the primate prefrontal cortex during associative learning. *Neuron* 21:1399–407

Audi R. 1993. Volition and agency. In *Action, Intention, and Reason*, ed. AR Mele, pp. 74–108. Ithaca, NY: Cornell Univ. Press

Badre D. 2008. Cognitive control, hierarchy, and the rostro-caudal organization of the frontal lobes. *Trends Cogn. Sci.* 12:193–200

Banks WP, ed. 2002. *Consciousness and Cognition*, Vol. 11. New York: Academic Press

Banks WP, Isham EA. 2009. We infer rather than perceive the moment we decided to act. *Psychol. Sci.* 20:17–21

Banks WP, Pockett S. 2007. Benjamin Libet's work on the neuroscience of free will. In *Blackwell Companion to Consciousness*, ed. M Velmans, S Schinder, pp. 657–70. Malden, MA: Blackwell

Barch DM, Braver TS, Sabb FW, Noll DC. 2000. Anterior cingulate and the monitoring of response conflict: evidence from an fMRI study of overt verb generation. *J. Cogn. Neurosci.* 12:298–309

Bittner T. 1996. Consciousness and the act of will. *Philos. Stud.* 81:331–41

Bode S, Haynes JD. 2009. Decoding sequential stages of task preparation in the human brain. *NeuroImage* 45:606–13

Brass M, Haggard P. 2007. To do or not to do: the neural signature of self-control. *J. Neurosci.* 27:9141–45

Brass M, Haggard P. 2008. The what, when, whether model of intentional action. *Neuroscientist* 14:319–25

Britten KH, Newsome WT, Shadlen MN, Celebrini S, Movshon JA. 1996. A relationship between behavioral choice and the visual responses of neurons in macaque MT. *Vis. Neurosci.* 13:87–100

Britten KH, Shadlen MN, Newsome WT, Movshon JA. 1992. The analysis of visual motion: a comparison of neuronal and psychophysical performance. *J. Neurosci.* 12:4745–65

Brown JW, Hanes DP, Schall JD, Stuphorn V. 2008. Relation of frontal eye field activity to saccade initiation during a countermanding task. *Exp. Brain Res.* 190:135–51

Bunge SA. 2004. How we use rules to select actions: a review of evidence from cognitive neuroscience. *Cogn. Affect. Behav. Neurosci.* 4:564–79

Bunge SA, Kahn I, Wallis JD, Miller EK, Wagner AD. 2003. Neural circuits subserving the retrieval and maintenance of abstract rules. *J. Neurophysiol.* 90:3419–28

Bunge SA, Wallis JD, Parker A, Brass M, Crone EA, et al. 2005. Neural circuitry underlying rule use in humans and nonhuman primates. *J. Neurosci.* 25:10347–50

Celebrini S, Newsome WT. 1994. Neuronal and psychophysical sensitivity to motion signals in extrastriate area MST of the macaque monkey. *J. Neurosci.* 14:4109–24

Chiu Y-C, Yantis S. 2009. A domain-independent source of cognitive control for task sets: shifting spatial attention and switching categorization rules. *J. Neurosci.* 29:3930–38

Churchland AK, Kiani R, Shadlen MN. 2008. Decision-making with multiple alternatives. *Nat. Neurosci.* 11:693–702

Ciaramidaro A, Adenzato M, Enrici I, Erk S, Pia L, et al. 2007. The intentional network: how the brain reads varieties of intentions. *Neuropsychologia* 45:3105–13

Corbetta M, Miezin FM, Dobmeyer S, Shulman GL, Petersen SE. 1990. Attentional modulation of neural processing of shape, color, and velocity in humans. *Science* 248:1556–59

Coxon JP, Stinear CM, Byblow WD. 2006. Intracortical inhibition during volitional inhibition of prepared action. *J. Neurophysiol.* 95:3371–83

Cui H, Andersen RA. 2007. Posterior parietal cortex encodes autonomously selected motor plans. *Neuron* 56:552–59

Cunnington R, Windischberger C, Deecke L, Moser E. 2002. The preparation and execution of self-initiated and externally-triggered movement: a study of event-related fMRI. *NeuroImage* 15:373–85

Cunnington R, Windischberger C, Robinson S, Moser E. 2006. The selection of intended actions and the observation of others' actions: a time-resolved fMRI study. *NeuroImage* 29:1294–302

Deecke L, Kornhuber HH. 1978. An electrical sign of participation of the mesial 'supplementary' motor cortex in human voluntary finger movement. *Brain Res.* 159:473–76

Deiber M-P, Honda M, Ibanez V, Sadato N, Hallett M. 1999. Mesial motor areas in self-initiated versus externally triggered movements examined with fMRI: effect of movement type and rate. *J. Neurophysiol.* 81:3065–77

de Lange FP, Spronk M, Willems RM, Toni I, Bekkering H. 2008. Complementary systems for understanding action intentions. *Curr. Biol.* 18:454–57

den Ouden HE, Frith U, Frith C, Blakemore SJ. 2005. Thinking about intentions. *NeuroImage* 28:787–96

Desmurget M, Epstein CM, Turner RS, Prablanc C, Alexander GE, Grafton ST. 1999. Role of the posterior parietal cortex in updating movements to a visual target. *Nat. Neurosci.* 2:563–67

Desmurget M, Reilly KT, Richard N, Szathmari A, Mottolese C, Sirigu A. 2009. Movement intention after parietal cortex stimulation in humans. *Science* 324:811–13

Desmurget M, Sirigu A. 2009. A parietal-premotor network for movement intention and motor awareness. *Trends Cogn. Sci.* 13:411–19

Donohue SE, Wendelken C, Bunge SA. 2008. Neural correlates of preparation for action selection as a function of specific task demands. *J. Cogn. Neurosci.* 20:694–706

Dorris MC, Glimcher PW. 2004. Activity in posterior parietal cortex is correlated with the relative subjective desirability of action. *Neuron* 44:365–78

Dorval AD. 2006. The rhythmic consequences of ion channel stochasticity. *Neuroscientist* 12:442–48

Dosenbach NUF, Fair DA, Cohen AL, Schlaggar BL, Petersen SE. 2008. A dual-networks architecture of top-down control. *Trends Cogn. Sci.* 12:99–105

Dosenbach NUF, Fair DA, Miezin FM, Cohen AL, Wenger KK, et al. 2007. Distinct brain networks for adaptive and stable task control in humans. *Proc. Natl. Acad. Sci. USA* 104:11073–78

Engbert K, Wohlschlager A, Haggard P. 2008. Who is causing what? The sense of agency is relational and efferent-triggered. *Cognition* 107:693–704

Farrer C, Frey SH, Van Horn JD, Tunik E, Turk D, et al. 2008. The angular gyrus computes action awareness representations. *Cereb. Cortex* 18:254–61

Fischer J, Ravizza M. 1998. *Responsibility and Control: A Theory of Moral Responsibility*. New York: Cambridge Univ. Press

Fleming SM, Mars RB, Gladwin TE, Haggard P. 2009. When the brain changes its mind: flexibility of action selection in instructed and free choices. *Cereb. Cortex* 19:2352–50

Fogassi L, Ferrari PF, Gesierich B, Rozzi S, Chersi F, Rizzolatti G. 2005. Parietal lobe: from action organization to intention understanding. *Science* 308:662–67

Franks KM, Stevens CF, Sejnowski TJ. 2003. Independent sources of quantal variability at single glutamatergic synapses. *J. Neurosci.* 23:3186–95

Fried I, Katz A, McCarthy G, Sass K, Williamson P, et al. 1991. Functional organization of human supplementary motor cortex studied by electrical stimulation. *J. Neurosci.* 11:3656–66

Frith CD, Friston K, Liddle PF, Frackowiak RSJ. 1991. Willed action and the prefrontal cortex in man: a study with PET. *Proc. R. Soc. Lond. Ser. B* 244:241–46

Glimcher PW. 2001. Making choices: the neurophysiology of visual-saccadic decision making. *Trends Neurosci.* 24:654–59

Glimcher PW. 2003. The neurobiology of visual-saccadic decision making. *Annu. Rev. Neurosci.* 26:133–79

Glimcher PW. 2005. Indeterminacy in brain and behavior. *Annu. Rev. Psychol.* 56:25–56

Gold JI, Shadlen MN. 2000. Representation of a perceptual decision in developing oculomotor commands. *Nature* 404:390–94

Gold JI, Shadlen MN. 2007. The neural basis of decision making. *Annu. Rev. Neurosci.* 30:535–74

Haggard P. 2005. Conscious intention and motor cognition. *Trends Cogn. Sci.* 9:290–95

Haggard P. 2008. Human volition: towards a neuroscience of will. *Nat. Rev. Neurosci.* 9:934–46

Haggard P, Eimer M. 1999. On the relation between brain potentials and the awareness of voluntary movements. *Exp. Brain Res.* 126:128–33

Hallett M. 2007. Volitional control of movement: the physiology of free will. *Clin. Neurophysiol.* 118:1179–92

Hamilton AF, Grafton ST. 2006. Goal representation in human anterior intraparietal sulcus. *J. Neurosci.* 26:1133–37

Hanks TD, Ditterich J, Shadlen MN. 2006. Microstimulation of macaque area LIP affects decision-making in a motion discrimination task. *Nat. Neurosci.* 9:682–89

Haynes JD, Sakai K, Rees G, Gilbert S, Frith C, Passingham RE. 2007. Reading hidden intentions in the human brain. *Curr. Biol.* 17:323–28

Heekeren HR, Marrett S, Ruff DA, Bandettini PA, Ungerleider LG. 2006. Involvement of human left dorsolateral prefrontal cortex in perceptual decision making is independent of response modality. *Proc. Natl. Acad. Sci. USA* 103:10023–28

Heekeren HR, Marrett S, Ungerleider LG. 2008. The neural systems that mediate human perceptual decision making. *Nat. Rev. Neurosci.* 9:467–79

Hesse MD, Thiel CM, Stephan KE, Fink GR. 2006. The left parietal cortex and motor intention: an event-related functional magnetic resonance imaging study. *Neuroscience* 140:1209–21

Huk AC, Shadlen MN. 2005. Neural activity in macaque parietal cortex reflects temporal integration of visual motion signals during perceptual decision making. *J. Neurosci.* 25:10420–36

Hyder F, Phelps EA, Wiggins CJ, Labar KS, Blamire AM, Shulman RG. 1997. "Willed action": a functional MRI study of the human prefrontal cortex during a sensorimotor task. *Proc. Natl. Acad. Sci. USA* 94:6989–94

Iacoboni M, Molnar-Szakacs I, Gallese V, Buccino G, Mazziotta JC, Rizzolatti G. 2005. Grasping the intentions of others with one's own mirror neuron system. *PLoS Biol.* 3:e79

Jahanshahi M, Jenkins IH, Brown RG, Marsden CD, Passingham RE, Brooks DJ. 1995. Self-initiated versus externally triggered movements: I. An investigation using measurement of regional cerebral blood flow with PET and movement-related potentials in normal and Parkinson's disease subjects. *Brain* 118:913–33

Jenkins IH, Jahanshahi M, Jueptner M, Passingham RE, Brooks DJ. 2000. Self-initiated versus externally triggered movements: II. The effect of movement predictability on regional cerebral blood flow. *Brain* 123:1216–28

Kennerley SW, Sakai K, Rushworth MFS. 2004. Organization of action sequences and the role of the Pre-SMA. *J. Neurophysiol.* 91:978–93

Kerns JG, Cohen JD, MacDonald AWI, Cho RY, Stenger VA, Carter CS. 2004. Anterior cingulate conflict monitoring and adjustments in control. *Science* 303:1023–26

Kiani R, Shadlen MN. 2009. Representation of confidence associated with a decision by neurons in the parietal cortex. *Science* 324:759–64

Kim J-N, Shadlen MN. 1999. Neural correlates of a decision in the dorsolateral prefrontal cortex of the macaque. *Nat. Neurosci.* 2:176–85

Knoch D, Fehr E. 2007. Resisting the power of temptations: the right prefrontal cortex and self-control. *Ann. NY Acad. Sci.* 1104:123–34

Kühn S, Brass M. 2009a. Retrospective construction of the judgement of free choice. *Conscious Cogn.* 18:12–21

Kühn S, Brass M. 2009b. When doing nothing is an option: the neural correlates of deciding whether to act or not. *NeuroImage* 46:1187–93

Kühn S, Gevers W, Brass M. 2009a. The neural correlates of intending not to do something. *J. Neurophysiol.* 101:1913–20

Kühn S, Haggard P, Brass M. 2009b. Intentional inhibition: how the "veto-area" exerts control. *Hum. Brain Mapp.* 30:2834–43

Lamm C, Fischer MH, Decety J. 2007. Predicting the actions of others taps into one's own somatosensory representations—a functional MRI study. *Neuropsychologia* 45:2480–91

Lau HC, Passingham RE. 2007. Unconscious activation of the cognitive control system in the human prefrontal cortex. *J. Neurosci.* 27:5805–11

Lau HC, Rogers RD, Haggard P, Passingham RE. 2004a. Attention to intention. *Science* 303:1208–10

Lau HC, Rogers RD, Passingham RE. 2006. On measuring the perceived onsets of spontaneous actions. *J. Neurosci.* 26:7265–71

Lau HC, Rogers RD, Passingham RE. 2007. Manipulating the experienced onset of intention after action execution. *J. Cogn. Neurosci.* 19:81–90

Lau HC, Rogers RD, Ramnani N, Passingham RE. 2004b. Willed action and attention to the selection of action. *NeuroImage* 21:1407–15

Leon MI, Shadlen MN. 1999. Effect of expected reward magnitude on the response of neurons in the dorsolateral prefrontal cortex of the macaque. *Neuron* 24:415–25

Libet B. 1985. Unconscious cerebral initiative and the role of conscious will in voluntary action. *Behav. Brain Sci.* 8:529–66

Libet B, Gleason CA, Wright EW, Pearl DK. 1983a. Time of conscious intention to act in relation to onset of cerebral activity (readiness-potential): the unconscious initation of a freely voluntary act. *Brain* 106:623–42

Libet B, Gleason CA, Wright EW, Pearl DK. 1983b. Time of conscious intention to act in relation to onset of cerebral activity (readiness-potential): the unconscious initation of a freely voluntry act. *Brain* 106:623–42

Libet B, Wright EW Jr, Gleason CA. 1982. Readiness-potentials preceding unrestricted "spontaneous" vs preplanned voluntary acts. *Electroencephalogr. Clin. Neurophysiol.* 54:322–35

Libet B, Wright EW Jr, Gleason CA. 1983c. Preparation or intention-to-act, in relation to pre-event potentials recorded at the vertex. *Electroencephalogr. Clin. Neurophysiol.* 56:367–72

Linser K, Goschke T. 2007. Unconscious modulation of the conscious experience of voluntary control. *Cognition* 104:459–75

Mainen ZF, Sejnowski TJ. 1995. Reliability of spike timing in neocortical neurons. *Science* 268:1503–6

Matsuhashi M, Hallett M. 2008. The timing of the conscious intention to move. *Eur. J. Neurosci.* 28:2344–51

Matsumoto K, Suzuki W, Tanaka K. 2003. Neuronal correlates of goal-based motor selection in the prefrontal cortex. *Science* 301:229–32

Mazurek ME, Roitman JD, Ditterich J, Shadlen MN. 2003. A role for neural integrators in perceptual decision making. *Cereb. Cortex* 13:1257–69

Mele AR. 2006. *Free Will and Luck.* Oxford: Oxford Univ. Press

Mele AR. 2009. *Effective Intentions: The Power of Conscious Will.* New York: Oxford Univ. Press

Monterosso J, Royzman EB, Schwartz B. 2005. Explaining away responsibility: effects of scientific explanation on perceived culpability. *Ethn. Behav.* 15:139–58

Moore J, Haggard P. 2008. Awareness of action: Inference and prediction. *Conscious Cogn.* 17:136–44

Moore JW, Lagnado D, Deal DC, Haggard P. 2009. Feelings of control: Contingency determines experience of action. *Cognition* 110:279–83

Mueller VA, Brass M, Waszak F, Prinz W. 2007. The role of the preSMA and the rostral cingulate zone in internally selected actions. *NeuroImage* 37:1354–61

Nachev P, Kennard C, Husain M. 2008. Functional role of the supplementary and presupplementary motor areas. *Nat. Rev. Neurosci.* 9:856–69

Nachev P, Rees G, Parton A, Kennard C, Husain M. 2005. Volition and conflict in human medial frontal cortex. *Curr. Biol.* 15:122–28

Nahmias E, Coates DJ, Kvaran T. 2007. Free will, moral responsibility, and mechanism: experiments on folk intuitions. *Midwest Stud. Philos.* 31:215–42

Newsome WT, Britten KH, Movshon JA. 1989. Neuronal correlates of a perceptual decision. *Nature* 341:52–54

O'Craven KM, Rosen BR, Kwong KK, Treisman AM, Savoy RL. 1997. Voluntary attention modulates fMRI activity in human MT-MST. *Neuron* 18:591–98

O'Doherty JEA. 2001. Abstract reward and punishment representations in the human orbitofrontal cortex. *Nat. Neurosci.* 4:95–102

Pacherie E. 2006. Toward a dynamic theory of intentions. In *Does Consciousness Cause Behavior? An Investigation of the Nature of Volition*, ed. S Pockett, WP Banks, S Gallagher, pp. 145–67. Cambridge, MA: MIT Press

Pacherie E. 2008. The phenomenology of action: a conceptual framework. *Cognition* 107:179–217

Palmer J, Huk AC, Shadlen MN. 2005. The effect of stimulus strength on the speed and accuracy of a perceptual decision. *J. Vis.* 5:376–404

Pesaran B, Nelson MJ, Andersen RA. 2008. Free choice activates a decision circuit between frontal and parietal cortex. *Nature* 453:406–9

Picard N, Strick PL. 1996. Motor areas of the medial wall: a review of their location and functional activation. *Cereb. Cortex* 6:342–53

Picard N, Strick PL. 2001. Imaging the premotor areas. *Curr. Opin. Neurobiol.* 11:663–72

Platt ML, Glimcher PW. 1999. Neural correlates of decision variables in parietal cortex. *Nature* 400:233–38

Praamstra P, Boutsen L, Humphreys GW. 2005. Frontoparietal control of spatial attention and motor intention in human EEG. *J. Neurophysiol.* 94:764–74

Quian Quiroga R, Snyder LH, Batista AP, Cui H, Andersen RA. 2006. Movement intention is better predicted than attention in the posterior parietal cortex. *J. Neurosci.* 26:3615–20

Ramnani N, Miall RC. 2004. A system in the human brain for predicting the actions of others. *Nat. Neurosci.* 7:85–90

Ridderinkhof KR, Ullsperger M, Crone EA, Nieuwenhuis S. 2004. The role of the medial frontal cortex in cognitive control. *Science* 306:443–47

Rizzolatti G, Sinigaglia C. 2007. Mirror neurons and motor intentionality. *Funct. Neurol.* 22:205–10

Roitman JD, Shadlen MN. 2002. Response of neurons in the lateral intraparietal area during a combined visual discrimination reaction time task. *J. Neurosci.* 22:9475–89

Roskies AL. 2006. Neuroscientific challenges to free will and responsibility. *Trends Cogn. Sci.* 10:419–23

Roskies AL. 2010. Why Libet's studies don't pose a threat to free will. In *Conscious Will and Responsibility*, ed. W Sinnott-Armstrong, L Nadel. New York: Oxford Univ. Press. In press

Rossi AF, Pessoa L, Desimone R, Ungerleider LG. 2009. The prefrontal cortex and the executive control of attention. *Exp. Brain Res.* 192:489–97

Rowe J, Hughes L, Eckstein D, Owen AM. 2008. Rule-selection and action-selection have a shared neuroanatomical basis in the human prefrontal and parietal cortex. *Cereb. Cortex* 18:2275–85

Rushworth MF. 2008. Intention, choice, and the medial frontal cortex. *Ann. NY Acad. Sci.* 1124:181–207

Schall JD, Boucher L. 2007. Executive control of gaze by the frontal lobes. *Cogn. Affect. Behav. Neurosci.* 7:396–412

Schall JD, Stuphorn V, Brown JW. 2002. Monitoring and control of action by the frontal lobes. *Neuron* 36:309–22

Schultz W, Tremblya L, Hollerman JR. 2000. Reward processing in primate orbitofrontal cortex and basal ganglia. *Cereb. Cortex* 10:272–83

Searle JR. 1983. *Intentionality: An Essay in the Philosophy of Mind*. Cambridge, UK: Cambridge Univ. Press

Serences JT, Yantis S. 2007. Spatially selective representations of voluntary and stimulus-driven attentional priority in human occipital, parietal, and frontal cortex. *Cereb. Cortex* 17:284–93

Shadlen MN, Britten KH, Newsome WT, Movshon JA. 1996. A computational analysis of the relationship between neuronal and behavioral responses to visual motion. *J. Neurosci.* 16:1486–510

Shadlen MN, Newsome WT. 1996. Motion perception: seeing and deciding. *Proc. Natl. Acad. Sci. USA* 93:628–33

Shadlen MN, Newsome WT. 2001. Neural basis of a perceptual decision in the parietal cortex (area LIP) of the rhesus monkey. *J. Neurophysiol.* 86:1916–36

Shibasaki H, Hallett M. 2006. What is the Bereitschaftspotential? *Clin. Neurophysiol.* 117:2341–56

Sinnott-Armstrong W, Nadel L, eds. 2010. *Conscious Will and Responsibility*. New York: Oxford Univ. Press. In press

Sirigu A, Daprati E, Ciancia S, Giraux P, Nighoghossian N, et al. 2004. Altered awareness of voluntary action after damage to the parietal cortex. *Nat. Neurosci.* 7:80–84

Soon CS, Brass M, Heinze H-J, Haynes J-D. 2008. Unconscious determinants of free decisions in the human brain. *Nat. Neurosci.* 11:543–45

Sumner P, Nachev P, Morris P, Peters AM, Jackson SR, et al. 2007. Human medial frontal cortex mediates unconscious inhibition of voluntary action. *Neuron* 54:697–711

Thaler D, Chen YC, Nixon PD, Stern CE, Passingham RE. 1995. The functions of the medial premotor cortex. I. Simple learned movements. *Exp. Brain Res.* 102:445–60

Thoenissen D, Zilles K, Toni I. 2002. Differential involvement of parietal and precentral regions in movement preparation and motor intention. *J. Neurosci.* 22:9024–34

Trevena JA, Miller J. 2002. Cortical movement preparation before and after a conscious decision to move. *Conscious Cogn.* 11:162–90

Tsakiris M, Haggard P, Franck N, Mainy N, Sirigu A. 2005. A specific role for efferent information in self-recognition. *Cognition* 96:215–31

Voss M, Ingram JN, Wolpert DM, Haggard P. 2008. Mere expectation to move causes attenuation of sensory signals. *PloS ONE* 3:e2866

Wegner D. 2002. *The Illusion of Conscious Will*. Cambridge, MA: MIT Press

Wegner D, Wheatley T. 1999. Apparent mental causation: sources of the experience of will. *Am. Psychol.* 54:480–92

Yang T, Shadlen MN. 2007. Probabilistic reasoning by neurons. *Nature* 447:1075–80

Young G. 2006. Preserving the role of conscious decision making in the initiation of intentional action. *J. Conscious Stud.* 13:51–68

Zhu J. 2004a. Intention and volition. *Can. J. Philos.* 34:175–93

Zhu J. 2004b. Understanding volition. *Philos. Psychol.* 17:247–73

Watching Synaptogenesis in the Adult Brain

Wolfgang Kelsch,[1,*] Shuyin Sim,[1,*] and Carlos Lois[2]

[1]Picower Institute of Learning and Memory, Departments of Biology and Brain and Cognitive Sciences, Massachussetts Institute of Technology, Cambridge, Massachusetts 02139; email: kelsch@mit.edu, shuyin@mit.edu

[2]Department of Neurobiology, University of Massachusetts Medical School, Aaron Lazare Medical Research Building, Worcester, Massachusetts 01605-2324; email: loisC@umassmed.edu

Annu. Rev. Neurosci. 2010. 33:131–49

The *Annual Review of Neuroscience* is online at neuro.annualreviews.org

This article's doi:
10.1146/annurev-neuro-060909-153252

*These authors contributed equally to this review

Key Words

synapse development, synaptic plasticity, activity-dependent synaptic formation, olfactory bulb, dentate gyrus

Abstract

Although the lifelong addition of new neurons to the olfactory bulb and dentate gyrus of mammalian brains is by now an accepted fact, the function of adult-generated neurons still largely remains a mystery. The ability of new neurons to form synapses with preexisting neurons without disrupting circuit function is central to the hypothesized role of adult neurogenesis as a substrate for learning and memory. With the development of several new genetic labeling and imaging techniques, the study of synapse development and integration of these new neurons into mature circuits both in vitro and in vivo is rapidly advancing our insight into their structural plasticity. Investigators' observation of synaptogenesis occurring in the adult brain is beginning to shed light on the flexibility that adult neurogenesis offers to mature circuits and the potential contribution of the transient plasticity that new neurons provide toward circuit refinement and adaptation to changing environmental demands.

Contents

OB: olfactory bulb

DG: dentate gyrus

INTRODUCTION

Adult neurogenesis in mammals was first described in the early 1960s (Altman & Das 1965),
but it was not until much later that investigators broadly accepted that the addition of new neurons occurs throughout life in both the olfactory bulb (OB) (Lois & Alvarez-Buylla 1993, Luskin 1993) and the dentate gyrus (DG) (Bayer 1980, Bayer et al. 1982, Gage et al. 1995). Although most neurons in the brain are added to immature circuits during assembly, neurons generated in adulthood face an additional challenge as they integrate into mature, fully functional circuits. Relatively little is known about the mechanisms that regulate the synaptic development of adult-born neurons and their connectivity within mature circuits. This review aims to present key aspects of the emerging understanding of synaptogenesis in adult-born neurons, as well as how activity in the brain modulates this process.

Synapse formation during adult neurogenesis raises several intriguing questions. Does synapse formation in adult-born neurons simply recapitulate the steps that occur during embryonic and neonatal development, or is it regulated by specific mechanisms specialized for integration into functioning circuits? How do new neurons make synapses with mature circuits without disrupting existing connectivity? An understanding of synaptogenesis in the adult brain will not only shed light on the putative function of adult neurogenesis in information processing and storage, but also provide new insights to develop strategies for successful neuronal replacement therapies to treat brain injury and neurodegenerative conditions.

We begin this review by discussing some of the techniques currently being used to study synaptogenesis in adult-born neurons. Next, we proceed to critically examine current literature on how the various types of adult-born neurons develop synaptic connections with their respective circuits and how this process is modulated by activity. We also discuss the functional properties of new neurons and their potential contribution toward refining the existing circuit. Finally, we conclude by reflecting on recent trends and discoveries in this dynamic field and exploring future directions toward understanding the integration of new neurons into adult circuits and the role of adult-born neurons in brain function.

TECHNIQUES FOR OBSERVING SYNAPTOGENESIS

Recent technical advances have accelerated the study of synapse formation in adult neurogenesis. In particular, two genetic methods that facilitate the selective labeling of new neurons with fluorescent proteins have been especially useful. First, oncoretroviral vectors can be used to label new neurons genetically (Jessberger et al. 2007, Kelsch et al. 2007), as they exclusively infect actively dividing cells, such as neuronal progenitors, but are unable to infect postmitotic cells, such as neurons (Roe et al. 1993). Second, investigators have developed several transgenic mouse lines that enable selective labeling of adult-born neurons. In two of these transgenic lines, expression of a fluorescent protein, either green fluorescent protein (GFP) or red fluorescent protein (dsRed), is driven by promoters that are active only in immature neurons, namely doublecortin (Wang et al. 2007) and nestin (Mignone et al. 2004). This process results in specific labeling of immature neurons in both the OB and the DG (Brown et al. 2003, Yamaguchi et al. 2000). In another line, the proopiomelanocortin (POMC) promoter drives GFP expression, which, unexpectedly, labels new neurons in the DG (Overstreet et al. 2004).

Genetic Labeling of Synapses of Adult-Born Neurons

Synapse formation has traditionally been studied in three main ways. First, in some cases, synapses can be easily identified on the basis of their association with neuronal structural specializations. For instance, many excitatory input synapses are located in dendritic spines, in which case spines may be used as a morphological substitute for synapses. One limitation of this method is that a large proportion of synapses, such as excitatory input synapses on cell somata and inhibitory synapses, are simply not associated with spines (Price & Powell 1970, Woolf et al. 1991). In addition, it is not possible to accurately quantify the density and measure the size of output synapses by simple morphological analyses. Second, antibody labeling against synaptic markers is a powerful method to label synapses in cultured neurons. However, this method is suboptimal in brain sections because the large number of synapses present severely complicates the attribution of synapses to individual new neurons. Emerging imaging techniques, however, may be able to overcome some of these problems soon (Micheva & Smith 2007). Third, synapses can be unambiguously identified by electron microscopy, but this technique is labor intensive and has yet to be sufficiently developed for high-throughput analysis.

Labeling synapses with genetically encoded markers addresses some of the limitations of the above-mentioned techniques and significantly simplifies the quantification of synaptic development in new neurons (Kelsch et al. 2008, Livneh et al. 2009, Meyer & Smith 2006, Niell et al. 2004). The visualization of pre- and postsynaptic terminals can be achieved via expression of fluorescent proteins fused to proteins specifically located in synapses. For instance, synaptophysin is a protein located in neurotransmitter vesicles that is selectively localized at presynaptic terminals (Sudhof & Jahn 1991) and can be used to identify release sites on axon terminals (**Figures 1d, 2b**). To identify postsynaptic terminals, PSD95, a scaffolding protein restricted to clusters in the postsynaptic density of most glutamatergic synapses (Ebihara et al. 2003, Gray et al. 2006, Niell et al. 2004, Sassoe-Pognetto et al. 2003, Sheng 2001, Washbourne et al. 2002), can be used (**Figures 1c,d and 2c**). When introducing these genetically encoded markers, it is critical to ensure that only modest levels of overexpression are achieved because excessive levels of these proteins may interfere with synaptic development or function (El-Husseini et al. 2000). Fortunately,

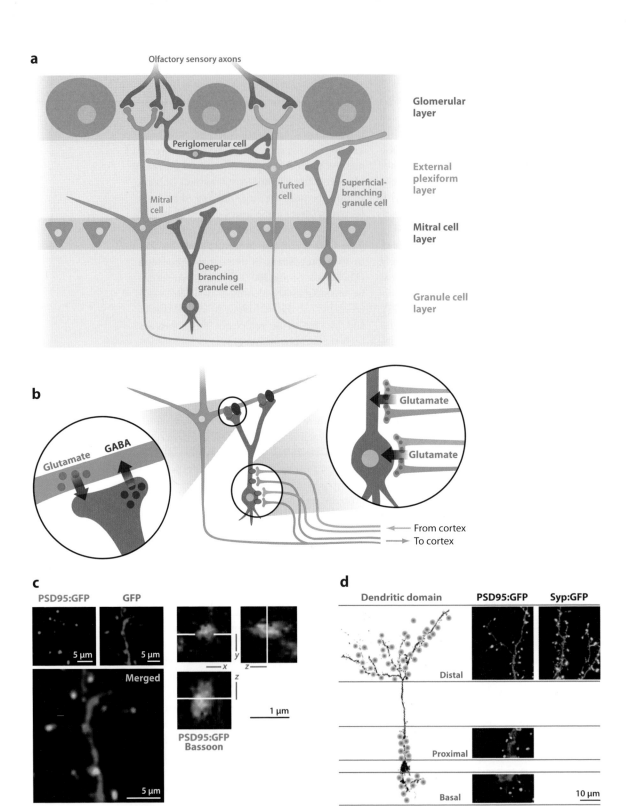

a

Olfactory sensory axons

Glomerular layer

Periglomerular cell

External plexiform layer

Mitral cell

Tufted cell

Superficial-branching granule cell

Mitral cell layer

Deep-branching granule cell

Granule cell layer

b

Glutamate   GABA

Glutamate

Glutamate

From cortex
To cortex

c

PSD95:GFP   GFP

5 μm   5 μm

Merged

5 μm

y
x   z
z

1 μm

PSD95:GFP
Bassoon

d

Dendritic domain   PSD95:GFP   Syp:GFP

Distal

Proximal

Basal

10 μm

retroviral vectors, which deliver single copies of transgenes into their target cells, produce sufficiently low levels of expression that leave synapse number and strength unperturbed (Kelsch et al. 2008). With genetically encoded markers, one can, in principle, analyze the complete set of a neuron's excitatory input synapses and output synapses, including those not associated with spines or axon terminals, respectively. This method also allows for software-aided quantification of synapses (Kelsch et al. 2008). Furthermore, viral vectors can be engineered such that in addition to synaptic markers other genes of interest, such as those coding for ion channels, growth factors, or cell adhesion molecules, can be introduced in the same vector to assess the effects of various manipulations on synapse formation and dynamics.

The labeling methods mentioned above allow investigators to observe synapse development in combination with a variety of imaging and recording techniques. Here we briefly discuss the strengths and limitations of four major techniques currently used in the field.

Two-photon laser scanning fluorescence microscopy in vivo. Two-photon microscope technology has taken off considerably in recent years and is still the only technique that allows for synapse imaging in vivo. This technique is extremely useful for observing real-time changes to experimental manipulations and allows investigators to visualize synapse dynamics. Owing to detection limits, only neurons up to ~800 μm below the brain's surface can be imaged (Helmchen & Denk 2005), largely restricting this technology to the study of adult-born periglomerular neurons (PGNs) in the glomerular layer of the OB and distal dendrites of OB granule neurons, which are located in the external plexiform layer (**Figure 1a**) (Mizrahi 2007). By removing part of the neocortex and white matter above the hippocampus, superficial dendrites of neurons in CA1 of the hippocampus can be imaged with two-photon technology (Mizrahi et al. 2004). However, adult-born granule cells in the DG cannot be imaged in the same manner without damaging a substantial part of the hippocampus. There

PGN: periglomerular neuron

Figure 1

Adult-born olfactory bulb granule cells and their synaptic wiring with the surrounding circuit. (*a*) Synaptic organization in the olfactory bulb. In the glomeruli of the olfactory bulb, olfactory sensory axons synapse on the apical dendrites of principal neurons, the mitral and tufted cells, as well as on periglomerular neurons (PGNs), which line these glomeruli. PGNs also form additional synaptic connections with the apical dendrites of principal neurons, whereas the lateral dendrites of principal neurons form synapses with granule cells (GC$_{OB}$). Two independent microcircuits may exist in the olfactory bulb, with GC$_{OB}$ with either deep or superficial dendritic branching connecting exclusively to mitral or tufted cells, respectively. (*b*) Synaptic connectivity of olfactory bulb granule neurons. GC$_{OB}$ form dendro-dendritic synapses with lateral dendrites of principal neurons in the bulb. These atypical synapses consist of a glutamatergic input synapse from the principal neuron onto the GC$_{OB}$ and a GABAergic output synapse onto the same lateral dendrite of the principal neuron, both located in a single spine. In addition, GC$_{OB}$ receive glutamatergic inputs onto their basal and proximal apical dendrites from centrifugal cortical axons and possibly also from axon collaterals of principal neurons. GC$_{OB}$ are also contacted by GABAergic input synapses from local interneurons in the olfactory bulb as well as cholinergic and monoaminergic inputs. (*c*) Genetic labeling of synapses. *Left*: Progenitors of GC$_{OB}$ were infected in the subventricular zone (SVZ) with retroviral vectors carrying genetic constructs encoding PSD95:GFP, a marker for postsynaptic glutamatergic sites, to generate PSD95:GFP-expressing GC$_{OB}$. PSD95:GFP-positive clusters can be detected by direct visualization of GFP (*shown as green puncta*). The dendritic morphology of the GC$_{OB}$ was revealed by amplifying the low levels of PSD95:GFP in the cytoplasm (that could not be detected by intrinsic fluorescence) by immunostaining against GFP with a red fluorescent secondary antibody. The merged images of PSD95:GFP positive clusters (*green*) and dendritic morphology (*red*) allow investigators to attribute clusters to specific dendritic domains of individual GC$_{OB}$ (scale bar, 5 μm). *Right*: Confocal three-dimensional image showing a PSD95:GFP positive cluster in a new GC$_{OB}$ that is contacted by the presynaptic marker, bassoon (scale bar, 1 μm). (*d*) Synaptic distribution in the dendritic domains of granule cells. GC$_{OB}$ have a basal dendrite and an apical dendrite, which consist of proximal and distal synaptic domains. The proximal domain is a specialized sector of the unbranched apical dendrite that emerges directly from the soma of GC$_{OB}$, which contains a high density of glutamatergic input synapses. The branched dendritic segment of the apical dendrite is known as the distal domain. Examples of genetic labeling of synapses in the dendritic domains of granule cells are given for postsynaptic glutamatergic synapses, PSD95:GFP, and for the presynaptic genetic marker, Synaptophysin:GFP (Syp:GFP).

has been great interest in using two-photon microscopy associated with endoscope lenses to image deep within the brain, but several technical obstacles need to be solved before this method can be routinely used (Barretto et al. 2009).

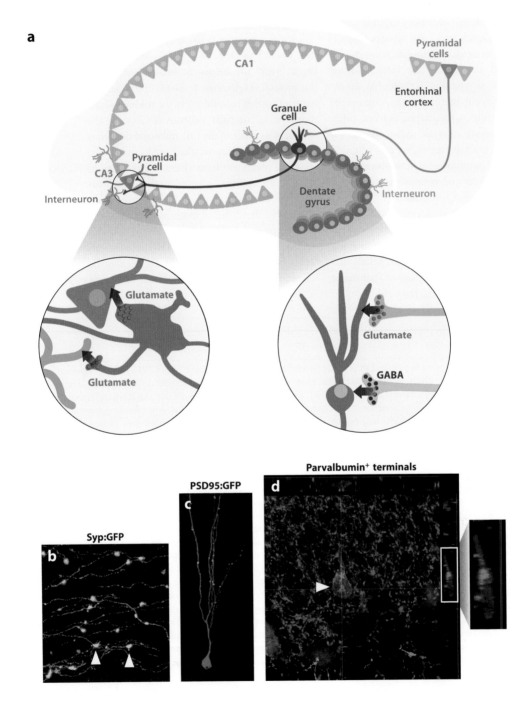

Confocal laser scanning microscopy. Because deep structures of the brain, such as the DG where adult-born granule cells are added, are beyond the depth limitation of two-photon microscopy in vivo, in vitro time-lapse confocal imaging of brain slices is sometimes carried out to study the dynamics of synapse formation (Galimberti et al. 2006, Toni et al. 2007). This technique is not widely used because of concerns about the integrity of cultured adult brain slices over long time periods with currently available culture techniques (Berdichevsky et al. 2009), as well as the possibility of abnormal synapse rearrangement due to fluctuations in culture conditions (Kirov et al. 2004). Confocal imaging of fixed slices is much more commonly used to study synaptogenesis, especially because this method of observation is technically straightforward and enables investigators to analyze many neurons at one time. Time course experiments can be performed to observe spine formation over days and months, but because only an instantaneous snapshot of the synapses can be obtained in fixed slices, this technique cannot be used to analyze the short-term dynamics of synapse formation in real time.

Electron microscopy. Electron microscopy is more technically challenging than confocal imaging but allows for simultaneous analysis of pre- and postsynaptic sites by visualizing synaptic vesicles and postsynaptic sites, respectively. Electron microscopy can also be used for three-dimensional high-resolution analysis of individual synapses on adult-born neurons and their synaptic partners (Toni et al. 2007). Recent developments within the past few years hint at the possibility of semiautomated sectioning and imaging of large neuropil volumes (Briggman & Denk 2006), thus facilitating high-throughput ultrastructural analyses of synapses.

Electrophysiology. Electrophysiological recording provides a functional confirmation of structural observations in studies of adult-born neurons. The frequency and amplitude of excitatory and inhibitory synaptic inputs help scientists understand changes in connectivity of new neurons during their maturation and the effects of diverse manipulations. Carleton et al. (2003) and van Praag et al. (2002) have used electrophysiology to describe synaptic properties of new neurons as they mature and integrate into their circuits. One of the most significant contributions of electrophysiology to the study of adult neurogenesis is the demonstration that new neurons display enhanced synaptic plasticity compared with fully mature neurons in both the OB and the DG (Nissant et al. 2009, Schmidt-Hieber et al. 2004). Scharfman et al. (2000) also used electrophysiological methods

---

Figure 2

Adult-born granule cells in the dentate gyrus and their synaptic wiring with the hippocampal and other circuits. (*a*) Synaptic organization of dentate granule cells. Adult-born granule cells in the dentate gyrus (GC_{DG}) receive excitatory glutamatergic input onto their apical dendrites from projection neurons in the entorhinal cortex and mossy cells in the hippocampus, as well as inhibitory GABAergic input from local interneurons. GC_{DG} project axons solely to the CA3 region of the hippocampus. At the CA3 region, these axons constitute two forms of specialized contact sites: large mossy terminals and en passant boutons. Large mossy terminals are compartmentalized release sites: The central portion of these terminals forms complex interdigitating synapses with proximal dendrites of CA3 pyramidal cells while the emanating filopodia of the terminals synapse on GABAergic interneurons in CA3. En passant boutons are smaller synaptic swellings along the axon collaterals that exclusively contact CA3 interneurons. (*b*) Genetic labeling of output synapses along axon collaterals of dentate granule cells. The release sites from GC_{DG} onto CA3 neurons, at large mossy terminals (*arrowheads*) and en passant boutons on the axons of adult-born GC_{DG}, can be visualized by a genetic presynaptic marker, synaptophysin:GFP (Syp:GFP, *yellow*). (*c*) Genetic labeling of input synapses in the apical dendrites of granule cells. Adult-born GC_{DG} develop glutamatergic input sites as visualized by PSD95:GFP (*yellow*) in their apical dendrite during their differentiation. Note the absence of PSD95:GFP-positive sites on the soma. (*d*) Identification of inhibitory innervation on the soma of adult-born dentate granule cells. Parvalbumin-positive inhibitory terminals (*red*) contacting the soma of an adult-born GC_{DG} (*arrowhead*) can be visualized by immunohistochemistry against parvalbumin. *Left*: Confocal image of a GFP-positive GC_{DG} labeled by retroviral infection of progenitors in the subgranular zone of the dentate gyrus. *Right*: Magnified z-stack cross-section of image on left.

to study how different manipulations affect neuronal integration into the adult brain, such as the effects of seizures on synaptic properties of adult-born dentate granule neurons.

SYNAPTOGENESIS IN NEURONS ADDED TO THE ADULT MAMMALIAN BRAIN

The three main types of neurons added to the adult brain are the granule cells (GC$_{OB}$) and PGNs in the OB and the granule cells in the DG (GC$_{DG}$). GC$_{OB}$ constitute the largest population of adult-born neurons. They are GABAergic interneurons that connect to the lateral dendrites of the OB's principal neurons (mitral and tufted cells; **Figure 1a**). PGNs are GABAergic and/or dopaminergic interneurons that modulate incoming information from olfactory sensory axons that connect to the apical dendrites of the olfactory bulb's principal neurons (**Figure 1a**). Granule neurons in the DG are excitatory neurons that receive input from the entorhinal cortex and project to the CA3 region of the hippocampal formation (**Figure 2a**).

SYNAPTOGENESIS IN ADULT-BORN OLFACTORY BULB GRANULE NEURONS

Stages of Synaptic Development

Adult-born GC$_{OB}$ arise from neural progenitors in the subventricular zone (SVZ), which lines the walls of the lateral ventricles (Lois & Alvarez-Buylla 1994). Neuroblasts travel long distances via the rostral migratory stream (RMS) into the OB where they migrate radially into the granule cell layer (Lois & Alvarez-Buylla 1994). Of the ∼30,000 new neurons produced daily in an adult mouse, more than 97% differentiate into GC$_{OB}$ while the remaining develop into PGNs (Lois & Alvarez-Buylla 1994, Winner et al. 2002).

GC$_{OB}$ are axonless neurons that have a basal and an apical dendrite (**Figure 1d**). Their apical dendrite is composed of an unbranched

segment emerging from the soma followed by a branched segment and can be divided into proximal and distal synaptic domains. The proximal synaptic domain is a region on the unbranched dendrite segment with a high concentration of glutamatergic input synapses. The distal domain consists of the branched dendritic segment and possesses spines containing bidirectional dendro-dendritic synapses, where input and output synapses are colocalized and functionally coupled. These bidirectional synapses receive glutamatergic input synapses from the lateral dendrites of principal neurons and release GABA back onto the same neurons (**Figure 1b**) (Mori 1987). Dendro-dendritic synapses are the only output of GC$_{OB}$ and are responsible for local inhibition of principal neurons in the OB (Chen et al. 2000, Halabisky & Strowbridge 2003, Mori 1987). The basal dendrite, or basal domain, and unbranched apical dendrite receive glutamatergic input from axon collaterals of the OB's principal neurons and olfactory cortex (Balu et al. 2007, Davis & Macrides 1981, Luskin & Price 1983, Mori et al. 1983).

The developmental stages of adult-born GC$_{OB}$ have been defined according to morphological criteria (Petreanu & Alvarez-Buylla 2002). Stage one neurons are those in the process of migration in the RMS. At stage two, new GC$_{OB}$ reach the granule cell layer and begin to extend their first neurites. At stage three, about ten days after the birth of GC$_{OB}$ in the SVZ, the main dendritic arbor of new GC$_{OB}$ continues to grow and cells start to receive inhibitory synaptic input (Carleton et al. 2003). At stage four, or about two weeks after their birth, new GC$_{OB}$ start receiving excitatory synaptic input (Carleton et al. 2003). In adult-born GC$_{OB}$, excitatory synapses appear first on the proximal segment of the apical dendrite at this stage (Kelsch et al. 2008). At this time there are few spines and synaptic sites on the distal branches of the apical dendrite (Petreanu & Alvarez-Buylla 2002). Finally, at stage five, between three and four weeks after birth of GC$_{OB}$, the distal branches of their apical dendrites develop dense spines, achieving full spine density in these dendrites by four weeks of development (Petreanu & Alvarez-Buylla 2002). During this final stage of maturation, new GC_{OB} acquire the ability to fire fast action potentials (Carleton et al. 2003) and form most of the input and output synapses on their distal branches (Kelsch et al. 2008). Although synaptic development is mostly complete by four weeks after the generation of adult-born GC_{OB} (Carleton et al. 2003, Kelsch et al. 2008, Mizrahi 2007, Petreanu & Alvarez-Buylla 2002, Whitman & Greer 2007a), Mizrahi (2007) has observed rearrangement of spines after this time, which suggests that GC_{OB} may maintain some capacity for synaptic modification even when they are mature.

Adult-born granule neurons first develop input synapses in their proximal dendritic domain, which lacks output synapses, before developing most of their dendro-dendritic output synapses and prior to acquiring the ability to fire action potentials (Kelsch et al. 2008), i.e., they "listen" before they can "speak." This sequential pattern of synaptic development of adult-born GC_{OB} sharply contrasts with the maturation of cells born during neonatal development, which is when most GC_{OB} are generated (Lemasson et al. 2005). First, neonatal-born GC_{OB} develop the ability to fire action potentials early in their development, during stage three, around the same time they start receiving synaptic inputs (Carleton et al. 2003). Second, neurons added to the neonatal brain also develop input and output synapses on the distal and proximal regions of apical dendrites simultaneously (Kelsch et al. 2008).

The different modes of synaptic development between adult and neonatal neurons could be due to intrinsic properties of new GC_{OB} already determined in their respective precursors. Alternatively, local cues in the neonatal and adult OB environment may be responsible for these differences. Heterochronic transplantation of postmitotic precursors could be helpful to clarify which aspects of synaptic development are governed by cell-autonomous versus external cues.

Synapse Connectivity Within Olfactory Circuits

The lateral dendrites of mitral and tufted cells, which form dendro-dendritic synapses with GC_{OB}, are located in the deep and superficial external plexiform layers, respectively (**Figure 1a**). Most GC_{OB} ramify distal dendritic branches only in one location within the external plexiform layer, not in both (Kelsch et al. 2007, Mori et al. 1983), and this phenomenon is genetically predetermined in neuronal progenitors as demonstrated by fate-mapping and transplantation studies (Kelsch et al. 2007, Merkle et al. 2007). The OB may have "independent microcircuits" such that specific populations of GC_{OB} target only one class of principal neurons (Mori 1987); GC_{OB} with superficial or deep dendrites may exclusively form synapses with tufted or mitral cells, respectively. However, the existence of these independent microcircuits has yet to be proven functionally. The concept of microcircuit-specific targeting of new neurons in the adult brain is consistent with the protomap model of circuit assembly (Rakic et al. 2009) and raises the possibility of genetically engineering stem cells to generate specific neuronal types to replace those lost to disease or injury.

Neuronal Addition and Turnover

The functional differences between deep and superficial GC_{OB} also extend to neuronal survival. Whereas neonatal-born GC_{OB} often reside in the superficial granule cell layers, adult-born neurons tend to localize within the deep layers (Imayoshi et al. 2008, Lemasson et al. 2005). Although most superficial and neonatal-born GC_{OB} survive for long periods approaching the animal's lifetime, deep and adult-born GC_{OB} tend to be short-lived (Imayoshi et al. 2008, Lemasson et al. 2005). A recent study using a transgenic labeling technique suggests that almost all deep, adult-born neurons are turned over and thus continuously replaced (Imayoshi et al. 2008), which supports Bayer's (1980) original observations. Two long-term studies suggest that cell death in adult-born GC_{OB} is limited to the first month after neuron birth (Lemasson et al. 2005, Winner et al. 2002), whereas another study suggests there is a further drop in cell survival after the first two months (Petreanu & Alvarez-Buylla 2002). However, the latter study has low temporal resolution after the two-month time point, and this result may also be caused by the high variability between samples. At least some of the adult-born GC_{OB} that persist throughout life maintain a synaptic density similar to the one they displayed a month after their birth (W. Kelsch & C. Lois, unpublished observations). In summary, the question of neuronal addition or turnover of adult-born GC_{OB} remains unresolved and warrants further clarification, especially in light of the implications on the potential role of adult neurogenesis for long-term memory storage.

Activity-Dependent Neuronal Survival

Only 50% of new GC_{OB} generated in the adult successfully integrate into the bulb's circuits and survive for extended time periods, and abundant evidence indicates that neuronal activity is important in determining their survival. Synaptic maturation in GC_{OB} occurs mostly in the third and fourth week of development (Carleton et al. 2003, Kelsch et al. 2008, Whitman & Greer 2007b). This period coincides with a time window during which the survival of GC_{OB} is most sensitive to sensory deprivation, when the proportion of surviving neurons is further reduced by half in a deprived bulb (Petreanu & Alvarez-Buylla 2002, Yamaguchi et al. 2000, Yamaguchi & Mori 2005). Silencing the circuit with pharmacologically enhanced inhibition also reduces survival of adult-born neurons during this critical period (Yamaguchi & Mori 2005). Rochefort et al. (2002) reported that exposure to an odor-enriched environment increases survival, particularly when the animal is rewarded for performing an odor-discrimination task (Alonso et al. 2006). However, the enhanced survival reported in these works has not been observed in other studies (Magavi et al. 2005). Although the source of this disparity is unclear, the enhanced survival reported may not be due solely to exposure to enriched odors, but also to the fact that the behavioral demand of the task may raise animals' attention levels (Alonso et al. 2006). These observations suggest that odor information processing via nascent synapses plays a critical role in the stable integration of new neurons in the adult olfactory system.

Activity-Dependent Synaptogenesis

Activity in the OB not only influences the survival of adult-born GC_{OB}, but also regulates their synaptic connectivity. When postnatal-born GC_{OB} are subject to sensory deprivation during the critical period, they display fewer synaptic spines (Saghatelyan et al. 2005). Kelsch et al. (2009) recently confirmed this finding using genetically encoded markers for excitatory synapses in adult-born GC_{OB}. The loss of input and output synapses triggered by sensory deprivation occurs only during early synaptic development and is not seen when sensory deprivation is performed after synaptic development is completed. This observation suggests that the critical period during which the survival of new neurons is dependent on sensory input coincides with a stage in which neurons have a high degree of plasticity, which facilitates the shaping of their synaptic organization. Similarly, a recent study by Nissant et al. (2009) has demonstrated that long-term potentiation can be induced in adult-born neurons during early stages of their maturation, but not after this period. It will be interesting to examine whether this critical period also applies to other forms of plasticity in adult-born neurons.

The effects of olfactory deprivation on synaptic development are complex: Adult-born GC_{OB} that survive after sensory deprivation display an increased density of proximal input synapses in the unbranched apical dendrite (Kelsch et al. 2009). This observation suggests that neurons may compensate for the absence of sensory input by receiving additional excitatory drive, which elevates their activity level above the threshold required for survival.

The relationship between cell-intrinsic excitability and synapse formation is not well understood. Recent experiments indicate that increasing the intrinsic excitability of adult-born GC_{OB} by expressing a voltage-gated bacterial sodium channel does not affect synapse formation or maintenance (Kelsch et al. 2009) and promotes stable integration of adult-born GC_{OB} (Lin et al. 2010). However, genetically increased excitability blocks sensory deprivation–triggered synaptic changes. GC_{OB} expressing this sodium channel that are born in a bulb deprived of sensory input develop a normal organization of glutamatergic input synapses, as measured by the density of their PSD95:GFP-positive clusters (Kelsch et al. 2009). Similarly, dampening the excitability of new GC_{OB} by overexpressing the potassium channel Kir2.1 does not affect the synapse numbers of the surviving neurons (Lin et al. 2010). Taken together with the finding that increased inhibition in the circuit decreases the survival of adult-born GCs (Yamaguchi & Mori 2005), these observations demonstrate that synaptogenesis in adult-born GC_{OB} is sensitive to changes in synaptic input and suggest that both survival selection and synaptic development are driven by a minimum threshold excitation level, to which several factors can contribute: first, local glutamatergic excitatory inputs from mitral or tufted cells, whose activity is regulated primarily by sensory experience; second, centrifugal glutamatergic inputs originating in other regions of the brain, such as the olfactory cortex, which act on the olfactory bulb; third, centrifugal inputs of neuromodulators such as acetylcholine, noradrenaline, or neuropeptides, which modulate neuronal activity on a longer timescale. For instance, cholinergic stimulation, which causes sustained depolarizations in GC_{OB} (Pressler et al. 2007), enhances the survival of new neurons both in the dentate gyrus and in the OB (Kaneko et al. 2006) and may be responsible for the reported enhanced survival of adult-born neurons when olfactory tasks involved

increased attention levels (Alonso et al. 2006). Hence, phasic excitation provided by synaptic input from mitral and tufted cells is only one of the many determinants of survival and integration of new GCs into the bulb's circuits.

We hope future studies will resolve the ambiguities surrounding the regulation of neuronal survival by centrifugal and sensory-driven inputs. Meanwhile, the transient synaptic plasticity displayed during synaptic development of new GC_{OB} may be an attractive model with which to study how neuronal connectivity during circuit assembly is regulated by activity.

SYNAPTOGENESIS IN ADULT-BORN OLFACTORY BULB PERIGLOMERULAR NEURONS

The second class of neurons generated throughout life in the OB is the PGN, which surrounds glomeruli where olfactory sensory axons connect to the apical dendrites of the OB's principal neurons (**Figure 1a**). PGNs receive excitatory synaptic input both from olfactory sensory axons as well as from the apical dendrites of principal neurons via dendro-dendritic synapses. The outputs of PGNs occur both through dendro-dendritic synapses and axonal output synapses with principal neurons, although not all PGNs have axons (Pinching & Powell 1971). PGNs are a highly diverse group of neurons and can be broadly divided into two groups: those whose dendrites synapse on only one or two glomeruli and those that synapse on many (Whitman & Greer 2007a). These neurons can be dopaminergic and/or GABAergic, and different subsets of GABAergic neurons also express different calcium-binding proteins such as calbindin and calretinin (Whitman & Greer 2007a). Different subtypes of PGNs are generated in the OB during embryogenesis, neonatal development, and adulthood (Batista-Brito et al. 2008, De Marchis et al. 2007).

Much less is known about the synaptic development of adult-born PGNs as compared with GC_{OB}. Given the diversity of adult-born

PGNs, one would expect heterogeneity in their synaptic development and organization as well. Indeed, the maturational sequence of spontaneous inputs is not stereotypical for PGNs: Some neurons develop GABAergic inputs first, whereas others develop glutamatergic inputs first (Grubb et al. 2008). Excitatory inputs to PGNs appear early during their development, and their frequency continues to increase until six weeks after their birth (Grubb et al. 2008).

During the first six weeks after their birth, adult-born PGNs develop a full dendritic arbor. In vivo two-photon imaging has shown that the spines of adult-born PGNs become more stable as they mature (Livneh et al. 2009). During the maturation of PGNs, strong functional changes occur in the synapses between sensory neurons and PGNs. These changes appear to be mostly restricted to the postsynaptic sites on the PGNs, whereas the characteristically high release probability at olfactory sensory neuron terminals (Murphy et al. 2004) is already present as soon as functional synapses are formed (Grubb et al. 2008). The highly dynamic rearrangement of input synapses in PGNs may be attributed to the continuous turnover of olfactory sensory axons (Zou et al. 2004), from which they receive their primary input, or could simply be an intrinsic property of these neurons.

Activity-Dependent Neuronal Survival and Synaptogenesis

Because PGNs are the first relay station of olfactory sensory input, it is hardly surprising that, akin to GC_{OB}, adult-born PGNs also display activity-dependent survival. Investigators have reported that sensory deprivation decreases (Mandairon et al. 2006), whereas olfactory enrichment (Rochefort et al. 2002) and olfactory discrimination tasks (Alonso et al. 2006) increase, adult-born PGN survival. Also, sensory enrichment accelerates the development of their glutamatergic input synapses as visualized by genetic synaptic markers (Livneh et al. 2009).

PGNs may be an attractive model with which to study the formation of synaptic connections and how they are affected by activity in real time, since their superficial location in the olfactory bulb makes them the only adult-born neurons amenable to in vivo two-photon imaging with sufficient spatial resolution to visualize these changes (Livneh et al. 2009).

SYNAPTOGENESIS IN ADULT-BORN DENTATE GYRUS GRANULE NEURONS

Stages of Synaptic Development

The third type of adult-born neurons in mammals is the DG granule cell (GC_{DG}), which arises from progenitors in the subgranular zone just beneath the granular layer where mature neurons eventually reside. About 9000 new granule cells are produced daily in the DG of young rats (Cameron & McKay 2001). These neurons receive their main excitatory input from the entorhinal cortex and provide glutamatergic input primarily to the excitatory pyramidal neurons and inhibitory interneurons in the CA3 region of the hippocampus (**Figure 2a**). In this manner, the DG acts as the main entry point for entorhinal cortex input into the hippocampus, relaying the information to CA3 for further processing before it is returned to the entorhinal cortex via CA1.

The distinct stages of neuronal maturation of adult-born GC_{DG} largely recapitulate those that occur during perinatal development, but at a slower pace (Overstreet-Wadiche et al. 2006a, Zhao et al. 2006). This observation could be due to the upregulation of DISC1 protein in the adult DG, which slows the increase in spine density of GC_{DG} during their development (Duan et al. 2007). The new GC_{DG} first receive GABAergic input to their dendrites approximately one week after they are generated. This innervation is initially depolarizing until two to four weeks, when it becomes hyperpolarizing (Esposito et al. 2005), owing to the transient expression of the inward chloride transporter

NKCC1 in immature neurons, which results in an elevated intracellular chloride concentration as compared with mature neurons (Ge et al. 2006). Expression of this transporter is necessary for normal development because its ablation leads to severely delayed neuronal maturation (Ge et al. 2006). In the second week after their birth, dendrites of GC_{DG} start to form spines and to receive glutamatergic input, and their GABAergic input becomes predominantly perisomatic (**Figure 2d**) (Esposito et al. 2005, Ge et al. 2006). Concurrently, axonal projections from new neurons reach the CA3 region and begin to form contacts that continue to mature for months (Toni et al. 2008). By two months of age, adult-born neurons have similar morphological and electrophysiological properties to those formed during perinatal development (Ge et al. 2007; Laplagne et al. 2006, 2007).

Activity-Dependent Neuronal Survival

Akin to GCs in the OB, 50% of new GC_{DG} born in the adult die by four weeks of age (Kempermann et al. 1997a), and their survival is most sensitive to environmental influences between the first and third weeks of development (Tashiro et al. 2006).

Levels of neurogenesis and subsequent survival of GC_{DG} are strongly influenced by neuronal activity. Increased levels of adult neurogenesis in the DG accompany changes in experiences through exercise or enriched environments (Kempermann et al. 1997b, van Praag et al. 1999). New neurons that are activated during learning are preferentially selected for incorporation into active DG circuits (Kee et al. 2007). Conversely, new GC_{DG} whose N-methyl-D-aspartate (NMDA) receptor-mediated input is eliminated experience a drastic reduction in survival rates (Tashiro et al. 2006). These observations illustrate that, as described in the OB, neuronal activity plays a role in selecting new neurons that eventually survive and integrate into the DG circuits.

Activity-Dependent Synaptogenesis and Pathology: Excitability-Induced Rewiring of Adult-Born Neurons in the Dentate Gyrus

The functional maturation of adult-born GC_{DG} is highly sensitive to changes in activity, and the strongest perturbations of synapse formation in new GC_{DG} are caused by seizures (Parent et al. 1997). Experimentally induced seizures accelerate synaptic development of new neurons such that new GC_{DG} added to an epileptic brain start receiving GABAergic input to their dendrites prior to two weeks after their birth, significantly earlier than in the unperturbed DG (Overstreet-Wadiche et al. 2006b).

In experimental seizure models, the DG exhibits network changes that resemble those observed in human pathology of temporal lobe epilepsy. This reorganization of connectivity may be attributed to anomalous integration of new neurons, in addition to other changes in preexisting neurons. Differentiating neurons are most susceptible to develop aberrant connectivity, and some morphological changes are seen only in new neurons generated within days of the onset of seizure, but not in neurons born a week before (Jessberger et al. 2007). Seizure-induced synaptic alterations to GC_{DG} in animal models include increased numbers of mushroom spines (spines with characteristically large heads) and spiny, branched basal dendrites that extend into the polymorphic cell layer (Jessberger et al. 2007). Seizures also perturb the migration of new GC_{DG}. The cell bodies of new GC_{DG} born during seizures aberrantly localize within the hilus and these neurons fire in synchrony with CA3 pyramidal neurons, which suggest that they contribute to increased excitability within the hippocampus (Scharfman et al. 2000). Seizures can also result in mossy fiber sprouting, axonal projection by GC_{DG} to the supragranular molecular layer. The consequences of sprouting are controversial; electrophysiological studies have proposed that this aberrant connectivity produces either recurrent excitatory circuits and subsequent hippocampal hyperexcitability (Okazaki et al. 1995), or conversely, recurrent inhibition by preferentially targeting inhibitory neurons in the molecular layer (Sloviter 1992).

Preliminary analyses of the addition of individual new neurons with genetically enhanced excitability into the adult dentate gyrus in vivo suggest that these neurons experience increased perisomatic inhibition as well as a reduction in the frequency of excitatory inputs and density of glutamatergic input synapses (S. Sim, C.W. Lin, and C. Lois, unpublished observations). Notably, these neurons display many of the seizure-induced alterations such as larger dendritic spines and basal dendrites, but they lack mossy fiber sprouting. Investigators have also observed synaptic rearrangements in these neurons' axon terminals. Normal GC_{DG} have axonal specializations known as large mossy terminals, where they form complex synapses with pyramidal neurons in the CA3 region of the hippocampus. Neurons with genetically enhanced excitability lose many of their large mossy terminals, suggesting a reduction in inputs to excitatory CA3 neurons. These results support previous findings in hippocampal cultures showing that these axonal connections are fairly dynamic, since synaptic rearrangements as well as changes in the size of large mossy terminals have been documented in response to changes in spiking activity (Galimberti et al. 2006).

Synaptic Plasticity During a Critical Period

Similar to the situation described for GC_{OB}, there is a critical period within which new neurons in the adult DG display increased synaptic plasticity compared with mature neurons as demonstrated by an enhanced propensity for long-term potentiation (Schmidt-Hieber et al. 2004). This enhanced synaptic plasticity lasts until the second month after new neurons are generated and then decreases to levels comparable to those of the surrounding mature neurons (Ge et al. 2007). Long-term potentiation during this critical period possesses several defining characteristics: It is dependent on the presence of the NR2B subunit of the NMDA

receptor and can be induced in the presence of intact inhibition (Ge et al. 2007). Similarly, a low-threshold calcium spike can boost fast sodium action potentials and contribute to long-term potentiation during this critical period (Schmidt-Hieber et al. 2004). Investigating how the flow of information through the hippocampus can shape the synaptic organization of new neurons will help investigators elucidate the role of adult neurogenesis in learning and memory.

PERSPECTIVES ON THE FUTURE

We conclude this review by raising several open questions that, we hope, may inspire future research.

First, adult neurogenesis is a widespread phenomenon in most vertebrates. It is interesting to note that mammals appear to be an exception among vertebrates: Their brains are composed mostly of long-lived, nonrenewable neurons born during the embryonic development. Why has a phenomenon that is common in so many classes of animals become less prevalent in mammals? Why is the human cerebellum or neocortex capable of processing, acquiring, and storing information for decades using a single set of neurons, whereas the dentate gyrus and olfactory bulb require the continuous addition of neurons into their circuits throughout life to perform their functions?

Second, in contrast with circuit assembly during embryonic development, which involves integration of new neurons in a mostly constant environment in utero, adult-born neurons integrate into mature, functioning circuits. This observation poses additional challenges because adult brain activity is constantly modulated by the ever-changing conditions of the outside world. Furthermore, because adult-generated neurons integrate into the brains of behaving animals, these neurons must form new synapses with minimal disruption to existing connectivity so that behavior is unperturbed. Do specialized mechanisms exist for synaptic integration in adult-born neurons, which differ from those used during embryonic brain development?

Third, synaptic plasticity of adult-born neurons is likely restricted to a specific time window early during their maturation. This phenomenon supports the idea that new neurons provide the mature circuit with a transient form of heightened plasticity, acting as a substrate for circuit refinement and adaptation. After this critical window, the neurons mature, become stably integrated into the brain, and partially lose their activity-dependent plasticity. The addition of cells endowed with such an initial short-lived flexibility and long-term stability enables information processing in the brain to be both versatile and reliable while faced with changing behavioral demands. The transient plasticity of new cells generated during adult neurogenesis may explain the requirement for additional new neurons to facilitate lifelong plasticity and reshaping of memory circuits. Which molecular mechanisms are responsible for this transient plasticity? Most mammalian brain regions, such as the thalamus, striatum, and neocortex, do not receive any new neurons after birth. Do neurons in these brain areas maintain their plasticity for longer periods of time compared with adult-born neurons added into the OB and DG?

Investigating these unanswered questions will shed some light on the mystery of why mammalian brain circuits are composed of two classes of neurons: those that live as long as the individual harboring them and those that are continuously added throughout life.

DISCLOSURE STATEMENT

The authors are not aware of any affiliations, memberships, funding, or financial holdings that might be perceived as affecting the objectivity of this review.

LITERATURE CITED

Alonso M, Viollet C, Gabellec MM, Meas-Yedid V, Olivo-Marin JC, Lledo PM. 2006. Olfactory discrimination learning increases the survival of adult-born neurons in the olfactory bulb. *J. Neurosci.* 26:10508–13

Altman J, Das GD. 1965. Autoradiographic and histological evidence of postnatal hippocampal neurogenesis in rats. *J. Comp. Neurol.* 124:319–35

Balu R, Pressler RT, Strowbridge BW. 2007. Multiple modes of synaptic excitation of olfactory bulb granule cells. *J. Neurosci.* 27:5621–32

Barretto RP, Messerschmidt B, Schnitzer MJ. 2009. In vivo fluorescence imaging with high-resolution microlenses. *Nat. Methods* 6:511–12

Batista-Brito R, Close J, Machold R, Fishell G. 2008. The distinct temporal origins of olfactory bulb interneuron subtypes. *J. Neurosci.* 28:3966–75

Bayer SA. 1980. Quantitative 3H-thymidine radiographic analyses of neurogenesis in the rat amygdala. *J. Comp. Neurol.* 194:845–75

Bayer SA, Yackel JW, Puri PS. 1982. Neurons in the rat dentate gyrus granular layer substantially increase during juvenile and adult life. *Science* 216:890–92

Berdichevsky Y, Sabolek H, Levine JB, Staley KJ, Yarmush ML. 2009. Microfluidics and multielectrode array-compatible organotypic slice culture method. *J. Neurosci. Methods* 178:59–64

Briggman KL, Denk W. 2006. Towards neural circuit reconstruction with volume electron microscopy techniques. *Curr. Opin. Neurobiol.* 16:562–70

Brown JP, Couillard-Despres S, Cooper-Kuhn CM, Winkler J, Aigner L, Kuhn HG. 2003. Transient expression of doublecortin during adult neurogenesis. *J. Comp. Neurol.* 467:1–10

Cameron HA, McKay RD. 2001. Adult neurogenesis produces a large pool of new granule cells in the dentate gyrus. *J. Comp. Neurol.* 435:406–17

Carleton A, Petreanu LT, Lansford R, Alvarez-Buylla A, Lledo PM. 2003. Becoming a new neuron in the adult olfactory bulb. *Nat. Neurosci.* 6:507–18

Chen WR, Xiong W, Shepherd GM. 2000. Analysis of relations between NMDA receptors and GABA release at olfactory bulb reciprocal synapses. *Neuron* 25:625–33

Davis BJ, Macrides F. 1981. The organization of centrifugal projections from the anterior olfactory nucleus, ventral hippocampal rudiment, and piriform cortex to the main olfactory bulb in the hamster: an autoradiographic study. *J. Comp. Neurol.* 203:475–93

De Marchis S, Bovetti S, Carletti B, Hsieh YC, Garzotto D, et al. 2007. Generation of distinct types of periglomerular olfactory bulb interneurons during development and in adult mice: implication for intrinsic properties of the subventricular zone progenitor population. *J. Neurosci.* 27:657–64

Duan X, Chang JH, Ge S, Faulkner RL, Kim JY, et al. 2007. Disrupted-In-Schizophrenia 1 regulates integration of newly generated neurons in the adult brain. *Cell* 130:1146–58

Ebihara T, Kawabata I, Usui S, Sobue K, Okabe S. 2003. Synchronized formation and remodeling of postsynaptic densities: long-term visualization of hippocampal neurons expressing postsynaptic density proteins tagged with green fluorescent protein. *J. Neurosci.* 23:2170–81

El-Husseini AE, Schnell E, Chetkovich DM, Nicoll RA, Bredt DS. 2000. PSD-95 involvement in maturation of excitatory synapses. *Science* 290:1364–68

Esposito MS, Piatti VC, Laplagne DA, Morgenstern NA, Ferrari CC, et al. 2005. Neuronal differentiation in the adult hippocampus recapitulates embryonic development. *J. Neurosci.* 25:10074–86

Gage FH, Coates PW, Palmer TD, Kuhn HG, Fisher LJ, et al. 1995. Survival and differentiation of adult neuronal progenitor cells transplanted to the adult brain. *Proc. Natl. Acad. Sci. USA* 92:11879–83

Galimberti I, Gogolla N, Alberi S, Santos AF, Muller D, Caroni P. 2006. Long-term rearrangements of hippocampal mossy fiber terminal connectivity in the adult regulated by experience. *Neuron* 50:749–63

Ge S, Goh EL, Sailor KA, Kitabatake Y, Ming GL, Song H. 2006. GABA regulates synaptic integration of newly generated neurons in the adult brain. *Nature* 439:589–93

Ge S, Yang CH, Hsu KS, Ming GL, Song H. 2007. A critical period for enhanced synaptic plasticity in newly generated neurons of the adult brain. *Neuron* 54:559–66

Gray NW, Weimer RM, Bureau I, Svoboda K. 2006. Rapid redistribution of synaptic PSD-95 in the neocortex in vivo. *PLoS Biol.* 4:e370

Grubb MS, Nissant A, Murray K, Lledo PM. 2008. Functional maturation of the first synapse in olfaction: development and adult neurogenesis. *J. Neurosci.* 28:2919–32

Halabisky B, Strowbridge BW. 2003. Gamma-frequency excitatory input to granule cells facilitates dendrodendritic inhibition in the rat olfactory Bulb. *J. Neurophysiol.* 90:644–54

Helmchen F, Denk W. 2005. Deep tissue two-photon microscopy. *Nat. Methods* 2:932–40

Imayoshi I, Sakamoto M, Ohtsuka T, Takao K, Miyakawa T, et al. 2008. Roles of continuous neurogenesis in the structural and functional integrity of the adult forebrain. *Nat. Neurosci.* 11:1153–61

Jessberger S, Zhao C, Toni N, Clemenson GD Jr, Li Y, Gage FH. 2007. Seizure-associated, aberrant neurogenesis in adult rats characterized with retrovirus-mediated cell labeling. *J. Neurosci.* 27:9400–7

Kaneko N, Okano H, Sawamoto K. 2006. Role of the cholinergic system in regulating survival of newborn neurons in the adult mouse dentate gyrus and olfactory bulb. *Genes Cells* 11:1145–59

Kee N, Teixeira CM, Wang AH, Frankland PW. 2007. Preferential incorporation of adult-generated granule cells into spatial memory networks in the dentate gyrus. *Nat. Neurosci.* 10:355–62

Kelsch W, Lin CW, Mosley CP, Lois C. 2009. A critical period for activity-dependent synaptic development during olfactory bulb adult neurogenesis. *J. Neurosci.* 29:11852–58

Kelsch W, Lin CW, Lois C. 2008. Sequential development of synapses in dendritic domains during adult neurogenesis. *Proc. Natl. Acad. Sci. USA* 105:16803–8

Kelsch W, Mosley CP, Lin CW, Lois C. 2007. Distinct mammalian precursors are committed to generate neurons with defined dendritic projection patterns. *PLoS Biol.* 5:e300

Kempermann G, Kuhn HG, Gage FH. 1997a. Genetic influence on neurogenesis in the dentate gyrus of adult mice. *Proc. Natl. Acad. Sci. USA* 94:10409–14

Kempermann G, Kuhn HG, Gage FH. 1997b. More hippocampal neurons in adult mice living in an enriched environment. *Nature* 386:493–95

Kirov SA, Petrak LJ, Fiala JC, Harris KM. 2004. Dendritic spines disappear with chilling but proliferate excessively upon rewarming of mature hippocampus. *Neuroscience* 127:69–80

Laplagne DA, Esposito MS, Piatti VC, Morgenstern NA, Zhao C, et al. 2006. Functional convergence of neurons generated in the developing and adult hippocampus. *PLoS Biol.* 4:e409

Laplagne DA, Kamienkowski JE, Esposito MS, Piatti VC, Zhao C, et al. 2007. Similar GABAergic inputs in dentate granule cells born during embryonic and adult neurogenesis. *Eur. J. Neurosci.* 25:2973–81

Lemasson M, Saghatelyan A, Olivo-Marin JC, Lledo PM. 2005. Neonatal and adult neurogenesis provide two distinct populations of newborn neurons to the mouse olfactory bulb. *J. Neurosci.* 25:6816–25

Lin CW, Sim S, Ainsworth A, Okada M, Kelsch W, Lois C. 2010. Genetically increased cell-intrinsic excitability enhances neuronal integration into adult brain circuits. *Neuron* 65:32–39

Livneh Y, Feinstein N, Klein M, Mizrahi A. 2009. Sensory input enhances synaptogenesis of adult-born neurons. *J. Neurosci.* 29:86–97

Lois C, Alvarez-Buylla A. 1993. Proliferating subventricular zone cells in the adult mammalian forebrain can differentiate into neurons and glia. *Proc. Natl. Acad. Sci. USA* 90:2074–77

Lois C, Alvarez-Buylla A. 1994. Long-distance neuronal migration in the adult mammalian brain. *Science* 264:1145–48

Luskin MB. 1993. Restricted proliferation and migration of postnatally generated neurons derived from the forebrain subventricular zone. *Neuron* 1:173–89

Luskin MB, Price JL. 1983. The topographic organization of associational fibers of the olfactory system in the rat, including centrifugal fibers to the olfactory bulb. *J. Comp. Neurol.* 216:264–91

Magavi SS, Mitchell BD, Szentirmai O, Carter BS, Macklis JD. 2005. Adult-born and preexisting olfactory granule neurons undergo distinct experience-dependent modifications of their olfactory responses in vivo. *J. Neurosci.* 25:10729–39

Mandairon N, Sacquet J, Jourdan F, Didier A. 2006. Long-term fate and distribution of newborn cells in the adult mouse olfactory bulb: influences of olfactory deprivation. *Neuroscience* 141:443–51

Merkle FT, Mirzadeh Z, Alvarez-Buylla A. 2007. Mosaic organization of neural stem cells in the adult brain. *Science* 317:381–84

Meyer MP, Smith SJ. 2006. Evidence from in vivo imaging that synaptogenesis guides the growth and branching of axonal arbors by two distinct mechanisms. *J. Neurosci.* 26:3604–14

Micheva KD, Smith SJ. 2007. Array tomography: a new tool for imaging the molecular architecture and ultrastructure of neural circuits. *Neuron* 55:25–36

Mignone JL, Kukekov V, Chiang AS, Steindler D, Enikolopov G. 2004. Neural stem and progenitor cells in nestin-GFP transgenic mice. *J. Comp. Neurol.* 469:311–24

Mizrahi A. 2007. Dendritic development and plasticity of adult-born neurons in the mouse olfactory bulb. *Nat. Neurosci.* 10:444–52

Mizrahi A, Crowley JC, Shtoyerman E, Katz LC. 2004. High-resolution in vivo imaging of hippocampal dendrites and spines. *J. Neurosci.* 24:3147–51

Mori K. 1987. Membrane and synaptic properties of identified neurons in the olfactory bulb. *Prog. Neurobiol.* 29:275–320

Mori K, Kishi K, Ojima H. 1983. Distribution of dendrites of mitral, displaced mitral, tufted, and granule cells in the rabbit olfactory bulb. *J. Comp. Neurol.* 219:339–55

Murphy GJ, Glickfeld LL, Balsen Z, Isaacson JS. 2004. Sensory neuron signaling to the brain: properties of transmitter release from olfactory nerve terminals. *J. Neurosci.* 24:3023–30

Niell CM, Meyer MP, Smith SJ. 2004. In vivo imaging of synapse formation on a growing dendritic arbor. *Nat. Neurosci.* 7:254–60

Nissant A, Bardy C, Katagiri H, Murray K, Lledo PM. 2009. Adult neurogenesis promotes synaptic plasticity in the olfactory bulb. *Nat. Neurosci.* 12:728–30

Okazaki MM, Evenson DA, Nadler JV. 1995. Hippocampal mossy fiber sprouting and synapse formation after status epilepticus in rats: visualization after retrograde transport of biocytin. *J. Comp. Neurol.* 352:515–34

Overstreet LS, Hentges ST, Bumaschny VF, de Souza FS, Smart JL, et al. 2004. A transgenic marker for newly born granule cells in dentate gyrus. *J. Neurosci.* 24:3251–59

Overstreet-Wadiche LS, Bensen AL, Westbrook GL. 2006a. Delayed development of adult-generated granule cells in dentate gyrus. *J. Neurosci.* 26:2326–34

Overstreet-Wadiche LS, Bromberg DA, Bensen AL, Westbrook GL. 2006b. Seizures accelerate functional integration of adult-generated granule cells. *J. Neurosci.* 26:4095–103

Parent JM, Yu TW, Leibowitz RT, Geschwind DH, Sloviter RS, Lowenstein DH. 1997. Dentate granule cell neurogenesis is increased by seizures and contributes to aberrant network reorganization in the adult rat hippocampus. *J. Neurosci.* 17:3727–38

Petreanu L, Alvarez-Buylla A. 2002. Maturation and death of adult-born olfactory bulb granule neurons: role of olfaction. *J. Neurosci.* 22:6106–13

Pinching AJ, Powell TP. 1971. The neuropil of the periglomerular region of the olfactory bulb. *J. Cell Sci.* 9:379–409

Pressler RT, Inoue T, Strowbridge BW. 2007. Muscarinic receptor activation modulates granule cell excitability and potentiates inhibition onto mitral cells in the rat olfactory bulb. *J. Neurosci.* 27:10969–81

Price JL, Powell TP. 1970. The synaptology of the granule cells of the olfactory bulb. *J. Cell Sci.* 7:125–55

Rakic P, Ayoub AE, Breunig JJ, Dominguez MH. 2009. Decision by division: making cortical maps. *Trends Neurosci.* 32:291–301

Rochefort C, Gheusi G, Vincent JD, Lledo PM. 2002. Enriched odor exposure increases the number of newborn neurons in the adult olfactory bulb and improves odor memory. *J. Neurosci.* 22:2679–89

Roe T, Reynolds TC, Yu G, Brown PO. 1993. Integration of murine leukemia virus DNA depends on mitosis. *EMBO J.* 12:2099–108

Saghatelyan A, Roux P, Migliore M, Rochefort C, Desmaisons D, et al. 2005. Activity-dependent adjustments of the inhibitory network in the olfactory bulb following early postnatal deprivation. *Neuron* 46:103–16

Sassoe-Pognetto M, Utvik JK, Camoletto P, Watanabe M, Stephenson FA, et al. 2003. Organization of postsynaptic density proteins and glutamate receptors in axodendritic and dendrodendritic synapses of the rat olfactory bulb. *J. Comp. Neurol.* 463:237–48

Scharfman HE, Goodman JH, Sollas AL. 2000. Granule-like neurons at the hilar/CA3 border after status epilepticus and their synchrony with area CA3 pyramidal cells: functional implications of seizure-induced neurogenesis. *J. Neurosci.* 20:6144–58

Schmidt-Hieber C, Jonas P, Bischofberger J. 2004. Enhanced synaptic plasticity in newly generated granule cells of the adult hippocampus. *Nature* 429:184–87

Sheng M. 2001. Molecular organization of the postsynaptic specialization. *Proc. Natl. Acad. Sci. USA* 98:7058–61

Sloviter RS. 1992. Possible functional consequences of synaptic reorganization in the dentate gyrus of kainate-treated rats. *Neurosci. Lett.* 137:91–96

Sudhof TC, Jahn R. 1991. Proteins of synaptic vesicles involved in exocytosis and membrane recycling. *Neuron* 6:665–77

Tashiro A, Sandler VM, Toni N, Zhao C, Gage FH. 2006. NMDA-receptor-mediated, cell-specific integration of new neurons in adult dentate gyrus. *Nature* 442:929–33

Toni N, Laplagne DA, Zhao C, Lombardi G, Ribak CE, et al. 2008. Neurons born in the adult dentate gyrus form functional synapses with target cells. *Nat. Neurosci.* 11:901–7

Toni N, Teng EM, Bushong EA, Aimone JB, Zhao C, et al. 2007. Synapse formation on neurons born in the adult hippocampus. *Nat. Neurosci.* 10:727–34

van Praag H, Kempermann G, Gage FH. 1999. Running increases cell proliferation and neurogenesis in the adult mouse dentate gyrus. *Nat. Neurosci.* 2:266–70

van Praag H, Schinder AF, Christie BR, Toni N, Palmer TD, Gage FH. 2002. Functional neurogenesis in the adult hippocampus. *Nature* 415:1030–34

Wang X, Qiu R, Tsark W, Lu Q. 2007. Rapid promoter analysis in developing mouse brain and genetic labeling of young neurons by doublecortin-DsRed-express. *J. Neurosci. Res.* 85:3567–73

Washbourne P, Bennett JE, McAllister AK. 2002. Rapid recruitment of NMDA receptor transport packets to nascent synapses. *Nat. Neurosci.* 5:751–59

Whitman MC, Greer CA. 2007a. Adult-generated neurons exhibit diverse developmental fates. *Dev. Neurobiol.* 67:1079–93

Whitman MC, Greer CA. 2007b. Synaptic integration of adult-generated olfactory bulb granule cells: basal axodendritic centrifugal input precedes apical dendrodendritic local circuits. *J. Neurosci.* 27:9951–61

Winner B, Cooper-Kuhn CM, Aigner R, Winkler J, Kuhn HG. 2002. Long-term survival and cell death of newly generated neurons in the adult rat olfactory bulb. *Eur. J. Neurosci.* 16:1681–89

Woolf TB, Shepherd GM, Greer CA. 1991. Serial reconstructions of granule cell spines in the mammalian olfactory bulb. *Synapse* 7:181–92

Yamaguchi M, Mori K. 2005. Critical period for sensory experience-dependent survival of newly generated granule cells in the adult mouse olfactory bulb. *Proc. Natl. Acad. Sci. USA* 102:9697–702

Yamaguchi M, Saito H, Suzuki M, Mori K. 2000. Visualization of neurogenesis in the central nervous system using nestin promoter-GFP transgenic mice. *Neuroreport* 11:1991–96

Zhao C, Teng EM, Summers RG Jr, Ming GL, Gage FH. 2006. Distinct morphological stages of dentate granule neuron maturation in the adult mouse hippocampus. *J. Neurosci.* 26:3–11

Zou DJ, Feinstein P, Rivers AL, Mathews GA, Kim A, et al. 2004. Postnatal refinement of peripheral olfactory projections. *Science* 304:1976–79

Neurological Channelopathies

Dimitri M. Kullmann

Institute of Neurology, University College London, Queen Square, London WC1N 3BG, United Kingdom; email: d.kullmann@ion.ucl.ac.uk

Annu. Rev. Neurosci. 2010. 33:151–72

First published online as a Review in Advance on March 23, 2010

The *Annual Review of Neuroscience* is online at neuro.annualreviews.org

This article's doi: 10.1146/annurev-neuro-060909-153122

Key Words

epilepsy, ataxia, migraine, myotonia, periodic paralysis, hyperekplexia

Abstract

Inherited ion channel mutations can affect the entire nervous system. Many cause paroxysmal disturbances of brain, spinal cord, peripheral nerve or skeletal muscle function, with normal neurological development and function in between attacks. To fully understand how mutations of ion channel genes cause disease, we need to know the normal location and function of the channel subunit, consequences of the mutation for biogenesis and biophysical properties, and possible compensatory changes in other channels that contribute to cell or circuit excitability. Animal models of monogenic channelopathies increasingly help our understanding. An important challenge for the future is to determine how more subtle derangements of ion channel function, which arise from the interaction of genetic and environmental influences, contribute to common paroxysomal disorders, including idiopathic epilepsy and migraine, that share features with rare monogenic channelopathies.

Contents

INTRODUCTION

Although rare, monogenic neurological channelopathies provide unparalleled insight into the mechanisms of neurological diseases, particularly idiopathic epilepsy and migraine, with which they share clinical features. Common forms of these diseases show high, albeit usually non-Mendelian, heritability, but the genetic mechanisms underlying them are unknown; and in most cases, the nature of the disorder of brain excitability remains unclear. Thus, neurological channelopathies are experiments of nature that potentially identify signaling pathways and circuits involved in diseases that account for a large burden to society. More broadly, they open a window onto the fundamental workings of the human brain, spinal cord, nerve, and muscle.

This review considers the challenges to understanding the mechanisms of neurological channelopathies, summarizes the main breakthroughs, and proposes directions for future research.

CLINICAL MANIFESTATIONS

A common feature of many channelopathies is that they cause discrete attacks, while the development of the nervous system is generally unperturbed, and affected individuals function normally between attacks. Similar to most genetic diseases, ion channel mutations tend to present early in life. Some manifest in a restricted developmental interval only, for instance, as self-limiting neonatal seizures. Although other disorders may not resolve, discrete attacks often give way to a fixed or progressive impairment, as seen in some muscle or cerebellar channelopathies. Among possible explanations for the developmental evolution of symptoms are changes in the expression of individual channels, maturation and compensatory alterations in circuit function, and excitotoxic damage to affected organs.

Most currently known channelopathies are dominantly inherited. This may not be a reflection of the true balance of different inheritance patterns because in many cases, the presence of several affected individuals in a family has led to the positional cloning or sequencing of candidate genes. Indeed, because some of the manifestations of individual channelopathies, such as migraine or seizures, are common and occur far more frequently as idiopathic disorders without clear Mendelian inheritance (Helbig et al. 2008, Wessman et al. 2007), it is difficult to justify the extensive sequencing of candidate genes in a patient without affected first-degree relatives.

Another obstacle to establishing the true prevalence of channelopathies is that many exhibit phenotypic heterogeneity. Some channelopathies are only partially penetrant or have highly diverse manifestations (even when the same mutation is inherited by different

Epilepsy and migraine

	Na+	K+	Ca2+	GABAA	Nicotinic
Epilepsy	SCN1A	KCNQ2	CACNA1H	GABRA1	CHRNA2
	SCN1B	KCNQ3		GABRB3	CHNRA4
	SCN2A	KCNMA1		GABRG2	CHRNB2
Migraine	SCN1A		CACNA1A		

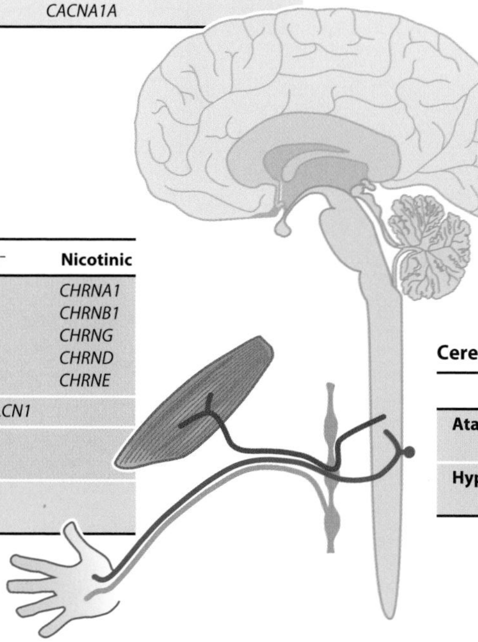

Neuromuscular disorders

	Na+	K+	Ca2+	Cl−	Nicotinic
Myasthenia					CHRNA1
Fetal akinesia					CHRNB1
					CHRNG
					CHRND
					CHRNE
Myotonia	SCN4A			CLCN1	
Periodic paralysis	SCN4A	KCNJ2	CACNA1S		
Pain Erythema	SCN9A				

Cerebellar ataxia and excessive startle

	K+	Ca2+	Glycine
Ataxia	KCNA1	CACNA1A	
	KCNC3		
Hyperexplexia			GLRA1
			GLRB

Figure 1

Channelopathies. Summary of the main ion channel classes, genes, and manifestations.

members of a kindred), consistent with an interaction with other genes or environmental factors.

Channelopathies can present in any area of clinical neurology (**Figure 1**).

(a) Epilepsy. Ion channel mutations can cause focal and generalized seizure disorders, which range in severity from benign neonatal convulsions to epileptic encephalopathies that lead to developmental regression and premature death. An important breakthrough came with the realization that an autosomal dominant pattern of inheritance occurs in some families in which multiple individuals have febrile and other types of seizures. Febrile seizures are generally considered a benign self-limiting condition that affects approximately 5%

of young children. However, in families with generalized epilepsy with febrile seizures plus (GEFS+), such seizures may fail to resolve by the age of six years, or individuals may experience generalized seizures without a febrile illness (Scheffer & Berkovic 1997). Some individuals in GEFS+ families may experience myoclonic or astatic seizures, or even develop temporal lobe epilepsy. Mutations of at least four ion channel genes cause GEFS+ (**Figure 2**).

(b) Migraine. Ion channel mutations account for some families with severe dominantly inherited migraine, in which attacks can be accompanied by reversible unilateral weakness [hence the term familial hemiplegic migraine (FHM)].

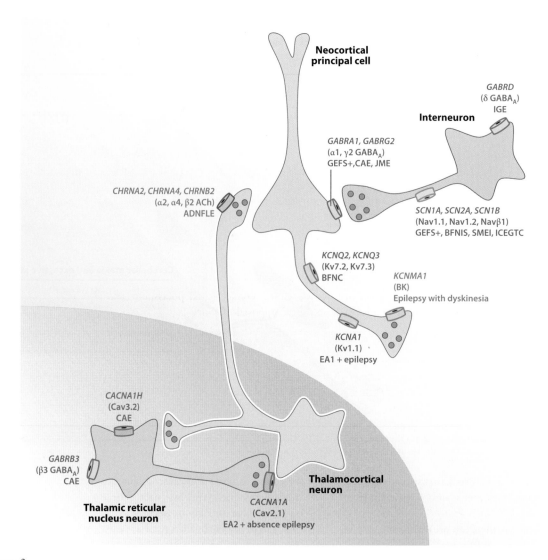

Figure 2

Epileptic channelopathies. The tentative roles of different ion channels implicated in epilepsy are indicated schematically, with gain and loss of function in red and blue, respectively. The locations of several channels are inferred mainly from rodent studies. Abbreviations: ADNFLE, autosomal dominant nocturnal frontal lobe epilepsy; BFNIS, benign familial neonatal-infantile seizures; BFNC, benign familial neonatal convulsions; CAE, childhood absence epilepsy; EA1, episodic ataxia type 1; EA2, episodic ataxia type 2; GEFS+, generalized epilepsy with febrile seizures plus; ICEGTC, intractable childhood epilepsy with generalized tonic-clonic seizures; IGE, idiopathic generalized epilepsy; JME, juvenile myoclonic epilepsy; SMEI, severe myoclonic epilepsy of infancy.

(c) Movement disorders. Channelopathies can cause either episodic or progressive cerebellar dysfunction. Other ion channel disorders manifest as attacks of dystonic posturing, jerking, or hyperekplexia, an exaggerated startle reaction.

(d) Disorders of peripheral nerve and autonomic function. Distinct mutations of one sodium-channel gene cause several disorders of nociception and of sympathetic tone. Some potassium channel mutations lead to spontaneous activity in motor axons. Although not considered further in this review, Charcot-Marie-Tooth type III (an X-linked hereditary neuropathy) is caused by mutations of Connexin32 (Bergoffen et al. 1993).

(e) Muscle disease. Channelopathies can result in either loss of muscle fiber excitability, which causes periodic paralysis, or excessive excitability, which manifests as myotonia (**Figure 3**). Although myotonic dystrophy, the most common myotonic disorder, is not usually considered a channelopathy, it may have similar mechanisms because mutations of the *DMPK* gene in this disorder alter the processing of mRNA that encodes the muscle chloride channel (Charlet-B et al. 2002, Mankodi et al. 2002). Another class of diseases, also beyond the scope of this review, is exemplified by malignant hyperthermia and central core disease, which can be caused by mutations of the ryanodine receptor (Priori 2005), an intracellular channel that gates calcium release from the sarcoplasmic reticulum.

(f) Psychiatric and cognitive disorders. Despite their strong heritability and frequently episodic course, psychiatric diseases have not been definitively attributed to ion channel mutations. Nevertheless, polymorphisms in ion channel genes have been associated with neurodevelopmental, cognitive, and psychiatric disorders (Huffaker et al. 2009, Pickard et al. 2008), and alcoholism (Edenberg et al. 2004).

UNDERSTANDING CHANNELOPATHIES

Explaining the clinical features of a channelopathy requires knowledge of where and when the mutated ion channel subunit is normally expressed and of the normal composition of the complete multimeric channel. Many missense mutations alter voltage- or ligand-dependent gating. However, other nucleotide changes generally result in stop codons, or frameshifts or altered splice sites that lead to aberrant

sequences before a premature termination. Some of these probably do not result in peptide synthesis because the mRNA is directed to the nonsense-mediated decay pathway. Larger-scale deletions and duplications of entire exons or genes have recently been detected in channelopathies (Suls et al. 2006, Labrum et al. 2009). Further complicating attempts to understand the effects of mutations, many ion channel genes undergo alternate splicing, thus a sequence change can have different consequences for different transcripts.

CNS Potassium Channelopathies

Potassium channels are the largest group of ion channels, the majority of which are expressed in the CNS. Voltage-gated potassium channels are composed of four homologous pore-forming subunits, which are encoded by more than 70 genes. With a few exceptions, each subunit contains six trans-membrane α-helices, of which the S4 segment acts as a voltage sensor. To date, human mutations have been identified in only a few of these genes. The salient features of these channelopathies are summarized below.

KCNA1. *KCNA1* encodes the potassium channel α subunit Kv1.1, which is widely expressed in axons, in the CNS and in peripheral nerve. Dominantly inherited mutations cause episodic ataxia type 1 (EA1) (Browne et al. 1994), characterized by brief paroxysms of cerebellar incoordination and interictal myokymia (spontaneous motor-unit activity). All the known mutations are missense, except for a premature stop codon (Eunson et al. 2000). Mutations impair channel function through different effects on activation and deactivation kinetics (Adelman et al. 1995), and many have been shown to exert a dominant-negative effect in heterologous expression systems, which implies that they coassemble with wild-type subunits. Interestingly, some mutations impair the rate of fast inactivation, arguably a paradoxical gain-of-function effect (Maylie et al. 2002). Because fast inactivation depends on other

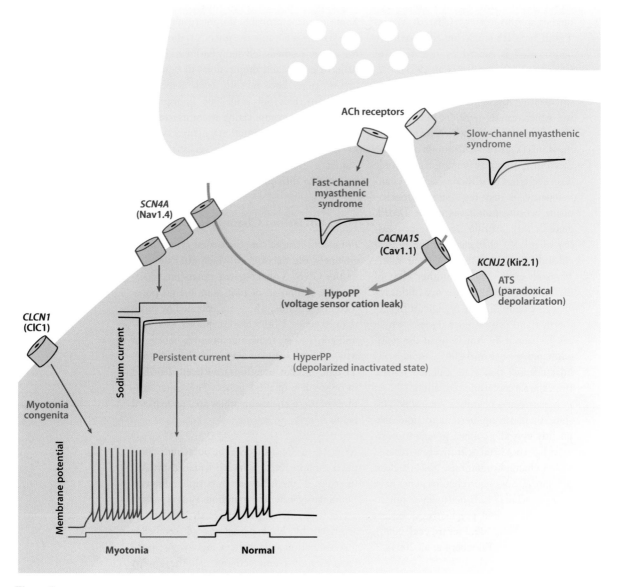

Figure 3

Muscle channelopathies. Gain- and loss-of-function mechanisms that underlie muscle channelopathies are indicated in red and blue, respectively. The slowed end-plate current and persistent sodium current that occur in slow-channel myasthenic syndrome and HyperPP, respectively, can lead to a vacuolar myopathy. The abnormal cation leak via the voltage-sensing pore of Cav1.1 or Nav1.4 that occurs in HypoPP represents a gain of abnormal function (*green*). ATS, Andersen-Tawil syndrome; HyperPP, hyperkalemic periodic paralysis; HypoPP, hypokalemic periodic paralysis.

coassembling subunits that occur only in some neurons, this finding may explain why many brain circuits that express Kv1.1 are unaffected. However, some *KCNA1* mutations can cause epilepsy (Zuberi et al. 1999). Interestingly,

the homozygous *KCNA1* null mutant mouse has seizures (Smart et al. 1998). In contrast, a heterozygous knock-in mouse that expresses an EA1 mutation exhibits stress-induced impaired coordination and evidence of increased GABA release from inhibitory neurons in the cerebellar cortex (Herson et al. 2003). However, the homozygous knock-in mutation is lethal, consistent with a promiscuous dominant negative effect on other Kv1 family subunits.

KCNQ2 and KCNQ3. Two other genes that encode potassium channels were initially identified through positional and homology cloning in families with dominantly inherited benign familial neonatal convulsions (BFNC). This disorder is characterized by brief generalized seizures that typically resolve spontaneously by six weeks of age. *KCNQ2* and *KCNQ3* encode Kv7.2 and Kv7.3, respectively (Singh et al. 1998, Charlier et al. 1998, Biervert et al. 1998), which coassemble to form a slowly activated channel (Wang et al. 1998) that is closed by the activation of M1 muscarinic receptors (Jentsch 2000). Some mutations interfere with channel assembly (Maljevic et al. 2008). The disease mechanism thus appears to be haploinsufficiency. Indeed, because only one allele is mutated, only a relatively modest loss of M current is sufficient to give rise to BFNC (Schroeder et al. 1998). Mice with engineered BFNC-associated heterozygous *Kcnq2* or *Kcnq3* mutations have a lower threshold for electrically induced seizures, whereas homozygous knock-in mice have spontaneous tonic-clonic seizures and a reduction in M-current density (Singh et al. 2008). Because Kv7.2 and Kv7.3 are preferentially expressed in the initial segments of axons (Shah et al. 2002, 2008), the disorder is likely to reflect a lowered threshold for action potential initiation in central neurons. Developmental upregulation of the channel may explain why seizures resolve (Tinel et al. 1998).

The spectrum of *KCNQ2* channelopathies has broadened with the identification of dominant-negative mutations that neutralize voltage-sensing residues in the S4 segment of Kv7.2. This results in a shift in activation threshold to more depolarized potentials (Dedek et al. 2001, Wuttke et al. 2007). These mutations are associated with spontaneous motor-unit activity (that manifests as

myokymia) in addition to neonatal seizures, consistent with the expression of Kv7.2 in motoneurons.

Mutations of two related genes, *KCNQ1* and *KCNQ4*, are associated with inherited forms of cardiac arrhythmia and deafness.

KCNMA1. A dominant missense mutation in the *KCNMA1* gene is associated with paroxysmal dyskinesia and epilepsy in one large kindred (Du et al. 2005). This gene encodes the ubiquitous neuronal BK Ca^{2+}-activated potassium channel, which opens in response to cytoplasmic Ca^{2+} and helps to repolarize neurons after action potentials. Surprisingly, the mutation increases Ca^{2+} sensitivity and allows the channel to open at more negative potentials. It is thus an exception to the general principle that only loss-of-function potassium channel mutations cause seizures.

KCNC3. Missense mutations of *KCNC3*, which encodes the Kv3.3 channel subunit, cause a dominantly inherited progressive ataxia (spinocerebellar ataxia type 13 or SCA13) (Waters et al. 2006). Paroxysmal symptoms are not prominent, which is unusual for a channelopathy. Although the channel is expressed in the cerebellum, the two identified mutations appear to have opposite effects on channel function, which precludes a simple explanation for the disease.

Muscle and Multisystem Potassium Channelopathies

Inwardly rectifying potassium channels lack several transmembrane segments, including S4. However, they preferentially mediate potassium flux at relatively negative membrane potentials.

KCNJ2. Dominant mutations of *KCNJ2*, which encodes the Kir2.1 subunit of inwardly rectifying channels, underlie Andersen-Tawil syndrome (ATS) (Andelfinger et al. 2002, Tristani-Firouzi et al. 2002, Plaster et al. 2001). Affected individuals experience attacks of skeletal muscle weakness (periodic paralysis), and cardiac arrhythmias, and can have relatively subtle cranio-facial, skeletal, and cognitive abnormalities. Missense mutations in ATS invariably have a dominant-negative effect (Davies et al. 2005, Tristani-Firouzi et al. 2002). Normal muscle fibers exhibit membrane potential bistability, with a depolarized inexcitable state favored by blockade of inwardly rectifying potassium channels, especially in low potassium (Foppen et al. 2002). Consistent with such a mechanism, the serum potassium concentration is often low during attacks of periodic paralysis in ATS.

Mutations of *KCNJ18*, which encodes the related inwardly rectifying channel Kir2.6, have increased susceptibility to hypokalemic periodic paralysis in patients suffering from thyrotoxicosis (Ryan et al. 2010). Mutations of two other inwardly rectifying potassium channel genes, *KCNJ10* and *KCNJ11*, cause multisystem disorders that include epilepsy, ataxia, or muscle weakness among their manifestations (Bockenhauer et al. 2009, Gloyn et al. 2004, Proks et al. 2004).

CNS Sodium Channelopathies

Sodium channels underlie the fast depolarization that typifies action potentials in mammalian neurons. They are structurally homologous to voltage-gated potassium channels, although they contain only a single pore-forming α subunit with four repeated domains. The main subtypes in the CNS are Nav1.1, Nav1.2, Nav1.3, and Nav1.6, encoded by *SCN1A*, *SCN2A*, *SCN3A*, and *SCN8A*, respectively. Mutations of the related sodium channel gene *SCN5A* are associated with congenital cardiac arrhythmias.

SCN1A. More than 300 dominant *SCN1A* mutations have been identified in a broad range of seizure disorders. Some missense mutations cause GEFS+ (Baulac et al. 1999, Escayg et al. 2000). Many other missense, nonsense, or frameshift mutations cause severe myoclonic

epilepsy of infancy (SMEI), an aggressive condition that includes multiple intractable seizure types and developmental regression (Claes et al. 2001). SMEI is often caused by de novo mutations. Idiopathic childhood epilepsy with generalized tonic-clonic seizures (ICEGTC) is a disorder of intermediate severity that can also be caused by missense *SCN1A* mutations (Fujiwara et al. 2003). Because of the high frequency of mutations, *SCN1A* sequencing is appropriate in any infant that presents with a cryptogenic epileptic encephalopathy. A few *SCN1A* mutations have also been identified in a variant of familial hemiplegic migraine (FHM3; Dichgans et al. 2005).

SCN1A encodes the pore-forming α subunit of Nav1.1 channels, which also include two β subunits. This fast sodium channel occurs widely in the CNS, and is especially densely expressed in the axon initial segments of parvalbumin-positive interneurons (Ogiwara et al. 2007). Impaired excitability of inhibitory neurons provides a compelling explanation for the paradoxical finding that the most severe epilepsy phenotype (SMEI) is typically caused by *SCN1A* haploinsufficiency (**Figure 2**). Indeed, heterozygous deletion of the mouse ortholog leads to spontaneous seizures (Yu et al. 2006). Although sodium current density is preserved in cultured excitatory neurons from such animals, it is reduced in interneurons. Missense mutations alter several aspects of Nav1.1 kinetics (Alekov et al. 2000, Spampanato et al. 2001), and some of them impair fast inactivation (Lossin et al. 2002), but a simple unifying theme is lacking. Interestingly, antiepileptic drugs that facilitate sodium channel inactivation are ineffective in SMEI and may even worsen seizures (Korff et al. 2007), consistent with a functional impairment of sodium channels. A relationship between *SCN1A* genotype and response to antiepileptic drugs may extend to patients without pathogenic mutations: a common polymorphism has been associated with differences in dosage of phenytoin and carbamazepine (Tate et al. 2005). The mechanistic basis for this association remains to be determined.

Missense mutations associated with FHM3 have also been examined biophysically. Although a common pattern has yet to emerge, at least some of them facilitate recovery from inactivation, which implies a gain of function (Dichgans et al. 2005, Kahlig et al. 2008, Cestèle et al. 2008). Why this should manifest as migraine is far from clear. Indeed, one mutation has been reported to cause migraine and epilepsy (Castro et al. 2009).

SCN2A. Far fewer mutations have been identified in *SCN2A*, which encodes the α subunit of another fast sodium channel, Nav1.2. However, a similar range of phenotypes occurs, which spans benign familial neonatal-infantile seizures (BFNIS) associated with missense mutations (Berkovic et al. 2004) and a SMEI-like disorder associated with a truncation (Kamiya et al. 2004). Missense mutations in BFNIS appear to confer a partial loss of function (Misra et al. 2008); thus, the disease mechanisms may mirror those of *SCN1A* mutations. Less direct evidence implicates *SCN3A* in epilepsy (Holland et al. 2008).

SCN1B. This gene encodes a small β1 auxiliary sodium-channel subunit and was the first to be linked to GEFS+ (Wallace et al. 1998). The originally described mutation disrupts a disulfide bond and probably reduces the trafficking of sodium channels to the membrane. Evidence for this includes a noninvasive study of peripheral nerve excitability in affected individuals (Kiernan et al. 2005b), which reveals changes similar to those seen in poisoning by tetrodotoxin (Kiernan et al. 2005a).

Peripheral Nerve Sodium Channelopathies

Although CNS-expressed sodium channels also occur in peripheral nerves, several additional channels occur mainly in dorsal root ganglion cells. Complementary roles for Na1.7, Nav1.8, and Nav1.9 (encoded by *SCN9A*, *SCN10A*, and *SCN11A*, respectively) in sensory transduction and nociception are beginning to emerge.

SCN9A. This gene encodes the α subunit of Nav1.7, which occurs in a subset of dorsal root ganglion (DRG) neurons, as well as in sympathetic ganglia. Missense mutations cause two distinct dominantly inherited disorders (Yang et al. 2004, Fertleman et al. 2006). Primary erythermalgia (PE) is characterized by episodes of burning pain and redness of extremities, triggered or exacerbated by high ambient temperatures and relieved by mexiletine. An allelic disorder, paroxysmal extreme pain disorder (PEPD) presents in early childhood with autonomic instability (especially skin flushing and syncope) and attacks of severe deep burning pain in a perineal, periorbital, or mandibular distribution (Fertleman et al. 2007). Carbamazepine, which promotes inactivation, is partially effective. In vitro studies show that PE mutations lower the activation threshold (Han et al. 2006), whereas PEPD mutations impair inactivation (Fertleman et al. 2006). A clue to the mechanisms of PE is that Nav1.7 undergoes relatively little closed state inactivation in response to slow subthreshold depolarizations and is therefore especially suited to amplifying small stimuli. These may be exaggerated by PE-associated mutations because of their left-shifted activation kinetics. Boosting of subthreshold depolarization in DRG neurons by mutant Nav1.7 predisposes to the generation of action potentials, which are partly mediated by colocalized Nav1.8 channels. The redness (erythema) however may result from simultaneous failure of action potential generation in sympathetic efferent neurons, because these do not express Nav1.8, and Nav1.7-mediated slow depolarization may instead lead to the inactivation of other sodium channels (Rush et al. 2006).

A third class of *SCN9A* mutations, which behave in a recessive manner, causes complete insensitivity to pain (Cox et al. 2006). Although originally identified in consanguineous families, compound heterozygous cases also occur (Nilsen et al. 2009). Affected children frequently self-mutilate or suffer the consequences of unhealed bone fractures, and associated anosmia implies a role for Nav1.7 in olfaction. All mutations reported to date are nonsense or frameshift, or affect splice sites giving rise to a truncated sequence (Goldberg et al. 2007).

Muscle Sodium Channelopathies

Nav1.4 is the main sodium channel of skeletal muscle. In common with fast sodium channels expressed in neurons, it inactivates rapidly. Impairment of fast inactivation underlies several channelopathies.

SCN4A. Dominant mutations of SCN4A, which encodes the α subunit of the muscle sodium channel Nav1.4, underlie a spectrum of disorders, which ranges from myotonia to periodic paralysis (Fontaine et al. 1990, Ptácek et al. 1991, Ebers et al. 1991, Rojas et al. 1991, McClatchey et al. 1992, Miller et al. 2004). In paramyotonia congenita (PC), patients experience muscle stiffness, especially noticeable in cold temperatures, that typically worsens with repetitive exercise (which distinguishes paramyotonia from other forms of myotonia). In potassium-aggravated myotonia (PAM), stiffness fluctuates with serum potassium levels and can be triggered during periods of rest after exercise, during fasting, or after eating potassium-rich foods. In hyperkalemic periodic paralysis (HyperPP) patients experience variable limb weakness, which lasts up to one or two days, associated with a high serum potassium (**Figure 3**). A distinct disorder, hypokalemic periodic paralysis (HypoPP), can also be caused by SCN4A mutations (Jurkat-Rott et al. 2000), although it is more commonly associated with CACNA1S mutations (see below). Attacks typically occur after a carbohydrate load, which leads to a fall in serum potassium. Most of these conditions present in childhood, and attacks of periodic paralysis can later fade, although a vacuolar myopathy sometimes develops.

Recordings from human myotubes in culture and heterologous expression studies suggest that PC, PAM, and HyperPP reflect variable degrees of gain of function due to impaired fast inactivation (Lerche et al. 1993, Cannon et al. 1991). In particular, HyperPP mutations tend to cause a small noninactivating component of the depolarization-evoked current. PAM and PC mutations also impair fast inactivation through variable changes in its rate, in its voltage dependency, and/or in the kinetics of recovery following repolarization. Mutations cluster around the domain III-IV linker, which contributes to the inactivation gate, and around other parts of the channel that interact with it. However, a few mutations do not fall neatly in this scheme and affect activation threshold instead, also in a manner consistent with a gain of function (Cummins et al. 1993).

Experimental and modeling studies converge on the following mechanistic account of the myotonia and paralysis (Cannon et al. 1993): With intermediate impairment of fast inactivation, sodium channels mediate a small, persistent depolarizing current that facilitates repetitive discharges (manifesting as myotonia). More severe impairment of inactivation leads to sustained depolarization, and sodium channels undergo a transition to a second inactivated state, which renders the muscle fibers inexcitable (corresponding to paralysis). Because depolarization causes potassium efflux, the essential components are in place for a positive feedback loop in which the extracellular potassium concentration builds and exacerbates the depolarization, which accounts for the association of paralysis with hyperkalemia. HypoPP represents a different disease mechanism and is considered below.

CNS Calcium Channelopathies

Calcium channels in the CNS can be classified into high-threshold Cav1 and Cav2 channels, which generally exhibit relatively slow inactivation, and low-threshold Cav3 channels, which inactivate relatively quickly. Their properties are determined principally by the identity of the pore-forming α1 subunit, which is structurally homologous to the α subunit of sodium channels. Congenital stationary night blindness is a retinal channelopathy associated with mutations of the gene encoding the α1 subunit of Cav1.4, whereas mutations affecting Cav1.3 are associated with cardiac arrhythmias and a multisystem disorder that includes autism and mental retardation.

CACNA1A. The *CACNA1A* gene encodes the pore-forming α1A subunit of Cav2.1, also known as P/Q-type, calcium channels, which contribute to neurotransmitter release at synapses widely in the nervous system. Cav2.1 channels also occur abundantly in cerebellar granule and Purkinje cells. Heterozygous mutations cause a spectrum of disorders, which include a form of FHM (FHM1), a paroxysmal cerebellar disorder [episodic ataxia type 2 (EA2)] and a progressive cerebellar degeneration [spinocerebellar ataxia type 6 (SCA6) (Zhuchenko et al. 1997, Ophoff et al. 1996). FHM1 is caused by missense mutations, although some mutations are associated with additional features, which include a mild cerebellar ataxia (Ducros et al. 2001) or cerebral edema after mild trauma (Stam et al. 2009). EA2 can also be caused by missense mutations (Denier et al. 2001), although it is more commonly caused by nonsense, frame-shift, or splice-site mutations (Denier et al. 1999). EA2 arises from loss of function (Guida et al. 2001), but it is not clear why this leads to episodic ataxia. FHM1 mutations have variable effects on kinetics (Tottene et al. 2002). Cultured neurons from *Cacna1a*-null mice, transfected with FHM mutant channels, exhibit alterations in synaptic transmission, consistent with impaired Cav2.1 function (Cao et al. 2004). However, analysis of a homozygous knock-in mouse model of FHM1 instead points to increased synaptic transmission at the neuromuscular junction (van den Maagdenberg et al. 2004) and at excitatory neocortical synapses (Tottene et al. 2009). FHM1 knock-in mice also show a lowered threshold and increased velocity of cortical spreading depression (CSD), and more severe post-CSD

deficits (van den Maagdenberg et al. 2004, Eikermann-Haerter et al. 2009). CSD underlies the aura that accompanies severe migraine. Why increased neurotransmitter release at glutamatergic synapses predisposes to CSD, rather than seizures, is not clear. Interestingly, several in-bred strains of mice with loss-of-function *Cacna1a* mutations have ataxia and spike-wave seizures. Some human mutations can also result in this combination (Imbrici et al. 2004, Jouvenceau et al. 2001). *CACNB4*, which encodes the β4 auxiliary subunit, has also been implicated in epilepsy and ataxia (Escayg et al. 2000).

CACNA1H. Another gene associated with absence seizures encodes the α1H subunit of Cav3.2, a transient or T-type channel abundantly expressed in thalamic reticular neurons (**Figure 2**). Rare genetic variants have been identified in some individuals with childhood absence epilepsy, although they do not cosegregate with the disorder (Chen et al. 2003). Functional expression studies have shown that several of these confer gain of function through various mechanisms (Khosravani et al. 2005, Vitko et al. 2005, Heron et al. 2007), which implies that they contribute to the risk of developing epilepsy and are not rare functionally neutral polymorphisms.

Muscle Calcium Channelopathies

The main calcium channel of skeletal muscle couples sarcolemmal depolarization to calcium release from the sarcoplasm via an interaction with the ryanodine receptor and is therefore essential for excitation-contraction coupling.

CACNA1S. Missense mutations in *CACNA1S*, which encodes the α1S subunit of the Cav1.1 muscle calcium channel, account for most cases of hypokalemic periodic paralysis (HypoPP). This disorder manifests as attacks of limb and neck muscle weakness associated with low serum potassium. All known mutations in HypoPP neutralize positively charged arginine residues in the voltage-sensing S4 segments of either *CACNA1S* or *SCN4A* (Ptáček et al. 1994, Jurkat-Rott et al. 1994, Matthews et al. 2009). Several such mutations allow a cation leak through a pore formed by the voltage-sensing segment (Sokolov et al. 2007, Struyk & Cannon 2007). The attacks of paralysis may share mechanisms in common with those in ATS (**Figure 3**): A reduction in extracellular potassium causes the inward rectifier Kir2.1 to stop conducting, which allows the inward cation leak through the voltage sensor to depolarize the muscle fiber and render it inexcitable (Jurkat-Rott et al. 2009).

Chloride Channelopathies

Voltage-dependent chloride channels are unrelated to the voltage-gated potassium channel family. Different members of this family show distinct patterns of activation. Two renal channelopathies, Dent disease and Bartter syndrome, are caused by mutations of *CLCN5* and *CLCNKB*, respectively.

CLCN1. Numerous mutations of the skeletal muscle chloride channel gene *CLCN1* underlie myotonia congenita (George et al. 1993, Koch et al. 1992). This disorder can show either dominant or recessive inheritance (Thomsen and Becker disease, respectively), and genetic counseling is complicated by the frequent occurrence of compound heterozygosity (Fialho et al. 2007). Patients experience fluctuating stiffness of facial and limb muscles, with a worsening in cold weather but improvement with muscle use. The dimeric ClC1 channel is only partly deactivated at resting membrane potential. It thus accounts for much of the muscle fiber resting membrane conductance but activates further upon depolarization. Pathogenic mutations reduce the current density and/or shift the activation curve to more depolarized potentials (**Figure 3**). Coexpression with wild-type *CLCN1* can shed light on the likely inheritance pattern of individual mutations because those that exert a dominant-negative effect in vitro are more likely to cause disease in the heterozygous state (Kubisch et al. 1998).

The status of *CLCN2*, which encodes the brain-expressed ClC2 channel, in epilepsy is uncertain. Both gain- and loss-of-function mutations have been identified, but incomplete segregation within affected kindreds implies that they may contribute to, rather than determine independently, an individual's risk of developing epilepsy (Saint-Martin et al. 2009).

LIGAND-GATED CHANNELOPATHIES

CNS Nicotinic Receptor Mutations

Nicotinic receptors in the CNS are either homopentameric, composed of five $\alpha 7$ subunits, or heteropentameric, consisting of $\beta 2$ together with $\alpha 2$, $\alpha 3$, or $\alpha 4$. The conditions upon which they are activated by synaptically released acetylcholine (ACh) remain poorly understood.

CHNRA4, CHRNA2, and CHRNB2. These genes encode the CNS nicotinic receptor subunits $\alpha 4$, $\alpha 2$, and $\beta 2$, which coassemble in an $(\alpha x)_2(\beta 2)_3$ stoichiometry and occur widely in the forebrain where they mediate a cholinergic facilitation of release of other neurotransmitters. Mutations of any of these genes are associated with autosomal dominant nocturnal frontal lobe epilepsy (ADNFLE) (De Fusco et al. 2000, Steinlein et al. 1995), which manifests with limb posturing or jerking and automatisms that arise during sleep-wake transitions especially. Although the disorder is dominantly inherited, it can show partial penetrance. The cellular mechanisms remain unclear, as well as the tendency for seizures to arise in the frontal lobes (**Figure 2**). Mutations tend to occur in the M2 pore-lining segment and alter various kinetic parameters, which include desensitization rate, Ca^{2+} permeability, and current density. However, other than a shift of the ACh dose-response curve to lower concentrations (Bertrand et al. 2002) and lower sensitivity to extracellular calcium (Rodrigues-Pinguet et al. 2003), no single parameter is consistently altered in the same way for all mutations. Several mouse models have been generated. Some knock-in mice exhibit EEG abnormalities and increase in nicotine-evoked release of GABA onto cortical neurons when studied in vitro (Klaassen et al. 2006). Excessive synchronous inhibition in thalamocortical circuits occurs in spike-wave epilepsy, and an analogous phenomenon could potentially underlie ADNFLE.

Although mutations of *CHRNA7*, which encodes the α7 subunit, have not been identified, this gene is implicated indirectly in idiopathic generalized epilepsy, mental retardation, autism, and schizophrenia, based on statistical associations with chromosomal deletions that encompass its locus (Sharp et al. 2008, Helbig et al. 2009, International Schizophrenia Consortium 2008, Stefansson et al. 2008).

Muscle Nicotinic Receptor Mutations

The skeletal muscle nicotinic receptor is among the most intensely studied ligand-gated ion channels and is the target of auto-antibodies in the acquired disease myasthenia gravis. Inherited channelopathies affecting the subunits of this receptor share many features in common with myasthenia gravis, although they present earlier in life.

CHRNA1, CHRNB1, CHRNG, CHRND, and CHRNE. Together, these genes encode the subunits that constitute the ACh receptor at the neuromuscular junction (α1, β1, γ, δ, and ε, respectively). Fetal receptors have a $(\alpha 1)_2 \beta 1 \gamma \delta$ stoichiometry, and after birth, ε replaces the γ subunit. Disease mechanisms associated with ACh receptors can be grouped into several main types, although the occurrence of compound heterozygous cases complicates matters (Michalk et al. 2008, Engel & Sine 2005). Homozygous null mutations of *CHNRA1, CHNRG,* or *CHRND* are prenatally lethal, owing to fetal akinesia. Homozygous nonsense mutations of *CHRNB1* or *CHRNE,* in contrast, are compatible with life but lead to a deficit of ACh receptors, which manifests as a congenital myasthenic syndrome (CMS). Affected children present soon after birth with hypotonia and muscle weakness affecting limb, trunk, ocular, and facial muscles. These can cause difficulties feeding and secondary craniofacial deformities. A further group of missense mutations that also cause CMS alter the kinetics of ACh receptors, such that the mean duration of opening is reduced, which corresponds to abnormally brief end-plate potentials (Engel et al. 1993) (**Figure 3**). These so-called fast current mutations can occur in any of the subunits expressed postnatally (α1, β1, δ, and ε). A final group of missense mutations, which affect any of these subunits, has the opposite effect and prolongs the duration of end-plate currents. This so-called slow current CMS can be associated with a slowly progressive myopathy, which probably represents cumulative excitotoxic damage to muscle fibers (Engel et al. 1982). CMS can thus arise from either loss or gain of ACh receptor function. However, this distinction has implications for drug treatment because acetylcholinesterase inhibition, which is effective in many acquired and inherited myasthenic disorders, can worsen slow-channel CMS.

GABA$_A$ Receptor Channelopathies

GABA$_A$ receptors are structurally homologous to nicotinic receptors. Although 16 subunits are known, only a small subset of the possible pentameric combinations are thought to occur in the brain.

GABRG2. GABA$_A$ receptors that underlie fast inhibition in the forebrain usually contain the γ2 subunit, which is encoded by *GABRG2* and is necessary for synaptic targeting. Several dominant mutations of this gene have been identified in association with GEFS+ and related generalized epilepsy syndromes (Baulac et al. 2001, Wallace et al. 2001) (**Figure 2**). Missense and nonsense mutations occur. Missense mutations impair GABA$_A$ receptor function through highly diverse mechanisms, which include defects in trafficking, impaired synaptic localization, and altered kinetics, as well as indirect effects on tonic inhibition mediated by α5-containing receptors (Bianchi et al. 2002,

Eugène et al. 2007). Dominant-negative effects of truncating mutations have also been documented (Kang et al. 2009). One missense mutation leads to epilepsy when knocked into the mouse ortholog, with evidence of reduced cortical inhibition (Tan et al. 2007). Interestingly, examination of a conditional mouse mutant suggests that the suppression of expression during a critical developmental window lowers subsequent seizure susceptibility (Chiu et al. 2008).

GABRA1. Mutations of *GABRA1* appear to be an infrequent cause of juvenile myoclonic epilepsy, with only one confirmed affected kindred known (Cossette et al. 2002). This gene encodes the α1 subunit of GABA$_A$ receptors, which contribute to fast inhibition in the forebrain. The dominantly inherited missense mutation disrupts protein folding and insertion into the membrane (Gallagher et al. 2007). A sporadic frameshift mutation has also been reported in an individual with childhood absence epilepsy (Maljevic et al. 2006).

GABRB3. Dominantly inherited mutations in *GABRB3*, which encodes the β3 subunit, have been identified in a few families with childhood absence epilepsy (Tanaka et al. 2008). They decrease surface expression of GABA$_A$ receptors in a non-neuronal expression system. Interestingly, *GABRB3* falls in the chromosome 15 region deleted in Angelman syndrome, a maternally inherited neurodevelopmental disorder associated with generalized seizures (Wagstaff et al. 1991). The β3 subunit is expressed in the reticular nucleus of the thalamus (RTN). Genetic deletion in the mouse results in the loss of fast GABA$_A$ receptor-mediated transmission in the RTN and hypersynchrony of thalamocortical neuron firing (Huntsman et al. 1999). The human mutations of both *GABRB3* and of the calcium channel gene *CACNA1H* thus provide direct supportive evidence that implicates the thalamocortical loop in this form of epilepsy (**Figure 2**). However, because the mutations of either gene were only identified in a few families and did not cosegregate fully, they are neither necessary nor sufficient to develop the disease.

GABRD. GABA receptors in the extrasynaptic membrane tend to contain the δ subunit, encoded by *GABRD*, rather than a γ subunit. Such receptors exhibit a high affinity for GABA and contribute to tonic inhibition (Semyanov et al. 2004). Two variants in *GABRD* have been suggested to increase the risk of generalized epilepsy (Dibbens et al. 2004). Although they do not cosegregate with the disease, a role for GABRD as a susceptibility locus is supported by evidence that they impair channel function (Feng et al. 2006).

Glycine Receptor Channelopathies

Glycine receptors are also members of the nicotinic superfamily of pentameric ligand-gated channels. They contribute to fast inhibition in the spinal cord and brain stem.

GLRA1 and GLRB. Disorders of glycinergic transmission underlie familial hyperekplexia. This disorder manifests as exaggerated startle in infancy. A variety of stimuli trigger stereotypical spasms consisting of abduction of the arms, trunk rigidity, and apnea, which can be life threatening. The spasms can be attenuated with benzodiazepines and gradually decrease in severity with development. *GLRA1* and *GLRB* encode α$_1$ and β subunits, respectively, which coassemble to form the main glycine receptor isoform in the spinal cord and brain stem, which exists in an (α1)$_2$β$_3$ stoichiometry and contributes to fast inhibition. Many familial cases can be attributed to dominantly inherited missense mutations of *GLRA1* (Shiang et al. 1993), although recessive mutations also occur, sometimes in combination in compound heterozygous cases (Harvey et al. 2008). *GLRB* mutations are much less common (Rees et al. 2002). The mutations act by impairing glycinergic transmission, some via a dominant-negative effect.

FROM CHANNELOPATHY TO SYNAPTOPATHY

Most of the channelopathies listed above cause paroxysmal symptoms. Conversely, the genes known to cause paroxysmal neurological disorders with Mendelian inheritance almost invariably code for ion channels. There are some exceptions, although these are also implicated in channel or synapse function. Thus, the only two non ion channel genes currently known to cause relatively pure monogenic human epilepsy (without developmental or cognitive impairment) encode proteins that interact with calcium channels (*EFHC1*) (Suzuki et al. 2004) and potassium channels and/or glutamate receptors (*LGI1*) (Fukata et al. 2006, Suzuki et al. 2004, Kalachikov et al. 2002). Another gene that causes familial hyperekplexia (*SLC6A5*) encodes a presynaptic glycine transporter that supplies glycinergic terminals with the inhibitory transmitter (Rees et al. 2006). And another gene that causes FHM (*ATP1A2*) encodes a subunit of the sodium-potassium ATPase (De Fusco et al. 2003). It is unclear why mutations of other proteins involved in regulating neuronal excitability, such as G-protein-coupled receptors, have not been identified in epilepsy or migraine.

Undoubtedly, many other ion channel gene mutations remain to be identified. However, an important challenge for future research is to understand the role played by ion channel genes in common forms of diseases that show high heritability but do not obey Mendelian patterns of transmission. Ion channel genes are excellent candidate susceptibility factors for idiopathic epilepsy and migraine; and indeed, some of the genes listed above (especially *CACNA1H*, *GABRB3*, and *GABRD*) appear to have such a role. Because association studies have failed to reveal a few common polymorphisms, a compelling model of the genetic architecture of such diseases is that many individually rare variants in multiple genes interact to determine the inherited disease risk to the individual. Exhaustive biophysical characterization of such variants may be required to understand their contribution to the disease risk (Rajakulendran et al. 2010).

DISCLOSURE STATEMENT

The author is not aware of any affiliations, memberships, funding, or financial holdings that might be perceived as affecting the objectivity of this review.

ACKNOWLEDGMENTS

Work in D.M.K.'s laboratory has been supported by the Medical Research Council, Action Medical Research, the European Research Council, the Wellcome Trust, and the Myasthenia Gravis Association. The author is grateful to S. Schorge for critical insights and comments on the manuscript.

LITERATURE CITED

Adelman JP, Bond CT, Pessia M, Maylie J. 1995. Episodic ataxia results from voltage-dependent potassium channels with altered functions. *Neuron* 15:1449–54

Alekov A, Rahman MM, Mitrovic N, Lehmann-Horn F, Lerche H. 2000. A sodium channel mutation causing epilepsy in man exhibits subtle defects in fast inactivation and activation in vitro. *J. Physiol. (Lond.)* 529(3):533–39

Andelfinger G, Tapper AR, Welch RC, Vanoye CG, George AL, Benson DW. 2002. KCNJ2 mutation results in Andersen syndrome with sex-specific cardiac and skeletal muscle phenotypes. *Am. J. Hum. Genet.* 71:663–68

Baulac S, Gourfinkel-An I, Picard F, Rosenberg-Bourgin M, Prud'homme JF, et al. 1999. A second locus for familial generalized epilepsy with febrile seizures plus maps to chromosome 2q21-q33. *Am. J. Hum. Genet.* 65:1078–85

Baulac S, Huberfeld G, Gourfinkel-An I, Mitropoulou G, Beranger A, et al. 2001. First genetic evidence of GABA$_A$ receptor dysfunction in epilepsy: a mutation in the gamma2-subunit gene. *Nat. Genet.* 28:46–48

Bergoffen J, Scherer SS, Wang S, Scott MO, Bone LJ, et al. 1993. Connexin mutations in X-linked Charcot-Marie-Tooth disease. *Science* 262:2039–42

Berkovic SF, Heron SE, Giordano L, Marini C, Guerrini R, et al. 2004. Benign familial neonatal-infantile seizures: characterization of a new sodium channelopathy. *Ann. Neurol.* 55:550–57

Bertrand D, Picard F, Le Hellard S, Weiland S, Favre I, et al. 2002. How mutations in the nAChRs can cause ADNFLE epilepsy. *Epilepsia* 43(5):S112–22

Bianchi MT, Song L, Zhang H, Macdonald RL. 2002. Two different mechanisms of disinhibition produced by GABAA receptor mutations linked to epilepsy in humans. *J. Neurosci.* 22:5321–27

Biervert C, Schroeder BC, Kubisch C, Berkovic SF, Propping P, et al. 1998. A potassium channel mutation in neonatal human epilepsy. *Science* 279:403–406

Bockenhauer D, Feather S, Stanescu HC, Bandulik S, Zdebik AA, et al. 2009. Epilepsy, ataxia, sensorineural deafness, tubulopathy, and KCNJ10 mutations. *N. Engl. J. Med.* 360:1960–70

Browne DL, Gancher ST, Nutt JG, Brunt ER, Smith EA, et al. 1994. Episodic ataxia/myokymia syndrome is associated with point mutations in the human potassium channel gene, KCNA1. *Nat. Genet.* 8:136–40

Cannon SC, Brown RH, Corey DP. 1991. A sodium channel defect in hyperkalemic periodic paralysis: potassium-induced failure of inactivation. *Neuron* 6:619–26

Cannon SC, Brown RH, Corey DP. 1993. Theoretical reconstruction of myotonia and paralysis caused by incomplete inactivation of sodium channels. *Biophys. J.* 65:270–88

Cao Y, Piedras-Rentería ES, Smith GB, Chen G, Harata NC, Tsien RW. 2004. Presynaptic Ca^{2+} channels compete for channel type-preferring slots in altered neurotransmission arising from Ca^{2+} channelopathy. *Neuron* 43:387–400

Castro M, Stam AH, Lemos C, deVries B, Vanmolkot KRJ, et al. 2009. First mutation in the voltage-gated Nav1.1 subunit gene SCN1A with co-occurring familial hemiplegic migraine and epilepsy. *Cephalalgia* 29:308–13

Cestele S, Scalmani P, Rusconi R, Terragni B, Franceschetti S, Mantegazza M. 2008. Self-limited hyper-excitability: functional effect of a familial hemiplegic migraine mutation of the Nav1.1 (SCN1A) Na$^+$ Channel. *J. Neurosci.* 28:7273–83

Charlet-B N, Savkur RS, Singh G, Philips AV, Grice EA, Cooper TA. 2002. Loss of the muscle-specific chloride channel in type 1 myotonic dystrophy due to misregulated alternative splicing. *Mol. Cell* 10:45–53

Charlier C, Singh NA, Ryan SG, Lewis TB, Reus BE, et al. 1998. A pore mutation in a novel KQT-like potassium channel gene in an idiopathic epilepsy family. *Nat. Genet.* 18:53–55

Chen Y, Lu J, Pan H, Zhang Y, Wu H, et al. 2003. Association between genetic variation of CACNA1H and childhood absence epilepsy. *Ann. Neurol.* 54:239–43

Chiu C, Reid CA, Tan HO, Davies PJ, Single FN, et al. 2008. Developmental impact of a familial GABAA receptor epilepsy mutation. *Ann. Neurol.* 64:284–93

Claes L, Del-Favero J, Ceulemans B, Lagae L, Van Broeckhoven C, De Jonghe P. 2001. De novo mutations in the sodium-channel gene SCN1A cause severe myoclonic epilepsy of infancy. *Am. J. Hum. Genet.* 68:1327–32

Cossette P, Liu L, Brisebois K, Dong H, Lortie A, et al. 2002. Mutation of GABRA1 in an autosomal dominant form of juvenile myoclonic epilepsy. *Nat. Genet.* 31:184–89

Cox JJ, Reimann F, Nicholas AK, Thornton G, Roberts E, et al. 2006. An SCN9A channelopathy causes congenital inability to experience pain. *Nature* 444:894–98

Cummins TR, Zhou J, Sigworth FJ, Ukomadu C, Stephan M, et al. 1993. Functional consequences of a Na$^+$ channel mutation causing hyperkalemic periodic paralysis. *Neuron* 10:667–78

Davies NP, Imbrici P, Fialho D, Herd C, Bisland LG, et al. 2005. Andersen-Tawil syndrome: new potassium channel mutations and possible phenotypic variation. *Neurology* 65:1083–89

De Fusco M, Becchetti A, Patrignani A, Annesi G, Gambardella A, et al. 2000. The nicotinic receptor beta 2 subunit is mutant in nocturnal frontal lobe epilepsy. *Nat. Genet.* 26:275–76

De Fusco M, Marconi R, Silvestri L, Atorino L, Rampoldi L, et al. 2003. Haploinsufficiency of ATP1A2 encoding the Na$^+$/K$^+$ pump alpha2 subunit associated with familial hemiplegic migraine type 2. *Nat. Genet.* 33:192–96

Dedek K, Kunath B, Kananura C, Reuner U, Jentsch TJ, Steinlein OK. 2001. Myokymia and neonatal epilepsy caused by a mutation in the voltage sensor of the KCNQ2 K$^+$ channel. *Proc. Natl. Acad. Sci. USA* 98:12272–77

Denier C, Ducros A, Durr A, Eymard B, Chassande B, Tournier-Lasserve E. 2001. Missense CACNA1A mutation causing episodic ataxia type 2. *Arch. Neurol.* 58:292–95

Denier C, Ducros A, Vahedi K, Joutel A, Thierry P, et al. 1999. High prevalence of CACNA1A truncations and broader clinical spectrum in episodic ataxia type 2. *Neurology* 52:1816–21

Dibbens LM, Feng H, Richards MC, Harkin LA, Hodgson BL, et al. 2004. GABRD encoding a protein for extra- or peri-synaptic GABAA receptors is a susceptibility locus for generalized epilepsies. *Hum. Mol. Genet.* 13:1315–19

Dichgans M, Freilinger T, Eckstein G, Babini E, Lorenz-Depiereux B, et al. 2005. Mutation in the neuronal voltage-gated sodium channel SCN1A in familial hemiplegic migraine. *Lancet* 366:371–77

Du W, Bautista JF, Yang H, Diez-Sampedro A, You S, et al. 2005. Calcium-sensitive potassium channelopathy in human epilepsy and paroxysmal movement disorder. *Nat. Genet.* 37:733–38

Ducros A, Denier C, Joutel A, Cecillon M, Lescoat C, et al. 2001. The clinical spectrum of familial hemiplegic migraine associated with mutations in a neuronal calcium channel. *N. Engl. J. Med.* 345:17–24

Ebers GC, George AL, Barchi RL, Ting-Passador SS, Kallen RG, et al. 1991. Paramyotonia congenita and hyperkalemic periodic paralysis are linked to the adult muscle sodium channel gene. *Ann. Neurol.* 30:810–16

Edenberg HJ, Dick DM, Xuei X, Tian H, Almasy L, et al. 2004. Variations in GABRA2, encoding the alpha 2 subunit of the GABA$_A$ receptor, are associated with alcohol dependence and with brain oscillations. *Am. J. Hum. Genet.* 74:705–14

Eikermann-Haerter K, Dileköz E, Kudo C, Savitz SI, Waeber C, et al. 2009. Genetic and hormonal factors modulate spreading depression and transient hemiparesis in mouse models of familial hemiplegic migraine type 1. *J. Clin. Invest.* 119:99–109

Engel AG, Lambert EH, Mulder DM, Torres CF, Sahashi K, et al. 1982. A newly recognized congenital myasthenic syndrome attributed to a prolonged open time of the acetylcholine-induced ion channel. *Ann. Neurol.* 11:553–69

Engel AG, Uchitel OD, Walls TJ, Nagel A, Harper CM, Bodensteiner J. 1993. Newly recognized congenital myasthenic syndrome associated with high conductance and fast closure of the acetylcholine receptor channel. *Ann. Neurol.* 34:38–47

Engel AG, Sine SM. 2005. Current understanding of congenital myasthenic syndromes. *Curr. Opin. Pharmacol.* 5:308–21

Escayg A, De Waard M, Lee DD, Bichet D, Wolf P, et al. 2000. Coding and noncoding variation of the human calcium-channel beta4-subunit gene CACNB4 in patients with idiopathic generalized epilepsy and episodic ataxia. *Am. J. Hum. Genet.* 66:1531–39

Escayg A, MacDonald BT, Meisler MH, Baulac S, Huberfeld G, et al. 2000. Mutations of SCN1A, encoding a neuronal sodium channel, in two families with GEFS+2. *Nat. Genet.* 24:343–45

Eugène E, Depienne C, Baulac S, Baulac M, Fritschy JM, et al. 2007. GABA$_A$ receptor gamma 2 subunit mutations linked to human epileptic syndromes differentially affect phasic and tonic inhibition. *J. Neurosci.* 27:14108–16

Eunson LH, Rea R, Zuberi SM, Youroukos S, Panayiotopoulos CP, et al. 2000. Clinical, genetic, and expression studies of mutations in the potassium channel gene KCNA1 reveal new phenotypic variability. *Ann. Neurol.* 48:647–56

Feng H, Kang J, Song L, Dibbens L, Mulley J, Macdonald RL. 2006. Delta Subunit susceptibility variants E177A and R220H associated with complex epilepsy alter channel gating and surface expression of {alpha}4beta2{delta} GABAA receptors. *J. Neurosci.* 26:1499–506

Fertleman CR, Ferrie CD, Aicardi J, Bednarek N, Eeg-Olofsson O, et al. 2007. Paroxysmal extreme pain disorder (previously familial rectal pain syndrome). *Neurology* 69:586–95

Fertleman CR, Baker MD, Parker KA, Moffatt S, Elmslie FV, et al. 2006. SCN9A mutations in paroxysmal extreme pain disorder: Allelic variants underlie distinct channel defects and phenotypes. *Neuron* 52:767–74

Fialho D, Schorge S, Pucovska U, Davies NP, Labrum R, et al. 2007. Chloride channel myotonia: exon 8 hot-spot for dominant-negative interactions. *Brain* 130:3265–74

Fontaine B, Khurana TS, Hoffman EP, Bruns GA, Haines JL, et al. 1990. Hyperkalemic periodic paralysis and the adult muscle sodium channel alpha-subunit gene. *Science* 250:1000–2

Foppen RJG, van Mil HGJ, van Heukelom JS. 2002. Effects of chloride transport on bistable behavior of the membrane potential in mouse skeletal muscle. *J. Physiol.* 542:181–91

Fujiwara T, Sugawara T, Mazaki-Miyazaki E, Takahashi Y, Fukushima K, et al. 2003. Mutations of sodium channel alpha subunit type 1 (SCN1A) in intractable childhood epilepsies with frequent generalized tonic-clonic seizures. *Brain* 126:531–46

Fukata Y, Adesnik H, Iwanaga T, Bredt DS, Nicoll RA, Fukata M. 2006. Epilepsy-related ligand/receptor complex LGI1 and ADAM22 regulate synaptic transmission. *Science* 313:1792–95

Gallagher MJ, Ding L, Maheshwari A, Macdonald RL. 2007. The GABAA receptor alpha1 subunit epilepsy mutation A322D inhibits transmembrane helix formation and causes proteasomal degradation. *Proc. Natl. Acad. Sci. USA* 104:12999–3004

George AL, Crackower MA, Abdalla JA, Hudson AJ, Ebers GC. 1993. Molecular basis of Thomsen's disease (autosomal dominant myotonia congenita). *Nat. Genet.* 3:305–10

Gloyn AL, Pearson ER, Antcliff JF, Proks P, Bruining GJ, et al. 2004. Activating mutations in the gene encoding the ATP-sensitive potassium-channel subunit Kir6.2 and permanent neonatal diabetes. *N. Engl. J. Med.* 350:1838–49

Goldberg YP, MacFarlane J, MacDonald ML, Thompson J, Dube M-P, et al. 2007. Loss-of-function mutations in the Na$_V$1.7 gene underlie congenital indifference to pain in multiple human populations. *Clin. Genet.* 71:311–19

Guida S, Trettel F, Pagnutti S, Mantuano E, Tottene A, et al. 2001. Complete loss of P/Q calcium channel activity caused by a CACNA1A missense mutation carried by patients with episodic ataxia type 2. *Am. J. Hum. Genet.* 68:759–64

Han C, Rush AM, Dib-Hajj SD, Li S, Xu Z, et al. 2006. Sporadic onset of erythermalgia: a gain-of-function mutation in Nav1.7. *Ann. Neurol.* 59:553–58

Harvey RJ, Topf M, Harvey K, Rees MI. 2008. The genetics of hyperekplexia: more than startle! *Trends Genet.* 24:439–47

Haug K, Warnstedt M, Alekov AK, Sander T, Ramirez A, et al. 2003. Mutations in CLCN2 encoding a voltage-gated chloride channel are associated with idiopathic generalized epilepsies. *Nat. Genet.* 33:527–32

Helbig I, Mefford HC, Sharp AJ, Guipponi M, Fichera M, et al. 2009. 15q13.3 microdeletions increase risk of idiopathic generalized epilepsy. *Nat. Genet.* 41:160–62

Helbig I, Scheffer IE, Mulley JC, Berkovic SF. 2008. Navigating the channels and beyond: unraveling the genetics of the epilepsies. *Lancet Neurol.* 7:231–45

Heron SE, Khosravani H, Varela D, Bladen C, Williams TC, et al. 2007. Extended spectrum of idiopathic generalized epilepsies associated with CACNA1H functional variants. *Ann. Neurol.* 62:560–68

Herson PS, Virk M, Rustay NR, Bond CT, Crabbe JC, et al. 2003. A mouse model of episodic ataxia type-1. *Nat. Neurosci.* 6:378–83

Holland KD, Kearney JA, Glauser TA, Buck G, Keddache M, et al. 2008. Mutation of sodium channel SCN3A in a patient with cryptogenic pediatric partial epilepsy. *Neurosci. Lett.* 433:65–70

Huffaker SJ, Chen J, Nicodemus KK, Sambataro F, Yang F, et al. 2009. A primate-specific, brain isoform of KCNH2 affects cortical physiology, cognition, neuronal repolarization and risk of schizophrenia. *Nat. Med.* 15:509–18

Huntsman MM, Porcello DM, Homanics GE, DeLorey TM, Huguenard JR. 1999. Reciprocal inhibitory connections and network synchrony in the mammalian thalamus. *Science* 283:541–43

Imbrici P, Jaffe SL, Eunson LH, Davies NP, Herd C, et al. 2004. Dysfunction of the brain calcium channel CaV2.1 in absence epilepsy and episodic ataxia. *Brain* 127:2682–92

International Schizophrenia Consortium. 2008. Rare chromosomal deletions and duplications increase risk of schizophrenia. *Nature* 455:237–41

Jentsch TJ. 2000. Neuronal KCNQ potassium channels: physiology and role in disease. *Nat. Rev. Neurosci.* 1:21–30

Jouvenceau A, Eunson LH, Spauschus A, Ramesh V, Zuberi SM, et al. 2001. Human epilepsy associated with dysfunction of the brain P/Q-type calcium channel. *Lancet* 358:801–7

Jurkat-Rott K, Lehmann-Horn F, Elbaz A, Heine R, Gregg RG, et al. 1994. A calcium channel mutation causing hypokalemic periodic paralysis. *Hum. Mol. Genet.* 3:1415–19

Jurkat-Rott K, Mitrovic N, Hang C, Kouzmekine A, Iaizzo P, et al. 2000. Voltage-sensor sodium channel mutations cause hypokalemic periodic paralysis type 2 by enhanced inactivation and reduced current. *Proc. Natl. Acad. Sci. USA* 97:9549–54

Jurkat-Rott K, Weber M, Fauler M, Guo X, Holzherr BD, et al. 2009. K^+-dependent paradoxical membrane depolarization and Na^+ overload, major and reversible contributors to weakness by ion channel leaks. *Proc. Natl. Acad. Sci. USA* 106:4036–41

Kahlig KM, Rhodes TH, Pusch M, Freilinger T, Pereira-Monteiro JM, et al. 2008. Divergent sodium channel defects in familial hemiplegic migraine. *Proc. Natl. Acad. Sci. USA* 105:9799–804

Kalachikov S, Evgrafov O, Ross B, Winawer M, Barker-Cummings C, et al. 2002. Mutations in LGI1 cause autosomal-dominant partial epilepsy with auditory features. *Nat. Genet.* 30:335–41

Kamiya K, Kaneda M, Sugawara T, Mazaki E, Okamura N, et al. 2004. A nonsense mutation of the sodium channel gene SCN2A in a patient with intractable epilepsy and mental decline. *J. Neurosci.* 24:2690–98

Kang J, Shen W, Macdonald RL. 2009. The GABRG2 Mutation, Q351X, Associated with generalized epilepsy with febrile seizures plus, has both loss of function and dominant-negative suppression. *J. Neurosci.* 29:2845–56

Khosravani H, Bladen C, Parker DB, Snutch TP, McRory JE, Zamponi GW. 2005. Effects of Cav3.2 channel mutations linked to idiopathic generalized epilepsy. *Ann. Neurol.* 57:745–49

Kiernan MC, Isbister GK, Lin CS, Burke D, Bostock H. 2005a. Acute tetrodotoxin-induced neurotoxicity after ingestion of puffer fish. *Ann. Neurol.* 57:339–48

Kiernan MC, Krishnan AV, Lin CS, Burke D, Berkovic SF. 2005b. Mutation in the Na^+ channel subunit SCN1B produces paradoxical changes in peripheral nerve excitability. *Brain* 128:1841–46

Klaassen A, Glykys J, Maguire J, Labarca C, Mody I, Boulter J. 2006. Seizures and enhanced cortical GABAergic inhibition in two mouse models of human autosomal dominant nocturnal frontal lobe epilepsy. *Proc. Natl. Acad. Sci. USA* 103:19152–57

Koch M, Steinmeyer K, Lorenz C, Ricker K, Wolf F, et al. 1992. The skeletal muscle chloride channel in dominant and recessive human myotonia. *Science* 257:797–800

Korff C, Laux L, Kelley K, Goldstein J, Koh S, Nordli D. 2007. Dravet syndrome (severe myoclonic epilepsy in infancy): a retrospective study of 16 patients. *J. Child Neurol.* 22:185–94

Kubisch C, Schmidt-Rose T, Fontaine B, Bretag AH, Jentsch TJ. 1998. ClC-1 chloride channel mutations in myotonia congenita: variable penetrance of mutations shifting the voltage dependence. *Hum. Mol. Genet.* 7:1753–60

Labrum RW, Rajakulendran S, Graves TD, Eunson LH, Bevan R, et al. 2009. Large-scale calcium channel gene rearrangements in episodic ataxia and hemiplegic migraine: implications for diagnostic testing. *J Med Genet.* 46(11):786–91

Lerche H, Heine R, Pika U, George AL, Mitrovic N, et al. 1993. Human sodium channel myotonia: slowed channel inactivation due to substitutions for a glycine within the III-IV linker. *J. Physiol. (Lond.)* 470:13–22

Lossin C, Wang DW, Rhodes TH, Vanoye CG, George AL. 2002. Molecular basis of an inherited epilepsy. *Neuron* 34:877–84

Maljevic S, Krampfl K, Cobilanschi J, Tilgen N, Beyer S, et al. 2006. A mutation in the GABAA receptor α1-subunit is associated with absence epilepsy. *Ann. Neurol.* 59:983–87

Maljevic S, Wuttke TV, Lerche H. 2008. Nervous system KV7 disorders: breakdown of a subthreshold brake. *J. Physiol.* 586:1791–801

Mankodi A, Takahashi MP, Jiang H, Beck CL, Bowers WJ, et al. 2002. Expanded CUG repeats trigger aberrant splicing of ClC-1 chloride channel premRNA and hyperexcitability of skeletal muscle in myotonic dystrophy. *Mol. Cell* 10:35–44

Matthews E, Labrum R, Sweeney MG, Sud R, Haworth A, et al. 2009. Voltage sensor charge loss accounts for most cases of hypokalemic periodic paralysis. *Neurology* 72:1544–47

Maylie B, Bissonnette E, Virk M, Adelman JP, Maylie JG. 2002. Episodic ataxia type 1 mutations in the human Kv1.1 potassium channel alter hKvbeta 1-induced N-type inactivation. *J. Neurosci.* 22:4786–93

McClatchey AI, Van den Bergh P, Pericak-Vance MA, Raskind W, Verellen C, et al. 1992. Temperature-sensitive mutations in the III-IV cytoplasmic loop region of the skeletal muscle sodium channel gene in paramyotonia congenita. *Cell* 68:769–74

Michalk A, Stricker S, Becker J, Rupps R, Pantzar T, et al. 2008. Acetylcholine receptor pathway mutations explain various fetal akinesia deformation sequence disorders. *Am. J. Hum. Genet.* 82:464–76

Miller TM, Dias da Silva MR, Miller HA, Kwiecinski H, Mendell JR, et al. 2004. Correlating phenotype and genotype in the periodic paralyses. *Neurology* 63:1647–55

Misra SN, Kahlig KM, George AL. 2008. Impaired NaV1.2 function and reduced cell surface expression in benign familial neonatal-infantile seizures. *Epilepsia* 49:1535–45

Nilsen KB, Nicholas AK, Woods CG, Mellgren SI, Nebuchennykh M, Aasly J. 2009. Two novel SCN9A mutations causing insensitivity to pain. *Pain* 143:155–58

Ogiwara I, Miyamoto H, Morita N, Atapour N, Mazaki E, et al. 2007. Nav1.1 localizes to axons of parvalbumin-positive inhibitory interneurons: a circuit basis for epileptic seizures in mice carrying an Scn1a gene mutation. *J. Neurosci.* 27:5903–14

Ophoff RA, Terwindt GM, Vergouwe MN, van Eijk R, Oefner PJ, et al. 1996. Familial hemiplegic migraine and episodic ataxia type-2 are caused by mutations in the Ca^{2+} channel gene CACNL1A4. *Cell* 87:543–52

Pickard BS, Knight HM, Hamilton RS, Soares DC, Walker R, et al. 2008. A common variant in the 3'UTR of the GRIK4 glutamate receptor gene affects transcript abundance and protects against bipolar disorder. *Proc. Natl. Acad. Sci. USA* 105:14940–45

Plaster NM, Tawil R, Tristani-Firouzi M, Canún S, Bendahhou S, et al. 2001. Mutations in Kir2.1 cause the developmental and episodic electrical phenotypes of Andersen's syndrome. *Cell* 105:511–19

Priori SG. 2005. Cardiac and skeletal muscle disorders caused by mutations in the intracellular Ca^{2+} release channels. *J. Clin. Investigation* 115:2033–38

Proks P, Antcliff JF, Lippiat J, Gloyn AL, Hattersley AT, Ashcroft FM. 2004. Molecular basis of Kir6.2 mutations associated with neonatal diabetes or neonatal diabetes plus neurological features. *Proc. Natl. Acad. Sci. USA* 101:17539–44

Ptácek LJ, George AL, Griggs RC, Tawil R, Kallen RG, et al. 1991. Identification of a mutation in the gene causing hyperkalemic periodic paralysis. *Cell* 67:1021–27

Ptácek LJ, Tawil R, Griggs RC, Engel AG, Layzer RB, et al. 1994. Dihydropyridine receptor mutations cause hypokalemic periodic paralysis. *Cell* 77:863–68

Rajakulendran S, Graves TD, Labrum RW, Kotzadimitriou D, Eunson L, et al. 2010. Genetic and functional characterisation of the P/Q calcium channel in episodic ataxia with epilepsy. *J. Physiol. (Lond.)* doi: 10.1113/jphysiol.2009.186437

Rees MI, Harvey K, Pearce BR, Chung S, Duguid IC, et al. 2006. Mutations in the gene encoding GlyT2 (SLC6A5) define a presynaptic component of human startle disease. *Nat. Genet.* 38:801–6

Rees MI, Lewis TM, Kwok JBJ, Mortier GR, Govaert P, et al. 2002. Hyperekplexia associated with compound heterozygote mutations in the beta-subunit of the human inhibitory glycine receptor (GLRB). *Hum. Mol. Genet.* 11:853–60

Rodrigues-Pinguet N, Jia L, Li M, Figl A, Klaassen A, et al. 2003. Five ADNFLE mutations reduce the Ca^{2+} dependence of the mammalian alpha4beta2 acetylcholine response. *J. Physiol. (Lond.)* 550:11–26

Rojas CV, Wang JZ, Schwartz LS, Hoffman EP, Powell BR, Brown RH. 1991. A Met-to-Val mutation in the skeletal muscle Na^+ channel alpha-subunit in hyperkalaemic periodic paralysis. *Nature* 354:387–89

Rush AM, Dib-Hajj SD, Liu S, Cummins TR, Black JA, Waxman SG. 2006. A single sodium channel mutation produces hyper- or hypoexcitability in different types of neurons. *Proc. Natl. Acad. Sci. USA* 103:8245–50

Ryan DP, daSilva MRD, Soong TW, Fontaine B, Donaldson MR, et al. 2010. Mutations in potassium channel Kir2.6 cause susceptibility to thyrotoxic hypokalemic periodic paralysis. *Cell* 140:88–98

Saint-Martin C, Gauvain G, Teodorescu G, Gourfinkel-An I, Fedirko E, et al. 2009. Two novel CLCN2 mutations accelerating chloride channel deactivation are associated with idiopathic generalized epilepsy. *Hum. Mutat.* 30:397–405

Scheffer IE, Berkovic SF. 1997. Generalized epilepsy with febrile seizures plus. A genetic disorder with heterogeneous clinical phenotypes. *Brain* 120(3):479–90

Schroeder BC, Kubisch C, Stein V, Jentsch TJ. 1998. Moderate loss of function of cyclic-AMP-modulated KCNQ2/KCNQ3 K$^+$ channels causes epilepsy. *Nature* 396:687–90

Semyanov A, Walker MC, Kullmann DM, Silver RA. 2004. Tonically active GABA$_A$ receptors: modulating gain and maintaining the tone. *Trends Neurosci.* 27:262–69

Shah MM, Mistry M, Marsh SJ, Brown DA, Delmas P. 2002. Molecular correlates of the M-current in cultured rat hippocampal neurons. *J. Physiol. (Lond.)* 544:29–37

Shah MM, Migliore M, Valencia I, Cooper EC, Brown DA. 2008. Functional significance of axonal Kv7 channels in hippocampal pyramidal neurons. *Proc. Natl. Acad. Sci. USA* 105:7869–74

Sharp AJ, Mefford HC, Li K, Baker C, Skinner C, et al. 2008. A recurrent 15q13.3 microdeletion syndrome associated with mental retardation and seizures. *Nat. Genet* 40:322–28

Shiang R, Ryan SG, Zhu YZ, Hahn AF, O'Connell P, Wasmuth JJ. 1993. Mutations in the alpha 1 subunit of the inhibitory glycine receptor cause the dominant neurologic disorder, hyperekplexia. *Nat. Genet.* 5:351–58

Singh NA, Charlier C, Stauffer D, DuPont BR, Leach RJ, et al. 1998. A novel potassium channel gene, KCNQ2, is mutated in an inherited epilepsy of newborns. *Nat. Genet.* 18:25–29

Singh NA, Otto JF, Dahle EJ, Pappas C, Leslie JD, et al. 2008 . Mouse models of human KCNQ2 and KCNQ3 mutations for benign familial neonatal convulsions show seizures and neuronal plasticity without synaptic reorganization. *J. Physiol. (Lond.)* 586:3405–23

Smart SL, Lopantsev V, Zhang CL, Robbins CA, Wang H, et al. 1998. Deletion of the K(V)1.1 potassium channel causes epilepsy in mice. *Neuron* 20:809–19

Sokolov S, Scheuer T, Catterall WA. 2007. Gating pore current in an inherited ion channelopathy. *Nature* 446:76–78

Spampanato J, Escayg A, Meisler MH, Goldin AL. 2001. Functional effects of two voltage-gated sodium channel mutations that cause generalized epilepsy with febrile seizures plus type 2. *J. Neurosci.* 21:7481–90

Stam AH, Luijckx GJ, Poll-The BT, Ginjaar I, Frants RR, et al. 2009. Early seizures and cerebral edema after trivial head trauma associated with the CACNA1A S218L mutation. *J. Neurol. Neurosurg. Psychiatr.* 80:1125–29

Stefansson H, Rujescu D, Cichon S, Pietiläinen OPH, Ingason A, et al. 2008. Large recurrent microdeletions associated with schizophrenia. *Nature* 455:232–36

Steinlein OK, Mulley JC, Propping P, Wallace RH, Phillips HA, et al. 1995. A missense mutation in the neuronal nicotinic acetylcholine receptor [alpha]4 subunit is associated with autosomal dominant nocturnal frontal lobe epilepsy. *Nat. Genet.* 11:201–203

Struyk AF, Cannon SC. 2007. A Na$^+$ channel mutation linked to hypokalemic periodic paralysis exposes a proton-selective gating pore. *J. Gen. Physiol.* 130:11–20

Suls A, Claeys KG, Goossens D, Harding B, Van Luijk R, et al. 2006. Microdeletions involving the SCN1A gene may be common in SCN1A-mutation-negative SMEI patients. *Hum. Mutat.* 27:914–20

Suzuki T, Delgado-Escueta AV, Aguan K, Alonso ME, Shi J, et al. 2004. Mutations in EFHC1 cause juvenile myoclonic epilepsy. *Nat. Genet.* 36:842–49

Tan HO, Reid CA, Single FN, Davies PJ, Chiu C, et al. 2007. Reduced cortical inhibition in a mouse model of familial childhood absence epilepsy. *Proc. Natl. Acad. Sci. USA* 104:17536–41

Tanaka M, Olsen RW, Medina MT, Schwartz E, Alonso ME, et al. 2008. Hyperglycosylation and reduced GABA currents of mutated GABRB3 polypeptide in remitting childhood absence epilepsy. *Am. J. Hum. Genet.* 82:1249–61

Tate SK, Depondt C, Sisodiya SM, Cavalleri GL, Schorge S, et al. 2005. Genetic predictors of the maximum doses patients receive during clinical use of the antiepileptic drugs carbamazepine and phenytoin. *Proc. Natl. Acad. Sci. USA* 102:5507–12

Tinel N, Lauritzen I, Chouabe C, Lazdunski M, Borsotto M. 1998. The KCNQ2 potassium channel: splice variants, functional and developmental expression. Brain localization and comparison with KCNQ3. *FEBS Lett.* 438:171–76

Tottene A, Conti R, Fabbro A, Vecchia D, Shapovalova M, et al. 2009. Enhanced excitatory transmission at cortical synapses as the basis for facilitated spreading depression in Ca(v)2.1 knockin migraine mice. *Neuron* 61:762–73

Tottene A, Fellin T, Pagnutti S, Luvisetto S, Striessnig J, et al. 2002. Familial hemiplegic migraine mutations increase Ca^{2+} influx through single human CaV2.1 channels and decrease maximal CaV2.1 current density in neurons. *Proc. Natl. Acad. Sci. USA* 99:13284–89

Tristani-Firouzi M, Jensen JL, Donaldson MR, Sansone V, Meola G, et al. 2002. Functional and clinical characterization of KCNJ2 mutations associated with LQT7 (Andersen syndrome). *J. Clin. Invest.* 110:381–88

van den Maagdenberg AMJM, Pietrobon D, Pizzorusso T, Kaja S, Broos LAM, et al. 2004. A Cacna1a knockin migraine mouse model with increased susceptibility to cortical spreading depression. *Neuron* 41:701–710

Vitko I, Chen Y, Arias JM, Shen Y, Wu X, Perez-Reyes E. 2005. Functional characterization and neuronal modeling of the effects of childhood absence epilepsy variants of CACNA1H, a T-type calcium channel. *J. Neurosci.* 25:4844–55

Wagstaff J, Knoll JH, Fleming J, Kirkness EF, Martin-Gallardo A, et al. 1991. Localization of the gene encoding the GABAA receptor beta 3 subunit to the Angelman/Prader-Willi region of human chromosome 15. *Am. J. Hum. Genet.* 49:330–37

Wallace RH, Marini C, Petrou S, Harkin LA, Bowser DN, et al. 2001. Mutant GABA$_A$ receptor gamma2-subunit in childhood absence epilepsy and febrile seizures. *Nat. Genet.* 28:49–52

Wallace RH, Wang DW, Singh R, Scheffer IE, George AL, et al. 1998. Febrile seizures and generalized epilepsy associated with a mutation in the Na$^+$-channel beta1 subunit gene SCN1B. *Nat. Genet.* 19:366–70

Wang HS, Pan Z, Shi W, Brown BS, Wymore RS, et al. 1998. KCNQ2 and KCNQ3 potassium channel subunits: molecular correlates of the M-channel. *Science* 282:1890–93

Waters MF, Minassian NA, Stevanin G, Figueroa KP, Bannister JPA, et al. 2006. Mutations in voltage-gated potassium channel KCNC3 cause degenerative and developmental central nervous system phenotypes. *Nat. Genet.* 38:447–51

Wessman M, Terwindt GM, Kaunisto MA, Palotie A, Ophoff RA. 2007. Migraine: a complex genetic disorder. *Lancet Neurol.* 6:521–32

Wuttke TV, Jurkat-Rott K, Paulus W, Garncarek M, Lehmann-Horn F, Lerche H. 2007. Peripheral nerve hyperexcitability due to dominant-negative KCNQ2 mutations. *Neurology* 69:2045–53

Yang Y, Wang Y, Li S, Xu Z, Li H, et al. 2004. Mutations in SCN9A, encoding a sodium channel alpha subunit, in patients with primary erythermalgia. *J. Med. Genet.* 41:171–74

Yu FH, Mantegazza M, Westenbroek RE, Robbins CA, Kalume F, et al. 2006. Reduced sodium current in GABAergic interneurons in a mouse model of severe myoclonic epilepsy in infancy. *Nat. Neurosci.* 9:1142–49

Zhuchenko O, Bailey J, Bonnen P, Ashizawa T, Stockton DW, et al. 1997. Autosomal dominant cerebellar ataxia (SCA6) associated with small polyglutamine expansions in the alpha 1A-voltage-dependent calcium channel. *Nat. Genet.* 15:62–69

Zuberi SM, Eunson LH, Spauschus A, De Silva R, Tolmie J, et al. 1999. A novel mutation in the human voltage-gated potassium channel gene (Kv1.1) associates with episodic ataxia type 1 and sometimes with partial epilepsy. *Brain* 122(5):817–25

Emotion, Cognition, and Mental State Representation in Amygdala and Prefrontal Cortex

C. Daniel Salzman[1,2,3,4,5,6] and Stefano Fusi[1]

[1]Department of Neuroscience, [2]Department of Psychiatry, [3]W.M. Keck Center on Brain Plasticity and Cognition, [4]Kavli Institute for Brain Sciences, [5]Mahoney Center for Brain and Behavior, Columbia University, New York, NY 10032; email: cds2005@columbia.edu, sf2237@columbia.edu, [6]New York State Psychiatric Institute, New York, NY 10032

Annu. Rev. Neurosci. 2010. 33:173–202

First published online as a Review in Advance on March 23, 2010

The *Annual Review of Neuroscience* is online at neuro.annualreviews.org

This article's doi:
10.1146/annurev.neuro.051508.135256

Key Words

neurophysiology, orbitofrontal cortex, value, reward, aversive, reinforcement learning

Abstract

Neuroscientists have often described cognition and emotion as separable processes implemented by different regions of the brain, such as the amygdala for emotion and the prefrontal cortex for cognition. In this framework, functional interactions between the amygdala and prefrontal cortex mediate emotional influences on cognitive processes such as decision-making, as well as the cognitive regulation of emotion. However, neurons in these structures often have entangled representations, whereby single neurons encode multiple cognitive and emotional variables. Here we review studies using anatomical, lesion, and neurophysiological approaches to investigate the representation and utilization of cognitive and emotional parameters. We propose that these mental state parameters are inextricably linked and represented in dynamic neural networks composed of interconnected prefrontal and limbic brain structures. Future theoretical and experimental work is required to understand how these mental state representations form and how shifts between mental states occur, a critical feature of adaptive cognitive and emotional behavior.

Contents

INTRODUCTION

The past century has witnessed a debate concerning the nature of emotion. When the brain is confronted with a stimulus that evokes emotion, does it first respond by activating a range of visceral and behavioral responses, which are only then followed by the conscious experience of emotion? For example, when we encounter a threatening snake, does autonomic reactivity, as well as behaviors such as freezing or fleeing, emerge prior to the feeling of fear? This view, championed by the psychologists William James and Carle Lange around the turn of the twentieth century (James 1884, 1894; Lange 1922), has attracted renewed interest because of the influential work of Damasio and colleagues (Damasio 1994). Alternatively, do visceral and behavioral responses occur as a result of central processing in the brain—processing that gives rise to emotional feelings—which then regulates or controls a variety of bodily responses [a possibility raised decades ago by Walter Cannon (1927) and Philip Bard (1928)]?

Neuroscientists have often sidestepped this debate by operationally defining a particular aspect of emotion—e.g., learning about fear—and using a specific behavioral or physiological assay—e.g., freezing—to investigate the neural basis of the process (Salzman et al. 2005). This approach is agnostic about which response comes first: the visceral and behavioral expression of emotion or the feeling of emotion. But it has proven powerful in helping to identify and characterize the neural circuitry responsible for specific aspects of emotional expression and regulation. These investigations have shown that one brain area, the amygdala, plays a vital role in many emotional processes (Baxter & Murray 2002, Lang & Davis 2006, LeDoux 2000, Phelps & LeDoux 2005) and that the amygdala and its interconnections with the prefrontal cortex (PFC) likely underlie many aspects of the interactions between emotion and cognition (Barbas & Zikopoulos 2007, Murray & Izquierdo 2007, Pessoa 2008, Price 2007).

Today, we still lack a resolution to the original debate concerning the relationship between emotional feelings and the bodily expression of emotions, in large part because both viewpoints appear to be supported in some circumstances. Emotional feelings do not necessarily involve visceral and behavioral components and vice versa (Lang 1994). But neurobiological advances—in particular, emerging data on the intimate relationship between the PFC and

PFC: prefrontal cortex

limbic areas such as the amygdala—begin to suggest a solution. As discussed below, the amygdala is essential for many of the visceral and behavioral expressions of emotion; meanwhile, the PFC—especially its medial and orbital regions—appears to be responsible for many of the cognitive aspects of emotional responses. However, recent studies suggest that both the functional and the electrophysiological characteristics of the amygdala and the PFC overlap and intimately depend on each other. Thus, the neural circuits mediating cognitive, emotional, physiological, and behavioral responses may not truly be separable and instead are inextricably linked. Moreover, we lack a unifying conceptual framework for understanding how the brain links these processes and how these processes change in unison.

MENTAL STATES: SYNTHESIZING COGNITION AND EMOTION

Here, we propose a theoretical foundation for understanding emotion in the context of its intimate relation to the cognitive, physiological, and behavioral responses that constitute emotional expression. We review recent neurobiological data concerning the amygdala and the PFC and discuss how these data fit into a proposed framework for understanding interactions between emotion and cognition.

The concept of a mental state plays a central role in our theoretical framework. We define a mental state as a disposition to action—i.e., every aspect of an organism's inner state that could contribute to its behavior or other responses—which may comprise all the thoughts, feelings, beliefs, intentions, active memories, and perceptions, etc., that are present at a given moment. Thus mental states can be described by a large number of variables, and the set of all mental state variables could provide a quantitative description of one's disposition to behavior. Of note, the identification of mental state variables is constrained by the language we use to describe them. Conse-

quently, mental state variables are not necessarily unique, and they are not necessarily independent from each other. Mental state variables need not be conscious or unconscious because both types of variables can predispose one to action. Overall, an organism's mental state incorporates internal variables, such as hunger or fear, as well as the representation of a set of environmental stimuli present at a given moment, and the temporal context of stimuli and events. Any given mental state predisposes an organism to respond in certain ways; these actions may be cognitive (e.g., making a decision), behavioral (e.g., freezing or fleeing), or physiological (e.g., increasing heart rate). Mental state variables are useful theoretical constructs because they provide quantitative metrics for analyzing and understanding behavioral and brain processes.

The concept of a mental state is intimately related to, but distinct from, what we call a brain state. Each mental state corresponds to one or more states of the dynamic variables—firing rates, synaptic weights, etc.—that describe the neural circuits of the brain; the full set of values of these variables constitutes a brain state. How are the variables characterizing a mental state represented at the neural circuit level—i.e., the current brain state? This is one way to phrase a fundamental and long-standing question for neuroscientists. At one end of the spectrum is the possibility that each neuron encodes only one variable. For example, a neuron may respond only to the pleasantness of a sensory stimulus, and not to its identity, to its meaning, or to the context in which the stimulus appears. When neurons encode only one variable, other neurons may easily read out the information represented, and the representation can, in principle, be modified without affecting other mental state variables.

One of the disadvantages of the type of representation described immediately above is well illustrated by what is known as the "binding problem" (Malsburg 1999). If each neuron represents only one mental state variable, then it is difficult to construct representations of complex situations. For example, consider a scene with two visual stimuli, one associated with reward and the other with punishment. The brain state should contain the information that pleasantness is associated with the first stimulus and not with the other. If neurons represent only one mental state variable at a time, like stimulus identity, or stimulus valence, then "binding" the information about different variables becomes a substantial challenge. In this case, there must be an additional mechanism that links the activation of the neuron representing pleasantness to the activation of the neuron representing the first stimulus. One simple and efficient way to solve this problem is to introduce neurons with mixed selectivity to conjunctions of events, such as a neuron that responds only when the first stimulus is pleasant. In this scheme, the representations of pleasantness and stimulus identity would be entangled and more difficult to decode, but the number of situations that could be represented would be significantly larger. As discussed below, different brain areas may contain representations with different degrees of entanglement.

How do emotions fit into the conceptual framework of mental states arising from brain states? One influential schema for characterizing emotion posits that emotions can vary along two axes: valence (pleasant versus unpleasant or positive versus negative) and intensity (or arousal) (Lang et al. 1990, Russell 1980). These two variables can simply be conceived as components of the current mental state. Two mental states correspond to different emotions when at least one of the two mental state variables—valence or intensity—is significantly different. Thus, variables describing emotions have the same ontological status as do variables that describe cognitive processes such as memory, attention, decision-making, language, and rule-based problem-solving. Below, we describe neurophysiological data documenting that variables such as valence and arousal are strongly encoded in the amygdala–prefrontal circuit, along with variables related to other cognitive processes. We suggest that neural representations in the amygdala may be more biased toward encoding mental state variables characterizing emotions (valence and intensity). PFC neurons may encode a broader range of variables in an entangled fashion, reflecting the complexity of the behavior and cognition that are its putative outputs.

The concept of a mental state unites cognition and emotion as part of a common framework. How does this framework contribute to the debate about the relationship between emotions and bodily responses? We argue that the issues raised in the debate essentially dissolve when one conceptualizes emotions as part of mental states: Neither emotional feelings nor bodily responses necessarily come first or second. Rather, both of these aspects of emotion are outputs of the neural networks that represent mental states. Furthermore, all the thoughts, physiological responses, and behaviors that constitute emotion are part of an ongoing feedback loop that alters the dynamic, ever-fluctuating brain state and generates new mental states from moment to moment.

How do mental states that integrate emotion and cognition arise from the activity of neural circuits? Below, we describe a potential anatomical substrate—the amygdala–prefrontal circuit—for emotional-cognitive interactions in the brain and how neurons in these areas could dynamically contribute to a subject's mental state. First, we review the bidirectional connections between the amygdala and the PFC that could form the basis of many interactions between cognition and emotion. Second, we review neurobiological studies that used lesions and pharmacological inactivation to investigate the function of the amygdala–PFC circuitry. Third, we review neurophysiological data from the amygdala and the PFC that reveal encoding of variables critical for the representation of mental states and for the learning algorithms—specifically, reinforcement learning (RL)—that emphasize the importance of encoding these parameters for adaptive behavior. For all these topics, we focus on data collected from nonhuman primates. Compared with their rodent counterparts, nonhuman primates are much more similar to humans in terms of both behavioral repertoire and anatomical development.

Finally, we describe a theoretical proposal that explains how mental states might emerge in the brain and how the amygdala and the PFC could play an integral role in this process. In particular, we propose that the interactions between emotion and cognition may be understood in the context of mental states that can be switched or gated by internal or external events.

AN ANATOMICAL SUBSTRATE FOR INTERACTIONS BETWEEN EMOTION AND COGNITION

This review focuses on interactions between the amygdala and the PFC because of their long-established roles in mediating emotional and cognitive processes (Holland & Gallagher 2004, Lang & Davis 2006, LeDoux 2000, Miller & Cohen 2001, Ochsner & Gross 2005, Wallis 2007). The amygdala is most often discussed in the context of emotional processes; yet it is extensively interconnected with the PFC, especially the posterior orbitofrontal cortex (OFC), and the anterior cingulate cortex (ACC). Here we provide a brief overview of amygdala and PFC anatomy, with an emphasis on the potential anatomical basis of interactions between cognitive and emotional processes.

Amygdala

The amygdala is a structurally and functionally heterogeneous collection of nuclei lying in the anterior medial portion of each temporal lobe. Sensory information enters the amygdala from advanced levels of visual, auditory, and somatosensory cortices, from the olfactory system, and from polysensory brain areas such as the perirhinal cortex and the parahippocampal gyrus (Amaral et al. 1992, McDonald 1998, Stefanacci & Amaral 2002). Within the lateral nucleus, the primary target of projections from unimodal sensory cortices, different sensory modalities are segregated anatomically. But, owing in part to intrinsic connections, multimodal encoding subsequently emerges in the lateral, basal, accessory basal, and other nuclei of the amygdala (Pitkanen & Amaral 1998,

Stefanacci & Amaral 2000). Output from the amygdala is directed to a wide range of target structures, including the PFC, the striatum, sensory cortices (including primary sensory cortices, connections which are probably unique to primates), the hippocampus, the perirhinal cortex, the entorhinal cortex, and the basal forebrain, and to subcortical structures responsible for aspects of physiological responses related to emotion, such as autonomic responses, hormonal responses, and startle (Davis 2000). In general, subcortical projections originate from the central nucleus, and projections to cortex and the striatum originate from the basal, accessory basal, and in some cases the lateral nuclei (Amaral et al. 1992, 2003; Amaral & Dent 1981; Carmichael & Price 1995a; Freese & Amaral 2005; Ghashghaei et al. 2007; Stefanacci et al. 1996; Stefanacci & Amaral 2002; Suzuki & Amaral 1994).

Prefrontal Cortex

The PFC, located in the anterior portion of the cerebral cortex and defined by projections from the mediodorsal nucleus of the thalamus (Fuster 2008), is composed of a group of interconnected brain areas. The distinctive feature of primate PFC is the emergence of dysgranular and granular cortices, which are completely absent in the rodent. In rodents, prefrontal cortex is entirely agranular (Murray 2008, Preuss 1995, Price 2007, Wise 2008). Therefore, much of the primate PFC does not have a clear-cut homolog in rodents. The PFC is often grouped into different subregions; Petrides & Pandya (1994) have described these as dorsal and lateral areas (Walker areas 9, 46, and 9/46), ventrolateral areas (47/12 and 45), medial areas (32 and 24), and orbitofrontal areas (10, 11, 13, 14, and 47/12). Of note, there are extensive interconnections between different PFC areas, allowing information to be shared within local networks (Barbas & Pandya 1989, Carmichael & Price 1996, Cavada et al. 2000), and information also converges from sensory cortices in multiple modalities (Barbas et al. 2002). In general, dorsolateral areas receive input from earlier

ACC: anterior cingulate cortex

OFC: orbitofrontal cortex

sensory areas (Barbas et al. 2002). Orbitofrontal areas receive inputs from advanced stages of sensory processing from every modality, including gustatory and olfactory (Carmichael & Price 1995b, Cavada et al. 2000, Romanski et al. 1999). Thus, extrinsic and intrinsic connections make the PFC a site of multimodal convergence of information about the external environment.

In addition, the PFC receives inputs that could inform it about internal mental state variables, such as motivation and emotions. Orbital and medial PFC are closely connected with limbic structures such as the amygdala (see below) and also have direct and indirect connections with the hippocampus and rhinal cortices (Barbas & Blatt 1995, Carmichael & Price 1995a, Cavada et al. 2000, Kondo et al. 2005, Morecraft et al. 1992). Medial and part of orbital PFC has connections to the hypothalamus and other subcortical targets that could mediate autonomic responses (Ongur et al. 1998). Neuromodulatory input to the PFC from dopaminergic, serotonergic, noradrenergic, and cholinergic systems could also convey information about internal state (Robbins & Arnsten 2009). Finally, outputs from the PFC, especially from dorsolateral PFC, are directed to motor systems, consistent with the notion that the PFC may form, represent, and/or transmit motor plans (Bates & Goldman-Rakic 1993, Lu et al. 1994). Altogether, the PFC receives inputs that provide information about many external and internal variables, including those related to emotions and to cognitive plans, providing a potential anatomical substrate for the representation of mental states.

Anatomical Interactions Between the PFC and Amygdala

Although there are diffuse bidirectional projections between amygdala and much of the PFC [see, e.g., figure 4 of Ghashghaei et al. (2007)], the densest interconnections are between the amygdala and orbital areas (e.g., caudal area 13) and medial areas (e.g., areas 24 and 25). The extensive anatomical connections among the amygdala, the PFC, and related structures are summarized in **Figure 1**. Amygdala input

to the PFC often terminates in both superficial and deep layers. OFC output to the amygdala originates in deep layers, and in some cases also in superficial layers, suggesting both feedforward and feedback modes of information transmission (Ghashghaei et al. 2007).

Previous work has established that the OFC output to the amygdala is complex and segregated, targeting multiple systems in the amygdala (Ghashghaei & Barbas 2002). Some OFC output is directed to the intercalated masses, a ribbon of inhibitory neurons in the amygdala that inhibits activity in the central nucleus (Ghashghaei et al. 2007, Pare et al. 2003). In addition, the OFC projects directly to the central nucleus, providing a means by which the OFC can activate this output structure in addition to inhibiting it (Ghashghaei & Barbas 2002, Stefanacci & Amaral 2000, 2002). Finally, the OFC projects to the basal, accessory basal, and lateral nuclei, where it may influence computations occurring within the amygdala (Ghashghaei & Barbas 2002, Stefanacci & Amaral 2000, 2002). Overall, the bidirectional communication between the amygdala and the OFC, as well as the connections with the rest of the PFC, provides a potential basis for the integration of cognitive, emotional, and physiological processes into a unified representation of mental states.

THE ROLE OF THE AMYGDALA AND THE PFC IN REPRESENTING MENTAL STATES: LESION STUDIES

Recent studies using lesions or pharmacological inactivation combined with behavioral studies in monkeys have begun to reveal the specific roles of the primate amygdala and various regions of the PFC in cognitive and emotional processes. We focus here on studies that have helped demonstrate the roles of these brain structures in processes such as valuation, rule-based actions, emotional processes, attention, goal-directed behavior, and working memory—processes that are likely to set some of the variables that constitute a subject's mental state.

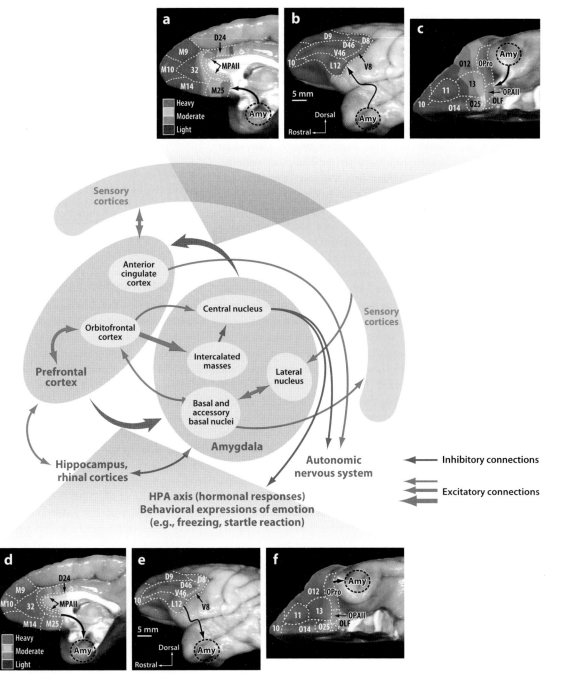

Figure 1

Overview of anatomical connections of the amygdala and the prefrontal cortex (PFC). Schematic showing some (but not all) the main projections of the amygdala and the PFC. The interconnections of the amygdala and the PFC (and especially the OFC) are emphasized. (*a–c*) Summary of projections from the amygdala to the PFC (density of projections is color coded). (*d–f*) Summary of projections from the PFC to the amygdala (projection density is color coded). The complex circuitry between the amygdala and the OFC is also highlighted (*red arrows connect the structures*). Medial amygdala nuclei not shown. Many additional connections of both amygdala and PFC are not shown. Panels *a–f* were adapted with permission from figures 5 and 6 of Ghashghaei et al. (2007).

Amygdala

Historically, lesions of primate amygdala produced a wide range of behavioral and emotional effects (Aggleton & Passingham 1981, Jones & Mishkin 1972, Kluver & Bucy 1939, Spiegler & Mishkin 1981, Weiskrantz 1956); but in recent years, scientists have increasingly recognized the importance of using anatomically precise lesions that spare fibers of passage. Many older studies had employed aspiration or radiofrequency lesions, which destroy both gray and white matter. By contrast, recent studies using excitotoxic chemical injections, which specifically kill cell bodies, have revised our understanding of cognitive and emotional functions that require the amygdala (Baxter & Murray 2000, Izquierdo & Murray 2007). Some conclusions, however, have been confirmed over many studies using both old and new techniques. In particular, scientists have most prominently used two types of behavioral tasks to establish the amygdala's role in forming or updating associations between sensory stimuli and reinforcement. First, consistent with findings from rodents, the primate amygdala is required for fear learning induced by Pavlovian conditioning (Antoniadis et al. 2009). Second, the amygdala is required for updating the value of a rewarding reinforcer during a devaluation procedure (Machado & Bachevalier 2007, Malkova et al. 1997, Murray & Izquierdo 2007). In this type of task, experimenters satiate an animal on a particular type of reward and test whether satiation changes subsequent choice behavior such that the animal chooses the satiated food type less often; amygdala lesions eliminate the effect of satiation. Pharmacological inactivation of the amygdala has confirmed the amygdala's role in updating a representation of a reinforcer's value; however, once this updating process finishes, the amygdala does not appear to be required (Wellman et al. 2005). In addition, the amygdala is important for other aspects of appetitive conditioned reinforcement (Parkinson et al. 2001) and for behavioral and physiological responding to emotional stimuli such as snakes and intruders in a manner consistent with its playing a role in processing both emotional valence and intensity (Izquierdo et al. 2005, Kalin et al. 2004, Machado et al. 2009). Finally, experiments using ibotenic acid instead of aspiration lesions in the amygdala have led to revisions in our understanding of the amygdala for reversal-learning task performance (during which stimulus-reinforcement contingencies are reversed). Recent evidence indicates that the amygdala is not required for reversal learning on tasks involving only rewards, unlike previous accounts (Izquierdo & Murray 2007). Overall, these data link the amygdala to functions that rely on neural processing related to both emotional valence and intensity.

Prefrontal Cortex

A long history of studies have used lesions to establish the importance of the PFC in goal-directed behavior, rule-guided behavior, and executive functioning more generally (Fuster 2008, Miller & Cohen 2001, Wallis 2007). These complex cognitive processes form an integral part of our mental state. In addition, lesions of orbitofrontal cortex (OFC) cause many emotional and cognitive deficits reminiscent of amygdala lesions, including deficits in reinforcer devaluation and in behavioral and hormonal responses to emotional stimuli (Izquierdo et al. 2004, 2005; Kalin et al. 2007; Machado et al. 2009; Murray & Izquierdo 2007). Recently, investigators have employed detailed trial-by-trial data analysis to enhance the understanding of the effects of lesions; this work led investigators to propose that ACC and OFC are more involved in the valuation of actions and stimuli, respectively (Kennerley et al. 2006, Rudebeck et al. 2008, Rushworth & Behrens 2008).

In addition, a recent study separately examined lesions of the dorsolateral PFC, the ventrolateral PFC, the principal sulcus (PS), the ACC, and the OFC on a task analogous to the Wisconsin Card Sorting Test (Buckley et al. 2009) used to assay PFC function in humans (Stuss et al. 2000). In the authors' version of the task, monkeys must discover by trial and error the current rule that is in effect; subjects needed to employ working memory for the rule, as well as to utilize information about recent reward history to guide behavior. Lesions in different PFC regions caused distinct deficit profiles: Deficits in working memory, reward-based updating of value representations, and active utilization of recent choice-outcome values were ascribed primarily to PS, OFC, and ACC lesions, respectively (Buckley et al. 2009). Of note, this study used aspiration lesions of the targeted brain regions, which almost certainly damaged fibers of passage located nearby.

A classic finding following aspiration lesions of the OFC is a deficit in learning about reversals of stimulus-reward contingencies (Jones & Mishkin 1972). However, a recent study used ibotenic acid to place a discrete lesion in OFC areas 11 and 13 (Kazama & Bachevalier 2009) and failed to find a deficit in reversal learning. We therefore may need to revise our understanding of how OFC contributes to reversal learning (similar to revisions made with reference to amygdala function, see above); however, the lack of an effect in the recent study may have been due to the anatomically restricted nature of the lesion. This issue will require further investigation.

Prefrontal-Amygdala Interactions

The amygdala is reciprocally connected with the PFC, primarily OFC and ACC, but also diffusely to other parts of the PFC (**Figure 1**). Studies have begun to examine possible functional interactions between the amygdala and the OFC in mediating different aspects of reinforcement-based and emotional behavior. In one powerful set of experiments, Baxter and colleagues (2000) performed a crossed surgical disconnection of the amygdala and the OFC by lesioning amygdala on one side of the brain and the OFC in the other hemisphere [connections between the amygdala and the OFC are ipsilateral (Ghashghaei & Barbas 2002)]. As noted above, bilateral lesions of monkey amygdala or the OFC impair reinforcer devaluation;

consistent with this finding, the authors found that surgical disconnection also impaired reinforcer devaluation, indicating that the amygdala and the OFC must interact to update the value of a reinforcer. Notably, in humans, neuroimaging studies on rare patients with focal amygdala lesions have revealed that the BOLD signal related to reward expectation in the ventromedial PFC is dependent on a functioning amygdala (Hampton et al. 2007). Investigators have also described functional interactions between the amygdala and the OFC in rodents (Saddoris et al. 2005, Schoenbaum et al. 2003); however, as noted above, rodent OFC may not necessarily correspond to any part of the primate granular/dysgranular PFC (Murray 2008, Preuss 1995, Wise 2008).

The lesion studies described above support the notion that the PFC and the amygdala, often in concert with each other, participate in executive functions such as attention, rule representation, working memory, planning, and valuation of stimuli and actions. In addition, these structures mediate aspects of emotional processing, including processing related to emotional valence and intensity. Together these variables form an integral part of what we have termed a mental state. However, one must exercise some caution when interpreting the results of lesion studies: Owing to potential redundancy in neural coding among brain circuits, a negative result does not necessarily imply that the lesioned area is not normally involved in the function in question. As discussed in the next section, neurons in many parts of the PFC have complex, entangled physiological properties. Given redundancy in encoding, it is therefore not surprising that lesions in these parts of the PFC often do not impair functioning related to the full range of response properties.

NEUROPHYSIOLOGICAL COMPONENTS OF MENTAL STATES

We have defined mental states as action dispositions, where actions are broadly defined to

CS: conditioned
stimulus

US: unconditioned
stimulus

include cognitive, physiological, or behavioral responses. Here, we focus on neural signals in the PFC and the amygdala that may encode key cognitive and emotional features of a mental state: the valuation of stimuli, the valence and intensity of emotional reactions to stimuli, our knowledge of the context of sensory stimuli and the requisite rules in that context, and our plans for interacting with stimuli in the environment. We review recent neurophysiological recordings from behaving nonhuman primates that demonstrate coding of all these variables, and they often feature entangled encoding of multiple variables.

Neural Representations of Emotional Valence and Arousal in the Amygdala and the OFC

In recent years, a number of physiological experiments have been directed at understanding the coding properties of neurons in the amygdala and the OFC. The amygdala has long been investigated with respect to aversive processing and its prominent role in fear conditioning, primarily in rodents (Davis 2000, LeDoux 2000, Maren 2005). However, a number of scientists have recognized that the amygdala also plays a role in appetitive processing (Baxter & Murray 2002). Early neurophysiological experiments in monkeys established the amygdala as a potential locus for encoding the affective properties of stimuli (Fuster & Uyeda 1971; Nishijo et al. 1988a, b; Sanghera et al. 1979; Sugase-Miyamoto & Richmond 2005).

To determine whether neurons in the primate amygdala preferentially encoded rewarding or aversive associations, Paton and colleagues (2006) recorded single neuron activity while monkeys learned that visual stimuli—novel abstract fractal images—predicted liquid rewards or aversive air puffs directed at the face, respectively. The experiments employed a Pavlovian procedure called trace conditioning, in which there is a brief temporal gap (the trace interval) between the presentation of a conditioned stimulus (CS) and an unconditioned stimulus (US)

(**Figure 2a**). Monkeys exhibited two behaviors that demonstrated their learning of the stimulus-outcome contingencies: anticipatory licking (an approach behavior) and anticipatory blinking (a defensive behavior). After monkeys learned the initial CS-US associations, reinforcement contingencies were reversed. Neurophysiological recordings revealed that the amygdala contained some neurons that respond more strongly when a CS is paired with a reward (positive value-coding neurons), and other neurons respond more strongly when the same CS is paired with an aversive stimulus (negative value-coding neurons). Although individual neurons exhibited this differential response during different time intervals (e.g., during the CS interval or parts of the trace interval), across the population of neurons, the value-related signal was temporally extended across the entire trial (**Figure 2b,c**). Positive and negative value-coding neurons appeared to be intermingled in the amygdala; both types of neurons dispersed within (and perhaps beyond) the basolateral complex (Belova et al. 2008, Paton et al. 2006).

Theoretical accounts of reinforcement learning often posit a neural representation of the value of the current situation as a whole (state value). Data from Belova et al. (2008) suggest that the amygdala could encode the value of the state instantiated by the CS presentation. Neural responses to the fixation point, which appeared at the beginning of trials, were consistent with a role of the amygdala in encoding state value. One can argue that the fixation point is a mildly positive stimulus because monkeys choose to look at it to initiate trials; and indeed, positive value-coding neurons tend to increase their firing in response to fixation point presentation, and negative value-coding neurons tend to decrease firing (**Figure 3**). Neural signaling after reward or air-puff presentation also indicates that amygdala neurons track state value, as differential levels of activity, on a population level, extending well beyond the termination of USs (**Figure 2b,c**). All these signals related to reinforcement contingencies could be used to coordinate physiological and behavioral

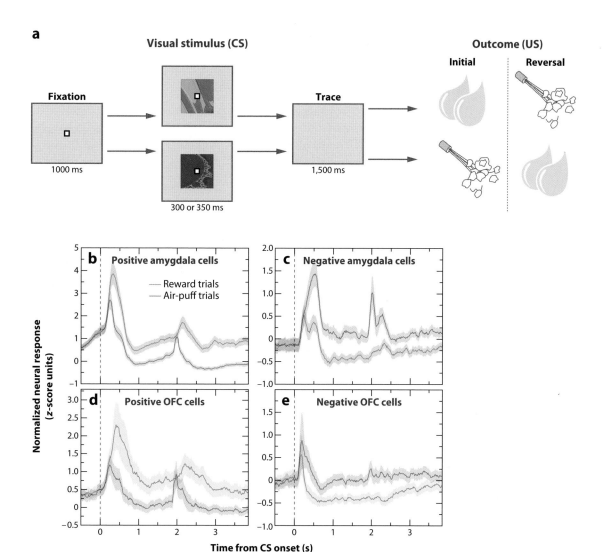

Figure 2

Neural representation of positive and negative valence in the amygdala and the OFC. (*a*) Trace-conditioning task involving both appetitive and aversive conditioning. Monkeys first centered gaze at a fixation point. Each experiment used novel abstract images as conditioned stimuli (CS). After fixating for 1 s, monkeys viewed a CS briefly, and following a 1.5-ms trace interval, unconditioned stimulus (US) delivery occurred. One CS predicted liquid reward, and a second CS predicted an aversive air puff directed at the face. After monkeys learned these initial associations, as indicated by anticipatory licking and blinking, the reinforcement contingencies were reversed. A third CS appeared on one-third of the trials, and it predicted either nothing or a much smaller reward throughout the experiment (not depicted in the figure). (*b–e*) Normalized and averaged population peri-stimulus time histograms (PSTHs) for positive and negative encoding amygdala (*b,c*) and OFC (*d,e*) neurons.

responses specific to appetitive and aversive systems; therefore, they form a potential neural substrate for positive and negative emotional variables.

As discussed earlier, however, valence is only one dimension of emotion; a second dimension is emotional intensity, or arousal. Recent data also link the amygdala to this second

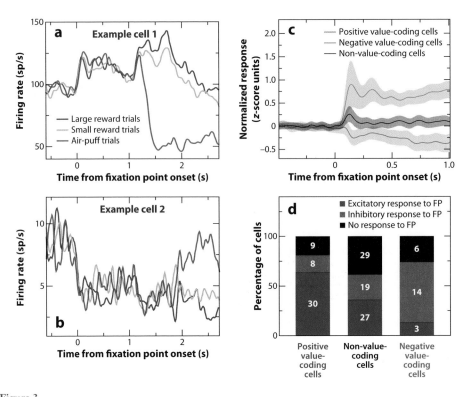

Figure 3

Amygdala neurons track state value during the fixation interval. (*a,b*) PSTHs aligned on fixation point onset from two example amygdala neurons (*a*, positive encoding; *b*, negative encoding) revealing responses to the fixation point consistent with their encoding state value. (*c*) Averaged and normalized responses to the fixation point for positive, negative, and nonvalue-coding amygdala neurons. (*d*) Histograms showing the number of cells that increased, decreased, or did not change their firing rates as a function of which valence the neuron encoded. Note that the fixation point may be understood as a mildly positive stimulus, so positive neurons tend to increase their response to it and negative neurons decrease their response. Adapted with permission from Belova et al. (2008, figure 2).

dimension. Belova and colleagues (2007) measured responses to rewards and aversive air puffs when they were either expected or unexpected. Surprising reinforcement is generally experienced as more arousing than when the same reinforcements occur predictably; consistent with this notion, expectation often modulated responses to reinforcement in the amygdala—in general, neural responses were enhanced when reinforcement was surprising. For some neurons, this modulation occurred only for rewards or for air puffs, but not for both (**Figure 4*a–d***). These neurons therefore could participate in valence-specific emotional and cognitive processes. However, many neurons modulated their responses to both rewards and air puffs (**Figure 4*e,f***). These neurons could underlie processes such as arousal or enhanced attention, which occur in response to intense emotional stimuli of both valences. Consistent with this role, neural correlates of skin conductance responses, which are mediated by the sympathetic nervous system, have been reported in the amygdala (Laine et al. 2009). Moreover, this type of valence-insensitive modulation of reinforcement responses by expectation could be appropriate for driving reinforcement learning through attention-based learning algorithms (Pearce & Hall 1980).

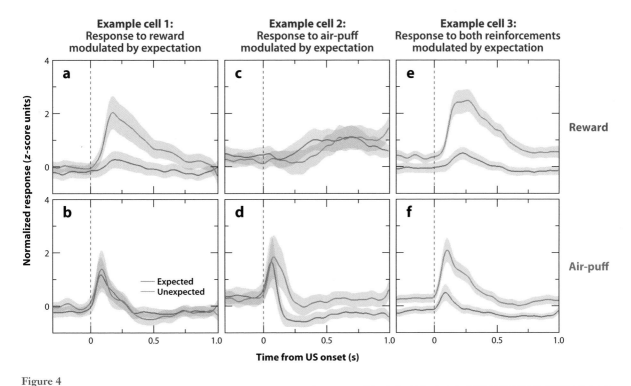

Example cell 1:
Response to reward
modulated by expectation

Example cell 2:
Response to air-puff
modulated by expectation

Example cell 3:
Response to both reinforcements
modulated by expectation

Reward

Air-puff

Expected
Unexpected

Normalized response (z-score units)

Time from US onset (s)

Figure 4

Valence-specific and valence-nonspecific encoding in the amygdala. (*a–f*). Normalized and averaged neural responses to reinforcement when it was expected (*magenta*) and unexpected (*cyan*) for reward (*a,c,e*) and air puff (*b,d,f*). Expectation modulated reinforcement responses for only one valence of reinforcement in some cells (*a–d*) but modulated reinforcement responses for both valences in many cells (*e,f*). These responses are consistent with a role of the amygdala in valence-specific processes, as well as valence-nonspecific processes, such as attention, arousal, and motivation. Adapted with permission from Belova et al. (2007, figure 3).

Of course, the amygdala does not operate in isolation; in particular, its close anatomical connectivity and functional overlap with the OFC raises the question of how OFC processing compares with and interacts with amygdala processing. Using a paradigm similar to that described above, Morrison and Salzman discovered that the OFC contains neurons that prefer rewarding or aversive associations, as in the amygdala, and that, across the population, the signals extend from shortly after CS onset until well after US offset (**Figure 2d,e**; data largely collected from area 13) (Morrison & Salzman 2009, Salzman et al. 2007). OFC responses to the fixation point are also modulated according to whether a cell has a positive and negative preference, in a manner similar to the amygdala (S. Morrison & C.D. Salzman, unpublished data). Together, these data

suggest that the OFC could also participate in a representation of state value.

Both positive and negative valences are represented in the amygdala and the OFC, but how might OFC and amygdala interact with each other? Unpublished data indicate that the appetitive system—composed of cells that prefer positive associations—updates more quickly in the OFC, adapting to changes in reinforcement contingencies faster than the appetitive system in the amygdala (S. Morrison & C.D. Salzman, personal communication, 2009). However, the opposite is true for the aversive system: Negative-preferring amygdala neurons adapt to changes in reinforcement contingencies more rapidly than do their counterparts in the OFC. Thus, the computational steps that update representations in appetitive and aversive systems are not the same in the amygdala and the OFC, even though the neurons appear to be anatomically interspersed in both structures. In contrast, after reinforcement contingencies are well learned in this task, the OFC signals upcoming reinforcement more rapidly than does the amygdala in both appetitive and aversive cells. This finding is consistent with a role for the OFC in rapidly signaling stimulus values and/or expected outcomes once learning is complete—a signal that could be used to exert prefrontal control over limbic structures such as the amygdala or to direct behavioral responding more generally.

The studies described above used Pavlovian conditioning—a procedure in which no action is required of the subject to receive reinforcement—to characterize neural response properties in relation to appetitive and aversive processing. However, many other studies have used decision-making tasks to quantify the extent to which neural response properties are related to reward values (Dorris & Glimcher 2004; Kennerley et al. 2008; Kim et al. 2008; Lau & Glimcher 2008; McCoy & Platt 2005; Padoa-Schioppa & Assad 2006, 2008; Platt & Glimcher 1999; Roesch & Olson 2004; Samejima et al. 2005; Sugrue et al. 2004; Wallis 2007; Wallis & Miller 2003); moreover, similar tasks are often used to examine human valuation processes using fMRI (Breiter et al. 2001, Gottfried et al. 2003, Kable & Glimcher 2007, Knutson et al. 2001, Knutson & Cooper 2005, McClure et al. 2004, Montague et al. 2006, O'Doherty et al. 2001, Rangel et al. 2008, Seymour et al. 2004). The strength of decision-making tasks is that the investigator can directly compare the subjects' preferences, on a fine scale, with neuron signaling. For example, Padoa-Schioppa & Assad (2006) trained monkeys to indicate which of two possible juice rewards they wanted; they offered the juices in different amounts by presenting visual tokens that indicated both juice type and juice amount (**Figure 5a**). Using this task, they discovered that multiple signals were present in different populations of neurons in the OFC. Some OFC neurons encoded what the authors termed "chosen value": Firing was correlated

with the value of the chosen reward; some neurons preferred higher and lower values, respectively (**Figure 5b,c**). These cell populations are reminiscent of the positive and negative value-coding neurons uncovered using the Pavlovian procedure described above. However, because negative valences were not explored in these experiments, these neurons may represent motivation, arousal, or attention, which are correlated with reward value (Maunsell 2004, Roesch & Olson 2004). Other OFC neurons encoded the value of one of the rewards offered (offer value cells; **Figure 5d**) and others still simply encoded the type of juice offered (taste neurons; **Figure 5e**), consistent with previous identification of taste-selective neurons in the OFC (Pritchard et al. 2007, Wilson & Rolls 2005). Further data suggested that the OFC responses were menu-invariant—i.e., if a cell prefers A to B, and B to C, it will also prefer A to C (Padoa-Schioppa & Assad 2008). This characteristic is called transitivity; it implies the ability to use the representation of value as a context-independent economic currency that could support decision-making. However, this finding may depend on the exact design of the task because other studies, focusing on partially overlapping regions of the OFC, have reported neural responses that reflect relative reward preferences, i.e., responses that vary with context and do not meet the standard of transitivity (Tremblay & Schultz 1999).

This rich variety of response properties in the OFC and the amygdala still represents only a subset of the types of encoding that have been observed in these brain areas. For example, amygdala neurons recorded during trace conditioning often exhibited image selectivity (Paton et al. 2006), and similar signals have been observed in the OFC (S. Morrison & C.D. Salzman, personal communication). Moreover, investigators have also described amygdala neural responses to faces, vocal calls, and combinations of faces and vocal calls (Gothard et al. 2007, Kuraoka & Nakamura 2007, Leonard et al. 1985). Meanwhile, the OFC neurons also encode gustatory working memory and modulate their responses depending on

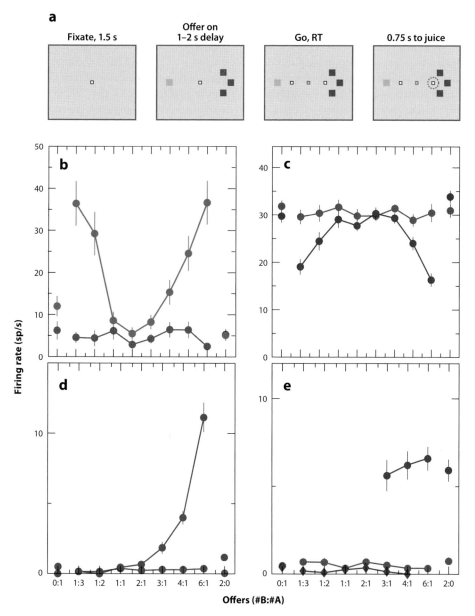

Figure 5

OFC neural responses during economic decision-making. (*a*) Behavioral task. Monkeys centered gaze at a fixation point and then viewed two visual tokens that indicate the type and quantity of juice reward being offered for potential saccades to each location (*tokens, yellow and blue squares*). After fixation point extinction, the monkey is free to choose which reward it wants by making a saccade to one of the targets. The amounts of juices offered of each type are titrated against each other to develop a full psychometric characterization of the monkey's preferences as a function of the two juice types offered. (*b–e*) Activity of four neurons revealing different types of response profiles. X-axis shows the quantity of each offer type. Chosen value neurons increased (*b*) or decreased (*c*) their firing when the value of their chosen option increased. Offer value neurons (*d*) increased their firing when the value of one of the juices offered increased. Juice neurons (*e*) increased their firing for trials with a particular juice type offered, independent of the amount of juice offered. Adapted from Padoa-Schioppa & Assad (2006, figures 1 and 3) with permission.

reward magnitude, reward probability, and the time and effort required to obtain a reward (Kennerley et al. 2008). Overall, in addition to encoding variables related to valence and arousal/intensity—two variables central to the representation of emotion—amygdala and OFC neurons encode a variety of other variables in an entangled fashion (Paton et al. 2006, Rigotti et al. 2010a).

Neural Representations of Cognitive Processes in the PFC

We have reviewed briefly the encoding of valence and arousal in the amygdala and the OFC, and now we turn our attention to the encoding of other mental state variables in the PFC. Our goal is not to discuss systematically every aspect of PFC neurophysiology, but instead to highlight response properties that may play an especially vital role in setting the variables that constitute a mental state: encoding of rules, which are essential for appropriately contextualizing environmental stimuli and other variables; flexible encoding of stimulus-stimulus associations across time and sensory modality; and encoding of complex motor plans.

Encoding of rules in the PFC. Understanding rules for behavior forms the basis for much of our social interaction; therefore, rules must routinely be represented in our brains. A critical feature of our cognitive ability is the ability to apply abstract, as opposed to concrete, rules, i.e., rules that can be generalized and flexibly applied to new situations. In a striking demonstration of this type of rule encoding in the PFC, Wallis and Miller recorded from three parts of the PFC (dorsolateral, ventrolateral, and the OFC) while monkeys performed a task requiring them to switch flexibly between two abstract rules (**Figure 6**) (Wallis et al. 2001). In this task, monkeys viewed two sequentially presented visual cues that could be either matching or nonmatching. In different blocks of trials, monkeys had to apply either a match rule or a nonmatch rule—indicated by the presentation of another cue at the start of the

trial—to guide their responding. The visual stimuli utilized in the blocks were identical; thus, the only difference between the blocks was the rule in effect, and this information must be a part of the monkey's mental state. Many neurons in all three parts of the PFC exhibited selective activity depending on the rule in effect; some neurons preferred match and others nonmatch (**Figure 6**). Of note, rule-selective activity was only one type of selectivity that was present: Neurons often responded selectively to the stimuli themselves, as well as to interactions between the stimuli and the rules. Therefore, it appears that these neurons represent abstract rules along with other variables in an entangled manner.

In the work by Wallis and Miller, the rule in effect was cued on every trial, and the monkeys switched from one rule to the other on a trial-by-trial basis. In contrast, Mansouri and colleagues (2006) used a task in which the rule switched in an uncued manner on a block-by-block basis, and monkeys had to discover the rule in effect in a given block (an analog of the Wisconsin Card Sorting Task). In one block of trials, monkeys had to apply a color-match rule to match two stimuli, and in the other block, monkeys had to apply a shape-match rule. The authors discovered that neural activity in the dorsolateral PFC encoded the rule in effect; different neurons encoded color and shape rules (**Figure 7a,b**). Rule encoding occurred during the trial itself but also during the fixation interval, and even during the intertrial interval (ITI) (**Figure 7c**). This observation implies that a neural signature of the rule in effect was maintained throughout a block of trials—even when the monkey was not performing a trial—as if the monkey had to keep the rule in mind. We suggest that this representation of rules therefore represents a distinctive component of a mental state.

Temporal integration of sensory stimuli and actions. One's current situation is defined not only in terms of the stimuli currently present, but also by the temporal context in which those stimuli appear, as well as by the

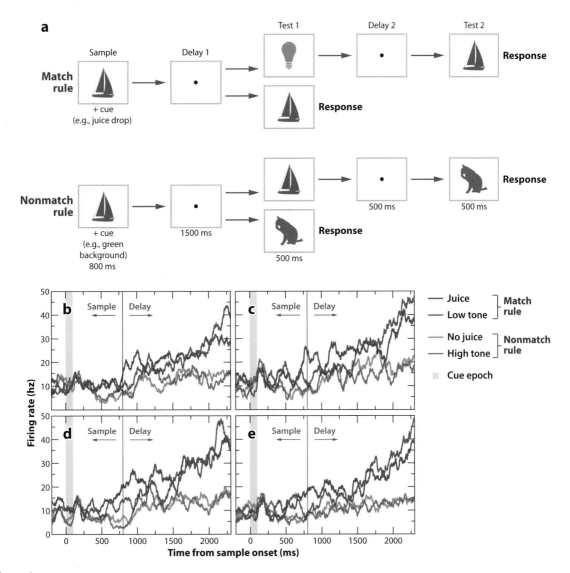

Figure 6

Single neurons encode rules in PFC. (*a*) Behavioral task. Monkeys grasped a lever to initiate a trial. They then had to center gaze at a fixation point while viewing a sample object, wait during a brief delay, and then view a test object. Two types of trials are depicted (*double horizontal arrows*). On match rule trials, monkeys had to release the lever if the test object matched the sample object. On nonmatch rule trials, monkeys had to release the lever if the test object did not match the sample. Otherwise, they had to hold the lever until a third object appeared that always required lever release. The rules in effect varied trial-by-trial by virtue of a different sensory cue (e.g., tones or juice) presented during viewing of the sample object. (*b,c*) PFC neurons encoding match (*b*) or nonmatch (*c*) rules. Activity was higher in relation to the rule in effect regardless of the stimuli shown. Adapted with permission from Wallis et al. (2001, figures 1 and 2).

associations those stimuli have with other stimuli. Fuster (2008) proposed that a cardinal function of the PFC is to provide a representation that reflects the temporal integration of relevant sensory information. Indeed, Fuster and colleagues (2000) have demonstrated this type of encoding in areas 6, 8, and 9/46 of the dorsolateral PFC. In this study, monkeys performed a task in which they had to associate an auditory tone (high or low) with a

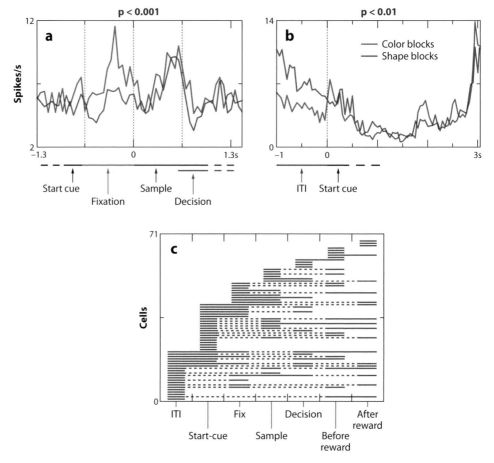

Figure 7

PFC neurons encode rules in effect across time within a trial. Monkeys performed a task in which they had to match either the shape or color of two simultaneously presented objects with a sample object viewed earlier in the trial. Monkeys learned by trial and error whether a shape or color rule was in effect within a block of trials, and block switches were uncued to the monkey. (*a,b*) Two PFC cells that fired differentially depending on the rule in effect; activity differences emerged during the fixation (*a*) and intertrial intervals (ITI) (*b*). Activity is aligned on a start cue, which occurs before fixation on every trial. During the sample interval, one stimulus is presented over the fovea. During the decision interval, two stimuli are presented to the left and right; one matched the sample stimulus in color, and the other matched in shape. The correct choice can be chosen only if one has learned the rule in effect for the current block. (*c*) Distribution of activity differences between shape and color rules for each cell studied in each time interval of a trial. Each line corresponds to a single cell, and the solid parts of a line indicate when the cell fired differentially between color and shape blocks. Encoding of rules occurred in all time epochs, indicating that PFC neurons encode the rule in effect across time within a trial. Adapted with permission from Mansouri et al. (2006, figures 2 and 3).

subsequently presented colored target (red or green). The authors discovered cells that responded selectively to associated tones and colors, e.g., cells that fired strongly only for the high tone and its associated target. Meanwhile, failure to represent the correct association ac-curately was correlated with behavioral errors; thus, PFC neurons' ability to form and represent cross-temporal and cross-modality representations was linked to subsequent actions.

The integration of sensory stimuli in the environment, as described above, is key for setting mental state variables; moreover, if we recall that a mental state can be defined as a disposition to action, any representation of planned actions must clearly be an important element of our mental state. Neural signals related to planned actions have been reported in numerous parts of the PFC in several tasks (Fuster 2008, Miller & Cohen 2001). In recent years, scientists have used more complex motor tasks to explore encoding of sequential movement plans. In the dorsolateral PFC, Tanji and colleagues have described activity related to cursor movements that will result from a series of planned arm movements (Mushiake et al. 2006). This activity therefore reflects future events that occur as a result of planned movements. Other studies of PFC neurons have discovered neural ensembles that predict a sequence of planned movements (Averbeck et al. 2006); when the required sequence of movements changes from block to block, the neural ensemble coding changes, too. In a manner reminiscent of rule encoding, this coding of planned movements was also present during the ITI, as if these cells were keeping note of the planned movement sequence throughout the block of trials (Averbeck & Lee 2007). Thus, the PFC not only tracks stimuli across time, but also represents the temporal integration of planned actions and the events that hinge on them. Dorsolateral PFC may well interact with the OFC and the ACC, and, via these areas, the amygdala, to make decisions based on the values of both environmental stimuli and internal variables and then to execute these decisions via planned action sequences.

NEURAL NETWORKS AND MENTAL STATES: A CONCEPTUAL AND THEORETICAL FRAMEWORK FOR UNDERSTANDING INTERACTIONS BETWEEN COGNITION AND EMOTION

We have so far reviewed how neurons in the amygdala and the PFC may encode neural signals representing variables—some more closely tied to emotional processes, and others to cognitive processes—that are components of mental states and how these representations are often entangled (i.e., more than one variable is encoded by a single neuron). But how do these neurons interact within a network to represent mental states in their entirety? Moreover, how can cognitive processes regulate emotional processes?

A central element of emotional regulation involves developing the ability to alter one's emotional response to a stimulus. In general, one can consider at least two basic ways in which this can occur. First, learning mechanisms may operate to change the representation of the emotional meaning of a stimulus. Indeed, one could simply forget or overwrite a previously stored association. Moreover given a stimulus previously associated with a particular reinforcement, such that the stimulus elicits an emotional response, re-experiencing the stimulus in the absence of the associated reinforcement can induce extinction. Extinction is thought to be a learning process whereby previously acquired responses are inhibited. In the case of fear extinction, scientists currently believe that original CS-US associations continue to be stored in the brain (so that they are not forgotten or overwritten), and inhibitory mechanisms develop that suppress the fear response (Quirk & Mueller 2008). Second, mechanisms must exist that can change or switch one's emotional responses depending on one's knowledge of his/her context or situation. A simple example of this phenomenology occurs when playing the game of blackjack. Here, the same card, such as a jack of clubs, can be rewarding, if it makes a total of 21 in your hand, or upsetting, if it makes a player go bust. Emotional responses to the jack of clubs can thereby vary on a moment-to-moment basis depending on the player's knowledge of the situation (e.g., his/her understanding of the rules of the game and of the cards already dealt). Emotional variables here depend critically on the cognitive variables representing one's understanding of the game and one's current hand of cards. Although mechanisms for this type of emotional regulation remain poorly understood, it presumably involves PFC-amygdala neural circuitry.

What type of theoretical framework could describe these different types of emotional regulation? Are there qualitative differences between the neural mechanisms that underlie them? Here we briefly describe one possible approach for explaining this phenomenology. Our proposal is built on the assumption that each mental state corresponds to a large number of states of dynamic variables that describe neurons, synapses, and other constituents of neural circuits. These components must interact such that neural circuit dynamics can actively maintain a representation of the current disposition of behavior, i.e., the current mental state. Complex interactions between these components must therefore correspond to the interactions between mental state variables such as emotional and cognitive parameters. Indeed, when brain states change, these changes typically and inherently involve correlated modifications of multiple mental state variables. In this section, we discuss how a class of neural mechanisms could underlie the representation of mental states and the potential interaction between cognition and emotion. We construct a conceptual framework whereby cognition-emotion interactions can occur via two sorts of mechanisms: associative learning and switching between mental states representing different contexts or situations.

A natural candidate mechanism for representing mental states is the reverberating activity that has been observed at the single neuron level in the form of selective persistent firing rates, such as that which has been described in the PFC (e.g., **Figure 7**) and other structures (Miyashita & Chang 1988, Yakovlev et al. 1998). Each mental state could be represented by a self-sustained, stable pattern of reverberating activity. Small perturbations of these activity patterns are damped by the interactions between neurons so that the state of the network is attracted toward the closest pattern of persistent activity representing a particular mental state. For this reason, these patterns

are called attractors of the neural dynamics. Attractor networks have been proposed as models for associative and working memory (Amit 1989, Hopfield 1982), for decision-making (Wang 2002), and for rule-based behavior (O'Reilly & Munakata 2000, Rolls & Deco 2002). Here, we suggest a scenario in which attractors represent stable mental states and every external or internal event encountered by an organism may steer the activity from one attractor to a different one. This type of mechanism could provide stable yet modifiable representations for the mental states, just like the on and off states of a switch. Thus mental states could be maintained over relatively long timescales but could also rapidly change in response to brief events.

Attractor networks can be utilized to model associative learning. Consider again the experiment performed by Paton and colleagues (2006), described in the section on neural representation of emotional variables. In a simple model, one can assume that learning involves modifying connections from neurons representing the CS (for simplicity, called external neurons) to some of the neurons representing the mental state, in particular those that represent the value of the CS in relation to reinforcement (called internal valence neurons). When the CSs are novel, a monkey does not know what to expect (reward or air puff). The monkey may know that it will be one of the two outcomes. Therefore, the CS in that particular context could induce a transition into one of the preexistent attractors representing the possible states. Some of these states correspond to the expectation of positive or negative reinforcement, and other states could correspond to neutral valence states. The external input starts a biased competition between all these different states. If the reinforcement received differs from the expected one, then the synapses connecting external and internal neurons will be modified such that the competition between mental states will generate a bias toward the correct association (see e.g., Fusi et al. 2007). This learning process is typical of situations in which there are one-to-one associations and, for example, the same CS always has the same value. The monkey can simply learn the stimulus-value associations by trial and error; with appropriate synaptic learning rules, the external connections are modified as needed.

In the situation described above, one CS always predicts reward, and the other punishment. But such conditions do not always exist. For example, Paton and colleagues reversed reinforcement contingencies after learning had occurred. In principle, learning these reversed contingencies could involve modifying the external connections to the neural circuit, thereby having new associations overwrite or override the previous associations. However, reversal tasks may not simply erase or unlearn associations; instead, reversal tasks may rely on processes similar to those invoked during extinction (Bouton 2002, Myers & Davis 2007). Increasing evidence implicates the amygdala-PFC circuit as playing a fundamental role in extinction (Gottfried & Dolan 2004, Izquierdo & Murray 2005, Likhtik et al. 2005, Milad & Quirk 2002, Olsson & Phelps 2004, Pare et al. 2004, Quirk et al. 2000).

For the second type of emotional regulation, during which emotional responses to stimuli depend on knowledge of one's situation or context, we need a qualitatively different learning mechanism. Consider a hypothetical variant of the experiment by Paton et al. (2006), in which the associations are reversed and changed multiple times. For example, stimulus A may initially be associated with a small reward and B with a small punishment. Then, in a second block of trials, A becomes associated with a large reward, and B with a large punishment. Assume that as the experiment proceeds, subjects go back and forth between these two types of blocks of trials so that the two contexts are alternated many times. In this case, if we can store a representation of both the two alternating contexts, we can adopt a significantly more efficient computational strategy. Instead of learning and forgetting associations, we can simply switch from one context to the other. For example, on the first trial of a block, if a large punishment follows B, the monkey can predict that

seeing A on subsequent trials will result in its receiving a large reward. Overall, in the first context, A and B can lead to only small rewards and punishments, respectively. In the second context, A and B always predict large rewards and punishments. To implement this switching type of computational strategy, internal synaptic connections within the neural network must be modified to create the neural representations of the mental states corresponding to the two contexts (Rigotti et al. 2010a).

To illustrate how a model employing an attractor neural network can describe the case of mental state switching described above, we employ an energy landscape metaphor [see e.g., Amit (1989) and **Figure 8**]. Each network state can be described as a vector containing the activation states of all neurons, and it can be characterized by its energy value. If we know the energy for each state, then we can predict the network behavior because the network state will evolve toward the state corresponding to the closest minimum of the energy. In **Figure 8**, we represent each network state as a point on a plane and the corresponding energy as a surface that resembles a hilly landscape. To describe the hypothetical experiment under discussion, which involves switching between contexts, we assume that two variables, context and valence (with two contexts and five different valences represented), characterize each mental state and that only one brain state corresponds to each mental state. As a consequence, for each point on the context-valence plane there is only one energy value (**Figure 8**, red surface). The network naturally relaxes toward the bottom of the valleys (minima of the energy), which represent different mental states. At the neural level, each of these points corresponds to a particular pattern of persistent activity. The six valleys in **Figure 8** correspond to six potential mental states created after modifying internal connections to represent the two different contexts and the related CS-US associations. As a result of the interactions due to recurrent connections, the cognitive variable corresponding to context constrains the set of accessible emotional states. In both contexts, interactions

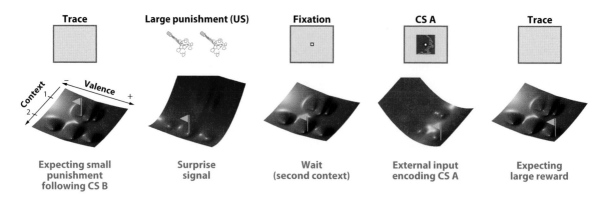

Trace	Large punishment (US)	Fixation	CS A	Trace

| Expecting small punishment following CS B | Surprise signal | Wait (second context) | External input encoding CS A | Expecting large reward |

Figure 8

The dynamics of context-dependent values. We consider a hypothetical variation of the experiment by Paton et al. (2006) in which there are two contexts. In the first context, CS A and B predict small rewards and punishments, respectively. In the second context, CS A and B are associated with large rewards and punishments. Six mental states now correspond to six valleys in the energy landscape. In panel 1, we consider the first trial of context 2, immediately after switching from context 1. Stimulus B has just been presented (not shown), and it is believed to predict small punishment. However a large punishment is delivered (not shown), and a surprise signal tilts the energy function (panel 2), inducing a transition to the neutral mental state of context 2 at the beginning of the next trial (panel 3; we assume for this example that the fixation interval has a neutral value). Now the system has already registered that it is in context 2. Consequently, the appearance of CS A tilts the energy landscape, and the mental state settles at a large positive value (panel 4). After CS disappearance, the network relaxes into the high positive value mental state of context 2. Thus the network does not need to relearn that CS A predicts a large reward because the network has already formed a representation for all the mental states contained within this simple experiment. Just knowing that the context has changed is sufficient for subjects to make an accurate prediction about impending reinforcement. For a detailed attractor model implementing this form of context dependency see Rigotti et al. (2010b).

with external neurons representing CS A or B can tilt temporarily the energy surface and bring the neural network into a different valley, corresponding to a different mental state. The final destination depends on the initial mental state representing the context. The valences of the states differ in the two contexts because valences associated with large rewards and punishments exist only in the second context.

This example illustrates the cognitive regulation of emotion because changes in a cognitive variable (context) cause a change in the possible associated emotional parameters (valence). Analogous mechanisms could underlie how other cognitive variables can influence emotional responses. For example, different social situations can demand different emotional responses to similar sensory stimuli, and knowledge of the social situation (essentially a context variable) can thereby constrain the emotional responses possible.

We based the forgoing discussion on the assumption that the mental states are represented by attractors of the neural dynamics. Alternative and complementary solutions are based on neural representations of mental states that change in time (Buonomano & Maass 2009, Jaeger & Hass 2004). For these neural systems, every trajectory or set of trajectories in the space of all possible brain states represents a particular mental state. These dynamic systems can generate complex temporal sequences that are important for motor planning (Susillo & Abbott 2009). However, they cannot instantaneously generalize to situations in which events are timed differently and they can be difficult to decode. Generally speaking, all known models of mental states provide a useful conceptual framework for understanding the principles of the dynamics of neural circuits, but they fall short of capturing the richness and complexity of real biological neural networks. For example, brain states are not encoded solely in the neuronal spiking activity. Investigators only now have begun to study interactions among dynamic variables operating on diverse timescales, all contributing to a particular brain state [see e.g., Mongillo et al. (2008) for a working memory model based on short-term synaptic facilitation].

The theoretical framework we propose provides a means for representing mental states in the distributed activity of networks of neurons encoding entangled representations. Because we have defined mental states as action dispositions, it is natural to wonder how mental states are linked to the selection and execution of actions. In recent years, reinforcement learning (RL) algorithms have provided an elegant framework for understanding both how subjects choose their actions to maximize reward and minimize punishment and how the brain may represent modeled parameters during this process (Daw et al. 2006, Dayan & Abbott 2001, Sutton & Barto 1998). In particular, scientists have attempted to link two types of RL algorithms onto specific neural structures: a model-based algorithm and a model-free algorithm (Daw et al. 2005). Model-based algorithms are suited for goal-directed actions and likely involve the PFC, whereas model-free algorithms could mediate the generation of habitual behavior (Graybiel 2008) and may involve the striatum. Of note, both types of RL algorithms actually require an already formed representation of states (i.e., of the relevant variables of one's current situation) to enable one to assign values to them to guide action selection. If one drives in an unfamiliar neighborhood packed with restaurants and needs to choose a restaurant, one must first build a mental map of the environment as it is experienced before assigning values to possible destinations. This process of creating mental state representations is not provided for by RL algorithms, which involve only the assignment and updating of values to already created states.

Creating a representation of mental states involves forging links between the many mental state variables that neurons represent. Recent work on the neural basis of object perception represents an initial step toward understanding how variables may be combined to support the formation of a mental state representation. The perception of objects requires one to develop a representation that is invariant for many viewing conditions, such as the precise retinal position of the object, the size, scale, or pose of the object, or the amount of clutter the object appears within the visual field. Di Carlo and colleagues have now provided evidence that unsupervised learning arising from the temporal contiguity of stimuli experienced during natural viewing leads to the formation of an invariant representation of a visual object (Li & DiCarlo 2008). Object perception corresponds to only one component of a mental state, but the scientific approach pioneered by Di Carlo's group may provide a path for understanding how other mental state variables also become linked to create a unified representation of a particular state. Indeed, some theoreticians have proposed that simple mechanisms such as temporal contiguity might underlie new mental state formation (O'Reilly & Munakata 2000, Rigotti et al. 2010a).

CONCLUSIONS

The conceptual framework we have put forth posits that mental states are composed of many variables that together correspond to an action disposition. These variables include parameters, such as valence and arousal, which are often ascribed to emotional processes, as well as parameters ascribed to cognitive processes, such as perceptions, memories, and plans. Of course, mental state parameters also include variables encoding our visceral state. Much in the way that Wittgenstein argued that philosophical controversies dissolve once one carefully disentangles the different ways in which language is being used (Wittgenstein 1958), we argue that the debate between scientists about the origin of emotional feelings—whether visceral processes precede or follow emotional feeling—dissolves. Instead, all these parameters may be linked and together form the representation of our mental state.

This conceptual framework has broad implications for understanding interactions between cognition and emotion in the brain. On the one hand, emotional processes can influence cognitive processes; on the other hand, cognitive processes can regulate or modify our emotions. Both of these interactions can be implemented by changing mental state variables (either emotional or cognitive ones); emotions and thoughts shift together, corresponding to the new mental state. Of course, different mechanisms may exist for implementing these interactions between cognition and emotion, such as mechanisms involving learning and

extinction, as well as mechanisms that support the creation of new mental state representations, such as when one learns a new rule. The process of understanding the complex encoding properties of the amygdala, the PFC, and related brain structures, as well as understanding their functional interactions, is in its infancy. Somehow the intricate connectivity of these brain structures gives rise to mental states and accounts for interactions between cognition and emotion that are fundamental to our well-being and our existence.

DISCLOSURE STATEMENT

The authors are not aware of any affiliations, memberships, funding, or financial holdings that might be perceived as affecting the objectivity of this review.

ACKNOWLEDGMENTS

We thank B. Lau and S. Morrison for helpful discussion, as well as comments on the manuscript, and H. Cline for invaluable support. C.D.S. gratefully acknowledges funding support from NIH (R01 MH082017, R01 DA020656, and RC1 MH088458) and the James S. McDonnell Foundation. S.F. receives support from DARPA SyNAPSE, the Gatsby Foundation, and the Sloan-Swartz Foundation.

LITERATURE CITED

Aggleton J, ed. 2000. *The Amygdala—A Functional Analysis*. Oxford: Oxford Univ. Press

Aggleton JP, Passingham RE. 1981. Syndrome produced by lesions of the amygdala in monkeys (*Macaca mulatta*). *J. Comp. Physiol. Psychol.* 95:961–77

Amaral D, Price J, Pitkanen A, Carmichael S. 1992. Anatomical organization of the primate amygdaloid complex. In *The Amygdala: Neurobiological Aspects of Emotion, Memory, and Mental Dysfunction*, ed. J Aggleton, pp. 1–66. New York: Wiley-Liss

Amaral DG, Behniea H, Kelly JL. 2003. Topographic organization of projections from the amygdala to the visual cortex in the macaque monkey. *Neuroscience* 118:1099–120

Amaral DG, Dent JA. 1981. Development of the mossy fibers of the dentate gyrus: I. A light and electron microscopic study of the mossy fibers and their expansions. *J. Comp. Neurol.* 195:51–86

Amit D. 1989. *Modeling Brain Function—The World of Attractor Neural Networks*. New York: Cambridge Univ. Press

Antoniadis EA, Winslow JT, Davis M, Amaral DG. 2009. The nonhuman primate amygdala is necessary for the acquisition but not the retention of fear-potentiated startle. *Biol. Psychiatry* 65:241–48

Averbeck BB, Lee D. 2007. Prefrontal neural correlates of memory for sequences. *J. Neurosci.* 27:2204–11

Averbeck BB, Sohn JW, Lee D. 2006. Activity in prefrontal cortex during dynamic selection of action sequences. *Nat. Neurosci.* 9:276–82

Barbas H, Blatt GJ. 1995. Topographically specific hippocampal projections target functionally distinct prefrontal areas in the rhesus monkey. *Hippocampus* 5:511–33

Barbas H, Ghashghaei H, Rempel-Clower N, Xiao D. 2002. Anatomic basis of functional specialization in prefrontal cortices in primates. In *Handbook of Neuropsychology*, ed. J Grafman, pp. 1–27. Amersterdam: Elsevier Science B.V.

Barbas H, Pandya DN. 1989. Architecture and intrinsic connections of the prefrontal cortex in the rhesus monkey. *J. Comp. Neurol.* 286:353–75

Barbas H, Zikopoulos B. 2007. The prefrontal cortex and flexible behavior. *Neuroscientist* 13:532–45

Bard P. 1928. A diencephalic mechanism for the expression of rage with special reference to the sympathetic nervous system. *Am. J. Physiol.* 84:490–515

Bates JF, Goldman-Rakic PS. 1993. Prefrontal connections of medial motor areas in the rhesus monkey. *J. Comp. Neurol.* 336:211–28

Baxter M, Murray E. 2000. Reinterpreting the behavioural effects of amygdala lesions in nonhuman primates. See Aggleton 2000, pp. 545–68

Baxter M, Murray EA. 2002. The amygdala and reward. *Nat. Rev. Neurosci.* 3:563–73

Baxter M, Parker A, Lindner CC, Izquierdo AD, Murray EA. 2000. Control of response selection by reinforcer value requires interaction of amygdala and orbital prefrontal cortex. *J. Neurosci.* 20:4311–19

Belova MA, Paton JJ, Morrison SE, Salzman CD. 2007. Expectation modulates neural responses to pleasant and aversive stimuli in primate amygdala. *Neuron* 55:970–84

Belova MA, Paton JJ, Salzman CD. 2008. Moment-to-moment tracking of state value in the amygdala. *J. Neurosci.* 28:10023–30

Bouton ME. 2002. Context, ambiguity, and unlearning: sources of relapse after behavioral extinction. *Biol. Psychiatry* 52:976–86

Breiter H, Aharon I, Kahneman D, Dale A, Shizgal P. 2001. Functional imaging of neural responses to expectancy and experience of monetary gains and losses. *Neuron* 30:619–39

Buckley MJ, Mansouri FA, Hoda H, Mahboubi M, Browning PG, et al. 2009. Dissociable components of rule-guided behavior depend on distinct medial and prefrontal regions. *Science* 325:52–58

Buonomano DV, Maass W. 2009. State-dependent computations: spatiotemporal processing in cortical networks. *Nat. Rev. Neurosci.* 10:113–25

Cannon W. 1927. The James-Lange theory of emotions: a critical examination and an alternative theory. *Am. J. Psychol.* 39:106–24

Carmichael ST, Price JL. 1995a. Limbic connections of the orbital and medial prefrontal cortex in macaque monkeys. *J. Comp. Neurol.* 363:615–41

Carmichael ST, Price JL. 1995b. Sensory and premotor connections of the orbital and medial prefrontal cortex of macaque monkeys. *J. Comp. Neurol.* 363:642–64

Carmichael ST, Price JL. 1996. Connectional networks within the orbital and medial prefrontal cortex of macaque monkeys. *J. Comp. Neurol.* 371:179–207

Cavada C, Company T, Tejedor J, Cruz-Rizzolo RJ, Reinoso-Suarez F. 2000. The anatomical connections of the macaque monkey orbitofrontal cortex. A review. *Cerebral. Cortex* 10:220–42

Damasio A. 1994. *Descartes's Error: Emotion, Reason, and the Human Brain.* New York: Harcourt Brace

Davis M. 2000. The role of the amygdala in conditioned and unconditioned fear and anxiety. See Aggleton 2000, pp. 213–87

Daw ND, Niv Y, Dayan P. 2005. Uncertainty-based competition between prefrontal and dorsolateral striatal systems for behavioral control. *Nat. Neurosci.* 8:1704–11

Daw ND, O'Doherty JP, Dayan P, Seymour B, Dolan RJ. 2006. Cortical substrates for exploratory decisions in humans. *Nature* 441:876–79

Dayan P, Abbott LF. 2001. *Theoretical Neuroscience.* Cambridge, MA: MIT Press

Dorris MC, Glimcher PW. 2004. Activity in posterior parietal cortex is correlated with the relative subjective desirability of action. *Neuron* 44:365–78

Freese JL, Amaral DG. 2005. The organization of projections from the amygdala to visual cortical areas TE and V1 in the macque monkey. *J. Comp. Neurol.* 486:295–317

Fusi S, Asaad WF, Miller EK, Wang XJ. 2007. A neural circuit model of flexible sensorimotor mapping: learning and forgetting on multiple timescales. *Neuron* 54:319–33

Fuster J. 2008. *The Prefrontal Cortex.* London: Elsevier

Fuster JM, Bodner M, Kroger JK. 2000. Cross-modal and cross-temporal association in neurons of frontal cortex. *Nature* 405:347–51

Fuster JM, Uyeda AA. 1971. Reactivity of limbic neurons of the monkey to appetitive and aversive signals. *Electroencephalogr. Clin. Neurophysiol.* 30:281–93

Ghashghaei H, Barbas H. 2002. Pathways for emotion: interactions of prefrontal and anterior temporal pathways in the amygdala of the rhesus monkey. *Neuroscience* 115:1261–79

Ghashghaei HT, Hilgetag CC, Barbas H. 2007. Sequence of information processing for emotions based on the anatomic dialogue between prefrontal cortex and amygdala. *Neuroimage* 34:905–23

Gothard KM, Battaglia FP, Erickson CA, Spitler KM, Amaral DG. 2007. Neural responses to facial expression and face identity in the monkey amygdala. *J. Neurophysiol.* 97:1671–83

Gottfried J, O'Doherty J, Dolan RJ. 2003. Encoding predictive reward value in human amygdala and orbitofrontal cortex. *Science* 301:1104–7

Gottfried JA, Dolan RJ. 2004. Human orbitofrontal cortex mediates extinction learning while accessing conditioned representations of value. *Nat. Neurosci.* 7:1144–52

Graybiel AM. 2008. Habits, rituals, and the evaluative brain. *Annu. Rev. Neurosci.* 31:359–87

Hampton AN, Adolphs R, Tyszka MJ, O'Doherty JP. 2007. Contributions of the amygdala to reward expectancy and choice signals in human prefrontal cortex. *Neuron* 55:545–55

Holland PC, Gallagher M. 2004. Amygdala-frontal interactions and reward expectancy. *Curr. Opin. Neurobiol.* 14:148–55

Hopfield JJ. 1982. Neural networks and physical systems with emergent collective computational abilities. *Proc. Natl. Acad. Sci. USA* 79:2554–58

Izquierdo A, Murray EA. 2005. Opposing effects of amygdala and orbital prefrontal cortex lesions on the extinction of instrumental responding in macaque monkeys. *Eur. J. Neurosci.* 22:2341–46

Izquierdo A, Murray EA. 2007. Selective bilateral amygdala lesions in rhesus monkeys fail to disrupt object reversal learning. *J. Neurosci.* 27:1054–62

Izquierdo A, Suda RK, Murray EA. 2004. Bilateral orbital prefrontal cortex lesions in rhesus monkeys disrupt choices guided by both reward value and reward contingency. *J. Neurosci.* 24:7540–48

Izquierdo A, Suda RK, Murray EA. 2005. Comparison of the effects of bilateral orbital prefrontal cortex lesions and amygdala lesions on emotional responses in rhesus monkeys. *J. Neurosci.* 25:8534–42

Jaeger H, Haas H. 2004. Harnessing nonlinearity: predicting chaotic systems and saving energy in wireless communication. *Science* 304:78–80

James W. 1884. What is an emotion? *Mind* 9:188–205

James W. 1894. The physical basis of emotion. *Psychol. Rev.* 1:516–29

Jones B, Mishkin M. 1972. Limbic lesions and the problem of stimulus-reinforcement associations. *Exp. Neurol.* 36:362–77

Kable JW, Glimcher PW. 2007. The neural correlates of subjective value during intertemporal choice. *Nat. Neurosci.* 10:1625–33

Kalin NH, Shelton SE, Davidson RJ. 2004. The role of the central nucleus of the amygdala in mediating fear and anxiety in the primate. *J. Neurosci.* 24:5506–15

Kalin NH, Shelton SE, Davidson RJ. 2007. Role of the primate orbitofrontal cortex in mediating anxious temperament. *Biol. Psychiatry* 62:1134–39

Kazama A, Bachevalier J. 2009. Selective aspiration or neurotoxic lesions of orbital frontal areas 11 and 13 spared monkeys' performance on the object discrimination reversal task. *J. Neurosci.* 29:2794–804

Kennerley SW, Dahmubed AF, Lara AH, Wallis JD. 2008. Neurons in the frontal lobe encode the value of multiple decision variables. *J. Cogn. Neurosci.* 21:1162–78

Kennerley SW, Walton ME, Behrens TE, Buckley MJ, Rushworth MF. 2006. Optimal decision making and the anterior cingulate cortex. *Nat. Neurosci.* 9:940–47

Kim S, Hwang J, Lee D. 2008. Prefrontal coding of temporally discounted values during intertemporal choice. *Neuron* 59:161–72. Erratum *Neuron* 59(3):522

Kluver H, Bucy P. 1939. Preliminary analysis of functions of the temporal lobes in monkeys. *Arch. Neurol. Psychiatry* 42:979–1000

Knutson B, Adams CM, Fong GW, Hommer D. 2001. Anticipation of increasing monetary reward selectively recruits nucleus accumbens. *J. Neurosci.* 21:RC159

Knutson B, Cooper JC. 2005. Functional magnetic resonance imaging of reward prediction. *Curr. Opin. Neurol.* 18:411–17

Kondo H, Saleem KS, Price JL. 2005. Differential connections of the perirhinal and parahippocampal cortex with the orbital and medial prefrontal networks in macaque monkeys. *J. Comp. Neurol.* 493:479–509

Kuraoka K, Nakamura K. 2007. Responses of single neurons in monkey amygdala to facial and vocal emotions. *J. Neurophysiol.* 97:1379–87

Laine CM, Spitler KM, Mosher CP, Gothard KM. 2009. Behavioral triggers of skin conductance responses and their neural correlates in the primate amygdala. *J. Neurophysiol.* 101:1749–54

Lang PJ. 1994. The varieties of emotional experience: a meditation on James-Lange theory. *Psychol. Rev.* 101:211–21

Lang PJ, Bradley MM, Cuthbert BN. 1990. Emotion, attention, and the startle reflex. *Psychol. Rev.* 97:377–95

Lang PJ, Davis M. 2006. Emotion, motivation, and the brain: reflex foundations in animal and human research. *Prog. Brain Res.* 156:3–29

Lange C. 1922. *The Emotions*. Baltimore, MD: Williams & Wilkins

Lau B, Glimcher PW. 2008. Value representations in the primate striatum during matching behavior. *Neuron* 58:451–63

LeDoux JE. 2000. Emotion circuits in the brain. *Annu. Rev. Neurosci.* 23:155–84

Leonard CM, Rolls ET, Wilson FA, Baylis GC. 1985. Neurons in the amygdala of the monkey with responses selective for faces. *Behav. Brain Res.* 15:159–76

Li N, DiCarlo JJ. 2008. Unsupervised natural experience rapidly alters invariant object representation in visual cortex. *Science* 321:1502–7

Likhtik E, Pelletier JG, Paz R, Pare D. 2005. Prefrontal control of the amygdala. *J. Neurosci.* 25:7429–37

Lu MT, Preston JB, Strick PL. 1994. Interconnections between the prefrontal cortex and the premotor areas in the frontal lobe. *J. Comp. Neurol.* 341:375–92

Machado CJ, Bachevalier J. 2007. The effects of selective amygdala, orbital frontal cortex or hippocampal formation lesions on reward assessment in nonhuman primates. *Eur. J. Neurosci.* 25:2885–904

Machado CJ, Kazama AM, Bachevalier J. 2009. Impact of amygdala, orbital frontal, or hippocampal lesions on threat avoidance and emotional reactivity in nonhuman primates. *Emotion* 9:147–63

Malkova L, Gaffan D, Murray EA. 1997. Excitotoxic lesions of the amygdala fail to produce impairment in visual learning for auditory secondary reinforcement but interfere with reinforcer devaluation effects in rhesus monkeys. *J. Neurosci.* 17:6011–20

Mansouri FA, Matsumoto K, Tanaka K. 2006. Prefrontal cell activities related to monkeys' success and failure in adapting to rule changes in a Wisconsin Card Sorting Test analog. *J. Neurosci.* 26:2745–56

Maren S. 2005. Synaptic mechanisms of associative memory in the amygdala. *Neuron* 47:783–86

Maunsell JH. 2004. Neuronal representations of cognitive state: reward or attention? *Trends Cogn. Sci.* 8:261–65

McClure SM, Laibson DI, Loewenstein G, Cohen JD. 2004. Separate neural systems value immediate and delayed monetary rewards. *Science* 306:503–7

McCoy AN, Platt ML. 2005. Risk-sensitive neurons in macaque posterior cingulate cortex. *Nat. Neurosci.* 8:1220–27

McDonald AJ. 1998. Cortical pathways to the mammalian amygdala. *Prog. Neurobiol.* 55:257–332

Milad MR, Quirk GJ. 2002. Neurons in medial prefrontal cortex signal memory for fear extinction. *Nature* 420:70–74

Miller EK, Cohen JD. 2001. An integrative theory of prefrontal cortex function. *Annu. Rev. Neurosci.* 24:167–202

Miyashita Y, Chang H. 1988. Neuronal correlate of pictorial short-term memory in the primate temporal cortex. *Nature* 331:68–70

Mongillo G, Barak O, Tsodyks M. 2008. Synaptic theory of working memory. *Science* 319:1543–46

Montague PR, King-Casas B, Cohen JD. 2006. Imaging valuation models in human choice. *Annu. Rev. Neurosci.* 29:417–48

Morecraft RJ, Geula C, Mesulam MM. 1992. Cytoarchitecture and neural afferents of orbitofrontal cortex in the brain of the monkey. *J. Comp. Neurol.* 323:341–58

Morrison S, Salzman C. 2009. The convergence of information about rewarding and aversive stimuli in single neurons. *J. Neurosci.* 29:11471–83

Murray EA. 2008. Neuropsychology of primate reward processes. In *New Encyclopedia of Neuroscience*, Vol. 6, ed. LR Squire, pp. 993–99. Oxford: Academic Press

Murray EA, Izquierdo A. 2007. Orbitofrontal cortex and amygdala contributions to affect and action in primates. *Ann. N.Y. Acad. Sci.* 1121:273–96

Mushiake H, Saito N, Sakamoto K, Itoyama Y, Tanji J. 2006. Activity in the lateral prefrontal cortex reflects multiple steps of future events in action plans. *Neuron* 50:631–41

Myers KM, Davis M. 2007. Mechanisms of fear extinction. *Mol. Psychiatry* 12:120–50

Nishijo H, Ono T, Nishino H. 1988a. Single neuron responses in amygdala of alert monkey during complex sensory stimulation with affective significance. *J. Neurosci.* 8:3570–83

Nishijo H, Ono T, Nishino H. 1988b. Topographic distribution of modality-specific amygdalar neurons in alert monkey. *J. Neurosci.* 8:3556–69

Ochsner KN, Gross JJ. 2005. The cognitive control of emotion. *Trends Cogn. Sci.* 9:242–49

O'Doherty J, Kringelbach ML, Rolls ET, Hornak J, Andrews C. 2001. Abstract reward and punishment representations in the human orbitofrontal cortex. *Nat. Neurosci.* 4:95–102

Olsson A, Phelps EA. 2004. Learned fear of "unseen" faces after Pavlovian, observational, and instructed fear. *Psychol. Sci.* 15:822–28

Ongur D, An X, Price JL. 1998. Prefrontal cortical projections to the hypothalamus in macaque monkeys. *J. Comp. Neurol.* 401:480–505

O'Reilly R, Munakata Y. 2000. *Computational Explorations in Cognitive Neuroscience.* Cambridge, MA: MIT Press

Padoa-Schioppa C, Assad JA. 2006. Neurons in the orbitofrontal cortex encode economic value. *Nature* 441:223–26

Padoa-Schioppa C, Assad JA. 2008. The representation of economic value in the orbitofrontal cortex is invariant for changes of menu. *Nat. Neurosci.* 11:95–102

Pare D, Quirk GJ, Ledoux JE. 2004. New vistas on amygdala networks in conditioned fear. *J. Neurophysiol.* 92:1–9

Pare D, Royer S, Smith Y, Lang EJ. 2003. Contextual inhibitory gating of impulse traffic in the intra-amygdaloid network. *Ann. N. Y. Acad. Sci.* 985:78–91

Parkinson J, Crofts HS, McGuigan M, Tomic DL, Everitt BJ, Roberts AC. 2001. The role of the primate amygdala in conditioned reinforcement. *J. Neurosci.* 21:7770–80

Paton J, Belova M, Morrison S, Salzman C. 2006. The primate amygdala represents the positive and negative value of visual stimuli during learning. *Nature* 439:865–70

Pearce J, Hall G. 1980. A model for Pavlovian conditioning: variations in the effectiveness of conditioned but not unconditioned stimuli. *Psychol. Rev.* 87:532–52

Pessoa L. 2008. On the relationship between emotion and cognition. *Nat. Rev. Neurosci.* 9:148–58

Petrides M, Pandya D. 1994. Comparative architectonic analysis of the human and macaque frontal cortex. In *Handbook of Neuropsychology*, ed. F Boller, J Grafman, pp. 17–57. New York: Elsevier

Phelps EA, LeDoux JE. 2005. Contributions of the amygdala to emotion processing: from animal models to human behavior. *Neuron* 48:175–87

Pitkanen A, Amaral DG. 1998. Organization of the intrinsic connections of the monkey amygdaloid complex: projections originating in the lateral nucleus. *J. Comp. Neurol.* 398:431–58

Platt ML, Glimcher PW. 1999. Neural correlates of decision variables in parietal cortex. *Nature* 400:233–38

Preuss TM. 1995. Do rats have prefrontal cortex? The Rose-Woolsey-Akert program reconsidered. *J. Cogn. Neurosci.* 7:1–24

Price JL. 2007. Definition of the orbital cortex in relation to specific connections with limbic and visceral structures and other cortical regions. *Ann. N. Y. Acad. Sci.* 1121:54–71

Pritchard TC, Schwartz GJ, Scott TR. 2007. Taste in the medial orbitofrontal cortex of the macaque. *Ann. N. Y. Acad. Sci.* 1121:121–35

Quirk GJ, Mueller D. 2008. Neural mechanisms of extinction learning and retrieval. *Neuropsychopharmacology* 33:56–72

Quirk GJ, Russo GK, Barron JL, Lebron K. 2000. The role of ventromedial prefrontal cortex in the recovery of extinguished fear. *J. Neurosci.* 20:6225–31

Rangel A, Camerer C, Montague PR. 2008. A framework for studying the neurobiology of value-based decision making. *Nat. Rev. Neurosci.* 9:545–56

Rigotti M, Ben Dayan Rubin D, Morrison SE, Salzman CD, Fusi S. 2010a. Attractor concretion as a mechanism for the formation of context representations. *NeuroImage* doi: 10.1016/j.neuroimage.2010.01.047. In press

Rigotti M, Ben-Dayan Rubin D, Wang X-J, Fusi S. 2010b. The importance of the diversity in neural responses in context-dependent tasks. Submitted

Robbins T, Arnsten A. 2009. The neuropsychopharmacology of fronto-executive function: monoaminergic modulation. *Annu. Rev. Neurosci.* 32:267–87

Roesch MR, Olson CR. 2004. Neuronal activity related to reward value and motivation in primate frontal cortex. *Science* 304:307–10

Rolls E, Deco G. 2002. *Computational Neuroscience of Vision*. Oxford: Oxford Univ. Press

Romanski LM, Bates JF, Goldman-Rakic PS. 1999. Auditory belt and parabelt projections to the prefrontal cortex in the rhesus monkey. *J. Comp. Neurol.* 403:141–57

Rudebeck PH, Bannerman DM, Rushworth MF. 2008. The contribution of distinct subregions of the ventromedial frontal cortex to emotion, social behavior, and decision making. *Cogn. Affect. Behav. Neurosci.* 8:485–97

Rushworth MF, Behrens TE. 2008. Choice, uncertainty and value in prefrontal and cingulate cortex. *Nat. Neurosci.* 11:389–97

Russell JA. 1980. A circumplex model of affect. *J. Pers. Soc. Psychol.* 39:1161–78

Saddoris MP, Gallagher M, Schoenbaum G. 2005. Rapid associative encoding in basolateral amygdala depends on connections with orbitofrontal cortex. *Neuron* 46:321–31

Salzman CD, Belova MA, Paton JJ. 2005. Beetles, boxes and brain cells: neural mechanisms underlying valuation and learning. *Curr. Opin. Neurobiol.* 15:721–29

Salzman CD, Paton JJ, Belova MA, Morrison SE. 2007. Flexible neural representations of value in the primate brain. *Ann. N. Y. Acad. Sci.* 1121:336–54

Samejima K, Ueda Y, Doya K, Kimura M. 2005. Representation of action-specific reward values in the striatum. *Science* 310:1337–40

Sanghera MK, Rolls ET, Roper-Hall A. 1979. Visual responses of neurons in the dorsolateral amygdala of the alert monkey. *Exp. Neurol.* 63:610–26

Schoenbaum G, Setlow B, Saddoris MP, Gallagher M. 2003. Encoding predicted outcome and acquired value in orbitofrontal cortex during cue sampling depends upon input from basolateral amygdala. *Neuron* 39:855–67

Seymour B, O'Doherty JP, Dayan P, Koltzenburg M, Jones AK, et al. 2004. Temporal difference models describe higher-order learning in humans. *Nature* 429:664–67

Spiegler BJ, Mishkin M. 1981. Evidence for the sequential participation of inferior temporal cortex and amygdala in the acquisition of stimulus-reward associations. *Behav. Brain Res.* 3:303–17

Stefanacci L, Amaral DG. 2000. Topographic organization of cortical inputs to the lateral nucleus of the macaque monkey amygdala: a retrograde tracing study. *J. Comp. Neurol.* 421:52–79

Stefanacci L, Amaral DG. 2002. Some observations on cortical inputs to the macaque monkey amygdala: an anterograde tracing study. *J. Comp. Neurol.* 451:301–23

Stefanacci L, Suzuki WA, Amaral DG. 1996. Organization of connections between the amygdaloid complex and the perirhinal and parahippocampal cortices in macaque monkeys. *J. Comp. Neurol.* 375:552–82

Stuss DT, Levine B, Alexander MP, Hong J, Palumbo C, et al. 2000. Wisconsin Card Sorting Test performance in patients with focal frontal and posterior brain damage: effects of lesion location and test structure on separable cognitive processes. *Neuropsychologia* 38:388–402

Sugase-Miyamoto Y, Richmond BJ. 2005. Neuronal signals in the monkey basolateral amygdala during reward schedules. *J. Neurosci.* 25:11071–83

Sugrue LP, Corrado GS, Newsome WT. 2004. Matching behavior and the representation of value in the parietal cortex. *Science* 304:1782–87

Susillo D, Abbott L. 2009. Generating coherent patterns of activity from chaotic neural networks. *Neuron.* 63:544–57

Sutton R, Barto A. 1998. *Reinforcement Learning*. Cambridge, MA: MIT Press

Suzuki WA, Amaral DG. 1994. Perirhinal and parahippocampal cortices of the macaque monkey: cortical afferents. *J. Comp. Neurol.* 350:497–533

Tremblay L, Schultz W. 1999. Relative reward preference in primate orbitofrontal cortex. *Nature* 398:704–8

von der Malsburg C. 1999. The what and why of binding: the modeler's perspective. *Neuron* 24:95–104

Wallis JD. 2007. Orbitofrontal cortex and its contribution to decision-making. *Annu. Rev. Neurosci.* 30:31–56

Wallis JD, Anderson KC, Miller EK. 2001. Single neurons in prefrontal cortex encode abstract rules. *Nature* 411:953–56

Wallis JD, Miller EK. 2003. Neuronal activity in primate dorsolateral and orbital prefrontal cortex during performance of a reward preference task. *Eur. J. Neurosci.* 18:2069–81

Wang XJ. 2002. Probabilistic decision making by slow reverberation in cortical circuits. *Neuron* 36:955–68

Weiskrantz L. 1956. Behavioral changes associated with ablation of the amygdaloid complex in monkeys. *J. Comp. Neurol.* 49:381–91

Wellman LL, Gale K, Malkova L. 2005. GABAA-mediated inhibition of basolateral amygdala blocks reward devaluation in macaques. *J. Neurosci.* 25:4577–86

Wilson FA, Rolls ET. 2005. The primate amygdala and reinforcement: a dissociation between rule-based and associatively-mediated memory revealed in neuronal activity. *Neuroscience* 133:1061–72

Wise SP. 2008. Forward frontal fields: phylogeny and fundamental function. *Trends Neurosci.* 31:599–608

Wittgenstein W. 1958. *Philosophical Investigations.* Oxford: Blackwell

Yakovlev V, Fusi S, Berman E, Zohary E. 1998. Inter-trial neuronal activity in inferior temporal cortex: a putative vehicle to generate long-term visual associations. *Nat. Neurosci.* 1:310–17

Category Learning in the Brain

Carol A. Seger[1] and Earl K. Miller[2]

[1] Department of Psychology and Program in Molecular, Cellular, and Integrative Neurosciences, Colorado State University, Fort Collins, Colorado 80523; email: Carol.Seger@colostate.edu

[2] The Picower Institute for Learning and Memory and Department of Brain and Cognitive Sciences, Massachusetts Institute of Technology, Cambridge, Massachusetts 02139; email: ekmiller@mit.edu

Annu. Rev. Neurosci. 2010. 33:203–19

The *Annual Review of Neuroscience* is online at neuro.annualreviews.org

This article's doi: 10.1146/annurev.neuro.051508.135546

Key Words

classification, concept learning, memory systems

Abstract

The ability to group items and events into functional categories is a fundamental characteristic of sophisticated thought. It is subserved by plasticity in many neural systems, including neocortical regions (sensory, prefrontal, parietal, and motor cortex), the medial temporal lobe, the basal ganglia, and midbrain dopaminergic systems. These systems interact during category learning. Corticostriatal loops may mediate recursive, bootstrapping interactions between fast reward-gated plasticity in the basal ganglia and slow reward-shaded plasticity in the cortex. This can provide a balance between acquisition of details of experiences and generalization across them. Interactions between the corticostriatal loops can integrate perceptual, response, and feedback-related aspects of the task and mediate the shift from novice to skilled performance. The basal ganglia and medial temporal lobe interact competitively or cooperatively, depending on the demands of the learning task.

Contents

INTRODUCTION

Although our brains can store specific experiences, it is not always advantageous for us to be too literal. A brain limited to storing an independent record of each experience would require a prodigious amount of storage and bog us down with details. We have instead evolved the ability to detect the higher-level structure of experiences, the commonalities across them that allow us to group experiences into meaningful categories and concepts. This process imbues the world with meaning. We instantly recognize and respond appropriately to objects, situations, expressions, etc., even though we have never encountered those exact examples before. It stimulates proactive, goal-directed thought by allowing us to generalize about (to imagine) future situations that share fundamental elements with past experience. Imagine the mental cacophony without this ability. The world would lack any deeper meaning. Experiences would be fragmented and unrelated. Things would seem strange and unfamiliar if they differed even trivially from previous examples. This situation describes many of the cognitive characteristics of neuropsychiatric disorders such as autism.

Here, we review how categories are learned by the brain. We begin with a brief definition of categories and describe how category learning is studied. We argue that categorization is not dependent on any single neural system, but rather results from the recruitment of a variety of neural systems depending on task demands. We then describe the primary brain areas involved in categorization learning: the visual cortex, the prefrontal and parietal cortices, the basal ganglia, and the medial temporal lobe. This leads to a discussion and hypotheses about how neural systems interact during category acquisition, which focus on interactions within and between corticostriatal loops connecting cortex and basal ganglia and between the basal ganglia and the medial temporal lobe. We end by summarizing principles by which the brain learns categories and other abstractions.

Categories

Categories represent our knowledge of groupings and patterns that are not explicit in the bottom-up sensory inputs. A simple example is crickets sharply dividing a range of pure tones into mate versus bat (a predator) (Wyttenbach et al. 1996). A wide range of tones on either side of a sharp boundary (16 kHz) are treated equivalently, whereas nearby tones that straddle it are treated differently. This grouping of experience by functional relevance occurs at many levels of processing and for a wide range of phenomena from more literal (e.g., color) to abstract (e.g., peace, love, and understanding). Many categorical distinctions are innate or result from many years of experience (e.g., faces), but key to human intelligence is our ability to learn new categories quickly, even when they are multivariate and abstract (e.g., Free Jazz, gastropub). Mahon & Caramazza (2009) and Martin (2007), among others, have published excellent reviews of innate or well-learned categories. We focus on category learning.

Examples of category learning tasks are shown in **Figure 1**. Many tasks use novel stimuli formed according to a particular perceptual manipulation and then grouped according to an experimenter-defined boundary. Some examples of stimuli used in tasks include prototypes, information integration, and stimuli morphed along a continuum (**Figure 1a–c**, respectively). We can also group events and actions into categories by more abstract properties or rules, which can range from simple deterministic rules based on a single easily identified dimension to more complex situations in which rules may be probabilistic, or complex (e.g., a conjunctive or disjunctive rule), or require identification of an abstract feature not actually present in the physical item (e.g., the rule "same" or "different" (**Figure 1e**). Categorization can even be completely arbitrary (**Figure 1d**). For example, imagine a group of students, half of whom are enrolled in one section of a course and half in the other section. The students within each section likely do not share any particular perceptual characteristics that are not shared by students in the other section. However, this categorization scheme has great utility for predicting which students are likely to attend class in a particular room at a particular time.

BRAIN AREAS INVOLVED IN CATEGORY LEARNING

Not surprisingly given the variety of above examples, category learning likely involves many brain systems including most of the neocortex, the hippocampus, and the basal ganglia. We review which types of category tasks recruit each region of the brain and describe each region's putative role. We make no claims to an exhaustive treatment; categorical representations are likely in many domains and their respective neural systems. For example, evidence indicates that the amygdala participates in generalization of knowledge about fearful or aversive types of stimuli (Barot et al. 2008).

Visual Cortex

We focus on the visual system, the best-studied modality. However, similar processes are likely present in other sensory modalities, including the auditory (Vallabha et al. 2007), the somatosensory (Romo & Salinas 2001), and the olfactory (Howard et al. 2009) systems.

Likely candidates for visual categorization are areas at the highest levels of visual processing. One is the inferior temporal cortex (ITC), whose neurons have complex shape selectivity (Desimone et al. 1984, Logothetis & Sheinberg 1996, Tanaka 1996). Investigators have known about neurons with category-like tuning properties since the seminal work on "face cells" by Gross and colleagues (Desimone et al. 1984). The human fusiform face area (Kanwisher et al. 1997), an ITC area with a preponderance of face cells, is recruited during learning of new face categories (DeGutis & D'Esposito 2007). Inferior temporal neurons in trained monkeys are specifically activated by trees or fish and show relatively little differentiation within those categories (Vogels 1999). Microstimulation of monkey ITC can facilitate visual classification of novel images (Kawasaki & Sheinberg 2008).

However, the ITC may play less of a role in learning explicit representations of category membership and more of a role in high-level analysis of features that contribute to categorization. ITC neurons often do not completely generalize among category members; they retain selectivity for underlying perceptual similarity between individuals (Freedman

ITC: inferotemporal cortex

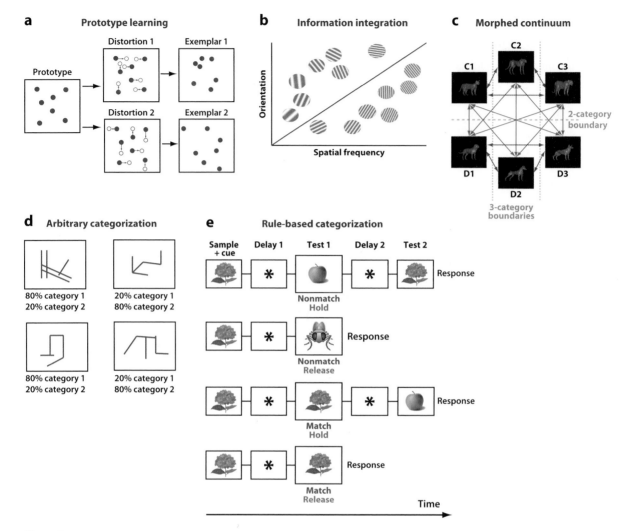

Figure 1

Categorization tasks. (*a*) Dot pattern prototype learning. A prototypical stimulus is selected (*left*), and category exemplars (*right*) are formed by randomly moving dots. Large amounts of movement (*bottom*) result in high distortion stimuli; smaller amounts of movement (*top*) result in low distortion stimuli. (*b*) Information integration task. Stimuli are formed by varying two incommensurate features: angle from vertical and width of the bars. Illustrated is a diagonal decision bound between categories; to learn the categorization successfully, subjects must integrate the knowledge of angle and width. (*c*) Cat-dog categorization task. Stimuli are formed as continuous morphs along each of the lines between prototype stimuli. The categorical decision bound arbitrarily divides the continuous perceptual space into two or three domains, or categories. (*d*) Arbitrary categorization task. Each stimulus is individually probabilistically associated with the categories; stimuli within a category do not share identifying common features. (*e*) "Same - different" rule task. Monkeys responded on the basis of whether novel pairs of images matched or did not match, depending on which rule was in effect.

et al. 2003, Jiang et al. 2007). They also emphasize certain critical stimuli or diagnostic features for the categories and show greater activity for stimuli near category boundaries (DeGutis & D'Esposito 2007, Freedman et al. 2003, Sigala & Logothetis 2002, Baker et al., 2002).

Simple shape-based perceptual categories may be acquired in earlier visual areas. A commonly used task is the dot pattern prototype learning task (see **Figure 1a**); subjects learn a single category (e.g., "A" versus "not A") via simply observing category members. This type of relatively simple category learning may depend on plasticity in the early visual system locus. fMRI studies show activity changes after dot pattern learning in the extrastriate visual cortex, typically around BA 18/19 and roughly corresponding to visual area V2 (Aizenstein et al. 2000; Reber et al. 1998, 2003). Performance on this task is preserved in persons with amnesia (Knowlton & Squire 1993; for a comprehensive review, see Smith 2008), indicating independence from the medial temporal lobe memory system, and is preserved in Parkinson disease (Reber & Squire 1999), indicating independence from corticostriatal systems. However, patients with moderate-severity Alzheimer disease, which can include damage to extrastriate visual cortex, are impaired (Keri et al. 1999). Other categorization tasks, however, do recruit corticostriatal and/or medial temporal lobe systems, especially more complex category learning that involves learning via trial-and-error feedback and learning of multiple categories (Casale & Ashby 2008, Little & Thulborn 2005, Vogels et al. 2002).

Plasticity in visual cortex likely involves local changes in the strength of cortical synapses owing to Hebbian learning (McClelland 2006) subserved by mechanisms of long-term potentiation. Sensory cortex typically emphasizes stability over plasticity, particularly in adults. Thus, perceptual categories, especially in early sensory cortex, do not usually result from just casual or limited amounts of passive experience with a stimulus.

Prefrontal Cortex

The prefrontal cortex occupies a far greater proportion of the human cerebral cortex than it does in other animals, which suggests that it might contribute to those cognitive capacities that separate humans from animals (Fuster 1995, Miller & Cohen 2001). It seems more readily modifiable by experience than does the sensory cortex.

For example, Freedman and colleagues (2001, 2002, 2003) trained monkeys to categorize stimuli along a morphing continuum of different blends of "cats" and "dogs" (see **Figure 1c**) and found a large proportion of randomly selected lateral prefrontal cortex (PFC) neurons with hallmarks of category representations: sharp differences in activity to similar-looking stimuli across a discrete category boundary yet similar activity to different-looking members of the same category. Simultaneous recording from the PFC and anterior-ventral ITC revealed weaker category effects in the ITC (they retained more selectivity for individual members) and that category signals appeared with a shorter latency in the PFC than in the ITC, as if it were fed back from the PFC (Freedman et al. 2003, Meyers et al. 2008). Human imaging studies found that ITC is sensitive to perceptual features of stimuli and perceptual distance between stimuli, but only PFC represents the boundary between actual categories or crucial conjunctions between features (Jiang et al. 2007, Li et al. 2009).

PFC neurons also reflect abstract rule-based categorical distinctions. For example, Wallis and Miller (Muhammad et al. 2006, Wallis et al. 2001, Wallis & Miller 2003,) trained monkeys to apply either a "same" or "different" rule to novel pairs of pictures (see **Figure 1e**). Many PFC neurons conveyed which rule was in effect independent of which specific cue signaled the rule, was not linked to the behavioral response, and was unaffected by the exact pictures the monkeys were judging. By contrast, the rules had relatively little effect in the ITC, even though the ITC is directly connected with the lateral PFC and it is critical for visual analysis of the pictures (Muhammad et al. 2006).

Parietal Cortex

The parietal cortex seems to emphasize visuospatial functions and linking information from perceptual cortex with potential responses. Many studies have examined its neural selectivity by having subjects discriminate the direction of motion of moving dots. Many

PFC: prefrontal cortex

direction-selective neurons are in extrastriate area V5/MT (Newsome et al. 1986) and project to the lateral inferior parietal lobe and insula, which integrate overall movement pattern (Ho et al. 2009, Rorie & Newsome 2005). Freedman & Assad (2006) trained monkeys to classify 360° of motion direction into two categories and found that category membership was strongly reflected in the lateral inferior parietal region, but much less so in V5/MT. The respective roles of the parietal and frontal cortices in categorization and visual cognition in general remain to be determined, but several studies indicate a close functional link between the lateral inferior parietal lobe and the PFC (Buschman & Miller 2007, Chafee & Goldman-Rakic 2000).

Premotor and Motor Cortex

Categorical decision tasks also involve selection and execution of an appropriate behavior. This recruits premotor cortex (PMC) and primary motor cortex within the frontal lobe. Category learning can also result in plasticity in brain systems involved in attention and eye movements (Blair et al. 2009). Little & Thulborn (2005) found changes in frontal eye field and supplementary eye field activity across training in a dot pattern categorization task that likely reflected improved visual scanning of the stimuli.

As expertise is developed, reliance on motor systems increases and reliance on other systems decreases. Indeed, PFC damage preferentially affects new learning: Animals and humans can still engage in complex behaviors as long as they were well learned before the damage (Dias et al. 1997, Murray et al. 2000, Shallice 1982). PFC neurons are more strongly activated during new learning than during execution of familiar tasks (Asaad et al. 1998). There are stronger signals in the dorsal PMC than in the PFC when humans performed familiar versus novel classifications (Boettiger & D'Esposito 2005) and when monkeys performed familiar abstract rules (Muhammad et al. 2006). Thus, the PFC may acquire new categories, but other areas such as

the PMC may execute them once they become familiar.

Hippocampus and the Medial Temporal Lobe

The medial temporal lobe (MTL) has anatomical and functional connections with cortex and seems specialized for rapid learning of individual instances (O'Reilly & Munakata 2000). The circuitry of the MTL and cortex forms a loop: Information from broad neocortical regions across the parietal, frontal, and temporal cortices projects to the entorhinal region of the parahippocampal gyrus. From the entorhinal cortex, the primary projections pass to the dentate gyrus, the CA3 field of the hippocampus, the CA1 field, and back to the entorhinal cortex. The CA3 field contains autoassociative recurrent links, which allow association formation during encoding and pattern completion during recall (Becker & Wojtowicz 2007, Gluck et al. 2003, O'Reilly & Munakata 2000).

Several lines of evidence suggest multiple roles for the MTL in categorization. Categorization can make use of the MTL's ability to learn individual instances. One task that requires instance learning is the arbitrary categorization task (**Figure 1d**), in which the category membership of each item must be remembered individually. fMRI studies find that MTL (among other systems, including corticostriatal systems) is often recruited during these tasks (Poldrack et al. 1999, 2001; Seger & Cincotta 2005). Likewise, monkey neurophysiology studies found that neurons in the hippocampus and temporal cortex show category-specific activity after training monkeys to group arbitrary stimuli (Hampson et al. 2004). Kreiman et al. (2000) found neurons in the human MTL that were selective for diverse pictures of familiar concepts such as Bill Clinton. The MTL's instance-learning capacity may also be invoked to store exceptions to rules and other categorical regularities (Love et al. 2004). Some degree of instance memory may be required in all categorization tasks that use novel stimuli; the MTL may be required to set up a memory representation of each stimulus that can then be accessed by other systems (Meeter et al. 2008).

Another important potential contribution of the MTL follows from observations that information acquired via the MTL can be transferred to new situations. One example is acquired equivalence. For example, if a subject learns that stimulus A is in categories 1 and 2, and stimulus B is in category 1, they can reasonably infer that stimulus B might also be in category 2. The MTL is involved in these tasks (Myers et al. 2003, Shohamy & Wagner 2008).

The Basal Ganglia and Corticostriatal Loops

The basal ganglia are a collection of subcortical nuclei that interact with cortex in corticostriatal loops. Cortical inputs arrive largely via the striatum and ultimately are directed back into the cortex via the thalamus. The basal ganglia maintain a degree of topographical separation in different loops, ensuring that the output is largely to the same cortical areas that gave rise to the initial inputs to the basal ganglia (Alexander et al. 1986, Hoover & Strick 1993, Kelly & Strick 2004, Parthasarathy et al. 1992). The frontal cortex receives the largest portion of BG outputs, suggesting some form of close collaboration between these structures (Middleton & Strick 1994, 2002). However, almost all cortical regions participate in corticostriatal loops (Flaherty & Graybiel 1991, Kemp & Powell 1970). Although there is overlap between the loops at their boundaries, it is useful to talk of four loops: executive, motivational, visual, and motor (Lawrence et al. 1998, Seger 2008), as illustrated in **Figure 2**. The basal ganglia exert a tonic inhibition on cortex; they selectively and phasically release the cortex to allow for selection of a movement (Humphries et al. 2006) or cognitive strategy (Frank 2005). In categorization tasks, this function may be recruited to help with selection of both an appropriate category representation and related strategies or behaviors (Seger 2008).

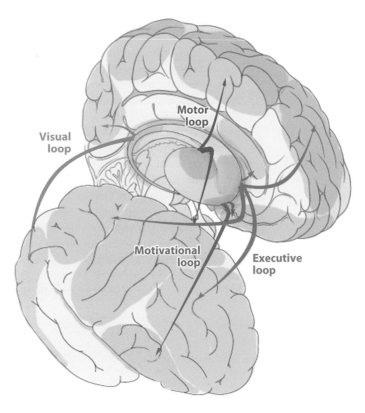

Figure 2

Corticostriatal loops. The motor loop (*blue*) connecting the motor cortex with the posterior putamen. Executive loop (*green*) connects the prefrontal cortex and the parietal cortex with the anterior caudate nucleus. The motivational loop (*red*) connects the ventral striatum with the orbitofrontal cortex. The visual loop (*orange*) connects extrastriate and inferotemporal cortices with the posterior caudate nucleus.

The basal ganglia are active in a wide variety of categorization tasks (Nomura et al. 2007; Poldrack et al. 1999, 2001; Seger & Cincotta 2005; Zeithamova et al. 2008), particularly those that require subjects to learn via trial and error (Cincotta & Seger 2007, Merchant et al. 1997). Performance on these tasks is impaired in patients with compromised basal ganglia functions owing to Parkinson and Huntington disease (Ashby & Maddox 2005, Knowlton et al. 1996, Shohamy et al. 2004). The roles of individual corticostriatal loops and their interactions during categorization are discussed further below.

Midbrain Dopaminergic System and Reinforcement Learning Mechanisms

Any form of supervised (reward-based) learning, including category learning, depends on the midbrain dopaminergic brain systems (the ventral tegmental area and the substantia nigra, pars compacta) (Schultz et al. 1992). Neurons in these areas show activity that seems to correspond to the reward prediction error signals suggested by animal learning models (Hollerman & Schultz 1998, Montague et al. 2004; but see Redgrave & Gurney 2006). They activate and release dopamine widely throughout the basal ganglia and cortex (especially in the frontal lobe) whenever animals are unexpectedly rewarded, and they pause when an expected reward is withheld. Over time the cells learn to respond to an event that directly predicts a reward: The event stands in for the reward (Schultz et al. 1993). Functional imaging has found that the basal ganglia, a primary target of dopamine neurons, are also sensitive to prediction error (Seymour et al. 2007).

Cortical inputs converge onto the dendrites of striatal spiny cells along with a strong input from midbrain dopaminergic neurons. Dopamine is required for synapse strengthening or weakening in the striatum by long-term depression or potentiation, respectively (Calabresi et al. 1992, Kerr & Wickens 2001, Otani et al. 1998). These anatomical and neurophysiological properties suggest that the striatum has an ideal infrastructure for rapid, reward-gated, supervised learning that quickly forms representations of the patterns of cortical connections that predict reward (Houk & Wise 1995, Miller & Buschman 2007). Functional imaging, neuropsychological, and computational studies suggest that feedback-based category learning via trial and error depends on both dopamine and the basal ganglia (Shohamy et al. 2008).

INTERACTION BETWEEN NEURAL SYSTEMS DURING CATEGORY LEARNING

Above, we discussed how categorization learning relies on multiple neural systems. For example, a visual categorization task may recruit the visual cortex and the MTL to represent and memorize the individual stimuli and facilitate processing of relevant features, the prefrontal cortex to learn and represent categorization rules and strategies, and the basal ganglia, parietal lobe, and motor cortices to make decisions and select behavioral responses on the basis of categorical information. In this section we discuss several ways that these neural systems may interact during category learning.

Interactions Between Fast Subcortical Plasticity and Slower Cortical Plasticity

A key issue in learning is the need to balance the advantages and disadvantages of fast versus slow plasticity (see sidebar, Computational Factors in Category Learning). Fast

---

COMPUTATIONAL FACTORS IN CATEGORY LEARNING

Generalized knowledge versus memory for specific instances. The complementary memory systems framework notes that generalized knowledge (e.g., the overall concept of a chair) conflicts with specific memories (e.g., one's own office chair) (O'Reilly & Munakata, 2000). Categorization learning emphasizes the acquisition of generalized knowledge about the world but also requires some specific representations, for example in the situation of arbitrary categories or in representing exceptions to general rules.

Fast versus slow learning. Fast learning has obvious advantages: One can learn to acquire resources and avoid obstacles faster and better than competitors. But fast learning comes at a cost; it does not allow the benefits that come from generalizing over multiple experiences, so by necessity it tends to be specific and error prone. For example, consider conditioned taste aversion: a one-trial and often erroneous aversion for a particular food. Extending learning across multiple episodes allows organisms to pick up on the regularities of predictive relationships and leave behind spurious associations and coincidences. This allows category formation by allowing learning mechanisms to identify the commonalities across different category members. We suggest that the brain balances the advantages and disadvantages of fast versus slow learning by having fast plasticity mechanisms (large changes in synaptic weights) in subcortical structures train slower plasticity (small weight changes) in cortical networks.

plasticity (large changes in synaptic weights with each episode) in a neural network has advantages in rapid storage of relevant activity patterns (and quick learning). But slow plasticity (small weight changes) allows networks to generalize; gradual changes result in neural ensembles that are not tied to specific inputs but instead store what is common among them. One possible solution is to have fast plasticity and slow plasticity systems interact (McClelland et al. 1995, O'Reilly & Munakata, 2000). For example, McClelland et al. (1995) suggested that long-term memory consolidation results from fast plasticity in the hippocampus, the output of which trains slower plasticity cortical networks that gradually elaborate the memories and link them to others. A similar relationship between the cerebellum and the cortex could underlie motor learning (Houk & Wise 1995). We suggest that an interaction between fast plasticity in the basal ganglia and slow plasticity in the cortex underlies many forms of category learning and abstraction (Miller & Buschman 2007).

Fast and slow plasticity may arise from different applications of the dopaminergic teaching signal. Both the cortex and the basal ganglia receive projections from midbrain dopaminergic neurons, but dopamine input to cortex is much lighter than that into the striatum (Lynd-Balta & Haber 1994). Dopamine projections also show a gradient in connectivity with heavier inputs in the PFC that drop off posteriorly (Goldman-Rakic et al. 1989, Thierry et al. 1973). This observation may explain why the PFC seems to show a greater deal of experience-dependent selectivity than does the visual cortex. In the striatum, the dopamine influence may be greater still. Dopamine neurons terminate near the synapse between a cortical axon and striatal spiny cell, a good position to gate plasticity between the cortex and the striatum. DA neurons synapse on the dendrites of cortical neurons, and therefore may have a lesser influence. Thus, whereas plasticity in the striatum may be fast and reward-gated in the cortex, it may be slower and reward-shaded. The striatum may be better suited to learn

details, the specific cues, responses, etc. that predict rewards, whereas the cortex acquires the commonalities among them that result in categories and abstractions (see Daw et al. 2005).

Some evidence suggests this notion. Pasupathy & Miller (2005) found that during conditional visuomotor learning in monkeys, striatal neural activity showed rapid, almost bistable, changes compared with a much slower trend in the PFC. Seger & Cincotta (2006) found that as humans learn rules, changes in striatal activity precede those in the frontal cortex. Abstract rules are more strongly represented (more neurons and stronger effects) and appear with a shorter latency in the frontal cortex than in the dorsal striatum (Muhammad et al. 2006), which is consistent with a greater cortical involvement in abstraction.

Under this view, normal learning depends on balance between the fast and the slow plasticity systems. An imbalance between these systems that causes basal ganglia plasticity to become abnormally strong and overwhelm the cortex might result in an autistic-like brain that is overwhelmed with details and cannot generalize. Recent work by Bear and colleagues may provide a molecular link (Dolen et al. 2007). They found that many psychiatric and neurological symptoms of Fragile X, including autism, can be explained by abnormally high activation of metabotropic glutamate receptor mGluR5. MGluR5 colocalizes with dopamine receptors in striatal neurons and is thought to regulate dopamine-dependent plasticity. The idea is that too much mGluR5 boosts dopaminergic plasticity mechanisms in striatum and overwhelms the cortex, resulting in an inability to generalize and fractionated, piecemeal cognition.

Interactions Within Corticostriatal Loops: Recursive Processing and Bootstrapping

As noted above, the cortex forms closed anatomical loops with the basal ganglia: Channels within the basal ganglia return outputs, via the thalamus, to the same cortical areas that gave rise to their initial cortical input (Hoover & Strick 1993, Kelly & Strick 2004). Closed loops suggest recursivity, bootstrapping operations in which the results from one iteration are fed back through the loop for further processing and elaboration. Some form of recursive processing must underlie the open-ended nature of human memory and thought. We suggest that recursive interactions between basal ganglia fast plasticity and slow cortical plasticity underlie construction of categories and abstractions. This idea may be reflected in a hallmark of human intelligence: It is easiest for us to understand new categories and concepts if they can be grounded first in familiar ones. We learn to multiply through serial addition, and we understand quantum mechanics by constructing analogies to waves and particles.

Interactions Between Corticostriatal Loops

Although basal ganglia–PFC connections are particularly prominent, the basal ganglia interact with all cortical regions. **Figure 2** illustrates the major patterns of projection, broken into loops. Functional imaging has shown that all four loops are recruited during categorization learning, albeit in different roles (Seger 2008, Seger & Cincotta 2005). The visual loop receives information from visual cortex; this information feeds forward to the executive and motor loops, providing a potential mechanism for selection of appropriate responses (Ashby et al. 1998, 2007), as well as back to visual cortex where it may assist in refinement of visual processing. The executive loop is associated with functions necessary for categorization learning, including feedback processing, working memory updating, and set shifting. The motor loop is involved in selecting and executing appropriate motor behavior, including selection of the motor response used to indicate category membership. The motivational loop is involved in processing reward and feedback.

The loops interact during learning. Seger and colleagues (2010) examined interactions between corticostriatal loops during categorization using Granger causality modeling and found patterns consistent with directed influence from the visual loop to the motor loop, and from the motor loop to the executive loop. This pattern is consistent with the processes required during each step of a typical categorization trial: processing the visual stimulus, preparing and executing the motor response indicating category membership, and receiving and processing feedback.

Corticostriatal loops also interact across many experiences or trials as subjects progress from being novices to experts in a categorization domain. The executive and motivational loops are most important early, when acquisition of information is fastest and feedback processing is the most useful, whereas the motor loop rises in importance as expertise is acquired (Williams & Eskandar 2006). The anterior caudate (executive loop) is sensitive to learning rate; activity is greatest when learning is occurring most rapidly (Williams & Eskandar 2006) and there is the greatest amount of prediction error (difference between expected outcome and actual outcome) to serve as a learning signal (Haruno & Kawato 2006). In contrast, the putamen (motor loop) is more engaged late in learning, when the category membership (and associated reward or feedback) is well learned. (Seger et al. 2010, Williams & Eskandar 2006). This idea is consistent with observations that the rodent dorsomedial striatum (equivalent to primate anterior caudate) is important for initial goal-oriented learning, whereas dorsolateral striatum (equivalent to primate posterior putamen) is important for later habit formation (Yin & Knowlton 2006).

Finally, corticostriatal loops can compete depending on the material being learned. Categories that can be learned via explicit rule-based processes tend to rely on the PFC and anterior caudate regions involved in the executive loop. Other category structures (such as information integration categories; **Figure 1b**) that are learned via more implicit processes rely on the visual loop. The COVIS model (Ashby et al. 1998, 2007) proposes that the executive and visual loops compete for dominance in controlling categorization. This proposal is supported by studies examining individual differences in prefrontal capacity: Subjects with high capacity tend to favor the rule-learning system and are relatively impaired at learning an information integration task that requires the more implicit strategy to achieve optimal performance (Decaro et al. 2008).

Interactions Between the Medial Temporal Lobe and Basal Ganglia

Both MTL and BG systems can form relationships between stimuli and categories. As described above, the MTL does so via explicit representation of the stimulus and its arbitrary category membership, whereas the basal ganglia map perceptual commonalities of categories to their associated behaviors. Human imaging studies suggest competition between MTL and BG systems during category learning: As BG activity increases, MTL activity decreases (Poldrack et al. 1999, 2001). However, relative decreases in MTL activity are difficult to interpret in functional imaging studies; apparent suppression of the MTL may simply be due to lower activity during categorization than during the comparison tasks (Law et al. 2005). Stronger evidence for competition between the two systems comes from lesion and pharmacological manipulations. When MTL is damaged or inhibited, the BG can take over a larger role in the control of behavior (Frank et al. 2006). Subjects with basal ganglia damage due to Parkinson disease recruit MTL to a larger extent than do controls during probabilistic classification category learning (Moody et al. 2004).

The BG and MTL may not invariably compete during categorization learning. Some studies show parallel recruitment of both systems, implying independent or cooperative contributions (Cincotta & Seger 2007). The MTL may be required initially to set up new individual item representations of stimuli (Meeter et al. 2008). These stimulus representations may then be accessible to BG systems for forming associations between stimuli and categories. Consistent with this theory,

Poldrack et al. (2001) found transient MTL activity at the beginning of a probabilistic classification task, which was then followed by a relative decrease in MTL activity and increase in BG activity.

It is unclear how interaction between MTL and BG is mediated. Some evidence indicates that the relationship is bilateral: Increases in BG activity lead to decreases in MTL activity and vice versa (Lee et al. 2008). Some research indicates that PFC is involved in this process (Poldrack & Rodriguez 2004). During distraction with PFC demanding dual tasks, categorization performance becomes more strongly related to striatal activity and less related to MTL (Foerde et al. 2006). In emotional situations, the amygdala can likely mediate the balance between systems (Wingard & Packard 2008).

CONCLUSION: PRINCIPLES OF CATEGORY LEARNING IN THE BRAIN

We are only beginning to understand how the brain learns categories. But we can posit some potential principles and hypotheses.

- Categorization involves both stimulus representations (e.g., of features, central tendencies, and degree of variability) and processes (e.g., decision-making processes establishing a criterion or rule for category membership) that recruit different neural systems depending on the type of category and how it is used.
- The brain does not have one single "categorization area." Categories are represented in a distributed fashion across the brain, and multiple neural systems are involved. Many of the systems involved in categorization have been identified in the multiple memory systems framework (Ashby & O'Brien 2005, Poldrack & Foerde 2008, Smith & Grossman 2008). Categorization tasks are not process-pure: Multiple systems may be recruited to solve any given categorization problem.

- Category learning withstands fundamental computational constraints. A trade-off exists between generalizing across previous experience and remembering specific items and events. This trade-off may be solved by having fast plasticity (large synaptic weight changes) in subcortical systems (e.g., basal ganglia and hippocampus) train slower plasticity (smaller weight changes) in the cortex, the latter of which builds the category representations by finding the commonalities across the specifics learned by the former. Normal learning depends on balance between these mechanisms. The balance can change depending on task demands. Certain neuropsychiatric disorders, such as autism, may result from an imbalance that causes the faster plasticity mechanisms in the subcortex to overwhelm the slower cortical plasticity, which could result in a brain that has great difficulty generalizing.

- Category learning may depend on recursive, bootstrapping interactions within corticostriatal loops. The open-ended nature of human thought likely depends on some form of recursive processing, and the closed anatomical loops the basal ganglia form with the cortex seem well suited. Different phases of learning and different aspects of a categorization task may also involve interactions across different corticostriatal loops.

- Category learning cuts across distinctions between implicit and explicit systems and declarative and nondeclarative memory systems. Explicit systems are those that are associated with some degree of conscious penetrability (Seger 1994). In categorization, these include PFC systems recruited in explicit rule-learning tasks, as well as MTL systems that result in consciously accessible episodic memories. Most other systems are typically considered to be implicit or unconscious (e.g., perceptual cortex); however, some (notably the corticostriatal loops) can be recruited in both explicit and implicit tasks. The declarative–nondeclarative distinction differs from the explicit–implicit distinction because it separates MTL-dependent memory processes (declarative) from other learning systems (nondeclarative). Categorization tasks may recruit various combinations of implicit and/or explicit, declarative and/or nondeclarative systems. For example, simple dot pattern learning is largely implicit (it occurs without intention to learn or awareness of learning) and nondeclarative (it is independent of MTL systems). Rule learning is explicit because subjects intend to learn and have awareness of what they have learned and is nondeclarative because it largely recruits prefrontal cortex and does not require the MTL for acquisition.

- A major challenge in understanding category learning is determining which category-learning systems are recruited in particular situations, and whether the systems function independently, cooperatively, or antagonistically. What is ultimately learned is an interaction between the structure of the information in the environment and the neural systems recruited to process the information (Reber et al. 2003, Zeithamova et al. 2008). Which systems are recruited can also depend on factors that can vary across individuals and situations, such as cognitive capacity (Decaro et al. 2008) and motivational state (Grimm et al. 2007).

DISCLOSURE STATEMENT

The authors are not aware of any affiliations, memberships, funding, or financial holdings that might be perceived as affecting the objectivity of this review.

ACKNOWLEDGMENTS

Preparation of this chapter was supported by the National Institute of Mental Health (R01-MH079182-05 to C.A.S.; 2-R01-MH065252-06 to E.K.M.) and Richard and Linda Hardy (to E.K.M.). We thank Timothy Buschman, Jason Cromer, Jefferson Roy, Brian Spiering, and Marlene Wicherski for valuable comments and Dan Lopez-Paniagua for preparing the figures.

LITERATURE CITED

Aizenstein HJ, MacDonald AW, Stenger VA, Nebes RD, Larson JK, et al. 2000. Complementary category learning systems identified using event-related functional MRI. *J. Cogn. Neurosci.* 12:977–87

Alexander GE, DeLong MR, Strick PL. 1986. Parallel organization of functionally segregated circuits linking basal ganglia and cortex. *Annu. Rev. Neurosci.* 9:357–81

Asaad WF, Rainer G, Miller EK. 1998. Neural activity in the primate prefrontal cortex during associative learning. *Neuron* 21:1399–407

Ashby FG, Alfonso-Reese LA, Turken AU, Waldron EM. 1998. A neuropsychological theory of multiple systems in category learning. *Psychol. Rev.* 105:442–81

Ashby FG, Ennis JM, Spiering BJ. 2007. A neurobiological theory of automaticity in perceptual categorization. *Psychol. Rev.* 114:632–56

Ashby FG, Maddox WT. 2005. Human category learning. *Annu. Rev. Psychol.* 56:149–78

Ashby FG, O'Brien JB. 2005. Category learning and multiple memory systems. *Trends Cogn. Sci.* 9:83–89

Baker CI, Behrmann M, Olson CR. 2002. Impact of learning on representation of parts and wholes in monkey inferotemporal cortex. *Nat. Neurosci.* 5:1210–16

Barot SK, Kyono Y, Clark EW, Bernstein IL. 2008. Visualizing stimulus convergence in amygdala neurons during associative learning. *Proc. Natl. Acad. Sci. USA* 105:20959–63

Becker S, Wojtowicz JM. 2007. A model of hippocampal neurogenesis in memory and mood disorders. *Trends Cogn. Sci.* 11:70–76

Blair MR, Watson MR, Walshe RC, Maj F. 2009. Extremely selective attention: eye-tracking studies of the dynamic allocation of attention to stimulus features in categorization. *J. Exp. Psychol. Learn. Mem. Cogn.* 35:1196–206

Boettiger CA, D'Esposito M. 2005. Frontal networks for learning and executing arbitrary stimulus-response associations. *J. Neurosci.* 25:2723–32

Buschman TJ, Miller EK. 2007. Top-down versus bottom-up control of attention in the prefrontal and posterior parietal cortices. *Science* 315:1860–62

Calabresi P, Maj R, Pisani A, Mercuri NB, Bernardi G. 1992. Long-term synaptic depression in the striatum: physiological and pharmacological characterization. *J. Neurosci.* 12:4224–33

Casale MB, Ashby FG. 2008. A role for the perceptual representation memory system in category learning. *Percept. Psychophys.* 70:983–99

Chafee MV, Goldman-Rakic PS. 2000. Inactivation of parietal and prefrontal cortex reveals interdependence of neural activity during memory-guided saccades. *J. Neurophysiol.* 83:1550–66

Cincotta CM, Seger CA. 2007. Dissociation between striatal regions while learning to categorize via feedback and via observation. *J. Cogn. Neurosci.* 19:249–65

Daw ND, Niv Y, Dayan P. 2005. Uncertainty-based competition between prefrontal and dorsolateral striatal systems for behavioral control. *Nat. Neurosci.* 8:1704–11

Decaro MS, Thomas RD, Beilock SL. 2008. Individual differences in category learning: sometimes less working memory capacity is better than more. *Cognition* 107:284–94

DeGutis J, D'Esposito M. 2007. Distinct mechanisms in visual category learning. *Cogn. Affect. Behav. Neurosci.* 7:251–59

Desimone R, Albright TD, Gross CG, Bruce C. 1984. Stimulus-selective properties of inferior temporal neurons in the macaque. *J. Neurosci.* 4:2051–62

Dias R, Robbins TW, Roberts AC. 1997. Dissociable forms of inhibitory control within prefrontal cortex with an analog of the Wisconsin Card Sort Test: restriction to novel situations and independence from "on-line" processing. *J. Neurosci.* 17:9285–97

Dolen G, Osterweil E, Rao BS, Smith GB, Auerbach BD, et al. 2007. Correction of fragile X syndrome in mice. *Neuron* 56:955–62

Flaherty AW, Graybiel AM. 1991. Corticostriatal transformations in the primate somatosensory system. Projections from physiologically mapped body-part representations. *J. Neurophysiol.* 66:1249–63

Foerde K, Knowlton BJ, Poldrack RA. 2006. Modulation of competing memory systems by distraction. *Proc. Natl. Acad. Sci. USA* 103:11778–83

Frank MJ. 2005. Dynamic dopamine modulation in the basal ganglia: a neurocomputational account of cognitive deficits in medicated and nonmedicated parkinsonism. *J. Cogn. Neurosci.* 17:51–72

Frank MJ, O'Reilly RC, Curran T. 2006. When memory fails, intuition reigns: Midazolam enhances implicit inference in humans. *Psychol. Sci.* 17:700–7

Freedman DJ, Assad JA. 2006. Experience-dependent representation of visual categories in parietal cortex. *Nature* 443:85–88

Freedman DJ, Riesenhuber M, Poggio T, Miller EK. 2001. Categorical representation of visual stimuli in the primate prefrontal cortex. *Science* 291:312–16

Freedman DJ, Riesenhuber M, Poggio T, Miller EK. 2002. Visual categorization and the primate prefrontal cortex: neurophysiology and behavior. *J. Neurophysiol.* 88:914–28

Freedman DJ, Riesenhuber M, Poggio T, Miller EK. 2003. A comparison of primate prefrontal and inferior temporal cortices during visual categorization. *J. Neurosci.* 23:5235–46

Fuster JM. 1995. *Memory in the Cerebral Cortex.* Cambridge, MA: MIT Press

Gluck MA, Meeter M, Myers CE. 2003. Computational models of the hippocampal region: linking incremental learning and episodic memory. *Trends Cogn. Sci.* 7:269–76

Goldman-Rakic PS, Leranth C, Williams SM, Mons N, Geffard M. 1989. Dopamine synaptic complex with pyramidal neurons in primate cerebral cortex. *Proc. Natl. Acad. Sci. USA* 86:9015–19

Grimm LR, Markman AB, Maddox WT, Baldwin GC. 2007. Differential effects of regulatory fit on category learning. *J. Exp. Soc. Psychol.* 44:920–27

Hampson RE, Pons TP, Stanford TR, Deadwyler SA. 2004. Categorization in the monkey hippocampus: a possible mechanism for encoding information into memory. *Proc. Natl. Acad. Sci. USA* 101:3184–89

Haruno M, Kawato M. 2006. Different neural correlates of reward expectation and reward expectation error in the putamen and caudate nucleus during stimulus-action-reward association learning. *J. Neurophysiol.* 95:948–59

Ho TC, Brown S, Serences JT. 2009. Domain general mechanisms of perceptual decision making in human cortex. *J. Neurosci.* 29:8675–87

Hollerman JR, Schultz W. 1998. Dopamine neurons report an error in the temporal prediction of reward during learning. *Nat. Neurosci.* 1:304–9

Hoover JE, Strick PL. 1993. Multiple output channels in the basal ganglia. *Science* 259:819–21

Houk JC, Wise SP. 1995. Distributed modular architectures linking basal ganglia, cerebellum, and cerebral cortex: their role in planning and controlling action. *Cereb. Cortex* 5:95–110

Howard JD, Plailly J, Grueschow M, Haynes JD, Gottfried JA. 2009. Odor quality coding and categorization in human posterior piriform cortex. *Nat. Neurosci.* 12(7):932–38

Humphries MD, Stewart RD, Gurney KN. 2006. A physiologically plausible model of action selection and oscillatory activity in the basal ganglia. *J. Neurosci.* 26:12921–42

Jiang X, Bradley E, Rini RA, Zeffiro T, Vanmeter J, Riesenhuber M. 2007. Categorization training results in shape- and category-selective human neural plasticity. *Neuron* 53:891–903

Kanwisher N, McDermott J, Chun MM. 1997. The fusiform face area: a module in human extrastriate cortex specialized for face perception. *J. Neurosci.* 17:4302–11

Kawasaki K, Sheinberg DL. 2008. Learning to recognize visual objects with microstimulation in inferior temporal cortex. *J. Neurophysiol.* 100:197–211

Kelly RM, Strick PL. 2004. Macro-architecture of basal ganglia loops with the cerebral cortex: use of rabies virus to reveal multisynaptic circuits. *Prog. Brain Res.* 143:449–59

Kemp JM, Powell TP. 1970. The cortico-striate projection in the monkey. *Brain* 93:525–46

Keri S, Kalman J, Rapcsak SZ, Antal A, Benedek G, Janka Z. 1999. Classification learning in Alzheimer's disease. *Brain* 122:1063–68

Kerr JND, Wickens JR. 2001. Dopamine D-1/D-5 receptor activation is required for long-term potentiation in the rat neostriatum in vitro. *J. Neurophysiol.* 85:117–24

Knowlton BK, Mangels JA, Squire LR. 1996. A neostriatal habit learning system in humans. *Science* 273:1399–402

Knowlton BK, Squire LR. 1993. The learning of categories: parallel brain systems for item memory and category knowledge. *Science* 262:1747–49

Kreiman G, Koch C, Fried I. 2000. Category-specific visual responses of single neurons in the human medial temporal lobe. *Nat. Neurosci.* 3:946–53

Law JR, Flanery MA, Wirth S, Yanike M, Smith AC, et al. 2005. Functional magnetic resonance imaging activity during the gradual acquisition and expression of paired-associate memory. *J. Neurosci.* 25:5720–29

Lawrence AD, Sahakian BJ, Robbins TW. 1998. Cognitive functions and corticostriatal circuits: insights from Huntington's disease. *Trends Cogn. Sci.* 2:379–88

Lee AS, Duman RS, Pittenger C. 2008. A double dissociation revealing bidirectional competition between striatum and hippocampus during learning. *Proc. Natl. Acad. Sci. USA* 105:17163–68

Li S, Mayhew SD, Kourtzi Z. 2009. Learning shapes the representation of behavioral choice in the human brain. *Neuron* 62:441–52

Little DM, Thulborn KR. 2005. Correlations of cortical activation and behavior during the application of newly learned categories. *Brain Res. Cogn. Brain Res.* 25:33–47

Logothetis NK, Sheinberg DL. 1996. Visual object recognition. *Annu. Rev. Neurosci.* 19:577–621

Love BC, Medin DL, Gureckis TM. 2004. SUSTAIN: a network model of category learning. *Psychol. Rev.* 111:309–32

Lynd-Balta E, Haber SN. 1994. The organization of midbrain projections to the ventral striatum in the primate. *Neuroscience* 59:609–23

Mahon BZ, Caramazza A. 2009. Concepts and categories: a cognitive neuropsychological perspective. *Annu. Rev. Psychol.* 60:27–51

Martin A. 2007. The representation of object concepts in the brain. *Annu. Rev. Psychol.* 58:25–45

McClelland JL. 2006. How far can you go with Hebbian learning, and when does it lead you astray? In *Processes of Change in Brain and Cognitive Development: Attention and Performance XXI*, ed. Y Munakata, MH Johnson, pp. 33–69. Oxford: Oxford Univ. Press

McClelland J, McNaughton B, O'Reilly R. 1995. Why there are complementary learning systems in the hippocampus and neocortex: insights from the successes and failures of connectionist models of learning and memory. *Psychol. Rev.* 102:419–57

Meeter M, Radics G, Myers CE, Gluck MA, Hopkins RO. 2008. Probabilistic categorization: How do normal participants and amnesic patients do it? *Neurosci. Biobehav. Rev.* 32:237–48

Merchant H, Zainos A, Hernández A, Salinas E, Romo R. 1997. Functional properties of primate putamen neurons during the categorization of tactile stimuli. *J. Neurophysiol.* 77:1132–54

Meyers EM, Freedman DJ, Kreiman G, Miller EK, Poggio T. 2008. Dynamic population coding of category information in inferior temporal and prefrontal cortex. *J. Neurophysiol.* 100:1407–19

Middleton FA, Strick PL. 1994. Anatomical evidence for cerebellar and basal ganglia involvement in higher cognitive function. *Science* 266:458–61

Middleton FA, Strick PL. 2002. Basal-ganglia 'projections' to the prefrontal cortex of the primate. *Cereb. Cortex* 12:926–35

Miller EK, Buschman TJ. 2007. Rules through recursion: how interactions between the frontal cortex and basal ganglia may build abstract, complex rules from concrete, simple ones. In *The Neuroscience of Rule-Guided Behavior*, ed. SB, JD Wallis, pp. 419–40. Oxford: Oxford Univ. Press

Miller EK, Cohen JD. 2001. An integrative theory of prefrontal function. *Annu. Rev. Neurosci.* 24:167–202

Montague PR, Hyman SE, Cohen JD. 2004. Computational roles for dopamine in behavioural control. *Nature* 431:760–67

Moody TD, Bookheimer SY, Vanek Z, Knowlton BJ. 2004. An implicit learning task activates medial temporal lobe in patients with Parkinson's disease. *Behav. Neurosci.* 118:438–42

Muhammad R, Wallis JD, Miller EK. 2006. A comparison of abstract rules in the prefrontal cortex, premotor cortex, inferior temporal cortex, and striatum. *J. Cogn. Neurosci.* 18:974–89

Murray EA, Bussey TJ, Wise SP. 2000. Role of prefrontal cortex in a network for arbitrary visuomotor mapping. *Exp. Brain Res.* 133:114–29

Myers CE, Shohamy D, Gluck MA, Grossman S, Kluger A, et al. 2003. Dissociating hippocampal versus basal ganglia contributions to learning and transfer. *J. Cogn. Neurosci.* 15:185–93

Newsome WT, Mikami A, Wurtz RH. 1986. Motion selectivity in macaque visual cortex. III. Psychophysics and physiology of apparent motion. *J. Neurophysiol.* 55:1340–51

Nomura EM, Maddox WT, Filoteo JV, Ing AD, Gitelman DR, et al. 2007. Neural correlates of rule-based and information-integration visual category learning. *Cereb. Cortex* 17:37–43

O'Reilly RC, Munakata Y. 2000. *Computational Explorations in Cognitive Neuroscience: Understanding the Mind.* Cambridge, MA: MIT Press

Otani S, Blond O, Desce JM, Crépel F. 1998. Dopamine facilitates long-term depression of glutamatergic transmission in rat prefrontal cortex. *Neuroscience* 85:669–76

Parthasarathy HB, Schall JD, Graybiel AM. 1992. Distributed but convergent ordering of corticostriatal projections—analysis of the frontal eye field and the supplementary eye field in the macaque monkey. *J. Neurosci.* 12:4468–88

Pasupathy A, Miller EK. 2005. Different time courses of learning-related activity in the prefrontal cortex and striatum. *Nature* 433:873–76

Poldrack RA, Clark J, Pare-Blagoev EJ, Shohamy D, Creso MJ, et al. 2001. Interactive memory systems in the human brain. *Nature* 414:546–50

Poldrack RA, Foerde K. 2008. Category learning and the memory systems debate. *Neurosci. Biobehav. Rev.* 32:197–205

Poldrack RA, Prabhakaran V, Seger CA, Gabrieli JDE. 1999. Striatal activation during acquisition of a cognitive skill. *Neuropsychology* 13:564–74

Poldrack RA, Rodriguez P. 2004. How do memory systems interact? Evidence from human classification learning. *Neurobiol. Learn. Mem.* 82:324–32

Reber PJ, Gitelman DR, Parrish TB, Mesulam MM. 2003. Dissociating explicit and implicit category knowledge with fMRI. *J. Cogn. Neurosci.* 15:574–83

Reber PJ, Squire LR. 1999. Intact learning of artificial grammars and intact category learning by patients with Parkinson's disease. *Behav. Neurosci.* 113:235–42

Reber PJ, Stark CE, Squire LR. 1998. Cortical areas supporting category learning identified using functional MRI. *Proc. Natl. Acad. Sci. USA* 95:747–50

Redgrave P, Gurney K. 2006. The short-latency dopamine signal: a role in discovering novel actions? *Nat. Rev. Neurosci.* 7:967–75

Romo R, Salinas E. 2001. Touch and go: decision-making mechanisms in somatosensation. *Annu. Rev. Neurosci.* 24:107–37

Rorie AE, Newsome WT. 2005. A general mechanism for decision-making in the human brain? *Trends Cogn Sci.* 9(2):41–43

Schultz W, Apicella P, Ljungberg T. 1993. Responses of monkey dopamine neurons to reward and conditioned stimuli during successive steps of learning a delayed response task. *J. Neurosci.* 13:900–13

Schultz W, Apicella P, Scarnati E, Ljungberg T. 1992. Neuronal activity in monkey ventral striatum related to the expectation of reward. *J. Neurosci.* 12:4595–610

Seger CA. 1994. Implicit learning. *Psychol. Bull.* 115:163–96

Seger CA. 2008. How do the basal ganglia contribute to categorization? Their roles in generalization, response selection, and learning via feedback. *Neurosci. Biobehav. Rev.* 32:265–78

Seger CA, Cincotta CM. 2005. The roles of the caudate nucleus in human classification learning. *J. Neurosci.* 25:2941–51

Seger CA, Cincotta CM. 2006. Dynamics of frontal, striatal, and hippocampal systems during rule learning. *Cereb. Cortex* 16:1546–55

Seger CA, Peterson E, Lopez-Paniagua D, Cincotta CM, Anderson CM. 2010. Dissociating the contributions of independent corticostriatal systems to visual categorization learning through the use of reinforcement learning modeling and Granger causality modeling. *NeuroImage* 50:644–56

Seymour B, Daw N, Dayan P, Singer T, Dolan R. 2007. Differential encoding of losses and gains in the human striatum. *J. Neurosci.* 27:4826–31

Shallice T. 1982. Specific impairments of planning. *Philos. Trans. R. Soc. Lond. B Biol. Sci.* 298:199–209

Shohamy D, Myers CE, Grossman S, Sage J, Gluck MA, Poldrack RA. 2004. Cortico-striatal contributions to feedback-based learning: converging data from neuroimaging and neuropsychology. *Brain* 127:851–59

Shohamy D, Myers CE, Kalanithi J, Gluck MA. 2008. Basal ganglia and dopamine contributions to probabilistic category learning. *Neurosci. Biobehav. Rev.* 32:219–36

Shohamy D, Wagner AD. 2008. Integrating memories in the human brain: hippocampal-midbrain encoding of overlapping events. *Neuron* 60:378–89

Sigala N, Logothetis NK. 2002. Visual categorization shapes feature selectivity in the primate temporal cortex. *Nature* 415:318–20

Smith EE. 2008. The case for implicit category learning. *Cogn. Affect. Behav. Neurosci.* 8:3–16

Smith EE, Grossman M. 2008. Multiple systems of category learning. *Neurosci. Biobehav. Rev.* 32:249–64

Tanaka K. 1996. Inferotemporal cortex and object vision. *Annu. Rev. Neurosci.* 19:109–39

Thierry AM, Blanc G, Sobel A, Stinus L, Glowinski J. 1973. Dopaminergic terminals in the rat cortex. *Science* 182:499–501

Vallabha GK, McClelland JL, Pons F, Werker JF, Amano S. 2007. Unsupervised learning of vowel categories from infant-directed speech. *Proc. Natl. Acad. Sci. USA* 104:13273–78

Vogels R. 1999. Categorization of complex visual images by rhesus monkeys. Part 2: single-cell study. *Eur J. Neurosci.* 11:1239–55

Vogels R, Sary G, Dupont P, Orban GA. 2002. Human brain regions involved in visual categorization. *Neuroimage* 16:401–14

Wallis JD, Anderson KC, Miller EK. 2001. Single neurons in the prefrontal cortex encode abstract rules. *Nature* 411:953–56

Wallis JD, Miller EK. 2003. From rule to response: neuronal processes in the premotor and prefrontal cortex. *J. Neurophysiol.* 90:1790–806

Williams ZM, Eskandar EN. 2006. Selective enhancement of associative learning by microstimulation of the anterior caudate. *Nat. Neurosci.* 9:562–68

Wingard JC, Packard MG. 2008. The amygdala and emotional modulation of competition between cognitive and habit memory. *Behav. Brain Res.* 193:126–31

Wyttenbach RA, May ML, Hoy RR. 1996. Categorical perception of sound frequency by crickets. *Science* 273:1542–44

Yin HH, Knowlton BJ. 2006. The role of the basal ganglia in habit formation. *Nat. Rev. Neurosci.* 7:464–76

Zeithamova D, Maddox WT, Schnyer DM. 2008. Dissociable prototype learning systems: evidence from brain imaging and behavior. *J. Neurosci.* 28:13194–201

Molecular and Cellular Mechanisms of Learning Disabilities: A Focus on NF1

C. Shilyansky, Y.S. Lee, and A.J. Silva

Department of Neurobiology, Psychology, Psychiatry and Biobehavioral Sciences, Semel Institute, University of California, Los Angeles, California 90095; email: silvaa@mednet.ucla.edu

Annu. Rev. Neurosci. 2010. 33:221–43

First published online as a Review in Advance on March 26, 2010

The *Annual Review of Neuroscience* is online at neuro.annualreviews.org

This article's doi: 10.1146/annurev-neuro-060909-153215

Key Words

Ras, GABA, LTP, animal model, neurodevelopmental disorder, ADHD

Abstract

Neurofibromatosis Type I (NF1) is a single-gene disorder characterized by a high incidence of complex cognitive symptoms, including learning disabilities, attention deficit disorder, executive function deficits, and motor coordination problems. Because the underlying genetic cause of this disorder is known, study of NF1 from a molecular, cellular, and systems perspective has provided mechanistic insights into the etiology of higher-order cognitive symptoms associated with the disease. In particular, studies of animal models of NF1 indicated that disruption of Ras regulation of inhibitory networks is critical to the etiology of cognitive deficits associated with NF1. Animal models of Nf1 identified mechanisms and pathways that are required for cognition, and represent an important complement to the complex neuropsychological literature on learning disabilities associated with this condition. Here, we review findings from NF1 animal models and human populations affected by NF1, highlighting areas of potential translation and discussing the implications and limitations of generalizing findings from this single-gene disease to idiopathic learning disabilities.

Contents

INTRODUCTION

Neurofibromatosis Type I (NF1) is one of a number of genetically determined neurocutaneous disorders that cause clusters of somatic and behavioral symptoms. Originally, these disorders were described according to their somatic symptoms, for example café au lait spots, Lisch nodules, and neurofibromas in NF1.

Occurrence of these symptoms in a consistent pattern helped to identify a patient population affected by a single disease. Heading the human genetics revolution, linkage studies mapped the disease-causing gene in NF1 to chromosome 17 (Barker et al. 1987, Seizinger et al. 1987). A small group of patients with NF1 were found to carry translocation breakpoints on chromosome 17, which greatly accelerated identification of the disease-causing gene (Fountain et al. 1989, Leach et al. 1989, Ledbetter et al. 1989). Further gene mapping within this region identified the *Nf1* gene (Viskochil et al. 1990, Wallace et al. 1990). Although early on, these studies focused on the somatic symptoms of NF1, additional work within the NF1 patient population also revealed behavioral and learning problems (Hyman et al. 2005, North et al. 1997). These cognitive symptoms occur often, and have a significant impact on quality of life in individuals affected by NF1 (North et al. 1997).

NF1 is a single gene–determined disorder characterized by multiple complex cognitive symptoms. Disease-causing mutations in the *Nf1* gene (Viskochil et al. 1990, Wallace et al. 1990) result in loss of function of its protein product, Neurofibromin. Individuals affected by NF1 are heterozygous for the *Nf1* gene mutation, as homozygous mutations appear to be lethal (Friedman 1999). Further study of these cognitive symptoms revealed that even in this single-gene disease, a complex picture of variable deficits and types of learning disabilities occurs. Here, we review human and animal studies that together have provided insight into the nature of cognitive symptoms associated with NF1 and the mechanisms underlying their expression and variability. Since NF1 and other neurocutaneous disorders are likely part of the broader cluster of learning disability disorders, we also discuss the current understanding of the evolution of cognitive symptoms in NF1 in the context of hypotheses regarding other learning disabilities with unknown etiology and poorly defined affected populations.

NF1:
Neurofibromatosis
Type I

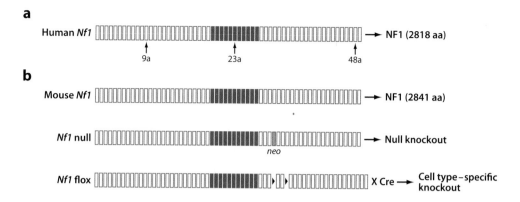

Figure 1

Schematic representation of human and mouse *Nf1*. (*a*) Human *Nf1* on chromosome 17 encodes a 2818–amino acid protein called neurofibromin. Small boxes represent exons. Filled boxes indicate exons 21 to 27a that encode the GAP-related domain (GRD). Alternatively spliced exons are marked by arrows. Disease-causing mutations are widely distributed throughout the entire *Nf1* gene and most of those result in loss of function of its protein product. (*b*) *Top*: the mouse *Nf1* gene on chromosome 11 has a structure very similar to that of human *Nf1* gene and encodes a protein with 98% identity to human NF1. *Middle*: *Nf1* null mutant mice were generated by inserting a *neomycin* cassette in exon 31 (Jacks et al. 1994). Light blue box indicates exon 31 including the *neo* gene. *Bottom*: conditional mutants were engineered by inserting loxP sites flanking exons 31 and 32 (Zhu et al. 2001). loxP sites are indicated by triangles. Delivery of Cre recombinase by crossing with Cre-expressing transgenic line or viral vector enables cell type–specific deletion of *Nf1*.

HUMAN GENETICS OF NF1

The *Nf1* gene, located on chromosome 17q, is one of the largest in the genome, encompassing 60 exons (**Figure 1**) (Li et al. 1995, Marchuk et al. 1991). It encodes multiple biochemical domains, including a Ras-GAP domain, which requires exon 23a for its activity. The *Nf1* gene has four splice variants, two of which are expressed in the CNS. The type I isoform of the *Nf1* gene contains exon 23a, and as a result encodes a version of Neurofibromin with efficient Ras-GAP activity. This variant is predominantly expressed in neurons. The type II isoform of the gene is a splice variant without exon 23a, and encodes a protein with ten times less Ras-GAP activity than type I. This variant is highly expressed in glia. Therefore, differential splicing of the *Nf1* gene confers different functions to its protein product across cell types.

Additionally, *Nf1* expression occurs across multiple brain systems that participate in a broad variety of behaviors. Within the CNS, *Nf1* transcription is seen in cortex, striatum, substantia nigra, brainstem, hippocampus, and cerebellum. In cortex and hippocampus, *Nf1* is expressed in pyramidal neurons, interneurons, and glia. In cortex, *Nf1* expression spans all cortical layers. In striatum, expression of *Nf1* is sparse, occurring in a pattern suggestive of interneuronal expression (Gutmann et al. 1995). It is also highly expressed in the Purkinje neurons of the cerebellum. This broad range of brain systems in which *Nf1* is expressed is also likely to contribute to the range of cognitive symptoms associated with its loss of function.

ROLE OF NEUROFIBROMIN IN CELL SIGNALING

The *Nf1* protein product, Neurofibromin, has multiple biochemical roles (**Figure 2**). This protein acts as a Ras-GAP (GTPase activating protein) to negatively regulate Ras signaling (Weiss et al. 1999), and it can also serve as an activator of adenylate cyclase (Tong et al. 2002). Its role as a Ras-GAP has been most clearly implicated in regulating neuronal function in mammals.

The Ras protein is part of a large superfamily of GTPases that mediate signaling from membrane receptors to intracellular cascades of kinases. Ras cycles between the inactive

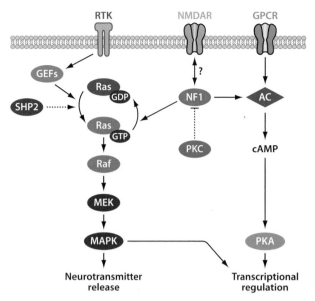

Figure 2

Neurofibromin and cell signaling. Neurofibromin (NF1) is a GTPase activating protein (GAP) that functions as a negative regulator of Ras-MAPK signaling cascade. Guanine-nucleotide exchange factors (GEFs), such as SOS, may counteract the GAP function of NF1. On the other hand, NF1 acts as an activator of adenylate cyclase (AC). Note that the key pathways are grossly simplified in this diagram. For example, the phosphatidylinositol 3-kinase (PI3K)-AKT-mTOR cascade is also modulated by NF1, but is omitted in the diagram. Arrows and barred lines indicate activation and suppression, respectively. MEK, mitogen-activated protein kinase or extracellular signal-regulated kinase kinase; NMDAR, N-methyl-D-aspartate receptor; GPCR, G protein-coupled receptor; RTK, receptor tyrosine kinase; PKA, protein kinase A; PKC, protein kinase C; SHP2, Src homology 2-containing tyrosine phosphatase.

GDP-bound state and the active GTP-bound state. The activity state of Ras is determined by a balance of activating proteins (GEFs, guanine nucleotide exchange factors, which allow bound GDP to be released so that GTP can bind) and inactivating proteins (GAPs, GTPase activating proteins, which increase the endogenous GTP hydrolyzing activity of Ras) (Bernards & Settleman 2004). Receptor tyrosine kinases, including the growth factor receptors (e.g., Trk B, the receptor for BDNF), are common upstream activators of Ras signaling (Bernards & Settleman 2005). This family of transmembrane receptors is characterized by phosphorylation of intracellular tyrosines upon ligand binding. Tyrosine phosphorylation then leads either to

direct binding and inactivation of Ras-GAPs such as Neurofibromin, or to indirect activation of Ras-GEFs (Patapoutian & Reichardt 2001, Weiss et al. 1999). Since this complex of receptors, Ras-GAPs, and Ras-GEFs resides at the membrane, Ras must also be tethered to the cell membrane to be activated. At the membrane, Ras can interact with downstream effectors such as Raf kinase, which then activate signaling cascades such as the MEK/MAPK pathway (Boguski & McCormick 1993, Weiss et al. 1999). These kinases then induce a number of downstream processes, including transcription of immediate early genes, which lead to short- and long-term changes in neuronal function.

In neurons, loss of Neurofibromin results in constitutive increases in Ras intracellular signaling (Li et al. 2005). Further, Neurofibromin loss of function can allow Ras signaling to become decoupled from extracellular triggers such as growth factors. For example, $Nf1^{-/-}$ sensory neurons no longer require the extracellular growth factor BDNF to survive and mature. This is presumably due to constitutively increased levels of activity of the Ras pathway when it is no longer subject to negative regulation by Neurofibromin (Vogel et al. 1995). It is currently unclear exactly which receptors are upstream of Neurofibromin regulation of Ras signaling during behavior. However, Neurofibromin is known to be quickly regulated by PKC in response to growth factors that utilize the tyrosine kinase receptors (EGF), and by ubiquitin-dependent proteolysis in response to both G-protein (LPA) and tyrosine kinase receptor ligands (EGF, PDGF).

The Ca^{2+}-sensitive PKC-alpha isoform quickly (within 1 min) phosphorylates the N terminal region of Neurofibromin in neuronal cultures. This phosphorylation increases Neurofibromin association with actin and maximizes its activity. Thus, this phosphorylation could also make the Ras-GAP activity of Neurofibromin activity dependent (Leondaritis et al. 2009). In this context, Neurofibromin may normally act to downregulate the Ras response to growth factors in an activity-dependent manner. The same growth factors can also quickly (by 5 min) but transiently induce ubiquitin-dependent proteolysis of Neurofibromin in an activity-independent manner. Thus, following growth factor release, Neurofibromin may play a key role in limiting and narrowing the time of Ras activation (Cichowski et al. 2003). As both forms of regulation of Neurofibromin were identified in embryonic culture systems, it remains to be seen whether similar regulation of Neurofibromin occurs in adult neuronal networks, whether it occurs downstream of growth factor receptors such as the BDNF receptors, and how either one or both mechanisms regulate neuronal networks during behavior.

Finally, although the receptors that utilize Neurofibromin to regulate Ras signaling in adult neurons have not been identified, one possible upstream receptor outside the growth factor receptors is the NMDA receptor. Neurofibromin is part of the NMDA receptor complex in the mouse forebrain (Husi et al. 2000). The NMDA receptor does interact with intracellular signaling cascades through its C terminus, in a manner that is independent of Ca^{2+} influx and critical to normal performance of delayed alternation, a behavior related to working memory (Bannerman et al. 2008). Although the functional relevance of Neurofibromin's localization to the NMDA receptor is unknown, it suggests that this protein plays a role in NMDA receptor dependent regulation of Ras signaling.

In addition to its role as a Ras-GAP, Neurofibromin acts as an activator of adenylate cyclase (Tong et al. 2002). Learning defects in *Nf1* null *Drosophila melanogaster* can be rescued by expression of a constitutively active form of PKA (Guo et al. 2000). These and other data suggested that the associative learning impairments in *Nf1* null flies are due to decreased activation of adenylate cyclase. Neurofibromin can increase adenylate cyclase activity in a Ras-dependent manner. This form of signaling occurs in response to growth factors such as EGF. Neurofibromin can also directly increase adenylate cyclase activity in a Ras-independent manner, through its C-terminal domain. This function of *Nf1* is required for stimulation of adenylate cyclase by neurotransmitters

such as serotonin or histamine (Hannan et al. 2006). *Nf1* loss in *Drosophila* causes specific phenotypes, such as small body size, through a direct decrease of adenylate cyclase activity (Tong et al. 2002). Other behavioral functions of *Nf1*, such as circadian rhythm modulation in *Drosophila*, are MAPK dependent and require the GAP domain (Williams et al. 2001). Still unclear is whether these biochemical effects of homozygous *Nf1* deletion are relevant to the behavioral and cognitive deficits associated with NF1, which are caused by heterozygous mutations. In both mouse and *Drosophila* models, heterozygous deletion of *Nf1* does not grossly affect regulation of the adenylate cyclase pathway (Tong et al. 2002). Nevertheless, it is possible that *Nf1* plays a role in maintaining the relative balance between Ras- and cAMP-dependent signaling (Weeber & Sweatt 2002).

The Neurofibromin protein interacts with a number of upstream regulators of Ras signaling, and has the potential to play multiple roles within neurons as part of various intracellular pathways. This may contribute to the range of phenotypes that have been observed in $Nf1^{+/-}$.

COGNITIVE PROFILE OF NF1 PATIENTS

NF1 is characterized by widespread symptoms affecting multiple organ systems (Lynch & Gutmann 2002, Williams et al. 2009). Among these symptoms are prominent cognitive impairments, which pose one of the most significant sources of lifetime morbidity for patients (North et al. 1997). NF1 specifically affects executive and other higher-order cognitive functions. In contrast, NF1 does not cause global cognitive impairments. In measures of global cognitive function, performance of affected individuals is comparable to unaffected siblings (Kayl & Moore 2000).

Executive function impairments are prevalent and can be severe in NF1. When specific cognitive domains are probed in neuropsychological batteries, 80%–90% of individuals affected by NF1 show impairments (Hyman et al. 2005, Krab et al. 2008a). In particular, NF1

Working memory: a flexible, continually updated system that maintains and manipulates information across short delay periods affects planning, visuospatial function, reading/vocabulary, and motor coordination (Hofman et al. 1994, Hyman et al. 2005). Additional executive function deficits are prominently seen in working memory, cognitive flexibility, and inhibitory control (Rowbotham et al. 2009). There is also a high comorbidity between NF1 and attention deficit disorder (Hofman et al. 1994, Hyman et al. 2005, Mautner et al. 2002). In up to 40% of children with NF1 who are identified as academic underachievers, underachievement is likely due to deficits in attention, planning, and organizational skills (Dilts et al. 1996, Hyman et al. 2005, Kayl & Moore 2000, Koth et al. 2000, North 2000). Comorbid attention deficit disorder affects not only academic but also social development for children with NF1 (Barton & North 2004). Children with NF1 tend to have social problems and appear socially awkward and withdrawn, in comparison to their siblings and to children with other chronic, life-threatening illnesses (Kayl & Moore 2000). These social deficits could be related to poor interpersonal skills resulting from decreased attention to social cues.

Developmental learning disabilities are also highly associated with NF1. Overall, individuals with NF1 are fourfold more likely to require special education (Krab et al. 2008a), and learning disabilities are diagnosed in up to 65% of individuals affected by NF1(Rosser & Packer 2003). Learning disabilities are characterized by impairments restricted to specific domains of mental function, leading to a discrepancy between tests of intellectual capability and actual achievement (Kelly 2004, Kronenberger & Dunn 2003). Both intellectual capability and achievement are typically assessed using standardized, normalized tests. Learning disability is diagnosed by a discrepancy of one to two standard deviations between achievement and intelligence test scores.

DSM-IV places learning disabilities into five major categories: reading disorders, mathematical disorders, disorders of written expression, nonverbal learning disorders, and learning disorders not otherwise specified (Kelly 2004, Kronenberger & Dunn 2003, Palumbo &

Lynch 2006). In addition, learning disabilities may be diagnosed according to specific symptoms, such as in dyslexia (reading disorder) or dyscalculia (math learning disorder). Thus, the criteria for diagnosis of learning disabilities are largely based on different systems of representing symptoms. However, a single learning disability category is often insufficient to describe all the symptoms of a given individual with idiopathic learning disability (Lagae 2008), or a group of individuals with an inherited learning disability such as NF1.

In fact, learning disabilities caused by NF1 show some characteristics of both nonverbal- and verbal-type learning disability. Components of nonverbal learning disability seen in NF1 include consistently poor performance in tests of visuospatial functioning and spatial learning (Kelly 2004), notable impairments in the ability to perceive social cues, poor organizational skills, and increased impulsiveness (Kayl & Moore 2000, North 2000). NF1 patients also show aspects of verbal learning disorder. Specifically, patients with NF1 have deficits in expressive and receptive language, vocabulary, visual naming, and phonologic awareness. In fact, reading and spelling are repeatedly found to be impaired more severely than predicted by IQ in NF1 patients (North 2000, North et al. 1997). Consistent with these impairments in language-based learning, NF1 patients show poorer academic achievement in reading and writing compared to their unaffected siblings. The pattern of learning disabilities seen in NF1 does not fall cleanly into one of the DSM-IV defined categories, but rather shares features with multiple learning disability categories. This shows that a single disorder, in this case NF1, can cause learning disabilities of varying phenotypes and presentations. Therefore, the various classifications and subdivisions of learning disabilities (for example, verbal versus nonverbal) may not actually represent different disease entities but rather different manifestations of a common disease cluster caused by a common set of underlying factors. In NF1, the drastic variability of symptom expression appears to be determined largely by independently inherited genetic modifiers.

GENETIC MODIFIERS OF BEHAVIORAL EXPRESSION OF NF1

Inheritance of NF1 shows an interesting pattern of complete genetic penetrance but variable expressivity (Ward & Gutmann 2005). Most people with an inactivating mutation in one allele of the *Nf1* gene show some symptoms of NF1. However, the clinical presentation of NF1 ranges from minimal symptom load to extremely high severity across many types of symptoms. One important factor underlying variable expressivity of cognitive symptoms in the NF1 population is the contribution of modifying genes, some of which may significantly alter behavior in the presence of mutant *Nf1*, but not in a wild-type (WT) background. Studies in both mouse models and human patient populations suggest that inheritance of genetic modifiers accounts for the majority of variability in NF1 expression (Easton et al. 1993, Sabbagh et al. 2009).

In NF1 mouse models, the phenotypic effect of the *Nf1* mutation depends on the background strain on which that mutation is expressed. This has been shown for both behavioral and somatic symptoms of NF1. Different strains of mice (for example, C57Bl/6J versus DBA/2J) carrying the $Nf1^{+/-}$ mutation have been shown to have different susceptibility to the formation of astrocytomas (Hawes et al. 2007). Similar studies have shown that the background strain also affects the behavioral phenotype of the *Nf1* mutation (Costa 2002). Since these strains are engineered with the same *Nf1* gene mutation, phenotypic differences across strains are attributed to differential expression of modifier genes across mouse strains that were inbred from genetically different founders. Such modifiers may interact with the gene directly, altering its level of expression, or, more likely, may interact on a more functional level to exacerbate or compensate for the signaling changes caused by loss of *Nf1*. The latter is likely to be most relevant as studies quantifying differences in levels of *Nf1* expression across strains find that although expression levels can vary dramatically, they do not always correlate with phenotype expression (Hawes et al. 2007). On the other hand, introducing functional modifiers of learning pathways can exacerbate the learning phenotype of $Nf1^{+/-}$ mice. This was demonstrated using a heterozygous null mutation of a receptor (NMDA; $NR1^{+/-}$) critical for learning and memory. Although the $NR1^{+/-}$ mutation alone does not have a spatial learning phenotype, it exacerbates the spatial learning phenotype of the $Nf1^{+/-}$ mutant mice (Silva et al. 1997). Therefore, mouse studies indicate that background genetic modifiers alter the expression of $Nf1^{+/-}$-related phenotypes, likely through functional interactions with the cellular pathways altered by the *Nf1* mutation.

In human patient studies, heritable genetic modifiers have also been found to affect expression of NF1, and current studies are in progress to identify specific loci encoding these modifiers. Analyses of large groups of extended families affected by NF1 suggest that the expression of NF1 symptoms is heritable. Variation in symptom expression between family members is consistent with a model of inheritance of modifier genes with little contribution from specific mutations to the *Nf1* gene itself or from variations in the WT *Nf1* allele (Easton et al. 1993, Sabbagh et al. 2009). Among multiple clinical symptoms of NF1, including learning disabilities and referral for remedial education, symptom expression shows high correlation among first-degree family members (who share 50% of their genes). Similarly, monozygotic twins affected by NF1 have a high correlation in symptom severity (Easton et al. 1993). As second- and third-degree family members are examined (who share 25% and 12.5% of their genes, respectively), the correlation in symptom expression decreases sharply. The strong genetic component in symptom expression in NF1 is consistent with modulation of disease symptoms by inherited genetic modifiers (Easton et al. 1993, Sabbagh et al. 2009). For

Genetic penetrance: the proportion of individuals carrying a genetic variant that also show phenotypic/symptomatic manifestations associated with that variant

Variable genetic expression: indicates the range of phenotypic/ symptomatic severity associated with the same genetic variant across phenotypes and different individuals; frequently seen in dominant genetic conditions

WT: wild type

example, proteins encoded by genetic modifiers may interact with the Neurofibromin signaling pathway to exacerbate effects of *Nf1* mutations or confer protection by compensating for the loss of function.

These studies of NF1 highlight the regulation of clinical presentation and symptom severity by inheritable genetic modifiers, and suggest that these modifiers may have a more dramatic impact in the presence of the *Nf1* mutation than in a WT background. Similar findings have been made in other genetically determined learning disabilities, such as Noonan syndrome (NS). NS is another common autosomal dominant genetic disorder with an incidence of 1 in ~2500 live births. It is characterized by facial abnormalities, short stature, webbed neck, motor delay, and cardiac disease (Noonan 1994, Tartaglia & Gelb 2005). Importantly, NS patients also show increased rates of learning disabilities and mental retardation (Lee et al. 2005, Money & Kalus 1979). Recently, Araki and colleagues generated NS mouse models by engineering NS-associated PTPN11 mutants. They demonstrated that the gain-of-function mutants show phenotypes similar to that found in NS patients (Araki et al. 2009, Araki et al. 2004). The penetrance of NS phenotypes is dependent on the specific *Ptpn11* allele studied as well as the genetic background of the mutant mice (Araki et al. 2009). *Ptpn11* mutations that modify tyrosine phosphatase SHP2 activity result in different heart defects. Half of the NS *Ptpn11$^{D61G/+}$* mutant mice are embryonic lethal on a mixed 129S4/SvJae X C57BL/6 genetic background (Araki et al. 2004). However, the *Ptpn11$^{D61G/+}$* mutation backcrossed to C57BL/6 revealed high lethality. In contrast, when crossed into 129S6/SvEv genetic background, this mutation showed normal viability, suggesting that there are modifier allele(s) that affect the viability of the NS *Ptpn11$^{D61G/+}$* mutation (Araki et al. 2009). The examples described above illustrate the dramatic influence of modifier genes on the clinical presentation of a single-gene disorder.

NF1 AND PHENOTYPES OF ANIMAL MODELS

Mice with a heterozygous null mutation of the *Nf1* gene (*Nf1$^{+/-}$*) have been the dominant rodent model used to study NF1. The *Nf1$^{+/-}$* mice show compelling genetic and behavioral parallels with human NF1, making them a useful system in which to study the mechanisms of behavioral phenotypes associated with NF1. Nevertheless, a word of caution is in order: There is a very large evolutionary gulf between mice and humans, and therefore the behavioral phenotypic parallels between animal models and NF1 must not be overinterpreted. Much needs to be done to fully elucidate the apparent parallels between rodent and human behavior.

The sequence, transcriptional regulation, and downstream targets of *Nf1* are conserved across species, including mouse and human (Bernards et al. 1993, Hajra et al. 1994). The majority of mutations (70%) of the *Nf1* gene in humans affected by NF1 lead to synthesis of a truncated, nonfunctional version of the encoded Neurofibromin protein (Shen et al. 1996, Thomson et al. 2002). Accordingly, the *Nf1$^{+/-}$* mouse model was made by inserting a *neo* gene in exon 31 of the *Nf1* gene, which leads to an unstable, quickly degraded transcript (**Figure 1**) (Jacks et al. 1994). Like patients with NF1, the *Nf1$^{+/-}$* mouse model is heterozygous for this loss-of-function mutation. Behavioral phenotypes of the *Nf1$^{+/-}$* mouse show a pattern of specific impairments in certain domains and preserved function in others. This pattern of behavioral phenotypes is reminiscent of the pattern of behavioral and cognitive symptoms seen in humans affected by NF1. Based on the genetic, biochemical, and behavioral parallels between the *Nf1$^{+/-}$* mouse model and human NF1, it is thought that this mouse offers a useful model of the behavioral and cognitive symptoms associated with the disorder. The *Nf1$^{+/-}$* mouse has therefore been utilized to identify electrophysiological and molecular mechanisms that contribute critically to cognitive and behavioral changes associated with NF1.

Conditional mutants of the *Nf1* gene have also been extensively studied to identify the effects of *Nf1* deletion within specific neuronal types. Conditional mutations can be created in mice using the Cre-loxP system, a powerful tool widely used for restricting gene deletions to specific time frames, cell types, or areas. A mouse line was engineered with loxP sites flanking exons 31–32 of the *Nf1* gene (*Nf1*$^{\text{flox/flox}}$) (**Figure 1**). The floxed *Nf1* gene acts like a WT allele prior to expression of Cre recombinase. Mice carrying one floxed *Nf1* allele and one deleted *Nf1* allele (*Nf1*$^{\text{flox/−}}$) show the same phenotypes as *Nf1*$^{+/−}$ mice. For example, they have the same survival profile: Deletion of the floxed *Nf1* allele following delivery of Cre recombinase caused embryonic lethality, similar to the *Nf1*$^{−/−}$ mutation (Zhu et al. 2001). Conditional *Nf1* mutants have revealed cell-specific roles of *Nf1* and interesting interactions between neighboring cells with different *Nf1* gene doses. Such interactions are critical to the development of some of the somatic events associated with NF1 such as neurofibromas (Zhu et al. 2002).

Finally, mouse models targeting specific biochemical domains of the *Nf1* gene have been examined to demonstrate the relative importance of the various regulatory functions of the *Nf1* protein product, Neurofibromin. For example, exon 23a of the *Nf1* gene is spliced out in peripheral tissue, but expressed in neuronal *Nf1*. Presence of exon 23a increases the Ras regulatory activity of Neurofibromin, suggesting that this represents a uniquely important aspect of *Nf1* loss in the brain. Indeed, many of the behavior deficits identified in *Nf1*$^{+/−}$ mice were also found in Nf1$^{23a−/−}$ mice, which carry a homozygous deletion restricted to exon 23 of the *Nf1* gene (Costa et al. 2001). This deletion creates a version of Neurofibromin that is functional but has diminished Ras regulatory activity. Importantly, the Nf1$^{23a−/−}$ mice show learning disability–related behavioral deficits, but none of the other somatic consequences of *Nf1* mutation such as increased tumor predisposition (Costa et al. 2001). This not only demonstrates that Neurofibromin has

a neuron-specific role as a Ras regulator, but also shows that altered behavioral performance in these mutant mice is a direct function of their *Nf1* mutation, rather than a secondary effect of tumor formation. Detailed studies have characterized the behavioral consequences of *Nf1* mutation.

TASKS USED TO ASSESS BEHAVIOR IN ANIMAL MODELS OF NF1

The behavioral tasks used to study disease mechanisms in *Nf1*$^{+/−}$ mice were chosen according to the following criteria: (*a*) face/construct validity to tasks performed less accurately by NF1 patients, (*b*) dependency on brain areas that contribute critically to cognitive functions affected by NF1, and (*c*) behaviors with well-established molecular and cellular underpinnings. To date, the behavioral paradigms used in the study of *Nf1*$^{+/−}$ mice were predominantly tests sensitive to hippocampal and parietal/prefrontal cortical function. This result reflects the prominence of symptoms thought to stem from dysfunction in these brain areas in NF1 patients. The Morris water maze was initially utilized in *Nf1*$^{+/−}$ mice as a test of spatial learning and memory that is highly sensitive to hippocampal function (Morris et al. 1982). Additionally, performance in the Morris water maze requires specific synaptic and cellular mechanisms, such as long-term potentiation (LTP) (Moser et al. 1998; Silva et al. 1992a,b). As such, this behavioral paradigm provided a bridge between functional/mechanistic changes in *Nf1*$^{+/−}$ mice and a mouse behavioral phenotype that could be related to the spatial learning deficits seen in NF1 patients. In the Morris water maze, mice are placed in a pool of opaque water in which there is an escape platform. The test subjects must use the spatial cues surrounding the pool to learn the location of the platform and navigate to it, a process that usually takes multiple trials over several days. Following training, memory of the spatial location of the platform is tested in probe trials where the platform is

Long-term potentiation (LTP): a form of synaptic plasticity where high-frequency stimulation leads to increased strength of synaptic transmission. In hippocampus, LTP is thought to be required for learning and memory

removed, and the experimenter measures the amount of time mice spend searching for the missing platform in the appropriate quadrant of the pool. In a control task (unperturbed by the $Nf1^{+/-}$ mutation), the position of the escape platform is directly marked by an object, and the mice just have to learn to go to it to escape from the water.

Extensive studies with the water maze and other hippocampal-dependent tasks helped to define the behavioral deficits in $Nf1$ mice and unravel underlying mechanisms (Costa et al. 2001, 2002; Cui et al. 2008; Silva et al. 1997). Subsequent behavioral studies have begun to address some of the prominent cognitive symptoms of NF1. Specific behavioral paradigms were utilized that have been developed in rodents to model core cognitive functions commonly affected in human disease. For example, the lateralized reaction time task was used in $Nf1^{+/-}$ mice to test attention deficit/hyperactivity (Li et al. 2005), as attention deficits are highly associated with NF1 in humans. Although attention, like memory and other behaviors modeled in mice, has important differences across species, there are also conserved elements that the behavioral tasks used in rodent models utilize (see Sidebar: Animal Models of Human Behaviors). In this case, the lateralized reaction time task was designed to test attention processes in a manner that captures key features of tasks used to probe attention in humans (Robbins 2002). Further, performance of the rodent task requires the prefrontal cortex, an area thought to contribute to attention in humans. Performance of rats in the lateralized reaction time task is also improved by stimulants used to treat ADHD (attention deficit and hyperactivity disorder) in humans (Jentsch et al. 2009), supporting common mechanistic underpinnings. The lateralized reaction time task is implemented in an operant chamber and requires mice or rats to fixate on a central hole while attending to two potential cues, one on either side of the center. In multiple, serial, self-initiated trials, one of the cues is lit for varying lengths of time, and the rodent must maintain attention divided across both sides of space in order to notice and respond to the lit cues. Beyond attention, similar tasks have also been adapted to examine working memory in rodents (Aarde & Jentsch 2006), in a manner analogous to the Sternberg style working memory tests used in humans (Cannon et al. 2005).

ANIMAL MODELS OF HUMAN BEHAVIORS

A key question often raised by animal models of human disease is whether the model captures all of the essential features of the disease. Despite the fact that some animal models only address a subset of the features of a disease, they can still be very useful for understanding the disease and developing treatments. To understand the complexity of the human condition a variety of different models is needed, not necessarily one with all of the features of the disorder. Another challenge is to extrapolate findings from animal models across species, despite differences between human and rodent control over cognition, timing, lifespan, and organization and development of brain structures such as prefrontal cortex. Currently, several approaches are being used in rodent models of disease. Behavioral batteries have been developed for animals to test for disease-relevant clusters of phenotypes. Appropriate and high-throughput behavioral testing has been developed with demonstrated face, construct, and some predictive validity with human disease symptoms (Chadman et al. 2009). Also, specific behavioral tasks have been developed for animals to model cognitive functions commonly affected in human disease. For example, the 5-choice serial reaction time task (5CSRTT) is one example of a behavioral paradigm that is designed to test multiple attention processes in a manner highly analogous to the probing of those processes in humans (Robbins 2002). Mechanisms contributing to normal performance of this task have been extensively characterized, including the neuromodulatory systems and brain areas that mediate normal performance of the 5CSRTT in rodents (Robbins 2002). This paradigm has also been adapted to examine working memory in rodents (Aarde & Jentsch 2006), in a manner analogous to the Sternberg style working memory tests used in humans (Cannon et al. 2005). Tasks are also being developed for use in humans to parallel key features of commonly used rodent tasks (Demeter et al. 2008). The use of these tasks to obtain convergent evidence from multiple approaches offers exciting new prospects for understanding disease through the integration of the cognitive, clinical, and basic neurosciences (Chadman et al. 2009, Fossella & Casey 2006).

These tasks provide very useful methods that allow for parallel studies in rodents and humans of core symptoms of NF1, including attention deficit and the working memory impairments often found in learning disabilities. As more evidence accumulates from the use of these tasks in rodents and humans, we will be able to better understand which behavioral mechanisms have been conserved across species and which tasks are best able to probe conserved mechanisms in the context of disease. This offers exciting new prospects for understanding diseases, such as NF1, through an integration of cognitive, clinical, and basic neuroscience approaches (Chadman et al. 2009, Fossella & Casey 2006).

PHENOTYPES OF NF1 ANIMAL MODELS AND PARALLELS TO HUMAN COGNITIVE SYMPTOMS

The phenotypes identified in Nf1 animal models in the tasks described above demonstrate that mutation of the *Nf1* gene in animal models causes impairments with striking analogy to that seen in the learning disabilities associated with NF1 in humans (**Table 1**). These phenotypes of the $Nf1^{+/-}$ mice can be improved with extended training, a feature also observed in human learning disabilities.

$Nf1^{+/-}$ mice show spatial learning deficits in the hidden version of the Morris water maze (Costa et al. 2001, 2002; Cui et al. 2008; Silva et al. 1997). Probe trials given early on during water maze training revealed that $Nf1^{+/-}$ mice require more training trials than controls to learn the position of the hidden platform. Thus, when searching for the missing platform in probe trials, the mutant mice spent less time in the appropriate quadrant compared to their WT littermates. Additional training led to continued improvement in spatial memory in $Nf1^{+/-}$ mice, such that probe trials given after additional days of training no longer revealed differences between genotypes. This sensitivity to overtraining is also a feature reported in NF1-associated learning disabilities in patients. Similarly, $Nf1^{+/-}$ mice also show deficits in contextual conditioning, a test where mice

ADHD: attention deficit and hyperactivity disorder

Table 1   Cognitive phenotypes of NF1 mouse models

Mouse Model	Behavioral phenotypes	Parallel human symptoms	References

$Nf1^{+/-}$	*Impaired in*
	  Morris water maze (hidden platform)
	  Contextual fear conditioning
	  Lateralized reaction time task
	  Pre-pulse inhibition
	Normal in
	  Open field and visible platform water maze
	  Cued fear conditioning
		Learning deficits in specific domain (visual-spatial)
		Attention deficits
			Silva et al. (1997)
			Cui et al. (2008)
			Li et al. (2005)

$Nf1^{23a-/-}$	*Impaired in*
	  Morris water maze (hidden platform)
	  Contextual discrimination
	  Rota-rod
	Normal in
	  Open field and visible platform water maze
	  Social transmission of food preference
		Learning deficits in specific domain (visual-spatial)
		Motor deficits
			Costa et al. (2001)

$Nf1^{flox/+}$; *SynI-cre*
$Nf1^{flox/+}$; *Dlx5/6-cre*
	Impaired in
	  Morris water maze (hidden platform)
	Normal in
	  Visible platform water maze
		Learning deficits in specific domain (visual-spatial)	Cui et al. (2008)

associated a novel chamber with a mild foot-shock (Cui et al. 2008). As with the water maze, contextual conditioning has a spatial learning component and requires hippocampal function.

In contrast to the significantly impaired spatial learning seen in $Nf1^{+/-}$ mice, neither visual learning nor simple associative learning is prominently disrupted in these mice (Costa et al. 2001, 2002; Cui et al. 2008; Silva et al. 1997). Performance of the $Nf1^{+/-}$ mice in the visual version of the Morris water maze is indistinguishable from WT. Additionally, $Nf1^{+/-}$ mice show no deficits in tone fear conditioning, where mice learn to associate a tone with a mild footshock. Unlike the Morris water maze, tone fear conditioning is a single-trial, simpler associative learning paradigm that requires amygdala function. Therefore, $Nf1$ mutation has very specific effects on spatial learning while sparing simple associative learning, possibly due to a greater effect on hippocampal function (Costa et al. 2002, Silva et al. 1997). This pattern of specific impairments is also seen in learning disabilities associated with NF1 in patients.

Finally, as NF1 subjects, $Nf1^{+/-}$ mice show attention deficits in the lateralized reaction time task (Li et al. 2005). ADHD is highly associated with NF1, and thought to contribute to academic and social changes in this patient population. In the lateralized reaction time task, $Nf1^{+/-}$ mice perform as accurately as WT mice when cues are lit for longer intervals (up to 1 s). However, as attention is taxed with shorter light presentations (0.5 s), a failure to maintain constant attention in the $Nf1^{+/-}$ mice is revealed. The $Nf1^{+/-}$ mice make significantly more omissions than do WT (Li et al. 2005). Humans with NF1 often show a type of ADHD characterized by increased omissions and lapses of attention (Hyman et al. 2005, Mautner et al. 2002).

These animal data demonstrate that mutation to the $Nf1$ gene is sufficient to directly cause a specific pattern of complex behavioral changes in mice. The behavioral phenotypes of the $Nf1^{+/-}$ mice in these tasks show analogy to the learning disabilities and other cognitive symptoms that characterize NF1.

However, various differences clearly exist between mouse and human. For example, the anatomical organization and connectivity of prefrontal cortex and other regions likely to be critical for the NF1 phenotype are different between the two species. Additionally, the complexity of the human NF1 phenotype can only be partially captured in animal models. This and other factors make direct translation from mouse to human studies difficult and unpredictable. In the context of NF1, where a large literature of both animal and human studies exists, it will be interesting to follow future experiments evaluating which specific molecular mechanisms, cellular effects, and behavioral paradigms carry translational value.

MULTIPLE PHENOTYPES LINKED TO THE SAME MECHANISM

Cognitive effects of the $Nf1^{+/-}$ mutation in mouse have been seen in diverse processes, from hippocampal memory to quick, flexible, prefrontal/parietal cortex-dependent attention processes. Although distinct underlying circuits and brain mechanisms mediate these cognitive functions, in NF1 both memory and attention are impaired by the same genetic change. In $Nf1^{+/-}$ mice, both deficits in memory and in attention are caused by increased Ras signaling, resulting from the loss of regulation by Neurofibromin (Costa et al. 2002, Cui et al. 2008, Li et al. 2005). To demonstrate the Ras dependency of deficits in the Morris water maze, $Nf1^{+/-}$ mice were crossed to null Ras mutants (K-$Ras^{+/-}$ or N-$Ras^{-/-}$ mice). As a result of the decreased levels of Ras activity in the Ras mutants, the $Nf1/Ras$ double mutants performed at the same level as did WT mice (Costa et al. 2002), even though each mutant individually ($Nf1^{+/-}$ and K-$Ras^{+/-}$) showed deficits in this task. Similarly, performance of $Nf1^{+/-}$ mice in the Morris water maze was rescued with farnesyl transferase inhibitors under conditions that did not enhance the performance of controls. These inhibitors pharmacologically decrease levels of Ras signaling (Costa et al. 2002) by blocking the posttranslational modification that is necessary for Ras to associate with the cellular membrane and therefore be active. This mechanism of action can also be targeted with a class of drugs, the statins, which also decrease isoprenylation (Li et al. 2005). Statins are relatively safe and are widely used to control cholesterol levels. In both the Morris water maze and in the lateralized reaction time task, lovastatin, a drug in the statin class, normalized the performance of $Nf1^{+/-}$ mice to levels indistinguishable from those of WT (Li et al. 2005). Importantly, statins also normalized the signaling and the neurophysiological deficits of the $Nf1^{+/-}$ mice. Hence, regardless of method used, decreases in Ras signaling rescue both the learning impairments and the attention deficits of the $Nf1^{+/-}$ mice. These results also confirm the wide range of cognitive dimensions affected by alterations of Ras signaling. The observation that multiple symptoms are caused by disruption of a single biochemical process is not specific to NF1, but is also seen following mutations to other Ras-regulatory genes. For example, SPRED1, another negative regulator of Ras-MEK/MAPK signaling, is responsible for a NF1-like syndrome (Legius syndrome) (Brems et al. 2007). Similar to $Nf1$, loss-of-function mutations of SPRED1 also result in hyperactivity in MEK/MAPK signaling. The disorder associated with $Spred1$ mutations involves some of the same symptom dimensions (other than neurofibromas) seen in NF1 (Pasmant et al. 2009). This includes NF1-like cognitive deficits in affected individuals (Brems et al. 2007, Pasmant et al. 2009, Spurlock et al. 2009). $Spred1$ knockout mice ($Spred1^{-/-}$) have been generated (Brems et al. 2007, Denayer et al. 2008) and found to show deficits in hippocampus-dependent learning and memory tasks, which may model cognitive deficits in patients with $Spred1$ mutation (Denayer et al. 2008). Therefore, disruption of the Ras signaling pathway through SPRED1 can also lead to a complex, multifaceted disorder. These disorders demonstrate that multiple types of symptoms can all be caused by alterations in a single gene/biochemical component.

INCREASED GABA RELEASE FROM INTERNEURONS UNDERLIES *Nf1* BEHAVIORAL PHENOTYPES

Nf1 deletion and increased Ras signaling have cell type–specific physiological effects that contribute to behavioral symptoms. In particular, *Nf1* expression in interneurons seems to be critical for behavior, whereas its role in pyramidal neurons appears to be nonessential for hippocampal learning in the strains of mice and behaviors tested (Cui et al. 2008). In the Morris water maze, heterozygous deletion of *Nf1* from inhibitory interneurons is sufficient to cause behavioral impairments. Conversely, deletion of *Nf1* from either pyramidal neurons or glia does not alter behavior in this task under the conditions tested. Thus, regulation of Ras signaling by *Nf1* is particularly critical within interneurons, but perhaps less so in pyramidal neurons (Cui et al. 2008). Within the hippocampus of $Nf1^{+/-}$ mice, increased interneuronal Ras signaling causes an increase in activity-dependent GABA release (**Figure 3**). This increased activity-dependent release of GABA (Cui et al. 2008) leads to larger evoked inhibitory currents in CA1, shifting the balance between inhibitory and excitatory processes within hippocampal networks of the mutant mice. As a result, LTP is impaired in the $Nf1^{+/-}$ mice, perhaps because the increased inhibition prevents sufficient depolarization of NMDARs during learning (Cui et al. 2008). In the hippocampus, there is strong evidence that LTP is required for learning of spatial and contextual information (Lee & Silva 2009, Martin et al. 2000, Martin & Morris 2002, Richter-Levin et al. 1995). Additionally, the very manipulations that reverse the learning deficits of the $Nf1^{+/-}$ mice (decreases in Ras signaling with statins or farnesyl transferase inhibitors), also reverse their LTP impairments. As additional training rescues the learning impairments of the mutants, additional synaptic stimulation (e.g., higher frequency) also rescues their LTP deficits. Therefore, the LTP deficits are thought to underlie the learning

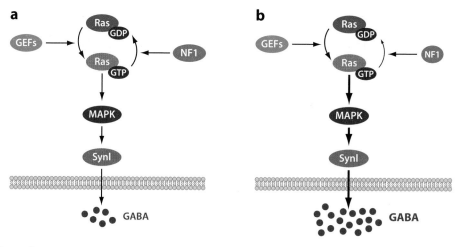

Figure 3

Proposed cellular mechanism underlying learning deficits of *Nf1* mutant mice. (*a*) Learning triggers interneuronal Ras signaling leading to increased GABA release. MAPK-dependent phosphorylation of synapsin I (SynI) plays a critical role in GABA release. Wild-type NF1 restricts the increase in GABA release within an appropriate range that modulates learning. (*b*) In *Nf1* mutants, reduced NF1 activity leads to abnormal hyperactivation of Ras signaling in inhibitory interneurons during learning, resulting in abnormally high GABA release. This increased activity-dependent GABA release shifts the balance between excitatory and inhibitory processes in neuronetworks of the mutant mice and impairs synaptic plasticity needed for learning and memory.

deficits seen in the *Nf1*$^{+/-}$ mice. Supporting this assertion, both the *Nf1*$^{+/-}$ LTP and water maze deficits can be improved using picrotoxin, a GABA$_A$ receptor antagonist (Cui et al. 2008), at concentrations that do not affect these phenomena in controls. These data demonstrate that Neurofibromin is an important regulator of inhibitory tone in hippocampal neuronal networks. This regulation of inhibition is functionally important as its disruption impairs behaviors, such as the acquisition of spatial and contextual information. Again, beyond manipulations of inhibition, Lovastatin, farnesyl transferase inhibitors, and the *N-ras* and *K-ras* mutations, all reverse both the LTP and the learning deficits of the *Nf1*$^{+/-}$ mutant mice, strengthening the link between the hippocampal LTP deficits of these mice and their hippocampal-dependent learning deficits (Costa et al. 2002, Li et al. 2005).

In other brain areas such as the prefrontal cortex and striatum, inhibition also plays an important role in regulating network phenomena that are critical for cognitive functions

affected in NF1, including working memory and attention. In prefrontal cortex, working memory–related cellular activity involves a balance between a high degree of excitation (allowing inputs to induce persistent activity) and a coordinated set of inhibitory mechanisms (imposing specificity to the activity). This balance between excitation and inhibition allows prefrontal cortex to encode information across delays (Compte et al. 2000). Activity in prefrontal cortex is regulated by both feedforward and feedback inhibitory microcircuits (Constantinidis et al. 2002, Hasenstaub et al. 2005, Sanchez-Vives & McCormick 2000), which can affect the onset and duration of persistent activity (Compte et al. 2000, Fellous & Sejnowski 2003). In addition, inhibition acts to tune persistent activity of prefrontal neurons to specifically represent cue location, and to organize temporal interactions between prefrontal neurons (Constantinidis & Goldman-Rakic 2002, Constantinidis et al. 2002). Decreasing inhibition within prefrontal cortex leads to loss of spatial tuning, which results in inaccurate working memory performance. Decreasing inhibition also unmasks spatial tuning in previously silent neurons. During a working memory task, this spatial tuning results in inappropriate, pre-emptive behavioral responses (Rao et al. 2000). Therefore, inhibitory interneurons in the prefrontal cortex have multiple functions during behavior, an observation with interesting implications for attention and executive deficits in both *Nf1* mice and NF1 subjects.

Similarly, striatal inhibitory interneurons modulate the activity of local medium spiny neurons. Inhibitory currents induced at the soma of medium spiny neurons by striatal interneurons can abolish or delay the onset of spiking (Koos & Tepper 1999, Tepper et al. 2008), making GABAergic interneurons a primary modulator of activity flow in the striatum. Importantly, striatal inteneurons are driven by glutamatergic inputs, including those descending from the cortex. Therefore, abnormally high levels of inhibition seen in $Nf1^{+/-}$ mice would be expected to significantly alter activity within the striatum, the prefrontal cortex, and the timing and flow of activity between striatum and cortex. Increased inhibition resulting from *Nf1* mutation likely underlies not only learning and memory deficits but also other behavioral phenotypes of the $Nf1^{+/-}$ mice that are more closely related to dysfunction in frontal/cortical brain areas (e.g., attention, executive function).

Increased inhibition can lead to behavioral deficits by altering either the ongoing function of neuronal networks in the adult brain, or by affecting neuronal development. In the $Nf1^{+/-}$ mice, increased inhibition in the adult brain contributes critically to behavioral deficits, since these deficits can be rescued in adult $Nf1^{+/-}$ mice by decreasing inhibition (e.g., using picrotoxin), or by normalizing Ras signaling (e.g., with statins). This finding suggests that these behavioral deficits occur as a result of a reversible increase in inhibition. However, it is important to also consider potential effects of increased inhibition on development of neuronal networks, since most *Nf1* gene mutations in patients and mice are present from birth. GABAergic inhibition plays an important role in the developmental patterning of neuronal networks, and so the *Nf1* mutation may cause developmental defects that correlate with symptoms of NF1 such as learning disabilities. The distinction between developmental and adult causes of NF1 is relevant for deciding which features of NF1 could be targeted for treatment.

During development, experience-dependent plasticity mediates organization of neuronal networks using mechanisms shared with adult experience-dependent plasticity. Therefore developmental plasticity is likely to be disrupted in NF1 by the same increase in inhibition that disrupts LTP in adult $Nf1^{+/-}$ mice. Further, GABAergic inhibitory networks play a role in regulating the developmental patterning of cortical networks by modulating the length of developmental sensitive periods. Sensitive periods are windows of time during development when experience has a greater effect on the organization of neuronal networks than in adulthood (Hensch 2004, Knudsen 2004). In primary sensory areas, the sensitive period is regulated by the maturation and activity of GABAergic interneurons. For example, tonic increases in GABAergic activity correlate with the closure of the critical period for ocular dominance in the visual cortex (Fagiolini & Hensch 2000, Hensch 2005). In NF1, increased inhibition may cause inappropriate, early closure of sensitive periods leading to altered patterning in cortical areas. Possibly related to increased inhibition, or to other effects of *Nf1* deletion, cortical barrels fail to form in mice where *Nf1* is homozygously deleted from neurons and astrocytes of the somtatosensory cortex (Lush et al. 2008). NF1 may be associated with changes in primary sensory processing that are caused by increased inhibition, but may not directly cause the higher-order cognitive symptoms such as learning disabilities. Consistent with this increased inhibition, learning disabilities are often found in association, but not correlated with, alterations in primary sensory processing (Ramus 2003). Because the severity of learning disabilities does not correlate well with the severity of alterations in sensory processing, such sensory processing deficits are not likely to directly cause learning disabilities. In NF1, such associations may simply arise from a common underlying mechanism. Alternatively, it is possible that the increase in inhibition characterized in the $Nf1^{+/-}$ mutant mice is sufficient to cause learning deficits, but not irreversible developmental changes. Further studies are needed to better understand interactions between developmental effects of NF1, any associated sensory processing deficits, and behavioral effects of constitutively altered Ras signaling/interneuronal regulation.

LEARNING DISABILITIES AS A DISTINCT PATHOLOGICAL ENTITY: LESSONS FROM NF1

A key question concerning learning disabilities is whether they represent a distinct disease cluster or rather the low end of a spectrum of performance in the human population. This is a question that is common to discussions of many disorders of cognitive function. For example, similar discussions in schizophrenia, autism, and Alzheimer's have concluded that these conditions are not part of a behavioral and cognitive phenotypic continuum in the human population, but that instead represent separate pathological clinical entities. Accordingly, individuals suffering from each of these conditions account for a very small percentage of the population (1%–2%). With respect to learning disabilities, there is little agreement in the literature on the prevalence of learning disabilities, but estimates on incidence range from 1% to 17% (Altarac & Saroha 2007, Lagae 2008). It is argued that an incidence of 10% is consistent with the idea that learning disability is not a single clinical entity, but reflects instead the wide distribution of cognitive traits in the population (normal spectrum hypothesis). Furthermore, the classification criteria used for these estimates are often broad, and these high frequencies of learning disability are also argued to be consistent with the normal spectrum hypothesis.

The studies of learning disabilities associated with single-gene disorders contradict these arguments. These studies demonstrate that the cognitive deficits defined in these individuals have unique pathological causes and can be classified into discrete clinical conditions with specific diagnoses. Despite their broad clinical manifestations, the studies with NF1, Fragile X, Tuberous Sclerosis, Noonan and Rett's syndrome, to name only a few, indicate that a wide range of cognitive symptoms can be caused by single-gene mutations. Thus, broad behavioral and cognitive profiles do not necessarily reflect either the low end of the cognitive continuum of the general population, or the presence of an equally broad and diffuse pathological etiology. Rather, in these diseases, it is clear that a broad range of cognitive phenotypes are caused by specific genetic mutations leading to distinct clinical entities.

The results from single-gene disorders also demonstrate that a wide range of severities can be caused by the same mutation in a single gene. For example, learning disability and attention deficit symptoms occur in NF1 in a continuum of severity, and they are known to be caused by a single underlying genetic pathology. In fact, some individuals affected by NF1 show very few cognitive manifestations, whereas others reveal profound, albeit often specific, cognitive deficits. In NF1, genetic factors modify the severity and expression of symptoms, such that the variability in clinical presentation does not reflect the uniformity of the principal genetic cause (i.e., mutation in the NF1 gene). Among idiopathic learning disabilities, it is also likely that variability in expression between individuals occurs as a result of other unknown mutations in the genetic background of affected individuals (i.e., modifier genes), as well as environmental factors. Therefore, variability in expression of learning disabilities between individuals cannot be taken to imply variability in, or lack of, a distinct underlying clinical pathology. These considerations also lend support to the model of learning disabilities as a distinct syndrome with underlying pathological mechanisms that alter learning, rather than being a low end of the normal cognitive spectrum of the general population.

As in schizophrenia, autism, and Alzheimer's, a small percentage of individuals with learning disabilities have known genetic mutations (e.g., mutations in the *Nf1* gene), whereas the genetic etiology of most individuals is unknown. As in these other disorders, the severity and range of symptoms manifested in any one affected individual can vary, perhaps reflecting the interplay between genetic and environmental factors. Again, as in these well-described disorders, there is a cluster of phenotypes that represent the clinical diagnostic criteria, as well as other clinically important comorbidities that are not part of the disorder itself. For example, epilepsy and mental retardation are often associated with autism, but they are not an intrinsic part of the diagnosis of this condition, just as attention problems are often associated with learning disabilities. Thus, learning disabilities share these and other features with other complex mental health problems. This complexity and diversity are not the exception but the rule in current classifications of disorders of brain function.

Identifying learning disabilities as a disorder, rather than a variation of "normal" learning, has more than purely academic implications, as it bears on approaches to treatment. Learning disabilities have been identified as one of the primary factors affecting quality of life in the NF1 population, even given the high incidence of tumors and serious somatic symptoms that are also associated with this condition. Additionally, without intervention, individuals affected with NF1 more often than not have difficulties in effectively compensating for their deficits in learning and attention. From the perspective of affected individuals, family members, and professional care-givers, the cognitive and behavioral challenges associated with NF1 require professional intervention and are both unique and distinguishable from the normal cognitive spectrum in the general population. The difficulty in compensating for significant cognitive impairments also

differentiates the learning problems associated with these disorders from those seen in "normal" populations. This need for specialized care, as well as the considerable negative repercussions of underlying pathologies on the life of affected individuals, further emphasizes the necessity to recognize learning disabilities as a distinct disorder that requires early detection, intervention, and treatment.

Importantly, early treatment in NF1 and other learning disabilities can be effective in improving cognition, as measured by neuropsychological testing as well as academic and social achievement. In the NF1 population, low doses of methylphenidate, which may address their attention problems, appear to improve performance in school as well as attention (Mautner et al. 2002). In idiopathic learning disabilities, current interventions include unified programs, involving phonology training, compensatory strategy training, as well as involvement of school teachers and parents. Such cognitive and behavioral therapy-based programs are also effective in improving academic outcomes, especially when initiated early (Lagae 2008). The benefits of therapeutic intervention for learning disabilities seem clearly to outweigh the risks of treatment. Therefore, it is critical to recognize learning disabilities as a disease entity that requires early diagnosis and intervention to improve quality of life in affected individuals.

ADULT TREATMENT OF NEURODEVELOPMENTAL DISORDERS: THE CASE FOR NF1

There is a reasonable and well-entrenched assumption that neurodevelopmental disorders such as NF1 disrupt developmental processes that cannot be fully reversed in adult individuals, and that hope for treatment rests in early (fetal stages) diagnosis and intervention (Ehninger et al. 2008). Indeed, there are compelling data that many of the genes that affect adult function also affect important developmental processes. For example, the *Nf1* gene affects trophic function during rodent development (Luikart et al. 2008, Vogel et al.

1995). It is reasonable to propose that disruption of these developmental processes could lead to neurophysiological and neuroanatomical abnormalities that are irreversible in adult individuals. Additionally, there is compelling evidence that the development of the brain has critical periods, where key neurodevelopmental events must take place, outside of which these same processes no longer can be fully realized. For example, developmental studies of vision identified critical periods with narrow temporal windows, where high levels of sensory plasticity are important for the large-scale organization of specific areas of the visual cortex (Hensch 2005). The idea of developmental critical periods has further strengthened the bias that it may not be possible to reverse the pathologies associated with neurodevelopmental disorders in adult individuals.

Recent results with a number of animal models of neurodevelopmental disorders, including NF1, Fragile X, Tuberous Sclerosis, Rett syndrome, Lhermitte-Duclos disease/ Cowden disease, Rubinstein-Taybi syndrome, Down syndrome, etc, demonstrate that it is possible to dramatically improve, if not fully reverse, the cognitive phenotypes associated with these neurodevelopmental disorders in adult animals (Ehninger et al. 2008). When the underlying biochemical and physiological pathologies are reversed in adult individuals, the cognitive phenotypes could be either dramatically improved or even fully reversed without necessarily intervening during development (Ehninger et al. 2008)! For example, studies reviewed above showed that brief treatments in adult mutants with either Lovastatin, a farnesyl transferase inhibitor, or picrotoxin could reverse the physiological and behavioral phenotypes of the $Nf1^{+/-}$ mutant mice. More importantly, recent clinical trials suggested that statins may be effective at reversing at least some of the cognitive phenotypes of NF1 patients (Krab et al. 2008b). Similarly, findings in Fragile X also suggest that treating a key physiological deficit in adult mice could reverse clinically relevant phenotypes in mouse models of this disease (Dolen et al. 2007). Additionally, results from pilot clinical trials are also starting to suggest that similar treatments may also be effective in Fragile X patients (Berry-Kravis et al. 2009, Hagerman et al. 2009).

It is too early to gauge the efficacy of adult treatments for neurodevelopmental disorders, since the interpretation of the few pilot clinical trials carried out so far is confounded by design flaws and limited numbers of subjects. Nevertheless, studies in animal models are starting to uncover the molecular, cellular, and systems mechanisms disrupted by neurodevelopmental disorders, as well as to use this information to develop targeted treatments. Importantly, these studies have led to recent results raising the exciting possibility of dramatically improving the life of individuals afflicted with neurodevelopmental disorders, even when treatments are started in adulthood.

DISCLOSURE STATEMENT

The authors are not aware of any affiliations, memberships, funding, or financial holdings that might be perceived as affecting the objectivity of this review.

ACKNOWLEDGMENTS

This work is supported by the CTF Young Investigators Award, ARCS Foundation, and NIH MCNB Training Grant (2T32MH019384-11A2) to C.S. This work was supported by grants from the NIH (R01 NS38480), Neurofibromatosis Inc., the Children's Tumor Foundation, and United States Army (W81XWH-06-1-0174) to A.J.S.

LITERATURE CITED

Aarde SM, Jentsch JD. 2006. Haploinsufficiency of the arginine-vasopressin gene is associated with poor spatial working memory performance in rats. *Horm. Behav.* 49:501–8

Altarac M, Saroha E. 2007. Lifetime prevalence of learning disability among US children. *Pediatrics* 119(Suppl. 1):S77–83

Araki T, Chan G, Newbigging S, Morikawa L, Bronson RT, Neel BG. 2009. Noonan syndrome cardiac defects are caused by PTPN11 acting in endocardium to enhance endocardial-mesenchymal transformation. *Proc. Natl. Acad. Sci. USA* 106:4736–41

Araki T, Mohi MG, Ismat FA, Bronson RT, Williams IR, et al. 2004. Mouse model of Noonan syndrome reveals cell type- and gene dosage-dependent effects of Ptpn11 mutation. *Nat. Med.* 10:849–57

Bannerman DM, Niewoehner B, Lyon L, Romberg C, Schmitt WB, et al. 2008. NMDA receptor subunit NR2A is required for rapidly acquired spatial working memory but not incremental spatial reference memory. *J. Neurosci.* 28:3623–30

Barker D, Wright E, Nguyen K, Cannon L, Fain P, et al. 1987. Gene for von Recklinghausen neurofibromatosis is in the pericentromeric region of chromosome 17. *Science* 236:1100–2

Barton B, North K. 2004. Social skills of children with neurofibromatosis type 1. *Dev. Med. Child Neurol.* 46:553–63

Bernards A, Settleman J. 2004. GAP control: regulating the regulators of small GTPases. *Trends Cell Biol.* 14:377–85

Bernards A, Settleman J. 2005. GAPs in growth factor signaling. *Growth Factors* 23:143–49

Bernards A, Snijders AJ, Hannigan GE, Murthy AE, Gusella JF. 1993. Mouse neurofibromatosis type 1 cDNA sequence reveals high degree of conservation of both coding and noncoding mRNA segments. *Hum. Mol. Genet.* 2:645–50

Berry-Kravis E, Hessl D, Coffey S, Hervey C, Schneider A, et al. 2009. A pilot open label, single dose trial of fenobam in adults with Fragile X syndrome. *J. Med. Genet.* 46:266–71

Boguski MS, McCormick F. 1993. Proteins regulating Ras and its relatives. *Nature* 366:643–54

Brems H, Chmara M, Sahbatou M, Denayer E, Taniguchi K, et al. 2007. Germline loss-of-function mutations in SPRED1 cause a neurofibromatosis 1-like phenotype. *Nat. Genet.* 39:1120–26

Cannon TD, Glahn DC, Kim J, Van Erp TG, Karlsgodt K, et al. 2005. Dorsolateral prefrontal cortex activity during maintenance and manipulation of information in working memory in patients with schizophrenia. *Arch. Gen. Psychiatry* 62:1071–80

Chadman KK, Yang M, Crawley JN. 2009. Criteria for validating mouse models of psychiatric diseases. *Am. J. Med. Genet. B Neuropsychiatr. Genet.* 150B:1–11

Cichowski K, Santiago S, Jardim M, Johnson BW, Jacks T. 2003. Dynamic regulation of the Ras pathway via proteolysis of the NF1 tumor suppressor. *Genes Dev.* 17:449–54

Compte A, Brunel N, Goldman-Rakic PS, Wang XJ. 2000. Synaptic mechanisms and network dynamics underlying spatial working memory in a cortical network model. *Cereb. Cortex* 10:910–23

Constantinidis C, Goldman-Rakic PS. 2002. Correlated discharges among putative pyramidal neurons and interneurons in the primate prefrontal cortex. *J. Neurophysiol.* 88:3487–97

Constantinidis C, Williams GV, Goldman-Rakic PS. 2002. A role for inhibition in shaping the temporal flow of information in prefrontal cortex. *Nat. Neurosci.* 5:175–80

Costa R, Elgersma Y, Silva AJ. 2003. Modeling cognitive disorders: from genes to therapies. In *Genetics and Genomics of Neurobehavioral Disorders*, ed. G Fisch, pp. 39–68. Totowa, NJ: Humana Press

Costa RM, Federov NB, Kogan JH, Murphy GG, Stern J, et al. 2002. Mechanism for the learning deficits in a mouse model of neurofibromatosis type 1. *Nature* 415:526–30

Costa RM, Yang T, Huynh DP, Pulst SM, Viskochil DH, et al. 2001. Learning deficits, but normal development and tumor predisposition, in mice lacking exon 23a of Nf1. *Nat. Genet.* 27:399–405

Cui Y, Costa RM, Murphy GG, Elgersma Y, Zhu Y, et al. 2008. Neurofibromin regulation of ERK signaling modulates GABA release and learning. *Cell* 135:549–60

Demeter E, Sarter M, Lustig C. 2008. Rats and humans paying attention: cross-species task development for translational research. *Neuropsychology* 22:787–99

Denayer E, Ahmed T, Brems H, Van Woerden G, Borgesius NZ, et al. 2008. Spred1 is required for synaptic plasticity and hippocampus-dependent learning. *J. Neurosci.* 28:14443–49

Dilts CV, Carey JC, Kircher JC, Hoffman RO, Creel D, et al. 1996. Children and adolescents with neurofibromatosis 1: a behavioral phenotype. *J. Dev. Behav. Pediatr.* 17:229–39

Dolen G, Osterweil E, Rao BS, Smith GB, Auerbach BD, et al. 2007. Correction of Fragile X syndrome in mice. *Neuron* 56:955–62

Easton DF, Ponder MA, Huson SM, Ponder BA. 1993. An analysis of variation in expression of neurofibromatosis (NF) type 1 (NF1): evidence for modifying genes. *Am. J. Hum. Genet.* 53:305–13

Ehninger D, Li W, Fox K, Stryker MP, Silva AJ. 2008. Reversing neurodevelopmental disorders in adults. *Neuron* 60:950–60

Fagiolini M, Hensch TK. 2000. Inhibitory threshold for critical-period activation in primary visual cortex. *Nature* 404:183–86

Fellous JM, Sejnowski TJ. 2003. Regulation of persistent activity by background inhibition in an in vitro model of a cortical microcircuit. *Cereb. Cortex* 13:1232–41

Fossella JA, Casey BJ. 2006. Genes, brain, and behavior: bridging disciplines. *Cogn. Affect. Behav. Neurosci.* 6:1–8

Fountain JW, Wallace MR, Bruce MA, Seizinger BR, Menon AG, et al. 1989. Physical mapping of a translocation breakpoint in neurofibromatosis. *Science* 244:1085–7

Friedman JM. 1999. Epidemiology of neurofibromatosis type 1. *Am. J. Med. Genet.* 89:1–6

Guo HF, Tong J, Hannan F, Luo L, Zhong Y. 2000. A neurofibromatosis-1-regulated pathway is required for learning in *Drosophila*. *Nature* 403:895–98

Gutmann DH, Geist RT, Wright DE, Snider WD. 1995. Expression of the neurofibromatosis 1 (NF1) isoforms in developing and adult rat tissues. *Cell Growth Differ.* 6:315–23

Hagerman RJ, Berry-Kravis E, Kaufmann WE, Ono MY, Tartaglia N, et al. 2009. Advances in the treatment of Fragile X syndrome. *Pediatrics* 123:378–90

Hajra A, Martin-Gallardo A, Tarle SA, Freedman M, Wilson-Gunn S, et al. 1994. DNA sequences in the promoter region of the NF1 gene are highly conserved between human and mouse. *Genomics* 21:649–52

Hannan F, Ho I, Tong JJ, Zhu Y, Nurnberg P, Zhong Y. 2006. Effect of neurofibromatosis type I mutations on a novel pathway for adenylyl cyclase activation requiring neurofibromin and Ras. *Hum. Mol. Genet.* 15:1087–98

Hasenstaub A, Shu Y, Haider B, Kraushaar U, Duque A, McCormick DA. 2005. Inhibitory postsynaptic potentials carry synchronized frequency information in active cortical networks. *Neuron* 47:423–35

Hawes JJ, Tuskan RG, Reilly KM. 2007. Nf1 expression is dependent on strain background: implications for tumor suppressor haploinsufficiency studies. *Neurogenetics* 8:121–30

Hensch TK. 2004. Critical period regulation. *Annu. Rev. Neurosci.* 27:549–79

Hensch TK. 2005. Critical period plasticity in local cortical circuits. *Nat. Rev. Neurosci.* 6:877–88

Hofman KJ, Harris EL, Bryan RN, Denckla MB. 1994. Neurofibromatosis type 1: the cognitive phenotype. *J. Pediatr.* 124:S1–8

Husi H, Ward MA, Choudhary JS, Blackstock WP, Grant SG. 2000. Proteomic analysis of NMDA receptor-adhesion protein signaling complexes. *Nat. Neurosci.* 3:661–69

Hyman SL, Shores A, North KN. 2005. The nature and frequency of cognitive deficits in children with neurofibromatosis type 1. *Neurology* 65:1037–44

Jacks T, Shih TS, Schmitt EM, Bronson RT, Bernards A, Weinberg RA. 1994. Tumor predisposition in mice heterozygous for a targeted mutation in Nf1. *Nat. Genet.* 7:353–61

Jentsch JD, Aarde SM, Seu E. 2009. Effects of atomoxetine and methylphenidate on performance of a lateralized reaction time task in rats. *Psychopharmacology* 202:497–504

Kayl AE, Moore BD, 3rd. 2000. Behavioral phenotype of neurofibromatosis, type 1. *Ment. Retard. Dev. Disabil. Res. Rev.* 6:117–24

Kelly DP. 2004. Neurodevelopmental dysfunction in the school-aged child. In *Nelson Textbook of Pediatrics*, ed. RE Behrman, RM Kliegman, HB Jenson, pp. 110–12. Philadelphia: Saunders. 17th ed.

Knudsen EI. 2004. Sensitive periods in the development of the brain and behavior. *J. Cogn. Neurosci.* 16:1412–25

Koos T, Tepper JM. 1999. Inhibitory control of neostriatal projection neurons by GABAergic interneurons. *Nat. Neurosci.* 2:467–72

Koth CW, Cutting LE, Denckla MB. 2000. The association of neurofibromatosis type 1 and attention deficit hyperactivity disorder. *Child Neuropsychol.* 6:185–94

Krab LC, Aarsen FK, de Goede-Bolder A, Catsman-Berrevoets CE, Arts WF, et al. 2008a. Impact of neurofibromatosis type 1 on school performance. *J. Child Neurol.* 23:1002–10

Krab LC, de Goede-Bolder A, Aarsen FK, Pluijm SM, Bouman MJ, et al. 2008b. Effect of simvastatin on cognitive functioning in children with neurofibromatosis type 1: a randomized controlled trial. *JAMA* 300:287–94

Kronenberger WG, Dunn DW. 2003. Learning disorders. *Neurol. Clin.* 21:941–52

Lagae L. 2008. Learning disabilities: definitions, epidemiology, diagnosis, and intervention strategies. *Pediatr. Clin. North Am.* 55:1259–68, vii

Leach RJ, Thayer MJ, Schafer AJ, Fournier RE. 1989. Physical mapping of human chromosome 17 using fragment-containing microcell hybrids. *Genomics* 5:167–76

Ledbetter DH, Rich DC, O'Connell P, Leppert M, Carey JC. 1989. Precise localization of NF1 to 17q11.2 by balanced translocation. *Am. J. Hum. Genet.* 44:20–24

Lee DA, Portnoy S, Hill P, Gillberg C, Patton MA. 2005. Psychological profile of children with Noonan syndrome. *Dev. Med. Child Neurol.* 47:35–38

Lee YS, Silva AJ. 2009. The molecular and cellular biology of enhanced cognition. *Nat. Rev.* 10:126–40

Leondaritis G, Petrikkos L, Mangoura D. 2009. Regulation of the Ras-GTPase activating protein neurofibromin by C-tail phosphorylation: implications for protein kinase C/Ras/extracellular signal-regulated kinase 1/2 pathway signaling and neuronal differentiation. *J. Neurochem.* 109(2):573–83

Li W, Cui Y, Kushner SA, Brown RA, Jentsch JD, et al. 2005. The HMG-CoA reductase inhibitor lovastatin reverses the learning and attention deficits in a mouse model of neurofibromatosis type 1. *Curr. Biol.* 15:1961–67

Li Y, O'Connell P, Breidenbach HH, Cawthon R, Stevens J, et al. 1995. Genomic organization of the neurofibromatosis 1 gene (NF1). *Genomics* 25:9–18

Luikart BW, Zhang W, Wayman GA, Kwon CH, Westbrook GL, Parada LF. 2008. Neurotrophin-dependent dendritic filopodial motility: a convergence on PI3K signaling. *J. Neurosci.* 28:7006–12

Lush ME, Li Y, Kwon CH, Chen J, Parada LF. 2008. Neurofibromin is required for barrel formation in the mouse somatosensory cortex. *J. Neurosci.* 28:1580–87

Lynch TM, Gutmann DH. 2002. Neurofibromatosis 1. *Neurol. Clin.* 20:841–65

Marchuk DA, Saulino AM, Tavakkol R, Swaroop M, Wallace MR, et al. 1991. cDNA cloning of the type 1 neurofibromatosis gene: complete sequence of the NF1 gene product. *Genomics* 11:931–40

Martin SJ, Grimwood PD, Morris RG. 2000. Synaptic plasticity and memory: an evaluation of the hypothesis. *Annu. Rev. Neurosci.* 23:649–711

Martin SJ, Morris RG. 2002. New life in an old idea: the synaptic plasticity and memory hypothesis revisited. *Hippocampus* 12:609–36

Mautner VF, Kluwe L, Thakker SD, Leark RA. 2002. Treatment of ADHD in neurofibromatosis type 1. *Dev. Med. Child Neurol.* 44:164–70

Money J, Kalus ME Jr. 1979. Noonan's syndrome. IQ and specific disabilities. *Am. J. Dis. Child.* 133:846–50

Morris RG, Garrud P, Rawlins JN, O'Keefe J. 1982. Place navigation impaired in rats with hippocampal lesions. *Nature* 297:681–83

Moser EI, Krobert KA, Moser MB, Morris RG. 1998. Impaired spatial learning after saturation of long-term potentiation. *Science* 281:2038–42

Noonan JA. 1994. Noonan syndrome. An update and review for the primary pediatrician. *Clin. Pediatr.* 33:548–55

North K. 2000. Neurofibromatosis type 1. *Am. J. Med. Genet.* 97:119–27

North KN, Riccardi V, Samango-Sprouse C, Ferner R, Moore B, et al. 1997. Cognitive function and academic performance in neurofibromatosis. 1: consensus statement from the NF1 Cognitive Disorders Task Force. *Neurology* 48:1121–27

Palumbo D, Lynch PA. 2006. Psychological testing in adolescent medicine. *Adolesc. Med. Clin.* 17:147–64

Pasmant E, Sabbagh A, Hanna N, Masliah-Planchon J, Jolly E, et al. 2009. SPRED1 germline mutations caused a neurofibromatosis type 1 overlapping phenotype. *J. Med. Genet.* 46:425–30

Patapoutian A, Reichardt LF. 2001. Trk receptors: mediators of neurotrophin action. *Curr. Opin. Neurobiol.* 11:272–80

Ramus F. 2003. Developmental dyslexia: specific phonological deficit or general sensorimotor dysfunction? *Curr. Opin. Neurobiol.* 13:212–18

Rao SG, Williams GV, Goldman-Rakic PS. 2000. Destruction and creation of spatial tuning by disinhibition: GABA$_A$ blockade of prefrontal cortical neurons engaged by working memory. *J. Neurosci.* 20:485–94

Richter-Levin G, Canevari L, Bliss TV. 1995. Long-term potentiation and glutamate release in the dentate gyrus: links to spatial learning. *Behav. Brain Res.* 66:37–40

Robbins TW. 2002. The 5-choice serial reaction time task: behavioural pharmacology and functional neuro-chemistry. *Psychopharmacology* 163:362–80

Rosser TL, Packer RJ. 2003. Neurocognitive dysfunction in children with neurofibromatosis type 1. *Curr. Neurol. Neurosci. Rep.* 3:129–36

Rowbotham I, Pit-ten Cate IM, Sonuga-Barke EJ, Huijbregts SC. 2009. Cognitive control in adolescents with neurofibromatosis type 1. *Neuropsychology* 23:50–60

Sabbagh A, Pasmant E, Laurendeau I, Parfait B, Barbarot S, et al. 2009. Unraveling the genetic basis of variable clinical expression in neurofibromatosis 1. *Hum. Mol. Genet.* 18:2779–90

Sanchez-Vives MV, McCormick DA. 2000. Cellular and network mechanisms of rhythmic recurrent activity in neocortex. *Nat. Neurosci.* 3:1027–34

Seizinger BR, Rouleau GA, Ozelius LJ, Lane AH, Faryniarz AG, et al. 1987. Genetic linkage of von Reck-linghausen neurofibromatosis to the nerve growth factor receptor gene. *Cell* 49:589–94

Shen MH, Harper PS, Upadhyaya M. 1996. Molecular genetics of neurofibromatosis type 1 (NF1). *J. Med. Genet.* 33:2–17

Silva AJ, Frankland PW, Marowitz Z, Friedman E, Laszlo GS, et al. 1997. A mouse model for the learning and memory deficits associated with neurofibromatosis type I. *Nat. Genet.* 15:281–84

Silva AJ, Paylor R, Wehner JM, Tonegawa S. 1992a. Impaired spatial learning in alpha-calcium-calmodulin kinase II mutant mice. *Science* 257:206–11

Silva AJ, Stevens CF, Tonegawa S, Wang Y. 1992b. Deficient hippocampal long-term potentiation in alpha-calcium-calmodulin kinase II mutant mice. *Science* 257:201–6

Spurlock G, Bennett E, Chuzhanova N, Thomas N, Jim HP, et al. 2009. SPRED1 mutations (Legius syn-drome): another clinically useful genotype for dissecting the neurofibromatosis type 1 phenotype. *J. Med. Genet.* 46:431–37

Tartaglia M, Gelb BD. 2005. Noonan syndrome and related disorders: genetics and pathogenesis. *Annu. Rev. Genomics Hum. Genet.* 6:45–68

Tepper JM, Wilson CJ, Koos T. 2008. Feedforward and feedback inhibition in neostriatal GABAergic spiny neurons. *Brain Res. Rev.* 58:272–81

Thomson SA, Fishbein L, Wallace MR. 2002. NF1 mutations and molecular testing. *J. Child Neurol.* 17:555–61; discussion 71–72, 646–51

Tong J, Hannan F, Zhu Y, Bernards A, Zhong Y. 2002. Neurofibromin regulates G protein-stimulated adenylyl cyclase activity. *Nat. Neurosci.* 5:95–96

Viskochil D, Buchberg AM, Xu G, Cawthon RM, Stevens J, et al. 1990. Deletions and a translocation interrupt a cloned gene at the neurofibromatosis type 1 locus. *Cell* 62:187–92

Vogel KS, Brannan CI, Jenkins NA, Copeland NG, Parada LF. 1995. Loss of neurofibromin results in neurotrophin-independent survival of embryonic sensory and sympathetic neurons. *Cell* 82:733–42

Wallace MR, Marchuk DA, Andersen LB, Letcher R, Odeh HM, et al. 1990. Type 1 neurofibromatosis gene: identification of a large transcript disrupted in three NF1 patients. *Science* 249:181–86

Ward BA, Gutmann DH. 2005. Neurofibromatosis 1: from lab bench to clinic. *Pediatr. Neurol.* 32:221–28

Weeber EJ, Sweatt JD. 2002. Molecular neurobiology of human cognition. *Neuron* 33:845–48

Weiss B, Bollag G, Shannon K. 1999. Hyperactive Ras as a therapeutic target in neurofibromatosis type 1. *Am. J. Med. Genet.* 89:14–22

Williams JA, Su HS, Bernards A, Field J, Sehgal A. 2001. A circadian output in *Drosophila* mediated by neurofibromatosis-1 and Ras/MAPK. *Science* 293:2251–56

Williams VC, Lucas J, Babcock MA, Gutmann DH, Korf B, Maria BL. 2009. Neurofibromatosis type 1 revisited. *Pediatrics* 123:124–33

Zhu Y, Ghosh P, Charnay P, Burns DK, Parada LF. 2002. Neurofibromas in NF1: Schwann cell origin and role of tumor environment. *Science* 296:920–22

Zhu Y, Romero MI, Ghosh P, Ye Z, Charnay P, et al. 2001. Ablation of NF1 function in neurons induces abnormal development of cerebral cortex and reactive gliosis in the brain. *Genes Dev.* 15:859–76

Wallerian Degeneration, Wld^S, and Nmnat

Michael P. Coleman[1] and Marc R. Freeman[2]

[1]Laboratory of Molecular Signaling, The Babraham Institute, Cambridge CB22 3AT, United Kingdom

[2]Department of Neurobiology, Howard Hughes Medical Institute, University of Massachusetts Medical School, Worcester, Massachusetts 01605-2324

Annu. Rev. Neurosci. 2010. 33:245–67

First published online as a Review in Advance on March 26, 2010

The *Annual Review of Neuroscience* is online at neuro.annualreviews.org

This article's doi: 10.1146/annurev-neuro-060909-153248

Key Words

axon, dying back disorder, axonal transport, Wallerian-like degeneration

Abstract

Traditionally, researchers have believed that axons are highly dependent on their cell bodies for long-term survival. However, recent studies point to the existence of axon-autonomous mechanism(s) that regulate rapid axon degeneration after axotomy. Here, we review the cellular and molecular events that underlie this process, termed Wallerian degeneration. We describe the biphasic nature of axon degeneration after axotomy and our current understanding of how Wld^S—an extraordinary protein formed by fusing a Ube4b sequence to Nmnat1—acts to protect severed axons. Interestingly, the neuroprotective effects of Wld^S span all species tested, which suggests that there is an ancient, Wld^S-sensitive axon destruction program. Recent studies with Wld^S also reveal that Wallerian degeneration is genetically related to several dying back axonopathies, thus arguing that Wallerian degeneration can serve as a useful model to understand, and potentially treat, axon degeneration in diverse traumatic or disease contexts.

Contents

INTRODUCTION

Axons are huge cellular structures. If a Volkswagen Beetle sprouted a tail proportional to the length of a human motor axon, it would be ~20 miles (or 30 km) long. Maintaining such an enormous cellular outgrowth is a major challenge for the nervous system, and it is accomplished through the combined support of neuronal cell bodies and axon-associated glial cells. Without the delivery of materials from cell bodies by axonal transport (De Vos et al. 2008), axons undergo Wallerian degeneration (Coleman 2005); and without glial support in vivo, axons also degenerate (Nave & Trapp 2008). Consequently, axon and synapse loss is increasingly recognized as a major contributor to neurodegenerative disease.

Despite the importance of extra-axonal support for long-term survival, axons are now thought to initiate their own degeneration when these systems fail. Moreover, the trigger is not a general lack of nutrients. Injured axons appear to self-destruct through a regulatable or active process (Buckmaster et al. 1995, Raff et al. 2002) that is distinct from apoptosis (Burne et al. 1996, Deckwerth & Johnson 1994, Finn et al. 2000). Related mechanisms are triggered in some dying back axonopathies that raise the prospect of intervention (Ferri et al. 2003, Mi et al. 2005, Samsam et al. 2003). All of this has come to light because of the fortuitous discovery of the *slow Wallerian degeneration* mutant mouse (*Wld[S]*) (Lunn et al. 1989), in which axon stumps that are distal to an injury survive ten times longer than normal. Over the past decade, advances in this intriguing and sometimes controversial field have begun to shed light on a novel form of neuroprotection (Araki et al. 2004, Avery et al. 2009, Beirowski et al. 2009, Conforti et al. 2009, Mack et al. 2001, Sasaki et al. 2009b, Yahata et al. 2009).

In this review, we highlight the basic cell biology of Wallerian degeneration and our current understanding of the mechanism by which Wld[S] delays injury-induced axon degeneration. Then, we review the relationship between Wallerian degeneration and central and peripheral neuropathies as defined by the use of Wld[S] as a tool to block Wallerian-like degeneration. Next, we examine new opportunities created by transferring the Wld[S] phenotype to other species. Finally, we discuss several outstanding questions, which include the identity of the molecular trigger for Wallerian degeneration, and we discuss future steps in understanding how Wld^S protects axons.

WALLERIAN DEGENERATION

Wallerian degeneration is classically defined as the degeneration of axons distal to an injury, following Augustus Waller's original nerve transection experiments (Waller 1850). Here, we focus primarily on axonal events, which culminate in the granular disintegration of the axonal cytoskeleton and axon fragmentation that leaves characteristic myelin ovoids behind (**Figure 1**). However, the glial and macrophage clearance of degenerating axons, which we touch on briefly, is also an important part of Wallerian degeneration (Vargas & Barres 2006). Similar processes occur in unmyelinated axons in mammals and invertebrates (Avery et al. 2009, Ayaz et al. 2008, Macdonald et al. 2006) and in the mammalian central nervous system (CNS). CNS axons exhibit focal swellings that are many times wider than a normal axon many hours before axons fragment (Beirowski et al. 2010, Cajal 1928) and the slower clearance of axonal debris may contribute to the poor regenerative environment of the CNS (Vargas & Barres 2006).

SLOW WALLERIAN DEGENERATION

The extended survival of Wld^S axons without their cell bodies has fundamentally changed our

Axonal transport: the bidirectional active transport of cargoes between axons and cell bodies and along axons

Wallerian degeneration: the degeneration of an axon distal to a site of injury

Wld^S: the slow Wallerian degeneration protein, an aberrant fusion protein that delays degeneration of injured axons by tenfold

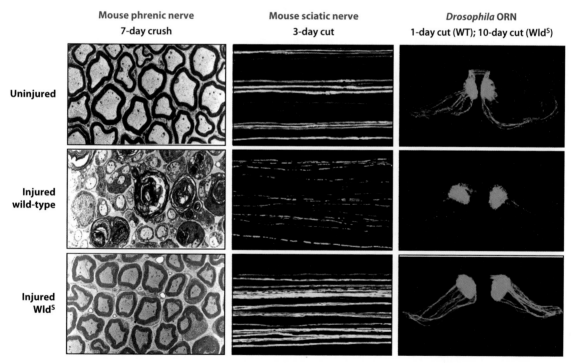

Figure 1

Wallerian degeneration in wild-type axons and preservation in Wld^S. Injured wild-type axons (middle row) exhibit granular disintegration of the cytoskeleton seen in electron microscopy (left column) and fragmentation visualized by fluorescence microscopy (middle and right columns). Cytoskeletal integrity, unswollen mitochondria, and axon continuity are preserved by the Wld^S gene in each case (bottom row). Note the remarkable consistency of Wallerian degeneration and the neuroprotective Wld^S phenotype between mice and *Drosophila*. ORN, olfactory receptor neuron. Left column from Brown et al. (1994); reprinted with permission from Wiley-Blackwell. Middle column from Conforti et al. (2007b); reprinted by permission from Macmillan Publishers Ltd., Nature Publishing Group.

view of axon degeneration. Wld^S transected axons in the sciatic nerve survive for over two weeks, compared to approximately 1.5 days in wild-type mice, and conduct evoked action potentials when stimulated (Lunn et al. 1989) **(Figure 1)**. Transected Wld^S axons in the CNS survive for similar extended periods (Perry et al. 1991). Surprisingly, *Wld^S* mice are viable and show normal motor function, although they exhibit a secondary delay in axon regeneration (Brown et al. 1994).

Transected *Wld^S* axons eventually degenerate in a process that is more atrophic and gradual than the sudden fragmentation that characterizes wild-type axons (Beirowski et al. 2005). This may reflect the gradual depletion of structural proteins from long-term anucleated axons. Thus, rapid Wallerian degeneration in wild-type nerves may be an active or at least regulated process, similar to apoptosis in principle.

Wld^S is a dose-dependent, semidominant phenotype that is inherited through a single locus (Mack et al. 2001, Perry et al. 1990b). It arose by spontaneous mutation at Harlan UK (then Harlan Olac, hence the original name C57BL/6/Ola) and was discovered by chance after it became homozygous (Lunn et al. 1989). The precise genetic background for *Wld^S* is uncertain (Lyon et al. 1993; V.H. Perry, personal communication), and there is further genomic divergence from C57BL/6 (A.L. Wilbrey, J.W. Tsao, and M.P. Coleman, manuscript in preparation).

The phenotype is intrinsic to nerves rather than macrophages (Perry et al. 1990a) and to axons rather than glia (Glass et al. 1993). In Schwann cell grafts between Wld^S and C57BL/6, host axons rather than donor Schwann cells determine the rate of degeneration (Glass et al. 1993); and primary neuronal cultures that lack glia show a remarkably similar delay in Wallerian degeneration after neurite transection, although neurites of both genotypes degenerate faster than in vivo (Buckmaster et al. 1995, Glass et al. 1993). Moreover, neuron-specific, but not glial, expression of the Wld^S gene confers the phenotype in *Drosophila* (Hoopfer et al. 2006,

Macdonald et al. 2006). Such an axon-specific effect on Wallerian degeneration is quite unique. Other mutations have been reported to influence Wallerian degeneration but seem to act on Schwann cell or macrophage responses rather than on axons (Keilhoff et al. 2002, Levy et al. 2001, Lopez-Vales et al. 2008, Narciso et al. 2009, Ramaglia et al. 2007).

The use of *Wld^S* mice as a genetic tool to explore the basis of cellular destruction pathways shows that neurodegenerative mechanisms are highly compartmentalized. Despite its robust effect on axon degeneration, *Wld^S* has no effect on apoptotic death of the cell soma, either in NGF-deprived sympathetic neuronal cultures or in axotomized motor neurons (Adalbert et al. 2006, Deckwerth & Johnson 1994), and no phenotypic change in any other cell type has been reported. Conversely, neither Bcl-2 overexpression nor Bax and Bak deletion alters Wallerian degeneration (Burne et al. 1996, Whitmore et al. 2003), and caspase 3 activation is neither detected in nor required for rapid Wallerian degeneration (Finn et al. 2000). Similar experiments established that axons in several disease models also die by nonapoptotic mechanisms. Bcl-2 overexpression and Bax deletion, respectively, rescue cell bodies in *pmn* mice and the DBA/2J glaucoma model but have no effect on axon degeneration (Libby et al. 2005, Sagot et al. 1995). Wld^S rescues axons in both cases (Ferri et al. 2003, Howell et al. 2007).

Synaptic terminals are also protected by *Wld^S* but act as another, partially-distinct compartment with respect to the timing of degeneration after injury (Gillingwater et al. 2002). Transected *Wld^S* motor axons support evoked neurotransmitter release at intact neuromuscular junctions for approximately five days compared to the usual 12–20 h (Ribchester et al. 1995), and CNS synapses are also protected (Gillingwater et al. 2006a). However, NMJ denervation occurs far sooner in wild-type and *Wld^S* animals than degeneration of the axon trunk. Moreover, neuromuscular synapse preservation is lost in young adult Wld^S mice without any change in *Wld^S*

expression, whereas Wld^S continues to preserve injured axon trunks (Gillingwater et al. 2002). Thus axonal and synaptic survival are both enhanced by Wld^S, but either the rate or the nature of the pathways involved differs.

New ENU mutant mice with enhanced synapse protection (Wong et al. 2009) and targeting of Nmnat1 to nerve terminals (E. Babetto, B. Beirowski, L. Janeckova, R. Brown, D. Thomson, R.R. Ribchester, M.P. Coleman, manuscript submitted) should shed more light on events at synapses. Interestingly, developmental synapse elimination is also unaltered in Wld^S mice (Parson et al. 1997), one of several findings that now distinguish developmental axon and synapse loss from injury-induced loss (Bishop et al. 2004, Hoopfer et al. 2006).

WALLERIAN-LIKE DEGENERATION

One central question is whether Wallerian degeneration is relevant to neurodegenerative disease. This occurred immediately to Waller, who stated: "It is particularly with reference to nervous diseases that it will be most desirable to extend these researches" (Waller 1850, p. 423). Decades later, dying-back-type axon degeneration in some peripheral nerve disorders was termed Wallerian-like degeneration based on morphological similarities (Cavanagh 1979, Griffin et al. 1996).

However, axon transection is rare in clinical neuroscience. Spinal injury and traumatic brain injury usually contuse and stretch axons respectively, which result in secondary axon interruption hours or days later. Peripheral nerves may be cut during surgery or wounding; but nerve damage by chronic pressure or metabolic, toxic, or inherited disorders is far more common. Axon transection has been observed directly in animal and cellular models of multiple sclerosis (Neumann et al. 2002; M. Kerschensteiner, personal communication), but whether this is the main mechanism in patients remains unclear. Axon endbulbs (Ferguson et al. 1997, Trapp et al. 1998) could alternatively begin as en passant swellings that

precipitate distal axon degeneration (Coleman 2005).

Although axon transection is rare, the disruption of axonal transport is extremely common and also isolates distal axons from cell bodies. As a tool to block Wallerian degeneration, Wld^S mice made it possible to test the hypothesis that similar mechanisms are triggered in noninjury disorders. Dying back follows a Wallerian-like mechanism in some motor neuron disease and peripheral neuropathy models and in nitric oxide damage (Alvarez et al. 2008, Ferri et al. 2003, Samsam et al. 2003, Wang et al. 2002). CNS studies extended this mechanism to models of Parkinson's disease, glaucoma, multiple sclerosis, and gracile axonal dystrophy (Beirowski et al. 2008, Hasbani & O'Malley 2006, Howell et al. 2007, Kaneko et al. 2006, Mi et al. 2005, Sajadi et al. 2004); and in primary culture, the neurotoxin vincristine and protein synthesis blockade also trigger Wallerian-like degeneration (Gilley & Coleman 2010, Wang et al. 2000). Thus, similar to using Bcl-2 to define apoptotic cell death, Wld^S sensitivity now provides a genetic definition for Wallerian-like degeneration.

Wld^S does not substantially block pathology in some models of amyotrophic lateral sclerosis or spinal muscular atrophy (Fischer et al. 2005, Kariya et al. 2009, Rose et al. 2008, Velde et al. 2004), whereas axonal swellings precede fragmentation by months in mouse models of familial Alzheimer's disease (Adalbert et al. 2009, Spires et al. 2005). Thus, Wallerian-like degeneration may not be the only outcome when transport is impaired. This may reflect a loss of different transport cargoes in different disorders or the reversion of synapse degeneration to wild type in older animals (Gillingwater et al. 2002). New transgenic mice with stronger synapse protection could help distinguish these possibilities (Beirowski et al. 2009).

THE WLD^S GENE AND PROTEIN

The Wld^S gene was mapped to mouse chromosome 4 (Lyon et al. 1993), in which an unusual genomic rearrangement brings two

Nmnat: nicotinamide mononucleotide adenylyltransferase; an enzyme that catalyzes NAD^+ synthesis in the salvage pathway. There are three isoforms in mammals

Dying back disorder: a neurodegenerative disorder in which axons die before cell bodies and/or in a retrograde pattern that begins with their distal ends

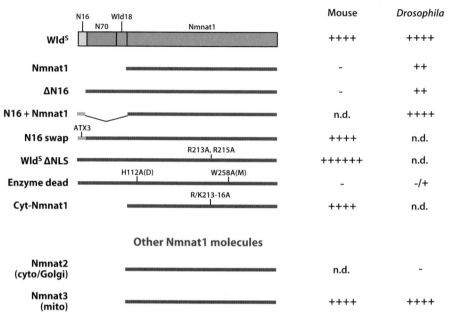

Figure 2

Axon protection mediated by Wld^S, Wld^S domains, and Nmnat molecules. The in vivo protective effects have been studied extensively in mouse and *Drosophila*. (*Top*) Wld^S-derived sequences represent constructs where specific domains of Wld^S were deleted or mutated. (*Bottom*) Other Nmnat molecules are mouse Nmnat2 and Nmnat3. Protection in either mouse or *Drosophila* is shown to the right. Protection and its relative strength are indicated by (+), a lack of protection is indicated by (-), and those not determined *in vivo* are indicated by n.d. Point mutations or domain swaps are shown above the diagram of each molecule, with point mutation positions that refer to their relative position in Wld^S.

N70: the N-terminal 70 amino acids of Wld^S and Ube4b

Ube4b: ubiquitin ligase E4b, also known as Ufd2a. It carries out multiubiquitination of substrates in the ubiquitin fusion degradation pathway

endogenous genes together. Their mRNAs splice to encode an in-frame fusion protein that is absent in wild-type mice (**Figure 2**) (Coleman et al. 1998, Conforti et al. 2000). The expression of this protein in transgenic mice replicates the Wld^S phenotype that identifies it as the Wld^S protein (Mack et al. 2001) (**Figure 2**), and the murine cDNA delays axon degeneration in rat, fly, and cell culture models (Adalbert et al. 2005, Araki et al. 2004, Hoopfer et al. 2006, Macdonald et al. 2006, Wang et al. 2001).

The C-terminal 285 amino acids comprise the complete protein sequence of nicotinamide mononucleotide adenylyltransferase 1 (Nmnat1), a key protein of the NAD$^+$ salvage pathway in mammals. Nmnat1 normally resides predominantly or exclusively in nuclei (Magni et al. 2004), where Wld^S is also abundant (Mack et al. 2001), and new roles for NAD$^+$ are emerging (Pollak et al. 2007). Other Nmnat isoforms synthesize NAD$^+$ in mitochondria and the Golgi apparatus (Berger et al. 2005, Raffaelli et al. 2002). The Wld^S protein synthesizes NAD$^+$ from nicotinamide mononucleotide and ATP but does not alter basal NAD$^+$ levels (Araki et al. 2004, Mack et al. 2001), probably owing to rapid NAD$^+$ catabolism (Pollak et al. 2007). The N-terminal 70 amino acids (N70) of Wld^S are derived from the N-terminus of Ube4b (or Ufd2a), an E4-type ubiquitin ligase that can add multiubiquitin chains to substrates of the ubiquitin fusion degradation (Ufd) pathway (Hatakeyama et al. 2001, Koegl et al. 1999). Only 6% of the Ube4b sequence is incorporated into Wld^S, which excludes the catalytic U box. Thus, Wld^S probably lacks ligase activity, but there is shared protein binding activity (Laser et al. 2006, Morreale et al. 2009). Finally, between the N70 and Nmnat1 sequences lies a unique 18 amino acid sequence generated by a read-through of Nmnat1 5′ UTR, the epitope of the specific Wld18 antibody (Samsam et al. 2003).

GAIN OF FUNCTION

Gain of function appears to be the most likely mechanism for Wld^S function. The genomic rearrangement retains the endogenous Nmnat1 and Ube4b genes, and the corresponding proteins are expressed at normal levels (Conforti et al. 2007b, Gillingwater et al. 2002). Thus, there is no obvious loss of function of either protein. Regarding a dominant negative mechanism, Nmnat activity is increased in Wld^S tissue (Mack et al. 2001), and deleting one allele of Nmnat1 does not alter Wallerian degeneration (L. Conforti, N. Smyth, and M.P. Coleman, manuscript in preparation), whereas haploinsufficiency for Ube4b causes axon pathology rather than axon protection (Kaneko-Oshikawa et al. 2005). For a gain of function mechanism, the key question remains whether this is an entirely novel function or whether Wld^S strengthens or mimics the function of an endogenous protein. Recent data support the latter model, indicating that Wld^S substitutes for Nmnat2, an essential axonal protein that is rapidly lost after axon injury (Gilley & Coleman 2010).

DEFINING PROTECTIVE DOMAINS IN THE MOLECULE AND THE CELL

Which domains of Wld^S are essential for the protection of severed axons? This question has been a central focus of the field for the past five years, but attempts to answer it have raised several controversies. Studies in primary culture have generally produced less consistent results regarding whether, where, and how Wld^S and its constituent domains protect severed axons (Araki et al. 2004, Conforti et al. 2007b, Wang et al. 2005) than studies in vivo. Structure-function analyses in mice and *Drosophila* (Avery et al. 2009; Beirowski et al. 2009; Conforti et al. 2007b, 2009; Sasaki et al. 2009a; Yahata et al. 2009) are now converging on a model in which the combinatorial activity of two key domains of Wld^S acts somewhere outside the nucleus to confer maximal axon protection (**Figure 2**).

Theoretically, Wld^S could promote axon protection through N70, Wld18, Nmnat1, or some combination of these domains. In vitro data argued that Nmnat1 could protect severed neurites in primary culture, although to a significantly lower degree than Wld^S (Araki et al. 2004, Sasaki et al. 2009a, Wang et al. 2005). However, overexpressed Nmnat1 is not sufficient to protect severed axons in transgenic mice (Conforti et al. 2007b), whether they are driven by the same β *actin* promoter used to identify Wld^S previously (Mack et al. 2001) or they express up to threefold higher levels using the *Prp* promoter (Yahata et al. 2009).

Can Nmnat1 protect axons in vivo in any context? In *Drosophila*, the expression of mouse Nmnat1 in ORNs using the Gal4/UAS binary expression system resulted in the strong protection of severed axons (Macdonald et al. 2006), but, importantly, the protection it afforded was consistently weaker and lasted for a shorter period when compared to Wld^S (Avery et al. 2009). Thus, Nmnat1 is likely an important part of Wld^S neuroprotective function, but why is there a discrepancy in Nmnat1-mediated protection between flies and mice? The most plausible explanation is that Gal4/UAS expresses extremely high levels of Nmnat1 in *Drosophila* ORNs. The Gal4 driver line used (*OR22a-Gal4*) is quite strong, which results in high levels of Gal4, and the expression of Nmnat1 (i.e., *UAS-Nmnat1*) is further amplified because Gal4 is an efficient transcriptional activator of *UAS*-regulated target genes. Perhaps experiments aimed at dramatically increasing the levels of Nmnat1 in mouse axons might ultimately provide some level of axon protection in the mouse model. Alternatively, these discrepancies may reflect differences in axon length, diameter, or other characteristics among *Drosophila* and mice that affect the initiation and execution of Wallerian degeneration.

Although wild-type Nmnat1 is not sufficient for the robust protection of severed axons in mice, its activity is clearly an essential part of the protective action of Wld[S]. Three groups recently generated animals that express Wld[S] variants in which the enzymatic activity of Nmnat1 was disrupted by mutation (Conforti et al. 2009, Yahata et al. 2009). In each case, the neuroprotective effects of Wld[S] were severely reduced (Avery et al. 2009) or abolished (Conforti et al. 2009, Yahata et al. 2009), this time consistent with primary culture results (Araki et al. 2004, Jia et al. 2007). These results are important because they implicate Nmnat1 enzymatic activity in Wld[S]-dependent axon protection and because they begin to address a second possible role, that of a chaperone, for Nmnat1 in axon protection. An interesting recent study in *Drosophila* identified mutants in the sole fly Nmnat gene (*dnmnat*), whose neurons appear to develop normally, which extend axons to the appropriate targets but then exhibit age- and activity-dependent degeneration (Zhai et al. 2006). These data raised the intriguing possibility that basic axon integrity in the mature nervous system might be regulated by constitutive dNmnat-dependent suppression of neuronal degeneration.

Surprisingly, in rescue experiments, dNmnat enzymatic activity was found to be dispensable for the rescue of *dnmnat* mutant neurodegenerative phenotypes (Zhai et al. 2006). A subsequent study proposed a novel chaperone-like role for dNmnat and mammalian Nmnat3 (e.g., in the refolding of denatured proteins), independent of NAD[+] biosynthetic activity; and the authors in turn proposed that this novel Nmnat chaperone activity might explain some aspects of the Wld[S]-dependent protection of severed axons (Zhai et al. 2008). Such a chaperone-like role, which could stabilize axonal proteins, might fit nicely with the neuroprotective effects

observed in Wld[S] mice, which essentially express a modified Nmnat1.

However, as mentioned above, the mutation of either the ATP binding site of Nmnat1 (Avery et al. 2009) or the NMN[+] binding site (Conforti et al. 2009, Yahata et al. 2009), which would be expected to leave chaperone function intact, potently blocked the ability of Wld[S] to protect severed axons. Is Nmnat-dependent chaperone activity important for Wld[S]-mediated axon protection? One possibility is that dNmnat and Nmnat3 chaperone-like functions have nothing to do with Wld[S]-mediated axon protection and are more specific to the neurodegenerative phenotypes observed in *dnmnat* mutants. Alternatively, Wld[S] may have two essential functions: to generate NAD[+] or another biosynthetic product that would require enzymatic activity and to act as a chaperone. If both activities are essential, then the disruption of either should suppress the axon protective effects of Wld[S], as found in transgenic mice and flies. We can conclude that Nmnat-dependent chaperone activity cannot be the sole role for the Nmnat1 domain of Wld[S].

The weaker axon protective effect of Nmnat1 in flies and mice suggests that other portions of Wld[S] are essential for Wld[S]-like levels of axon protection, through affecting either its localization or activity. A protein interaction site within the N-terminal 16 amino acids (N16), which is derived from Ube4b, coimmunoprecipitates valosin containing protein (VCP/p97) from mouse brain homogenates (Laser et al. 2006). VCP is a AAA-ATPase with key roles in the UPS and membrane fusion (Wang et al. 2004) and is the only abundant, direct binding partner precipitated by this region.

Subsequent work has shown that N16 is necessary and sufficient to explain the differences in protective effects between Wld[S] and Nmnat1 in vivo. The deletion of N16 from Wld[S] completely suppresses the axon protection afforded by Wld[S] in mice (Conforti et al. 2009) and greatly weakens the protection of axons in *Drosophila* to a level found by expressing

Nmnat1 alone (**Figure 2**) (Avery et al. 2009). Moreover, fusing N16 directly to Nmnat1 results in levels of axon protection that are indistinguishable from Wld^S (Avery et al. 2009). Does N16 exert its effects on axon preservation through VCP? Two in vivo experiments support this notion. First, replacing N16 in Wld^S with a well-characterized VCP-binding motif from ataxin 3 (Morreale et al. 2009), which shares only five amino acids with N16, restores Wld^S-like axon protection in mice (Conforti et al. 2009). Second, RNAi knockdown of fly VCP is sufficient to suppress axon protection by Wld^S to levels indistinguishable from those afforded by Nmnat1 alone. Together, these data argue strongly that N16, likely working through VCP, and an enzymatically active Nmnat1, are the critical domains and essential activities for Wld^S-like levels of axon protection.

A second major question regarding Wld^S function relates to its site of action. Is it functioning in the nucleus or elsewhere in the cell? Clarifying this point is critical for understanding precisely how Wld^S can so potently suppress Wallerian degeneration. The striking nuclear localization of Wld^S has led to several studies of expression patterns in Wld^S versus wild-type neurons (Gillingwater et al. 2006b, Simonin et al. 2007), in the hope of identifying changes in the expression of key genes that modulate axon autodestruction (see below). However, more recent careful analysis of the localization of Wld^S and potential extranuclear sites of action proposes that Wld^S exerts its neuroprotective effects outside the nucleus.

Nmnat1 contains a strong nuclear localization sequence (NLS), which may account for the nuclear localization of Wld^S in vivo. If Wld^S is required in the nucleus, then deleting the NLS from Nmnat1 should weaken its protective effects. In striking contrast, the mutation of the Nmnat1 NLS from Wld^S doubled the latest timepoint when surviving axons could be identified and greatly enhanced synaptic protection especially in older mice (Beirowski et al. 2009). Moreover, although the expression of Nmnat1 alone in mice fails to suppress

Wallerian degeneration (Conforti et al. 2007b, Yahata et al. 2009), the generation of mice that harbor a cytoplasmic mutant of Nmnat1 resulted in robust axon protection (Sasaki et al. 2009a), although the relative contributions of cytoplasmic targeting and high expression levels in these mice remain unclear. Thus, excluding Wld^S and Nmnat1 from the nucleus makes them more protective of severed axons. This observation hints at a possible role for N16 in relocalizing Nmnat1 activity outside the nucleus to another cellular location. Indeed, a careful analysis of Wld^S expression in peripheral nerves revealed low but detectable levels of expression (Beirowski et al. 2009, Yahata et al. 2009), consistent with a potential non-nuclear site of action for Wld^S in mice.

OUTSTANDING QUESTIONS ON THE WLD^S AXON PROTECTIVE MECHANISM

What is the Cellular Site of WLD^S Action?

The discussion above establishes a nonnuclear site of action for Wld^S; now, we need to identify the site. One approach is to distinguish between roles in the axon and soma, and another approach is to identify the appropriate organelle(s) or protein complex. This may be less straightforward than resolving nuclear or cytoplasmic actions. Wld^S localizes to multiple internal membranous organelles (Beirowski et al. 2009, Yahata et al. 2009). Mitochondria have attracted particular interest because injured axons are protected by overexpressed Nmnat3, the mitochondrial isoform (Avery et al. 2009, Yahata et al. 2009). However, overexpressed proteins may have ectopic locations, and the experience with nuclei reminds us that an observable location does not identify the site of action (Beirowski et al. 2009). Thus, other locations still need to be considered.

Other possible sites include the Golgi apparatus and the ER, particularly because these are sites where VCP is abundant. VCP binding redistributes Wld^S within nuclei, which suggests that cytoplasmic Wld^S is likely to migrate to sites where VCP is abundant (Wilbrey et al. 2008), although this also includes mitochondria (Braun et al. 2006). To complicate matters further, mitochondria are rapidly transported in axons (Misgeld et al. 2007), have close interactions with the ER that may influence axon survival (Merkwirth & Langer 2008), and exchange NAD$^+$ with their surroundings (Todisco et al. 2006). Thus, any NAD$^+$ produced within mitochondria could act on nearby structures to protect axons or vice versa. Wld^S becomes ineffective when targeted to the internal surface of the plasma membrane, so this may not be the site of action (Avery et al. 2009), but distinguishing between other sites could be far more complex.

Is NAD$^+$ a Protective WLD^S Product?

An alternative route forward is to identify functions downstream of Wld^S Nmnat activity. In addition to its long established role in bioenergetic metabolism, NAD$^+$ is also a substrate for protein deacetylation by sirtuins, for synthesis of cyclic ADP ribose and ADP ribose, both regulators of internal calcium stores, and for mono- and poly-ADP ribosylation of proteins (Hassa et al. 2006, Pollak et al. 2007). It also potentiates the response of sodium-activated potassium channels to sodium (Tamsett et al. 2009), although an action of Wld^S at plasma membranes seems unlikely (see above). NAD$^+$ is also used to synthesize NADP$^+$, whose reduced form has roles in detoxification and oxidative defense (Pollak et al. 2007). Identifying one of these as an important downstream step could help resolve the site of action because rapid NAD$^+$ catabolism limits its long-range diffusion (Pollak et al. 2007).

However, despite agreement that Wld^S needs Nmnat activity to protect axons (above), it is less clear whether NAD$^+$ is the protective enzyme product. No increase in NAD$^+$ is detectable with Wld^S (Araki et al. 2004, Mack et al. 2001) or when Nmnat activity is raised more than 15-fold (Sasaki et al. 2009a). Knockdown or inhibition of Nampt, the rate-limiting enzyme in the NAD$^+$ salvage pathway, failed to revert Wld^S phenotype in one study (Sasaki et al. 2009b) and reverted it only partially in another (Conforti et al. 2009). Conversely, increasing cellular NAD$^+$ by blocking NAD$^+$ catabolizing enzyme CD38 does not confer a Wld^S phenotype (Sasaki et al. 2009b; A.L. Wilbrey & M.P. Coleman, unpublished observations). One proposal is that Nmnat activity decreases reactive oxygen species in mitochondria because it protects axons from rotenone-induced damage without restoring normal ATP levels (Press & Milbrandt 2008). Another proposal is that Nmnat catalyzes the reverse reaction under stress conditions, which generates an emergency supply of ATP (Yahata et al. 2009), although this appears unlikely to preserve axons for several weeks. Many of these observations are also consistent with NAD$^+$ acting at a highly localized site, but clearly, alternative substrates and products of Nmnat isoforms (Hassa et al. 2006, Sorci et al. 2007) need to be considered alongside NAD$^+$ as candidates for the axon protective mechanism.

THE MOLECULAR TRIGGER FOR WALLERIAN DEGENERATION

To fully understand how Wld^S, or some Nmnats, delay Wallerian degeneration, we must understand the process they delay. Remarkably, we still do not know the molecular pathway for Wallerian degeneration, a question that far predates Wld^S mice (Lubinska 1977). Almost all recent studies have focused on overexpressed or modified proteins, including Wld^S itself, which is absent in wild-type organisms. The results are exciting but do not tell us directly how endogenous proteins behave when a wild-type axon is injured or sick. The "potential to throw light on the normal processes of nerve degeneration" was a major driving force for identifying Wld^S in the first place (Lyon et al. 1993, p. 9717), and realizing this goal remains a major gap in the field.

However, there are some clues from studying how Wld^S alters disease models. Two alternative models for how injury triggers

Wallerian degeneration are a prodegeneration signal generated at the lesion site (e.g., calcium influx through the cut end) and the absence of a prosurvival signal derived from the cell body. The disease studies clearly show that Wallerian-like processes can be triggered in the complete absence of physical injury. In contrast, there is a strong correlation with disorders of axonal transport (Beirowski et al. 2008, Ferri et al. 2003, Howell et al. 2007, Wang et al. 2005). Non-lethal impairment of protein synthesis also triggers Wallerian-like degeneration (Gilley & Coleman 2010). These observations support the survival factor model and argue against the need for any signal derived from the injury site.

SURVIVAL AND CATASTROPHE

We have much to learn from the biphasic degeneration of wild-type axons. An initial latent phase lasts 36–44 hours after injury in a mouse sciatic nerve, followed by a sudden and catastrophic fragmentation phase (Beirowski et al. 2005, Kerschensteiner et al. 2005, Lubinska 1977). During the latent phase, nodal gaps widen and motor axons lose their terminals (Conforti et al. 2007a, Miledi & Slater 1970), but axon trunks remain continuous and can conduct evoked action potentials (Lunn et al. 1989, Moldovan et al. 2009). Fly axons, zebrafish axons (S. Martin & A. Sagasti, personal communication), and transected neurites in primary culture also show a latent phase, albeit shorter than in mice. Both between and within species, there is a correlation with axon stump length.

After surviving 1.5 days without a cell body, several centimeters of mouse wild-type axon undergoes catastrophic fragmentation, possibly in one hour (**Figure 3**) (Beirowski et al. 2005). An equally rapid process, captured in live in vivo imaging, occurs in CNS axons, where there is also more axon swelling (Kerschensteiner et al. 2005). The onset of this fragmentation phase is heterogeneous among axons in the same nerve and depends on intrinsic properties such as axon diameter and extrinsic properties such as temperature (Gamble & Jha 1958, Lubinska 1977).

The model posited to explain these data is strikingly similar to the survival signal discussed above. Lubinska suggested that distal axons require the constant delivery of a trophic factor, anterogradely transported from cell bodies (Lubinska 1977) but partially redistributed by retrograde transport (Lubinska 1982). Injured axons degenerate when this putative factor falls below a threshold level. This can explain why cold temperatures and proteasome inhibition extend the latent phase because these are likely to increase the trophic factor half-life (Gamble & Jha 1958, Macinnis & Campenot 2005). This also explains reports that fragmentation begins at the proximal end of a transected axon stump and progresses distally (Beirowski et al. 2005, Lubinska 1977) because steady degradation of the putative survival factor in distal axons will produce a net anterograde flux, which depletes the survival factor in regions closer to the injury sooner than in distal axons. However, the directionality remains controversial (Beirowski et al. 2005). The most consistent, and perhaps the most important, point is the sudden switch.

Lubinska's reasons for suggesting a single trophic factor are unclear but probably relate to the simple biphasic kinetics. A sudden switch is best explained by a single event: the depletion of one axonal component. Multiple survival factors with slightly different half-lives and transport kinetics would be expected to produce a more gradual switch.

Any complete model for the molecular mechanism of Wallerian degeneration must explain what is happening during the latent phase and why it ends so suddenly. A complete model for the action of Wld[S] should also explain whether and how the longer latent phase in Wld[S] axons relates to the shorter latent phase in wild type.

WLD[S] AND SURVIVAL SIGNALING

The key to future progress on this question is to fit the Wld[S] protective mechanism into the survival signal, or trophic factor model, of wild-type axon degeneration. The ability of axons

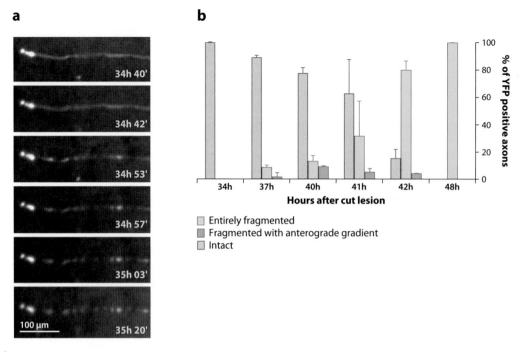

Figure 3

Wallerian degeneration in wild-type axons is a biphasic process. (*a*) Distal stumps of injured wild-type axons in the dorsal column of mouse spinal cord remain continuous for a latent phase of over 34 h before fragmenting over the course of a few minutes. From Kerschensteiner et al. (2005) reprinted by permission from Macmillan Publishers Ltd., Nature Publishing Group. (*b*) Quantification of fragmented, unfragmented, and partially fragmented wild-type axons in distal sciatic nerve after transection injury shows a similar latent phase followed by fragmentation of all axons in the nerve over the next few hours. The timing of fragmentation is heterogenous within the axon population, but once started, fragmentation is rapid, such that the percentage of partially fragmented axons never exceeds 10% at any one time. From Beirowski et al. (2005) reprinted with permission from BioMed Central.

to survive for 2–3 weeks originally cast doubt on the survival signal model and suggested that short-term axon survival is independent of cell bodies. However, these are mutant axons in which one or more axonal components must have been altered prior to any nerve lesion. Wld^S could not, and indeed does not (Wishart et al. 2007), cause major alterations to the axonal proteome, but if Lubinska's proposal of a single, endogenous trophic factor is correct, only one or a few changes may be sufficient to circumvent its loss. This might be achieved by the following:

(*a*) directly adding to the pool of a wild-type survival factor,

(*b*) delivering an endogenous survival factor in greater quantities to axons,

(*c*) depleting axons of a factor needed to execute degeneration,

(*d*) stabilizing a survival factor in injured axons,

(*e*) promoting local synthesis of an endogenous survival factor in axons, or

(*f*) substituting for an endogenous survival factor.

Model (*a*) is clearly not the case. Wld^S cannot be identical to a wild-type survival factor because it is absent from wild-type cells.

GENE EXPRESSION

Models (*b*) and (*c*) rely on altering the supply of other axonal components. This could be achieved by altering their axonal transport or their expression level. Wld^S has no obvious similarity to any axonal transport motor or regulator, but it is abundant in nuclei (Mack et al. 2001), and several transcripts and proteins are expressed at different levels in Wld^S and C57BL/6 mice (Chitnis et al. 2007; Gillingwater et al. 2006b; Simonin et al. 2007; Wishart et al. 2007, 2008). Moreover, sirtuin-1, a nuclear NAD$^+$ dependent gene silencing protein, was reported to be required for axon protection by exogenous NAD$^+$ (Araki et al. 2004).

However, no causative link has been established yet between gene expression or proteomic changes and axon protection by Wld^S, and sirtuins are not required for Wld^S to protect axons (Avery et al. 2009, Wang et al. 2005). Many of these gene expression changes are also not conserved in Wld^S rats, and some changes reflect the genomic divergence of Wld^S and C57BL/6 mice directly (A.L. Wilbrey, J.W. Tsao, M.R. Cookson, and M.P. Coleman, manuscript in preparation). Moreover, the finding that Wld^S has a cytoplasmic site of action (above) argues against the existence of a mechanism that involves changes in gene expression.

UPS IMPAIRMENT

Model (*d*), stabilizing a survival factor in injured axons, most clearly fits with a possible impairment of the ubiquitin proteasome system (UPS) (Coleman & Ribchester 2004, Ehlers 2003). Impairing the UPS prolongs the survival of injured axons (Hoopfer et al. 2006, Macinnis & Campenot 2005, Zhai et al. 2003), but it is less clear whether this is how Wld^S prolongs axon survival. General impairment of the UPS is more likely to make axons sick than protect them. Proteasome inhibition causes neurite degeneration in culture and peripheral neuropathy in humans (Kane et al. 2003, Laser et al. 2003), whereas several mouse axonopathies are caused by UPS impairment (Kaneko-Oshikawa et al. 2005, Saigoh et al. 1999). An efficient UPS appears to be essential for axon health (Coleman & Ribchester 2004); so the robust

health of Wld^S mice, rats, flies, and primary cultures argues against any general UPS impairment.

Interaction between Wld^S and VCP has the potential to impair specific UPS functions. In addition to Ube4b, several other ubiquitin ligases and the deubiquitinating enzyme ataxin 3 bind VCP at the same site, so Wld^S could compete for VCP binding with any or all of these proteins (Morreale et al. 2009). However, the VCP binding site is dispensable in the context of modified Nmnat1 or Nmnat3 (Avery et al. 2009, Sasaki et al. 2009a, Yahata et al. 2009); thus, these actions are not needed to protect axons.

LOCAL PROTEIN SYNTHESIS

Model (*e*) is based on the observation that injured axons elevate local protein synthesis on both sides of a lesion (Court et al. 2008, Perlson et al. 2005), so Wld^S might protect axons by stimulating the local synthesis of a survival factor. The local synthesis of Wld^S in axons has also been proposed (Fainzilber & Twiss 2006). However, the protein synthesis machinery appears at the same time in injured Wld^S and wild-type axons (Court et al. 2008), and the protective capacity of Wld^S is unabated when protein synthesis is suppressed (Gilley & Coleman 2010). Thus, local synthesis is not required for Wld^S to protect axons.

SUBSTITUTING FOR A SURVIVAL FACTOR

Model (*f*) would fit with Wld^S substituting for an endogenous Nmnat in injured axons. This concept seems implicit in the various Nmnat overexpression studies but has not been phrased this way, perhaps because the nucleus was previously seen as a likely site of action. The recent knowledge that Wld^S acts outside nuclei, and is present in axons, refocuses attention on what it may do there, consistent with an earlier report of a local site of action (Wang et al. 2005). A key requirement is that the respective Nmnat should also be present in axons. Interestingly, it now seems that

Nmnat2, like some other Golgi proteins (Merianda et al. 2009), is present in axons and that its depletion is necessary for rapid Wallerian degeneration in vitro of injured axons and sufficient for Wallerian-like degeneration of uninjured axons (Gilley & Coleman 2010). This observation supports a model in which Nmnat2 is an endogenous axon survival factor, and Wld[S], a far more stable protein, can substitute for a prolonged period when it is present.

ACTIVE OR PASSIVE

A related discussion is whether Wallerian degeneration is an active process. Active could have several meanings. Unlike apoptosis, Wallerian degeneration does not require de novo protein synthesis (Gilley & Coleman 2010). Proteases are required to execute it because calpains are involved in the later stages (Schlaepfer 1974), but more exciting would be whether the molecular trigger involves a cascade of regulatory proteases. The involvement of caspase 6 in axonal pruning shows that such cascades can regulate axon degeneration (Nikolaev et al. 2009), although no evidence links this to Wallerian degeneration yet. Kinase cascades may also be involved. Deleting dual leucine kinase or applying a partially specific JNK inhibitor modestly delays Wallerian degeneration (Miller et al. 2009), and proteasome inhibition fails to delay Wallerian degeneration if MEK activity is blocked (Macinnis & Campenot 2005). In both cases, it will be interesting to know whether and how this relates to the axon protective effect of Wld[S]. Finally, screens for loss-of-function mutations (below) could be informative. Even a single loss-of-function mutation in a neuronal gene that phenocopied Wld[S] would demonstrate that Wallerian degeneration is indeed an active process, driven by an underlying and definable genetic program.

THE WLD[S] ZOO: FROM MICE TO FLIES AND BEYOND

For 16 years, Wld[S] could only be studied in mice. In the past five years, this has changed dramatically with the generation of Wld[S] rats and flies (Adalbert et al. 2005, Hoopfer et al. 2006, Macdonald et al. 2006) and recently also zebra fish (S. Martin & A. Sagasti, personal comunication). This replication in diverse species using mouse cDNA raises intriguing evolutionary questions and provides new tools for mechanism and disease studies.

Wld[S] rats were generated for those surgical procedures and disease models in which rats have advantages over mice (Adalbert et al. 2005). For example, ventral root avulsion was used to show that Wld[S] neuroprotection is compartment-specific in vivo (Adalbert et al. 2006), and laser photocoagulation of the trabecular meshwork was used to show axon protection in an induced glaucoma model (Beirowski et al. 2008). Wld[S] rats have also made an unexpected contribution to mechanism studies. Changes that mediate Wld[S] action in mice must also be present in rats, which enables us to eliminate CD200 as a candidate despite its elevation in Wld[S] mice (Chitnis et al. 2007). Wld[S] rats also provide an abundant source of tissue for biochemical and proteomic studies and establish a precedent for a replicating Wld[S] phenotype in other mammals.

The serendipitous identification of *Wld[S]* mice by Perry and colleagues was a fortunate event that revolutionized how we think about Wallerian degeneration, but the probability of further spontaneous mutants is low. Moreover, beyond the fact that Wld[S] can block it, we know almost nothing about the molecular regulation of Wallerian degeneration. Is there a single or many genetic switches that must be thrown to activate axon autodestruction? What initiates and executes these events? And importantly, what is the mechanism by which Wld[S] impinges upon these pathways?

One powerful approach to these and other outstanding questions that has been lacking in the field is unbiased forward genetic screening for mutants that modify Wallerian degeneration or axon protection by Wld[S]. A recent screen for dominant ENU-induced mutants in mice has produced a new strain that strengthens protection of neuromuscular synapses (Wong et al. 2009), but screening for loss-of-function mutants in mice is labor intensive, expensive, and slow.

Two essential features to define an invertebrate system as tractable (and relevant) for the study of Wallerian degeneration are activation of an axon autodestruction program after axotomy that is morphologically similar to mammalian Wallerian degeneration and genetic regulation of this degenerative event by Wld^S. The recent development of a simple and reproducible approach for assaying axon degeneration in the adult *Drosophila* olfactory system allowed for the first detailed in vivo analysis of Wallerian degeneration and Wld^S function in invertebrate models (Macdonald et al. 2006) (**Figure 1**). Interestingly, this work revealed that the events that lead to axon destruction indeed appear morphologically similar to Wallerian degeneration in mammals: Severed axons remain intact for a defined latent phase of 6–8 h, subsequently show beading and cytoskeletal breakdown, and finally undergo wholesale fragmentation (Macdonald et al. 2006). Similarly, when PDF$^+$ CNS axons were severed in primary cultures of the adult *Drosophila* brain, these axons also underwent degeneration within a day (Ayaz et al. 2008). Thus, when *Drosophila* axons are severed, the distal fragments degenerate after a latent phase and ultimately disappear from the CNS.

Can these injury-induced degenerative events in invertebrate axons be regulated by Wld^S? Impressively, in the case of *Drosophila* adult ORNs, mouse Wld^S suppresses Wallerian degeneration for >3 weeks after injury (Avery et al. 2009, Macdonald et al. 2006). Moreover, a recent structure-function study of Wld^S, Nmnat1, Nmnat2, and Nmnat3 in *Drosophila* (Avery et al. 2009) led to results strikingly similar to those found in mammalian systems with similar molecules (Conforti et al. 2009, Yahata et al. 2009) (**Figure 2**). Together, these observations argue strongly that the cellular and molecular mechanisms that mediate axon autodestruction are ancient features of neuronal cell types and are well conserved in mice and flies.

Laser axotomy, and the use of mutants with fragile axons that spontaneously break in response to worm movement, are emerging as useful techniques to sever axons in vivo in *C. elegans*. Although the focus of this work is exploring mechanisms of regeneration of the proximal stump (Guo et al. 2008; Hammarlund et al. 2007, 2009; Wu et al. 2007; Yanik et al. 2004), we can glean some information regarding degeneration of the distal fragment, which beads, degenerates, and disappears in a manner similar to Wallerian degeneration in mice and flies (Hammarlund et al. 2007, Wu et al. 2007). Whether Wld^S can suppress this degeneration remains to be determined.

Drosophila and *C. elegans* appear poised to contribute in major ways to understanding the basic cellular and molecular mechanisms that drive Wallerian degeneration and to further our understanding of how Wld^S protects severed axons. Foremost, these organisms are highly amenable to rapid forward genetic analysis, which allows for straightforward genetic screens for Wallerian degeneration mutants or modifiers of Wld^S-function. Exploiting this opportunity, along with other tools available in these systems such as genetically-encoded whole-genome RNAi collections, and genetic mosaic approaches to study cell autonomous roles of essential genes should help tremendously in rapidly defining the pathways that promote Wallerian degeneration or Wld^S function.

Finally, although Wld^S has been an extremely useful tool to explore the molecular relationship between Wallerian degeneration and mouse models of neurodegenerative disease, forward genetic screens in invertebrates are expected to lead to the identification of new Wld^S-independent tools that modulate other steps in the Wallerian degeneration pathway. These, in turn, will represent a new battery of genetic reagents with which to reassess this central question and could ultimately lead to the characterization of a core set of axon destruction genes used in diverse degenerative settings.

LESSONS FROM NATURE: DEGENERATION NOT REQUIRED

Do all severed axons have to degenerate? Some invertebrate axons exhibit very slow Wallerian degeneration, even in wild-type organisms (Benbassat & Spira 1994, Nordlander & Singer 1972). This is particularly well studied in crustaceans in which an evoked transmitter release can occur up to one year after axon transection (Parnas et al. 1991). Similarly, in two species of crickets, *T. commodus* and *G. bimaculatus*, severed PDF-positive axons survive for up to 90 days after axotomy, and behaviors subject to their control appear to remain intact (Stengl 1995). A few wild-type vertebrate axons show a similar phenotype, usually in extremely large-diameter axons (Zottoli et al. 1987). Unlike injured Wld^S axons, these axons are typically invaded by hypertrophic adaxonal glia, which are thought to transfer proteins to axons. Similar events may occur in mammalian axons, but this does not appear to contribute to the Wld^S phenotype (Court et al. 2008). Thus, the means of resisting Wallerian degeneration may be different, although the identities of any proteins supplied by glia would be very interesting for understanding axon survival mechanisms.

All this leads naturally to the question of whether a similar pathway operates in humans. A repeat of the tandem triplication that gave rise to Wld^S in mice seems unlikely. However, because minor changes to Nmnat1 or Nmnat3 can delay axon degeneration, it will be interesting to determine whether polymorphisms in these proteins, or in homologs of other proteins identified in invertebrate screens, alter susceptibility to neurodegenerative disorders.

ROLES FOR GLIA IN AXON SURVIVAL AND WALLERIAN DEGENERATION

The role of glia in Wallerian degeneration is normally thought to be limited to clearance of axonal debris and myelin ovoids (Vargas & Barres 2006), but could glia also play an instructive role in axon degeneration? To date,

no clear genetic evidence links glial phagocytic activity to the destruction of target cell. For example, when glial engulfment activity is blocked by either mutations in the *draper* gene, which encodes an engulfment receptor required for clearance of axonal debris (Macdonald et al. 2006), or by suppressing endocytosis with a glial-expressed dominant temperature-sensitive dynamin (Doherty et al. 2009; J. Zeigenfuss and M.R. Freeman, unpublished observations), severed axons fragment on schedule. Likewise, although the glial clearance of fragmented axons and myelin debris is much slower in the CNS than in the PNS, CNS axons themselves fragment over a normal time frame (Vargas & Barres 2006). However, a potential role for Schwann cells has been suggested, based on the observation that Wallerian degeneration appears to be nucleated in the middle of each internode following the widening of nodes of Ranvier (Lubinska 1977). Future studies that directly address the precise sequence of these events and the neuron-glia signaling mechanisms involved in axon/myelin clearance should shed light on these important events.

What is the role of glia in supporting severed Wld^S-expressing axons during the many weeks they linger (without cell body support) in the CNS? It would be amazing if any axons survived on their own during this time without any contribution from glia. More likely, several key cellular components, including high energy metabolites, are passed from glia to surviving axon stumps (Court et al. 2008). Such a requirement might explain why Wld^S and wild-type axons degenerate more quickly in vitro than in vivo (Buckmaster et al. 1995). A major goal for the field should be to define precisely how axons are nourished by surrounding glia, and how these mechanisms impact axon survival or degeneration in Wallerian degeneration and neurodegenerative disease.

CONCLUSIONS

The past decade has seen a revolution in how we think about axon destruction after injury. We now understand that Wallerian degeneration is a highly regulated process, in which a poorly understood latent phase precedes the rapid and catastrophic destruction of the axon. Amazingly, Wallerian degeneration can be suppressed by a single protein, Wld^S, and this effect is robust even in diverse species. Two domains of Wld^S appear critical for its neuroprotective function and likely function at a highly-localized non-nuclear site, but precisely how and where they exert their effects remains unclear. There are several extremely interesting outstanding questions in the field, including the following:

1. What is the molecular trigger that activates Wallerian degeneration?

2. What other proteins regulate axon survival/destruction? Do these include executors of an active process as well as inhibitors?

3. What are the endogenous regulators of Wallerian degeneration in vivo, and what are their roles in axon degeneration disorders?

4. Where in the cytoplasm/axoplasm does Wld^S act?

5. Is NAD$^+$ the Nmnat product responsible for axon survival, and if so, what does it do?

6. Why does Wld^S protect axons robustly in some axonopathies but not in others?

7. Does a Wld^S phenotype occur in the human population, and does this influence neurodegeneration?

It seems that Waller was right and that studies of Wallerian degeneration are informative about the molecular bases of axon degeneration in diverse injury and disease contexts. Answering the above questions is the next critical step in revealing the pathway and will advance our understanding of axon biology in fundamental ways.

DISCLOSURE STATEMENT

M.C. discloses that the patent for the Nmnat2 modulator is pending.

ACKNOWLEDGMENTS

This work was supported by funding from the Biotechnology and Biological Sciences Research Council (M.P.C.) and the U.S. National Institutes of Health (grant NS059991 to M.R.F.). M.R.F. is an Early Career Scientist with the Howard Hughes Medical Institute. We would like to thank members of the Coleman laboratory for critical reading of the text and Michelle Avery for *Drosophila* images in **Figure 1**.

LITERATURE CITED

Adalbert R, Gillingwater TH, Haley JE, Bridge K, Beirowski B, et al. 2005. A rat model of slow Wallerian degeneration (WldS) with improved preservation of neuromuscular synapses. *Eur. J. Neurosci.* 21:271–77

Adalbert R, Nogradi A, Babetto E, Janeckova L, Walker SA, et al. 2009. Severely dystrophic axons at amyloid plaques remain continuous and connected to viable cell bodies. *Brain* 132:402–16

Adalbert R, Nogradi A, Szabo A, Coleman MP. 2006. The slow Wallerian degeneration gene in vivo protects motor axons but not their cell bodies after avulsion and neonatal axotomy. *Eur. J. Neurosci.* 24:2163–68

Alvarez S, Moldovan M, Krarup C. 2008. Acute energy restriction triggers Wallerian degeneration in mouse. *Exp. Neurol.* 212:166–78

Araki T, Sasaki Y, Milbrandt J. 2004. Increased nuclear NAD biosynthesis and SIRT1 activation prevent axonal degeneration. *Science* 305:1010–13

Avery MA, Sheehan A, Kerr KS, Wang J, Freeman MR. 2009. WldS requires Nmnat1 enzymatic activity and N16-VCP interactions to suppress Wallerian degeneration. *J. Cell Biol.* 184:501–13

Ayaz D, Leyssen M, Koch M, Yan J, Srahna M, et al. 2008. Axonal injury and regeneration in the adult brain of *Drosophila*. *J. Neurosci.* 28:6010–21

Beirowski B, Adalbert R, Wagner D, Grumme D, Addicks K, et al. 2005. The progressive nature of Wallerian degeneration in wild-type and slow Wallerian degeneration (WldS) nerves. *BMC Neurosci.* 6:6

Beirowski B, Babetto E, Coleman MP, Martin KR. 2008. The WldS gene delays axonal but not somatic degeneration in a rat glaucoma model. *Eur. J. Neurosci.* 28:1166–79

Beirowski B, Babetto E, Gilley J, Mazzola F, Conforti L, et al. 2009. Non-nuclear Wld(S) determines its neuroprotective efficacy for axons and synapses in vivo. *J. Neurosci.* 29:653–68

Beirowski B, Nógrádi A, Babetto E, Garcia-Alias G, Coleman MP. 2010. Mechanisms of axonal spheroid formation in central nervous system Wallerian degeneration. *J. Neuropath. Exp. Neurol.* In press

Benbassat D, Spira ME. 1994. The survival of transected axonal segments of cultured aplysia neurons is prolonged by contact with intact nerve cells. *Eur. J. Neurosci.* 6:1605–14

Berger F, Lau C, Dahlmann M, Ziegler M. 2005. Subcellular compartmentation and differential catalytic properties of the three human nicotinamide mononucleotide adenylyltransferase isoforms. *J. Biol. Chem.* 280:36334–41

Bishop DL, Misgeld T, Walsh MK, Gan WB, Lichtman JW. 2004. Axon branch removal at developing synapses by axosome shedding. *Neuron* 44:651–61

Braun RJ, Zischka H, Madeo F, Eisenberg T, Wissing S, et al. 2006. Crucial mitochondrial impairment upon CDC48 mutation in apoptotic yeast. *J. Biol. Chem.* 281:25757–67

Brown MC, Perry VH, Hunt SP, Lapper SR. 1994. Further studies on motor and sensory nerve regeneration in mice with delayed Wallerian degeneration. *Eur. J. Neurosci.* 6:420–28

Buckmaster EA, Perry VH, Brown MC. 1995. The rate of Wallerian degeneration in cultured neurons from wild type and C57BL/WldS mice depends on time in culture and may be extended in the presence of elevated K+ levels. *Eur. J. Neurosci.* 7:1596–602

Burne JF, Staple JK, Raff MC. 1996. Glial cells are increased proportionally in transgenic optic nerves with increased numbers of axons. *J. Neurosci.* 16:2064–73

Cajal SRy. 1928. *Degeneration and Regeneration of the Nervous System.* London: Oxford Univ. Press

Cavanagh JB. 1979. The 'dying back' process. A common denominator in many naturally occurring and toxic neuropathies. *Arch. Pathol. Lab Med.* 103:659–64

Chitnis T, Imitola J, Wang Y, Elyaman W, Chawla P, et al. 2007. Elevated neuronal expression of CD200 protects Wlds mice from inflammation-mediated neurodegeneration. *Am. J. Pathol.* 170:1695–712

Coleman M. 2005. Axon degeneration mechanisms: commonality amid diversity. *Nat. Rev. Neurosci.* 6:889–98

Coleman MP, Conforti L, Buckmaster EA, Tarlton A, Ewing RM, et al. 1998. An 85-kb tandem triplication in the slow Wallerian degeneration (Wlds) mouse. *Proc. Natl. Acad. Sci. USA* 95:9985–90

Coleman MP, Ribchester RR. 2004. Programmed axon death, synaptic dysfunction and the ubiquitin proteasome system. *Curr. Drug Targets CNS Neurol. Disord.* 3:227–38

Conforti L, A. W, Morreale G, Janeckova L, Beirowski B, et al. 2009. WldS protein requires Nmnat activity and a short N-terminal sequence to protect axons in mice. *J. Cell Biol.* 184:491–500

Conforti L, Adalbert R, Coleman MP. 2007a. Neuronal death: Where does the end begin? *Trends Neurosci.* 30:159–66

Conforti L, Fang G, Beirowski B, Wang MS, Sorci L, et al. 2007b. NAD(+) and axon degeneration revisited: Nmnat1 cannot substitute for Wld(S) to delay Wallerian degeneration. *Cell Death Differ.* 14:116–27

Conforti L, Tarlton A, Mack TG, Mi W, Buckmaster EA, et al. 2000. A Ufd2/D4Cole1e chimeric protein and overexpression of rbp7 in the slow Wallerian degeneration (WldS) mouse. *Proc. Natl. Acad. Sci. USA* 97:11377–82

Court FA, Hendriks WT, Macgillavry HD, Alvarez J, van Minnen J. 2008. Schwann cell to axon transfer of ribosomes: toward a novel understanding of the role of glia in the nervous system. *J. Neurosci.* 28:11024–29

De Vos KJ, Grierson AJ, Ackerley S, Miller CC. 2008. Role of axonal transport in neurodegenerative diseases. *Annu. Rev. Neurosci.* 31:151–73

Deckwerth TL, Johnson EM, Jr. 1994. Neurites can remain viable after destruction of the neuronal soma by programmed cell death (apoptosis). *Dev. Biol.* 165:63–72

Doherty J, Logan MA, Tasdemir OE, Freeman MR. 2009. Ensheathing glia function as phagocytes in the adult *Drosophila* brain. *J. Neurosci.* 29:4768–81

Ehlers MD. 2003. Eppendorf 2003 prize-winning essay. Ubiquitin and the deconstruction of synapses. *Science* 302:800–1

Fainzilber M, Twiss JL. 2006. Tracking in the Wld(s)—the hunting of the SIRT and the luring of the draper. *Neuron* 50:819–21

Ferguson B, Matyszak MK, Esiri MM, Perry VH. 1997. Axonal damage in acute multiple sclerosis lesions. *Brain* 120:393–99

Ferri A, Sanes JR, Coleman MP, Cunningham JM, Kato AC. 2003. Inhibiting axon degeneration and synapse loss attenuates apoptosis and disease progression in a mouse model of motoneuron disease. *Curr. Biol.* 13:669–73

Finn JT, Weil M, Archer F, Siman R, Srinivasan A, Raff MC. 2000. Evidence that wallerian degeneration and localized axon degeneration induced by local neurotrophin deprivation do not involve caspases. *J. Neurosci.* 20:1333–41

Fischer LR, Culver DG, Davis AA, Tennant P, Wang M, et al. 2005. The Wld(S) gene modestly prolongs survival in the SOD1(G93A) fALS mouse. *Neurobiol. Dis.* 19:293–300

Gamble HJ, Jha BD. 1958. Some effects of temperature upon the rate and progress of Wallerian degeneration in mammalian nerve fibers. *J. Anat.* 92:171–77

Gilley J, Coleman MP. 2010. Endogenous Nmnat2 is an essential survival factor for maintenance of health axons. *PLoS Biol.* 8:e1000200

Gillingwater TH, Ingham CA, Parry KE, Wright AK, Haley JE, et al. 2006a. Delayed synaptic degeneration in the CNS of Wlds mice after cortical lesion. *Brain* 129:1546–56

Gillingwater TH, Thomson D, Mack TG, Soffin EM, Mattison RJ, et al. 2002. Age-dependent synapse withdrawal at axotomised neuromuscular junctions in Wld(s) mutant and Ube4b/Nmnat transgenic mice. *J. Physiol.* 543:739–55

Gillingwater TH, Wishart TM, Chen PE, Haley JE, Robertson K, et al. 2006b. The neuroprotective WldS gene regulates expression of PTTG1 and erythroid differentiation regulator 1-like Gene in Mice and human cells. *Hum. Mol. Genet.* 15:625–35

Glass JD, Brushart TM, George EB, Griffin JW. 1993. Prolonged survival of transected nerve fibres in C57BL/Ola mice is an intrinsic characteristic of the axon. *J. Neurocytol.* 22:311–21

Griffin JW, George EB, Chaudhry V. 1996. Wallerian degeneration in peripheral nerve disease. *Baillieres Clin. Neurol.* 5:65–75

Guo SX, Bourgeois F, Chokshi T, Durr NJ, Hilliard MA, et al. 2008. Femtosecond laser nanoaxotomy lab-on-a-chip for in vivo nerve regeneration studies. *Nat. Methods* 5:531–33

Hammarlund M, Jorgensen EM, Bastiani MJ. 2007. Axons break in animals lacking beta-spectrin. *J. Cell Biol.* 176:269–75

Hammarlund M, Nix P, Hauth L, Jorgensen EM, Bastiani M. 2009. Axon regeneration requires a conserved MAP kinase pathway. *Science* 323:802–6

Hasbani DM, O'Malley KL. 2006. Wld(S) mice are protected against the Parkinsonian mimetic MPTP. *Exp. Neurol.* 202:93–99

Hassa PO, Haenni SS, Elser M, Hottiger MO. 2006. Nuclear ADP-ribosylation reactions in mammalian cells: Where are we today and where are we going? *Microbiol. Mol. Biol. Rev.* 70:789–829

Hatakeyama S, Yada M, Matsumoto M, Ishida N, Nakayama KI. 2001. U-Box proteins as a new family of ubiquitin-protein ligases. *J. Biol. Chem.* 276:33111–20

Hoopfer ED, McLaughlin T, Watts RJ, Schuldiner O, O'Leary DD, Luo L. 2006. Wld(s) protection distinguishes axon degeneration following injury from naturally occurring developmental pruning. *Neuron* 50:883–95

Howell GR, Libby RT, Jakobs TC, Smith RS, Phalan FC, et al. 2007. Axons of retinal ganglion cells are insulted in the optic nerve early in DBA/2J glaucoma. *J. Cell Biol.* 179:1523–37

Jia H, Yan T, Feng Y, Zeng C, Shi X, Zhai Q. 2007. Identification of a critical site in Wld(s): essential for Nmnat enzyme activity and axon-protective function. *Neurosci. Lett.* 413:46–51

Kane RC, Bross PF, Farrell AT, Pazdur R. 2003. Velcade: U.S. FDA approval for the treatment of multiple myeloma progressing on prior therapy. *Oncologist* 8:508–13

Kaneko S, Wang J, Kaneko M, Yiu G, Hurrell JM, et al. 2006. Protecting axonal degeneration by increasing nicotinamide adenine dinucleotide levels in experimental autoimmune encephalomyelitis models. *J. Neurosci.* 26:9794–804

Kaneko-Oshikawa C, Nakagawa T, Yamada M, Yoshikawa H, Matsumoto M, et al. 2005. Mammalian E4 is required for cardiac development and maintenance of the nervous system. *Mol. Cell Biol.* 25:10953–64

Kariya S, Mauricio R, Dai Y, Monani UR. 2009. The neuroprotective factor Wld(s) fails to mitigate distal axonal and neuromuscular junction (NMJ) defects in mouse models of spinal muscular atrophy. *Neurosci. Lett.* 449:246–51

Keilhoff G, Fansa H, Wolf G. 2002. Differences in peripheral nerve degeneration/regeneration between wild-type and neuronal nitric oxide synthase knockout mice. *J. Neurosci. Res.* 68:432–41

Kerschensteiner M, Schwab ME, Lichtman JW, Misgeld T. 2005. In vivo imaging of axonal degeneration and regeneration in the injured spinal cord. *Nat. Med.* 11:572–77

Koegl M, Hoppe T, Schlenker S, Ulrich HD, Mayer TU, Jentsch S. 1999. A novel ubiquitination factor, E4, is involved in multiubiquitin chain assembly. *Cell* 96:635–44

Laser H, Conforti L, Morreale G, Mack TG, Heyer M, et al. 2006. The slow Wallerian degeneration protein, WldS, binds directly to VCP/p97 and partially redistributes it within the nucleus. *Mol. Biol. Cell* 17:1075–84

Laser H, Mack TG, Wagner D, Coleman MP. 2003. Proteasome inhibition arrests neurite outgrowth and causes "dying-back" degeneration in primary culture. *J. Neurosci. Res.* 74:906–16

Levy D, Kubes P, Zochodne DW. 2001. Delayed peripheral nerve degeneration, regeneration, and pain in mice lacking inducible nitric oxide synthase. *J. Neuropathol. Exp. Neurol.* 60:411–21

Libby RT, Li Y, Savinova OV, Barter J, Smith RS, et al. 2005. Susceptibility to neurodegeneration in glaucoma is modified by Bax gene dosage. *PLoS Genet.* 1:e4

Lopez-Vales R, Navarro X, Shimizu T, Baskakis C, Kokotos G, et al. 2008. Intracellular phospholipase A(2) group IVA and group VIA play important roles in Wallerian degeneration and axon regeneration after peripheral nerve injury. *Brain* 131:2620–31

Lubinska L. 1977. Early course of Wallerian degeneration in myelinated fibres of the rat phrenic nerve. *Brain Res.* 130:47–63

Lubinska L. 1982. Patterns of Wallerian degeneration of myelinated fibres in short and long peripheral stumps and in isolated segments of rat phrenic nerve. Interpretation of the role of axoplasmic flow of the trophic factor. *Brain Res.* 233:227–40

Lunn ER, Perry VH, Brown MC, Rosen H, Gordon S. 1989. Absence of Wallerian degeneration does not hinder regeneration in peripheral nerve. *Eur. J. Neurosci.* 1:27–33

Lyon MF, Ogunkolade BW, Brown MC, Atherton DJ, Perry VH. 1993. A gene affecting Wallerian nerve degeneration maps distally on mouse chromosome 4. *Proc. Natl. Acad. Sci. USA* 90:9717–20

Macdonald JM, Beach MG, Porpiglia E, Sheehan AE, Watts RJ, Freeman MR. 2006. The *Drosophila* cell corpse engulfment receptor draper mediates glial clearance of severed axons. *Neuron* 50:869–81

Macinnis BL, Campenot RB. 2005. Regulation of Wallerian degeneration and nerve growth factor withdrawal-induced pruning of axons of sympathetic neurons by the proteasome and the MEK/Erk pathway. *Mol. Cell Neurosci.* 28:430–39

Mack TG, Reiner M, Beirowski B, Mi W, Emanuelli M, et al. 2001. Wallerian degeneration of injured axons and synapses is delayed by a Ube4b/Nmnat chimeric gene. *Nat. Neurosci.* 4:1199–206

Magni G, Amici A, Emanuelli M, Orsomando G, Raffaelli N, Ruggieri S. 2004. Structure and function of nicotinamide mononucleotide adenylyltransferase. *Curr. Med. Chem.* 11:873–85

Merianda TT, Lin AC, Lam JS, Vuppalanchi D, Willis DE, et al. 2009. A functional equivalent of endoplasmic reticulum and Golgi in axons for secretion of locally synthesized proteins. *Mol. Cell Neurosci.* 40:128–42

Merkwirth C, Langer T. 2008. Mitofusin 2 builds a bridge between ER and mitochondria. *Cell* 135:1165–67

Mi W, Beirowski B, Gillingwater TH, Adalbert R, Wagner D, et al. 2005. The slow Wallerian degeneration gene, WldS, inhibits axonal spheroid pathology in gracile axonal dystrophy mice. *Brain* 128:405–16

Miledi R, Slater CR. 1970. On the degeneration of rat neuromuscular junctions after nerve section. *J. Physiol.* 207:507–28

Miller BR, Press C, Daniels RW, Sasaki Y, Milbrandt J, DiAntonio A. 2009. A dual leucine kinase-dependent axon self-destruction program promotes Wallerian degeneration. *Nat. Neurosci.* 12:387–89

Misgeld T, Kerschensteiner M, Bareyre FM, Burgess RW, Lichtman JW. 2007. Imaging axonal transport of mitochondria in vivo. *Nat. Methods* 4:559–61

Moldovan M, Alvarez S, Krarup C. 2009. Motor axon excitability during Wallerian degeneration. *Brain* 132:511–23

Morreale G, Conforti L, Coadwell J, Wilbrey AL, Coleman MP. 2009. Evolutionary divergence of valosin-containing protein/cell division cycle protein 48 binding interactions among endoplasmic reticulum-associated degradation proteins. *FEBS J.* 276:1208–20

Narciso MS, Mietto Bde S, Marques SA, Soares CP, Mermelstein Cdos S, et al. 2009. Sciatic nerve regeneration is accelerated in galectin-3 knockout mice. *Exp. Neurol.* 217:7–15

Nave KA, Trapp BD. 2008. Axon-glial signaling and the glial support of axon function. *Annu. Rev. Neurosci.* 31:535–61

Neumann H, Medana IM, Bauer J, Lassmann H. 2002. Cytotoxic T lymphocytes in autoimmune and degenerative CNS diseases. *Trends Neurosci.* 25:313–19

Nikolaev A, McLaughlin T, O'Leary DD, Tessier-Lavigne M. 2009. APP binds DR6 to trigger axon pruning and neuron death via distinct caspases. *Nature* 457:981–89

Nordlander RH, Singer M. 1972. Electron microscopy of severed motor fibers in the crayfish. *Z. Zellforsch. Mikrosk. Anat.* 126:157–81

Parnas I, Dudel J, Atwood HL. 1991. Synaptic transmission in decentralized axons of rock lobster. *J. Neurosci.* 11:1309–15

Parson SH, Mackintosh CL, Ribchester RR. 1997. Elimination of motor nerve terminals in neonatal mice expressing a gene for slow Wallerian degeneration (C57Bl/Wlds). *Eur. J. Neurosci.* 9:1586–92

Perlson E, Hanz S, Ben-Yaakov K, Segal-Ruder Y, Seger R, Fainzilber M. 2005. Vimentin-dependent spatial translocation of an activated MAP kinase in injured nerve. *Neuron* 45:715–26

Perry VH, Brown MC, Lunn ER. 1991. Very slow retrograde and Wallerian degeneration in the CNS of C57BL/Ola mice. *Eur. J. Neurosci.* 3:102–5

Perry VH, Brown MC, Lunn ER, Tree P, Gordon S. 1990a. Evidence that very slow Wallerian degeneration in C57BL/Ola mice is an intrinsic property of the peripheral nerve. *Eur. J. Neurosci.* 2:802–8

Perry VH, Lunn ER, Brown MC, Cahusac S, Gordon S. 1990b. Evidence that the rate of Wallerian degeneration is controlled by a single autosomal dominant gene. *Eur. J. Neurosci.* 2:408–13

Pollak N, Dolle C, Ziegler M. 2007. The power to reduce: pyridine nucleotides–small molecules with a multitude of functions. *Biochem. J.* 402:205–18

Press C, Milbrandt J. 2008. Nmnat delays axonal degeneration caused by mitochondrial and oxidative stress. *J. Neurosci.* 28:4861–71

Raff MC, Whitmore AV, Finn JT. 2002. Axonal self-destruction and neurodegeneration. *Science* 296:868–71

Raffaelli N, Sorci L, Amici A, Emanuelli M, Mazzola F, Magni G. 2002. Identification of a novel human nicotinamide mononucleotide adenylyltransferase. *Biochem. Biophys. Res. Commun.* 297:835–40

Ramaglia V, King RH, Nourallah M, Wolterman R, de Jonge R, et al. 2007. The membrane attack complex of the complement system is essential for rapid Wallerian degeneration. *J. Neurosci.* 27:7663–72

Ribchester RR, Tsao JW, Barry JA, Asgari-Jirhandeh N, Perry VH, Brown MC. 1995. Persistence of neuromuscular junctions after axotomy in mice with slow Wallerian degeneration (C57BL/WldS). *Eur. J. Neurosci.* 7:1641–50

Rose FF, Jr, Meehan PW, Coady TH, Garcia VB, Garcia ML, Lorson CL. 2008. The Wallerian degeneration slow (Wld(s)) gene does not attenuate disease in a mouse model of spinal muscular atrophy. *Biochem. Biophys. Res. Commun.* 375:119–23

Sagot Y, Dubois-Dauphin M, Tan SA, de Bilbao F, Aebischer P, et al. 1995. Bcl-2 overexpression prevents motoneuron cell body loss but not axonal degeneration in a mouse model of a neurodegenerative disease. *J. Neurosci.* 15:7727–33

Saigoh K, Wang YL, Suh JG, Yamanishi T, Sakai Y, et al. 1999. Intragenic deletion in the gene encoding ubiquitin carboxy-terminal hydrolase in gad mice. *Nat. Genet.* 23:47–51

Sajadi A, Schneider BL, Aebischer P. 2004. Wld(s)-mediated protection of dopaminergic fibers in an animal model of parkinson disease. *Curr. Biol.* 14:326–30

Samsam M, Mi W, Wessig C, Zielasek J, Toyka KV, et al. 2003. The Wlds mutation delays robust loss of motor and sensory axons in a genetic model for myelin-related axonopathy. *J. Neurosci.* 23:2833–39

Sasaki Y, Vohra BP, Baloh RH, Milbrandt J. 2009a. Transgenic mice expressing the Nmnat1 protein manifest robust delay in axonal degeneration in vivo. *J. Neurosci.* 29:6526–34

Sasaki Y, Vohra BP, Lund FE, Milbrandt J. 2009b. Nicotinamide mononucleotide adenylyl transferase-mediated axonal protection requires enzymatic activity but not increased levels of neuronal nicotinamide adenine dinucleotide. *J. Neurosci.* 29:5525–35

Schlaepfer WW. 1974. Effects of energy deprivation on Wallerian degeneration in isolated segments of rat peripheral nerve. *Brain Res.* 78:71–81

Simonin Y, Perrin FE, Kato AC. 2007. Axonal involvement in the Wlds neuroprotective effect: analysis of pure motoneurons in a mouse model protected from motor neuron disease at a presymptomatic age. *J. Neurochem.* 101:530–42

Sorci L, Cimadamore F, Scotti S, Petrelli R, Cappellacci L, et al. 2007. Initial-rate kinetics of human NMN-adenylyltransferases: substrate and metal ion specificity, inhibition by products and multisubstrate analogues, and isozyme contributions to NAD$^+$ biosynthesis. *Biochemistry* 46:4912–22

Spires TL, Meyer-Luehmann M, Stern EA, McLean PJ, Skoch J, et al. 2005. Dendritic spine abnormalities in amyloid precursor protein transgenic mice demonstrated by gene transfer and intravital multiphoton microscopy. *J. Neurosci.* 25:7278–87

Stengl M. 1995. Pigment-dispersing hormone-immunoreactive fibers persist in crickets which remain rhythmic after bilateral transection of the optic stalks. *J. Comp. Physiol. A* 176:217–28

Tamsett TJ, Picchione KE, Bhattacharjee A. 2009. NAD$^+$ activates KNa channels in dorsal root ganglion neurons. *J. Neurosci.* 29:5127–34

Todisco S, Agrimi G, Castegna A, Palmieri F. 2006. Identification of the mitochondrial NAD$^+$ transporter in Saccharomyces cerevisiae. *J. Biol. Chem.* 281:1524–31

Trapp BD, Peterson J, Ransohoff RM, Rudick R, Mork S, Bo L. 1998. Axonal transection in the lesions of multiple sclerosis. *N. Engl. J. Med.* 338:278–85

Vargas ME, Barres BA. 2007. Why is Wallerian degeneration in the CNS so slow? *Annu. Rev. Neurosci.* 30:153–79

Velde CV, Garcia ML, Yin X, Trapp BD, Cleveland DW. 2004. The neuroprotective factor Wlds does not attenuate mutant SOD1-mediated motor neuron disease. *Neuromolecular Med.* 5:193–204

Waller A. 1850. Experiments on the section of glossopharyngeal and hypoglossal nerves of the frog and observations of the alternatives produced thereby in the structure of their primitive fibres. *Philos. Trans. R. Soc. Lond. B Biol. Sci.* 140:423–29

Wang J, Zhai Q, Chen Y, Lin E, Gu W, et al. 2005. A local mechanism mediates NAD-dependent protection of axon degeneration. *J. Cell Biol.* 170:349–55

Wang MS, Davis AA, Culver DG, Glass JD. 2002. WldS mice are resistant to paclitaxel (taxol) neuropathy. *Ann. Neurol.* 52:442–47

Wang MS, Fang G, Culver DG, Davis AA, Rich MM, Glass JD. 2001. The WldS protein protects against axonal degeneration: a model of gene therapy for peripheral neuropathy. *Ann. Neurol.* 50:773–79

Wang MS, Wu Y, Culver DG, Glass JD. 2000. Pathogenesis of axonal degeneration: parallels between Wallerian degeneration and vincristine neuropathy. *J. Neuropathol. Exp. Neurol.* 59:599–606

Wang Q, Song C, Li CC. 2004. Molecular perspectives on p97-VCP: progress in understanding its structure and diverse biological functions. *J. Struct. Biol.* 146:44–57

Whitmore AV, Lindsten T, Raff MC, Thompson CB. 2003. The proapoptotic proteins Bax and Bak are not involved in Wallerian degeneration. *Cell Death Differ.* 10:260–61

Wilbrey AL, Haley JE, Wishart TM, Conforti L, Morreale G, et al. 2008. VCP binding influences intracellular distribution of the slow Wallerian degeneration protein, Wld(S). *Mol. Cell Neurosci.* 38:325–40

Wishart TM, Paterson JM, Short DM, Meredith S, Robertson KA, et al. 2007. Differential proteomic analysis of synaptic proteins identifies potential cellular targets and protein mediators of synaptic neuroprotection conferred by the slow Wallerian degeneration (Wlds) gene. *Mol. Cell Proteomics* 6(8):1318–30

Wishart TM, Pemberton HN, James SR, McCabe CJ, Gillingwater TH. 2008. Modified cell cycle status in a mouse model of altered neuronal vulnerability (slow Wallerian degeneration; Wlds). *Genome Biol.* 9:R101

Wong F, Fan L, Wells S, Hartley R, Mackenzie FE, et al. 2009. Axonal and neuromuscular synaptic phenotypes in WldS, SOD1(G93A) and ostes mutant mice identified by fiber-optic confocal microendoscopy. *Mol. Cell Neurosci.* 42:296–307

Wu Z, Ghosh-Roy A, Yanik MF, Zhang JZ, Jin Y, Chisholm AD. 2007. *Caenorhabditis elegans* neuronal regeneration is influenced by life stage, ephrin signaling, and synaptic branching. *Proc. Natl. Acad. Sci. USA* 104:15132–37

Yahata N, Yuasa S, Araki T. 2009. Nicotinamide mononucleotide adenylyltransferase expression in mitochondrial matrix delays Wallerian degeneration. *J. Neurosci.* 29:6276–84

Yanik MF, Cinar H, Cinar HN, Chisholm AD, Jin Y, Ben-Yakar A. 2004. Neurosurgery: functional regeneration after laser axotomy. *Nature* 432:822

Zhai Q, Wang J, Kim A, Liu Q, Watts R, et al. 2003. Involvement of the ubiquitin-proteasome system in the early stages of wallerian degeneration. *Neuron* 39:217–25

Zhai RG, Cao Y, Hiesinger PR, Zhou Y, Mehta SQ, et al. 2006. *Drosophila* NMNAT maintains neural integrity independent of its NAD synthesis activity. *PLoS Biol.* 4:e416

Zhai RG, Zhang F, Hiesinger PR, Cao Y, Haueter CM, Bellen HJ. 2008. NAD synthase NMNAT acts as a chaperone to protect against neurodegeneration. *Nature* 452:887–91

Zottoli SJ, Marek LE, Agostini MA, Strittmatter SL. 1987. Morphological and physiological survival of goldfish Mauthner axons isolated from their somata by spinal cord crush. *J. Comp. Neurol.* 255:272–82

Neural Mechanisms for Interacting with a World Full of Action Choices

Paul Cisek and John F. Kalaska

Groupe de Recherche sur le Système Nerveux Central (FRSQ), Département de Physiologie, Université de Montréal, Montréal, Québec H3C 3J7 Canada; email: paul.cisek@umontreal.ca

Annu. Rev. Neurosci. 2010. 33:269–98

First published online as a Review in Advance on March 26, 2010

The *Annual Review of Neuroscience* is online at neuro.annualreviews.org

This article's doi: 10.1146/annurev.neuro.051508.135409

0147-006X/10/0721-0269$20.00

Key Words

neurophysiology, sensorimotor control, decision making, embodied cognition

Abstract

The neural bases of behavior are often discussed in terms of perceptual, cognitive, and motor stages, defined within an information processing framework that was originally inspired by models of human abstract problem solving. Here, we review a growing body of neurophysiological data that is difficult to reconcile with this influential theoretical perspective. As an alternative foundation for interpreting neural data, we consider frameworks borrowed from ethology, which emphasize the kinds of real-time interactive behaviors that animals have engaged in for millions of years. In particular, we discuss an ethologically-inspired view of interactive behavior as simultaneous processes that specify potential motor actions and select between them. We review how recent neurophysiological data from diverse cortical and subcortical regions appear more compatible with this parallel view than with the classical view of serial information processing stages.

Contents

INTRODUCTION

In this review, we discuss some potential implications of recent neurophysiological results to large-scale theories of behavior. We focus primarily on data from the cerebral cortex of nonhuman primates, data that are often interpreted in terms of a theoretical framework influenced by studies of human cognition. In that framework, the brain is seen as an information processing system that first transforms sensory information into perceptual representations, then uses these to construct knowledge about the world and make decisions, and finally implements decisions through action.

However, neurophysiological data we review below appear to be at odds with many of the assumptions of this influential view. For example, studies on the neural mechanisms of decision making have repeatedly shown that correlates of decision processes are distributed throughout the brain, notably including cortical and subcortical regions that are strongly implicated in the sensorimotor control of movement. Neural correlates of putative decision variables (such as payoff) appear to be expressed by the same neurons that encode the attributes (such as direction) of the potential motor responses used to report the decision, which reside within sensorimotor circuits that guide the on-line execution of movements. These data and their implications for the computational mechanisms of decision making have been the subject of several recent reviews (Glimcher 2003, Gold & Shadlen 2007, Schall 2004). Here, we consider more general questions of what these and other recent data imply for large-scale theories of the neural organization of behavior. Why should supposedly cognitive processes take place within sensorimotor circuits? Why should individual neurons appear to change in time from encoding sensory qualities to encoding motor parameters? What should be the time course of neural processing? Many recent results do not appear to be compatible with the classical distinctions between perceptual, cognitive, and motor systems (Lebedev & Wise 2002); and we consider whether an alternative framework can more readily account for these observations.

As an alternative perspective, we consider frameworks inspired by many decades of ethological research focused on natural animal behavior in the wild. Such behavior involves continuous sensorimotor interaction between an organism and its environment, in contrast with conditions often used in the laboratory in which time is divided into a series of individual trials. Because the brain's functional architecture originally evolved to serve the needs of interactive behavior, and was strongly conserved during phylogeny, we believe an ethological foundation may be more appropriate for understanding neurophysiological data about voluntary sensorimotor behavior compared to frameworks inspired by studies of advanced human abilities. Indeed, we believe that a wide variety of neurophysiological results are more readily interpreted within the perspective of ethologically-based theories.

COGNITIVE NEUROSCIENCE AND THE INFORMATION PROCESSING FRAMEWORK

Imagine yourself sitting at a computer, replying to a friend's email message. From the outside, the act looks simple: A person is seated, hardly moving their body, looking at a computer screen. After some time, fingers begin to tap some keys on the keyboard. Of course, on the inside, this behavior involves a large set of incredibly complex processes. These include visual recognition of the letters displayed on the screen, parsing of the words and sentences, analysis of their meaning, emotional reactions to it, consideration of the many factors that influence the nature of the answer and how to phrase the reply, and finally, the production of precise finger movements to produce a new set of letters. We can classify these processes into three general categories: perception—the processes that take information from the outside world to build knowledge about it; cognition—the internal processes of knowledge manipulation, including semantic analysis, decision making, etc.; and action—the control of the muscular contractions that produce our response. In tasks such as replying to an email, these processes are likely performed in a largely serial manner. We sense the world, think about it, and then act upon it.

This informal description of behavior is often reflected in how we study it in the lab. Perceptual scientists study how the brain processes information to produce internal representations of external phenomena. Cognitive scientists study, among other things, how knowledge is acquired, how memories are stored and retrieved, and how abstract decisions are made between distinct choices such as whether to try gamble A or B. Motor scientists like us study how a voluntary plan of action is transformed into patterns of muscular contraction that move our limbs or eyes or produce utterances. This division of labor among neuroscientists reflects how we think about behavior and influences how we teach brain science. Indeed, the view of the brain as an information processing system is formalized in a theoretical framework that has dominated psychological thinking and teaching for more than 50 years.

Information processing was established as the theoretical foundation of cognitive psychology during the mid-twentieth century, when it replaced the then-dominant paradigm of behaviorism. From the information processing perspective, perception involves the construction of increasingly sophisticated and abstract internal representations of the world (Marr 1982) that are used as the input to cognitive systems. These cognitive systems bring salient context-dependent information together in a temporary working memory buffer (Miller 1956), manipulate representations to build complex knowledge (Johnson-Laird 1988, Pylyshyn 1984), store and retrieve information from long-term memory (Newell & Simon 1972), perform deductive reasoning (Smith & Osherson 1995), and make decisions (Shafir & Tversky 1995, Tversky & Kahneman 1981). Finally, the motor systems are seen simply as tools that implement action plans chosen by cognitive processes. They are often conceived according to formalisms borrowed from engineering control theory, in which a predetermined motor program or desired trajectory (Keele 1968, Miller et al. 1960) is passed to a controller that executes it via feedforward and feedback control mechanisms.

This classical framework was originally proposed as an explanation of complex human abilities of abstract problem solving, such as chess playing (Newell & Simon 1972, Pylyshyn 1984)—the kinds of problems that require the subject to obtain information about the world and perform a great deal of computation before taking any external actions. It was not originally meant to be a general theory of all behavior. Eventually, however, the architecture of an information processing system was seen as so powerful that its basic concepts have come to influence nearly every domain of brain theory. For example, the concept of the bandwidth of transmission in information processing channels has provided a foundation for theories

Motor program:
a theoretical representation of planned movements

of attention (Broadbent 1958) and working memory (Miller 1956), as well as studies of motor control (Fitts 1954). The project of explaining complex behavior in terms of neural mechanisms is often called cognitive neuroscience (Albright et al. 2000, Gazzaniga 2000). It is an approach that inherits specific concepts from cognitive psychology and maps them onto particular regions of the brain. Today, even those of us who study sensorimotor control have a tendency to phrase the problem as one of transforming input representations to output representations through a series of intermediate processing stages.

However, attempts to interpret neural data from this perspective encounter several challenges. For example, functions that should be unified appear distributed throughout the brain, whereas those that should be distinct appear to involve the same regions, or even the same cells. Below, we discuss several examples of such challenges, which lead us to question whether the basic structure of sensing, thinking, and acting is indeed the optimal blueprint for understanding how the brain implements much of the real-time interactive behavior whose demands drove neural evolution.

Perceptual Processing

Psychological and computational theories often propose that our perception of the world is the result of a reconstruction process that uses sensory information to build and update an internal representation of the external world (Marr 1982, Riesenhuber & Poggio 1999, Riesenhuber & Poggio 2002). We usually assume that this internal representation must be unified (linking diverse information into a common form available to diverse systems) and stable (reflecting the stable nature of the physical world) to be useful for building knowledge and making decisions. To date, however, neural data do not support the existence of such an internal representation. Indeed, the representation of the external world generated by the most studied sensory modality, the visual system, appears to be neither unified nor

stable at the level of single neurons or neural populations.

For example, Ungerleider & Mishkin (1982) reviewed data indicating that visual information in the cerebral cortex diverges into two partially distinct streams of processing: (*a*) an occipitotemporal ventral stream in which cells are sensitive to information that pertains to the identity of objects and (*b*) an occipitoparietal dorsal stream in which cells are sensitive to spatial information. Within each of these, information diverges further. There are separate visual streams for processing color, shape, and motion (Felleman & Van Essen 1991); and there are multiple representations of space within the posterior parietal cortex (Colby & Goldberg 1999, Stein 1992). From the traditional cognitive perspective, the ventral stream builds a representation of what is in the environment, whereas the dorsal stream builds a representation of where things are. Presumably, all of these visual substreams must be bound together to form a unified representation of the world; but whether and how this binding occurs remain unresolved, despite vibrant research efforts (Engel et al. 2001, Shadlen & Movshon 1999, Singer 2001).

Furthermore, activity in much of the visual system appears to be strongly influenced by attentional modulation (Boynton 2005, Colby & Goldberg 1999, Moran & Desimone 1985, Treue 2001), even when a quiescent monkey spontaneously scans a familiar stable environment (Bushnell et al. 1981, Gottlieb et al. 1998, Mountcastle et al. 1981). This is usually exhibited as an enhancement of neural activity from the regions of space to which attention is directed and a suppression of activity from unattended regions. Such attentional modulation is found in both the ventral and dorsal streams and increases as one ascends the visual hierarchy (Treue 2001). Consequently, the neural representation of the visual world, at least in higher visual areas, appears "dominated by the behavioral relevance of the information, rather than designed to provide an accurate and complete description of it" (Treue 2001, p. 295). Furthermore, because the direction of attention is frequently shifting from one place to another, the activity in visual regions is anything but stable. It is constantly changing, even if one is fixating a completely motionless scene.

To summarize, the classical assumption of a unified and stable internal representation (an internal replica of the external world) does not appear to be well supported by the divergence of the visual system and the widespread influence of attentional and contextual modulation. If something that resembles a perception module exists, it overlaps so strongly with cognitive processes that the distinction between them becomes blurred.

Motor Control

According to the information processing view of voluntary behavior, the role of the motor system is to implement the course of action commanded by the cognitive system. This has led to the common assumption that by the time motor processing begins, cognitive processes have decided what to do, and only a single motor program is prepared before movement initiation (Keele 1968, Miller et al. 1960). However, neural data do not appear to support this assumption. First, many of the same regions that appear to be involved in movement planning are also active during movement execution (Alexander & Crutcher 1990b, Crammond & Kalaska 2000, Hoshi & Tanji 2007, Kalaska et al. 1998, Wise et al. 1997). Neural correlates of both planning and execution processes can be found even in the activity of individual cells, whose association with motor output changes in time from abstract aspects of the task to limb movement-related parameters (Cisek et al. 2003, Crammond & Kalaska 2000, Shen & Alexander 1997). Furthermore, whenever planning activity has been studied in tasks that present animals with choices, that same activity also appears related to decision making processes that should have been completed by the cognitive system (Cisek & Kalaska 2005, Gold & Shadlen 2007, Hoshi & Tanji 2007, Platt & Glimcher 1999, Romo et al. 2004, Wallis & Miller 2003). Such functional

heterogeneity at the level of single neurons is difficult to reconcile with the breakdown of behavior into perception, cognition, and action.

Instead of encoding the unique and detailed motor program predicted by classical models (Keele 1968, Miller et al. 1960), neural activity in motor regions appears to initially encode information about relevant stimuli and then changes to represent motor variables, such as the direction of movement. For example, during visual search tasks, cells in frontal eye fields (FEF) initially respond to all salient stimuli, including multiple distracters, but later reflect only the final selected target (Schall & Bichot 1998). During reach/antireach tasks, neural activities in the dorsal premotor cortex (PMd) first appear to encode the location of a stimulus and later reflect the movement direction instructed by that stimulus (Crammond & Kalaska 1994, Gail et al. 2009). Such findings have often been interpreted as early visual responses, which are followed by motor activity, but it is unclear how a traditional model could account for both of these being encoded in the same region, sometimes by the same neurons. Additionally, neural activity in motor regions appears to be modulated by a variety of putatively cognitive variables, as described below. In summary, if an action module exists, then it appears to be closely entwined with both perceptual and cognitive processes (Lebedev & Wise 2002).

Cognitive Functions

The search for the modules that lie between perception and action has been even more problematic. According to classical views (Fodor 1983, Pylyshyn 1984), cognition is separate from sensorimotor control. However, a hallmark executive function, decision making (Tversky & Kahneman 1981), does not appear to be localized within particular higher cognitive centers such as the primate prefrontal cortex. Instead, there is growing evidence that decisions, at least those reported through action, are found within the same sensorimotor circuits that are responsible for planning and executing the associated actions (Cisek &

FEF: frontal eye fields

PMd: dorsal premotor cortex

LIP: lateral intraparietal area

PPC: posterior parietal cortex

Salience map: a spatial representation of the most salient features of the environment

Kalaska 2005, Gold & Shadlen 2007, Pesaran et al. 2008, Romo et al. 2002, Romo et al. 2004, Scherberger & Andersen 2007). For example, Romo and colleagues (Hernandez et al. 2002, Romo et al. 2002, Romo et al. 2004) found that during tasks in which a nominally tactile perceptual decision is reported by an arm movement, correlates of all of the putative sensory encoding, memory, discrimination, and decision-making processes were much stronger within premotor regions than in classical somatic sensory areas. Similarly, when monkeys were required to decide whether to hold or release a lever in response to a sequence of visual stimuli, neural correlates of the behavioral rule (match/nonmatch) and the action decision (release/hold) were stronger and appeared earlier in the premotor regions related to hand movements than in the prefrontal cortex (Wallis & Miller 2003). Likewise, decisions about eye movements appear to involve the same circuits that execute eye movements, which include the lateral intraparietal area (LIP) (Dorris & Glimcher 2004, Gold & Shadlen 2007, Platt & Glimcher 1999, Sugrue et al. 2004, Yang & Shadlen 2007), the FEF (Coe et al. 2002, Schall & Bichot 1998), and the superior colliculus (Basso & Wurtz 1998, Carello & Krauzlis 2004, Horwitz et al. 2004, Thevarajah et al. 2009), which is a brainstem structure that is just two synapses away from the motor neurons that move the eye. In all of these cases, the same neurons appear to first reflect decision-related variables such as the quality of evidence in favor of a given choice and then later encode the metrics of the action used to report the decision (Cisek & Kalaska 2005, Kim & Basso 2008, Roitman & Shadlen 2002, Schall & Bichot 1998, Yang & Shadlen 2007).

Consequently, it has proven to be notoriously difficult to assign a specific perceptual, cognitive, or motor function to cortical associative regions such as the posterior parietal cortex (PPC), where cells appear to be related to all of these functions (Andersen & Buneo 2003, Colby & Duhamel 1996, Colby & Goldberg 1999, Culham & Kanwisher 2001, Kalaska & Crammond 1995). The PPC represents spatial sensory information on the location of behaviorally salient objects in the environment (Colby & Duhamel 1996, Colby & Goldberg 1999, Stein 1992), strongly modulated by attention and behavioral context (Colby & Duhamel 1996, Colby & Goldberg 1999, Kalaska 1996, Mountcastle et al. 1975). This has led to the hypothesis that the parietal cortex is involved in directing attention to different parts of space and in constructing a salience map of the environment (Constantinidis & Steinmetz 2001, Kusunoki et al. 2000). Presumably, this forms part of the perceptual representation that serves as input to the cognitive system. However, there is also strong evidence that parietal cortical activity contains representations of action intentions (Andersen & Buneo 2003, Colby & Duhamel 1996, Kalaska et al. 1997, Mazzoni et al. 1996, Platt & Glimcher 1997, Snyder et al. 2000), which include activity that specifies the direction of intended saccades (Snyder et al. 2000) and arm reaching movements (Andersen & Buneo 2003, Buneo et al. 2002, Kalaska & Crammond 1995), and different subregions of the PPC are specialized for different effectors (Calton et al. 2002, Cui & Andersen 2007). Because action representations are supposedly activated by the output of the cognitive system, it is difficult to reconcile these findings with the sensory properties of the PPC, which leads to persistent debates about its role. Furthermore, neural activity in the PPC is also modulated by a range of variables associated with decision making, such as expected utility (Platt & Glimcher 1999), local income (Sugrue et al. 2004), relative subjective desirability (Dorris & Glimcher 2004), and log-likelihood estimates (Yang & Shadlen 2007). In short, the PPC does not appear to fit neatly into any of the categories of perception, cognition, or action; or alternatively, the PPC reflects all categories at once without respecting those theoretical distinctions. Indeed, it is difficult to see how neural activity in the PPC can be interpreted using any of the concepts of classical cognitive psychology (Culham & Kanwisher 2001).

The data and resulting disagreements reviewed above have motivated us to reflect on some of our assumptions for interpreting neural activity. Perhaps specific functions such as perception are not implemented by particular cortical regions. Instead, they may be implemented by different layers or subnetworks of cells distributed within many parts of the nervous system. Perhaps distinct roles such as perceptual or motor representations can be performed by the same neurons at different times. These possibilities are worth considering and studying experimentally. Here, however, we explore a different possibility. We consider whether the distinctions among perceptual, cognitive, and motor systems may not reflect the natural categories of neural computations that underlie sensory-guided behavior (Hendriks-Jansen 1996, Lebedev & Wise 2002). The framework of serial information processing may not be the optimal blueprint for the global functional architecture of the brain. Instead, we consider whether alternative theoretical frameworks for the large-scale organization of behavior may facilitate interpretations of neural activity.

AN ECOLOGICAL PERSPECTIVE

One of the most important facts we know about the brain is that it evolved. This not only motivates our theories to describe mechanisms that confer selective advantage, but more importantly, it constrains theories to respect the brain's phylogenetic history. Contrary to popular belief, brain evolution has been remarkably conservative. Since the development of the telencephalon, the basic outline of the vertebrate nervous system has been strongly conserved (Butler & Hodos 2005, Holland & Holland 1999, Katz & Harris-Warrick 1999). Even recently elaborated structures such as the mammalian neocortex have homologs among nonmammals (Medina & Reiner 2000), and the topology of neural circuitry is analogous across diverse species (Karten 1969).

The conservative nature of brain evolution motivates us to think about large-scale theories of neural organization from the perspective of the kinds of behaviors that animals engaged in many millions of years ago, when that neural

organization was being laid down. Throughout evolutionary history, organisms and their nervous systems have been preoccupied by almost constant interaction with a complex and ever changing environment, which continuously offers a potentially bewildering variety of opportunities and demands for action. Interaction with such an environment cannot be broken down into a sequence of distinct and self-contained events that each start with a discrete stimulus and end with a specific response, similar to the isolated trials we typically use in many psychological or neurophysiological experiments. Instead, it involves the continuous modification of ongoing actions through feedback control, the continuous evaluation of alternative activities that may become available, and continuous tradeoffs between choosing to persist in a given activity and switching to a different one. The internal processes that are most useful for such behavior may not be those that first construct an accurate internal description of objective and abstract knowledge about the world and then reflect upon it with some introspective, intelligent circuits. Instead, pragmatic processes that mediate sensorimotor interaction in the here and now, on the basis of continuous streams of sensory inputs as well as prior knowledge and experiences, are much more useful for guiding interactive behavior (Gibson 1979).

An emphasis on real-time, natural behavior has been the foundation of ethological research for a long time (Hinde 1966). In the early twentieth century, researchers such as Von Uexküll, Tinbergen, and Lorenz focused their studies on the observation of animals in the wild rather than in the laboratory. Consequently, instead of focusing on how knowledge is represented or what variables are included in the motor program, they focused on how competition between potential actions is resolved, how ongoing behavior is fine tuned by feedback mechanisms that operate at multiple hierarchical levels, and how animals trade off activity against metabolic costs.

Some of the original founders of psychological science also emphasized the importance of

Affordances:
opportunities for
action defined by the
environment around
an animal

Embodied cognition:
a study of cognition
that emphasizes its
role in sensorimotor
control and action

interactions with the environment. For example, John Dewey (1896) criticized the view of behavior as a process of receiving a stimulus and producing a response, and wrote that "[w]hat we have is a circuit...the motor response determines the stimulus, just as truly as sensory stimulus determines movement." (p. 363). Similar emphases on sensorimotor control were made by Hughlings Jackson (1884) and Merleau-Ponty (1945), among many others. Perhaps the best known example is the work of the eminent psychologist Jean Piaget (1954), who suggested that the abstract cognitive abilities of adult humans are constructed upon the basis of the sensorimotor interactions experienced as a child. This is supported by a variety of neural studies, which include the classic experiments of Held & Hein (1963), who found that the visual behavior of newborn kittens did not develop properly unless they were allowed to exert their own active control upon their visual input.

The perceptual psychologist James Gibson was another well-known proponent of an ecological view of behavior. Similar to ethologists, Gibson viewed the constrained environment of a typical psychological experiment as concealing the true interactive nature of behavior. He argued that perception is not about passively constructing an internal representation of the world, but rather it is about actively picking up information of interest to one's behavior. Inspired by earlier work of Gestalt psychologists such as Koffka, he emphasized that the environment contains information relevant for an animal's activity and that a large part of perception is the accumulation of that information. He defined the concept of affordances (Gibson 1979) as the opportunities for action that the environment presents to an animal.

Ethological concepts have been very useful in research on autonomous robotics, which is increasingly abandoning classical serial architectures based on explicit representations of the environment in favor of hierarchical control systems in which the basic elements are sensorimotor feedback loops (Ashby 1965, Brooks 1991, Hendriks-Jansen 1996, Meyer 1995, Sahin et al. 2007). For a robot that

interacts in the real world, such architectures have simply proven to be more effective than serial information processing through distinct perception, cognition, and action modules. These concepts are also becoming increasingly influential in a branch of cognitive science that is sometimes called embodied cognition (Clark 1997, Klatzky et al. 2008, Núñez & Freeman 2000, Thelen et al. 2001).

Such concepts may also be useful for interpreting neurophysiological data. For example, Graziano & Aflalo (2007) proposed that the multiple motor areas in the precentral gyrus may not be organized on the basis of a sequential planning and execution architecture, as commonly assumed. Instead, the precentral gyrus may reflect the animal's natural behavioral repertoire, with different regions that are specialized for different actions such as bringing objects to the mouth, manipulating objects in central vision, climbing, or defensive behavior. Although controversial, this conjecture has intriguing similarities to theoretical proposals that evolution constructs complex behaviors by using simpler ones as building blocks (Brooks 1991, Hendriks-Jansen 1996). This has clear ecological advantages because it reflects the need for animals to partially plan many different classes of potential actions, such as grasping a piece of fruit while also being ready to scamper away in case of danger.

One particularly important and influential example of how a perspective of interactive behavior may shed light on neurophysiology is the work of Melvyn Goodale and David Milner (1992, Milner & Goodale 1995). As discussed above, visual processing diverges in the cerebral cortex into a ventral stream, where cells are sensitive to stimulus features, and a dorsal stream, where cells are sensitive to spatial relationships (Ungerleider & Mishkin 1982). Instead of describing these, respectively, as the what and where systems, Goodale & Milner suggested that the predominant role of the dorsal stream is to mediate visually guided behavior. They proposed that the dorsal stream (now often called the how system) is sensitive to spatial information, not to build a representation of the environment for central knowledge acquisition, but because spatial information is critical for specifying the parameters of potential and ongoing actions. This view explains many other properties of dorsal pathway processing, such as its emphasis on concrete and current information (Milner & Goodale 1995) and its intimate interconnection with frontal regions involved in movement control (Johnson et al. 1996, Wise et al. 1997). From this perspective, processing in the parietal cortex and reciprocally connected premotor regions is not exclusively concerned with descriptive representations of objects in the external world but primarily with pragmatic representations of the opportunities for action that those objects afford (Cisek 2007, Colby & Duhamel 1996, Fadiga et al. 2000, Kalaska et al. 1998, Rizzolatti & Luppino 2001). Indeed, parietal activity in both monkeys (Iriki et al. 1996, Mountcastle et al. 1975) and humans (Gallivan et al. 2009) is often stronger when objects are within reach.

Several groups have developed these ideas further. For example, Fagg & Arbib (1998) have suggested that the PPC represents a set of currently available potential actions, one of which is ultimately selected for overt execution. Similarly, we and others have suggested that the dorsal stream is involved in specifying the parameters of potential actions, whereas the ventral stream provides further information for their selection (Andersen & Buneo 2003, Cisek 2007, Kalaska et al. 1998, Passingham & Toni 2001, Sakagami & Pan 2007). This has much in common with a long history of proposals, made on the basis of EEG studies (Coles et al. 1985) and stimulus-response compatibility effects (Kornblum et al. 1990), that neural processing is continuous and not organized in distinct serial stages. It is also similar to the proposal that the brain begins to prepare several actions in parallel while collecting evidence for selecting between them (Shadlen et al. 2008), a view that is strongly supported by neurophysiological studies of decision making (Gold & Shadlen 2007, Kim & Basso 2008, Ratcliff et al. 2007).

Some of these ideas are summarized in **Figure 1**, which shows what we call the affordance competition hypothesis (Cisek 2007). This general hypothesis is directly inspired by the work of Gibson, Ashby, Goodale & Milner, Arbib, and many others mentioned above. It begins with a distinction between two types of problems that animals behaving in the natural environment continuously face: deciding what to do and how to do it. We can call these the problems of action selection and action specification. However, although traditional psychological theories assume that selection (decision making) occurs before specification (movement planning), we consider the possibility that, at least during natural interactive behavior, these processes operate simultaneously and in an integrated manner (Cisek 2007).

For the particular case of visually-guided movement, action specification (**Figure 1**, dark blue lines) may involve the dorsal visual stream and a distributed and reciprocally interconnected network of areas in the posterior parietal and caudal frontal cortex (Andersen & Buneo 2003; Andersen et al. 1997; Goodale & Milner 1992; Johnson et al. 1996; Kalaska 1996; Kalaska & Crammond 1995; Milner & Goodale 1995; Rizzolatti & Luppino 2001; Wise et al. 1996, 1997). These circuits perform transformations that convert information about objects in sensory coordinates into the parameters of actions (Andersen & Buneo 2003, Andersen et al. 1997, Wise et al. 1997). Along the way, each area can represent information that is pertinent to several potential actions simultaneously as patterns of tuned activity within distributed populations of cells. This forms a representation of possible movements that is conceptually similar to a probability density function (Sanger 2003). Importantly, these same brain regions ultimately guide the execution of those actions. Because multiple actions usually cannot be performed at the same time, there is competition between options, perhaps through mutual inhibition among cells with different tuning properties (Cisek 2006) and/or through differential selection in corticostriatal circuits (Leblois et al. 2006).

Action selection: the process of choosing an action from among many possible alternatives

Action specification: the process of specifying the spatiotemporal aspects of possible actions

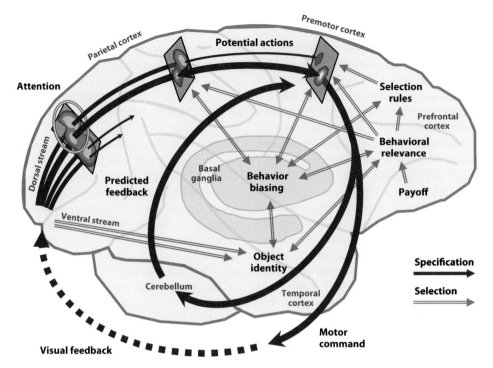

Figure 1

Sketch of the affordance competition hypothesis in the context of visually-guided movement. The primate brain is shown, emphasizing the cerebral cortex, cerebellum, and basal ganglia. Dark blue arrows represent processes of action specification, which begin in the visual cortex and proceed rightward across the parietal lobe, and which transform visual information into representations of potential actions. Polygons represent three neural populations along this route. Each population is depicted as a map where the lightest regions correspond to peaks of tuned activity, which compete for further processing. This competition is biased by input from the basal ganglia and prefrontal cortical regions that collect information for action selection (*red double-line arrows*). These biases modulate the competition in several loci, and because of reciprocal connectivity, their influences are reflected over a large portion of the cerebral cortex. The final selected action is released into execution and causes overt feedback through the environment (*dotted blue arrow*) as well as internal predictive feedback through the cerebellum. Modified with permission from Cisek (2007).

If a competition between representations of potential actions exists in frontoparietal circuits, then intelligent behavior requires a way to influence that competition by factors related to rewards, costs, risks, or any variable pertinent to making good choices. A variety of brain systems can contribute their votes into this selection process simply by biasing activity within the ongoing frontoparietal competition (**Figure 1**, red double-line arrows). This includes influences from subcortical structures such as the basal ganglia (Mink 1996, Redgrave et al. 1999, Schultz 2004) and cortical regions such as the prefrontal cortex (Miller 2000,

Sakagami & Pan 2007, Tanji & Hoshi 2001, Wise 2008). In turn, the prefrontal areas receive information pertinent to action selection that include object identity from the temporal lobe (Pasupathy & Connor 2002, Tanaka et al. 1991) and subjective value from the orbitofrontal cortex (Padoa-Schioppa & Assad 2008, Schultz et al. 2000, Wallis 2007). In summary, the hypothesis is that interaction with the environment involves continuous and simultaneous processes of sensorimotor control and action selection from among the distributed representations of a limited number of response options. This perspective is consistent with a large family of computational models of decisions, which suggest that neural activity related to different response choices builds up in separate accumulators as a function of the evidence for or against those choices until a threshold is reached that favors one over the others (Gold & Shadlen 2007, Ratcliff et al. 2007). Whereas classic decision models have treated the accumulators as separate modules that correspond to distinct choices, our hypothesis suggests that they emerge from a continuous population that represents parameters of potential actions (Cisek 2006, Erlhagen & Schöner 2002, Furman & Wang 2008, Tipper et al. 2000), at least in cases when decisions are reported through specific actions. Indeed, the neurons that have been implicated in the evidence accumulation process are always found within populations tuned for motor output parameters such as direction (Glimcher 2003, Gold & Shadlen 2007, Kim & Basso 2008, Ratcliff et al. 2007).

We propose that the kind of general theoretical architecture shown in **Figure 1**, although highly simplified, can nevertheless help us interpret many of the neural data briefly reviewed above, data that have proven difficult to interpret within the traditional view of serial information processing stages. The rest of this review is devoted to discussing data relevant to that claim.

REVISITING NEURAL DATA

In revisiting some of the neural data discussed at the beginning of this review, we emphasize two main conjectures: (*a*) The control of interactive behavior involves competition between parallel sensorimotor control loops, and (*b*) neural representations involved in this control are pragmatic—that is, they are adapted to produce good control as opposed to producing accurate descriptions of the sensory environment or a motor plan. Both of these proposals have been made repeatedly for over a hundred years of research from Dewey to Gibson to Goodale & Milner and others, but they have not often been used as the theoretical framework for interpreting neurophysiological data.

Frontoparietal Specification of Potential Actions

Following Goodale & Milner (1992, Milner & Goodale 1995), one can interpret the dorsal visual stream as part of the system for specifying the parameters of potential actions using visual information, a process that continues during movement execution (Resulaj et al. 2009). As mentioned above, the dorsal stream is not unified but progressively diverges into parallel subsystems, each specialized toward the demands of different sensorimotor functions and effectors (Andersen et al. 1997, Colby & Duhamel 1996, Colby & Goldberg 1999, Rizzolatti & Luppino 2001, Stein 1992, Wise et al. 1997). Area LIP represents space in an ego-centered reference frame (Colby & Duhamel 1996, Snyder et al. 1998), is involved in control of gaze (Snyder et al. 2000), and is interconnected with other parts of the gaze control system that includes the FEF and the superior colliculus (Paré & Wurtz 2001). The medial intraparietal area (MIP) is involved in the control of arm reaching movements (Cui & Andersen 2007, Kalaska & Crammond 1995, Pesaran et al. 2008, Scherberger & Andersen 2007, Snyder et al. 2000), represents target locations with respect to the direction of gaze and the position of the arm (Buneo et al. 2002), and is interconnected with frontal regions that are involved in reaching, such as PMd (Johnson et al. 1996, Wise et al. 1997). Neurons in the anterior intraparietal area (AIP) are involved in grasping (Baumann et al. 2009), their activity is sensitive to object size and orientation, and they are interconnected with the grasp-related ventral premotor cortex (PMv) (Nakamura et al. 2001, Rizzolatti & Luppino 2001). To summarize, the dorsal stream diverges into parallel subsystems, each of which specifies the spatial parameters of different kinds of potential actions and plays a direct role in guiding their execution during movement.

In such a distributed system, several actions can be specified simultaneously. For example, if a monkey is presented with a spatial target but not instructed about whether an arm or

MIP: medial intraparietal area

AIP: anterior intraparietal area

PMv: ventral premotor cortex

eye movement is required, neurons begin to discharge in both LIP and MIP (Calton et al. 2002, Cui & Andersen 2007). Later, if an arm movement is instructed (Calton et al. 2002) or autonomously chosen (Cui & Andersen 2007), the activity becomes stronger in MIP than LIP. Conversely, if a saccade is instructed or chosen, activity becomes stronger in LIP than MIP. This is consistent with the proposal that before the effector is selected, reach and saccade plans begin to be specified simultaneously by different parts of the PPC. Indeed, during natural activity, eye and hand movements are usually executed in unison. Similar findings have been reported for decisions regarding hand choice. For example, Hoshi & Tanji (2007) showed that when monkeys are first presented with a reach target location in a bimanual response-choice task without specifying which arm to use, neural activity in the premotor cortex reflects the potential movements of both hands until the monkey is instructed about which hand to use.

Simultaneous specification of multiple potential actions can occur even within the same effector system (Basso & Wurtz 1998, Bastian et al. 1998, Baumann et al. 2009, Cisek & Kalaska 2005, McPeek & Keller 2002, Platt & Glimcher 1997, Powell & Goldberg 2000, Schall & Bichot 1998, Scherberger & Andersen 2007). For example, behavioral data (McPeek et al. 2000) and neurophysiological data (McPeek & Keller 2002) suggest that the preparation of multiple sequential saccades can overlap in time. When two or more potential saccade targets are presented simultaneously, neural correlates for each are observed in area LIP (Platt & Glimcher 1997, Powell & Goldberg 2000) and even in the superior colliculus, where they are modulated by selection probability (Basso & Wurtz 1998, Kim & Basso 2008).

Likewise, behavioral studies of reaching have suggested that the brain simultaneously processes information about multiple potential actions. For example, the trajectory of a reaching movement to a target is influenced by the presence of distracters (Song & Nakayama 2008, Tipper et al. 2000, Welsh et al. 1999)

and veers away from regions of risk (Trommershauser et al. 2006). Patients with frontal lobe damage often cannot suppress actions associated with distracters even while they are planning actions directed elsewhere (Humphreys & Riddoch 2000), and such effects may be the result of competition among parallel simultaneous representations of potential actions, with a bias toward the actions with the highest stimulus-response compatibility (Castiello 1999).

Neurophysiological studies support this interpretation. For example, partial information on possible upcoming movements engages the activity of cells in reach-related regions before the animal selects the movement that will be made (Bastian et al. 1998, Kurata 1993, Riehle & Requin 1989). In particular, when a reach direction is initially specified ambiguously by sensory information, neural activity arises in the motor and premotor cortex that spans the entire angular range of potential directions. Later, when the direction is specified more precisely, the directional spread of population activity narrows to reflect this choice (Bastian et al. 1998). Neural correlates of multiple potential reaching actions have been reported in the dorsal premotor cortex (PMd) even when the choices are distinct and mutually exclusive (Cisek & Kalaska 2005). As shown in **Figure 2**, when a monkey was presented with two opposite potential reaching actions, only one of which would later be indicated as the correct choice by a nonspatial cue, neural activity in the premotor cortex specified both directions simultaneously. When information for selecting one action over the other became available, the representation of the chosen direction was strengthened while that of the unwanted direction was suppressed. The monkey used a strategy of preparing both movements simultaneously during the initial period of uncertainty despite the fact that the task design permitted the use of an alternative strategy (more consistent with traditional models of processing) in which target locations are stored in a general-purpose working memory buffer that is distinct from motor representations and only

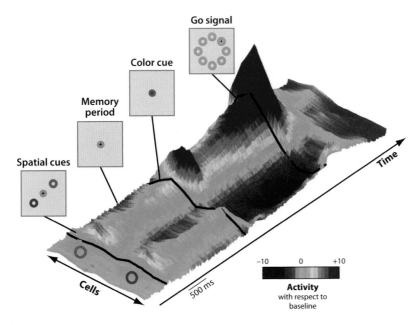

Figure 2

Population activity in the dorsal premotor cortex during a reach-selection task. The 3D colored surface depicts neural activity with respect to baseline, with cells sorted by their preferred direction along the bottom edge. Diagrams on the left show the stimuli presented to the monkey at different points during the trial (cross indicates the cursor). Note that during the period of ambiguity, even after stimuli vanished, the population encodes two potential directions. Data from Cisek & Kalaska (2005).

converted to a motor plan after the decision is made. In contrast, we propose that multiple movement options are specified within the same system that is used to prepare and guide the execution of the movement that is ultimately selected. The simultaneous specification of multiple actions can even occur when only a single object is viewed. For example, the multiple affordances offered by a single object can evoke neural activity in the grasp-related area AIP that can represent several potential grasps until one is instructed (Baumann et al. 2009), in agreement with the predictions of theoretical models (Fagg & Arbib 1998).

Evidence that the nervous system can simultaneously represent multiple potential actions suggests a straightforward interpretation of the finding, described above, that early responses in many premotor and parietal regions first appear to encode information about relevant stimuli and later change to encode motor variables. Perhaps the early activity,

time-locked to stimulus appearance, does not encode the stimuli themselves but rather the set of potential actions that are most strongly associated with those stimuli (Wise et al. 1996), such as actions with high stimulus-response compatibility (Crammond & Kalaska 1994). This would imply that the functional role of this activity does not change in time from sensory to motor encoding but simply reflects the arrival of selection influences from slower but more sophisticated mechanisms for deciding which action is most appropriate.

Recent computational models have proposed that whenever multiple potential targets are available, representations of potential actions emerge within several frontoparietal neural populations, each composed of a continuum of cells with different preferences for the potential parameters of movement (Cisek 2006, Erlhagen & Schöner 2002, Tipper et al. 2000). In each population, cells with similar preferences mutually excite each other (even if they are not physically adjacent), which leads to the activation of groups of cells with similar tuning. At the same time, cells with different preferences inhibit each other, thus implementing a competition between representations of actions that are mutually exclusive. Unlike classical models of decisions, in which the different choices are abstract and clearly distinct (e.g., choosing between gambles A or B), models in which decisions emerge within tuned populations suggest that the same mechanism—lateral inhibition—is responsible for defining the choices as well as for implementing the competition between them. Importantly, they suggest that decisions about actions emerge within the same populations of cells that define the physical properties of those actions and guide their execution.

This proposal can account for phenomena that cannot be explained using models in which the decision process occurs in an abstract space that is separate from a representation of the metrics of motor options. For example, although it is well-known that reaction time (RT) generally increases with the number of choices presented to a subject, it is less widely recognized that RT is also dependent upon the spatial separation of the response options (Bock & Eversheim 2000). As another example, Ghez et al. (1997) showed that forced rapid choices between precued options are dependent on target separation. If cues are close together, subjects initially move in between them (continuous mode) before deviating toward one or the other in mid-reach. If the cues are far apart, they choose one at random and move to it directly (discrete mode). To explain such results, models of decisions must capture how the choices themselves are defined in physical space and how the similarity of potential actions influences their interactions (Cisek 2006, Erlhagen & Schöner 2002). This is straightforward if the representations of choices exist within neural populations that encode the physical parameters of the movements used to report the choice.

From this perspective, it is not surprising that neural activity in the frontal and parietal cortex encodes information that appears to be sensory, motor, and cognitive in nature (Wise et al. 1996). The case of area LIP is particularly instructive. If LIP is involved in the specification of potential saccades, then its activity must correlate with the location of possible saccade targets (Mazzoni et al. 1996, Snyder et al. 2000), even when multiple potential saccades are processed simultaneously (Platt & Glimcher 1997, Powell & Goldberg 2000). At the same time, however, the ongoing selection of potential actions will modulate the strength of activities in LIP. Such modulation is influenced by target salience (Colby & Goldberg 1999, Kusunoki et al. 2000), reward size and selection probability (Platt & Glimcher 1999, Yang & Shadlen 2007), and other decision variables (Dorris & Glimcher 2004, Sugrue et al. 2004), as well as prior information on the type of action to be performed (Calton et al. 2002, Cui & Andersen 2007). The progressive elimination of potential saccade targets along the dorsal stream also explains why the representation of visual space in LIP is so sparse (Gottlieb et al. 1998): Only the most promising and salient targets make it to LIP.

If the presence of salient targets can engage the simultaneous specification of several potential actions in a variety of frontoparietal systems, then this process is closely related to the concept of a salience map (Kusunoki et al. 2000, Powell & Goldberg 2000). In particular, the front end of a system for action selection should enhance the most behaviorally salient information in the environment to bias sensorimotor systems toward the most behaviorally relevant potential actions. Thus, it will be action dependent (Snyder et al. 2000) but still influenced by the salience of stimuli, even while actions are instructed elsewhere (Kusunoki et al. 2000). In short, attention and intention may be different aspects of a common process that progressively narrows the set of potential actions that will be processed further downstream. This agrees with the proposal (Allport 1987, Neumann 1990) that attention is a mechanism for early action selection and not a solution to the purely internal problem of a computational bottleneck for processing sensory information

(Broadbent 1958), as is often assumed. Indeed, models of action selection compatible with the kind of framework shown on **Figure 1** (Cisek 2006, Erlhagen & Schöner 2002) are functionally equivalent to the biased competition model used to explain data on visual attention (Boynton 2005, Desimone & Duncan 1995, Treue 2001). In both cases, parallel representations of targets/actions compete through lateral inhibition; and in both cases, they are biased by top-down and or bottom-up influences.

The idea that decisions emerge within a continuum of tuned cells can even go beyond strictly action-related decisions (Song & Nakayama 2009) and may provide insights into mechanisms that have traditionally been studied using tasks in which subjects report choices by pressing distinct keys on a keyboard. For example, when subjects are asked to report on seemingly binary yes/no questions (such as, is the sky ever green?) by moving a cursor instead of just pressing one key or another, their movement trajectories reveal a wealth of phenomena that suggest the decision is represented in a continuous space that spills into overt movement (McKinstry et al. 2008). For example, features of the trajectory such as its endpoint and peak velocity correlate with the subject's confidence about their choice. Even linguistic decisions, such as those made while parsing an ambiguous sentence, appear to occur in a continuous parameter space that can influence movement trajectories (Farmer et al. 2007). These findings are difficult to reconcile with the idea that cognition is separate from sensorimotor control (Fodor 1983) but make good sense if the continuous nature of the representations that underlie the selection of actions has been retained as selection systems evolved to implement increasingly abstract decisions.

Action Selection Signals from Prefrontal and Subcortical Sources

From the perspective of the parallel architecture reviewed here, there may be no single central executive module that guides decisions. As shown by many of the studies reviewed above,

neural correlates of cognitive processes can be seen throughout the brain during sensorimotor and decision tasks. When we look for the neural correlates of cognition, it does not appear as an independent module (Fodor 1983) that receives input from perceptual modules and sends goal signals to motor centers. Instead, it appears as a process that is closely integrated with action selection, evaluation, and motor execution (Cisek 2007, Glimcher 2003, Gold & Shadlen 2007, Heekeren et al. 2008, Hoshi & Tanji 2007, Pesaran et al. 2008, Rizzolatti & Luppino 2001, Shadlen et al. 2008).

The recent evolution of primates is distinguished by advances in the ability to select actions based on increasingly abstract and arbitrary criteria. This kind of selection may have been made possible by the dramatic elaboration of the prefrontal cortex (Hauser 1999), especially the granular frontal cortex, which does not appear to have a homolog in rodents (Wise 2008). These frontal regions are strongly implicated in decision making and action selection (Fuster et al. 2000, Kim & Shadlen 1999, Miller 2000, Romo et al. 2004, Rowe et al. 2000, Tanji & Hoshi 2001). For example, neurons in the dorsolateral prefrontal cortex (DLPFC) are sensitive to various combinations of stimulus features, and this sensitivity is always related to the particular demands of the task at hand (Barraclough et al. 2004, Hoshi et al. 1998, Kim & Shadlen 1999, Quintana & Fuster 1999, Rainer et al. 1998). For example, an experiment on reach target selection (Hoshi et al. 2000) found that when arbitrary iconic stimuli were presented, activity in the DLPFC was sensitive to potentially relevant stimulus features, such as shape and location. After the presentation of a signal that indicated the correct selection rule (shape-match or location-match), rule-sensitive neurons briefly became active, selecting out the relevant memorized stimulus features needed to make the response choice. After this process was complete, the remaining activity reflected the intended movement choice. Prefrontal decisions appear to evolve through the collection of votes for categorically selecting one choice over others. For example, when monkeys were trained to report perceptual discriminations using saccades (Kim & Shadlen 1999), DLPFC activity initially reflected the quality of evidence in favor of a given target and later simply reflected the monkey's choice. Similar effects have been reported in PMv and prefrontal cortex during tactile vibration frequency discrimination tasks (Romo et al. 2004).

The information for visually-based action selection can also come from the ventral visual stream (Cisek 2007, Kalaska et al. 1998, Passingham & Toni 2001, Sakagami & Pan 2007), where cells are sensitive to object identity (Pasupathy & Connor 2002, Sugase et al. 1999, Tanaka et al. 1991) and modulated by attention (Treue 2001). The detection of stimulus features relevant for action selection is reminiscent of what ethologists referred to as the detection of a key stimulus that releases specific behaviors (Ewert 1997, Tinbergen 1950). The detection of key stimuli need not require full-fledged object recognition (and indeed may be its precursor) because often a fragment or specific feature that has consistent meaning within an animal's niche is all that is necessary to elicit behavior. Therefore, the properties of ventral stream processing may not have originally evolved for a role in pure perception, but may instead reveal its earlier and more fundamental role in collecting information useful for action selection. Indeed, the distinction between pure vision-for-perception versus vision-for-action systems is difficult to make at the neuroanatomical level (Lebedev & Wise 2002).

Wallis (2007) reviews evidence that the motivational value of potential actions is computed by the orbitofrontal cortex (OFC), which integrates sensory and affective information to estimate the value of a reward outcome. Single-neuron studies have shown that the OFC represents the value of goods in a manner that is not dependent on their relative value with respect to other available choices (Padoa-Schioppa & Assad 2008) but scales with the range of values in a given context (Padoa-Schioppa 2009). Sakagami & Pan (2007) suggest that this information is further integrated with sensory signals from the ventral visual stream to provide an estimate of the behavioral relevance of potential actions in the ventrolateral prefrontal cortex (VLPFC), which then projects to the DLPFC and premotor regions to influence action selection.

Because action selection is a fundamental problem faced by even the most primitive vertebrates, it likely involves structures that were prominently developed long ago and have been conserved in evolution. The basal ganglia are promising candidates (Kalivas & Nakamura 1999, Mink 1996, Redgrave et al. 1999). The basal ganglia may form a central locus in which excitation that arrives from different motor systems competes, and a winning behavior is selected while others are inhibited through projections back to the motor systems (Brown et al. 2004, Leblois et al. 2006, Mink 1996, Redgrave et al. 1999). Afferents to the input nuclei of the basal ganglia (the striatum and subthalamic nucleus) arrive from nearly the entire cerebral cortex and from the limbic system, converge onto the output nuclei (substantia nigra and globus pallidus), and project through the thalamus back to the cerebral cortex. This cortico-striatal-pallido-thalamo-cortical loop is organized into multiple parallel channels, which run through specific motor regions as well as through regions implicated in higher cognitive functions (Alexander & Crutcher 1990a, Middleton & Strick 2000). In agreement with the hypothesis of basal ganglia selection, cell activity in the input nuclei is related to movement parameters (Alexander & Crutcher 1990a) but is also influenced by the expectation of reward (Schultz et al. 2000, Takikawa et al. 2002). During learning of arbitrary visuomotor mappings, striatal activity evolves in concert with PMd activity to indicate the selected movement (Buch et al. 2006). The inactivation of cells in output nuclei disrupts movement speed in a manner consistent with the proposal that the inhibition of competing motor programs is disrupted (Wenger et al. 1999). Furthermore, the finding that the basal ganglia connect with prefrontal regions, in a manner similar to their connections with premotor cortex, suggests that basal ganglia innervation of prefrontal regions also mediates selection but on a more abstract level (Hazy et al. 2007). This is also consistent with motor and cognitive aspects of basal ganglia diseases (Mink 1996, Sawamoto et al. 2002).

Parallel Operation

Continuous interactive behavior often does not allow one to stop to think or to collect information to build a complete knowledge of one's surroundings. The demands of survival in an ever-changing environment drove evolution to endow animals with an architecture that allows them to partially prepare several courses of action simultaneously, so that alternatives can be ready for release at short notice. Such an ecological view of behavior suggests that the processes of action selection and specification normally occur simultaneously and continue even during the overt performance of movements, which allows animals to switch to another option if they change their mind (Resulaj et al. 2009). That is, sensory information that arrives from the world is continuously used to specify several currently available potential actions, in parallel, while other kinds of information are collected to select the one that will be released into execution at a given moment (Cisek & Kalaska 2005, Glimcher 2003, Gold & Shadlen 2007, Kalaska et al. 1998, Kim & Shadlen 1999, Shadlen et al. 2008). From this perspective, behavior is viewed as a constant competition between internal representations of conflicting demands and opportunities.

Suppose that an animal is endowed with this kind of parallel architecture, adapted for continuous real-time interaction with a natural environment. What would happen if we remove that animal from its natural environment and place it in a neurophysiological laboratory? In this highly controlled and impoverished setting, time is broken into discrete trials, each starting with the presentation of a stimulus and ending with the production of a response (and if the response is correct, a reward). Furthermore, unlike in natural behavior, most features of the sensory input are deliberately made independent from the animal's actions—the response

in a given trial usually does not determine the stimulus in the next trial. Of course, animals are capable of dealing with this artificial scenario and able to learn which responses yield the best rewards. The question addressed here is the following: What would a parallel architecture such as that of **Figure 1** predict about the time course of processing in such a situation?

When the stimulus is first presented, we should expect an initial fast feedforward sweep of activity along the dorsal stream, crudely representing the potential actions that are most directly specified by the stimulus. Indeed, Schmolesky et al. (1998) showed that neural responses to simple visual flashes appear quickly throughout the dorsal visual system and engage putatively motor-related areas such as FEF in as little as 50 ms. This is significantly earlier than some visual areas such as V2 and V4. In general, even within the visual system neural activation does not appear to follow a serial sequence from early to late areas (Paradiso 2002). In a reaching task, population activity in PMd responds to a learned visual cue within 50 ms of its appearance (Cisek & Kalaska 2005). Such fast responses are not purely visual because they reflect the context within which the stimulus is presented. For example, they reflect whether the monkey expects to see one or two stimuli (Cisek & Kalaska 2005), reflect anticipatory biases or priors (Coe et al. 2002, Takikawa et al. 2002), and can be entirely absent if the monkey already knows what action to take and can ignore the stimulus altogether (Crammond & Kalaska 2000) (**Figure 3**). In short, these phenomena are compatible with the notion of a fast dorsal specification system that quickly uses novel visual information to specify the potential actions most consistently associated with a given stimulus (Gibson 1979, Milner & Goodale 1995).

After the initial options are quickly specified, slower selection processes should begin to sculpt the neural activity patterns by introducing a variety of task-relevant biasing factors. Indeed, extrastriate visual areas MT and 7a respond to a stimulus in approximately 50 ms but begin to reflect the influence of attention in 100–120 ms (Constantinidis & Steinmetz 2001,

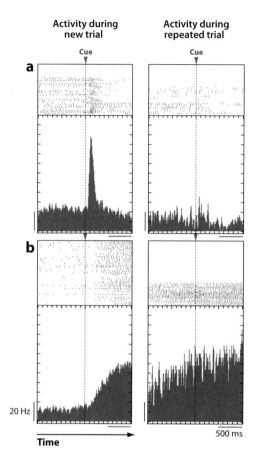

Activity during new trial

Cue

a

b

20 Hz

Time

Activity during repeated trial

Cue

500 ms

Figure 3

(*a*) Pooled activity of three phasic PMd neurons during trials in their preferred direction, aligned on cue onset. The left panel shows activity during trials in which a novel and unpredictable cue is presented, instructing the monkey about the required reaching movement. The right panel shows activity from trials that follow errors and repeat the same cue at the same location. Because the monkey has learned from experience that the same target will be presented in trials following an error, the presentation of the target stimulus provides it with no new salient information about movement and does not evoke a neural response. (*b*) Pooled activity of three tonic PMd neurons, same format. Note that in trials following errors, the directionally tuned activity is already present before the cue appears. This reflects the retention of prior knowledge about the imminent presentation of the same target after an error. Reprinted with permission from Crammond & Kalaska (2000).

Treue 2001). FEF neurons respond to the onset of a stimulus in 50 ms (Schmolesky et al. 1998), but detect the singleton of a visual-search array with a median of approximately 100 ms and discriminate pro- versus antisaccades in approximately 120 ms (Sato & Schall 2003). LIP neurons respond to stimulus onset in approximately

50 ms and discriminate targets from distracters in 138 ms (Thomas & Pare 2007). Neurons in dorsal premotor cortex respond to the locations of cues instructing two potential movements in 70 ms but begin to predict the monkey's choice in 110–130 ms (Cisek & Kalaska 2005).

A recent study by Ledberg et al. (2007) provides an overall picture of the time course observed in all of the experiments described above. These authors simultaneously recorded local field potentials (LFPs) from up to 15 cerebral cortical regions of monkeys that performed a conditional Go/NoGo task (**Figure 4*a***). Through an elegant experimental design, Ledberg and colleagues identified the first neural events that responded to the appearance of a stimulus, those which discriminated its identity, as well as those that predicted the monkey's chosen response (**Figure 4*b***). In agreement with earlier studies (c.f. Schmolesky et al. 1998), they observed a fast feedforward sweep of stimulus onset-related activity appearing within 50–70 ms in striate and extrastriate cortex and 55–80 ms in FEF and premotor cortex. Discrimination of different stimulus categories occurred later, within approximately 100 ms of onset in prestriate areas and 200 ms in prefrontal sites. The Go/NoGo decision appeared approximately 150 ms after stimulus onset, nearly simultaneously within a diverse mosaic of cortical sites including prestriate, inferotemporal, parietal, premotor, and prefrontal areas.

In summary, when behavior is experimentally isolated in the lab, the continuous and parallel processes critical for interaction appear as two waves of activation: an early wave crudely specifying a menu of options and a second wave that selects among them approximately 120–150 ms after stimulus onset (Ledberg et al. 2007). It appears that the brain can quickly specify multiple potential actions within its fast frontoparietal sensorimotor control system, but it takes approximately 150 ms (in the case of simple tasks) to integrate sufficient information to make a decision between them. However, the apparent serial order of these events is largely a result of the experimental strategy of

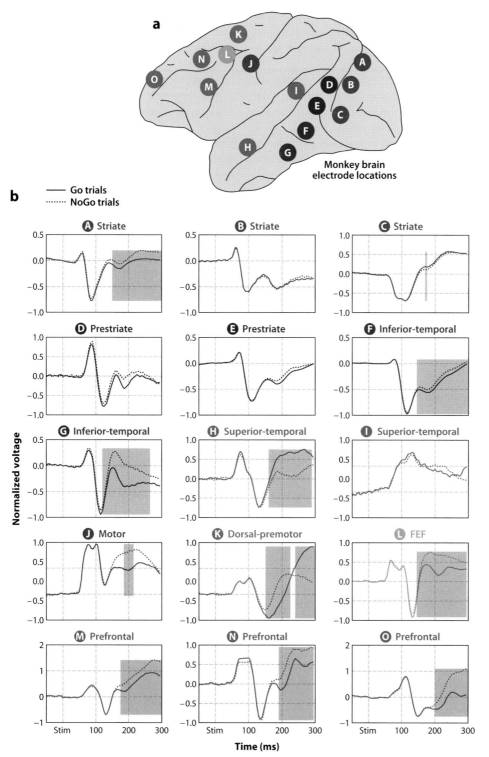

a

K N L J A

O M I D B

E C

F

H G

Monkey brain
electrode locations

b

—— Go trials
········ NoGo trials

Normalized voltage

Ⓐ Striate

Ⓑ Striate

Ⓒ Striate

Ⓓ Prestriate

Ⓔ Prestriate

Ⓕ Inferior-temporal

Ⓖ Inferior-temporal

Ⓗ Superior-temporal

Ⓘ Superior-temporal

Ⓙ Motor

Ⓚ Dorsal-premotor

Ⓛ FEF

Ⓜ Prefrontal

Ⓝ Prefrontal

Ⓞ Prefrontal

Stim 100 200 300

Time (ms)

Figure 4

(*a*) Anatomical location of electrodes measuring local-field potentials. (*b*) Time course of average normalized event-related potentials from the electrodes shown in (*a*). (*Solid line*) Potentials during Go trials. (*Dashed line*) Potentials during NoGo trials. Gray-shaded regions highlight epochs of time during which these differ. Note that the earliest responses to stimuli appear throughout the cerebral cortex after approximately 50 ms, and activity begins to predict the monkey's choice after approximately 150 ms in a distributed network that includes parietal and frontal regions. Reprinted with permission from Ledberg et al. (2007).

dividing behavior into a sequence of discrete and independent trials. During natural activity, all of these events presumably occur continuously.

These results are consistent with the hypothesis that decisions emerge through a distributed consensus achieved among reciprocally connected frontoparietal regions, each of which may contain representations of several potential actions. This makes a further prediction: The precise order in which decisions appear across the cerebral cortex will be highly task dependent (Cisek 2006). For example, if the factors that lead to a decision are bottom-up visual features such as stimulus salience, then neural correlates of that decision should appear first in the parietal cortex and then in frontal regions. In contrast, if the biasing factors require the kinds of complex stimulus-rule conjunctions that engage neurons in the prefrontal cortex, then the decision should emerge first in frontal regions before propagating back to the parietal cortex. Recent studies have supported that prediction. For example, when monkeys perform pop-out visual search tasks, neural activity in LIP reflects the choice before FEF, but if the task involves conjunction search then FEF reflects the choice before LIP (Buschman & Miller 2007). Interestingly, during a Go/NoGo task in which monkeys made decisions on the basis of cognitive rules, activity that predicted the response appeared in PMd even before the prefrontal cortex (PFC) (Wallis & Miller 2003). It is as if, at least in that kind of task, a decision may be influenced by noisy neural votes arriving from the PFC but is determined by a consensus that is reached in a frontoparietal network that includes the PMd (Pesaran et al. 2008).

FUTURE DIRECTIONS

We conclude this article by returning to several practical issues. If we hypothesize that the functional architecture of the brain consists of simultaneous competing sensorimotor control systems and distributed selection mechanisms—how should we proceed to study these processes? Clearly, neurophysiological experiments must be conducted in a careful and quantitative manner to allow the interpretation of resulting data. But how can one quantitatively study natural behavior, which is inherently variable and unconstrained?

One approach is to continue as before. There is no reason why data obtained in a classical laboratory setting cannot still be interpreted in the broader context of natural behavior. For example, the study of Ledberg et al. (2007) on the timing of cortical processes was done with head-fixed animals that were observing an impoverished stimulus and making a single Go/Nogo response, but its results can still be used to gain valuable insights into the organization of a flexible parallel system for interactive behavior. In fact, all of the studies discussed above are still relevant and amenable to interpretation in terms of interactive behavior. However, we should not mistake our experimental method for the outline of a theory.

Controlled laboratory experiments can also be designed to be inspired by natural behavior. For example, visual-search tasks capture many aspects of natural foraging activity, which requires animals to discriminate food (targets) within a cluttered environment (distracters). Recent studies by Michael Dorris and colleagues take the analogy further, by presenting monkeys with a visual foraging task in which they can explore their environment through unconstrained saccades, making tradeoffs between harvesting rewards by looking at one stimulus versus searching for better payoffs among other stimuli (Kan & Dorris 2009).

Finally, technical and mathematical advances are starting to make it possible to study truly unconstrained behavior while still yielding solid and interpretable data. For example, d'Avella & Bizzi (2005) studied motor control in frogs by allowing them to freely move around their environment—swimming, jumping, and walking without constraints—while EMG activity was chronically recorded from 13 hindlimb muscles. Through a careful analysis of muscular patterns, they extracted the motor synergies that appear to underlie these natural behaviors. A still more ambitious approach is to implant wireless multielectrode arrays in the brains of rats (Sodagar et al. 2007) or monkeys (Chestek et al. 2009) and record ensemble activity during free behavior. However, unconstrained behavior requires novel ways of analyzing and thinking about our data. In particular, we need methods that move away from the standard approach of averaging over similar trials and toward analyzing behavior and the neural firing patterns of many neurons during each individual action (Yu et al. 2009).

CONCLUSIONS

One of the major goals of neuroscience research is to reveal the brain's functional architecture to build a theoretical framework that bridges the brain and behavior (Schall 2004). In recent decades, an influential framework has been based on concepts developed in cognitive psychology, which were originally intended to explain human abstract problem solving behavior. These led to a functional architecture of information processing stages that can be roughly categorized as perceptual, cognitive, and motor control modules. However, as we review above, a growing number of neurophysiological studies in nonhuman primates appear difficult to interpret from this perspective. This motivates us to consider alternative theoretical frameworks. Above, we consider proposals developed for many decades in fields such as ethology, ecological psychology, and autonomous robotics research, which were designed to explain the original and still primary purpose of the brain—to endow organisms with the ability to adaptively interact with their environment. We discuss diverse neurophysiological data on the control of voluntary behavior that, in our opinion, appear to be more compatible with these alternative frameworks. This leads to the claim that perhaps they provide a better foundation for interpreting neural data and designing future experiments. We acknowledge that similar claims have been made many times in the past, but limitations of space and our own knowledge make it impossible to list all

of the contributors to these views. Our purpose here is to draw attention to how recent neurophysiological experiments lend strong support to these alternative ways of thinking about the brain.

The proposals we review above address interactive sensorimotor behavior and are not meant to constitute a general theory of all brain function. Clearly, tasks such as writing an email or playing chess involve processes that are far removed from simple sensorimotor control. Nevertheless, theories of embodied cognition have suggested, following Piaget, that cognitive abilities may have evolved within the context of ancestral abilities for interacting with the world (Hendriks-Jansen 1996, Pezzulo & Castelfranchi 2009, Powers 1973, Thelen et al. 2001, Toates 1998). For example, Toates (1998) proposed that primitive switching mechanisms, which mediate between different stimulus-response associations, have become elaborated through evolution into the cognitive systems that now allow us to make complex and sophisticated decisions. Pezzulo & Castelfranchi (2009) suggest that our ability to think about the world results from the internalization of the processes of predicting the consequences of actions. Their cognitive leverage hypothesis proposes that as the sensorimotor control system gradually evolved, it began to predict increasingly abstract consequences of behavior. This eventually allowed the mental rehearsal of entire sequences of acts and evaluation of their potential outcomes, without overt motor activity. Their hypothesis is consistent with the close relationship between mental imagery and the systems for motor preparation (Cisek & Kalaska 2004, di Pellegrino et al. 1992, Rizzolatti & Craighero 2004, Umilta et al. 2001) and potentially explains how an organism may go beyond merely reacting to properties of the immediate environment and act in a goal-directed manner. In conclusion, the neural systems that mediate the sensorimotor behavior of our ancient ancestors may have provided the foundations for modern cognitive abilities, and their consideration may shed light on the neural mechanisms that underlie human thought.

SUMMARY POINTS

1. Brains evolved for sensorimotor control and retained much of that architecture—even the neocortex is still part of that old circuit.

2. Natural interactive behavior requires sensorimotor control and selection systems to operate continuously and in parallel.

3. Distinctions between perceptual, cognitive, and motor processes, although descriptively useful, might not reflect the natural categories of the brain's functional organization.

4. Decisions appear to be made through a distributed consensus that emerges in competitive populations.

5. Neurophysiological data may be more readily interpreted from the perspective of interactive behavior than from the perspective of serial information processing.

FUTURE ISSUES

1. What are possible experimental approaches for studying behavior without constraining its interactive nature? Although some behavioral abilities are already studied in a relatively natural setting (e.g., locomotion), other systems demand new technologies for wireless multi-unit recording and new methods for analyzing data.

2. For a deeper understanding of the evolution of modern behavioral abilities, we would like to reconstruct the sequence of phylogenetic elaborations of a given system along a particular branch of the evolutionary tree. This calls for comparative neurophysiological studies of a diverse set of related species. However, this poses a significant challenge, not only because homologies are difficult to establish but also because practical matters motivate scientists to study particular species whose brains are already well mapped (e.g., rats, cats, and macaques).

3. From a theoretical standpoint, we need models that explain how sensorimotor control systems could have become elaborated to implement more sophisticated behavior. For example, if action selection takes place within a space defined by movement parameters (e.g., reach direction), what are the parameter spaces in which high-level decisions are made? Do these high-level spaces maintain topology, similar to the somatotopy and spatial maps useful for action selection?

4. How can an advanced agent discover the high-level opportunities afforded within its behavioral niche and link them with long-term goals?

DISCLOSURE STATEMENT

The authors are not aware of any affiliations, memberships, funding, or financial holdings that might be perceived as affecting the objectivity of this review.

ACKNOWLEDGMENTS

This article is dedicated to Steve Wise, a friend and colleague whose work has had a major influence on our thinking about the brain. We thank Ignasi Cos, Trevor Drew, Andrea Green, and Christopher Pack for their comments and their contributions to the development of the ideas described in this paper. We apologize in advance to all of the investigators whose relevant research could not be cited owing to space limitations. Our research is supported by grants from the Canadian Institutes of Health Research, the Fonds de la Recherche en Santé du Québec, the Natural Sciences and Engineering Research Council of Canada, the National Institutes of Health, the Canadian Foundation for Innovation, and the EJLB Foundation.

LITERATURE CITED

Albright TD, Kandel ER, Posner MI. 2000. Cognitive neuroscience. *Curr. Opin. Neurobiol.* 10(5):612–24

Alexander GE, Crutcher MD. 1990a. Functional architecture of basal ganglia circuits: neural substrates of parallel processing. *TINS* 13(7):266–71

Alexander GE, Crutcher MD. 1990b. Neural representations of the target (goal) of visually guided arm movements in three motor areas of the monkey. *J. Neurophysiol.* 64(1):164–78

Allport DA. 1987. Selection for action: some behavioral and neurophysiological considerations of attention and action. In *Perspectives on Perception and Action*, ed. H Heuer, AF Sanders, pp. 395–419. Hillsdale, NJ: Erlbaum

Andersen RA, Buneo CA. 2003. Sensorimotor integration in posterior parietal cortex. *Adv. Neurol.* 93:159–77

Andersen RA, Snyder LH, Bradley DC, Xing J. 1997. Multimodal representation of space in the posterior parietal cortex and its use in planning movements. *Annu. Rev. Neurosci.* 20:303–30

Ashby WR. 1965. *Design for a Brain: The Origin of Adaptive Behavior*. London: Chapman and Hall

Barraclough DJ, Conroy ML, Lee D. 2004. Prefrontal cortex and decision making in a mixed-strategy game. *Nat. Neurosci.* 7(4):404–10

Basso MA, Wurtz RH. 1998. Modulation of neuronal activity in superior colliculus by changes in target probability. *J. Neurosci.* 18(18):7519–34

Bastian A, Riehle A, Erlhagen W, Schöner G. 1998. Prior information preshapes the population representation of movement direction in motor cortex. *Neuroreport* 9:315–19

Baumann MA, Fluet MC, Scherberger H. 2009. Context-specific grasp movement representation in the macaque anterior intraparietal area. *J. Neurosci.* 29(20):6436–48

Bock O, Eversheim U. 2000. The mechanisms of movement preparation: a precuing study. *Behav. Brain Res.* 108(1):85–90

Boynton GM. 2005. Attention and visual perception. *Curr. Opin. Neurobiol.* 15(4):465–69

Broadbent DE. 1958. *Perception and Communication*. New York: Pergamon

Brooks R. 1991. Intelligence without representation. *Artif. Intell.* 47:139–59

Brown JW, Bullock D, Grossberg S. 2004. How laminar frontal cortex and basal ganglia circuits interact to control planned and reactive saccades. *Neural Networks* 17(4):471–510

Buch ER, Brasted PJ, Wise SP. 2006. Comparison of population activity in the dorsal premotor cortex and putamen during the learning of arbitrary visuomotor mappings. *Exp. Brain Res.* 169(1):69–84

Buneo CA, Jarvis MR, Batista AP, Andersen RA. 2002. Direct visuomotor transformations for reaching. *Nature* 416(6881):632–36

Buschman TJ, Miller EK. 2007. Top-down versus bottom-up control of attention in the prefrontal and posterior parietal cortices. *Science* 315(5820):1860–62

Bushnell MC, Goldberg ME, Robinson DL. 1981. Behavioral enhancement of visual responses in monkey cerebral cortex. I. Modulation in posterior parietal cortex related to selective visual attention. *J. Neurophysiol.* 46(4):755–72

Butler AB, Hodos W. 2005. *Comparative Vertebrate Neuroanatomy: Evolution and Adaptation*. New York: Wiley-Liss

Calton JL, Dickinson AR, Snyder LH. 2002. Non-spatial, motor-specific activation in posterior parietal cortex. *Nat. Neurosci.* 5(6):580–88

Carello CD, Krauzlis RJ. 2004. Manipulating intent: evidence for a causal role of the superior colliculus in target selection. *Neuron* 43(4):575–83

Castiello U. 1999. Mechanisms of selection for the control of hand action. *Trends. Cogn. Sci.* 3(7):264–71

Chestek CA, Gilja V, Nuyujukian P, Kier RJ, Solzbacher F et al. 2009. HermesC: low-power wireless neural recording system for freely moving primates. *IEEE Trans. Neural Syst. Rehabil. Eng.* 17(4):330–38

Cisek P. 2007. Cortical mechanisms of action selection: the affordance competition hypothesis. *Philos. Trans. R. Soc. Lond. B Biol. Sci.* 362(1485):1585–99

Cisek P. 2006. Integrated neural processes for defining potential actions and deciding between them: a computational model. *J. Neurosci.* 26(38):9761–70

Cisek P, Crammond DJ, Kalaska JF. 2003. Neural activity in primary motor and dorsal premotor cortex in reaching tasks with the contralateral versus ipsilateral arm. *J. Neurophysiol.* 89(2):922–42

Cisek P, Kalaska JF. 2004. Neural correlates of mental rehearsal in dorsal premotor cortex. *Nature* 431(7011):993–96

Cisek P, Kalaska JF. 2005. Neural correlates of reaching decisions in dorsal premotor cortex: Specification of multiple direction choices and final selection of action. *Neuron* 45(5):801–14

Clark A. 1997. *Being There: Putting Brain, Body, and World Together Again*. Cambridge, MA: MIT Press

Coe B, Tomihara K, Matsuzawa M, Hikosaka O. 2002. Visual and anticipatory bias in three cortical eye fields of the monkey during an adaptive decision-making task. *J. Neurosci.* 22(12):5081–90

Colby CL, Duhamel JR. 1996. Spatial representations for action in parietal cortex. *Brain Res. Cogn. Brain Res.* 5(1–2):105–15

Colby CL, Goldberg ME. 1999. Space and attention in parietal cortex. *Annu. Rev. Neurosci.* 22:319–49

Coles MG, Gratton G, Bashore TR, Eriksen CW, Donchin E. 1985. A psychophysiological investigation of the continuous flow model of human information processing. *J. Exp. Psychol. Hum. Percept. Perform.* 11(5):529–53

Constantinidis C, Steinmetz MA. 2001. Neuronal responses in area 7a to multiple-stimulus displays. I. Neurons encode the location of the salient stimulus. *Cereb. Cortex.* 11(7):581–91

Crammond DJ, Kalaska JF. 2000. Prior information in motor and premotor cortex: activity during the delay period and effect on premovement activity. *J. Neurophysiol.* 84(2):986–1005

Crammond DJ, Kalaska JF. 1994. Modulation of preparatory neuronal activity in dorsal premotor cortex due to stimulus-response compatibility. *J. Neurophysiol.* 71(3):1281–84

Cui H, Andersen RA. 2007. Posterior parietal cortex encodes autonomously selected motor plans. *Neuron* 56(3):552–59

Culham JC, Kanwisher NG. 2001. Neuroimaging of cognitive functions in human parietal cortex. *Curr. Opin. Neurobiol.* 11(2):157–63

d'Avella A, Bizzi E. 2005. Shared and specific muscle synergies in natural motor behaviors. *Proc. Natl. Acad. Sci. USA* 102(8):3076–81

Desimone R, Duncan J. 1995. Neural mechanisms of selective visual attention. *Annu. Rev. Neurosci.* 18:193–222

Dewey J. 1896. The reflex arc concept in psychology. *Psychol. Rev.* 3(4):357–70

di Pellegrino G, Fadiga L, Fogassi L, Gallese V, Rizzolatti G. 1992. Understanding motor events: a neurophysiological study. *Exp. Brain Res.* 91(1):176–80

Dorris MC, Glimcher PW. 2004. Activity in posterior parietal cortex is correlated with the relative subjective desirability of action. *Neuron* 44(2):365–78

Engel AK, Fries P, Singer W. 2001. Dynamic predictions: oscillations and synchrony in top-down processing. *Nat. Rev. Neurosci.* 2(10):704–16

Erlhagen W, Schöner G. 2002. Dynamic field theory of movement preparation. *Psychol. Rev.* 109(3):545–72

Ewert J-P. 1997. Neural correlates of key stimulus and releasing mechanism: a case study and two concepts. *TINS* 20(8):332–39

Fadiga L, Fogassi L, Gallese V, Rizzolatti G. 2000. Visuomotor neurons: ambiguity of the discharge or 'motor' perception? *Int. J. Psychophysiol.* 35(2–3):165–77

Fagg AH, Arbib MA. 1998. Modeling parietal-premotor interactions in primate control of grasping. *Neural Netw.* 11(7–8):1277–303

Farmer TA, Cargill SA, Spivey MJ. 2007. Gradiency and visual context in syntactic garden-paths. *J. Mem. Lang.* 57(4):570–95

Felleman DJ, Van Essen DC. 1991. Distributed hierarchical processing in the primate cerebral cortex. *Cereb. Cortex* 1(1):1–47

Fitts PM. 1954. The information capacity of the human motor system in controlling the amplitude of movement. *J. Exp. Psychol.* 47(6):381–91

Fodor JA. 1983. *Modularity of Mind.* Cambridge, MA: MIT Press

Furman M, Wang XJ. 2008. Similarity effect and optimal control of multiple-choice decision making. *Neuron* 60(6):1153–68

Fuster JM, Bodner M, Kroger JK. 2000. Cross-modal and cross-temporal association in neurons of frontal cortex. *Nature* 405(6784):347–51

Gail A, Klaes C, Westendorff S. 2009. Implementation of spatial transformation rules for goal-directed reaching via gain modulation in monkey parietal and premotor cortex. *J. Neurosci.* 29(30):9490–99

Gallivan JP, Cavina-Pratesi C, Culham JC. 2009. Is that within reach? fMRI reveals that the human superior parieto-occipital cortex encodes objects reachable by the hand. *J. Neurosci.* 29(14):4381–91

Gazzaniga MS. 2000. *The New Cognitive Neurosciences.* Cambridge, MA: MIT Press

Ghez C, Favilla M, Ghilardi MF, Gordon J, Bermejo R, Pullman S. 1997. Discrete and continuous planning of hand movements and isometric force trajectories. *Exp. Brain Res.* 115(2):217–33

Gibson JJ. 1979. *The Ecological Approach to Visual Perception.* Boston: Houghton Mifflin

Glimcher PW. 2003. The neurobiology of visual-saccadic decision making. *Annu. Rev. Neurosci.* 26:133–79

Gold JI, Shadlen MN. 2007. The neural basis of decision making. *Annu. Rev. Neurosci.* 30:535–74

Goodale MA, Milner AD. 1992. Separate visual pathways for perception and action. *TINS* 15(1):20–25

Gottlieb JP, Kusunoki M, Goldberg ME. 1998. The representation of visual salience in monkey parietal cortex. *Nature* 391(6666):481–84

Graziano MS, Aflalo TN. 2007. Mapping behavioral repertoire onto the cortex. *Neuron* 56(2):239–51

Hauser MD. 1999. Perseveration, inhibition and the prefrontal cortex: a new look. *Curr. Opin. Neurobiol.* 9(2):214–22

Hazy TE, Frank MJ, O'Reilly RC. 2007. Towards an executive without a homunculus: computational models of the prefrontal cortex/basal ganglia system. *Philos. Trans. R. Soc. Lond. B Biol. Sci.* 362(1485):1601–13

Heekeren HR, Marrett S, Ungerleider LG. 2008. The neural systems that mediate human perceptual decision making. *Nat. Rev. Neurosci.* 9(6):467–79

Held R, Hein A. 1963. Movement-produced stimulation in the development of visually guided behavior. *J. Comp. Physiol. Psychol.* 56(5):872–76

Hendriks-Jansen H. 1996. *Catching Ourselves in the Act: Situated Activity, Interactive Emergence, Evolution, and Human Thought.* Cambridge, MA: MIT Press

Hernandez A, Zainos A, Romo R. 2002. Temporal evolution of a decision-making process in medial premotor cortex. *Neuron* 33(6):959–72

Hinde RA. 1966. *Animal Behavior: A Synthesis of Ethology and Comparative Psychology.* New York: McGraw-Hill

Holland LZ, Holland ND. 1999. Chordate origins of the vertebrate central nervous system. *Curr. Opin. Neurobiol.* 9(5):596–602

Horwitz GD, Batista AP, Newsome WT. 2004. Representation of an abstract perceptual decision in macaque superior colliculus. *J. Neurophysiol.* 91(5):2281–96

Hoshi E, Tanji J. 2007. Distinctions between dorsal and ventral premotor areas: anatomical connectivity and functional properties. *Curr. Opin. Neurobiol.* 17(2):234–42

Hoshi E, Shima K, Tanji J. 1998. Task-dependent selectivity of movement-related neuronal activity in the primate prefrontal cortex. *J. Neurophysiol.* 80(6):3392–97

Hoshi E, Shima K, Tanji J. 2000. Neuronal activity in the primate prefrontal cortex in the process of motor selection based on two behavioral rules. *J. Neurophysiol.* 83(4):2355–73

Hughlings Jackson J. 1884. Evolution and dissolution of the nervous system. In *Selected Writings of John Hughlings Jackson*, ed. J Taylor, pp. 45–75. London: Staples

The definitive work on ecological psychology and the concept of affordances.

An excellent review of research on the neural mechanisms of decisions, particularly in the saccadic system.

An outstanding synthesis of ethological, evolutionary, and developmental literature and implications for large-scale brain theories.

Humphreys GW, Riddoch JM. 2000. One more cup of coffee for the road: object-action assemblies, response blocking and response capture after frontal lobe damage. *Exp. Brain Res.* 133:81–93

Iriki A, Tanaka M, Iwamura Y. 1996. Coding of modified body schema during tool use by macaque postcentral neurones. *Neuroreport* 7(14):2325–30

Johnson PB, Ferraina S, Bianchi L, Caminiti R. 1996. Cortical networks for visual reaching: physiological and anatomical organization of frontal and parietal arm regions. *Cereb. Cortex* 6(2):102–19

Johnson-Laird PN. 1988. *The Computer and the Mind: An Introduction to Cognitive Science.* Cambridge, MA: Harvard Univ. Press

Kalaska JF. 1996. Parietal cortex area 5 and visuomotor behavior. *Can. J. Physiol. Pharmacol.* 74:483–98

Kalaska JF, Crammond DJ. 1995. Deciding not to GO: neuronal correlates of response selection in a GO/NOGO task in primate premotor and parietal cortex. *Cereb. Cortex* 5:410–28

Kalaska JF, Scott SH, Cisek P, Sergio LE. 1997. Cortical control of reaching movements. *Curr. Opin. Neurobiol.* 7:849–59

Kalaska JF, Sergio LE, Cisek P. 1998. Cortical control of whole-arm motor tasks. In *Sensory Guidance of Movement, Novartis Foundation Symposium #218,* ed. M Glickstein, pp. 176–201. Chichester, UK: Wiley

Kalivas PW, Nakamura M. 1999. Neural systems for behavioral activation and reward. *Curr. Opin. Neurobiol.* 9(2):223–27

Kan JY, Dorris MC. 2009. Superior colliculus activity reflects saccade profitability during visual foraging task. *Progr. No. 285.12 2009 Neuroscience Meeting Planner.* Chicago IL: Soc. Neurosci. Online

Karten HJ. 1969. The organization of the avian telencephalon and some speculations on the phylogeny of the amniote telencephalon. *Ann. N. Y. Acad. Sci.* 167(1):164–79

Katz PS, Harris-Warrick RM. 1999. The evolution of neuronal circuits underlying species-specific behavior. *Curr. Opin. Neurobiol.* 9(5):628–33

Keele SW. 1968. Movement control in skilled motor performance. *Psychol. Bull.* 70:387–403

Kim B, Basso MA. 2008. Saccade target selection in the superior colliculus: a signal detection theory approach. *J. Neurosci.* 28(12):2991–3007

Kim J-N, Shadlen MN. 1999. Neural correlates of a decision in the dorsolateral prefrontal cortex of the macaque. *Nat. Neurosci.* 2(2):176–85

Klatzky R, Behrmann M, MacWhinney B. 2008. *Embodiment, Ego-Space and Action.* New York: Psychology Press

Kornblum S, Hasbroucq T, Osman A. 1990. Dimensional overlap: cognitive basis for stimulus-response compatibility—a model and taxonomy. *Psychol. Rev.* 97(2):253–70

Kurata K. 1993. Premotor cortex of monkeys: set- and movement-related activity reflecting amplitude and direction of wrist movements. *J. Neurophysiol.* 69(1):187–200

Kusunoki M, Gottlieb J, Goldberg ME. 2000. The lateral intraparietal area as a salience map: the representation of abrupt onset, stimulus motion, and task relevance. *Vision Res.* 40(10–12):1459–68

Lebedev MA, Wise SP. 2002. Insights into seeing and grasping: distinguishing the neural correlates of perception and action. *Behav. Cogn. Neurosci. Rev.* 1:108–29

Leblois A, Boraud T, Meissner W, Bergman H, Hansel D. 2006. Competition between feedback loops underlies normal and pathological dynamics in the basal ganglia. *J. Neurosci.* 26(13):3567–83

Ledberg A, Bressler SL, Ding M, Coppola R, Nakamura R. 2007. Large-scale visuomotor integration in the cerebral cortex. *Cereb. Cortex* 17(1):44–62

Marr DC. 1982. *Vision.* San Francisco: W. H. Freeman

Mazzoni P, Bracewell RM, Barash S, Andersen RA. 1996. Motor intention activity in the macaque's lateral intraparietal area. I. Dissociation of motor plan from sensory memory. *J. Neurophysiol.* 76:1439–57

McKinstry C, Dale R, Spivey MJ. 2008. Action dynamics reveal parallel competition in decision making. *Psychol. Sci.* 19(1):22–24

McPeek RM, Keller EL. 2002. Superior colliculus activity related to concurrent processing of saccade goals in a visual search task. *J. Neurophysiol.* 87(4):1805–15

McPeek RM, Skavenski AA, Nakayama K. 2000. Concurrent processing of saccades in visual search. *Vision Res.* 40(18):2499–516

Medina L, Reiner A. 2000. Do birds possess homologues of mammalian primary visual, somatosensory and motor cortices? *TINS* 23(1):1–12

An important electrophysiological study showing the time-course of neural processing across diverse regions of the primate cerebral cortex.

Merleau-Ponty M. 1945. *Phénoménologie de la Perception*. Paris: Gallimard

Meyer J-A. 1995. The animat approach to cognitive science. In *Comparative Approaches to Cognitive Science*, ed. HL Roitblat, J-A Meyer, pp. 27–44. Cambridge, MA: MIT Press

Middleton FA, Strick PL. 2000. Basal ganglia and cerebellar loops: motor and cognitive circuits. *Brain Res. Rev.* 31(2–3):236–50

Miller EK. 2000. The prefrontal cortex and cognitive control. *Nat. Rev. Neurosci.* 1(1):59–65

Miller GA. 1956. The magical number seven, plus or minus two: some limits on our capacity for processing information. *Psychol. Rev.* 63:81–97

Miller GA, Galanter E, Pribram KH. 1960. *Plans and the Structure of Behavior*. New York: Holt, Rinehart and Winston

Milner AD, Goodale MA. 1995. *The Visual Brain in Action*. Oxford, UK: Oxford Univ. Press

Mink JW. 1996. The basal ganglia: focused selection and inhibition of competing motor programs. *Prog. Neurobiol.* 50(4):381–425

Moran J, Desimone R. 1985. Selective attention gates visual processing in the extrastriate cortex. *Science* 229:782–84

Mountcastle VB, Andersen RA, Motter BC. 1981. The influence of attentive fixation upon the excitability of the light-sensitive neurons of the posterior parietal cortex. *J. Neurosci.* 1(11):1218–25

Mountcastle VB, Lynch JC, Georgopoulos AP, Sakata H, Acuna C. 1975. Posterior parietal association cortex of the monkey: command functions for operations within extrapersonal space. *J. Neurophysiol.* 38(4):871–908

Nakamura H, Kuroda T, Wakita M, Kusunoki M, Kato A et al. 2001. From three-dimensional space vision to prehensile hand movements: the lateral intraparietal area links the area V3A and the anterior intraparietal area in macaques. *J. Neurosci.* 21(20):8174–87

Neumann O. 1990. Visual attention and action. In *Relationships Between Perception and Action: Current Approaches*, ed. O Neumann, W Prinz, pp. 227–267. Berlin: Springer-Verlag

Newell A, Simon HA. 1972. *Human Problem Solving*. Englewood Cliffs, NJ: Prentice-Hall

Núñez R, Freeman WJ. 2000. *Reclaiming Cognition: The Primacy of Action, Intention and Emotion*. Thorverton: Imprint Academic

Padoa-Schioppa C. 2009. Range-adapting representation of economic value in the orbitofrontal cortex. *J. Neurosci.* 29(44):14004–14

Padoa-Schioppa C, Assad JA. 2008. The representation of economic value in the orbitofrontal cortex is invariant for changes of menu. *Nat. Neurosci.* 11(1):95–102

Paradiso MA. 2002. Perceptual and neuronal correspondence in primary visual cortex. *Curr. Opin. Neurobiol.* 12(2):155–61

Paré M, Wurtz RH. 2001. Progression in neuronal processing for saccadic eye movements from parietal cortex area LIP to superior colliculus. *J. Neurophysiol.* 85(6):2545–62

Passingham RE, Toni I. 2001. Contrasting the dorsal and ventral visual systems: guidance of movement versus decision making. *Neuroimage* 14(1 Pt. 2):S125–31

Pasupathy A, Connor CE. 2002. Population coding of shape in area V4. *Nat. Neurosci.* 5(12):1332–38

Pesaran B, Nelson MJ, Andersen RA. 2008. Free choice activates a decision circuit between frontal and parietal cortex. *Nature* 453(7193):406–9

Pezzulo G, Castelfranchi C. 2009. Thinking as the control of imagination: a conceptual framework for goal-directed systems. *Psychol. Res.* 73(4):559–77

Piaget J. 1954. *The Construction of Reality in the Child*. New York: Basic Books

Platt ML, Glimcher PW. 1999. Neural correlates of decision variables in parietal cortex. *Nature* 400(6741):233–38

Platt ML, Glimcher PW. 1997. Responses of intraparietal neurons to saccadic targets and visual distractors. *J. Neurophysiol.* 78(3):1574–89

Powell KD, Goldberg ME. 2000. Response of neurons in the lateral intraparietal area to a distractor flashed during the delay period of a memory-guided saccade. *J. Neurophysiol.* 84(1):301–10

Powers WT. 1973. *Behavior: The Control of Perception*. New York: Aldine Publ.

Pylyshyn ZW. 1984. *Computation and Cognition: Toward a Foundation for Cognitive Science*. Cambridge, MA: MIT Press

A detailed review of the anatomical circuits between the cerebral cortex, cerebellum, and the basal ganglia.

An influential synthesis of research on the dorsal and ventral visual pathways and their functional interpretation.

A control-theoretic perspective on how sensorimotor control systems may have evolved toward more sophisticated planning.

Quintana J, Fuster JM. 1999. From perceptions to actions: temporal integrative functions of prefrontal and parietal neurons. *Cereb. Cortex* 9(3):213–21

Rainer G, Asaad WF, Miller EK. 1998. Selective representation of relevant information by neurons in the primate prefrontal cortex. *Nature* 363(6885):577–79

Ratcliff R, Hasegawa YT, Hasegawa RP, Smith PL, Segraves MA. 2007. Dual diffusion model for single-cell recording data from the superior colliculus in a brightness-discrimination task. *J. Neurophysiol.* 97(2):1756–74

Redgrave P, Prescott TJ, Gurney K. 1999. The basal ganglia: a vertebrate solution to the selection problem? *Neuroscience* 89(4):1009–23

Resulaj A, Kiani R, Wolpert DM, Shadlen MN. 2009. Changes of mind in decision-making. *Nature* 461(7261):263–66

Riehle A, Requin J. 1989. Monkey primary motor and premotor cortex: single-cell activity related to prior information about direction and extent of an intended movement. *J. Neurophysiol.* 61(3):534–49

Riesenhuber M, Poggio T. 1999. Are cortical models really bound by the "binding problem"? *Neuron* 24(1):87–93

Riesenhuber M, Poggio T. 2002. Neural mechanisms of object recognition. *Curr. Opin. Neurobiol.* 12(2):162–68

Rizzolatti G, Craighero L. 2004. The mirror-neuron system. *Annu. Rev. Neurosci.* 27(1):169–92

Rizzolatti G, Luppino G. 2001. The cortical motor system. *Neuron* 31(6):889–901

Roitman JD, Shadlen MN. 2002. Response of neurons in the lateral intraparietal area during a combined visual discrimination reaction time task. *J. Neurosci.* 22(21):9475–89

Romo R, Hernandez A, Zainos A. 2004. Neuronal correlates of a perceptual decision in ventral premotor cortex. *Neuron* 41(1):165–73

Romo R, Hernandez A, Zainos A, Lemus L, Brody CD. 2002. Neuronal correlates of decision-making in secondary somatosensory cortex. *Nat. Neurosci.* 5(11):1217–25

Rowe JB, Toni I, Josephs O, Frackowiak RS, Passingham RE. 2000. The prefrontal cortex: response selection or maintenance within working memory? *Science* 288(5471):1656–60

Sahin E, Çakmak M, Dogar MR, Ugur E, Üçoluk G. 2007. To afford or not to afford: a new formalization of affordances toward affordance-based robot control. *Adaptive Behav.* 15(4):447–71

Sakagami M, Pan X. 2007. Functional role of the ventrolateral prefrontal cortex in decision making. *Curr. Opin. Neurobiol.* 17(2):228–33

Sanger TD. 2003. Neural population codes. *Curr. Opin. Neurobiol.* 13(2):238–49

Sato TR, Schall JD. 2003. Effects of stimulus-response compatibility on neural selection in frontal eye field. *Neuron* 38(4):637–48

Sawamoto N, Honda M, Hanakawa T, Fukuyama H, Shibasaki H. 2002. Cognitive slowing in Parkinson's disease: a behavioral evaluation independent of motor slowing. *J. Neurosci.* 22(12):5198–203

Schall JD. 2004. On building a bridge between brain and behavior. *Annu. Rev. Psychol.* 55:23–50

Schall JD, Bichot NP. 1998. Neural correlates of visual and motor decision processes. *Curr. Opin. Neurobiol.* 8:211–17

Scherberger H, Andersen RA. 2007. Target selection signals for arm reaching in the posterior parietal cortex. *J. Neurosci.* 27(8):2001–12

Schmolesky MT, Wang Y, Hanes DP, Thompson KG, Leutgeb S et al. 1998. Signal timing across the macaque visual system. *J. Neurophysiol.* 79(6):3272–78

Schultz W. 2004. Neural coding of basic reward terms of animal learning theory, game theory, microeconomics and behavioural ecology. *Curr. Opin. Neurobiol.* 14(2):139–47

Schultz W, Tremblay L, Hollerman JR. 2000. Reward processing in primate orbitofrontal cortex and basal ganglia. *Cereb. Cortex* 10(3):272–84

Shadlen MN, Kiani R, Hanks TD, Churchland AK. 2008. Neurobiology of decision making: An intentional framework. In *Better than Conscious? Decision Making, the Human Mind, and Implications for Institutions*, ed. C Engel, W Singer, pp. 71–101. Cambridge, MA: MIT Press

Shadlen MN, Movshon JA. 1999. Synchrony unbound: a critical evaluation of the temporal binding hypothesis. *Neuron* 24(1):67–25

A broad discussion of a general theoretical framework for decision-making, and implications for human cognitive function.

Shafir E, Tversky A. 1995. Decision making. In *Thinking: An Invitation to Cognitive Science*, ed. EE Smith, DN Osherson, pp. 77–100. Cambridge, MA: MIT Press

Shen L, Alexander GE. 1997. Neural correlates of a spatial sensory-to-motor transformation in primary motor cortex. *J. Neurophysiol.* 77:1171–94

Singer W. 2001. Consciousness and the binding problem. *Ann. N. Y. Acad. Sci.* 929:123–46

Smith EE, Osherson DN. 1995. *Thinking: An Invitation to Cognitive Science*. Cambridge, MA: MIT Press

Snyder LH, Batista AP, Andersen RA. 1998. Change in motor plan, without a change in the spatial locus of attention, modulates activity in posterior parietal cortex. *J. Neurophysiol.* 79(5):2814–19

Snyder LH, Batista AP, Andersen RA. 2000. Intention-related activity in the posterior parietal cortex: a review. *Vision Res.* 40(10–12):1433–41

Sodagar AM, Wise KD, Najafi K. 2007. A fully integrated mixed-signal neural processor for implantable multichannel cortical recording. *IEEE Trans. Biomed. Eng.* 54(6):1075–88

Song JH, Nakayama K. 2008. Target selection in visual search as revealed by movement trajectories. *Vision Res.* 48:853–61

Song JH, Nakayama K. 2009. Hidden cognitive states revealed in choice reaching tasks. *Trends Cogn Sci.* 13(8):360–66

Stein JF. 1992. The representation of egocentric space in the posterior parietal cortex. *Behav. Brain Sci.* 15:691–700

Sugase Y, Yamane S, Ueno S, Kawano K. 1999. Global and fine information coded by single neurons in the temporal visual cortex. *Nature* 400(6747):869–73

Sugrue LP, Corrado GS, Newsome WT. 2004. Matching behavior and the representation of value in the parietal cortex. *Science* 304(5678):1782–87

Takikawa Y, Kawagoe R, Hikosaka O. 2002. Reward-dependent spatial selectivity of anticipatory activity in monkey caudate neurons. *J. Neurophysiol.* 87(1):508–15

Tanaka K, Saito H-A, Fukada Y, Moriya M. 1991. Coding visual images of objects in the inferotemporal cortex of the macaque monkey. *J. Neurophysiol.* 66(1):170–89

Tanji J, Hoshi E. 2001. Behavioral planning in the prefrontal cortex. *Curr. Opin. Neurobiol.* 11(2):164–70

Thelen E, Schöner G, Scheier C, Smith LB. 2001. The dynamics of embodiment: a field theory of infant perseverative reaching. *Behav. Brain Sci.* 24(1):1–34

Thevarajah D, Mikulic A, Dorris MC. 2009. Role of the superior colliculus in choosing mixed-strategy saccades. *J. Neurosci.* 29(7):1998–2008

Thomas NW, Pare M. 2007. Temporal processing of saccade targets in parietal cortex area LIP during visual search. *J. Neurophysiol.* 97(1):942–47

Tinbergen N. 1950. The hierarchical organisation of nervous mechanisms underlying instinctive behavior. *Symp. Soc. Exp. Biol.* 4:305–12

Tipper SP, Howard LA, Houghton G. 2000. Behavioural consequences of selection from neural population codes. In *Control of Cognitive Processes: Attention and Performance XVIII*, ed. S Monsell, J Driver, pp. 223–45. Cambridge, MA: MIT Press

Toates F. 1998. The interaction of cognitive and stimulus-response processes in the control of behavior. *Neurosci. Biobehav. Rev.* 22(1):59–83

Treue S. 2001. Neural correlates of attention in primate visual cortex. *Trends Neurosci.* 24(5):295–300

Trommershauser J, Landy MS, Maloney LT. 2006. Humans rapidly estimate expected gain in movement planning. *Psychol. Sci.* 17(11):981–88

Tversky A, Kahneman D. 1981. The framing of decisions and the psychology of choice. *Science* 211:453–58

Umilta MA, Kohler E, Gallese V, Fogassi L, Fadiga L et al. 2001. I know what you are doing. a neurophysiological study. *Neuron* 31(1):155–65

Ungerleider LG, Mishkin M. 1982. Two cortical visual systems. In *Analysis of Visual Behavior*, ed. DJ Ingle, MA Goodale, RJW Mansfield, 18:549–586. Cambridge, MA: MIT Press

Wallis JD. 2007. Orbitofrontal cortex and its contribution to decision-making. *Annu. Rev. Neurosci.* 30:31–56

Wallis JD, Miller EK. 2003. From rule to response: neuronal processes in the premotor and prefrontal cortex. *J. Neurophysiol.* 90(3):1790–806

Welsh TN, Elliott D, Weeks DJ. 1999. Hand deviations toward distractors. Evidence for response competition. *Exp. Brain Res.* 127(2):207–12

Wenger KK, Musch KL, Mink JW. 1999. Impaired reaching and grasping after focal inactivation of Globus Pallidus pars interna in the monkey. *J. Neurophysiol.* 82(5):2049–60

Wise SP. 2008. Forward frontal fields: phylogeny and fundamental function. *Trends Neurosci.* 31(12):599–608

Wise SP, Boussaoud D, Johnson PB, Caminiti R. 1997. Premotor and parietal cortex: corticocortical connectivity and combinatorial computations. *Annu. Rev. Neurosci.* 20:25–42

Wise SP, di Pellegrino G, Boussaoud D. 1996. The premotor cortex and nonstandard sensorimotor mapping. *Can. J. Physiol. Pharmacol.* 74:469–82

Yang T, Shadlen MN. 2007. Probabilistic reasoning by neurons. *Nature* 447(7148):1075–80

Yu BM, Cunningham JP, Santhanam G, Ryu SI, Shenoy KV, Sahani M. 2009. Gaussian-process factor analysis for low-dimensional single-trial analysis of neural population activity. *J. Neurophysiol.* 102(1):614–35

The Role of the Human Prefrontal Cortex in Social Cognition and Moral Judgment*

Chad E. Forbes[1] and Jordan Grafman[2]

[1]Imaging Sciences Training Program, Radiology and Imaging Sciences, Clinical Center and National Institute of Biomedical Imaging and Bioengineering; Cognitive Neuroscience Section, National Institute of Neurological Disorders and Stroke, National Institutes of Health, Bethesda, Maryland 20892; email: forbesce@cc.nih.gov

[2]Cognitive Neuroscience Section, National Institute of Neurological Disorders and Stroke, National Institutes of Health, Bethesda, Maryland 20892; email: GrafmanJ@ninds.nih.gov

Annu. Rev. Neurosci. 2010. 33:299–324

First published online as a Review in Advance on March 29, 2010

The *Annual Review of Neuroscience* is online at neuro.annualreviews.org

This article's doi:
10.1146/annurev-neuro-060909-153230

0147-006X/10/0721-0299$20.00

Key Words

implicit and explicit social cognitive and moral judgment processing, frontal lobes, neural function, social cognitive neuroscience, structured event complex theory

Abstract

Results from functional magnetic resonance imaging and lesion studies indicate that the prefrontal cortex (PFC) is essential for successful navigation through a complex social world inundated with intricate norms and moral values. This review examines regions of the PFC that are critical for implicit and explicit social cognitive and moral judgment processing. Considerable overlap between regions active when individuals engage in social cognition or assess moral appropriateness of behaviors is evident, underscoring the similarity between social cognitive and moral judgment processes in general. Findings are interpreted within the framework of structured event complex theory, providing a broad organizing perspective for how activity in PFC neural networks facilitates social cognition and moral judgment. We emphasize the dynamic flexibility in neural circuits involved in both implicit and explicit processing and discuss the likelihood that neural regions thought to uniquely underlie both processes heavily interact in response to different contextual primes.

Contents

INTRODUCTION

Take a second to recall a recent interaction you had with someone you met for the first time at a social event. Hopefully the conversation went swimmingly and you devoted most of your attention to the verbal interaction. If probed, however, you could probably recall, in addition to the information learned verbally, the person's ethnicity, whether he/she was similar to you, and your impressions of which type of person you thought they were in general (and subsequently whether you liked them). You probably even processed all of this information within a few seconds. Thus within a matter of seconds, you encoded identifiable information about the person, predicted what would be best to say to facilitate a pleasant interaction, and formed an impression of him/her, all while your prefrontal cortex (PFC) simultaneously monitored and evaluated neural information from your five modalities, coordinated movements and actions, and kept you focused on adhering to social or personally relevant norms. Yet you likely processed such information with nary a thought, largely because your PFC can engage in most of the aforementioned functions outside of your conscious awareness. With this scenario in mind, you can begin to appreciate the complexity of interactions between multiple cognitive and sensory processes originating from neural regions both within and outside the PFC that provide the foundation for complex social behavior and moral judgment.

The PFC is anatomically organized in a way that allows for information from all five senses to be integrated [see the discussion of Banyas (1999), Fuster (1997) below]. The PFC is also critical for higher order functions such as focusing attention on goal-relevant stimuli and inhibiting distractions (Badre & Wagner 2004, Botvinick et al. 2004, Dolcos et al. 2007, MacDonald et al. 2000, Milham et al. 2001, Miller & D'Esposito 2005), while evaluating and interpreting information within the context of past experiences (Adolphs 1999, Damásio 1994, Rolls & Grabenhorst 2008), storing semantic information about the self and others (Johnson et al. 2002, Kelley et al. 2002, Schmitz et al. 2004), and temporally organizing actions or planned behaviors (A. Barbey, M. Koenigs, and J. Grafman, under review; Fuster 1997, Wager & Smith 2003). Given all the cognitive functions for which the PFC is necessary, it is not surprising that the PFC makes higher-order cognitive functions such as social cognition and

PFC: prefrontal cortex

moral judgment possible. As we describe here, the PFC processes all these functions at varying processing speeds, e.g., impressions and judgments of others and one's surroundings can be reached within milliseconds (termed implicit, fast, or automatic processing) or derived over longer periods of time (termed explicit, slow, or controlled processing).

Below, we review literature linking PFC functions to those of social cognition and moral judgment. We begin with a general overview of social cognition and moral judgment and highlight how the PFC is organized to enable such higher-order cognitive processes and the regions they require most. The existing literature is then interpreted within the framework of the structured event complex theory (Grafman 2002, Wood & Grafman 2003), which offers a broad organizing perspective for how activity in the PFC neural network provides the impetus for social cognition and moral judgment. We highlight how PFC activity underlying the implicit and explicit processes necessary for social cognition and moral judgment varies in different situational contexts, i.e., how situational cues change the way stimuli are processed psychologically and neurologically.

A BRIEF PRIMER ON SOCIAL COGNITION

Social cognition refers to the processes by which we make sense of ourselves, the social environment or culture in which we live, and the people around us (Fiske 1993, Macrae & Bodenhausen 2000). Although the term social cognition can encompass any cognitive process engaged to understand and interpret the self, others, and the self-in-relation-to-others within the social environment, it can be broken down into several primary categories that have both implicit and explicit components. These categories include social perceptual processes, attributional processes, and social categorization processes that use schemas or stereotypes. Research highlighting each category is discussed in turn.

Social Perceptual Processes

Research on social perception includes how one processes and identifies faces and categorical information such as gender and ethnicity, mimics others on a basic level, and interprets others' movements and intentions. Given that humans can process social markers on the order of milliseconds (Cunningham et al. 2004a), engage different neural systems in the processing of inanimate objects compared with people (Mitchell et al. 2002), and appear to be uniquely sensitive to human faces (Kouider et al. 2009), one can presume that humans are innately sensitive to social information (Adolphs 1999, Van Overwalle 2009). Using electroencephalographic (EEG) methodology, research has documented that faces can be consciously distinguished from non-face stimuli by 170 ms, and ethnic in-group and out-group faces are differentiated by 250 ms (Ito et al. 2004). Distinctions between ethnicity and facial recognition can occur as quickly as 30–50 ms postpresentation in neural regions such as the amygdala and middle fusiform gyrus as well (Cunningham et al. 2004a, Kouider et al. 2009). The ability to process social information rapidly and with little other information provides individuals with a perceptual blueprint or schema that allows them to identify and seek out goal-relevant stimuli and avoid potentially dangerous stimuli in a remarkably efficient manner.

Attributional Processes

Attributional processes refer to an innate drive to explain and understand others' actions and behaviors as well as our own (Gilbert & Malone 1995). Attributional processes are highly contingent on the perspective of the perceiver, i.e., whether one is interpreting others' actions from their own perspective or from that of another (Storms 1973). The default mode is for individuals to evaluate others' actions from their own perspective, which typically leads to a fundamental attribution error, i.e., the tendency for people to attribute others' behaviors to trait characteristics and attribute their own behavior to situational factors (Gilbert & Malone 1995, Ross 1977). Individuals are capable of taking the perspective of others, however (e.g., "walk a mile in another's shoes"), which has formed the basis for the perspective-taking literature in social psychology and a widely investigated theory in the cognitive neuroscience literature termed theory of mind (TOM). TOM refers to the general ability to infer the thoughts and beliefs of one's self and others (Carrington & Bailey 2009). Although the mechanisms involved in TOM are still unknown, i.e., whether it occurs via inferring others' intentions and beliefs on the basis of observed behaviors (the "theory" theory; Gallese & Goldman 1998) or via inferring others' intentions and beliefs by simulating how the observer might feel in similar situations (simulation theory; Gallese & Goldman 1998, Ramnani & Miall 2004), general consensus indicates that TOM is the fundamental basis for social interactions (for a review, see Carrington & Bailey 2009). As a whole, attributional processes serve as the foundation for developing self-knowledge (e.g., via self-perceptual processes; Bem 1967) in relation to others in our world and rely heavily on executive resources, and subsequently PFC resources, in general (Saxe et al. 2006).

Social Categorization Processes

Another line of social cognitive research examines how categorization processes such as schema and stereotype activation affect individuals' perception and information processing. In regard to social cognition, schemas represent a cognitive framework for social categories and the associations among them (Fiske & Taylor 1991). Stereotypes, defined as a general belief one has toward different groups or cognitive objects (Allport 1954, Fiske 1998, Macrae et al. 1994), can be considered a specific type of schema. Although schemas and stereotypes make information processing dramatically more efficient (Macrae et al. 1994), they come with a price. For example, research finds that exposure to racial out-group members makes negative stereotypes immediately cognitively

accessible (termed automatic stereotype activation; Devine 1989). Once activated, stereotypes can be cognitively taxing to downregulate or suppress (C. Forbes and T. Schmader, under review; Richeson et al. 2003, Schmader et al. 2008), can bias nonverbal behaviors (Dovidio et al. 2002), and can negatively bias explicit perceptions toward out-group members behaving ambiguously (Rudman & Lee 2002), all seemingly unbeknownst to the perceiver. Thus although stereotypes facilitate the processing of information that is consistent with expectations, they also increase the incidence of judgment errors when information contradicts expectations and biases perceptions to a greater extent than the stereotyping individual assumes.

THE ROLE OF IMPLICIT AND EXPLICIT PROCESSES IN SOCIAL COGNITION

The field of social cognition has devoted much effort to understand better the differential effects of implicit and explicit processes on the aforementioned cognitive processes. Implicit processes unfold rapidly, require little cognitive effort, occur outside individuals' conscious awareness, and involve posterior cortical and subcortical regions of the brain (Amodio & Devine 2006, Cunningham & Zelazo 2007). In addition to face recognition and stereotype activation, other implicit processes include conditioned, evaluative associations between ideas or categories and stimuli that fit those categories (termed implicit attitudes; Fazio & Olson 2003, Greenwald & Banaji 1995) and self-serving biases. Conversely, explicit processes are deliberative, cognitively taxing, consciously accessible, and largely rely on the PFC (Amodio & Devine 2006, Cunningham & Zelazo 2007). Examples of these processes include deliberative evaluations of objects (termed explicit attitudes; Oskamp & Schultz 2005), introspective perceptions of self and others, and attributions.

Although these processes have historically been treated as distinct from one another, given ambiguities in behavioral findings and the nature of neuroanatomical connections (e.g., extensive reciprocal connections between PFC and subcortical regions in the brain; see **Figure 1**), recent perspectives argue that implicit and explicit processes may interact at all stages of cognitive processing (Cunningham & Johnson 2007, Devine & Sharp 2009; C. Forbes, C. Cox, T. Schmader, and L. Ryan, under review). Failure to identify a unitary, rapidly unfolding interaction between implicit and explicit processes could be due largely to methodological constraints and the nature of the issue being investigated (Devine & Sharp 2009). Pertaining to the nature of issues being assessed, implicit attitudes can be highly predictive of explicit attitudes, but implicit attitudes can also account for variance that is unexplained by explicit attitudes when researchers assess socially sensitive subjects such as race or gender (Cunningham et al. 2004b, Devine 1989, Greenwald et al. 2009, Nosek et al. 2002).

For instance, women often readily express dislike toward the math domain and can quickly pair negative words with math-related words on an implicit associations test (a common measure used to assess implicit attitudes; Nosek et al. 2002), suggesting they have a negative implicit attitude toward math as well (Nosek et al. 2002). When asking white individuals about their attitudes toward black individuals, however, although they explicitly report positive attitudes toward blacks, they often demonstrate a negative implicit attitude toward them on the implicit associations test (Greenwald et al. 2009, Greenwald & Banaji 1995). These negative implicit associations in turn can predict decreased math effort in women and negative nonverbal behaviors toward blacks during interracial interactions among other things (Dovidio et al. 2002; C. Forbes & T. Schmader, under review).

In the case of women's attitudes toward math, attitudes resulting from implicit and explicit processes appear to represent two ends on a continuum (i.e., faster to slower) that can be generated either rapidly or slowly and result in a unitary attitude. In the example of whites' attitudes toward blacks, attitudes would appear to result from two separate systems (i.e., faster versus slower) that engender two distinct

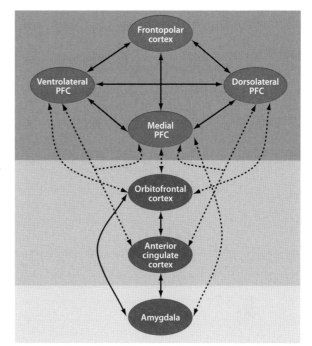

Explicit

Implicit

Figure 1

Diagram of direct neural connectivity between regions critical for implicit and explicit social cognitive and moral judgment processes. Solid bidirectional arrows denote direct reciprocal neural connectivity between two regions within a given processing system, i.e., implicit or explicit processing. Dashed bidirectional arrows represent direct reciprocal connectivity between two regions typically involved in either implicit or explicit processing. Blue circles denote neural regions that are typically involved in more implicit cognitive processing. Red circles signify neural regions that are typically involved in more explicit cognitive processing. These regions are not exclusively involved in implicit or explicit processing however. This conjecture is represented by the three shaded boxes and large arrow on the right. The lightest gray box represents neural regions, namely the amygdala here for the sake of simplicity, that are largely involved in implicit processes. Likewise, the darkest gray box highlights neural regions largely involved in explicit processes. The medium shaded box represents regions that are recruited during both implicit and explicit processing. PFC, prefrontal cortex.

attitudes. However, it is also possible that the latter scenario is reflective of an individual's ability to produce a rapid, visceral evaluation of a stimulus that can then be assessed within a given context for appropriateness and be down-regulated, suppressed, or altered accordingly.

This interpretation would suggest that the implicit attitude and explicit attitude are unitary, but situational demands (e.g., the desire to be politically correct in an interaction with a new acquaintance) necessitate alteration of the

fMRI: functional magnetic resonance imaging

overt expression of the explicit attitude, making them appear unique accordingly. In this instance, implicit and explicit processes likely interacted to engender the appearance of two distinct attitudes; however, given that these interactions can reach fruition within 500 ms, one can appreciate the difficulty inherent in examining them effectively [for detailed theoretical examinations of the conditions that may elicit implicit and explicit attitude overlap or differentiation, see Fazio & Olson (2003), Gawronski & Bodenhausen (2006), Petty et al. (2007), Wilson et al. (2000)]. Advances in EEG and functional magnetic resonance imaging (fMRI) methodologies, e.g., combining the exceptional temporal and spatial advantages of each respectively, could soon pave the way for successful assessments of these interactions.

The examination of implicit and explicit processes is not unique to social cognition. Given that visceral, emotional responses to stimuli can also be considered an implicit response, in recent years investigators have devoted much attention to understanding how emotional reactions can bias otherwise rational perceptions when people evaluate the appropriateness of others' behaviors. As you might expect, attributional products of implicit and explicit processes also play a prominent role in our next topic: the field of moral judgment.

A BRIEF PRIMER ON MORAL JUDGMENT

Moral judgments are broadly defined as evaluative judgments of the appropriateness of one's behavior within the context of socialized perceptions of right and wrong (Moll et al. 2005). **Table 1** provides an overview of traditional philosophical dilemmas as well as more recent, real-world examples that have been used to assess moral judgment processes in the literature. The study of morality and subsequent moral judgments has a long history in philosophy and more recently psychology, and as opposed to the fundamental top-down and bottom-up processes the field of social cognition usually tackles, moral judgments have long been thought

to rely solely on controlled, rational, and logical thought processes (Kohlberg 1969, Turiel 1983). Given the likely evolutionary origins of morality and the degree to which it permeates all facets of society and cognition from early childhood on, the idea that moral judgments are only a product of controlled cognitive processing has been convincingly challenged (Haidt 2001, Moll et al. 2005, Schulkin 2000).

More recent research on moral judgments has begun to incorporate the likelihood that moral judgments have an emotional component to them as well. On one end of the spectrum, some researchers have argued that moral judgments are largely direct products of intuitive or implicit emotional processes (Haidt 2001, Nichols 2002, van den Bos 2003). According to Haidt's social intuitionist model (Haidt 2001), moral behavior is predicated largely on implicit moral emotions such as guilt or compassion that compete and interact to guide morality outside of conscious awareness. These moral emotions were likely essential to our survival and evolution as a species and thus influence our perceptions and thoughts at all levels of cognitive processing in the form of "moral intuitions" (Moll et al. 2003). Controlled cognitive processes are likely to play a role in moral judgment only when situational demands necessitate them, e.g., situations that engender moral dilemmas (Moll et al. 2003). Although theories such as Haidt's rightfully identify a critical role for implicit emotional processes, they deemphasize the importance of explicit cognitive processing. As such, it is more difficult to explain findings demonstrating the importance of explicit processes, and the integral role that the PFC appears to play, in moral judgments overall (discussed below).

Dual Process and Interactionist Models of Moral Judgment

To bridge the gap between moral judgments and implicit and explicit processes, other researchers have incorporated the concept of a dual-process theory, positing that moral judgments can be derived via implicit, emotional

Table 1 A sample of traditional philosophically based (numbers 1 and 2; e.g., Greene et al. 2001, Koenigs et al. 2007) and more pragmatic (numbers 3, 4, and 5; Knutson et al. 2010) vignettes used to assess moral judgment processes

Type of moral judgment assessed	Typical examples	Moral judgment
1. Impersonal moral dilemmas	The Standard Trolley dilemma: You are at the wheel of a runaway trolley quickly approaching a fork in the tracks. On the tracks extending to the left is a group of five railway workmen. On the tracks extending to the right is a single railway workman. If you do nothing the trolley will proceed to the left, causing the deaths of the five workmen. The only way to avoid the deaths of these workmen is to hit a switch on your dashboard that will cause the trolley to proceed to the right, causing the death of the single workman.	Is it appropriate for you to hit the switch to avoid the deaths of the five workmen? Note: The decision to hit the switch is considered the utilitarian choice. The decison to not hit the switch is the non-utilitarian or deontological choice.
2. Personal moral dilemmas	The Crying Baby dilemma: Enemy soldiers have taken over your village. They have orders to kill all remaining civilians. You and some of your townspeople have sought refuge in the cellar of a large house. Outside you hear the voices of soldiers who have come to search the house for valuables. Your baby begins to cry loudly. You cover his mouth to block the sound. If you remove your hand from his mouth his crying will summon the attention of the soldiers who will kill you, your child, and the others hiding out in the cellar. To save yourself and the others you must smother your child to death.	Is it appropriate for you to smother your child to save yourself and the other townspeople?
3. Violations of social norms	As I was backing out of a parking lot, I bumped a parked car and left a minor dent. I did not even feel the impact when I hit the car, but it left a little bit of damage. I drove away without leaving a message or trying to contact the person.	Rating the vignette on dimensions of emotional intensity, emotional aversion, harm, self-benefit, other-benefit, premeditation, illegality, social norm violations, the extent to which other individuals were involved in the scenario, likelihood of event occurring in real life, personal familiarity, general familiarity, and moral appropriateness.
4. Social affective/ aversive situations	In the late nineties, I got my girlfriend pregnant. She told me she was pregnant and that she wanted me to be there for her. At first, I took her to her appointments and such, until I became nervous and ended the relationship.	Rating the vignette on dimensions of emotional intensity, emotional aversion, harm, self-benefit, other-benefit, premeditation, illegality, social norm violations, the extent to which other individuals were involved in the scenario, likelihood of event occurring in real life, personal familiarity, general familiarity, and moral appropriateness.
5. Intent involved in situations that engender self-benefit	I was taking a statistics class and the professor's instructions were very unclear. So, all the students in the class helped each other on homework assignments and tests. When it came time to take the test we would just give each other the answers.	Rating the vignette on dimensions of emotional intensity, emotional aversion, harm, self-benefit, other-benefit, premeditation, illegality, social norm violations, the extent to which other individuals were involved in the scenario, likelihood of event occurring in real life, personal familiarity, general familiarity, and moral appropriateness.

routes and explicit, controlled routes (Greene et al. 2001, 2004). Specific to moral dilemmas, Greene and colleagues have argued that emotional and controlled cognitive processes are key components of moral decisions involving utilitarian (e.g., choosing to sacrifice one to save many) and nonutilitarian choices (e.g., choosing to risk detection by enemy soldiers in lieu of smothering one's crying baby to death; see **Table 1**), and at times can play competitive roles (Greene et al. 2008). Whereas utilitarian judgments require controlled cognitive processing and are thus cognitively demanding, nonutilitarian or deontological judgments are implicit and predicated on emotional responses.

As opposed to implicit and explicit processes competing against each other, other researchers argue that the answer may lie somewhere in between (Moll & de Oliveira-Souza 2007). Moll et al. (2005) suggest that moral processes are products of the integration of social contextual knowledge, social semantic knowledge, and basic motivational and emotional drives. These three component representations interact to produce what Moll et al. term event–feature–emotion complexes (EFECs), which bind together via sequential, temporal, and third-party binding mechanisms and are influenced by one's situational and cultural context. Through these interactions, the EFEC framework makes specific hypotheses for moral judgments and moral emotions in different contexts. As discussed below, the model proposed by Moll et al. (2005) also makes specific predictions regarding patterns of neural activity in the PFC and subcortical regions engendered by situations that require moral judgments.

In sum, there is much debate regarding how moral judgments are cognitively derived and how they influence overt perceptions of others. Regardless of which theoretical conjecture is most plausible, moral judgments likely arise out of complex interactions between implicit and explicit processes. The field of cognitive neuroscience and the fundamental role the PFC plays in cognition can help shed light on the nature of social cognitive and moral judgment processes. Next we briefly describe the

structure and functions of the PFC that enable social cognition and moral judgments.

STRUCTURE AND FUNCTION OF THE PFC

The PFC can be parsed into dorsolateral (DLPFC), ventrolateral (VLPFC), dorsomedial (DMPFC), ventromedial (VMPFC), and orbitofrontal (OFC) regions. Whereas the VMPFC and OFC regions evolved from subcortical regions in the limbic system, the DLPFC likely evolved much later from motor regions such as the basal ganglia, the premotor cortex, and the supplementary motor area (Banyas 1999, Fuster 1997). Given that the motor areas of the cortex are thought to store motor programs, i.e., representations of well-learned mechanistic procedures, regions of the PFC that evolved more recently may be related to these evolutionarily older regions because they provide a representational basis for goal-directed action (Barbey et al. 2009, Wood & Grafman 2003). By examining the axonal projections distributed by and received by each major region of the PFC, we can learn more about the basic functions in which PFC regions are involved.

The medial and orbitofrontal regions of the PFC are hubs for integrating emotional, viscerally arousing information and relaying that information to the DLPFC (Fuster 1997). Specifically, medial and orbitofrontal regions of the PFC receive direct afferents from the amygdala, from most other limbic structures, from the striatum, and from temporal visual association areas. Medial and orbitofrontal regions also receive indirect afferents from the mesencephalic reticular formation and from the inferior temporal cortex [e.g., the fusiform face area (FFA)] via the magnocellular portion of the mediodorsal nucleus in the thalamus (Fuster 1997). Studies in a variety of animal species, including the monkey, indicate that the medial and orbitofrontal regions of the PFC, in turn, send a multitude of efferents to the DLPFC (Amaral & Price 1984, Ghashghaei & Barbas 2002, Ghashghaei et al. 2007), which suggests that the medial PFC regions (DMPFC, VMPFC, and OFC) are integral for monitoring individuals' internal states and motivations and for relaying that information to the DLPFC (Elliott & Deakin 2005, Wood & Grafman 2003).

The DLPFC is thought to be involved in the execution of movement and planned behaviors as well as the integration of sensory information (Barbey et al. 2009, Beauregard et al. 2001, MacDonald et al. 2000, Miller & Cohen 2001, Wood & Grafman 2003). This is particularly likely given that the DLPFC is reciprocally connected with the basal ganglia, the premotor cortex, the supplementary motor area, the cingulate cortex, and association areas and receives indirect afferents from the substantia nigra, the cerebellum, and the globus pallidus via mediodorsal and ventrolateral thalamic nuclei (Fuster 1997). Highlighting a fundamental yet pervasive role for the DLPFC in goal-directed behavior, the neural circuit that embodies the reciprocal connections between the DLPFC and anterior cingulate cortex (ACC) has been heavily implicated in the detection of and subsequent correction for behaviors that engender outcomes that differ from expectations (e.g., Cavanagh et al. 2009). Pyramidal cells in the DLPFC also exhibit the potential to fire over extended periods of time (Levy & Goldman-Rakic 2000) and across events (Bodner et al. 1996, Fuster & Alexander 1971), which suggests that the PFC is well suited for representing action and engaging in behaviors that allow humans to execute long-term goals (Barbey et al. 2009).

Overall, the PFC receives direct afferents from most brain regions, including the hypothalamus and hippocampus (at least in the rat, cat, and monkey), and is extensively interconnected with systems within the occipital, parietal, and temporal lobes that are involved in sensory processing for each modality (Fuster 1997). Often considered the "top" of a top-down hierarchically determined architecture, the PFC is critical for setting and achieving long-term goals (Koechlin et al. 2003) and for setting activation thresholds in nonfrontal

brain regions to detect goal-relevant stimuli. As such, in conjunction with subcortical regions, the PFC likely provides the key representations for implicit and explicit social cognitive and moral judgment processing. We next discuss the specific regions involved in said processes and findings supporting these conjectures.

ACC: anterior cingulate cortex

FPC: frontopolar cortex

PFC REGIONS CRITICAL FOR SOCIAL COGNITION AND MORAL JUDGMENTS

Many social cognitive processes are uniquely human and, as such, utilize distinct neural processes over and above those involved in memory, executive function, perception, and language in general (Adolphs 1999, Mitchell et al. 2006b, Van Overwalle 2009). Overall, implicit social cognitive and moral judgment processing typically involves the amygdala, insula, hypothalamus, ACC, OFC, and sensory cortex (Cunningham & Zelazo 2007, Moll et al. 2005). Conversely, slower explicit processing typically involves the VMPFC, the DLPFC, the VLPFC, and the anterior most portion of the PFC, termed the frontopolar cortex (FPC) (Cunningham & Zelazo 2007, Moll et al. 2005). Although these regions may be typically involved in implicit or explicit processing, they may not be exclusively involved in one form of processing or the other and, as noted above, likely interact at all levels of processing (see **Figure 1**). For instance, whereas regions such as the amygdala, ACC, and OFC may be typically recruited for implicit social cognitive processes, in the presence of contextual cues that prime different motivational or affective states, these regions may be recruited for, or remain particularly active during, explicit cognitive processing as well and exert influence over PFC regions accordingly (see the discussion of C. Forbes, C. Cox, T. Schmader, & L. Ryan, under review).

Activity in PFC regions associated with social cognition and moral judgment varies on the basis of the situational context and how stimuli are presented. For instance, studies have shown differential neural activity when individuals engage in self compared with other processing (Jenkins et al. 2008, Mitchell et al. 2006a, Ochsner et al. 2005), when they are presented with predictable or random patterns of stimuli (Dreher et al. 2002), and whether stimuli are presented for short or long durations (Cunningham et al. 2004a; C. Forbes, C. Cox, T. Schmader, & L. Ryan, under review). Using the primers on social cognition and moral judgments as guides, we review the literature linking brain regions with the social processes outlined in these sections.

The Role of the PFC in Social Perceptual Processes

The social neuroscience literature reveals that social perceptual processes are dynamic and sensitive to basic, even arbitrary distinctions among identifying features of others. Presenting individuals with subliminal faces of outgroup members elicits amygdala activation that can be regulated by the ACC, OFC, and DLPFC (Cunningham et al. 2004a; C. Forbes, C. Cox, T. Schmader, & L. Ryan, under review). Van Bavel and colleagues (2008) demonstrated that this typical neural response can be situationally manipulated on the basis of arbitrary social distinctions as well.

Utilizing the minimal group paradigm, a process known to establish novel in-groups and out-groups effectively using arbitrary information (e.g., tossing a coin or selecting a certain painting over another; Tajfel 1982), Van Bavel et al. (2008) randomly assigned white subjects to different arbitrary, mixed-race teams under the assumption that their team would be competing against another team later in the experiment. Subjects were first presented with the supposed faces of their team and the other team (which consisted of equal numbers of white and black faces), were asked to encode them, and then were presented with the faces later during a task that asked them to categorize faces by ethnicity or team membership. After the experiment, subjects provided likeability ratings for the different faces they were presented with throughout the task. Behavioral findings

replicated the basic minimal-group effect: Subjects reported liking members of their team, regardless of ethnicity, more than members of the other team. The neuroimaging results indicated that compared with out-group faces, exposure to in-group faces engendered greater activity in the amygdala, FFA, OFC, and dorsal striatum. Furthermore, activity in the OFC mediated the biased, in-group liking ratings. These findings suggest that the neural networks underlying social perceptual processes are quite sensitive to the malleability of social group membership and self-categorization. A situational shift in self-categorization can alter the way neural networks process and attend to social stimuli, such as an out-group face compared with a teammate's, in spite of otherwise well-learned negative associations linked to out-group members.

The Role of the PFC in Attributional Processes

An abundance of findings in the TOM literature suggests that specific social cognitive neural networks are involved when individuals engage in attributional processes in general (for a recent meta-analysis, see Van Overwalle 2009). Indeed, TOM would not be possible without large contributions from adequately functioning PFC regions and the consistent finding that individuals with autism have particular difficulties with TOM tasks is particularly persuasive. For instance, Castelli and colleagues (2002) asked adults with autism or Asperger syndrome and normal controls to watch animated sequences of triangles engaging in various movements, some of which imply intent. Subjects were asked to describe what was happening in each animation, and sequences that depicted some kind of intent among the triangles led normal controls to "mentalize," or make inferences that implied the triangles had mental states. During this task, autistic subjects demonstrated a lack of mentalizing compared with controls during animated sequences that implied intent; autistic subjects also demonstrated less activity in the medial PFC, the temporoparietal junction, and the temporal poles than did controls.

In addition to findings demonstrating abnormal functioning in autistic individuals' medial PFC during self-related and mentalizing tasks in general (Gilbert et al. 2009), these findings suggest the medial PFC plays a critical role in allowing individuals to identify themselves as unique agents in a social world that, in turn, allows them to infer intent in their behavior and in others'.

The VMPFC seems to be particularly important for perspective taking, TOM, and self and other processing. Research from patient studies supports this conjecture. In addition to the tamping-iron-through-the-medial-PFC-induced curmudgeonry incurred by Phineas Gage, individuals with VMPFC lesions and frontotemporal dementia are known to have particular difficulties with mental tasks such as inferring a person's psychological state by interpreting the directionality and expression of their eyes (i.e., the reading the mind in the eyes test; Baron-Cohen et al. 2001) and faux paus detection (Gregory et al. 2002). Patients with VMPFC as well as OFC lesions are also notorious for inappropriate or irrational social behaviors (Barrash et al. 2000, Beer et al. 2006, Grafman et al. 1996, Koenigs & Tranel 2007). Investigators have also found increased activity in the medial PFC when individuals engage in tasks that are introspective in nature, which suggests that the medial PFC may be a hub for self, other, and self-in-relation-to-other processing in general (Amodio & Frith 2006; Gusnard et al. 2001; Johnson et al. 2002, 2006).

The Role of the PFC in Social Categorization Processes

The medial PFC also plays a critical role in social categorization processes. Individuals with VMPFC lesions have exhibited reduced implicit stereotyping compared with normal controls and those with lesions in the DLPFC (Milne & Grafman 2001), which suggests that this region is important for the representation of social schemas and stereotypes. In light of the functional connectivity between the VMPFC and the DLPFC, stereotype primes likely engender increased activity in the VMPFC, which then relays information to the DLPFC. Not surprisingly then, many studies suggest the DLPFC is critical for regulating or suppressing stereotype activation in general (Knutson et al. 2007, Payne 2005, Richeson et al. 2003), and studies of patients with primarily VLPFC or VMPFC lesions (Gozzi et al. 2009) support this idea. The latter study also demonstrated that another important component in implicit stereotyping is conceptual social knowledge and that anterior temporal lobe lesions, particularly in the right hemisphere, can compromise this form of social representation.

Highlighting the dynamic flexibility of the neural substrates that underlie these social phenomena, including those involved in implicit and explicit processing, C. Forbes, C. Cox, T. Schmader, & L. Ryan (under review) investigated the effects of priming negative in-group and out-group stereotypes on individuals' motivation to regulate stereotype activation. In this study, white subjects who reported being explicitly nonprejudiced and motivated to remain so were presented with subliminal (30 ms) and supraliminal (525 ms) black and white faces. To prime negative in-group and out-group stereotypes, either a violent death metal or violent rap song, i.e., stimuli that prime negative stereotypes for whites or blacks respectively, was played in the background while subjects were exposed to the novel faces. Results revealed that when negative white stereotypes were primed, the typical amygdala response to subliminal black faces was not evident. When negative stereotypes of blacks were primed, however, amygdala activity was elicited in response to black faces at implicit processing speeds that persisted into explicit processing speeds, i.e. the amygdala response was evident in response to black faces presented at both 30 ms and 525 ms.

Furthermore, functional connectivity analyses revealed that the increased amygdala response to subliminal black faces covaried with increases in OFC and DLPFC activity among other regions. The amygdala response to supraliminal black faces covaried with decreases in ACC activity and again with increases in DLPFC activity, which suggested that the negative stereotype activation engendered increased processing demands in the DLPFC specifically at implicit and explicit processing speeds. These findings indicate that neural regions involved in implicit and explicit processing may interact at fast and slow cognitive processing speeds and that situational primes can alter how neural networks involved in social cognition perceive and react to social stimuli.

The PFC's Role in Moral Judgment Processes

The basic components of social cognitive and moral judgment processing appear similar, and neural regions involved in these types of processes reflect this. Many key neural regions involved in social cognition overlap with those involved in moral judgment, including the medial PFC and DLPFC. For instance, Greene and colleagues (Greene et al. 2001, Greene & Haidt 2002) suggest that when individuals consider personal moral dilemmas and/or make non-utilitarian judgments, areas involved in emotion processing are likely to exhibit increased activity, and this effect would be mediated by the medial PFC. Conversely, considering impersonal moral dilemmas and/or making utilitarian judgments are more likely to elicit increased activation in regions associated with cognitive control processes and the DLPFC specifically. Again, according to Greene and colleagues, these dual routes to decision making may often compete or create tension.

To test this hypothesis, Greene et al. (2004) asked subjects to read a series of moral dilemmas, ranging from impersonal (e.g., flipping a trolley track switch to save the lives of many compared to one) to personal (e.g., smothering your crying baby to death to avoid being detected by oppositional soldiers who will kill you and others) and indicate which alternative they would choose. Results revealed that personal moral judgments elicited activity in the medial PFC among other regions, replicating previous findings (Greene et al. 2001). In contrast, impersonal moral judgments elicited activity in the DLPFC, which suggested that individuals engaged in more effortful cognitive, as opposed to emotional, processing in analyzing these moral dilemmas. Increased activity in the DLPFC was also evident when individuals made utilitarian choices on personal moral dilemmas (e.g., indicating they would smother their baby to death to save the lives of many).

One set of findings inconsistent with Greene's dual process theory is that individuals with VMPFC lesions make more emotional choices in an ultimatum game and more utilitarian moral judgments in general compared with normal controls (Koenigs et al. 2007, Koenigs & Tranel 2007, Moll & de Oliveira-Souza 2007). Other important factors to consider when contrasting personal versus impersonal moral decisions include the frequency of exposure to the described scenario. Personal moral decision making as well as more real-world or pragmatic scenarios would be more likely to recruit familiar analogous personal memories, whereas impersonal moral decision making and those dilemmas grounded in traditional philosophy would be less likely to do so on the basis of frequency of experience alone. Controlling for these and other factors that could affect scenario processing is critical, and Knutson and colleagues have now established a normative database of brief, real-world moral scenarios for this purpose (Knutson et al. 2010).

Thus, Moll and colleagues argue for the EFEC framework described above and posit that a network of closely interconnected neural regions is responsible for integrating the diverse functions involved in moral appraisals and judgments. In addition to the VMPFC and DLPFC, other critical regions are the FPC, the anterior temporal cortex, the superior temporal sulcus region, and the limbic structures including the amygdala, the angular gyrus, and the posterior cingulate (Moll et al. 2005, Moll & de Oliveira-Souza 2007, Raine & Yang 2006). Moll and colleagues' EFEC framework (Moll et al. 2005, Moll & de Oliveira-Souza 2007) makes specific hypotheses regarding the neural regions involved in moral judgment, emotions, and values.

Specific to the PFC, the EFEC framework predicts that whereas the anterior PFC is integral for enabling humans to assess the possible long-term consequences of their behavior with respect to others, the DLPFC is integral for predicting outcomes of one's behavior in novel contexts. Consistent with the hypothesized role of the VMPFC in the representation of social knowledge, Moll et al. (2005) argue that the VMPFC plays an important role in allowing one to adhere to social norms and cultural values that individuals derive through the socialization process. Finally, the OFC is likely necessary for comparing social cues in a given context with one's preexisting representations of social knowledge to help one determine appropriate behaviors within a given context. Consistent with one theme of this review, the nature of the presenting stimuli and/or that which individuals are asked to deliberate can differentially activate regions involved in more implicit emotional processing (i.e., the limbic system) or slow explicit processing (e.g., the DLPFC) depending on the context. Although a host of evidence from lesion and experimental studies supports these conjectures, to date no investigators have directly compared the various models of moral judgment and processing in social neuroscience experiments.

Overall, the medial and lateral PFC as well as the ACC and OFC are clearly necessary for social cognitive and moral judgment processing (**Figure 2**). How these regions interact to facilitate everything from basic social cognitive processes to more complex processes such as moral judgment and in what capacity are still questions that are much debated in the literature. Next, we outline a theory that attempts to answer these questions by providing a framework for the types of information stored in the different regions of the PFC and how they may interact to enable social cognitive and moral judgment processing in general.

STRUCTURED EVENT COMPLEX THEORY DESCRIBES THE PFC'S ROLE IN SOCIAL COGNITION AND MORAL JUDGMENT

As mentioned above, on the basis of the functional connectivity between different PFC and subcortical regions and evolutionary and neurophysiological evidence, the primary role of the PFC is in the representation of action and guidance of behavior (Barbey et al. 2009). Any given behavior can be broken down into a series of recognizable events, which are semantic in nature and of a fixed temporal duration

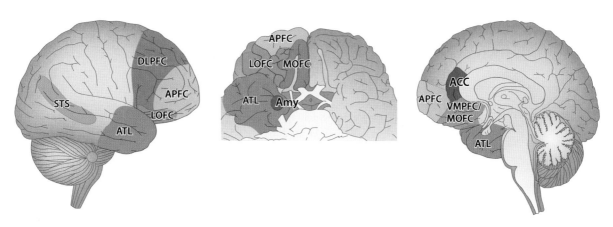

Figure 2

Neural regions identified as critical for social cognitive and moral judgment processing. Neural regions include the anterior prefrontal cortex (APFC), dorsolateral prefrontal cortex (DLPFC), medial and lateral orbitofrontal cortex (MOFC and LOFC), ventromedial regions of the prefrontal cortex (VMPFC), anterior cingulate cortex (ACC), anterior temporal lobes (ATL), amygdala (Amy), and the superior temporal sulcus (STS) region. Modified, with permission, from Moll et al. (2005), figure 1.

SEC: structured event complex

(Barbey et al. 2009, Zacks & Tversky 2001). In a given situation, a series of events can be primed and linked together to form a script that guides behavior and allows one to predict how the situation will unfold. The linking of events can, in turn, represent a set of goal-oriented events, one that is sequentially ordered and composed of social norms that guide behavior and perceptions. We refer to this goal-oriented set of events as a structured event complex (SEC) (Barbey et al. 2009, Grafman 2002, Wood & Grafman 2003).

Components of SECs can be semantically independent, but they are encoded and retrieved as an episode using simulation mechanisms or feature maps (Barsalou et al. 2003a,b; Damasio 1989). SECs provide goal-directed actions with semantic and temporal structure and are activated or primed by environmental cues, arming the organism with ammunition to predict how different social scenarios will unfurl. SECs ultimately represent myriad bits of knowledge that can be organized in predictable or unique manners to allow for increasingly complex behaviors and predictions. The type of knowledge a given SEC contains and the format of the behaviors they facilitate can be localized to specific neural regions.

SECs Are Composed of Multiple Dimensions

According to the SEC framework, SECs have multiple dimensions to them, including predictability, complexity, and category specificity, and the nature of the binding process is predicated on the hemisphere in which binding occurs (for a review, see Barbey et al. 2009). Overall, whereas the left PFC is hypothesized to integrate meaning and features between single adjacent events, the right PFC integrates meaning and information across events. In terms of degree of predictability, the medial PFC stores predictable SECs, or those SECs that are engrained in individuals and have structured, familiar goals and behaviors associated with them. Predictable SECs can be thought of as a schema or stereotype in general; e.g., they represent

how different events such as going to a party or a lecture typically unfold, but they are tailored toward the individual's goals. For instance, when going to a party, an introvert is likely to have different goals than an extrovert, and subsequently their SEC for attending a party, while similar, will uniquely vary on the basis of their individual goals for the evening (e.g., stand in a corner like a wall flower versus meet as many new people as possible).

Conversely, the lateral PFC has evolved to store adaptive SECs, which are more flexible in nature and allow for adaptations to unique or ambiguous situations. For instance, when meeting someone new, an individual is likely to activate SECs on the basis of the person's appearance, which allow an individual to predict which behaviors are required for a successful interaction; however, unexpected feedback will stimulate different SECs that, in turn, will update predictions, goals, and ultimately behavior. This conjecture is supported by past research indicating that the VLPFC is particularly active when individuals experience attitude ambivalence, i.e., a situational cue primes both positive and negative information, thus requiring the individual to resolve the ambiguity in a novel manner (Cunningham et al. 2004b).

SECs vary substantially in complexity as well. Given its proximity to phylogenetically older regions of the brain, the posterior PFC stores simple, well-learned SECs that consist of minimal information about event sequences (e.g., see Kruger et al. 2009a,b). This is likely where basic social cognitive SECs, such as neural networks sensitive to different facial expressions or body movements, are stored, which, when activated by a situational cue, in turn activate more complex SECs in the anterior PFC. The most anterior portions of the PFC store the most complex SECs, including long-term goals and integration of multistage event complexes, which is likely why the FPC region is so heavily involved in complex moral judgments (Berthoz et al. 2002; Moll et al. 2001, 2002, 2005).

The VMPFC and DLPFC regions enable categorical specificity in SECs. Specifically, the VMPFC stores SECs specific to social norms and scripts. In addition to Milne & Grafman's (2001) study demonstrating less stereotype activation by individuals with ventral PFC lesions (Milne & Grafman 2001), fMRI studies have demonstrated that stereotype activation is likely to engender increased activity in the VMPFC specifically (Knutson et al. 2007, Quadflieg et al. 2009). Violations of social norms also elicit activity in the VMPFC (Berthoz et al. 2002), which suggests that this region is upregulated for the purpose of retrieving stored social norms when contextual cues necessitate comparisons between others' behaviors and known norms. Together these findings support the notion that the VMPFC stores SECs specific to general beliefs about various social groups and social norms, which are necessary for one to navigate the social world in an efficient yet moral manner.

Not unlike other theories, the SEC framework posits that the DLPFC is involved in planning and action. The SEC framework specifically hypothesizes that the DLPFC stores event sequences that represent the planning and action necessary to achieve a primed goal state. Based on situational cues, overarching goals dictate how components from SECs activated in other brain regions are integrated and organized to formulate an action plan with a desirable outcome.

One way to assess the mechanisms underlying these assertions would be to examine individual differences between biased and nonbiased individuals interacting with an out-group member. For instance, we might expect interactions with an out-group member to elicit activation in the medial PFC in general, representing stereotypic activation associated with the out-group. We would also expect, however, for DLPFC activity to vary on the basis of an individual's goal to be nonbiased toward the out-group member, by which increased DLPFC activity could represent the manipulation and integration of SECs to allow the individual to actualize an egalitarian goal. One study by Rilling and colleagues (Rilling et al. 2008) provides a format to investigate these questions.

In one study, Rilling et al. (2008) randomly assigned subjects to a red or black team (purportedly based on results from a personality test) and asked them to complete a prisoner's dilemma task with a supposed in-group and out-group partner while brain activity was assessed via fMRI. Participants were classified as discriminators or nondiscriminators post-hoc on the basis of whether they reported feeling different when interacting with the in-group partner compared with the out-group partner. Results indicated that when playing with arbitrarily defined out-group members, both nondiscriminators and discriminators demonstrated increased activity in the medial PFC, suggesting they were activating SECs associated with social norms and perhaps representations of other known out-group members in hopes of predicting their partner's behaviors.

Activity in the DLPFC, however, was modulated by feelings toward the out-group partner. Nondiscriminators elicited greater activity in the DLPFC compared with discriminators. Given that typical minimal group paradigms such as these engender immediate in-group bias and disliking for out-group members (particularly during competitions), increased DLPFC activity in reported nondiscriminators could represent their attempts to organize activated SECs from the medial PFC in a manner consistent with an overarching egalitarian goal. These findings support the conjecture that a given situational cue activates stereotype-related SECs in the medial PFC, but the DLPFC monitors and temporally organizes these SECs within the context of meta-goal states, such as the desire to behave in a morally just way toward others to regulate behavior accordingly.

The SEC framework also provides a means for understanding how different PFC regions contribute to implicit and explicit processes and how these processes differentially affect PFC neural networks. Consistent with the role neural regions along the midline play in implicit processing in general (e.g., the amygdala and the ACC), the medial PFC likely contributes to implicit processes because this region stores predictable SECs associated with habituated sequences, schemas, and stereotypes (e.g., one cannot have stereotype activation without the stereotype). Likewise, posterior regions of the PFC are likely involved in implicit processing to the extent they store basic features associated with others and others' intentions, which are automatically activated upon encounter. Conversely, the lateral and anterior portions of the PFC are more involved with explicit processing, given that these regions store adaptive and complex SECs involved in explicit planning, action, and detailed sequences.

Any given behavior enacted in a specific context would necessarily involve an interaction between the various SECs that contribute to implicit and explicit processing (see **Table 2**). For example, let's say you run into a friend of a friend at a bar whom you had met a few times before and this second-degree friend is behaving strangely. This behavior prompts you to engage in attributional processing to try and explain your acquaintance's behavior. To do this, simple SECs associated with the individual's features and context, predictable SECs associated with how the individual behaved in the past, and adaptive and complex SECs accounting for the unique situation would all be activated. Your motivation to understand or explain your acquaintance's behavior would in turn dictate whether conclusions would be based more on implicit or on explicit processes. If you do not necessarily like the person, your explicit attributions for their behavior are likely to be more influenced by predictable SECs that reflect your lack of motivation to understand their behavior and your disliking of them in general (e.g., "this person is weird just like I thought"). If the person is someone you want to like, adaptive and complex SECs will dictate explicit attributions as you are motivated to find specific contextual factors that may be influencing their behavior (e.g., "maybe they had a rough day and one drink too many").

Thus the lateral PFC can regulate the medial PFC, i.e. adaptive SECs will be utilized and predictable SECs will be inhibited or restructured within the overarching SEC accordingly, when habituated sequences are not appropri-

ate or desired in light of contextual cues. When motivation is lacking, contributions of the medial PFC to implicit processing will have a greater influence on explicit perceptions, and predictable SECs will be utilized in lieu of adaptive SECs. In both instances, however, the construction of the overarching SEC would be the product of the interaction between multiple SECs varying in complexity associated with contextual primes, norms, values, and current goal states or plans of action.

Overall, the SEC framework provides a rationale for the myriad cognitive processes involved in social cognition and moral judgment, from the heuristic and efficient to the dynamically flexible and cognitively demanding. It also highlights the pivotal role that different neural regions of the PFC play in those processes and the necessity for these neural regions to interact at multiple speeds of cognitive processing (see **Table 2** for a mapping of social processes on to brain regions and SEC components). Together, utilizing knowledge of the phylogenetically hierarchical structure of the PFC in conjunction with physiological properties of the PFC, the SEC framework provides a comprehensive view of how a given social context can activate and integrate SECs throughout the PFC that enable individuals to assess their situations and make predictions that satisfy personal goals in socially and morally appropriate ways either extremely quickly or more deliberately.

SUMMARY AND CONCLUSIONS

The PFC is vital for human social cognitive and moral judgment processing. Because more anterior regions of the PFC serve as the last, integrative stop for all facets of perceptual and emotional processing, regions such as the OFC, the VMPFC, and the DLPFC are likely critical for evaluating current motivational and emotional states and situational cues and for integrating this information within the context of current goal states and past experience. Through interactions among these regions, possibly at both implicit and explicit levels, humans can build immediate impressions of others, infer what

Table 2 A summary of different SECs involved in social cognitive and moral judgment processes, key PFC regions involved, and the tendency for a given process to occur implicitly (fast) or explicitly (deliberative). BA, Brodmann's area; PT, perspective taking; SEC, structured event complex; TOM, theory of mind; labels following a given BA region denote whether fMRI studies (F), lesion studies (L), or transcranial magnetic stimulation (T) provided evidence for that brain region's involvement in the respective social process

	Underlying information processing components	Key PFC regions involved	Implicit or explicit process?
Social perceptual processes			
Other individual identification	Single-event processing: identify basic social features of individual	BA 9 (F), BA 10 (F), BA 11 (F), BA 47 (F)	Implicit
Social mimicry	Single-event processing: sequential dependencies based on partner behaviors that vary by single adjacent event	BA 6 (F), BA 44 (F)	Implicit
Movement intentions	Single-event processing: identify intent and meaning of basic biological movements	BA 9 (F), BA 11 (F)	Implicit
Attributional processes			
Self-related processing	Predictable SECs associated with self-concept; occasional use of adaptive SECs when behaviors contradict perceived self-concept	BA 9 (F), BA 11 (F, L), BA 12 (F, L), BA 32 (F)	Both
Other processing/ PT/TOM	Predictable SECs associated with knowledge of others; adaptive SECs to account for ambiguous behaviors; social category–specific SECs bias perceptions based on learned associations of others' group memberships	BA 6 (F), BA 9 (F, L), BA 11 (F, L), BA 12 (F, L), BA 32 (F), BA 46 (T)	Both
Moral judgments	Predictable SECs associated with self-perceived appropriate behaviors; adaptive SECs are applied to morally ambigous situations; social and nonsocial category–specific SECs consisting of socially and personally acceptable beliefs and norms	BA 9 (F), BA 10 (F), BA 11 (F, L), BA 12 (F, L), BA 32 (F), BA 46 (F), BA 47 (F)	Both
Social categorization processes			
Stereotypes/schemas/ scripts	Predictable SECs representing well-learned beliefs toward others and normal courses of behavior in general; social category–specific SECs representing well-learned beliefs, attitudes, and social norms	BA 9 (F), BA 11 (F, L), BA 12 (F, L), BA 46 (F, L), BA 47 (F)	Implicit
Impression formation	Predictable SECs involved with estimating others' intentions based on limited identifying information; social category–specific SECs associated with others' perceived social categorical information	BA 9 (F), BA 11 (F), BA 12 (F)	Implicit

(Continued)

Table 2 (*Continued*)

	Underlying information processing components	Key PFC regions involved	Implicit or explicit process?
Predictive processes			
Future planning	Long duration SECs involved in planning and action; predictable SECs associated with self-concept	BA 9 (F), BA 10 (F), BA 11 (F), BA 12 (F), BA 46 (F, L)	Explicit
Strategies in novel contexts	Nonsocial category–specific SECs involved in predicting behavior in novel context and planning and action	BA 46 (F), BA 47 (F)	Explicit
Strategies in learned contexts	Social category–specific SECs involved in well-learned social rules and scripts	BA 9 (F), BA 11 (F), BA 12 (F), BA 47 (F)	Implicit

others are thinking, and plan actions that are likely to facilitate a successful interaction with others, all within the context of social norms. These basic behaviors and impressions, in turn, form the foundation for more complex cognitive processes such as evaluating another's behavior in relation to culturally ascribed rules or to function successfully within large social groups.

The SEC framework provides a means to understand how regions within the PFC interact to enable social cognitive and moral judgment processing. Although outside the realm of this article, in light of our understanding of the functional connectivity between the PFC and most other subcortical neural regions, SECs underlying social cognitive and moral judgment processing likely integrate emotional and reward-related responses to contextual stimuli and basic perceptual processes (Wood & Grafman 2003). These SECs, in turn, are influenced by the socialization process and individual differences. The complex interaction between these processes can occur at multiple processing speeds to facilitate sophisticated implicit and explicit social cognition and moral judgment. Moll et al. (2005) posit in their EFEC framework that complex behaviors such as moral judgment likely involve the integration of SECs that represent social norms, basic perceptual features in one's environment such as facial expressions, visceral emotional responses to stimuli, and references to past experience

via autobiographical memory reconstruction, as well as SECs that represent meta or long-term goal states among other things. Such integration would require interplay among neural regions such as the anterior PFC, the DLPFC, the VMPFC, the OFC, the superior temporal sulcus, the anterior temporal lobe, and the limbic system, including the hypothalamus, the septal area, and the amygdala in a matter of milliseconds (**Figure 2**). In line with evolutionary perspectives, the order in which the aforementioned list of critical neural regions are listed may also represent a hierarchical structure that allows for increasingly complex social and moral behaviors.

In attempts to assess such complex neural and social interactions in a scientifically valid manner, social neuroscience experiments have begun to include an increasing array of tasks and problems reflecting the varied social experiences of real life. Social processes are not simply another brain activity worthy of attention but are key brain processes determining such things as outcomes after brain injury, the trajectory of human evolution, and the modulation of human impulses. This review has emphasized the PFC's important role in these behaviors. Whereas some brain regions, such as the ventral axis structures from the brain stem to cortex concerned with reward or limbic structures concerned with emotion and attachment, have been understandably linked to social behaviors, other brain areas, such as regions within the parietal cortex, which have been implicated (primarily via functional neuroimaging) in social behavior, have not been firmly established yet as being crucial to the examined social process. Although it has become common to attribute social processes to certain regions within the PFC, there is less certainty about how to conceptualize the spatial topography to understand better why certain social processes lie near each other in these brain areas and what might the underlying computational processes be that support these social (and presumably other kinds of cognitive) processes.

The PFC has evolved to represent the most complex aspects of knowledge and information processing. Recent research has shown that some areas within the PFC are part of an increasingly hierarchical system for processing information from single events to the linkage of sets of events. More dorsal regions of the PFC are more likely to process information pertaining to an agent's actions toward external stimuli (e.g., agents acting upon objects), whereas more ventral regions appear more likely to assess the relevance of information to the social agent (e.g., reward value attributed to external stimuli). Hemispheric asymmetries in information processing are also apparent; the left hemisphere codes a dominant meaning or characterization of stimuli, whereas the right hemisphere codes multiple concepts in parallel without committing to one concept or meaning. Although less efficient than left hemisphere processing, particularly when a single solution is optimal, such processing would be superior when more than one solution to a social problem is possible.

This very brief speculative characterization of the PFC's functional roles applies to all kinds of functional domains including the moral and social processes reviewed above. The fact that some of the most prominent deficits that occur following damage to the PFC are social-cognitive suggests that at least certain aspects of social behavior are tightly coupled to the processing constraints of the PFC. In recent reviews, our colleagues offer some examples of how the schema we sketched above is related to specific social and cognitive processing deficits in patients with lesions to specific areas of the PFC as well as which prefrontal cortical regions might be activated depending on specific processing demands (e.g., Barbey et al. 2009, Krueger et al. 2009a, Moll et al. 2005, Wood & Grafman 2003).

Computational forces within the PFC appear to allow for at least two hierarchical mechanisms to operate. One mechanism involves representational complexity, which links a deep search tree with parallel searches occurring simultaneously. The other mechanism involves temporal coding across events that merges apparently separable events into a single engram, allowing for streamlined forecasting and memory retrieval. These two mechanisms enable more detailed and elaborated conceptions of social behavior to be represented and utilized than would be available from a single event embedded in a stream of events. Although humans may benefit from the development of such elaborated conceptions in our behavior, constraints on resource utilization may bias us to rely more on attitudes, heuristics, and simplified personal vignettes when making decisions about beliefs.

If important aspects of social beliefs are represented in memory in the PFC, how similar is that representation to that which is seen in other forms of representational memory such as semantic memory concerned with the meaning of words and objects? Simpler forms of social representation, such as attitudes, may obey principles similar to that of semantic representations, including being sensitive to frequency of exposure, influenced by context, etc. Little evidence indicates whether the same constraints would apply to more complex beliefs such as religious or political beliefs (e.g., see Kapogiannis et al. 2009, Zamboni et al. 2009). Frequency of event exposure affects the activation site within the medial PFC as does complexity of event information (Krueger et al. 2009a,b). Less frequently exposed information is associated with the complexity of representation and activation of the FPC. This association suggests that more frequently exposed information can lead to sparser cue representation because the behavioral action sequence would be more predictable and procedurally rigid (also predicting storage sites more posterior within the frontal lobes than FPC), whereas less frequent information would require more complex cuing and deeper representational search and deliberation because outcomes would be less predictable, given less direct personal experience with the event sequence.

This level of complex representation should be slower to process and enact than other forms of representation such as object naming or even social concepts. What kind of advantage does it offer? This form of complex representation would offer an advantage because it could inhibit more impulsive behavior and have a supervening role in decision making. In particular, it would automatically activate potential or certain consequences in the future of an immediate action. Such foresight allows the individual to implement strategic action that is potentially costly in the present but advantageous in the long run and can overcome potential liabilities in physical capabilities or other attributes. In a brain that is built with inherent inhibitory pathways between competitive brain regions, a brain region that was composed of SECs would have the mechanisms to inhibit simple associative behaviors using temporal information encapsulated within engrams stored in the PFC. Compared with other recent evolutionary changes, this particular change would offer substantial advantages in many situations—both social and nonsocial—and could stimulate the evolution or development of other functional brain properties. For example, if our representational memories captured and integrated social information over long time durations, that evolutionary change may have stimulated the development of language content that conveyed the social consequences (e.g., allowing the verbal expression of foresight) associated with one's current actions and motives, thereby facilitating the expansion of language beyond its use for simple object identity and naming.

To make these ideas more concrete, let us take an example of one social event. You are with politically informed friends and are discussing whom you are going to vote for in the next presidential election. How might your cognitive and neural processes support that social interaction? Because the discussion is within a known social group and it is likely that at least one other person in the group will choose the same candidate as you will, your brain would activate regions within the PFC that are concerned with reading others' intentions, bonding, the heuristics of voting for a party's candidate (if favored, your reward system would also be active), narrative discussion of the candidates' qualifications versus his or her opponent, as well as the overall context of the election. Equal emphasis in information processing is not paid to all aspects of this scenario at any one time, so regions concerned with each of the above social processes are likely to be differentially activated at any one point in time.

Most of the social processes described above are explicit, but other implicit social processes may also be engaged. If you are much more familiar with your own candidate's background, you may be likely to activate stereotypes and biases about his opponent (who may be classified as being a candidate from an out-group, from your perspective). Expectation of how the conversation will go allows priming of future narratives to occur and also taxes cognitive structures concerned with foresight and action planning. Similar conversations with less informed people could wind up taxing more primitive and implicit cognitive and social representations and typically evoke more emotional, and less rationale, discourse. The dynamics of such discussions cause many regions and social processes to be simultaneously primed in preparation for retrieval of information (relevant or irrelevant).

It is difficult to capture the above scenario within a laboratory setting, and even compartmentalizing such a dynamic situation is challenging. Investigators have tended to isolate the major social process contributors to simple responses with various kinds of probes. What differentiates us from other species that have their own social structures is the human ability to use solutions and ideas that do not depend on the surface features of the subjects (e.g., size, attractiveness, progeny) nor their physical capabilities (e.g., strength, aggressive tendency). Despite these differences, under certain circumstances with unknown individuals, the selection of the person who is judged to be the most popular or whom you might vote for often corresponds with the physical features of the face, and a significant association exists between those attractive features and the candidate's likelihood to win an election (Spezio et al. 2008, Todorov et al. 2005). So although human social complexity has distanced us from related species, e.g., chimpanzees or gorillas, they are not out of view.

In this article, we have emphasized functional neuroanatomy and have not discussed the chemical and genetic features of human social behavior. Quantifying the anatomic, chemical, genetic, and behavioral components of social behavior in combination will allow researchers to account for as much variance in behavior as is possible in open societies. Social neuroscience will play a key role in this effort.

The social interactions and judgments of humans have been based on evolutionary pressures and environmental and social contingencies. They will continue to evolve in parallel with technological changes. The widespread use of devices that can provide almost instant information (e.g., face recognition and identification) and feedback (e.g., eliciting pleasurable sensations via individually tailored visual stimuli on Web sites) as well as the emergence of social networks based simply on user-provided input will change the way we interact with others and the way the brain evolves or devolves in the future. In addition, introducing such sophisticated technology at a young age may affect development and enhance brain systems concerned with more immediate results and gratification. Public discussion of these issues will become more important as we judiciously manage the benefits of new technology in balance with its effects on social behavior and on the development of the social brain.

DISCLOSURE STATEMENT

The authors are not aware of any affiliations, memberships, funding, or financial holdings that might be perceived as affecting the objectivity of this review.

ACKNOWLEDGMENTS

This work was supported by the Imaging Sciences Training Program, Radiology and Imaging Sciences, Clinical Center, and the National Institute of Biomedical Imaging and Bioengineering and the Intramural Research Program of the National Institute of Neurological Disorders and Stroke, National Institutes of Health. The authors thank Joshua Poore and an anonymous reviewer for helpful comments on earlier versions of this manuscript.

LITERATURE CITED

Adolphs R. 1999. Social cognition and the human brain. *Trends Cogn. Sci.* 3:469–79

Allport GW. 1954. *The Nature of Prejudice*. Oxford, UK: Addison-Wesley. xviii, 537 pp.

Amaral DG, Price JL. 1984. Amygdalo-cortical projections in the monkey (*Macaca fascicularis*). *J. Comp. Neurol.* 230:465–96

Amodio DM, Devine PG. 2006. Stereotyping and evaluation in implicit race bias: evidence for independent constructs and unique effects on behavior. *J. Pers. Soc. Psychol.* 91:652–61

Amodio DM, Frith CD. 2006. Meeting of minds: the medial frontal cortex and social cognition. *Nat. Rev. Neurosci.* 7:268–77

Badre D, Wagner AD. 2004. Selection, integration, and conflict monitoring: assessing the nature and generality of prefrontal cognitive control mechanisms. *Neuron* 41:473–87

Banyas CA. 1999. Evolution and phylogenetic history of the frontal lobes. In *The Human Frontal Lobes: Functions and Disorders*, ed. BL Miller, JL Cummings, pp. 83–106. New York: Guilford

Barbey AK, Krueger F, Grafman J. 2009. Structured event complexes in the medial prefrontal cortex support counterfactual representations for future planning. *Philos. Trans. R. Soc. Lond. B Biol. Sci.* 364:1291–300

Baron-Cohen S, Wheelwright S, Hill J, Raste Y, Plumb I. 2001. The "Reading the Mind in the Eyes" test revised version: a study with normal adults, and adults with Asperger syndrome or high-functioning autism. *J. Child Psychol. Psychiatry* 42:241–51

Barrash J, Tranel D, Anderson SW. 2000. Acquired personality disturbances associated with bilateral damage to the ventromedial prefrontal region. *Dev. Neuropsychol.* 18:355–81

Barsalou LW, Kyle Simmons W, Barbey AK, Wilson CD. 2003a. Grounding conceptual knowledge in modality-specific systems. *Trends Cogn. Sci.* 7:84–91

Barsalou LW, Niedenthal PM, Barbey AK, Ruppert JA. 2003b. Social embodiment. In *The Psychology of Learning and Motivation: Advances in Research and Theory*, Vol. 43, pp. 43–92: New York: Elsevier Sci.

Beauregard M, Levesque J, Bourgouin P. 2001. Neural correlates of conscious self-regulation of emotion. *J. Neurosci.* 21:RC165

Beer JS, John OP, Scabini D, Knight RT. 2006. Orbitofrontal cortex and social behavior: integrating self-monitoring and emotion-cognition interactions. *J. Cogn. Neurosci.* 18:871–79

Bem DJ. 1967. Self-perception: an alternative interpretation of cognitive dissonance phenomena. *Psychol. Rev.* 74:183–200

Berthoz S, Armony JL, Blair RJ, Dolan RJ. 2002. An fMRI study of intentional and unintentional (embarrassing) violations of social norms. *Brain* 125:1696–708

Bodner M, Kroger J, Fuster JM. 1996. Auditory memory cells in dorsolateral prefrontal cortex. *Neuroreport* 7:1905–8

Botvinick MM, Cohen JD, Carter CS. 2004. Conflict monitoring and anterior cingulate cortex: an update. *Trends Cogn. Sci.* 8:539–46

Carrington SJ, Bailey AJ. 2009. Are there theory of mind regions in the brain? A review of the neuroimaging literature. *Hum. Brain Mapp.* 30:2313–35

Castelli F, Frith C, Happe F, Frith U. 2002. Autism, Asperger syndrome and brain mechanisms for the attribution of mental states to animated shapes. *Brain* 125:1839–49

Cavanagh JF, Cohen MX, Allen JJ. 2009. Prelude to and resolution of an error: EEG phase synchrony reveals cognitive control dynamics during action monitoring. *J. Neurosci.* 29:98–105

Cunningham WA, Johnson MK. 2007. Attitudes and evaluation: toward a component process framework. In *Social Neuroscience: Integrating Biological and Psychological Explanations of Social Behavior*, ed. E Harmon-Jones, P Winkielman, pp. 227–45. New York: Guilford

Cunningham WA, Johnson MK, Raye CL, Chris Gatenby J, Gore JC, Banaji MR. 2004a. Separable neural components in the processing of black and white faces. *Psychol. Sci.* 15:806–13

Cunningham WA, Nezlek JB, Banaji MR. 2004b. Implicit and explicit ethnocentrism: revisiting the ideologies of prejudice. *Pers. Soc. Psychol. Bull.* 30:1332–46

Cunningham WA, Zelazo PD. 2007. Attitudes and evaluations: a social cognitive neuroscience perspective. *Trends Cogn. Sci.* 11:97–104

Damásio AR. 1989. Time-locked multiregional retroactivation: a systems-level proposal for the neural substrates of recall and recognition. *Cognition* 33:25–62

Damásio AR. 1994. *Descartes' Error: Emotion, Reason, and the Human Brain*. New York: Quill

Devine PG. 1989. Stereotypes and prejudice: their automatic and controlled components. *J. Personal. Soc. Psychol.* 56:5–18

Devine PG, Sharp LB. 2009. Automaticity and control in stereotyping and prejudice. In *Handbook of Prejudice, Stereotyping, and Discrimination*, ed. TD Nelson, pp. 61–87. New York: Psychol. Press

Dolcos F, Miller B, Kragel P, Jha A, McCarthy G. 2007. Regional brain differences in the effect of distraction during the delay interval of a working memory task. *Brain Res.* 1152:171–81

Dovidio JF, Kawakami K, Gaertner SL. 2002. Implicit and explicit prejudice and interracial interaction. *J. Pers. Soc. Psychol.* 82:62–68

Dreher JC, Koechlin E, Ali SO, Grafman J. 2002. The roles of timing and task order during task switching. *Neuroimage* 17:95–109

Elliott R, Deakin B. 2005. Role of the orbitofrontal cortex in reinforcement processing and inhibitory control: evidence from functional magnetic resonance imaging studies in healthy human subjects. *Int. Rev. Neurobiol.* 65:89–116

Fazio RH, Olson MA. 2003. Implicit measures in social cognition research: their meaning and use. *Annu. Rev. Psychol.* 54:297–327

Fiske ST. 1993. Social cognition and social perception. *Annu. Rev. Psychol.* 44:155–94

Fiske ST. 1998. Stereotyping, prejudice, and discrimination. In *The Handbook of Social Psychology*, Vols. 1, 2, ed. DT Gilbert, ST Fiske, G Lindzey, pp. 357–411. New York: McGraw-Hill. 4th ed.

Fiske ST, Taylor SE. 1991. *Social Cognition*. New York: McGraw-Hill. 2nd ed. xviii, 717 pp.

Fuster JM. 1997. *The Prefrontal Cortex: Anatomy, Physiology, and Neuropsychology of the Frontal Lobe*. New York: Raven

Fuster JM, Alexander GE. 1971. Neuron activity related to short-term memory. *Science* 173:652–54

Gallese V, Goldman A. 1998. Mirror neurons and the simulation theory of mind-reading. *Trends Cogn. Sci.* 2:493–501

Gawronski B, Bodenhausen GV. 2006. Associative and propositional processes in evaluation: an integrative review of implicit and explicit attitude change. *Psychol. Bull.* 132:692–731

Ghashghaei HT, Barbas H. 2002. Pathways for emotion: interactions of prefrontal and anterior temporal pathways in the amygdala of the rhesus monkey. *Neuroscience* 115:1261–79

Ghashghaei HT, Hilgetag CC, Barbas H. 2007. Sequence of information processing for emotions based on the anatomic dialogue between prefrontal cortex and amygdala. *Neuroimage* 34:905–23

Gilbert DT, Malone PS. 1995. The correspondence bias. *Psychol. Bull.* 117:21–38

Gilbert SJ, Meuwese JD, Towgood KJ, Frith CD, Burgess PW. 2009. Abnormal functional specialization within medial prefrontal cortex in high-functioning autism: a multi-voxel similarity analysis. *Brain* 132:869–78

Gozzi M, Raymont V, Solomon J, Koenigs M, Grafman J. 2009. Dissociable effects of prefrontal and anterior temporal cortical lesions on stereotypical gender attitudes. *Neuropsychologia* 47:2125–32

Grafman J. 2002. The structured event complex and the human prefrontal cortex. In *Principles of Frontal Lobe Function*, ed. DTH Stuss, RT Knight, p. 616. Oxford/New York: Oxford Univ. Press

Grafman J, Schwab K, Warden D, Pridgen A, Brown HR, Salazar AM. 1996. Frontal lobe injuries, violence, and aggression: a report of the Vietnam Head Injury Study. *Neurology* 46:1231–38

Greene J, Haidt J. 2002. How (and where) does moral judgment work? *Trends Cogn. Sci.* 6:517–23

Greene JD, Morelli SA, Lowenberg K, Nystrom LE, Cohen JD. 2008. Cognitive load selectively interferes with utilitarian moral judgment. *Cognition* 107:1144–54

Greene JD, Nystrom LE, Engell AD, Darley JM, Cohen JD. 2004. The neural bases of cognitive conflict and control in moral judgment. *Neuron* 44:389–400

Greene JD, Sommerville RB, Nystrom LE, Darley JM, Cohen JD. 2001. An fMRI investigation of emotional engagement in moral judgment. *Science* 293:2105–8

Greenwald AG, Banaji MR. 1995. Implicit social cognition: attitudes, self-esteem, and stereotypes. *Psychol. Rev.* 102:4–27

Greenwald AG, Poehlman TA, Uhlmann EL, Banaji MR. 2009. Understanding and using the Implicit Association Test: III. Meta-analysis of predictive validity. *J. Pers. Soc. Psychol.* 97:17–41

Gregory C, Lough S, Stone V, Erzinclioglu S, Martin L, et al. 2002. Theory of mind in patients with frontal variant frontotemporal dementia and Alzheimer's disease: theoretical and practical implications. *Brain* 125:752–64

Gusnard DA, Akbudak E, Shulman GL, Raichle ME. 2001. Medial prefrontal cortex and self-referential mental activity: relation to a default mode of brain function. *Proc. Natl. Acad. Sci. USA* 98:4259–64

Haidt J. 2001. The emotional dog and its rational tail: a social intuitionist approach to moral judgment. *Psychol. Rev.* 108:814–34

Ito TA, Thompson E, Cacioppo JT. 2004. Tracking the timecourse of social perception: the effects of racial cues on event-related brain potentials. *Pers. Soc. Psychol. Bull.* 30:1267–80

Jenkins AC, Macrae CN, Mitchell JP. 2008. Repetition suppression of ventromedial prefrontal activity during judgments of self and others. *Proc. Natl. Acad. Sci. USA* 105:4507–12

Johnson MK, Raye CL, Mitchell KJ, Touryan SR, Greene EJ, Nolen-Hoeksema S. 2006. Dissociating medial frontal and posterior cingulate activity during self-reflection. *Soc. Cogn. Affect. Neurosci.* 1:56–64

Johnson SC, Baxter LC, Wilder LS, Pipe JG, Heiserman JE, Prigatano GP. 2002. Neural correlates of self-reflection. *Brain* 125:1808–14

Kapogiannis D, Barbey AK, Su M, Zamboni G, Krueger F, Grafman J. 2009. Cognitive and neural foundations of religious belief. *Proc. Natl. Acad. Sci. USA* 106:4876–81

Kelley WM, Macrae CN, Wyland CL, Caglar S, Inati S, Heatherton TF. 2002. Finding the self? An event-related fMRI study. *J. Cogn. Neurosci.* 14:785–94

Knutson KM, Krueger F, Koenigs M, Hawley A, Escobedo JR, et al. 2010. Behavioral norms for condensed moral vignettes. *Soc. Cogn. Affect. Neurosci.* In press

Knutson KM, Mah L, Manly CF, Grafman J. 2007. Neural correlates of automatic beliefs about gender and race. *Hum. Brain Mapp.* 28:915–30

Koechlin E, Ody C, Kouneiher F. 2003. The architecture of cognitive control in the human prefrontal cortex. *Science* 302:1181–85

Koenigs M, Tranel D. 2007. Irrational economic decision-making after ventromedial prefrontal damage: evidence from the Ultimatum Game. *J. Neurosci.* 27:951–56

Koenigs M, Young L, Adolphs R, Tranel D, Cushman F, et al. 2007. Damage to the prefrontal cortex increases utilitarian moral judgements. *Nature* 446:908–11

Kohlberg L. 1969. Stage and sequence: the cognitive-developmental approach to socialization. In *Handbook of Socialisation Theory and Research*, ed. DA Goslin, pp. 347–480. Chicago: Rand McNally

Kouider S, Eger E, Dolan R, Henson RN. 2009. Activity in face-responsive brain regions is modulated by invisible, attended faces: evidence from masked priming. *Cereb. Cortex* 19:13–23

Krueger F, Barbey AK, Grafman J. 2009a. The medial prefrontal cortex mediates social event knowledge. *Trends Cogn. Sci.* 13:103–9

Krueger F, Spampinato MV, Barbey AK, Huey ED, Morland T, Grafman J. 2009b. The frontopolar cortex mediates event knowledge complexity: a parametric functional MRI study. *Neuroreport* 20:1093–97

Levy R, Goldman-Rakic PS. 2000. Segregation of working memory functions within the dorsolateral prefrontal cortex. *Exp. Brain Res.* 133:23–32

MacDonald AW 3rd, Cohen JD, Stenger VA, Carter CS. 2000. Dissociating the role of the dorsolateral prefrontal and anterior cingulate cortex in cognitive control. *Science* 288:1835–38

Macrae CN, Bodenhausen GV. 2000. Social cognition: thinking categorically about others. *Annu. Rev. Psychol.* 51:93–120

Macrae CN, Milne AB, Bodenhausen GV. 1994. Stereotypes as energy-saving devices: a peek inside the cognitive toolbox. *J. Personal. Soc. Psychol.* 66:37–47

Milham MP, Banich MT, Webb A, Barad V, Cohen NJ, et al. 2001. The relative involvement of anterior cingulate and prefrontal cortex in attentional control depends on nature of conflict. *Brain Res. Cogn. Brain Res.* 12:467–73

Miller BT, D'Esposito M. 2005. Searching for "the top" in top-down control. *Neuron* 48:535–38

Miller EK, Cohen JD. 2001. An integrative theory of prefrontal cortex function. *Annu. Rev. Neurosci.* 24:167–202

Milne E, Grafman J. 2001. Ventromedial prefrontal cortex lesions in humans eliminate implicit gender stereo-typing. *J. Neurosci.* 21:RC150

Mitchell JP, Heatherton TF, Macrae CN. 2002. Distinct neural systems subserve person and object knowledge. *Proc. Natl. Acad. Sci. USA* 99:15238–43

Mitchell JP, Macrae CN, Banaji MR. 2006a. Dissociable medial prefrontal contributions to judgments of similar and dissimilar others. *Neuron* 50:655–63

Mitchell JP, Mason MF, Macrae CN, Banaji MR. 2006b. Thinking about others: the neural substrates of social cognition. In *Social Neuroscience: People Thinking about Thinking People*, ed. JT Cacioppo, PS Visser, pp. 63–82. Cambridge, MA: MIT Press

Moll J, de Oliveira-Souza R. 2007. Moral judgments, emotions and the utilitarian brain. *Trends Cogn. Sci.* 11:319–21

Moll J, de Oliveira-Souza R, Bramati IE, Grafman J. 2002. Functional networks in emotional moral and nonmoral social judgments. *Neuroimage* 16:696–703

Moll J, de Oliveira-Souza R, Eslinger PJ. 2003. Morals and the human brain: a working model. *Neuroreport* 14:299–305

Moll J, Eslinger PJ, Oliveira-Souza R. 2001. Frontopolar and anterior temporal cortex activation in a moral judgment task: preliminary functional MRI results in normal subjects. *Arq. Neuropsiquiatr.* 59:657–64

Moll J, Zahn R, de Oliveira-Souza R, Krueger F, Grafman J. 2005. Opinion: the neural basis of human moral cognition. *Nat. Rev. Neurosci.* 6:799–809

Nichols S. 2002. Norms with feeling: towards a psychological account of moral judgment. *Cognition* 84:221–36

Nosek BA, Banaji MR, Greenwald AG. 2002. Math = male, me = female, therefore math not = me. *J. Pers. Soc. Psychol.* 83:44–59

Ochsner KN, Beer JS, Robertson ER, Cooper JC, Gabrieli JD, et al. 2005. The neural correlates of direct and reflected self-knowledge. *Neuroimage* 28:797–814

Oskamp S, Schultz PW. 2005. *Attitudes and Opinions*. Mahwah, NJ: Lawrence Erlbaum. 3rd. ed. xii, 578 pp.

Payne BK. 2005. Conceptualizing control in social cognition: how executive functioning modulates the expression of automatic stereotyping. *J. Pers. Soc. Psychol.* 89:488–503

Petty RE, Brinol P, DeMarree KG. 2007. The Meta-Cognitive Model (MCM) of attitudes: implications for attitude measurement, change, and strength. *Soc. Cogn.* 25:657–86

Quadflieg S, Turk DJ, Waiter GD, Mitchell JP, Jenkins AC, Macrae CN. 2009. Exploring the neural correlates of social stereotyping. *J. Cogn. Neurosci.* 21:1560–70

Raine A, Yang Y. 2006. Neural foundations to moral reasoning and antisocial behavior. *Soc. Cogn. Affect. Neurosci.* 1:203–13

Ramnani N, Miall RC. 2004. A system in the human brain for predicting the actions of others. *Nat. Neurosci.* 7:85–90

Richeson JA, Baird AA, Gordon HL, Heatherton TF, Wyland CL, et al. 2003. An fMRI investigation of the impact of interracial contact on executive function. *Nat. Neurosci.* 6:1323–28

Rilling JK, Dagenais JE, Goldsmith DR, Glenn AL, Pagnoni G. 2008. Social cognitive neural networks during in-group and out-group interactions. *Neuroimage* 41:1447–61

Rolls ET, Grabenhorst F. 2008. The orbitofrontal cortex and beyond: from affect to decision-making. *Prog. Neurobiol.* 86:216–44

Ross L, ed. 1977. *The Intuitive Psychologist and His Shortcomings*, Vol. 10. New York: Academic

Rudman LA, Lee MR. 2002. Implicit and explicit consequences of exposure to violent and misogynous rap music. *Group Process. Intergroup Relat.* 5:133–50

Saxe R, Schulz LE, Jiang YV. 2006. Reading minds versus following rules: dissociating theory of mind and executive control in the brain. *Soc. Neurosci.* 1:284–98

Schmader T, Johns M, Forbes C. 2008. An integrated process model of stereotype threat effects on performance. *Psychol. Rev.* 115:336–56

Schmitz TW, Kawahara-Baccus TN, Johnson SC. 2004. Metacognitive evaluation, self-relevance, and the right prefrontal cortex. *Neuroimage* 22:941–47

Schulkin J. 2000. *Roots of Social Sensibility and Neural Function*. Cambridge, MA: MIT Press. xviii, 206 pp.

Spezio M, Rangel A, Alvarez R, O'Doherty J, Mattes K, et al. 2008. A neural basis for the effect of candidate appearance on election outcomes. *Soc. Cogn. Affect. Neurosci.* 3:344–52

Storms MD. 1973. Videotape and the attribution process: reversing actors' and observers' points of view. *J. Pers. Soc. Psychol.* 27:165–75

Tajfel H. 1982. Social psychology of intergroup relations. *Annu. Rev. Psychol.* 33:1–39

Todorov A, Mandisodza A, Goren A, Hall C. 2005. Inferences of competence from faces predict election outcomes. *Science* 308:1623–26

Turiel E. 1983. *The Development of Social Knowledge: Morality and Convention*. Cambridge, UK: Cambridge Univ. Press. 252 pp.

Van Bavel JJ, Packer DJ, Cunningham WA. 2008. The neural substrates of in-group bias: a functional magnetic resonance imaging investigation. *Psychol. Sci.* 19:1131–39

van den Bos K. 2003. On the subjective quality of social justice: the role of affect as information in the psychology of justice judgments. *J. Personal. Soc. Psychol.* 85:482–98

Van Overwalle F. 2009. Social cognition and the brain: a meta-analysis. *Hum. Brain Mapp.* 30:829–58

Wager TD, Smith EE. 2003. Neuroimaging studies of working memory: a meta-analysis. *Cogn. Affect. Behav. Neurosci.* 3:255–74

Wilson TD, Lindsey S, Schooler TY. 2000. A model of dual attitudes. *Psychol. Rev.* 107:101–26

Wood JN, Grafman J. 2003. Human prefrontal cortex: processing and representational perspectives. *Nat. Rev. Neurosci.* 4:139–47

Zacks JM, Tversky B. 2001. Event structure in perception and conception. *Psychol. Bull.* 127:3–21

Zamboni G, Gozzi M, Krueger F, Duhamel JR, Sirigu A, Grafman J. 2009. Individualism, conservatism, and radicalism as criteria for processing political beliefs: a parametric fMRI study. *Soc. Neurosci.* 4:367–83

Sodium Channels in Normal and Pathological Pain

Sulayman D. Dib-Hajj,[1,2,3] Theodore R. Cummins,[4] Joel A. Black,[1,2,3] and Stephen G. Waxman[1,2,3]

[1]Department of Neurology and [2]Center for Neuroscience and Regeneration Research, Yale University School of Medicine, New Haven, Connecticut 06510

[3]Rehabilitation Research Center, Veterans Affairs Connecticut Healthcare System, West Haven, Connecticut 06516

[4]Department of Pharmacology and Toxicology, Stark Neurosciences Institute, Indiana University School of Medicine, Indianapolis, Indiana 46202; email: stephen.waxman@yale.edu

Annu. Rev. Neurosci. 2010. 33:325–47

First published online as a Review in Advance on April 1, 2010

The *Annual Review of Neuroscience* is online at neuro.annualreviews.org

This article's doi: 10.1146/annurev-neuro-060909-153234

Key Words

erythromelalgia, paroxysmal extreme pain disorder, congenital insensitivity to pain, sensory neurons, sympathetic neurons

Abstract

Nociception is essential for survival whereas pathological pain is maladaptive and often unresponsive to pharmacotherapy. Voltage-gated sodium channels, $Na_v1.1$–$Na_v1.9$, are essential for generation and conduction of electrical impulses in excitable cells. Human and animal studies have identified several channels as pivotal for signal transmission along the pain axis, including $Na_v1.3$, $Na_v1.7$, $Na_v1.8$, and $Na_v1.9$, with the latter three preferentially expressed in peripheral sensory neurons and $Na_v1.3$ being upregulated along pain-signaling pathways after nervous system injuries. $Na_v1.7$ is of special interest because it has been linked to a spectrum of inherited human pain disorders. Here we review the contribution of these sodium channel isoforms to pain.

Contents

INTRODUCTION

Nociception is the physiological system for the perception of pain, and thus it contributes to survival because it warns of impending harm. Signaling along the pain axis from peripheral receptors to higher-order brain centers optimally discriminates potentially harmful from innocuous stimuli. If pain is inappropriately magnified or prolonged, or occurs in the absence of appropriate external stimuli, it is pathological. The responses of nociceptors to stimuli are encoded by action potentials, whose genesis and propagation are dependent on voltage-gated sodium channels, and it is thus not surprising that aberrant expression patterns of channels and inherited sodium channelopathies have been linked to neuropathic and inflammatory pain. Here we review current knowledge of sodium channels that are preferentially expressed along the pain-signaling pathways.

Nine pore-forming sodium channel α-subunits ($Na_v1.1$–$Na_v1.9$, referred to as channels hereinafter), encoded by the *SCN1A-SCN5A* and *SCN8A-SCN11A* genes, have been identified in mammals, and their expression is spatially and temporally regulated (Catterall et al. 2005). These channels are large polypeptides (1700–2000 amino acids) that fold into four domains (DI–DIV), each domain including six transmembrane segments, linked by three loops (Catterall 2000). Different channels gate with different kinetics and voltage-dependent properties (Catterall et al. 2005), with six channels sensitive to block by nanomolar concentrations of tetrodotoxin (TTX-S), and three channels resistant to this blocker (TTX-R) (**Table 1**) (Catterall et al. 2005). Because channel properties are cell-type dependent and sodium channel properties can be modulated in a cell type–specific manner (for example, see Cummins et al. 2001, Choi et al. 2007), these channels should, whenever practicable, be studied within neurons in which they are normally expressed. Methods to study sodium channels within peripheral sensory neurons (Dib-Hajj et al. 2009b) have yielded important information about the contribution of individual sodium channels to electrogenesis within these neurons (Rush et al. 2006).

SODIUM CHANNELS IN DRG NEURONS

Dorsal root ganglia (DRG) house neurons of diverse sensory modalities that require precise electrogenic tuning. DRG neurons can express up to five sodium channels (Black et al. 1996, Dib-Hajj et al. 1998b), more than in any other neuronal cell type. Adult DRG neurons can express the TTX-S channels $Na_v1.1$, $Na_v1.6$, and $Na_v1.7$, and the TTX-R channels $Na_v1.8$

Nociceptors: pain- or damage-sensing neurons

Inherited sodium channelopathies: pathologies linked to mutations in sodium channels

Na_v1: voltage-gated sodium channel subfamily 1

TTX: tetrodotoxin

DRG: dorsal root ganglion

Table 1 Sodium channels preferentially expressed in peripheral neurons

Channel	Expression in peripheral sensory neurons	TTX sensitivity	Physiological attributes
$Na_v1.3$	Normally expressed during embryogenesis, but continues to be expressed in sympathetic neurons in adult Upregulated in DRG neurons after injury	S Kd = 1.8–4nM	Rapid repriming Produces large ramp current Produces persistent current Amplifies small depolarizing inputs
$Na_v1.7$	Preferentially expressed in DRG and sympathetic neurons	S Kd = 4.3–25nM	Slow repriming Produces large ramp current Amplifies small depolarizing inputs
$Na_v1.8$	Selectively expressed in DRG neurons	R Kd = 40–60 μM	Depolarized voltage-dependence for activation and inactivation. Rapid repriming Produces majority of current during AP upstroke Supports repetitive firing in response to depolarizing input Different slow-inactivation properties in IB4[+] and IB4[−] DRG neurons
$Na_v1.9$	Selectively expressed in small-diameter nonpeptidergic DRG neurons	R Kd = 40 μM	Hyperpolarized voltage-dependency of activation Slow activation kinetics Ultra slow inactivation Broad overlap between activation and fast-inactivation Amplifies and prolongs small depolarizations close to resting membrane potential Depolarizes resting potential of DRG neurons

and $Na_v1.9$ (as well as $Na_v1.5$ at low levels) (**Figure 1**). $Na_v1.1$ and $Na_v1.6$ expression is common to central nervous system (CNS) and peripheral nervous system (PNS) neurons, whereas $Na_v1.7$, $Na_v1.8$, and $Na_v1.9$ are specific to PNS neurons.

$Na_v1.3$ is the major channel in embryonic neurons; in rodents, it is significantly reduced in neonates and is undetectable in adult DRG neurons (Waxman et al. 1994) and is at low levels in adult brain (Beckh et al. 1989). However, levels of $Na_v1.3$, comparable to other channels, are present in adult sympathetic ganglion neurons (Rush et al. 2006). $Na_v1.3$ channel expression is upregulated in axotomized rodent DRG neurons (Waxman et al. 1994, Dib-Hajj et al. 1996), as is discussed in detail below. The recent discovery of a mutation in $Na_v1.3$ linked to childhood epilepsy (Holland et al. 2008), however, is consistent with higher $Na_v1.3$ expression levels in human brains after

birth (Whitaker et al. 2001), and suggests that $Na_v1.3$ may play a larger role in human than in rodent nociception.

Three channels, $Na_v1.7$, $Na_v1.8$, and $Na_v1.9$, appear to have evolved relatively recently because their sequences have not been reported from nonmammalian species thus far. $Na_v1.7$ is expressed in sensory and sympathetic (Toledo-Aral et al. 1997), and myenteric (Sage et al. 2007) neurons, whereas $Na_v1.9$ is expressed in sensory and myenteric neurons (Dib-Hajj et al. 1998b, Rugiero et al. 2003), and $Na_v1.8$ only in sensory neurons (Akopian et al. 1996, Rugiero et al. 2003). $Na_v1.7$, $Na_v1.8$, and $Na_v1.9$, produced by functionally identified nociceptive neurons (Fang et al. 2002; Djouhri et al. 2003a,b), may have evolved a specialized sensory role in mammals, including pain, and are attractive targets for the development of new pharmaceutical agents to treat pain.

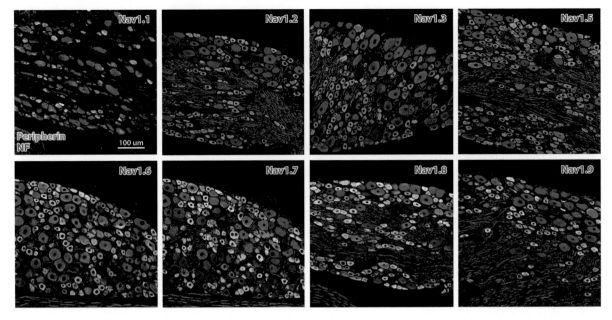

Figure 1

Five sodium channels are expressed in adult DRG neurons. Sodium channel isoforms $Na_v1.1$, $Na_v1.6$, $Na_v1.7$, $Na_v1.8$, and $Na_v1.9$ (*red*) are colocalized in DRG neurons expressing peripherin (*green*), a specific small neuron marker, and neurofilament (*blue*), a marker of medium and large neurons. $Na_v1.8$ is expressed preferentially in small and medium neurons, whereas $Na_v1.9$ is expressed exclusively in small neurons. $Na_v1.7$ is highly expressed in small neurons, but is also present in some large neurons. $Na_v1.6$ generally has expression in all size classes of neurons, whereas the limited $Na_v1.1$ expression is largely confined to large neurons. Colocalization of sodium channels with peripherin is depicted in yellow and with neurofilament in magenta.

NORMAL AND PATHOLOGICAL ELECTROGENESIS IN NOCICEPTORS

Neuropathic and inflammatory pain signals originate predominantly in peripheral sensory terminals, but are maintained by central sensitization. Sodium channel mutations or dysregulated expression within peripheral primary afferents and within CNS neurons along the pain-signaling axis have been shown to contribute to the establishment and maintenance of pain states (Waxman & Hains 2006; Dib-Hajj et al. 2007, 2009a). This pivotal role of sodium channels in pain has been empirically confirmed by symptomatic relief in patients treated with sodium channel blockers (Rice & Hill 2006, Dworkin et al. 2007), but the nonspecific nature and side effects of existing blockers (Sindrup & Jensen 2007, Gerner & Strichartz 2008) have limited their clinical utility,

Central sensitization: enhanced excitability of CNS neurons typically triggered by hyperexcitable peripheral nociceptors leading to an exaggerated response to a normal stimulus

providing an impetus for the search for isoform-specific sodium channel blockers.

SODIUM CHANNEL DYSREGULATION IN EXPERIMENTAL MODELS OF PAIN

Several animal models of nerve injury and inflammation have shown transcriptional regulation of sodium channel genes in DRG neurons, with transcription of some channels "turned off" and others "turned on" (Waxman 2001). For example, $Na_v1.3$ channel expression, which is undetected in adult rat DRG neurons, is upregulated in axotomized neurons (Waxman et al. 1994, Dib-Hajj et al. 1996, Black et al. 1999), whereas $Na_v1.8$ and $Na_v1.9$, which are abundant in small rat DRG neurons, are downregulated in axotomized neurons (Dib-Hajj et al. 1998b, Sleeper et al. 2000, Decosterd et al. 2002). Importantly, two studies have reported a reduction of $Na_v1.7$, $Na_v1.8$, and $Na_v1.9$ within injured human DRG neurons (Coward et al. 2000, 2001), but other studies have shown that $Na_v1.7$ and $Na_v1.8$ can accumulate within injured axons in painful human neuromas (Kretschmer et al. 2002, Bird et al. 2007, Black et al. 2008). The dysregulated expression of sodium channels is not a generalized injury-induced nonspecific response or a recapitulation of a developmental expression program, since transection of the centrally projecting axons of DRG neurons by dorsal rhizotomy does not alter $Na_v1.3$, $Na_v1.8$, or $Na_v1.9$ expression (Black et al. 1999, Sleeper et al. 2000) and severing peripheral axons does not lead to an upregulation of $Na_v1.2$, which is normally expressed during embryogenesis within DRG neurons (Waxman et al. 1994).

CONTRIBUTION OF INDIVIDUAL SODIUM CHANNELS TO ACQUIRED PAIN

$Na_v1.3$

Peripheral nerve injury triggers upregulated $Na_v1.3$ expression in DRG (Waxman et al. 1994, Dib-Hajj et al. 1996, Black et al. 1999, Kim et al. 2001, Lindia et al. 2005), dorsal horn (Hains et al. 2004), and thalamic (Zhao et al. 2006) neurons. Similarly, spinal cord injury (SCI) leads to $Na_v1.3$ expression within dorsal horn and thalamic neurons (Hains et al. 2003, 2005; Lampert et al. 2006b). Injury-induced $Na_v1.3$ upregulation within DRG neurons is reversed by administration of neurotrophic growth factor (NGF) and glial-derived neurotrophic factor (GDNF) (Boucher et al. 2000, Leffler et al. 2002), suggesting that loss of target-derived neurotrophic factors derepresses transcriptional silencing of $Na_v1.3$. The trigger for injury-induced $Na_v1.3$ upregulation within second- and third-order neurons is not yet understood, but may also involve changes regulated by alterations in neurotrophic factor levels.

$Na_v1.3$ channel activation produces a fast-inactivating, rapid-repriming (recovery from inactivation) TTX-S current, with slow closed-state inactivation that yields a substantial ramp current in response to small, slow depolarizations, and these properties are modulated in a cell type–dependent manner (Cummins et al. 2001). Contactin/F3, a cell-adhesion molecule that translocates to the cell surface in an activity-dependent manner (Pierre et al. 2001), interacts with $Na_v1.3$ and increases its current density (Shah et al. 2004). Similarities in the biophysical properties of $Na_v1.3$ and the TTX-S current within injured DRG neurons (Cummins & Waxman 1997) implicate $Na_v1.3$ in injury-induced neuron hyperexcitability. $Na_v1.3$ has also been linked to an increase in persistent current within dorsal horn neurons following SCI (Lampert et al. 2006b).

Ectopic firing within neuromas is now well established (Devor 2006). $Na_v1.3$ channels have been localized within distal axon tips in experimental neuromas in rats (Black et al. 1999) and in human neuromas (Black et al. 2008). Contactin, which is upregulated following axotomy, coaccumulates with $Na_v1.3$ in experimental neuromas (Shah et al. 2004), suggesting a positive feedback loop in which ectopic activity enhances trafficking of contactin to the plasma membrane, leading to increased $Na_v1.3$ expression at the axonal tips, and exacerbation of neuropathic pain (Shah et al. 2004). The attenuation of ectopic firing (Liu et al. 2001) and amelioration of pain behavior (Lyu et al. 2000) by 20nM TTX is consistent with a contribution of $Na_v1.3$ channels and other TTX-S channels to ectopic discharges within neuromas.

In support of a link between upregulated expression of $Na_v1.3$ and hyperexcitability of primary afferants and central neurons in the ascending pain pathway, enlargement of their peripheral receptive fields, and neuropathic pain, Hains et al. (2003, 2004) observed that intrathecal treatment with antisense oligonucleotide (ODN) targeting $Na_v1.3$ reduces levels of $Na_v1.3$ within dorsal horn neurons and ameliorates pain after sciatic nerve and spinal cord injury (**Table 2**). However, Lindia et al.

Table 2 Results of knock-down and knock-out studies

Channel	Knock-down	Knock-out
$Na_v1.3$	Attenuation of pain with antisense ODN after SCI and CCI in rat (Hains et al. 2003, 2004) No effect on pain behavior with antisense ODN in rat (Lindia et al. 2005)	Normal neuropathic pain behavior (Nassar et al. 2005)
$Na_v1.7$	Attenuation of pain with HSV-delivered antisense construct in mice (Yeomans et al. 2005)	Abrogated inflammatory response and thermal hyperalgesia (Nassar et al. 2004) Normal neuropathic pain behavior (Nassar et al. 2006)
$Na_v1.8$	Attenuation of pain with antisense ODN after CCI in rat (Lai et al. 2002, Porreca et al. 1999, Joshi et al. 2006, Gold et al. 2003, Yoshimura et al. 2001) Attenuation of pain following lentivirus-delivered siRNA (Dong et al. 2007)	Impaired thermal hyperalgesia and inflammatory (Akopian et al. 1999, Laird et al. 2002) and cold pain (Zimmermann et al. 2007)
$Na_v1.9$	No ODN effect on neuropathic pain in rat (Porreca et al. 1999)	Impaired inflammatory pain (Priest et al. 2005, Amaya et al. 2006), but see (Hillsley et al. 2006)

(2005) did not observe amelioration of neuropathic pain after peripheral nerve injury following knock-down of $Na_v1.3$ by different ODNs (**Table 2**). Additionally, global or DRG-specific knock-out of $Na_v1.3$ does not impair pain behavior after nerve injury (Nassar et al. 2006), a result suggesting that either this channel does not contribute to injured neuron hyperexcitability, that its function is redundant, or that compensatory changes obscure the effect of losing this channel. Despite clear upregulation of $Na_v1.3$ expression in DRG, dorsal horn, and thalamic neurons after axonal injury, sodium channels other than $Na_v1.3$ may mediate injury-induced hyperexcitability.

$Na_v1.7$

$Na_v1.7$ is preferentially expressed in DRG and sympathetic ganglion neurons (Sangameswaran et al. 1997, Toledo-Aral et al. 1997, Djouhri et al. 2003b). $Na_v1.7$ produces a fast-activating and -inactivating, slow-repriming, TTX-S current (Klugbauer et al. 1995). Slow closed-state inactivation of $Na_v1.7$ yields a substantial ramp current in response to small, slow depolarizations (Cummins et al. 1998, Herzog et al. 2003). Based on its biophysical properties, $Na_v1.7$ is poised to amplify generator potentials in neurons expressing it, including nociceptors, and to act as a threshold channel for firing

action potentials (Rush et al. 2007), thereby setting the gain in pain-signaling neurons (Waxman 2006, Dib-Hajj et al. 2007). The switch from slow-repriming to rapid-repriming TTX-S currents in injured DRG neurons (Cummins & Waxman 1997) is consistent with reduced $Na_v1.7$ mRNA levels following axotomy (Kim et al. 2002). However, the incomplete loss of $Na_v1.7$ channels in injured DRG neurons suggests that other factors influence repriming of residual TTX-S channels in these neurons, including modulation of $Na_v1.6$, the other TTX-S channel within small DRG neurons (Black et al. 2002).

In agreement with animal studies, reduced levels of $Na_v1.7$ in DRG neurons have been reported following peripheral nerve injury in humans (Coward et al. 2000, 2001). However, recent studies have demonstrated accumulation of $Na_v1.7$ within axons in painful human neuromas, including those in amputees with phantom limb pain (Kretschmer et al. 2002, Bird et al. 2007, Black et al. 2008). Activated p38 and ERK1/2 MAPK (mitogen-activated protein kinase) also accumulate within axons in human neuromas (Black et al. 2008) raising the possibility that modulation of $Na_v1.7$ by activated MAPKs (Stamboulian et al. 2010) may contribute to ectopic firing at neuromas.

Inflammation causes an upregulation of $Na_v1.7$ and TTX-S current in DRG neurons

Generator potentials: passively transmitted, typically small stimulus-evoked currents that depolarize cell membrane and, once it reaches a threshold, trigger an all-or-none action potential at the first trigger zone of the neuron

MAPK: mitogen-activated protein kinase

that project to the inflamed area (Black et al. 2004, Gould et al. 2004, Strickland et al. 2008). Inflammatory mediators, e.g., NGF, upregulate $Na_v1.7$ expression (Toledo-Aral et al. 1997, Gould et al. 2000), and increased levels of $Na_v1.7$ transcripts and phosphorylated $Na_v1.7$ protein have been reported in a rat model of painful diabetic neuropathy (Hong et al. 2004, Chattopadhyay et al. 2008). Activated p38 MAPK and PKC, which are signal transducers of inflammatory mediators, have been reported to regulate the expression of $Na_v1.7$ in diabetic neuropathy (Chattopadhyay et al. 2008). Peptide toxins that preferentially block $Na_v1.7$ have been identified (Middleton et al. 2002, Xiao et al. 2008); however, they do not ameliorate pain in nerve-injury animal models, perhaps owing to impaired accessibility (Schmalhofer et al. 2008). In contrast, an important role for $Na_v1.7$ in inflammatory pain is supported by the observations that $Na_v1.7$ knock-down in primary afferents ameliorates thermal hyperalgesia in mice following complete Freund's adjuvant injection into the paw (Yeomans et al. 2005).

A role for $Na_v1.7$ in inflammatory pain is confirmed in knock-out studies. Global knock-out of $Na_v1.7$ was neonatal lethal, but a conditional $Na_v1.7$ knock-out in $Na_v1.8$-expressing mouse DRG neurons abrogates inflammation-induced pain (Nassar et al. 2004). The loss of $Na_v1.7$ in $Na_v1.8$-expressing DRG neurons did not impair neuropathic pain. The presence of $Na_v1.7$ in $Na_v1.8$-negative DRG neurons may perhaps contribute to neuropathic pain-signaling, although this is a very limited population of cells (J.A. Black and S.G. Waxman, unpublished observations). Alternatively, $Na_v1.7$ may be dispensable for neuropathic pain signaling in animal models, or the animal models are not suitable to uncover the role of this channel (see discussion below on the limitations of animal models in pain testing).

In the aggregate, the preferential expression of $Na_v1.7$ in nociceptors, the functional role of $Na_v1.7$ in regulating neuronal excitability, and the results of knock-down studies support a critical role for $Na_v1.7$ in pain signaling, which is further supported by identification of $Na_v1.7$ mutations in human hereditary pain disorders (see below).

$Na_v1.8$

$Na_v1.8$ is a sensory neuron-specific channel that is preferentially expressed in DRG and trigeminal ganglia (Akopian et al. 1996, Sangameswaran et al. 1996), most of which are nociceptive (Djouhri et al. 2003a), and is also present along peripheral axons shafts (Rush et al. 2005) and free nerve terminals in skin (Zhao et al. 2008) and cornea (Black & Waxman 2002). $Na_v1.8$ produces a slow-inactivating, rapid-repriming TTX-R current with depolarized activation and inactivation voltage-dependency (Akopian et al. 1996, 1999). Activation and inactivation properties of the slow-inactivating TTX-R current are conserved in human DRG neurons (Dib-Hajj et al. 1999). $Na_v1.8$ trafficking to the cell membrane is enhanced by annexin II light-chain (Okuse et al. 2002), and by contactin in IB4$^+$ but not in IB4$^-$ DRG neurons (Rush et al. 2005). In contrast, SCLT1 (Sodium-Channel-CLaThrin-linker 1; previously known as CAP1A) internalizes $Na_v1.8$ in a clathrin-dependent manner (Liu et al. 2005), and Nedd4-2 ubiquitin ligase, but not Nedd4, induces a reduction in $Na_v1.8$ current density (Fotia et al. 2004).

$Na_v1.8$ contributes most of the sodium current underlying the action potential upstroke in neurons that expresses it (Renganathan et al. 2001, Blair & Bean 2002). Depolarized inactivation and rapid repriming may explain why $Na_v1.8$ accounts for most of the current in later spikes in a train (Blair & Bean 2003). $Na_v1.8$ slow-inactivation is differentially modulated in peptidergic (IB4$^-$) and nonpeptidergic (IB4$^+$) nociceptors (Choi et al. 2007), possibly contributing to different degrees of adaptation of action potential firing in response to sustained stimulation (Blair & Bean 2003, Tripathi et al. 2006, Choi et al. 2007).

The biophysical properties of $Na_v1.8$, its critical role in repetitive firing, and its presence in free nerve endings, where pain-signaling is initiated, suggest that $Na_v1.8$ can significantly influence nociceptor excitability, thus contributing to pain. The role of $Na_v1.8$ in neuropathic pain is, however, not well understood. Axonal transection in the sciatic nerve causes a downregulation of $Na_v1.8$ mRNA, protein and current in injured neurons (Dib-Hajj et al. 1996, Cummins & Waxman 1997, Sleeper et al. 2000, Decosterd et al. 2002). However, increased $Na_v1.8$ levels have been reported in spared axons and neuronal cell bodies in neuropathic pain models (Gold et al. 2003, Zhang et al. 2004), possibly in response to inflammatory cytokines produced during Wallerian degeneration. $Na_v1.8$-mediated hyperexcitability of uninjured neurons provides a plausible explanation for a contribution of $Na_v1.8$ to neuropathic pain in animal models. Human patients with chronic neuropathic pain show increased $Na_v1.8$ channel expression proximal to injury sites (Coward et al. 2000, Yiangou et al. 2000, Black et al. 2008). A role for $Na_v1.8$ in neuropathic pain is also suggested by studies of knockdown (Lai et al. 2002, Joshi et al. 2006, Dong et al. 2007), toxin-inhibition (Bulaj et al. 2006, Ekberg et al. 2006), and the small molecule inhibitor, A-803467 (Jarvis et al. 2007) (**Table 2**).

The contribution of $Na_v1.8$ in inflammatory pain is well documented in animal studies. $Na_v1.8$ levels in DRG neurons are increased following carrageenan injection into rat hindpaw (Tanaka et al. 1998, Black et al. 2004), and following treatment of cultured DRG neurons with inflammatory mediators (Gold et al. 1996, Jin & Gereau 2006, Binshtok et al. 2008). Injection of complete Freund's adjuvant into rat hindpaw does not increase $Na_v1.8$ levels in DRG neurons (Okuse et al. 1997), but does increase $Na_v1.8$ translocation to myelinated and unmyelinated axons in the sciatic nerve (Coggeshall et al. 2004). Increased $Na_v1.8$ current density was also reported in an animal model of colitis (Beyak et al. 2004). Further evidence for an important role of $Na_v1.8$ in inflammatory pain is provided by knock-down studies (Joshi et al. 2006) and inhibition by A-803467 (Jarvis et al. 2007).

Pro-inflammatory mediators, released by damaged tissue and infiltrating immune cells (Scholz & Woolf 2007), have been shown to modulate sodium currents through activation of protein kinases (Jin & Gereau 2006, Hucho & Levine 2007, Binshtok et al. 2008). NGF, an inflammatory cytokine, upregulates $Na_v1.8$ within DRG neurons in vivo (Dib-Hajj et al. 1998a, Leffler et al. 2002) and in vitro (Fjell et al. 1999b, Cummins et al. 2000); ceramide, a second messenger for NGF, increases $Na_v1.8$ current density (Zhang et al. 2002). PGE2 and other inflammatory mediators act through PKA and PKC kinases (England et al. 1996; Gold et al. 1998, 2002; Zhou et al. 2002; Hucho & Levine 2007) to increase $Na_v1.8$ current density and produce a hyperpolarizing shift in activation voltage-dependency, possibly via PKA/PKC phosphorylation of serine residues within L1 (Fitzgerald et al. 1999, Vijayaragavan et al. 2004). Patch-clamp studies of DRG neurons from diabetic rats show increased slow-inactivating TTX-R current amplitude and hyperpolarizing shifts of activation and steady-state inactivation, consistent with increased serine/threonine phosphorylation of $Na_v1.8$ (Hong et al. 2004).

In contrast, treatment of DRG neurons with proinflammatory cytokines TNF-α (Jin & Gereau 2006) and IL-1β (Binshtok et al. 2008) increases $Na_v1.8$ current density without altering its gating properties, via a p38 MAPK-mediated mechanism. The p38-mediated increase in $Na_v1.8$ current density results from phosphorylation of two serine residues within $Na_v1.8$-L1 that are distinct from the PKA/PKC phosphorylation sites (Hudmon et al. 2008). Similarly, inflammation of visceral organs causes an increase in $Na_v1.8$ current density without a hyperpolarizing shift in activation voltage-dependency (Yoshimura et al. 2001, Bielefeldt et al. 2002). Thus, multiple inflammatory modalities may differentially regulate the $Na_v1.8$ current.

Studies in $Na_v1.8$ knock-out mice have confirmed a role of $Na_v1.8$ in somatic inflammatory (Akopian et al. 1999, Kerr et al. 2001) and cold (Zimmermann et al. 2007) pain. $Na_v1.8$ is expressed in all DRG neurons that innervate the colon (Gold et al. 2002), and $Na_v1.8$ knock-out mice show deficits in visceral inflammatory pain (Laird et al. 2002, Hillsley et al. 2006). However, a role for $Na_v1.8$ in neuropathic pain was not observed in the $Na_v1.8$ knock-out mouse (Akopian et al. 1999) despite a report of 20-fold reduction of ectopic discharges in neuromas in these mice (Roza et al. 2003). Additionally, the double $Na_v1.7/Na_v1.8$ deletion did not attenuate neuropathic pain response in mice (Nassar et al. 2005). Increased $Na_v1.7$ expression in DRG from $Na_v1.8$ knock-out mice (Akopian et al. 1999) may contribute to, but does not totally explain, their normal neuropathic pain behavior, especially because the absence of $Na_v1.8$ has been shown to attenuate the excitability of neurons expressing a gain-of-function $Na_v1.7$ mutation (Rush et al. 2006).

$Na_v1.9$

$Na_v1.9$ is preferentially expressed in small-diameter, nonpeptidergic DRG neurons (Dib-Hajj et al. 1998b, 2002), which are largely nociceptors (Fang et al. 2002, 2006), and in trigeminal ganglion and myenteric neurons (Dib-Hajj et al. 2002, Rugiero et al. 2003), and has been found within free nerve terminals in skin and cornea (Black & Waxman 2002, Dib-Hajj et al. 2002). $Na_v1.9$ expression is regulated by the trophic factor GDNF but not NGF (Fjell et al. 1999a, Leffler et al. 2002). $Na_v1.9$ current density is significantly reduced in $IB4^+$ neurons from contactin-null mice, suggesting a role for contactin in $Na_v1.9$ trafficking (Rush et al. 2005). Thus, the expression of $Na_v1.9$ appears to be tightly regulated within DRG neurons, and may contribute to the functional specialization (Stucky & Lewin 1999, Braz et al. 2005) of $IB4^+$ and $IB4^-$ neurons.

$Na_v1.9$ current is TTX-R, with a hyperpolarized voltage-dependency of activation close to the resting membrane potential of neurons (-60 to -70 mV) and an ultraslow inactivation leading to a persistent current (Cummins

et al. 1999). Glycosylation of $Na_v1.9$ is developmentally regulated and hyperpolarizes inactivation voltage-dependency (Tyrrell et al. 2001). Recombinant $Na_v1.9$ produces a small current with similar properties in HEK 293 cell line (Dib-Hajj et al. 2002). The persistent TTX-R current is missing from DRG neurons of $Na_v1.9$ knock-out mice (Priest et al. 2005, Amaya et al. 2006, Ostman et al. 2007), and can be restored by expression of recombinant $Na_v1.9$ channels (Ostman et al. 2007), unequivocally confirming the identity of the current. Importantly, native human $Na_v1.9$ current activates at ~-80 mV, 10–20 mV more negative than $Na_v1.9$ current in rodent DRG neurons, likely owing to species-specific differences in primary protein sequence (Dib-Hajj et al. 1999). The ultraslow kinetics of $Na_v1.9$ suggest that it does not contribute to the action potential upstroke, but that it may enhance and prolong the response to subthreshold depolarizations (Cummins et al. 1999, Herzog et al. 2001), and lower the threshold for single action potentials and repetitive firing (Baker et al. 2003). Based on computer simulations (Herzog et al. 2001) and empirical evidence (Baker et al. 2003, Ostman et al. 2007, Copel et al. 2009), $Na_v1.9$ appears to act as a threshold channel.

$Na_v1.9$ is sensitive to intracellular fluoride (Coste et al. 2004), suggesting modulation by kinases/phosphatases. Additionally, direct activation of G proteins in DRG neurons increases $Na_v1.9$ current with a subsequent reduction in action potential threshold and an increase in spontaneous firing (Baker et al. 2003, Ostman et al. 2007). Recently, $Na_v1.9$ current density has been shown to increase because of a rapid, transient hyperpolarizing shift of activation and inactivation following neurokinin 3 receptor activation in enteric neurons (which is also mimicked by activation of PKC), reducing the threshold for action potential generation of these neurons (Copel et al. 2009).

Experimental evidence supports a role for $Na_v1.9$ in inflammatory and diabetic neuropathy pain. Expression of $Na_v1.9$ has been shown to increase in DRG neurons innervating inflamed rat hindpaw (Tate et al. 1998). PGE2, acting via G protein–coupled receptors, increases $Na_v1.9$ current density in DRG neurons in vitro, accompanied by hyperpolarized shifts of activation and inactivation (Rush & Waxman 2004), while treatment with IL-1β increases persistent TTX-R in a p38 MAPK-dependent manner (Binshtok et al. 2008). Although expression levels of $Na_v1.9$ do not appear to be altered in small DRG neurons from diabetic rats, increased $Na_v1.9$ levels in large-diameter neurons suggest a contribution to painful diabetic neuropathy (Craner et al. 2002). In contrast, $Na_v1.9$ mRNA and protein levels and current density are downregulated in several animal models of neuropathic pain (Cummins & Waxman 1997, Dib-Hajj et al. 1998b, Cummins et al. 2000, Sleeper et al. 2000, Decosterd et al. 2002). An early study using $Na_v1.9$ antisense ODN treatment did not report amelioration of neuropathic pain (Porreca et al. 1999). However, activation of neurokinin 3 receptor causes potentiation of $Na_v1.9$ leading to increased excitability of enteric neurons (Copel et al. 2009), which suggests an effect on nociceptive DRG neurons that coexpress neurokinin 3 receptor and $Na_v1.9$.

$Na_v1.9$ knock-out mice show impaired somatic inflammatory pain behavior (Priest et al. 2005, Amaya et al. 2006), but normal neuropathic pain (Amaya et al. 2006). Mice that are heterozygous for the $Na_v1.9$ null-allele manifested impaired inflammatory response (Priest et al. 2005), suggesting haploinsufficiency; however, similar findings have not been reported in another study with a different $Na_v1.9$ knock-out mouse (Amaya et al. 2006). $Na_v1.9$ knock-out mice did not manifest gastrointestinal or apparent nutritional deficits (Priest et al. 2005, Amaya et al. 2006). Another independently produced $Na_v1.9$ knock-out mouse strain displayed no deficits in visceral inflammatory pain (Hillsley et al. 2006), suggesting a differential role of this channel in somatic versus visceral pain.

Altogether, convincing evidence indicates a role for $Na_v1.9$ in inflammatory and diabetic neuropathy pain, although a role in neuropathic pain is less clear.

SODIUM CHANNELS IN HEREDITARY HUMAN PAIN

A compelling case can be made for a direct involvement of a target by establishing a monogenic link of mutations to disease. The recent discovery of a genetic link of $Na_v1.7$ to pain disorders in humans solidified the status of $Na_v1.7$ as central to pain-signaling. Dominant gain-of-function mutations in *SCN9A*, the gene that encodes sodium channel $Na_v1.7$, have been linked to two severe pain syndromes, inherited erythromelalgia (IEM) and paroxysmal extreme pain disorder (PEPD), and recessive loss-of-function mutations have been linked to congenital insensitivity to pain (CIP) (Dib-Hajj et al. 2007, Drenth & Waxman 2007). Electrophysiological characterization of these mutations has elucidated the molecular basis for altered excitability of DRG neurons that express these mutant channels, establishing a mechanistic link to human pain.

Pain in IEM is localized to the distal extremities (feet and hands) and has been reported as early as 1-year-old (early onset), in the second decade (delayed-onset), and in adults (adult-onset) (Dib-Hajj et al. 2007, Drenth & Waxman 2007). Mutations in $Na_v1.7$ have been identified in patients with early- and delayed-onset IEM (**Table 3**), but the molecular basis of adult-onset IEM remains unknown. Treatment for IEM, even with sodium channel blockers, e.g., lidocaine or mexiletine, is largely ineffective (Dib-Hajj et al. 2007, Drenth & Waxman 2007), and in one case may be the result of reduced affinity of the mutant channel to these drugs (Sheets et al. 2007). Recently, however, two cases of IEM were reported with favorable pain management with sodium channel blockers: V872G, controlled by lidocaine/mexiletine (Choi et al. 2009), and V400M, controlled by carbamazepine (Fischer et al. 2009). The V872G mutation shows enhanced use-dependent block by lidocaine (Choi et al. 2009), whereas V400M displays a

Table 3 Effects of IEM and PEPD mutations on gating properties of $Na_v1.7$, compared to wild-type channels

IEM mutations

Mutation	$\Delta V_{1/2}$ activation	$\Delta V_{1/2}$ inactivation	Slow inactivation	Deactivation	Ramp current	Repriming	Persistent current[a]	Reference
Q10R	−5.3 mV	0	Enhanced	Slower	Unchanged	Unchanged	0	Han et al. 2009
I136V	−5.7 mV	0	Enhanced	Slower	3.4X WT	Faster	0	Cheng et al. 2008
F216S	−11.8 mV	0	Enhanced	Slower	2X	Unchanged	0	Choi et al. 2006
S241T	−8.4 mV	0	Enhanced	Slower	5X	Unchanged	0	Lampert et al. 2006a
N395K	−7.7 mV	0	Impaired	Slower	ND	ND	0	Sheets et al. 2007
V400M	−6.5 mV	+7.3 mV	Unchanged	Slower	2X[b]	ND	0	Fischer et al. 2009
L823R	−14.6 mV	−9.8 mV	ND	Slower	ND	ND	0	Lampert et al. 2009
I848T	−13.8 mV	0	Impaired	Slower	3X	ND	0	Cummins et al. 2004
L858H	−13.3 mV	0	Enhanced	Slower	4.5X	ND	0	Cummins et al. 2004
L858F	−9.0 mV	+3.0 mV	Unchanged	Slower	4.5X	Faster	0	Han et al. 2006
A863P	−8 mV	+10 mV	Steep VD	Slower	10X	Faster	0	Harry et al. 2006
V872G	−9.3 mV	0	ND	Slower	2X	ND	0	Choi et al. 2009
F1449V	−7.6 mV	+4.3 mV	Unchanged	Unchanged	Unchanged	Faster	0	Dib-Hajj et al. 2005

PEPD mutations

Mutation	$\Delta V_{1/2}$ activation	$\Delta V_{1/2}$ inactivation	Slow inactivation	Deactivation	Ramp current	Repriming	Persistent current[a]	Reference
V1298F	+6.3 mV	+19.8 mV	Impaired	Unchanged	2X	Faster	6@8ms	Jarecki et al. 2008
V1299F	+4.5 mV	+21.0 mV	Impaired	Unchanged	5X	Faster	6@8ms	Jarecki et al. 2008
I1461T	+2.5 mV	+18.8 mV	Impaired	Unchanged	2X	Faster	6@8ms	Jarecki et al. 2008
I1461T	0	+29.3 mV	ND	ND	ND	ND	40@8ms	Fertleman et al. 2006
T1464I	+6.8 mV	+9.4 mV	ND	ND	ND	ND	30@8ms	Fertleman et al. 2006
M1627K	−2.9 mV	+19.3 mV	Unchanged	Unchanged	7X	Faster	0@8ms	Dib-Hajj et al. 2008
M1627K	0	+22.2 mV	ND	ND	ND	ND	7@8ms	Fertleman et al. 2006

Mixed phenotype mutations

Mutation	$\Delta V_{1/2}$ activation	$\Delta V_{1/2}$ inactivation	Slow inactivation	Deactivation	Ramp current	Repriming	Persistent current[a]	Reference
A1632E[c]	−7.0 mV	+17 mV	Impaired	ND	5X	Faster	4@8ms	Estacion et al. 2008

[a] Persistent current is presented as percentage of peak current, measured at 8 ms after initiation of pulse to compare all published results.
[b] ND: Not done.
[c] A1632E patients presented with symptoms of both IEM and PEPD.

depolarizing shift of activation by carbamazepine, suggesting an allosteric effect of the drug on the mutant channel (Fischer et al. 2009).

A second set of mutations of $Na_v1.7$ (**Table 3**) underlies many of the PEPD cases reported thus far. Some cases of PEPD do not show this linkage to $Na_v1.7$, suggesting involvement of another target (Fertleman et al. 2006). Severe perirectal pain in PEPD along with skin flushing can start in infancy and possibly in utero, but with no reported involvement of feet and hands (Fertleman et al. 2007). Although seizures and cardiac symptoms may accompany PEPD, a link to the expression of the mutant $Na_v1.7$ channel in sympathetic neurons has not yet been established. As patients age, pain extends to ocular and maxillary/mandibular areas and is triggered by cold, eating, or emotional state (Fertleman et al. 2007). PEPD symptoms, in contrast to IEM, are well controlled by the anticonvulsant sodium channel blocker carbamazepine (Fertleman et al. 2006, Dib-Hajj et al. 2008, Estacion et al. 2008).

All IEM mutations in $Na_v1.7$ characterized thus far shift activation voltage-dependency in a hyperpolarized direction, increase ramp current and slow deactivation (**Table 3**; **Figure 2**). In contrast, PEPD mutations shift the voltage-dependency of steady-state fast-inactivation in a depolarizing direction and, depending upon the specific mutation, may make inactivation incomplete resulting in a persistent current (**Table 3**; **Figure 2**). The A1632E mutation displays changes both in hyperpolarizing activation and depolarizing steady-state inactivation, and there is a mixed phenotype including IEM and PEPD symptoms in this patient (Estacion et al. 2008). Thus, IEM and PEPD mutations can be considered part of a physiological continuum that can produce a continuum of clinical phenotypes including IEM, PEPD, and overlap disorders with a characteristic of both (**Figure 2**). At the cellular level, IEM mutant $Na_v1.7$ channels lower threshold for single action potentials and increase firing frequency in DRG neurons, with all but one (F1449V)

causing a depolarizing shift in resting potential (**Figure 3**) (Dib-Hajj et al. 2005, Harty et al. 2006, Rush et al. 2006, Han et al. 2009). PEPD $Na_v1.7$ mutant channels lower threshold for single action potential and increase frequency of firing in DRG neurons, but without altering resting potential (Dib-Hajj et al. 2008, Estacion et al. 2008). Impaired inactivation of PEPD $Na_v1.7$ mutant channels could explain the favorable response of the patients to carbamazepine.

$Na_v1.7$-related CIP is caused by recessive loss-of-function mutations that truncate the channel protein or impair splicing signals to prevent the production of channel mRNA (Cox et al. 2006). Truncated $Na_v1.7$ mutant channels do not produce functional channels (Cox et al. 2006, Ahmad et al. 2007), or act as dominant negative proteins (Ahmad et al. 2007). Heterozygous parents are asymptomatic, indicating that loss of one *SCN9A* allele does not lead to haploinsufficiency, and the occurrence of $Na_v1.7$-related CIP in progeny of nonconsanguinous marriages (Goldberg et al. 2007, Nilsen et al. 2009) indicates a more common occurrence of carriers of nonfunctional *SCN9A* alleles than initially thought after the reporting of $Na_v1.7$-related CIP in consanguinous Pakistani families (Cox et al. 2006). Patients do not report any form of pain, but report intact sensory modalities except for impaired olfaction (Goldberg et al. 2007, Nilsen et al. 2009), and do not display motor, cognitive, sympathetic, or gastrointestinal deficits.

To study the effect of gain-of-function mutations of $Na_v1.7$ on sympathetic neurons, in which $Na_v1.7$ is normally expressed, Rush et al. (2006) expressed the L858H IEM $Na_v1.7$ channel mutant (Yang et al. 2004) in superior cervical ganglion (SCG) neurons. Current-clamp analysis showed that L858H mutant channels depolarize resting potential in both DRG and SCG neurons by 6 mV, but render DRG neurons hyperexcitable and SCG neurons hypoexcitable. Co-expression of $Na_v1.8$, which is normally present in DRG but not SCG, rescued electrogenesis in SCG

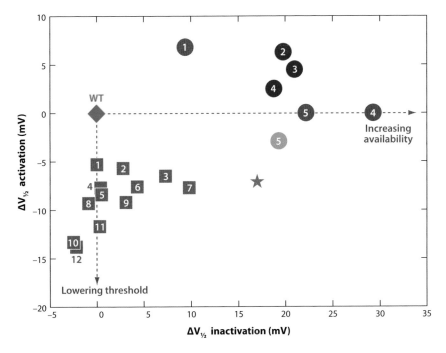

Figure 2

IEM and PEPD mutations are part of a physiological continuum linked to a continuum of clinical phenotypes. Shifts in the voltage-dependency of activation and fast-inactivation of each mutant compared to wild-type hNa$_V$1.7 are plotted for IEM mutants (*tan squares*) and PEPD mutants (*circles*) numbered to identify the specific mutation and reference from which the data were compiled. For PEPD mutations, same-colored symbols indicate mutations that were characterized electrophysiologically by the same group, while duplicate numbers indicate that the same mutation was profiled by different groups. The wild-type control is plotted as a green diamond at (0,0). The dotted lines through (0,0) demarcate between positive and negative shifts and indicate the outcome for the shifts. The A1632E mutation, from a patient with a mixed clinical phenotype, plotted with the red star symbol, shows shifts in activation and inactivation common to both IEM and PEPD mutants. The identities of the numbered IEM mutation are as follows (*shown as tan squares*): [1] Q10R (Han et al. 2009), [2] I136V (Cheng et al. 2008), [3] V400M (Fischer et al. 2009), [4] N395K (Sheets et al. 2007), [5] S241T (Lampert et al. 2006a), [6] F1449V (Dib-Hajj et al. 2005), [7] A863P (Harty et al. 2006), [8] V872G (Choi et al. 2009), [9] L858F (Han et al. 2006), [10] F216S (Choi et al. 2006), [11] L858H (Cummins et al. 2004), [12] I848T (Cummins et al. 2004). The identities of the numbered PEPD mutation are as follows (shown as colored circles): 1 (*gray*), T1464I (Fertleman et al. 2006); 2 (*blue*), V1298F (Jarecki et al. 2008); 3 (*blue*), V1299F (Jarecki et al. 2008); 4 (*gray*), I1461T (Fertleman et al. 2006); 4 (*blue*), I1461T (Jarecki et al. 2008); 5 (*gray*), M1627K (Fertleman et al. 2006); 5 (*orange*), M1627K (Dib-Hajj et al. 2008); red star, A1632E (Estacion et al. 2008). Adapted with permission from Estacion et al. (2008).

neurons that express the L858H mutant channels (Rush et al. 2006). Sympathetic neuron hypoexcitability may reduce tonic cutaneous vasoconstriction, thereby contributing to skin flushing in IEM. Why patients with gain-of-function mutations in Na$_v$1.7 do not suffer global sympathetic deficits, however, remains enigmatic.

LESSONS FROM DISCREPANCIES BETWEEN HUMAN AND ANIMAL STUDIES

Can we learn anything from the discrepancies (**Table 2**) between different knock-down and knock-out studies, and from the different phenotype in the murine global Na$_v$1.7 knock-out and Na$_v$1.7-related CIP in humans? Different

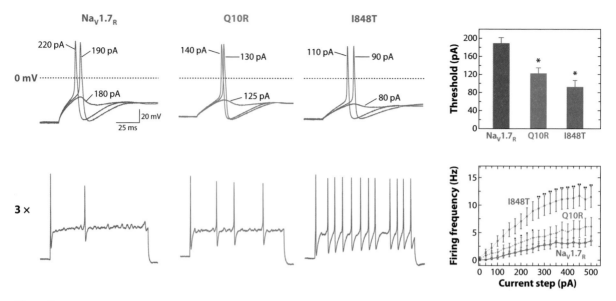

Figure 3

Both Q10R and I848T mutations decrease the action potential threshold in small DRG neurons, and increase firing frequency in small DRG neurons, but to different degrees. *Upper panels*: Representative traces from a cell expressing $Na_V1.7$ wild-type channels (*blue trace*), showing subthreshold response to 180 pA current injection and subsequent action potentials evoked by injections of 190 pA (current threshold for this neuron) and 220 pA. Representative traces from a cell expressing Q10R channels (*green trace*), showing a lower current threshold (130 pA for this cell) for action potential generation. Representative traces from a DRG neuron expressing I848T channels (*dark yellow trace*), showing a significantly lower current threshold (90 pA for this cell) for action potential generation. Histogram shows that threshold for action potential generation decreases significantly (*denotes $p < 0.05$) in those expressing Q10R channels and I848T channels. *Lower panels*: Response of cells expressing wild-type (*blue trace*), Q10R (*green traces*) and I848T (*dark yellow trace*) channels respectively to 1 s depolarizing current steps that are 3X the current threshold for action potential generation. Comparison of mean fire frequency among cells expressing wild-type, Q10R, and I848T channels across the range of current injections from 25 to 500 pA shows a quantitative difference between the effect on firing frequency of Q10R and I848T mutation, which is correlated with their hyperpolarized shifts of activation (*denotes $p < 0.05$, Q10R versus wild-type; **denotes $p < 0.05$, I848T versus Q10R). Adapted with permission from Han et al. (2009).

animal species, interstrain genetic differences, sex differences, and differences between responses to pain in rat (where most knock-down studies are performed) and mice (knock-out studies) may explain in part the apparently conflicting findings in different studies in experimental pain models (Mogil 2009). Additionally, multiple splice isoforms of $Na_V1.3$ (Thimmapaya et al. 2005) and $Na_V1.7$ (Raymond et al. 2004) exist, and it is possible that they may differentially contribute to hyperexcitability of neurons in which they are expressed. Compensatory changes during development may confound observations in channel-specific knock-out mice or after permanent ablation of a class of neurons (Abrahamsen et al. 2008), but

they are less likely to occur in transient (knock-down) experiments, after acute block in adults or ablation of a specific cell type in adults. For instance, ablation of $Mrpgrd^+$ small DRG neurons in adult mouse produced more profound effects on mechanosensitivity to noxious stimuli compared to Mrpgrd knock-out mice (Cavanaugh et al. 2009). Mismatches between knock-out and knock-down studies may also result from off-target effects of antisense reagents. These factors may have contributed to the narrow conclusion from knock-out studies that $Na_V1.7$, $Na_V1.8$, and $Na_V1.9$ are contributors to inflammatory but not neuropathic pain.

Humans tolerate total loss of $Na_V1.7$ with few physiological deficits other than insensitivity to pain and blunted olfaction (Cox et al. 2006, Goldberg et al. 2007, Nilsen et al. 2009). Global $Na_v1.7$ knock-out is, however, lethal in mice (Nassar et al. 2004). Ahmad et al. (2007) reported the expression of $Na_v1.7$ in the hypothalamus and several brainstem nuclei of rodents but not humans and suggested this as a basis for the species-specific effect of global $Na_v1.7$ knock-out. Impaired olfaction in patients with global $Na_v1.7$ (Goldberg et al. 2007, Nilsen et al. 2009) suggests a more likely alternative explanation of blunted olfaction in $Na_v1.7$ knock-out mice, which would be consistent with the report of Nassar et al. (2004) that neonatal $Na_v1.7$ knock-out mice die because they apparently are unable to feed.

Additionally, methods of pain assessment may contribute to the incongruent manifestation of pain behavior in experimental pain models and human pain symptoms. For example, allodynia and mechanical hyperalgesia, which are typically inferred from reduction in latency for paw withdrawal threshold using von Frey filaments, do not describe the response to suprathreshold stimuli. Electrophysiological recordings in $Na_v1.8$ knock-out animals demonstrated marked reductions in responses to suprathreshold mechanical stimuli, compared to wild-type, although behavioral assessments could not be used in this range (Matthews et al. 2006). Interpretation of data supporting the role of individual sodium channels in pain necessitates careful consideration of the model itself and the methods for pain assessment.

PROSPECTS FOR NEW PAIN THERAPEUTICS

There is a large set of potential targets for development of pain therapeutics. Each of the channels discussed here—$Na_v1.3$, $Na_v1.7$, $Na_v1.8$, and $Na_v1.9$—merits further study. The genetic linkage of $Na_v1.7$ to human pain disorders and the fact that total loss of this channel does not pose an immediate threat to life (no cardiac, cognitive, or motor deficit) have triggered substantial interest in $Na_v1.7$. Total loss of $Na_v1.7$ may predispose these patients to injuries and their complications. However, in a pharmacotherapeutic context, total block of this channel may not be needed. Indeed, characterization of Q10R, a delayed-onset IEM mutation that produces a small shift in $Na_v1.7$ activation, has demonstrated a quantitative difference in its effects on the gating properties of the channel and on its effects on DRG neuron excitability, compared to I848T, an early-onset IEM mutation that produced a larger shift in $Na_v1.7$ activation, providing evidence for a genotype-phenotype correlation (**Figure 3**) (Han et al. 2009). $Na_v1.7$ plays a definitive role in pain signaling, and current evidence suggests that it acts as a rheostat that sets the gain on pain. The physiological coupling of $Na_v1.7$ and $Na_v1.8$ (Rush et al. 2006) suggests that $Na_v1.8$, which is expressed exclusively in peripheral sensory neurons, may represent another especially opportune target.

Specific block of peripheral sodium channels may minimize risk of serious side effects. Robust $Na_v1.7$ expression within heterologous cells lends itself to high-throughput screening of small molecules and biological blockers. The discovery of a specific small molecule $Na_v1.8$ blocker (Jarvis et al. 2007) is encouraging and suggests that identification of isoform-specific, small molecule blockers is not unrealistic. Given the important role of the peripheral channels in pain states, and tolerance to loss of $Na_v1.7$ (humans, mice), $Na_v1.8$, and $Na_v1.9$ (mice), the pursuit of peripheral sodium channel blockers is an exciting prospect.

DISCLOSURE STATEMENT

The authors are not aware of any affiliations, memberships, funding, or financial holdings that might be perceived as affecting the objectivity of this review.

ACKNOWLEDGMENTS

We thank the members of our group for valuable discussions and Bart Toftness for technical assistance. Work in the S.G.W. laboratory is supported in part by grants from the National Multiple Sclerosis Society and the Rehabilitation Research and Development Service and Medical Research Service, Department of Veterans Affairs. T.R.C. was supported by research grant NS053422 from the National Institute of Health. The Center for Neuroscience and Regeneration Research is a Collaboration of the Paralyzed Veterans of America and the United Spinal Association with Yale University.

LITERATURE CITED

Abrahamsen B, Zhao J, Asante CO, Cendan CM, Marsh S, et al. 2008. The cell and molecular basis of mechanical, cold, and inflammatory pain. *Science* 321:702–5

Ahmad S, Dahllund L, Eriksson AB, Hellgren D, Karlsson U, et al. 2007. A stop codon mutation in SCN9A causes lack of pain sensation. *Hum. Mol. Genet.* 16:2114–21

Akopian AN, Sivilotti L, Wood JN. 1996. A tetrodotoxin-resistant voltage-gated sodium channel expressed by sensory neurons. *Nature* 379:257–62

Akopian AN, Souslova V, England S, Okuse K, Ogata N, et al. 1999. The tetrodotoxin-resistant sodium channel SNS has a specialized function in pain pathways. *Nat. Neurosci.* 2:541–48

Amaya F, Wang H, Costigan M, Allchorne AJ, Hatcher JP, et al. 2006. The voltage-gated sodium channel Na$_v$1.9 is an effector of peripheral inflammatory pain hypersensitivity. *J. Neurosci.* 26:12852–60

Baker MD, Chandra SY, Ding Y, Waxman SG, Wood JN. 2003. GTP-induced tetrodotoxin-resistant Na$^+$ current regulates excitability in mouse and rat small diameter sensory neurones. *J. Physiol.* 548:373–82

Beckh S, Noda M, Lubbert H, Numa S. 1989. Differential regulation of three sodium channel messenger RNAs in the rat central nervous system during development. *EMBO J.* 8:3611–16

Beyak MJ, Ramji N, Krol KM, Kawaja MD, Vanner SJ. 2004. Two TTX-resistant Na$^+$ currents in mouse colonic dorsal root ganglia neurons and their role in colitis-induced hyperexcitability. *Am. J. Physiol. Gastrointest. Liver Physiol.* 287:G845–55

Bielefeldt K, Ozaki N, Gebhart GF. 2002. Mild gastritis alters voltage-sensitive sodium currents in gastric sensory neurons in rats. *Gastroenterology* 122:752–61

Binshtok AM, Wang H, Zimmermann K, Amaya F, Vardeh D, et al. 2008. Nociceptors are interleukin-1beta sensors. *J. Neurosci.* 28:14062–73

Bird EV, Robinson PP, Boissonade FM. 2007. Na$_v$1.7 sodium channel expression in human lingual nerve neuromas. *Arch. Oral. Biol.* 52:494–502

Black JA, Cummins TR, Plumpton C, Chen YH, Hormuzdiar W, et al. 1999. Upregulation of a silent sodium channel after peripheral, but not central, nerve injury in DRG neurons. *J. Neurophysiol.* 82:2776–85

Black JA, Dib-Hajj S, McNabola K, Jeste S, Rizzo MA, et al. 1996. Spinal sensory neurons express multiple sodium channel alpha-subunit mRNAs. *Mol. Brain Res.* 43:117–31

Black JA, Liu S, Tanaka M, Cummins TR, Waxman SG. 2004. Changes in the expression of tetrodotoxin-sensitive sodium channels within dorsal root ganglia neurons in inflammatory pain. *Pain* 108:237–47

Black JA, Nikolajsen L, Kroner K, Jensen TS, Waxman SG. 2008. Multiple sodium channel isoforms and mitogen-activated protein kinases are present in painful human neuromas. *Ann. Neurol.* 64:644–53

Black JA, Renganathan M, Waxman SG. 2002. Sodium channel Na$_v$1.6 is expressed along nonmyelinated axons and it contributes to conduction. *Mol. Brain Res.* 105:19–28

Black JA, Waxman SG. 2002. Molecular identities of two tetrodotoxin-resistant sodium channels in corneal axons. *Exp. Eye Res.* 75:193–9

Blair NT, Bean BP. 2002. Roles of tetrodotoxin (TTX)-sensitive Na$^+$ current, TTX-resistant Na$^+$ current, and Ca^{2+} current in the action potentials of nociceptive sensory neurons. *J. Neurosci.* 22:10277–90

Blair NT, Bean BP. 2003. Role of tetrodotoxin-resistant Na$^+$ current slow inactivation in adaptation of action potential firing in small-diameter dorsal root ganglion neurons. *J. Neurosci.* 23:10338–50

Boucher TJ, Okuse K, Bennett DL, Munson JB, Wood JN, McMahon SB. 2000. Potent analgesic effects of GDNF in neuropathic pain states. *Science* 290:124–27

Braz JM, Nassar MA, Wood JN, Basbaum AI. 2005. Parallel "pain" pathways arise from subpopulations of primary afferent nociceptor. *Neuron* 47:787–93

Bulaj G, Zhang MM, Green BR, Fiedler B, Layer RT, et al. 2006. Synthetic mO-conotoxin MrVIB blocks TTX-resistant sodium channel Na$_V$1.8 and has a long-lasting analgesic activity. *Biochemistry* 45:7404–14

Catterall WA. 2000. From ionic currents to molecular mechanisms: the structure and function of voltage-gated sodium channels. *Neuron* 26:13–25

Catterall WA, Goldin AL, Waxman SG. 2005. Int. Union Pharmacol. XLVII. Nomenclature and structure-function relationships of voltage-gated sodium channels. *Pharmacol. Rev.* 57:397–409

Cavanaugh DJ, Lee H, Lo L, Shields SD, Zylka MJ, et al. 2009. Distinct subsets of unmyelinated primary sensory fibers mediate behavioral responses to noxious thermal and mechanical stimuli. *Proc. Natl. Acad. Sci. USA* 106:9075–80

Chattopadhyay M, Mata M, Fink DJ. 2008. Continuous delta-opioid receptor activation reduces neuronal voltage-gated sodium channel NaV1.7 levels through activation of protein kinase C in painful diabetic neuropathy. *J. Neurosci.* 28:6652–58

Cheng X, Dib-Hajj SD, Tyrrell L, Waxman SG. 2008. Mutation I136V alters electrophysiological properties of the NaV1.7 channel in a family with onset of erythromelalgia in the second decade. *Mol. Pain* 4:1

Choi JS, Dib-Hajj SD, Waxman SG. 2006. Inherited erythermalgia. Limb pain from an S4 charge-neutral Na channelopathy. *Neurology* 67:1563–67

Choi JS, Dib-Hajj SD, Waxman S. 2007. Differential slow inactivation and use-dependent inhibition of Nav1.8 channels contribute to distinct firing properties in IB^{4+} and IB^{4-} DRG neurons. *J. Neurophysiol.* 97:1258–65

Choi JS, Zhang L, Dib-Hajj SD, Han C, Tyrrell L, et al. 2009. Mexiletine-responsive erythromelalgia due to a new Na$_v$1.7 mutation showing use-dependent current fall-off. *Exp. Neurol.* 216:383–89

Coggeshall RE, Tate S, Carlton SM. 2004. Differential expression of tetrodotoxin-resistant sodium channels Na$_v$1.8 and Na$_v$1.9 in normal and inflamed rats. *Neurosci. Lett.* 355:45–48

Copel C, Osorio N, Crest M, Gola M, Delmas P, Clerc N. 2009. Activation of neurokinin 3 receptor increases Na$_v$1.9 current in enteric neurons. *J. Physiol.* 587:1461–79

Coste B, Osorio N, Padilla F, Crest M, Delmas P. 2004. Gating and modulation of presumptive NaV1.9 channels in enteric and spinal sensory neurons. *Mol. Cell Neurosci.* 26:123–34

Coward K, Aitken A, Powell A, Plumpton C, Birch R, et al. 2001. Plasticity of TTX-sensitive sodium channels PN1 and brain III in injured human nerves. *NeuroReport* 12:495–500

Coward K, Plumpton C, Facer P, Birch R, Carlstedt T, et al. 2000. Immunolocalization of SNS/PN3 and NaN/SNS2 sodium channels in human pain states. *Pain* 85:41–50

Cox JJ, Reimann F, Nicholas AK, Thornton G, Roberts E, et al. 2006. An SCN9A channelopathy causes congenital inability to experience pain. *Nature* 444:894–98

Craner MJ, Klein JP, Renganathan M, Black JA, Waxman SG. 2002. Changes of sodium channel expression in experimental painful diabetic neuropathy. *Ann. Neurol.* 52:786–92

Cummins TR, Aglieco F, Renganathan M, Herzog RI, Dib-Hajj SD, Waxman SG. 2001. Nav1.3 sodium channels: rapid repriming and slow closed-state inactivation display quantitative differences after expression in a mammalian cell line and in spinal sensory neurons. *J. Neurosci.* 21:5952–61

Cummins TR, Black JA, Dib-Hajj SD, Waxman SG. 2000. Glial-derived neurotrophic factor upregulates expression of functional SNS and NaN sodium channels and their currents in axotomized dorsal root ganglion neurons. *J. Neurosci.* 20:8754–61

Cummins TR, Dib-Hajj SD, Black JA, Akopian AN, Wood JN, Waxman SG. 1999. A novel persistent tetrodotoxin-resistant sodium current In SNS-null and wild-type small primary sensory neurons. *J. Neurosci.* 19:RC43

Cummins TR, Dib-Hajj SD, Waxman SG. 2004. Electrophysiological properties of mutant Nav1.7 sodium channels in a painful inherited neuropathy. *J. Neurosci.* 24:8232–36

Cummins TR, Howe JR, Waxman SG. 1998. Slow closed-state inactivation: a novel mechanism underlying ramp currents in cells expressing the hNE/PN1 sodium channel. *J. Neurosci.* 18:9607–19

Cummins TR, Waxman SG. 1997. Downregulation of tetrodotoxin-resistant sodium currents and upregulation of a rapidly repriming tetrodotoxin-sensitive sodium current in small spinal sensory neurons after nerve injury. *J. Neurosci.* 17:3503–14

Decosterd I, Ji RR, Abdi S, Tate S, Woolf CJ. 2002. The pattern of expression of the voltage-gated sodium channels $Na_V1.8$ and $Na_V1.9$ does not change in uninjured primary sensory neurons in experimental neuropathic pain models. *Pain* 96:269–77

Devor M. 2006. Sodium channels and mechanisms of neuropathic pain. *J. Pain* 7(Suppl. 1):S3–12

Dib-Hajj SD, Black JA, Cummins TR, Kenney AM, Kocsis JD, Waxman SG. 1998a. Rescue of alpha-SNS sodium channel expression in small dorsal root ganglion neurons following axotomy by in vivo administration of nerve growth factor. *J. Neurophysiol.* 79:2668–76

Dib-Hajj SD, Black JA, Cummins TR, Waxman SG. 2002. NaN/Nav1.9: a sodium channel with unique properties. *Trends Neurosci.* 25:253–59

Dib-Hajj SD, Black JA, Felts P, Waxman SG. 1996. Down-regulation of transcripts for Na channel alpha-SNS in spinal sensory neurons following axotomy. *Proc. Natl. Acad. Sci. USA* 93:14950–54

Dib-Hajj SD, Black JA, Waxman SG. 2009a. Voltage-gated sodium channels: therapeutic targets for pain. *Pain Med.* 10(7):1260–69

Dib-Hajj SD, Choi JS, Macala LJ, Tyrrell L, Black JA, et al. 2009b. Transfection of rat or mouse neurons by biolistics or electroporation. *Nat. Protoc.* 4:1118–26

Dib-Hajj SD, Cummins TR, Black JA, Waxman SG. 2007. From genes to pain: $Na_V1.7$ and human pain disorders. *Trends Neurosci.* 30: 555–63

Dib-Hajj SD, Estacion M, Jarecki BW, Tyrrell L, Fischer TZ, et al. 2008. Paroxysmal extreme pain disorder M1627K mutation in human Nav1.7 renders DRG neurons hyperexcitable. *Mol. Pain* 4:37

Dib-Hajj SD, Rush AM, Cummins TR, Hisama FM, Novella S, et al. 2005. Gain-of-function mutation in Nav1.7 in familial erythromelalgia induces bursting of sensory neurons. *Brain* 128:1847–54

Dib-Hajj SD, Tyrrell L, Black JA, Waxman SG. 1998b. NaN, a novel voltage-gated Na channel, is expressed preferentially in peripheral sensory neurons and down-regulated after axotomy. *Proc. Natl. Acad. Sci. USA* 95:8963–68

Dib-Hajj SD, Tyrrell L, Cummins TR, Black JA, Wood PM, Waxman SG. 1999. Two tetrodotoxin-resistant sodium channels in human dorsal root ganglion neurons. *FEBS Lett.* 462:117–20

Djouhri L, Fang X, Okuse K, Wood JN, Berry CM, Lawson S. 2003a. The TTX-resistant sodium channel Nav1.8 (SNS/PN3): expression and correlation with membrane properties in rat nociceptive primary afferent neurons. *J. Physiol.* 550:739–52

Djouhri L, Newton R, Levinson SR, Berry CM, Carruthers B, Lawson SN. 2003b. Sensory and electrophysiological properties of guinea-pig sensory neurones expressing $Na_V1.7$ (PN1) Na^+ channel alpha-subunit protein. *J. Physiol.* 546:565–76

Dong XW, Goregoaker S, Engler H, Zhou X, Mark L, et al. 2007. Small interfering RNA-mediated selective knockdown of $Na_V1.8$ tetrodotoxin-resistant sodium channel reverses mechanical allodynia in neuropathic rats. *Neuroscience* 146:812–21

Drenth JP, Waxman SG. 2007. Mutations in sodium-channel gene SCN9A cause a spectrum of human genetic pain disorders. *J. Clin. Invest.* 117:3603–9

Dworkin RH, O'Connor AB, Backonja M, Farrar JT, Finnerup NB, et al. 2007. Pharmacologic management of neuropathic pain: evidence-based recommendations. *Pain* 132:237–51

Ekberg J, Jayamanne A, Vaughan CW, Aslan S, Thomas L, et al. 2006. mO-conotoxin MrVIB selectively blocks $Na_V1.8$ sensory neuron specific sodium channels and chronic pain behavior without motor deficits. *Proc. Natl. Acad. Sci. USA* 103:17030–35

England S, Bevan S, Docherty RJ. 1996. PGE2 modulates the tetrodotoxin-resistant sodium current in neonatal rat dorsal root ganglion neurones via the cyclic AMP-protein kinase A cascade. *J. Physiol.* 495:429–40

Estacion M, Dib-Hajj SD, Benke PJ, Te Morsche RH, Eastman EM, et al. 2008. $Na_V1.7$ gain-of-function mutations as a continuum: A1632E displays physiological changes associated with erythromelalgia and paroxysmal extreme pain disorder mutations and produces symptoms of both disorders. *J. Neurosci.* 28:11079–88

Fang X, Djouhri L, Black JA, Dib-Hajj SD, Waxman SG, Lawson SN. 2002. The presence and role of the tetrodotoxin-resistant sodium channel $Na_V1.9$ (NaN) in nociceptive primary afferent neurons. *J. Neurosci.* 22:7425–33

Fang X, Djouhri L, McMullan S, Berry C, Waxman SG, et al. 2006. Intense isolectin-B4 binding in rat dorsal root ganglion neurons distinguishes C-fiber nociceptors with broad action potentials and high Nav1.9 expression. *J. Neurosci.* 26:7281–92

Fertleman CR, Baker MD, Parker KA, Moffatt S, Elmslie FV, et al. 2006. SCN9A mutations in paroxysmal extreme pain disorder: allelic variants underlie distinct channel defects and phenotypes. *Neuron* 52:767–74

Fertleman CR, Ferrie CD, Aicardi J, Bednarek NA, Eeg-Olofsson O, et al. 2007. Paroxysmal extreme pain disorder (previously familial rectal pain syndrome). *Neurology* 69:586–95

Fischer TZ, Gilmore ES, Estacion M, Eastman E, Taylor S, et al. 2009. A novel Na$_v$1.7 mutation producing carbamazepine-responsive erythromelalgia. *Ann. Neurol.* 65:733–41

Fitzgerald EM, Okuse K, Wood JN, Dolphin AC, Moss SJ. 1999. cAMP-dependent phosphorylation of the tetrodotoxin-resistant voltage-dependent sodium channel SNS. *J. Physiol.* 516:433–46

Fjell J, Cummins TR, Dib-Hajj SD, Fried K, Black JA, Waxman SG. 1999a. Differential role of GDNF and NGF in the maintenance of two TTX-resistant sodium channels in adult DRG neurons. *Mol. Brain Res.* 67:267–82

Fjell J, Cummins TR, Fried K, Black JA, Waxman SG. 1999b. In vivo NGF deprivation reduces SNS expression and TTX-R sodium currents in IB4-negative DRG neurons. *J. Neurophysiol.* 81:803–10

Fotia AB, Ekberg J, Adams DJ, Cook DI, Poronnik P, Kumar S. 2004. Regulation of neuronal voltage-gated sodium channels by the ubiquitin-protein ligases Nedd4 and Nedd4-2. *J. Biol. Chem.* 279:28930–35

Gerner P, Strichartz GR. 2008. Sensory and motor complications of local anesthetics. *Muscle Nerve* 37:421–25

Gold MS, Levine JD, Correa AM. 1998. Modulation of TTX-R/(Na) by PKC and PKA and their role in PGE$_2$-induced sensitization of rat sensory neurons in vitro. *J. Neurosci.* 18:10345–55

Gold MS, Reichling DB, Shuster MJ, Levine JD. 1996. Hyperalgesic agents increase a tetrodotoxin-resistant Na$^+$ current in nociceptors. *Proc. Natl. Acad. Sci. USA* 93:1108–12

Gold MS, Zhang L, Wrigley DL, Traub RJ. 2002. Prostaglandin E$_2$ modulates TTX-R I(Na) in rat colonic sensory neurons. *J. Neurophysiol.* 88:1512–22

Gold MS, Weinreich D, Kim CS, Wang R, Treanor J, et al. 2003. Redistribution of Na$_V$1.8 in uninjured axons enables neuropathic pain. *J. Neurosci.* 23:158–66

Goldberg Y, Macfarlane J, Macdonald M, Thompson J, Dube MP, et al. 2007. Loss-of-function mutations in the Na$_v$1.7 gene underlie congenital indifference to pain in multiple human populations. *Clin. Genet.* 71:311–19

Gould HJ 3rd, England JD, Soignier RD, Nolan P, Minor LD, et al. 2004. Ibuprofen blocks changes in Na$_v$ 1.7 and 1.8 sodium channels associated with complete freund's adjuvant-induced inflammation in rat. *J. Pain* 5:270–80

Gould HJ 3rd, Gould TN, England JD, Paul D, Liu ZP, Levinson SR. 2000. A possible role for nerve growth factor in the augmentation of sodium channels in models of chronic pain. *Brain Res.* 854:19–29

Hains BC, Klein JP, Saab CY, Craner MJ, Black JA, Waxman SG. 2003. Upregulation of sodium channel Na$_v$1.3 and functional involvement in neuronal hyperexcitability associated with central neuropathic pain after spinal cord injury. *J. Neurosci.* 23:8881–92

Hains BC, Saab CY, Klein JP, Craner MJ, Waxman SG. 2004. Altered sodium channel expression in second-order spinal sensory neurons contributes to pain after peripheral nerve injury. *J. Neurosci.* 24:4832–39

Hains BC, Saab CY, Waxman SG. 2005. Changes in electrophysiological properties and sodium channel Na$_v$1.3 expression in thalamic neurons after spinal cord injury. *Brain* 128:2359–71

Han C, Rush AM, Dib-Hajj SD, Li S, Xu Z, et al. 2006. Sporadic onset of erythermalgia: a gain-of-function mutation in Na$_v$1.7. *Ann. Neurol.* 59:553–58

Han C, Dib-Hajj SD, Lin Z, Li Y, Eastman EM, et al. 2009. Early- and late-onset inherited erythromelalgia: genotype-phenotype correlation. *Brain* 132:1711–22

Harty TP, Dib-Hajj SD, Tyrrell L, Blackman R, Hisama FM, et al. 2006. Na$_V$1.7 mutant A863P in erythromelalgia: effects of altered activation and steady-state inactivation on excitability of nociceptive dorsal root ganglion neurons. *J. Neurosci.* 26:12566–75

Herzog RI, Cummins TR, Ghassemi F, Dib-Hajj SD, Waxman SG. 2003. Distinct repriming and closed-state inactivation kinetics of Nav1.6 and Nav1.7 sodium channels in mouse spinal sensory neurons. *J. Physiol.* 551:741–50

Herzog RI, Cummins TR, Waxman SG. 2001. Persistent TTX-resistant Na^+ current affects resting potential and response to depolarization in simulated spinal sensory neurons. *J. Neurophysiol.* 86:1351–64

Hillsley K, Lin JH, Stanisz A, Grundy D, Aerssens J, et al. 2006. Dissecting the role of sodium currents in visceral sensory neurons in a model of chronic hyperexcitability using $Na_v1.8$ and $Na_v1.9$ null mice. *J. Physiol.* 576:257–67

Holland KD, Kearney JA, Glauser TA, Buck G, Keddache M, et al. 2008. Mutation of sodium channel SCN3A in a patient with cryptogenic pediatric partial epilepsy. *Neurosci. Lett.* 433:65–70

Hong S, Morrow TJ, Paulson PE, Isom LL, Wiley JW. 2004. Early painful diabetic neuropathy is associated with differential changes in tetrodotoxin-sensitive and -resistant sodium channels in dorsal root ganglion neurons in the rat. *J. Biol. Chem.* 279:29341–50

Hucho T, Levine JD. 2007. Signaling pathways in sensitization: toward a nociceptor cell biology. *Neuron* 55:365–76

Hudmon A, Choi JS, Tyrrell L, Black JA, Rush AM, et al. 2008. Phosphorylation of sodium channel $Na_v1.8$ by p38 mitogen-activated protein kinase increases current density in dorsal root ganglion neurons. *J. Neurosci.* 28:3190–201

Jarecki BW, Sheets PL, Jackson JO 2nd, Cummins TR. 2008. Paroxysmal extreme pain disorder mutations within the D3/S4-S5 linker of $Na_v1.7$ cause moderate destabilization of fast inactivation. *J. Physiol.* 586:4137–53

Jarvis MF, Honore P, Shieh CC, Chapman M, Joshi S, et al. 2007. A-803467, a potent and selective $Na_v1.8$ sodium channel blocker, attenuates neuropathic and inflammatory pain in the rat. *Proc. Natl. Acad. Sci. USA* 104:8520–25

Jin X, Gereau RW. 2006. Acute p38-mediated modulation of tetrodotoxin-resistant sodium channels in mouse sensory neurons by tumor necrosis factor-alpha. *J. Neurosci.* 26:246–55

Joshi SK, Mikusa JP, Hernandez G, Baker S, Shieh CC, et al. 2006. Involvement of the TTX-resistant sodium channel $Na_v1.8$ in inflammatory and neuropathic, but not post-operative, pain states. *Pain* 123:75–82

Kerr BJ, Souslova V, McMahon SB, Wood JN. 2001. A role for the TTX-resistant sodium channel $Na_v1.8$ in NGF-induced hyperalgesia, but not neuropathic pain. *NeuroReport* 12:3077–80

Kim CH, Oh Y, Chung JM, Chung K. 2001. The changes in expression of three subtypes of TTX sensitive sodium channels in sensory neurons after spinal nerve ligation. *Mol. Brain Res.* 95:153–61

Kim CH, Oh Y, Chung JM, Chung K. 2002. Changes in three subtypes of tetrodotoxin sensitive sodium channel expression in the axotomized dorsal root ganglion in the rat. *Neurosci. Lett.* 323:125–28

Klugbauer N, Lacinova L, Flockerzi V, Hofmann F. 1995. Structure and functional expression of a new member of the tetrodotoxin-sensitive voltage-activated sodium channel family from human neuroendocrine cells. *EMBO J.* 14:1084–90

Kretschmer T, Happel LT, England JD, Nguyen DH, Tiel RL, et al. 2002. Accumulation of PN1 and PN3 sodium channels in painful human neuroma—evidence from immunocytochemistry. *Acta Neurochir.* 144:803–10

Lai J, Gold MS, Kim CS, Bian D, Ossipov MH, et al. 2002. Inhibition of neuropathic pain by decreased expression of the tetrodotoxin-resistant sodium channel, NaV1.8. *Pain* 95:143–52

Laird JM, Souslova V, Wood JN, Cervero F. 2002. Deficits in visceral pain and referred hyperalgesia in Nav1.8 (SNS/PN3)-null mice. *J. Neurosci.* 22:8352–56

Lampert A, Dib-Hajj SD, Eastman EM, Tyrrell L, Lin Z, et al. 2009. Erythromelalgia mutation L823R shifts activation and inactivation of threshold sodium channel Nav1.7 to hyperpolarized potentials. *Biochem. Biophys. Res. Commun.* 390:319–24

Lampert A, Dib-Hajj SD, Tyrrell L, Waxman SG. 2006a. Size matters: Erythromelalgia mutation S241T in Nav1.7 alters channel gating. *J. Biol. Chem.* 281 36029–35

Lampert A, Hains BC, Waxman SG. 2006b. Upregulation of persistent and ramp sodium current in dorsal horn neurons after spinal cord injury. *Exp. Brain Res.* 174:660–66

Leffler A, Cummins TR, Dib-Hajj SD, Hormuzdiar WN, Black JA, Waxman SG. 2002. GDNF and NGF reverse changes in repriming of TTX-sensitive Na^+ currents following axotomy of dorsal root ganglion neurons. *J. Neurophysiol.* 88:650–58

Lindia JA, Kohler MG, Martin WJ, Abbadie C. 2005. Relationship between sodium channel $Na_v1.3$ expression and neuropathic pain behavior in rats. *Pain* 117:145–53

Liu C, Cummins TR, Tyrrell L, Black JA, Waxman SG, Dib-Hajj SD. 2005. CAP-1A is a novel linker that binds clathrin and the voltage-gated sodium channel Na$_V$1.8. *Mol. Cell Neurosci.* 28:636–49

Liu X, Zhou JL, Chung K, Chung JM. 2001. Ion channels associated with the ectopic discharges generated after segmental spinal nerve injury in the rat. *Brain Res.* 900:119–27

Lyu YS, Park SK, Chung K, Chung JM. 2000. Low dose of tetrodotoxin reduces neuropathic pain behaviors in an animal model. *Brain Res.* 871:98–103

Matthews EA, Wood JN, Dickenson AH. 2006. Nav 1.8-null mice show stimulus-dependent deficits in spinal neuronal activity. *Mol. Pain* 2:5

Middleton RE, Warren VA, Kraus RL, Hwang JC, Liu CJ, et al. 2002. Two tarantula peptides inhibit activation of multiple sodium channels. *Biochemistry* 41:14734–47

Mogil JS. 2009. Animal models of pain: progress and challenges. *Nat. Rev. Neurosci.* 10:283–94

Nassar MA, Stirling LC, Forlani G, Baker MD, Matthews EA, et al. 2004. Nociceptor-specific gene deletion reveals a major role for Nav1.7 (PN1) in acute and inflammatory pain. *Proc. Natl. Acad. Sci. USA* 101:12706–11

Nassar MA, Levato A, Stirling C, Wood JN. 2005. Neuropathic pain develops normally in mice lacking both Nav1.7 and Nav1.8. *Mol. Pain* 1:24

Nassar MA, Baker MD, Levato A, Ingram R, Mallucci G, et al. 2006. Nerve injury induces robust allodynia and ectopic discharges in Nav1.3 null mutant mice. *Mol. Pain* 2:33

Nilsen KB, Nicholas AK, Woods CG, Mellgren SI, Nebuchennykh M, Aasly J. 2009. Two novel SCN9A mutations causing insensitivity to pain. *Pain* 143:155–58

Okuse K, Chaplan SR, McMahon SB, Luo ZD, Calcutt NA, et al. 1997. Regulation of expression of the sensory neuron-specific sodium channel SNS in inflammatory and neuropathic pain. *Mol. Cell Neurosci.* 10:196–207

Okuse K, Malik-Hall M, Baker MD, Poon WY, Kong H, et al. 2002. Annexin II light chain regulates sensory neuron-specific sodium channel expression. *Nature* 417:653–56

Ostman JA, Nassar MA, Wood JN, Baker MD. 2007. GTP up-regulated persistent Na$^+$ current and enhanced nociceptor excitability require Na$_V$1.9. *J. Physiol.* 586:1077–87

Pierre K, Dupouy B, Allard M, Poulain DA, Theodosis DT. 2001. Mobilization of the cell adhesion glycoprotein F3/contactin to axonal surfaces is activity dependent. *Eur. J. Neurosci.* 14:645–56

Porreca F, Lai J, Bian D, Wegert S, Ossipov MH, et al. 1999. A comparison of the potential role of the tetrodotoxin-insensitive sodium channels, PN3/SNS and NaN/SNS2, in rat models of chronic pain. *Proc. Natl. Acad. Sci. USA* 96:7640–44

Priest BT, Murphy BA, Lindia JA, Diaz C, Abbadie C, et al. 2005. Contribution of the tetrodotoxin-resistant voltage-gated sodium channel NaV1.9 to sensory transmission and nociceptive behavior. *Proc. Natl. Acad. Sci. USA* 102:9382–87

Raymond CK, Castle J, Garrett-Engele P, Armour CD, Kan Z, et al. 2004. Expression of alternatively spliced sodium channel alpha-subunit genes: Unique splicing patterns are observed in dorsal root ganglia. *J. Biol. Chem.* 279:46234–41

Renganathan M, Cummins TR, Waxman SG. 2001. Contribution of Nav1.8 sodium channels to action potential electrogenesis in DRG neurons. *J. Neurophysiol.* 86:629–40

Rice AS, Hill RG. 2006. New treatments for neuropathic pain. *Annu. Rev. Med.* 57:535–51

Roza C, Laird JM, Souslova V, Wood JN, Cervero F. 2003. The tetrodotoxin-resistant Na$^+$ channel Nav1.8 is essential for the expression of spontaneous activity in damaged sensory axons of mice. *J. Physiol.* 550:921–26

Rugiero F, Mistry M, Sage D, Black JA, Waxman SG, et al. 2003. Selective expression of a persistent tetrodotoxin-resistant Na$^+$ current and NaV1.9 subunit in myenteric sensory neurons. *J. Neurosci.* 23:2715–25

Rush AM, Craner MJ, Kageyama T, Dib-Hajj SD, Waxman SG, Ranscht B. 2005. Contactin regulates the current density and axonal expression of tetrodotoxin-resistant but not tetrodotoxin-sensitive sodium channels in DRG neurons. *Eur. J. Neurosci.* 22:39–49

Rush AM, Cummins TR, Waxman SG. 2007. Multiple sodium channels and their roles in electrogenesis within dorsal root ganglion neurons. *J. Physiol.* 579(Part 1):1–14

Rush AM, Dib-Hajj SD, Liu S, Cummins TR, Black JA, Waxman SG. 2006. A single sodium channel mutation produces hyper- or hypoexcitability in different types of neurons. *Proc. Natl. Acad. Sci. USA* 103:8245–50

Rush AM, Waxman SG. 2004. PGE_2 increases the tetrodotoxin-resistant $Na_v1.9$ sodium current in mouse DRG neurons via G-proteins. *Brain Res.* 1023:264–71

Sage D, Salin P, Alcaraz G, Castets F, Giraud P, et al. 2007. $Na_v1.7$ and $Na_v1.3$ are the only tetrodotoxin-sensitive sodium channels expressed by the adult guinea pig enteric nervous system. *J. Comp. Neurol.* 504:363–78

Sangameswaran L, Delgado SG, Fish LM, Koch BD, Jakeman LB, et al. 1996. Structure and function of a novel voltage-gated, tetrodotoxin-resistant sodium channel specific to sensory neurons. *J. Biol. Chem.* 271:5953–56

Sangameswaran L, Fish LM, Koch BD, Rabert DK, Delgado SG, et al. 1997. A novel tetrodotoxin-sensitive, voltage-gated sodium channel expressed in rat and human dorsal root ganglia. *J. Biol. Chem.* 272:14805–9

Schmalhofer WA, Calhoun J, Burrows R, Bailey T, Kohler MG, et al. 2008. ProTx-II, a selective inhibitor of NaV1.7 sodium channels, blocks action potential propagation in nociceptors. *Mol. Pharmacol.* 74:1476–84

Scholz J, Woolf CJ. 2007. The neuropathic pain triad: neurons, immune cells and glia. *Nat. Neurosci.* 10:1361–68

Shah BS, Rush AM, Liu S, Tyrrell L, Black JA, et al. 2004. Contactin associates with sodium channel Nav1.3 in native tissues and increases channel density at the cell surface. *J. Neurosci.* 24:7387–99

Sheets PL, Jackson JO, Waxman SG, Dib-Hajj S, Cummins TR. 2007. A Nav1.7 channel mutation associated with hereditary erythromelalgia contributes to neuronal hyperexcitability and displays reduced lidocaine sensitivity. *J. Physiol.* 581:1019–31

Sindrup SH, Jensen TS. 2007. Are sodium channel blockers useless in peripheral neuropathic pain? *Pain* 128:6–7

Sleeper AA, Cummins TR, Dib-Hajj SD, Hormuzdiar W, Tyrrell L, et al. 2000. Changes in expression of two tetrodotoxin-resistant sodium channels and their currents in dorsal root ganglion neurons after sciatic nerve injury but not rhizotomy. *J. Neurosci.* 20:7279–89

Stamboulian S, Choi JS, Ahn HS, Chang YW, Tyrrell L, et al. 2010. ERK1/2 mitogen-activated protein kinase phosphorylates sodium channel Nav1.7 and alters its gating properties. *J. Neurosci.* 30:1637–47

Strickland IT, Martindale JC, Woodhams PL, Reeve AJ, Chessell IP, McQueen DS. 2008. Changes in the expression of $Na_V1.7$, $Na_V1.8$ and $Na_V1.9$ in a distinct population of dorsal root ganglia innervating the rat knee joint in a model of chronic inflammatory joint pain. *Eur. J. Pain* 12:564–72

Stucky CL, Lewin GR. 1999. Isolectin B_4-positive and -negative nociceptors are functionally distinct. *J. Neurosci.* 19:6497–505

Tanaka M, Cummins TR, Ishikawa K, Dib-Hajj SD, Black JA, Waxman SG. 1998. SNS Na^+ channel expression increases in dorsal root ganglion neurons in the carrageenan inflammatory pain model. *NeuroReport* 9:967–72

Tate S, Benn S, Hick C, Trezise D, John V, et al. 1998. Two sodium channels contribute to the TTX-R sodium current in primary sensory neurons. *Nat. Neurosci.* 1:653–55

Thimmapaya R, Neelands T, Niforatos W, Davis-Taber RA, Choi W, et al. 2005. Distribution and functional characterization of human Na1.3 splice variants. *Eur. J. Neurosci.* 22:1–9

Toledo-Aral JJ, Moss BL, He ZJ, Koszowski AG, Whisenand T, et al. 1997. Identification of PN1, a predominant voltage-dependent sodium channel expressed principally in peripheral neurons. *Proc. Natl. Acad. Sci. USA* 94:1527–32

Tripathi PK, Trujillo L, Cardenas CA, Cardenas CG, de Armendi AJ, Scroggs RS. 2006. Analysis of the variation in use-dependent inactivation of high-threshold tetrodotoxin-resistant sodium currents recorded from rat sensory neurons. *Neuroscience* 143:923–38

Tyrrell L, Renganathan M, Dib-Hajj SD, Waxman SG. 2001. Glycosylation alters steady-state inactivation of sodium channel Nav1.9/NaN in dorsal root ganglion neurons and is developmentally regulated. *J. Neurosci.* 21:9629–37

Vijayaragavan K, Boutjdir M, Chahine M. 2004. Modulation of Nav1.7 and Nav1.8 peripheral nerve sodium channels by protein kinase A and protein kinase C. *J. Neurophysiol.* 91:1556–69

Waxman SG. 2001. Transcriptional channelopathies: an emerging class of disorders. *Nat. Rev. Neurosci.* 2:652–59

Waxman SG. 2006. Neurobiology: a channel sets the gain on pain. *Nature* 444:831–32

Waxman SG, Hains BC. 2006. Fire and phantoms after spinal cord injury: Na$^+$ channels and central pain. *Trends Neurosci.* 29:207–15

Waxman SG, Kocsis JD, Black JA. 1994. Type III sodium channel mRNA is expressed in embryonic but not adult spinal sensory neurons, and is reexpressed following axotomy. *J. Neurophysiol.* 72:466–70

Whitaker WR, Faull RL, Waldvogel HJ, Plumpton CJ, Emson PC, Clare JJ. 2001. Comparative distribution of voltage-gated sodium channel proteins in human brain. *Mol. Brain Res.* 88:37–53

Xiao Y, Bingham JP, Zhu W, Moczydlowski E, Liang S, Cummins TR. 2008. Tarantula huwentoxin-IV inhibits neuronal sodium channels by binding to receptor site 4 and trapping the domain II voltage sensor in the closed configuration. *J. Biol. Chem.* 283:27300–13

Yang Y, Wang Y, Li S, Xu Z, Li H, et al. 2004. Mutations in SCN9A, encoding a sodium channel alpha subunit, in patients with primary erythermalgia. *J. Med. Genet.* 41:171–74

Yeomans DC, Levinson SR, Peters MC, Koszowski AG, Tzabazis AZ, et al. 2005. Decrease in inflammatory hyperalgesia by Herpes vector–mediated knockdown of Na$_v$1.7 sodium channels in primary afferents. *Hum. Gene Ther.* 16:271–77

Yiangou Y, Birch R, Sangameswaran L, Eglen R, Anand P. 2000. SNS/PN3 and SNS2/NaN sodium channel-like immunoreactivity in human adult and neonate injured sensory nerves. *FEBS Lett.* 467:249–52

Yoshimura N, Seki S, Novakovic SD, Tzoumaka E, Erickson VL, et al. 2001. The involvement of the tetrodotoxin-resistant sodium channel Na$_v$1.8 (PN3/SNS) in a rat model of visceral pain. *J. Neurosci.* 21:8690–96

Zhang XF, Zhu CZ, Thimmapaya R, Choi WS, Honore P, et al. 2004. Differential action potentials and firing patterns in injured and uninjured small dorsal root ganglion neurons after nerve injury. *Brain Res.* 1009:147–58

Zhang YH, Vasko MR, Nicol GD. 2002. Ceramide, a putative second messenger for nerve growth factor, modulates the TTX-resistant Na$^+$ current and delayed rectifier K$^+$ current in rat sensory neurons. *J. Physiol.* 544:385–402

Zhao P, Barr TP, Hou Q, Dib-Hajj SD, Black JA, et al. 2008. Voltage-gated sodium channel expression in rat and human epidermal keratinocytes: evidence for a role in pain. *Pain* 139:90–105

Zhao P, Waxman SG, Hains BC. 2006. Sodium channel expression in the ventral posterolateral nucleus of the thalamus after peripheral nerve injury. *Mol. Pain* 2:27

Zhou Z, Davar G, Strichartz G. 2002. Endothelin-1 (ET-1) selectively enhances the activation gating of slowly inactivating tetrodotoxin-resistant sodium currents in rat sensory neurons: a mechanism for the pain-inducing actions of ET-1. *J. Neurosci.* 22:6325–30

Zimmermann K, Leffler A, Babes A, Cendan CM, Carr RW, et al. 2007. Sensory neuron sodium channel Nav1.8 is essential for pain at low temperatures. *Nature* 447:855–58

RELATED RESOURCES

Costigan M, Scholz J, Woolf CJ. 2009. Neuropathic pain: a maladaptive response of the nervous system to damage. *Annu. Rev. Neurosci.* 32:1–32

Dib-Hajj SD, Binshtok AM, Cummins TR, Jarvis MF, Samad T, Zimmermann K. 2009. Voltage-gated sodium channels in pain states: role in pathophysiology and targets for treatment. *Brain Res. Rev.* 60:65–83

Ji RR, Strichartz G. 2004. Cell signaling and the genesis of neuropathic pain. *Sci. STKE* 2004:reE14

McCleskey EW, Gold MS. 1999. Ion channels of nociception. *Annu. Rev. Physiol.* 61:835–56

Momin A, Wood JN. 2008. Sensory neuron voltage-gated sodium channels as analgesic drug targets. *Curr. Opin. Neurobiol.* 18:383–88

Mechanisms of Synapse and Dendrite Maintenance and Their Disruption in Psychiatric and Neurodegenerative Disorders

Yu-Chih Lin[1] and Anthony J. Koleske[1,2,3]

[1]Department of Molecular Biophysics and Biochemistry, [2]Department of Neurobiology, and [3]Interdepartmental Neuroscience Program, Yale University, New Haven, Connecticut 06520-8024; email: yu-chih.lin@yale.edu, anthony.koleske@yale.edu

Annu. Rev. Neurosci. 2010. 33:349–78

First published online as a Review in Advance on April 5, 2010

The *Annual Review of Neuroscience* is online at neuro.annualreviews.org

This article's doi: 10.1146/annurev-neuro-060909-153204

Key Words

neurodegeneration, dendritic spine, actin cytoskeleton, psychiatric disease

Abstract

Emerging evidence indicates that once established, synapses and dendrites can be maintained for long periods, if not for the organism's entire lifetime. In contrast to the wealth of knowledge regarding axon, dendrite, and synapse development, we understand comparatively little about the cellular and molecular mechanisms that enable long-term synapse and dendrite maintenance. Here, we review how the actin cytoskeleton and its regulators, adhesion receptors, and scaffolding proteins mediate synapse and dendrite maintenance. We examine how these mechanisms are reinforced by trophic signals passed between the pre- and postsynaptic compartments. We also discuss how synapse and dendrite maintenance mechanisms are compromised in psychiatric and neurodegenerative disorders.

Contents

Filopodia: thin,
finger-like structures
with long, parallel
bundled actin
filaments that protrude
from the cell surface

PREFACE: SYNAPSES STABILIZE DENDRITIC ARBORS

Excitatory glutamatergic synapses form between presynaptic axon specializations and specialized protrusions, called dendritic spines, that extend from the dendritic shaft. Dendritic spines initiate as thin filopodial structures that emerge from the dendritic shaft (Fiala et al. 1998, Holtmaat et al. 2005, Ziv & Smith 1996, Zuo et al. 2005). These dendritic filopodia contain small clusters of adhesion and scaffolding proteins that serve as the building blocks of the postsynaptic specialization (Niell et al.

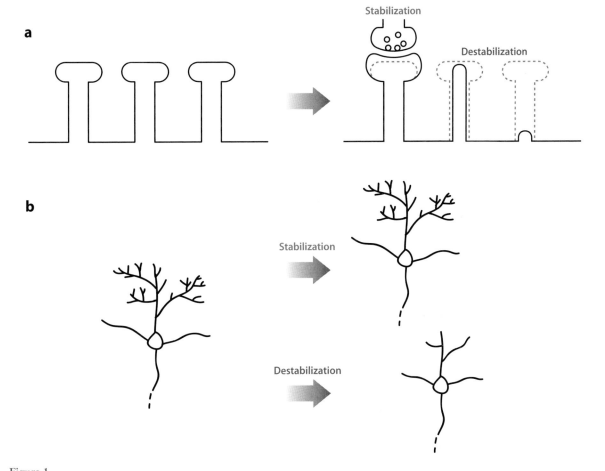

Figure 1

Stabilization of dendritic spines. (*a*) Synaptic activity, including adhesive contact between pre- and postsynaptic compartments and trophic signaling between the two compartments, enhances dendritic spine stability. (*b*) Synaptic support maintains the integrity of dendritic architecture over time. The loss of this support results in destabilization of dendritic spines followed by dendritic simplification.

2004). Contact of a filopodium with a presynaptic compartment promotes the stabilization and enlargement of the filopodium tip into a dendritic spine (Ziv & Smith 1996, Zuo et al. 2005).

Dendritic arbors are highly dynamic, exhibiting frequent branch additions and retractions, with only a subset stabilized and maintained (Cline 2001, Dailey & Smith 1996, Wong et al. 2000). Productive synapse formation and the accompanying activation of postsynaptic signaling mechanisms promote arbor stability (Niell et al. 2004, Rajan et al. 1999,

Wu & Cline 1998). These stabilizing influences likely explain the age-dependent slowing of dendritic spine and branch dynamics that follows the peak of synapse formation and refinement (Holtmaat et al. 2005, Trachtenberg et al. 2002, Wu et al. 1999). Conversely, a loss of, or reduction in, synaptic inputs leads to dendritic loss in vivo (**Figure 1**), which indicates that the maintenance of synaptic input is critical for dendritic stability (Coleman & Riesen 1968, Jones & Thomas 1962, Le Gros Clark 1957, Matthews & Powell 1962, Sfakianos et al. 2007).

INTRODUCTION: SYNAPSES AND DENDRITES MUST BE MAINTAINED FOR LONG PERIODS

The vast majority of stable dendritic spines and filopodia-like protrusions on mature dendrites have synapses (Arellano et al. 2007, Harris 1999). Long-term confocal microscopy imaging in adult rodents indicates that a large fraction of dendritic spines as well as complete dendritic arbors are stable for extended time periods of several months, and possibly years (Holtmaat et al. 2005, Majewska et al. 2006, Trachtenberg et al. 2002, Zuo et al. 2005). Together, these findings argue that individual synapses may last for the majority of an organism's lifetime, possibly decades in the case of humans.

Synapse and dendritic spine loss and dendritic atrophy are observed in the aging human brain (Uylings et al. 2000). Reductions in synapse number and dendritic arbor size are also associated with psychiatric illnesses, such as schizophrenia and major depressive disorder (MDD), as well as neurodegenerative diseases, such as Alzheimer's disease (AD) (Anderton et al. 1998; Broadbelt et al. 2002; Cotter et al. 2001, 2002; Flood 1991; Flood et al. 1987a, 1987b; Glantz & Lewis 1997, 2000; Kalus et al. 2000; Rajkowska et al. 1999; Woo et al. 1998; Hanks & Flood 1991; Law et al. 2004; Stockmeier et al. 2004). These reductions in synaptic connectivity are believed to be a major contributor to the altered mood and impaired perception and cognition that characterize these conditions.

Given the importance of synapse stability for human brain health, a major unmet goal in neuroscience is to understand how neurons achieve long-term synaptic stability. Elucidating these mechanisms is an essential prerequisite to developing therapies to prevent synapse loss in aging and disease. Here, we review the cellular and biochemical mechanisms that support long-term synaptic maintenance. We also discuss how disease pathology may undermine the mechanisms that maintain long-term synapse and dendrite stability.

EXPERIMENTAL APPROACHES TO STUDY SYNAPSE AND DENDRITE MAINTENANCE

Synapse and dendrite maintenance can be studied in long-term, established neuronal cultures, which are maintained long enough to allow synapses to form; in brain slices; and in whole animals. Each of these systems has both benefits and limitations. Cultured neuron and slice models offer more ease of manipulation, but may lack essential in vivo determinants of synapse and dendrite maintenance. For example, neuronal cultures grown on two-dimensional plastic dishes do not faithfully mimic the soft tissue properties or supportive (e.g., glia) or nutrient supply systems of the brain. Cultured neurons also do not establish normal neuronal wiring patterns. For these reasons, synapses in cultured neurons may be inherently less stable than synapses and dendrites in vivo. Although whole animal systems are more difficult to manipulate, they offer the only opportunity to investigate the importance of synapse and dendrite maintenance regulators in the physiological context. Ideally, investigations of synapse and dendrite maintenance would employ complementary techniques/approaches, using cultured systems to pose specific mechanistic questions and animal models to explore how manipulation of these mechanisms affects synapse and dendrite stability in vivo. Indeed, for most of the regulatory molecules we discuss below, loss of function leads to similar effects in both in vitro and in vivo systems.

Many key regulators of synapse and dendrite maintenance also have essential functions in neuronal migration, axon or dendrite targeting, or synapse formation. Thus, the general approach to study roles in synapse maintenance involves using genetic (e.g., RNA interference or gene knockout) or pharmacological tactics to inhibit a potential regulator after synapses and dendrites are properly formed. Excitatory synapse dissolution requires dendritic spine shrinkage or retraction. The effects of these manipulations on dendritic spine stability are assessed by monitoring both spine density and morphology, including spine length, head shape, and width or volume. Defects in spine maintenance can be reflected by a decrease in spine number and size, as well as a proportional increase in the smaller, thinner spines or filopodia-like protrusions. Dendrite arbor structure can be assessed by labeling the neurons via dye injection or fluorescent protein expression, and quantified using computer-assisted camera lucida tracing systems.

PROTEINS THAT PROVIDE SUPPORT FOR SYNAPSE MAINTENANCE

Long-term synapse and dendrite maintenance relies heavily on structural proteins that support the overall synaptic framework and organize the synaptic signaling machinery. We review here how the actin cytoskeleton, adhesion receptors, and scaffolding proteins provide structural support for long-term synapse and dendrite maintenance.

F-Actin Provides Long-Term Structural Support for the Presynaptic Compartment and Dendritic Spines

The actin cytoskeleton provides essential structural support for long-term synapse maintenance. A presynaptic filamentous (F-) actin meshwork acts as a scaffold to organize the neurotransmitter release machinery, tether regulators near sites of release, and facilitate vesicle trafficking and endocytosis (Dillon & Goda 2005, Halpain 2003, Zhang & Benson 2001). Postsynaptically, F-actin provides integrity for dendritic spines and organizes signaling machinery at the postsynaptic density (Dillon & Goda 2005, Schubert & Dotti 2007, Sekino et al. 2007). We highlight here features of the F-actin cytoskeleton and its regulators that are especially critical for long-term synapse maintenance. For a more comprehensive treatment of F-actin and its regulators at the synapse, we refer readers to an excellent recent review in this series (Dillon & Goda 2005).

Dendritic spines have diverse morphologies (e.g., thin, stubby, and mushroom shapes), which can be altered in response to activity or injury (Harris 1999, Sorra & Harris 2000). F-actin is highly enriched in dendritic spines compared with other parts of the neuron (Cohen et al. 1985, Fifkova & Delay 1982, Matus et al. 1982). The dendritic spine neck is composed of bundled actin similar to that found in filopodia and microvilli. The spine head contains a branched F-actin network analogous to that found in lamellipodia (Cheng et al. 2000, Landis & Reese 1983). The visualization of GFP-labeled actin in spines reveals that spines can undergo significant changes in shape (Fischer et al. 1998, Star et al. 2002). This dynamic behavior results from cycles of continuous actin polymerization/depolymerization involving up to 95% of the total spine actin (Honkura et al. 2008, Okamoto et al. 2004, Star et al. 2002, Zhang & Benson 2001). These observations reveal the importance of F-actin dynamics in supporting spine shape and its dynamic remodeling. This structural plasticity likely contributes to the ability of synapses to alter their synaptic signaling strength based on experience, a phenomenon known as synaptic plasticity.

Although the precise configuration of the actin cytoskeleton determines dendritic spine morphology, long-term spine maintenance requires perfectly balanced cycles of actin polymerization and depolymerization within the spine. Treatment with jasplakinolide, which induces actin polymerization, results in spine enlargement, whereas latrunculin-A or cytochalasin-D, which shift the actin equilibrium toward increased G-actin, reduce spine motility and cause spine shrinkage or loss (Allison et al. 1998, Fischer et al. 1998, Honkura et al. 2008, Okamoto et al. 2004, Star et al. 2002). Together, these experiments reveal that F-actin structure is critical for dendritic spine maintenance.

F-ACTIN REGULATORS CONTRIBUTE TO SYNAPSE AND DENDRITIC SPINE MAINTENANCE

F-actin structure is controlled by a large set of regulatory proteins, a growing number of which have been implicated in long-term synapse and dendritic spine maintenance (**Figure 2**). In many cases, disruption or altered activity of these regulators contributes to psychiatric or neurodegenerative disorders associated with defects in dendritic spine or synapse maintenance (**Table 1**). Here, we highlight key actin regulators that play essential roles in synapse and dendritic spine maintenance.

Rho Family GTPases: Rho and Rac

Rho family GTPases are a class of small GTPases that control actin dynamics and rearrangement to regulate numerous cellular functions, including spine morphogenesis and stability (Govek et al. 2004, Nadif Kasri & Van Aelst 2008). In particular, Rac1 (Rac) and RhoA (Rho) play important roles in spine maintenance (**Figure 3**) (Nakayama et al. 2000, Tashiro & Yuste 2004). In general, Rac activation stimulates F-actin polymerization and dendritic spine stability, whereas Rho activation in mature neurons leads to synapse and dendritic spine loss and dendritic regression (Govek et al. 2004, Lee et al. 2000, Ruchhoeft et al. 1999, Tashiro et al. 2000), although localized reductions in Rho activity are associated with spine collapse under some conditions (Schubert et al. 2006).

Rac and Rho regulate spine structure and maintenance by controlling actin cytoskeletal dynamics. Downstream effectors of Rac1 include p21-activated kinase 1 (Pak1) and insulin receptor substrate p53 (IRSp53) (Ethell & Pasquale 2005). Activation of Pak1 regulates spine stability by activating LIM kinase-mediated cofilin inhibition (Meng et al. 2002, Penzes et al. 2003). Activation of IRSp53 by Rac induces WASP family verprolin-homologous protein (WAVE) and Arp2/3-mediated actin polymerization, resulting in spine head enlargement (Ethell & Pasquale 2005, Miki et al. 2000). Rho activation stimulates Rho-associated protein kinase (ROCK), a major Rho effector, which stimulates actomyosin contractility, resulting in spine shortening and retraction (Ethell & Pasquale 2005, Fukata et al. 2001). Interestingly, Rac and Rho share some

downstream effectors, including myosin light chain kinase and LIMK (Maekawa et al. 1999). Exactly how they engage similar molecules to achieve opposite effects on spine stability needs to be further investigated. Finally, Rac and Rho are activated by guanine nucleotide-exchange factors (GEFs) and turned off by GTPase-activating proteins (GAPs), and both classes of regulators have been implicated in the regulation of synapse and dendritic spine stability.

Regulators of Rho Family GTPases

Oligophrenin-1. Mutations in oligophrenin-1 (OPHN1), a RhoGAP, are associated with X-linked mental retardation (Billuart et al. 1998, Zanni et al. 2005). OPHN1 is predominantly expressed in the brain and colocalizes with PSD-95 and F-actin in dendritic spines (Govek et al. 2004). OPHN1 overexpression in established neuronal cultures has no effect on spine density, but it enlarges spine head size, which is known to correlate with increased spine stability (Nadif Kasri et al. 2009). Conversely, OPHN1 knockdown in vitro increases RhoA activity and destabilizes dendritic spines by reducing spine density and length (Govek et al. 2004, Nadif Kasri et al. 2009). The OPHN1 knockdown spine phenotype mimics the phenotype resulting from RhoA activation, whereas inhibiting ROCK blocks the spine destabilization that accompanies OPHN1 knockdown in mature neurons (Govek et al. 2004).

Arg and p190RhoGAP. Our laboratory has identified a Rho inhibitory pathway that regulates synapse and dendrite stability. In non-neuronal cells, integrin-mediated adhesion activates the Abl-related-gene (Arg) nonreceptor tyrosine kinase to phosphorylate and activate the Rho inhibitor p190RhoGAP (Bradley et al. 2006, Hernandez et al. 2004, Peacock et al. 2007). p190RhoGAP is concentrated in the postsynaptic compartment and aids in maintaining dendritic spine stability (Moresco & Koleske 2003, Sfakianos et al. 2007). p190RhoGAP phosphorylation

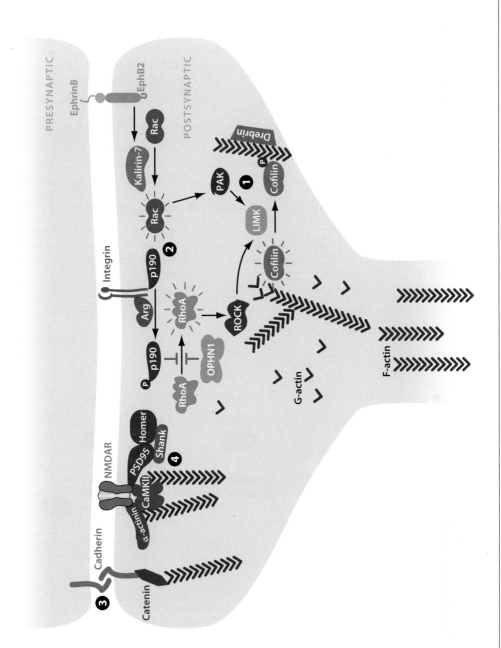

Figure 2

Molecular mechanisms that regulate dendritic spine stability. The actin cytoskeleton and its regulators, adhesion receptors, and scaffolding proteins provide physical support for long-term synaptic maintenance. Most signaling pathways regulate spine stability by either 1. directly or 2. indirectly (via RhoGTPase signaling pathways) regulating actin dynamics or its interactions with adhesion and scaffolding molecules. 3. Adhesion molecules mediate signaling from the presynaptic or extracellular compartments into dendritic spines to evoke changes in Rho GTPase and other signaling cascades that control F-actin structure and stability. 4. Scaffolding proteins both interact directly with F-actin to support cytoskeletal structure and organize signaling molecules that impinge on these cytoskeletal control mechanisms.

Table 1 Molecules associated with neurological disorders with reduced synapse or destabilized dendritic spine pathology. The identified molecules are known to regulate the synapse and/or dendritic spine stability, and their up- or downregulation, deletion, or loss-of-function (LOF) mutations have been implicated in psychiatric, neurological, or neurodegenerative disorders[a]

Molecule	Alteration	Related neurological disorders	References
Actin-related proteins			
Cofilin	↑	AD	Zhao et al. 2006
LIMK	+/−	Williams syndrome	Bellugi et al. 1999, Frangiskakis et al. 1996
Drebrin	↓	AD	Lacor et al. 2007
Rho-GTPase regulators			
Oligophrenin-1	LOF mutation	X-linked mental retardation	Billuart et al. 1998, Zanni et al. 2005
Kalirin-7	↓	AD, schizophrenia	Hill et al. 2006, Youn et al. 2007
Scaffolding protein			
Shank3	+/−	ASD	Durand et al. 2007
Adhesion molecule			
Neurexin:Neuroligin	−/−, LOF mutation	ASD, mental retardation, schizophrenia	Reviewed by Sudhof, 2008
Trophic factors			
BDNF	↓	MDD, Rett syndrome	Chang et al. 2006
Neuregulin-1:ErbB4	LOF mutation	Schizophrenia	Reviewed by Mei & Xiong, 2008
Corticosteroids	↑	MDD, chronic stress, Cushing's disease	Tata et al. 2006, Woolley et al. 1990, Patil et al. 2007

[a]Abbreviations: AD, Alzheimer disease; ASD, autism spectrum disorders; MDD, major depressive disorder.

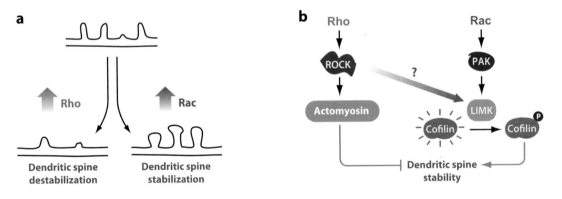

Figure 3

Rho family GTPases have opposite effects on dendritic spine stability. (*a*) Rho and Rac have opposite effects on dendritic spine maintenance. Rho activation causes dendritic spine destabilization characterized by spine retraction and collapse, resulting in dendritic segments that lack spines. Rac activation promotes dendritic spine enlargement and stabilization. (*b*) Activation of RhoA stimulates Rho-associated protein kinase (ROCK) and actomyosin contractility, which causes spine collapse and synapse loss. Conversely, Rac activates Pak1 and LIMK to phosphorylate and inhibit cofilin, thereby stabilizing F-actin and promoting dendritic spine stability. Rho signaling through ROCK can also activate LIMK, but whether this contributes to spine stability is unclear.

is significantly reduced in the $arg^{-/-}$ mouse hippocampus, and this is accompanied by elevated RhoA activity (Hernandez et al. 2004, Sfakianos et al. 2007). Axons, dendrites, and synapses develop normally in the $arg^{-/-}$ mouse hippocampus and cortex and reach their fully mature size and number by postnatal day 21. However, hippocampal synapses then destabilize, leading to 30% loss of synapses by postnatal day 31, followed by dendritic atrophy by early adulthood (Sfakianos et al. 2007). Biochemical studies suggest that the peak of Arg-mediated p190RhoGAP activation occurs just after the peak of synaptogenesis (Hernandez et al. 2004, Sfakianos et al. 2007), suggesting Arg and p190RhoGAP signaling is critical for long-term stabilization and maintenance of a subset of nascent synapses.

Kalirin-7 RacGEF. Kalirins (Kal) are RacGEFs, and Kal-7 is the predominant isoform in the adult rat brain (Johnson et al. 2000; Penzes et al. 2000, 2001). Kal-7 interacts and colocalizes with PSD-95 and F-actin in dendritic spines (Ma et al. 2003, Penzes et al. 2001). Knockdown or genetic ablation of Kal-7 in established neuronal cultures leads to dendritic spine simplification, leading to longer and thinner dendritic filopodia, and remaining synapses contain thinner postsynaptic densities (Ackermann & Matus 2003, Ma et al. 2008b). Several synaptic adhesion pathways act through Kal-7 to regulate F-actin rearrangement and spine stability. For example, ephrinB-EphB2 signaling across the synapse activates Kal-7 signaling through Rac and Pak1 to maintain dendritic spine stability (Penzes et al. 2003). Moreover, N-cadherin engagement and NMDA receptor activity stimulate Kal-7-dependent Rac1 activation, resulting in F-actin polymerization and enlargement of existing spines (Xie et al. 2008, 2007). Kal-7 expression is reduced in the brains of patients with schizophrenia and AD, consistent with the reduction of synapses and dendritic spines in these diseases (Hill et al. 2006, Youn et al. 2007).

ADF/Cofilin and LIM Kinase 1

The closely related actin depolymerizing factors (ADFs) and cofilin catalyze the severing of actin filaments. Depending on local cellular conditions (G- and F-actin concentration, ADP- versus ATP-bound F-actin, and other factors), cofilin can either disassemble actin filaments or sever filaments to provide new barbed ends to support F-actin assembly (Carlier et al. 1997, dos Remedios et al. 2003). Thus, cofilin controls the balance of G-actin versus F-actin locally. Cofilin localizes to the spine head in both developing and mature synapses where it associates with the spine periphery and with the postsynaptic density (Hotulainen et al. 2009, Racz & Weinberg 2006). Cofilin knockdown in established neuronal cultures reduces F-actin turnover in spines and decreases dendritic spine density, with a more pronounced effect on less stable thin spines than on other classes (Hotulainen et al. 2009).

LIM kinases phosphorylate cofilin at Serine 3, thereby inhibiting its function. LIM kinase 1 (LIMK-1) is a neuron-specific ADF/cofilin kinase highly enriched in mature synapses (Bernard et al. 1994, Proschel et al. 1995, Wang et al. 2000, Yang et al. 1998). Cultured hippocampal neurons from $limk$-$1^{-/-}$ mice exhibit reduced ADF/cofilin phosphorylation and altered F-actin accumulation in spines, resulting in abnormal spines with thicker necks and smaller heads (Meng et al. 2003, 2002). Synaptic GTPase activating protein (synGAP), a Rac inhibitor, attenuates this pathway to regulate cofilin activity in spines (Carlisle et al. 2008). Cultured hippocampal neurons from $syngap^{+/-}$ and $syngap^{-/-}$ mice exhibit increased phospho-cofilin and increased actin polymerization, leading to spine head enlargement, which demonstrates that disruption of cofilin regulation alters F-actin structure within spines.

LIMK-1 also has an important role in the presynaptic compartment to regulate *Drosophila* neuromuscular junction stability (Eaton & Davis 2005). In this context, LIMK-1 acts downstream of the Wit (*wishful thinking*) type II bone morphogenetic protein (BMP) receptor.

AMPA/NMDA receptors: Two glutamate-gated ion channels: AMPA (α-amino-3-hydroxy-5-methyl-4-isoxazolepropionic acid) receptors and NMDA (N-methyl D-aspartic acid) receptors. AMPA receptors mediate Na^+ influx to allow membrane depolarization. This depolarization, together with glutamate binding, activates NMDA receptors to mediate Ca^{2+} influx.

ADF: actin depolymerizing factor

Profilin: small actin-binding protein that exchanges ATP for ADP, allowing G-actin:ATP to be added to the barbed end of a growing actin filament

Loss of LIMK-1 function causes presynaptic motor neuron terminal regression, leaving behind unpaired postsynaptic membrane specializations. Overexpression of the ADF/cofilin phosphatase *slingshot*, which, like the *limk-1* mutant, should reduce ADF/cofilin phosphorylation, does not cause neuromuscular junction instability at the synapse. This finding suggests that LIMK-1 targets a presynaptic substrate other than ADF/cofilin to regulate synapse stability.

In humans, the LIMK-1 gene is among a microcluster of genes located on chromosome 7q11.23 and haploinsufficient in Williams syndrome, a developmental disorder associated with hypersociability and impairments in some cognitive tasks. These observations suggest that the altered synaptic cytoskeletal control may underlie some of these behavioral deficits, but more biochemical and cytoarchitectural work must be done to clearly establish this relationship (Bellugi et al. 1999, Frangiskakis et al. 1996).

Myosin IIB

Actomyosin contractility is an essential contributor to long-term spine maintenance. Class II myosins are cytoskeletal motor proteins that promote actomyosin contractility in nonmuscle cells. Myosin IIB localizes to dendritic spines, where it is enriched in the postsynaptic density (PSD) (Ryu et al. 2006). Inhibition of myosin IIB by siRNA or the myosin IIB inhibitor blebbistatin causes dendritic spine heads to elongate and become more filopodial in mature cultured neurons; coincident with this morphological transition, myosin IIB inhibition leads to a reduction in mini-excitatory postsynaptic currents, which reflects a loss of synapses (Ryu et al. 2006, Webb et al. 2007). These observations indicate that myosin IIB–based contractility is essential for synapse maintenance.

Drebrin

Drebrin is an F-actin binding protein expressed in neurons predominantly during early post-natal brain development (Ishikawa et al. 1994, Shirao et al. 1988). Drebrin colocalizes with F-actin in dendritic spine heads in mature cultured hippocampal neurons (Biou et al. 2008, Takahashi et al. 2003) and the amount of endogenous drebrin in dendritic spines positively correlates with the spine head size in the adult mouse cerebral cortex (Kobayashi et al. 2007). Overexpression of drebrin in immature neurons induces formation of long F-actin bundled filopodia and accumulation of synaptic proteins such as PSD-95 (Mizui et al. 2005). After synapses are properly formed, overexpression of drebrin causes destabilization of globular spines to polymorphic and very thin filopodia-like spines (Biou et al. 2008, Hayashi & Shirao 1999, Ivanov et al. 2009). These alterations of spine structure are dependent on the F-actin binding of drebrin (Biou et al. 2008, Hayashi & Shirao 1999, Ivanov et al. 2009). Together, these data suggest that drebrin levels must be perfectly balanced in the dendritic spine for long-term synapse stability, but drebrin-mediated regulation of spine stability is poorly understood. Drebrin may interact with profilin to facilitate actin polymerization and elongate dendritic spines (Mammoto et al. 1998). In support of this hypothesis, profilin IIa can regulate activity-dependent spine stabilization (Ackermann & Matus 2003).

CaMKII

CaMKII is a calcium-calmodulin protein kinase implicated in synaptic plasticity (Colbran & Brown 2004, Fink & Meyer 2002, Hudmon & Schulman 2002). CaMKII consists of two predominant isoforms, α and β, in the brain. CaMKIIβ is expressed from embryonic stages to adulthood, but CaMKIIα is only expressed after birth (Fink et al. 2003, Lin & Redmond 2008). Both isoforms are implicated in long-term dendritic spine stability. For example, CaMKIIα levels in spines correlate with overall spine size (Asrican et al. 2007), and blocking CaMKIIα activity in mature neuronal cultures can inhibit activity-induced spine enlargement and lead to spine destabilization

(Yamagata et al. 2009, Zha et al. 2009). In addition to its enzymatic role in dendritic spine signaling, CaMKIIβ uniquely contains an F-actin-binding domain that directly regulates F-actin stability (Fink et al. 2003, Lin & Redmond 2008, Okamoto et al. 2007, Shen et al. 1998). CaMKIIβ knockdown in hippocampal slices leads to a significant loss of mature spines, transforming them into immature dendritic filopodia (Okamoto et al. 2007). Intriguingly, CaMKIIβ's spine maintenance function does not require its kinase activity. In fact, CaMKIIβ activation dislocates it from the postsynaptic area, leading to dendritic spine destabilization (Okamoto et al. 2007).

NMDA Receptor

The actin cytoskeleton interacts functionally with NMDA receptors to support synaptic activity, which is essential for long-term synapse activity. Both depolymerization of the actin cytoskeleton and actomyosin inhibitors diminish NMDA receptor activity, indicating that NMDA receptor activity is tightly coupled to the configuration of the F-actin cytoskeleton (Lei et al. 2001, Rosenmund & Westbrook 1993). The NR1A, -1C, and -2B receptor subunits bind directly to the F-actin-crosslinking protein α-actinin, an interaction that can be blocked by calmodulin binding to the tails (Wyszynski et al. 1997). Calmodulin binds the NR1A tail and mediates Ca^{2+}-dependent inactivation of NMDA receptor currents (Ehlers et al. 1996, Zhang et al. 1998). Thus, α-actinin binding may protect NMDA receptors from this mode of inhibition. NMDA receptors also cluster at the postsynaptic density via interactions between their tails and postsynaptic scaffolding proteins (Feng & Zhang 2009), many of which couple directly or indirectly to the cytoskeleton. This may explain why synaptic NMDA receptor clustering can be maintained in established neuronal cultures even when F-actin or α-actinin is removed (Allison et al. 1998, Halpain et al. 1998).

RNAi-mediated knockdown of the NMDA receptor in vitro causes spine destabilization and loss of synapses, but NMDA channel blockers do not destabilize spines or synapses (Alvarez et al. 2007). Instead, structure/function studies indicate that specific NMDA C-terminal sequences are essential to mediating long-term synapse and dendrite maintenance. These findings indicate that, in addition to their activity-based roles, NMDA receptors have distinct structural roles in maintaining spines and synapses.

SYNAPTIC SCAFFOLDING PROTEINS PROVIDE STRUCTURAL SUPPORT FOR LONG-TERM MAINTENANCE

The PSD is an electron-dense submembranous region that directly apposes the presynaptic compartment. PSD scaffolding proteins organize and concentrate receptors and signaling molecules in this small area to optimize responsivity to presynaptic activity. PSD size contributes to the size of the spine head, which positively correlates with spine maintenance (Harris & Stevens 1989, Holtmaat et al. 2006, Kasai et al. 2003, Trachtenberg et al. 2002). It is therefore not surprising that scaffolding proteins have critical roles in synapse and dendritic spine stabilization (**Figure 2**).

PSD-95

PSD-95 is a major PSD scaffolding protein that belongs to the membrane-associated guanylate kinase (MAGUK) protein family (Cho et al. 1992). PSD-95 contains several PDZ and protein-protein interaction domains that stabilize interactions between neurotransmitter receptors, synaptic adhesion receptors, and actin-associated scaffolding molecules (Ethell & Pasquale 2005). Knockdown of PSD-95 in cultured neurons results in reductions in spine density and size (Ehrlich et al. 2007). PSD-95 appears to be particularly important for activity-dependent synapse stabilization. Stimuli that normally promote long-term enhancement of synaptic efficacy in hippocampal slices instead lead to spine destabilization in PSD-95 knockdown neurons (Ehrlich et al. 2007).

Shank/Homer

The Shank and Homer proteins are associated with PSD-95 and colocalize in the PSD (Naisbitt et al. 1999, Tu et al. 1999, Xiao et al. 2000). In addition to providing stabilizing support to PSD-95, Shank/Homer complexes recruit IP_3 receptors and F-actin to synapses, and therefore enlarge spine heads to stabilize dendritic spines (Sala et al. 2001, Xiao et al. 2000). Shank1B overexpression in mature neurons selectively enlarges the spine head, consistent with a primary function in spine stabilization (Sala et al. 2001). Accordingly, spine density is not affected in neurons from $shank1^{-/-}$ mice, although they have smaller dendritic spines and thinner PSDs (Hung et al. 2008). By contrast, knockdown of Shank3 in cultured hippocampal neurons results in the increase of dendritic filopodia, indicating destabilized dendritic spines (Roussignol et al. 2005). Mutations of *shank3* are linked to Autism spectrum disorders, suggesting that the failure of Shank/Homer-induced synapse stabilization may have significant consequences during brain development in humans (Durand et al. 2007).

STABILIZING INTERACTIONS AT THE SYNAPSE VIA ADHESION SYSTEMS

Once formed, synapses are maintained by adhesion systems that mediate stabilizing interactions between the pre- and postsynaptic compartments and between the synapse and supporting glia and extracellular matrix (ECM) (**Figure 2**).

Cadherins and Catenins

Dimerization of both pre- and postsynaptic cadherins mediates stable adhesion between the compartments. Cadherins also employ intracellular binding partners called catenins to regulate the activity of RhoA and Rac1 GTPases to coordinate F-actin remodeling within the spine and promote cadherin clustering at adhesion

sites (Arikkath & Reichardt 2008, Kwiatkowski et al. 2007, Nelson 2008). Cadherin-catenin complexes couple functionally to the actin cytoskeleton, thereby providing additional stabilization of the actin cytoskeleton. Blocking N-cadherin function in cultured hippocampal neurons alters the normal distribution of postsynaptic proteins and leads to increased numbers of filopodia, suggesting that N-cadherin helps scaffold the postsynaptic proteins that mediate spine stability (Togashi et al. 2002). Here, it is important to remember that diverse cadherin isoforms present at the synapse may play redundant roles in spine stabilization (Manabe et al. 2000).

Overexpression of αN-catenin increases spine head width and reduces dendritic spine turnover. αN-catenin-deficient neurons have more motile and destabilized spines (Abe et al. 2004). Deleting β-catenin after synaptogenesis destabilizes mushroom spines and increases the number of thin filopodia-like spines (Okuda et al. 2007). Cultured hippocampal neurons from β-catenin knockout mice exhibit greater spine density with reduced head width and length, suggesting that β-catenin also plays a stabilizing role in spines (Arikkath 2009). Interestingly, p120ctn deletion in mature neurons also causes a significant reduction in N-cadherin levels, decreased Rac1 activity, and increased RhoA activity. These alterations result in a significant loss of dendritic spines and synapses in vitro and in vivo (Elia et al. 2006), highlighting the role of catenins in maintaining dendritic spine stability.

Neurexin/Neuroligin Complexes

Neuroligins are postsynaptic transmembrane proteins that bind to presynaptic neurexins to stabilize synaptic contacts. Neuroligins form complexes with scaffolding proteins via their intracellular domain, thereby regulating synaptic function and stability (Craig & Kang 2007). Overexpression of Neuroligin-1 (NL-1) in cultured hippocampal neurons results in altered spine morphology, with an increase in spine-like protrusions and excitatory synaptic puncta (Chih et al. 2005). Knockdown of either neuroligin isoform reduces spine density and suppresses excitatory synapse formation (Chih et al. 2005). Although knockdown of neuroligins in cultured neurons shows reduced synapse formation, cultured NL1-3 triple knockout neurons have synapses that are indistinguishable from wild-type neurons (Chih et al. 2005, Varoqueaux et al. 2006). Although synapse numbers are not changed in NL1-3 knockout neurons, glutamatergic and GABAergic synaptic activity is significantly reduced. These studies suggest that neuroligins support long-term synaptic maintenance by recruiting receptors to the synapse and allowing proper synaptic function in vivo. In humans, deletions or mutations in neuroligins are associated with several cognitive disorders, including Autism spectrum disorders, mental retardation, and schizophrenia, but the precise genetic and biochemical mechanisms by which these mutations contribute to disease are still unknown (Sudhof 2008).

Ephrins and Eph Receptors

Eph receptors, including both EphA and EphB receptors, are receptor tyrosine kinases located on the postsynaptic compartment. EphAs (including EphA1-8 and A10) and EphBs (EphB1-4, and B6) interact with their presynaptic ligands, ephrinAs 1-5 and ephrinBs 1-3, respectively, to form bidirectional adhesion and signaling complexes that regulate synapse formation and maintenance (Kayser et al. 2008, Klein 2009). EphB1-3 knockout neurons exhibit an increased number of dendritic filopodia, whereas knockdown of EphB receptors reduces spine density (Kayser et al. 2008). Besides interacting with its ligand, the EphB receptor also interacts in *cis* with syndecan-2, a heparin sulfate proteoglycan enriched at postsynaptic sites (Ethell & Yamaguchi 1999, Irie & Yamaguchi 2004, Rapraeger & Ott 1998). Syndecan-2 regulates dendritic spine morphology via the actin cytoskeleton by signaling

through EphB2 receptors and the downstream Rho GTPase (Ethell et al. 2001, Irie & Yamaguchi 2004, Penzes et al. 2003, Shi et al. 2009, Tolias et al. 2007). Knockdown of syndecan-2 in mature hippocampal neurons results in reduced spine numbers (Lin et al. 2007).

Ephrin-A3 is the predominant ephrin-A isoform in the adult hippocampus. EphA4, the ephrin-A3 receptor, localizes on dendritic spines while ephrin-A3 localizes on astrocytes, together forming a bridge between the dendritic spine and surrounding glia cell. EphA4 activation by ephrin-A3 inhibits integrin adhesion pathways and therefore destabilizes dendritic spines. However, disrupting EphA4-ephrin interaction or deletion of EphA4 also results in aberrant spine morphology, although the overall dendritic structure is normal (Bourgin et al. 2007, Murai et al. 2003). These studies suggest that glial cells provide negative cues via ephrin-A3 and EphA4 signaling to regulate spine stability (Ethell & Pasquale 2005). EphA4 activation by ephrin-A1 recruits and activates cyclin-dependent kinase 5, resulting in activation of the RhoGEF ephexin1, leading to increased Rho activity and spine destabilization (Fu et al. 2007). In other contexts, EphA4 activation can contribute to Rac1-mediated spine stabilization (Inoue et al. 2009). A recent study indicates that EphA4 is processed by γ-secretase to enhance the formation and maintenance of dendritic spines (Inoue et al. 2009). Mutation of presenilin1 (PS1), a subunit of γ-secretase, impairs EphA4 processing, and this may contribute to the pathology of the familial AD resulting from PS1 mutations. Thus, EphA4 activation may generate opposite effects on spine maintenance that are dependent on the stimuli and downstream signaling pathways.

Integrins

Integrins are the major mediators of cell-ECM interactions. Functional integrin receptors are formed by heterodimers of α and β subunits (Juliano 2002). Integrins bind to ECM and

BDNF: brain-derived neurotrophic factor

TrkB: tyrosine receptor kinase B

relay intracellular signaling events that promote actin assembly via the recruitment of actin polymerization regulators or structural proteins (e.g., actin crosslinker proteins) that couple directly to the actin cytoskeleton (Blystone 2004, DeMali et al. 2003).

Inhibiting β1-integrin signaling using ligand-blocking peptides results in the transformation of existing dendritic spines to elongated dendritic filopodia (Bourgin et al. 2007, Shi & Ethell 2006). This destabilization of dendritic spines is mediated by NMDA receptor and CaMKII activation; blocking NMDAR and CaMKII activation reverts the destabilization phenotype. One major consequence of blocking β1-integrin signaling is dephosphorylation of Crk-associated substrate (Cas). Correspondingly, Cas knockdown decreases dendritic spine density and length, phenotypes similar to those observed when β1-integrin signaling is blocked (Bourgin et al. 2007). Blocking β1-integrin signaling with disrupting peptides leads to dramatic and rapid retraction of dendritic arbors of chick retinal ganglion cells, a phenotype that may result from loss of stabilizing synaptic contacts (Marrs et al. 2006).

Although it is diffusely expressed in neurons, integrin α5, which can pair with β1, localizes to synapses after synaptic stimulation, where it regulates spine stability via Src and Rac activation. This process depends on the activation of GIT1, a signaling adaptor that localizes Rac (Webb et al. 2007). Integrin α5 knockdown in cultured hippocampal neurons leads to an 80% decrease of synapse numbers and a reduced number of spines and dendritic protrusions (Webb et al. 2007). Several integrin ligands, such as laminin and reelin, have also been shown to affect dendritic spine stability (Liu et al. 2001, Seil 1998). Mutations or altered expression of these ligands have been linked to neurological disorders, including schizophrenia and AD, suggesting that altering integrin signaling may be involved in disease pathology (Costa et al. 2001, Huang et al. 1995, Liu et al. 2001, Rodriguez et al. 2000, Zhan et al. 1995).

SYNAPTIC ACTIVITY PROMOTES LONG-TERM SYNAPSE AND DENDRITE MAINTENANCE: THE IMPORTANCE OF TROPHIC SUPPORT PATHWAYS

Classic experiments demonstrated that destruction of afferent axons leads to the atrophy of target dendrites (Coleman & Riesen 1968, Jones & Thomas 1962, Le Gros Clark 1957, Matthews & Powell 1962). Postsynaptic degeneration could result simply from the physical loss of the presynaptic pairing partner; however, manipulations that reduce synaptic activity (e.g., odor or light deprivation) also compromise synapse and dendritic spine number and overall dendrite maintenance (Benson et al. 1984, Valverde 1967). These findings provided the first hints that synaptic activity supports long-term synapse maintenance. Subsequent work has revealed that the pairing of pre- and postsynaptic compartments is reinforced by the transmission of stabilizing signals between the two compartments. In this section, we review how trophic signaling pathways act to promote long-term synapse and dendrite stability.

BDNF and TrkB Mediate Bidirectional Stabilizing Cross-Talk Across the Synapse

Brain-derived neurotrophic factor (BDNF) was first identified as a factor that allowed survival of neurons that were unresponsive to nerve growth factor (NGF) (Barde et al. 1982). Subsequent work has demonstrated that BDNF is a member of the NGF family of neurotrophins, which includes NGF, NT-3, and NT-4/5 (Carvalho et al. 2008, Reichardt 2006). BDNF binds to both the P75^{NTR} receptor and the tyrosine receptor kinase B (TrkB) receptor, but we focus on TrkB binding and its role in promoting long-term synapse and dendrite maintenance (Carvalho et al. 2008).

The loss of BDNF or TrkB in the mouse brain leads to dramatic and widespread loss of dendritic spines and dendritic arbors in diverse neuronal populations (Baquet et al. 2004,

Gorski et al. 2003, Schober et al. 1998, Xu et al. 2000). These effects are especially prominent in the cerebral cortex, which is significantly smaller in both BDNF- and TrkB-deficient brains (Gorski et al. 2003, Xu et al. 2000). Both knockout animals also exhibit an increased neuronal packing density that results from a gross reduction in cortical neuron dendrite arbor size. Reductions in hippocampal spine density are reported in aged $trkB^{+/-}$ animals, suggesting that aging and reduced BDNF signaling may confer vulnerability to synapse loss (von Bohlen und Halbach et al. 2003).

Both TrkB and BDNF are expressed by pre- and postsynaptic neurons. BDNF is stored in vesicles in the pre- and postsynaptic compartments and can be released by depolarization and/or glutamatergic stimulation (Hartmann et al. 2001, Kohara et al. 2001, Kojima et al. 2001). Once released, BDNF exerts a stabilizing influence on both the pre- and postsynaptic compartments (**Figure 4a**). BDNF signaling through TrkB is essential to stabilize presynaptic release sites in Xenopus optic neuron axons (Hu et al. 2005, Marshak et al. 2007). TrkB is also expressed postsynaptically at the

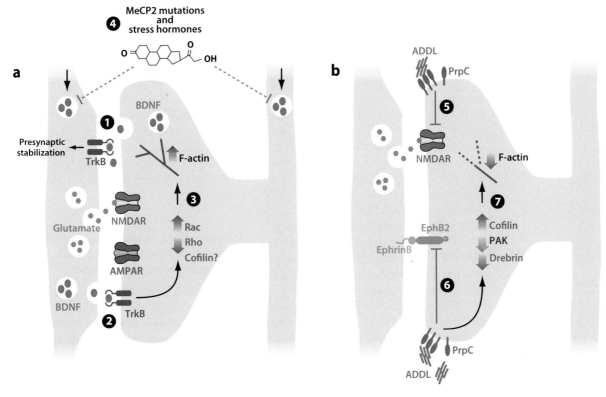

Figure 4

Targeting of synaptic stabilization mechanisms in disease. (*a*) Model for synapse stabilization by BDNF signaling through TrkB. 1. Synaptic activity stimulates BDNF release from the postsynaptic compartment, which binds to presynaptic TrkB receptors. This leads to stabilization of the presynaptic compartment. 2. Activity-based BDNF release from the presynaptic compartment activates postsynaptic TrkB. 3. Postsynaptic TrkB may activate Rac, inhibit Rho, and regulate cofilin to stimulate increased F-actin assembly to stabilize synapses. 4. Mutations in the MeCP2 gene, as in Rett syndrome, and elevated stress hormones likely destabilize synapses by reducing BDNF levels. (*b*) Model for synapse destabilization by Aβ-derived diffusible ligands. 5, 6. Binding of ADDL (Aβ-derived diffusible ligands) to its receptor PrpC on the dendritic spine leads to reduced amounts of both NMDA receptor (5) and EphB2 (6), both of which help stabilize synapses. 7. ADDL stimulation increases cofilin activity, decreases PAK activity, and reduces drebrin levels leading to reduced F-actin levels and spine shrinkage.

neuromuscular junction where it colocalizes in muscle with acetylcholine receptors (AChRs) (Gonzalez et al. 1999). A kinase-deficient splice TrkB isoform (TrkB.t1) can block AChR clustering and destabilize neuromuscular postsynaptic specializations (Gonzalez et al. 1999). Similarly, TrkB.t1 can also induce postsynaptic dendritic degeneration when expressed in facial motor neurons (De Wit et al. 2006).

BDNF acts through TrkB to stimulate several well-characterized downstream signaling cascades, including activation of Ras GTPase, Erk1/2 MAP kinase, phospholipase C gamma, and phosphatidylinositol-3-kinase (Carvalho et al. 2008, Reichardt 2006). Ongoing work should clarify which, if any, of these pathways mediate the stabilizing influence of BDNF on synapses. In addition, BDNF stimulates several cytoskeletal effector pathways that may be especially relevant for these processes. For example, BDNF stimulation promotes cofilin dephosphorylation in developing neuron growth cones, leading to increased growth cone filopodia (Fass et al. 2004, Gehler et al. 2004). Similar activation of BDNF signaling may promote cofilin-dependent F-actin network assembly at synapses. In support of this, bath application of BDNF to hippocampal slices synergizes with electrical stimulation to produce robust postsynaptic F-actin assembly (Rex et al. 2007). However, BDNF treatment also increases cofilin phosphorylation under these conditions. Determining the precise role of cofilin will require testing how nonphosphorylatable "activated" or phosphomimetic "inactive" cofilin affects synaptic F-actin assembly in response to activity.

BDNF signaling through TrkB also acts on the Tiam1, a GTPase exchange factor, to activate the Rac1 GTPase, stimulating local F-actin assembly (Miyamoto et al. 2006). Consistent with Rac1 as a downstream target, BDNF stimulation activates the PAK in hippocampal brain slices (Rex et al. 2007). In addition, the kinase-deficient TrkB.t1 isoform has a unique intracellular sequence that can bind RhoGDI, a Rho inhibitor. BDNF activation stimulates release of RhoGDI from TrkB.t1 to attenu-

ate Rho signaling (Ohira et al. 2006). Thus, BDNF may promote spine and synapse stability via TrkB-mediated activation of Rac and inhibition of Rho.

Loss of BDNF signaling in neurons may contribute to developmental and neurodegenerative disorders in humans. The majority of Rett syndrome (RS) cases are caused by mutations in the MeCP2 DNA binding protein (Van den Veyver & Zoghbi 2000), and BDNF expression is diminished in Mecp2-null mice (Chang et al. 2006). These mutants exhibit similar pathology to brain region-specific BDNF-knockout mice, and mice deficient for both proteins exhibit greatly reduced viability compared with either single knockout. BDNF overexpression in Mecp2-null mice increases survival rates and spontaneous synaptic activity. These observations suggest deficiencies in BDNF may contribute to the pathology of Rett syndrome. Furthermore, TrkB mRNA levels are known to decrease in the aging rat and human hippocampus and may contribute to reductions in synapses, dendritic spines, and dendritic complexity that occur in the aging brain (Silhol et al. 2005, Uylings et al. 2000, Webster et al. 2006).

Neuregulin-1 Stimulates ErbB4 to Enhance Excitement (and Stability) at the Synapse

Neuregulin-1 (Nrg1) is a growth factor that binds and activates the ErbB family of receptor tyrosine kinases. ErbB4 is enriched in the postsynaptic density, and its activity can be increased or decreased by treatments that elevate or inhibit synaptic activity, respectively (Li et al. 2007). Overexpression of ErbB4 in hippocampal slices leads to dendritic spine enlargement, whereas ErbB4 knockdown leads to spine shrinkage and loss. ErbB4 overexpression increases AMPA currents, an effect that requires ErbB4 kinase activity and activity of the NMDA receptors, prominent tyrosine kinase substrates. Knockdown of ErbB4 leads to relative reductions in both AMPA and NMDA currents, but this effect is blocked by Nrg1 knockdown in the presynaptic neuron. Exogenous Nrg1 can also promote increased numbers of spines and spine enlargement in cultured hippocampal neurons. Together, these data suggest that presynaptic Nrg1 signaling to postsynaptic ErbB4 supports long-term synapse maintenance by promoting synaptic activity and dendritic spine size. In support of this model, conditional knockout of both ErbB2 and ErbB4 in mice leads to a reduction in hippocampal and cortical dendritic spine density, despite the apparent absence of other gross neuronal defects (Barros et al. 2009).

SYNAPSE LOSS AND DENDRITIC ATROPHY ARE COMMON IN PSYCHIATRIC AND NEURODEGENERATIVE DISORDERS

Reduced synaptic connectivity and dendritic atrophy are hallmarks of several psychiatric disorders and neurodegenerative diseases, where they are associated with impairments in perception, affect, and cognition. Although defects in initial circuit development may contribute to these conditions, the impairments of brain function often only become manifest in late adolescence or adulthood. An emerging theme is that the pathophysiological changes responsible for these diseases act by specifically targeting mechanisms of synapse and dendrite maintenance. In this section, we review these pathological conditions, with a specific emphasis on the current view of how disease-specific mechanisms are believed to destabilize synapses and dendrites.

Major Depression, Chronic Stress, and Cushing's Disease: Synapse Destabilization by Corticosteroids

MDD is a debilitating disorder characterized by pervasive sadness; feelings of guilt or worthlessness; reduced engagement in pleasurable activities; and alterations in sleep, eating, and work patterns. Several anatomical features consistent with a reduction in synapses and dendritic arbor

extent have been observed in the hippocampi and anterior cingulate and orbitofrontal cortices of individuals with MDD, including reduced tissue volume, decreased cell soma size, increased neuronal packing density, and decreased staining for dendritic spine markers (Cotter et al. 2002, 2001; Law et al. 2004; Rajkowska et al. 1999; Stockmeier et al. 2004). These observations strongly suggest that reduced synaptic connectivity may contribute to MDD symptomology.

Chronic stress is a significant risk factor for MDD (Kendler et al. 1999, Pittenger & Duman 2008). In rats, chronic restraint stress leads to significant dendritic atrophy in the hippocampus and prefrontal cortex, which are areas affected in humans with MDD (Cook & Wellman 2004, Radley et al. 2008, Watanabe et al. 1992). Stress leads to an elevation of the corticosteroid stress hormones, which has been causally linked to synapse loss and dendritic regression. For example, daily injection of corticosteroids over three weeks leads to reduced synapse density in the rat hippocampal CA3 region (Tata et al. 2006), accompanied by regression of dendrite arbors (Woolley et al. 1990), and similar effects have been observed in the prefrontal cortex (Wellman 2001).

Elevated circulating corticosteroids are also found in patients with Cushing's disease, a multisymptom disease caused by pituitary tumors that result in increased levels of adrenocorticotropic hormone, which stimulates elevated cortisol production. A similar disorder, Cushing's syndrome, is sometimes observed in individuals undergoing excessive long-term glucocorticoid treatment for inflammatory and autoimmune diseases (Patil et al. 2007). Imaging studies reveal a significant loss of cortical and subcortical brain volume in these patients, likely reflecting a significant loss of synapses and dendrite arbors (Patil et al. 2007).

Long-term elevation of corticosteroids causes synapse loss and dendritic regression by interfering directly with synapse and dendrite maintenance mechanisms. Steroids act through their receptors, which are DNA-binding transcriptional regulators. Significant efforts have focused on identifying the relevant target genes regulated by corticosteroids. Both chronic stress and chronic corticosteroid exposure decrease hippocampal BDNF mRNA and protein (Gourley et al. 2009, Nibuya et al. 1999, Smith et al. 1995), and BDNF levels are also reduced in MDD patients (Schmidt & Duman 2007) (**Figure 4a**). This finding suggests that corticosteroids cause synapse loss and dendritic atrophy by attenuating trophic support provided by BDNF:TrkB signaling. In support of this hypothesis, direct infusion of BDNF into the brain can ameliorate the depressive symptoms that accompany synapse loss and dendritic atrophy in animal models of chronic stress or depression (Gourley et al. 2008, Shirayama et al. 2002). Also, antidepressant treatments lead to an elevation of BDNF levels in MDD patients and appear to block the inhibition of BDNF levels caused by chronic stress or corticosteroids in animal models (Schmidt & Duman 2007). Moreover, genetic disruption or blockade of BDNF:TrkB signaling can prevent the efficacy of antidepressant treatment in chronic stress animal models (Schmidt & Duman 2007).

Although these studies implicate reduced BDNF:TrkB signaling as a major contributor to the behavioral symptoms of chronic stress disorders and MDD, the cellular and molecular mechanisms by which antidepressants or BDNF infusion reduce depression-like symptoms are not yet known. It will be particularly interesting to examine whether these treatments act by restoring synapse loss and reducing dendritic atrophy associated with these disorders.

Alterations in Nrg1:ErbB4 Signaling: A Link to Schizophrenia

Schizophrenia is a brain disorder characterized by significant changes in perception, including hallucinations and delusions, but also by flat affect, reduced socialization, and cognitive impairment. Increased neuronal packing density in the prefrontal cortex of the schizophrenic brain provided the first hints that reduced synaptic connectivity may contribute to its

pathophysiology (Selemon & Goldman-Rakic 1999). Reduced immunostaining for presynaptic markers (Glantz & Lewis 1997, Woo et al. 1998), decreased dendritic spine density, and reductions in dendritic arbor size have been noted in schizophrenia patients (Broadbelt et al. 2002, Glantz & Lewis 2000, Kalus et al. 2000).

Recent studies revealed an association between polymorphisms in *nrg1* and *erbb4* and increased risk for schizophrenia (Mei & Xiong 2008). These mutations may compromise Nrg1:ErbB4-mediated trophic synapse support, thereby leading to reductions in synaptic connectivity. Such an erosion of synaptic connectivity would be consistent with the decrease in glutamatergic function observed in schizophrenia (Mei & Xiong 2008). However, Nrg1 stimulation increases ErbB4 signaling and heightens the attenuation of NMDA receptor phosphorylation in tissue isolated from schizophrenia patients (Hahn et al. 2006), suggesting ErbB4 signaling pathways may be hypersensitized in schizophrenia. Thus, a major unresolved issue is exactly how synaptic Nrg1:ErbB4 signaling is disrupted in schizophrenia and if this contributes to schizophrenia pathophysiology.

Aβ-Derived Diffusible Ligands in Alzheimer's Disease: Synaptic Superdestroyers?

AD is the most common form of human dementia and it leads to progressive impairments in memory, cognition, and behavior. The brains of affected individuals exhibit significant synapse loss and dendritic arbor regression, followed by neuronal death (Anderton et al. 1998; Flood 1991; Flood et al. 1987a,b; Hanks & Flood 1991). The degree of cognitive and memory impairment in AD patients correlates with the extent of synapse and dendritic arbor loss, and not with neuronal death (Falke et al. 2003, Terry et al. 1991). Thus, preventing synapse loss and dendritic regression represents a major therapeutic target for AD treatment.

One hallmark of AD that distinguishes it from the normal aging brain is the presence of elevated levels of amyloid beta (Aβ) peptide fragments derived from the amyloid precursor protein. Aβ peptides form soluble aggregates, termed Aβ-derived diffusible ligands (ADDLs), and insoluble aggregates, termed amyloid plaques (APs), in the AD brain. APs are often found in contact with dystrophic neurites, neuronal processes with abnormal trajectories that may reflect axon or dendrite degeneration adjacent to the plaque, but AP burden does not correlate well with the extent of disease symptomology (Haass & Selkoe 2007). ADDL levels correlate better than APs with the cognitive impairment in AD patients, and injection of patient-derived ADDLs into animals can induce acute memory loss, suggesting a direct link between ADDLs and AD symptomology (Haass & Selkoe 2007, Shankar et al. 2008).

Several recent studies characterized the potent ability of ADDLs to induce synapse destabilization (**Figure 4b**). Soluble Aβ oligomers can bind directly and preferentially to dendritic spines and promote spine destabilization and functional synapse loss (Lacor et al. 2007, Lauren et al. 2009, Ma et al. 2008a, Shankar et al. 2007). Aβ oligomer application leads to reductions in the EphB2 and NMDA receptors (Lacor et al. 2007). In addition, ADDL application leads to reductions in PAK levels and activity (Zhao et al. 2006). These reductions correlate with increased levels of the actin-severing protein cofilin, and decreased levels of the F-actin-stabilizing spine protein drebrin (Lacor et al. 2007, Zhao et al. 2006). These changes appear to have functional relevance as parallel changes in PAK, cofilin,

and drebrin are observed in AD patient samples (Zhao et al. 2006). Moreover, cofilin activity is essential for the spine-destabilizing effects of ADDLs (Shankar et al. 2007).

In addition to their effects on cytoskeletal signaling pathways, ADDLs induce cofilin aggregation (Zhao et al. 2006). Similar cofilin-rich insoluble aggregates, termed Hirano bodies, are found in AD brains. Cofilin aggregates disrupt the microtubular network within neurites, and may "choke off" synapses by preventing delivery of materials via the microtubule highway (Minamide et al. 2000). Together, these data suggest that ADDLs may be "synaptic superdestroyers" that target synapses for destabilization by interfering simultaneously with synaptic adhesive systems (EphB2), activity-based maintenance mechanisms (NMDA receptor), and cytoskeletal stabilization mechanisms (PAK, drebrin, and cofilin). The recent identification of prion precursor protein (PrP) as a functional ADDL receptor should greatly promote efforts to understand how ADDL triggers the chain of events that result in synapse destabilization (Lauren et al. 2009).

Alterations in Rho GTPase signaling may also contribute to synapse destabilization in AD. One recent study demonstrated an increase in Rho activity and a decrease in Rac activity in an APP transgenic AD mouse model (Petratos et al. 2008). The decreased Rac1 activity may underlie the reductions in PAK levels or activity noted in AD brains (Zhao et al. 2006). These pathways may represent important targets for therapies to block AD progression.

SUMMARY POINTS

1. Long-term stability of synapses, dendritic spines, and dendrites requires structural support from the actin cytoskeleton and adhesion and scaffolding molecules. Dysfunction of several cytoskeletal regulators is associated with synapse and dendrite instability.

2. Several activity-dependent trophic signaling pathways mediate cross-talk between the pre- and postsynaptic compartment. These pathways function by activating or reinforcing the synapse stabilization mechanisms.

3. Premature synapse and dendrite loss is associated with neurodegenerative and psychiatric disorders. In many cases, the pathology of these diseases selectively targets synapse and dendrite maintenance mechanisms.

FUTURE DIRECTIONS

1. Synapse maintenance is greatly affected by the quality and frequency of synaptic inputs. A major goal is elucidating how synaptic activity patterns differentially activate maintenance mechanisms.

2. Mutations or polymorphisms in several key synaptic regulators are associated with different pathological outcomes. A major challenge is understanding how these mutations lead to synapse instability and disease pathology.

3. Each of the synaptic stabilization mechanisms is potentially a gateway for therapeutic strategies to protect against synapse loss to arrest disease progression. Future efforts must aim to develop small molecules that can activate stabilizing pathways to reinforce synaptic stability or block destabilizing mechanisms.

DISCLOSURE STATEMENT

The authors are not aware of any affiliations, memberships, funding, or financial holdings that might be perceived as affecting the objectivity of this review.

ACKNOWLEDGMENTS

We apologize to our colleagues whose research could not be appropriately cited owing to space limitations. We thank Thomas Biederer, Ron Duman, Shannon Gourley, Charles Greer, Michael Koelle, Susumu Tomita, and Sloan Warren for thoughtful discussions and critical feedback on the review. Our research is supported by NIH Grants NS39475 and CA133346, an American Heart Association Established Investigator Award, awards from the Yale Interdisciplinary Research Consortium on Stress, Self-Control and Addiction, and an anonymous donor.

LITERATURE CITED

Abe K, Chisaka O, Van Roy F, Takeichi M. 2004. Stability of dendritic spines and synaptic contacts is controlled by alpha N-catenin. *Nat. Neurosci.* 7:357–63

Ackermann M, Matus A. 2003. Activity-induced targeting of profilin and stabilization of dendritic spine morphology. *Nat. Neurosci.* 6:1194–200

Allison DW, Gelfand VI, Spector I, Craig AM. 1998. Role of actin in anchoring postsynaptic receptors in cultured hippocampal neurons: differential attachment of NMDA versus AMPA receptors. *J. Neurosci.* 18:2423–36

Alvarez VA, Ridenour DA, Sabatini BL. 2007. Distinct structural and ionotropic roles of NMDA receptors in controlling spine and synapse stability. *J. Neurosci.* 27:7365–76

Anderton BH, Callahan L, Coleman P, Davies P, Flood D, et al. 1998. Dendritic changes in Alzheimer's disease and factors that may underlie these changes. *Prog. Neurobiol.* 55:595–609

Arellano JI, Espinosa A, Fairen A, Yuste R, DeFelipe J. 2007. Non-synaptic dendritic spines in neocortex. *Neuroscience* 145:464–69

Arikkath J. 2009. Regulation of dendrite and spine morphogenesis and plasticity by catenins. *Mol. Neurobiol.* 40:46–54

Arikkath J, Reichardt LF. 2008. Cadherins and catenins at synapses: roles in synaptogenesis and synaptic plasticity. *Trends Neurosci.* 31:487–94

Asrican B, Lisman J, Otmakhov N. 2007. Synaptic strength of individual spines correlates with bound Ca^{2+}-calmodulin-dependent kinase II. *J. Neurosci.* 27:14007–11

Baquet ZC, Gorski JA, Jones KR. 2004. Early striatal dendrite deficits followed by neuron loss with advanced age in the absence of anterograde cortical brain-derived neurotrophic factor. *J. Neurosci.* 24:4250–58

Barde YA, Edgar D, Thoenen H. 1982. Purification of a new neurotrophic factor from mammalian brain. *EMBO J.* 1:549–53

Barros CS, Calabrese B, Chamero P, Roberts AJ, Korzus E, et al. 2009. Impaired maturation of dendritic spines without disorganization of cortical cell layers in mice lacking NRG1/ErbB signaling in the central nervous system. *Proc. Natl. Acad. Sci. USA* 106:4507–12

Bellugi U, Lichtenberger L, Mills D, Galaburda A, Korenberg JR. 1999. Bridging cognition, the brain and molecular genetics: evidence from Williams syndrome. *Trends Neurosci.* 22:197–207

Benson TE, Ryugo DK, Hinds JW. 1984. Effects of sensory deprivation on the developing mouse olfactory system: a light and electron microscopic, morphometric analysis. *J. Neurosci.* 4:638–53

Bernard O, Ganiatsas S, Kannourakis G, Dringen R. 1994. Kiz-1, a protein with LIM zinc finger and kinase domains, is expressed mainly in neurons. *Cell Growth Differ.* 5:1159–71

Billuart P, Bienvenu T, Ronce N, des Portes V, Vinet MC, et al. 1998. Oligophrenin-1 encodes a rhoGAP protein involved in X-linked mental retardation. *Nature* 392:923–26

Biou V, Brinkhaus H, Malenka RC, Matus A. 2008. Interactions between drebrin and Ras regulate dendritic spine plasticity. *Eur. J. Neurosci.* 27:2847–59

Blystone SD. 2004. Integrating an integrin: a direct route to actin. *Biochim. Biophys. Acta* 1692:47–54

Bourgin C, Murai KK, Richter M, Pasquale EB. 2007. The EphA4 receptor regulates dendritic spine remodeling by affecting beta1-integrin signaling pathways. *J. Cell Biol.* 178:1295–307

Bradley WD, Hernandez SE, Settleman J, Koleske AJ. 2006. Integrin signaling through Arg activates p190RhoGAP by promoting its binding to p120RasGAP and recruitment to the membrane. *Mol. Biol. Cell* 17:4827–36

Broadbelt K, Byne W, Jones LB. 2002. Evidence for a decrease in basilar dendrites of pyramidal cells in schizophrenic medial prefrontal cortex. *Schizophr. Res.* 58:75–81

Carlier MF, Laurent V, Santolini J, Melki R, Didry D, et al. 1997. Actin depolymerizing factor (ADF/cofilin) enhances the rate of filament turnover: implication in actin-based motility. *J. Cell Biol.* 136:1307–22

Carlisle HJ, Manzerra P, Marcora E, Kennedy MB. 2008. SynGAP regulates steady-state and activity-dependent phosphorylation of cofilin. *J. Neurosci.* 28:13673–83

Carvalho AL, Caldeira MV, Santos SD, Duarte CB. 2008. Role of the brain-derived neurotrophic factor at glutamatergic synapses. *Br. J. Pharmacol.* 153(Suppl. 1):S310–24

Chang Q, Khare G, Dani V, Nelson S, Jaenisch R. 2006. The disease progression of Mecp2 mutant mice is affected by the level of BDNF expression. *Neuron* 49:341–48

Cheng XT, Hayashi K, Shirao T. 2000. Non-muscle myosin IIB-like immunoreactivity is present at the drebrin-binding cytoskeleton in neurons. *Neurosci. Res.* 36:167–73

Chih B, Engelman H, Scheiffele P. 2005. Control of excitatory and inhibitory synapse formation by neuroligins. *Science* 307:1324–28

Cho KO, Hunt CA, Kennedy MB. 1992. The rat brain postsynaptic density fraction contains a homolog of the *Drosophila* discs-large tumor suppressor protein. *Neuron* 9:929–42

Cline HT. 2001. Dendritic arbor development and synaptogenesis. *Curr. Opin. Neurobiol.* 11:118–26

Cohen RS, Chung SK, Pfaff DW. 1985. Immunocytochemical localization of actin in dendritic spines of the cerebral cortex using colloidal gold as a probe. *Cell Mol. Neurobiol.* 5:271–84

Colbran RJ, Brown AM. 2004. Calcium/calmodulin-dependent protein kinase II and synaptic plasticity. *Curr. Opin. Neurobiol.* 14:318–27

Coleman PD, Riesen AH. 1968. Environmental effects on cortical dendritic fields. I. Rearing in the dark. *J. Anat.* 102:363–74

Cook SC, Wellman CL. 2004. Chronic stress alters dendritic morphology in rat medial prefrontal cortex. *J. Neurobiol.* 60:236–48

Costa E, Davis J, Grayson DR, Guidotti A, Pappas GD, Pesold C. 2001. Dendritic spine hypoplasticity and downregulation of reelin and GABAergic tone in schizophrenia vulnerability. *Neurobiol. Dis.* 8:723–42

Cotter D, Mackay D, Chana G, Beasley C, Landau S, Everall IP. 2002. Reduced neuronal size and glial cell density in area 9 of the dorsolateral prefrontal cortex in subjects with major depressive disorder. *Cereb. Cortex* 12:386–94

Cotter D, Mackay D, Landau S, Kerwin R, Everall I. 2001. Reduced glial cell density and neuronal size in the anterior cingulate cortex in major depressive disorder. *Arch. Gen. Psychiatry* 58:545–53

Craig AM, Kang Y. 2007. Neurexin-neuroligin signaling in synapse development. *Curr. Opin. Neurobiol.* 17:43–52

Dailey ME, Smith SJ. 1996. The dynamics of dendritic structure in developing hippocampal slices. *J. Neurosci.* 16:2983–94

De Wit J, Eggers R, Evers R, Castren E, Verhaagen J. 2006. Long-term adeno-associated viral vector-mediated expression of truncated TrkB in the adult rat facial nucleus results in motor neuron degeneration. *J. Neurosci.* 26:1516–30

DeMali KA, Wennerberg K, Burridge K. 2003. Integrin signaling to the actin cytoskeleton. *Curr. Opin. Cell Biol.* 15:572–82

Dillon C, Goda Y. 2005. The actin cytoskeleton: integrating form and function at the synapse. *Annu. Rev. Neurosci.* 28:25–55

dos Remedios CG, Chhabra D, Kekic M, Dedova IV, Tsubakihara M, et al. 2003. Actin binding proteins: regulation of cytoskeletal microfilaments. *Physiol. Rev.* 83:433–73

Durand CM, Betancur C, Boeckers TM, Bockmann J, Chaste P, et al. 2007. Mutations in the gene encoding the synaptic scaffolding protein SHANK3 are associated with autism spectrum disorders. *Nat. Genet.* 39:25–27

Eaton BA, Davis GW. 2005. LIM Kinase1 controls synaptic stability downstream of the type II BMP receptor. *Neuron* 47:695–708

Ehlers MD, Zhang S, Bernhadt JP, Huganir RL. 1996. Inactivation of NMDA receptors by direct interaction of calmodulin with the NR1 subunit. *Cell* 84:745–55

Ehrlich I, Klein M, Rumpel S, Malinow R. 2007. PSD-95 is required for activity-driven synapse stabilization. *Proc. Natl. Acad. Sci. USA* 104:4176–81

Elia LP, Yamamoto M, Zang K, Reichardt LF. 2006. p120 catenin regulates dendritic spine and synapse development through Rho-family GTPases and cadherins. *Neuron* 51:43–56

Ethell IM, Irie F, Kalo MS, Couchman JR, Pasquale EB, Yamaguchi Y. 2001. EphB/syndecan-2 signaling in dendritic spine morphogenesis. *Neuron* 31:1001–13

Ethell IM, Pasquale EB. 2005. Molecular mechanisms of dendritic spine development and remodeling. *Prog. Neurobiol.* 75:161–205

Ethell IM, Yamaguchi Y. 1999. Cell surface heparan sulfate proteoglycan syndecan-2 induces the maturation of dendritic spines in rat hippocampal neurons. *J. Cell Biol.* 144:575–86

Falke E, Nissanov J, Mitchell TW, Bennett DA, Trojanowski JQ, Arnold SE. 2003. Subicular dendritic arborization in Alzheimer's disease correlates with neurofibrillary tangle density. *Am. J. Pathol.* 163:1615–21

Fass J, Gehler S, Sarmiere P, Letourneau P, Bamburg JR. 2004. Regulating filopodial dynamics through actin-depolymerizing factor/cofilin. *Anat. Sci. Int.* 79:173–83

Feng W, Zhang M. 2009. Organization and dynamics of PDZ-domain-related supramodules in the postsynaptic density. *Nat. Rev. Neurosci.* 10:87–99

Fiala JC, Feinberg M, Popov V, Harris KM. 1998. Synaptogenesis via dendritic filopodia in developing hippocampal area CA1. *J. Neurosci.* 18:8900–11

Fifkova E, Delay RJ. 1982. Cytoplasmic actin in neuronal processes as a possible mediator of synaptic plasticity. *J. Cell Biol.* 95:345–50

Fink CC, Bayer KU, Myers JW, Ferrell JE Jr, Schulman H, Meyer T. 2003. Selective regulation of neurite extension and synapse formation by the beta but not the alpha isoform of CaMKII. *Neuron* 39:283–97

Fink CC, Meyer T. 2002. Molecular mechanisms of CaMKII activation in neuronal plasticity. *Curr. Opin. Neurobiol.* 12:293–99

Fischer M, Kaech S, Knutti D, Matus A. 1998. Rapid actin-based plasticity in dendritic spines. *Neuron* 20:847–54

Flood DG. 1991. Region-specific stability of dendritic extent in normal human aging and regression in Alzheimer's disease. II. Subiculum. *Brain Res.* 540:83–95

Flood DG, Buell SJ, Horwitz GJ, Coleman PD. 1987a. Dendritic extent in human dentate gyrus granule cells in normal aging and senile dementia. *Brain Res.* 402:205–16

Flood DG, Guarnaccia M, Coleman PD. 1987b. Dendritic extent in human CA2–3 hippocampal pyramidal neurons in normal aging and senile dementia. *Brain Res.* 409:88–96

Frangiskakis JM, Ewart AK, Morris CA, Mervis CB, Bertrand J, et al. 1996. LIM-kinase1 hemizygosity implicated in impaired visuospatial constructive cognition. *Cell* 86:59–69

Fu WY, Chen Y, Sahin M, Zhao XS, Shi L, et al. 2007. Cdk5 regulates EphA4-mediated dendritic spine retraction through an ephexin1-dependent mechanism. *Nat. Neurosci.* 10:67–76

Fukata Y, Amano M, Kaibuchi K. 2001. Rho-Rho-kinase pathway in smooth muscle contraction and cytoskeletal reorganization of nonmuscle cells. *Trends Pharmacol. Sci.* 22:32–39

Gehler S, Shaw AE, Sarmiere PD, Bamburg JR, Letourneau PC. 2004. Brain-derived neurotrophic factor regulation of retinal growth cone filopodial dynamics is mediated through actin depolymerizing factor/cofilin. *J. Neurosci.* 24:10741–49

Glantz LA, Lewis DA. 1997. Reduction of synaptophysin immunoreactivity in the prefrontal cortex of subjects with schizophrenia. Regional and diagnostic specificity. *Arch. Gen. Psychiatry* 54:660–69

Glantz LA, Lewis DA. 2000. Decreased dendritic spine density on prefrontal cortical pyramidal neurons in schizophrenia. *Arch. Gen. Psychiatry* 57:65–73

Gonzalez M, Ruggiero FP, Chang Q, Shi YJ, Rich MM, et al. 1999. Disruption of Trkb-mediated signaling induces disassembly of postsynaptic receptor clusters at neuromuscular junctions. *Neuron* 24:567–83

Gorski JA, Zeiler SR, Tamowski S, Jones KR. 2003. Brain-derived neurotrophic factor is required for the maintenance of cortical dendrites. *J. Neurosci.* 23:6856–65

Gourley SL, Kedves AT, Olausson P, Taylor JR. 2009. A history of corticosterone exposure regulates fear extinction and cortical NR2B, GluR2/3, and BDNF. *Neuropsychopharmacology* 34:707–16

Gourley SL, Kiraly DD, Howell JL, Olausson P, Taylor JR. 2008. Acute hippocampal brain-derived neurotrophic factor restores motivational and forced swim performance after corticosterone. *Biol. Psychiatry* 64:884–90

Govek EE, Newey SE, Akerman CJ, Cross JR, Van der Veken L, Van Aelst L. 2004. The X-linked mental retardation protein oligophrenin-1 is required for dendritic spine morphogenesis. *Nat. Neurosci.* 7:364–72

Haass C, Selkoe DJ. 2007. Soluble protein oligomers in neurodegeneration: lessons from the Alzheimer's amyloid beta-peptide. *Nat. Rev. Mol. Cell Biol.* 8:101–12

Hahn CG, Wang HY, Cho DS, Talbot K, Gur RE, et al. 2006. Altered neuregulin 1-erbB4 signaling contributes to NMDA receptor hypofunction in schizophrenia. *Nat. Med.* 12:824–28

Halpain S. 2003. Actin in a supporting role. *Nat. Neurosci.* 6:101–2

Halpain S, Hipolito A, Saffer L. 1998. Regulation of F-actin stability in dendritic spines by glutamate receptors and calcineurin. *J. Neurosci.* 18:9835–44

Hanks SD, Flood DG. 1991. Region-specific stability of dendritic extent in normal human aging and regression in Alzheimer's disease. I. CA1 of hippocampus. *Brain Res.* 540:63–82

Harris KM. 1999. Structure, development, and plasticity of dendritic spines. *Curr. Opin. Neurobiol.* 9:343–48

Harris KM, Stevens JK. 1989. Dendritic spines of CA 1 pyramidal cells in the rat hippocampus: serial electron microscopy with reference to their biophysical characteristics. *J. Neurosci.* 9:2982–97

Hartmann M, Heumann R, Lessmann V. 2001. Synaptic secretion of BDNF after high-frequency stimulation of glutamatergic synapses. *EMBO J.* 20:5887–97

Hayashi K, Shirao T. 1999. Change in the shape of dendritic spines caused by overexpression of drebrin in cultured cortical neurons. *J. Neurosci.* 19:3918–25

Hernandez SE, Krishnaswami M, Miller AL, Koleske AJ. 2004. How do Abl family kinases regulate cell shape and movement? *Trends Cell Biol.* 14:36–44

Hill JJ, Hashimoto T, Lewis DA. 2006. Molecular mechanisms contributing to dendritic spine alterations in the prefrontal cortex of subjects with schizophrenia. *Mol. Psychiatry* 11:557–66

Holtmaat A, Wilbrecht L, Knott GW, Welker E, Svoboda K. 2006. Experience-dependent and cell-type-specific spine growth in the neocortex. *Nature* 441:979–83

Holtmaat AJ, Trachtenberg JT, Wilbrecht L, Shepherd GM, Zhang X, et al. 2005. Transient and persistent dendritic spines in the neocortex in vivo. *Neuron* 45:279–91

Honkura N, Matsuzaki M, Noguchi J, Ellis-Davies GC, Kasai H. 2008. The subspine organization of actin fibers regulates the structure and plasticity of dendritic spines. *Neuron* 57:719–29

Hotulainen P, Llano O, Smirnov S, Tanhuanpaa K, Faix J, et al. 2009. Defining mechanisms of actin polymerization and depolymerization during dendritic spine morphogenesis. *J. Cell Biol.* 185:323–39

Hu B, Nikolakopoulou AM, Cohen-Cory S. 2005. BDNF stabilizes synapses and maintains the structural complexity of optic axons in vivo. *Development* 132:4285–98

Huang DY, Weisgraber KH, Strittmatter WJ, Matthew WD. 1995. Interaction of apolipoprotein E with laminin increases neuronal adhesion and alters neurite morphology. *Exp. Neurol.* 136:251–57

Hudmon A, Schulman H. 2002. Neuronal CA^{2+}/calmodulin-dependent protein kinase II: the role of structure and autoregulation in cellular function. *Annu. Rev. Biochem.* 71:473–510

Hung AY, Futai K, Sala C, Valtschanoff JG, Ryu J, et al. 2008. Smaller dendritic spines, weaker synaptic transmission, but enhanced spatial learning in mice lacking Shank1. *J. Neurosci.* 28:1697–708

Inoue E, Deguchi-Tawarada M, Togawa A, Matsui C, Arita K, et al. 2009. Synaptic activity prompts gamma-secretase-mediated cleavage of EphA4 and dendritic spine formation. *J. Cell Biol.* 185:551–64

Irie F, Yamaguchi Y. 2004. EPHB receptor signaling in dendritic spine development. *Front. Biosci.* 9:1365–73

Ishikawa R, Hayashi K, Shirao T, Xue Y, Takagi T, et al. 1994. Drebrin, a development-associated brain protein from rat embryo, causes the dissociation of tropomyosin from actin filaments. *J. Biol. Chem.* 269:29928–33

Ivanov A, Esclapez M, Pellegrino C, Shirao T, Ferhat L. 2009. Drebrin A regulates dendritic spine plasticity and synaptic function in mature cultured hippocampal neurons. *J. Cell Sci.* 122:524–34

Johnson RC, Penzes P, Eipper BA, Mains RE. 2000. Isoforms of kalirin, a neuronal Dbl family member, generated through use of different 5′- and 3′-ends along with an internal translational initiation site. *J. Biol. Chem.* 275:19324–33

Jones WH, Thomas DB. 1962. Changes in the dendritic organization of neurons in the cerebral cortex following deafferentation. *J. Anat.* 96:375–81

Juliano RL. 2002. Signal transduction by cell adhesion receptors and the cytoskeleton: functions of integrins, cadherins, selectins, and immunoglobulin-superfamily members. *Annu. Rev. Pharmacol. Toxicol.* 42:283–323

Kalus P, Muller TJ, Zuschratter W, Senitz D. 2000. The dendritic architecture of prefrontal pyramidal neurons in schizophrenic patients. *Neuroreport* 11:3621–65

Kasai H, Matsuzaki M, Noguchi J, Yasumatsu N, Nakahara H. 2003. Structure-stability-function relationships of dendritic spines. *Trends Neurosci.* 26:360–68

Kayser MS, Nolt MJ, Dalva MB. 2008. EphB receptors couple dendritic filopodia motility to synapse formation. *Neuron* 59:56–69

Kendler KS, Karkowski LM, Prescott CA. 1999. Causal relationship between stressful life events and the onset of major depression. *Am. J. Psychiatry* 156:837–41

Klein R. 2009. Bidirectional modulation of synaptic functions by Eph/ephrin signaling. *Nat. Neurosci.* 12:15–20

Kobayashi C, Aoki C, Kojima N, Yamazaki H, Shirao T. 2007. Drebrin A content correlates with spine head size in the adult mouse cerebral cortex. *J. Comp. Neurol.* 503:618–26

Kohara K, Kitamura A, Morishima M, Tsumoto T. 2001. Activity-dependent transfer of brain-derived neurotrophic factor to postsynaptic neurons. *Science* 291:2419–23

Kojima M, Takei N, Numakawa T, Ishikawa Y, Suzuki S, et al. 2001. Biological characterization and optical imaging of brain-derived neurotrophic factor-green fluorescent protein suggest an activity-dependent local release of brain-derived neurotrophic factor in neurites of cultured hippocampal neurons. *J. Neurosci. Res* 64:1–10

Kwiatkowski AV, Weis WI, Nelson WJ. 2007. Catenins: playing both sides of the synapse. *Curr. Opin. Cell Biol.* 19:551–56

Lacor PN, Buniel MC, Furlow PW, Clemente AS, Velasco PT, et al. 2007. Abeta oligomer-induced aberrations in synapse composition, shape, and density provide a molecular basis for loss of connectivity in Alzheimer's disease. *J. Neurosci.* 27:796–807

Landis DM, Reese TS. 1983. Cytoplasmic organization in cerebellar dendritic spines. *J. Cell Biol.* 97:1169–78

Lauren J, Gimbel DA, Nygaard HB, Gilbert JW, Strittmatter SM. 2009. Cellular prion protein mediates impairment of synaptic plasticity by amyloid-beta oligomers. *Nature* 457:1128–32

Law AJ, Weickert CS, Hyde TM, Kleinman JE, Harrison PJ. 2004. Reduced spinophilin but not microtubule-associated protein 2 expression in the hippocampal formation in schizophrenia and mood disorders: molecular evidence for a pathology of dendritic spines. *Am. J. Psychiatry* 161:1848–55

Le Gros Clark W. 1957. Inquiries into the anatomical basis of olfactory discrimination. *Proc. R. Soc. Lond.* 146:299–319

Lee T, Winter C, Marticke SS, Lee A, Luo L. 2000. Essential roles of *Drosophila* RhoA in the regulation of neuroblast proliferation and dendritic but not axonal morphogenesis. *Neuron* 25:307–16

Lei S, Czerwinska E, Czerwinski W, Walsh MP, MacDonald JF. 2001. Regulation of NMDA receptor activity by F-actin and myosin light chain kinase. *J. Neurosci.* 21:8464–72

Li B, Woo RS, Mei L, Malinow R. 2007. The neuregulin-1 receptor erbB4 controls glutamatergic synapse maturation and plasticity. *Neuron* 54:583–97

Lin YC, Redmond L. 2008. CaMKIIbeta binding to stable F-actin in vivo regulates F-actin filament stability. *Proc. Natl. Acad. Sci. USA* 105:15791–96

Lin YL, Lei YT, Hong CJ, Hsueh YP. 2007. Syndecan-2 induces filopodia and dendritic spine formation via the neurofibromin-PKA-Ena/VASP pathway. *J. Cell Biol.* 177:829–41

Liu WS, Pesold C, Rodriguez MA, Carboni G, Auta J, et al. 2001. Down-regulation of dendritic spine and glutamic acid decarboxylase 67 expressions in the reelin haploinsufficient heterozygous reeler mouse. *Proc. Natl. Acad. Sci. USA* 98:3477–82

Ma QL, Yang F, Calon F, Ubeda OJ, Hansen JE, et al. 2008a. p21-activated kinase-aberrant activation and translocation in Alzheimer disease pathogenesis. *J. Biol. Chem.* 283:14132–43

Ma XM, Huang J, Wang Y, Eipper BA, Mains RE. 2003. Kalirin, a multifunctional Rho guanine nucleotide exchange factor, is necessary for maintenance of hippocampal pyramidal neuron dendrites and dendritic spines. *J. Neurosci.* 23:10593–603

Ma XM, Kiraly DD, Gaier ED, Wang Y, Kim EJ, et al. 2008b. Kalirin-7 is required for synaptic structure and function. *J. Neurosci.* 28:12368–82

Maekawa M, Ishizaki T, Boku S, Watanabe N, Fujita A, et al. 1999. Signaling from Rho to the actin cytoskeleton through protein kinases ROCK and LIM-kinase. *Science* 285:895–98

Majewska AK, Newton JR, Sur M. 2006. Remodeling of synaptic structure in sensory cortical areas in vivo. *J. Neurosci.* 26:3021–29

Mammoto A, Sasaki T, Asakura T, Hotta I, Imamura H, et al. 1998. Interactions of drebrin and gephyrin with profilin. *Biochem. Biophys. Res. Commun.* 243:86–89

Manabe T, Togashi H, Uchida N, Suzuki SC, Hayakawa Y, et al. 2000. Loss of cadherin-11 adhesion receptor enhances plastic changes in hippocampal synapses and modifies behavioral responses. *Mol. Cell Neurosci.* 15:534–46

Marrs GS, Honda T, Fuller L, Thangavel R, Balsamo J, et al. 2006. Dendritic arbors of developing retinal ganglion cells are stabilized by beta 1-integrins. *Mol. Cell Neurosci.* 32:230–41

Marshak S, Nikolakopoulou AM, Dirks R, Martens GJ, Cohen-Cory S. 2007. Cell-autonomous TrkB signaling in presynaptic retinal ganglion cells mediates axon arbor growth and synapse maturation during the establishment of retinotectal synaptic connectivity. *J. Neurosci.* 27:2444–56

Matthews MR, Powell TP. 1962. Some observations on transneuronal cell degeneration in the olfactory bulb of the rabbit. *J. Anat.* 96:89–102

Matus A, Ackermann M, Pehling G, Byers HR, Fujiwara K. 1982. High actin concentrations in brain dendritic spines and postsynaptic densities. *Proc. Natl. Acad. Sci. USA* 79:7590–94

Mei L, Xiong WC. 2008. Neuregulin 1 in neural development, synaptic plasticity and schizophrenia. *Nat. Rev. Neurosci.* 9:437–52

Meng Y, Zhang Y, Tregoubov V, Falls DL, Jia Z. 2003. Regulation of spine morphology and synaptic function by LIMK and the actin cytoskeleton. *Rev. Neurosci.* 14:233–40

Meng Y, Zhang Y, Tregoubov V, Janus C, Cruz L, et al. 2002. Abnormal spine morphology and enhanced LTP in LIMK-1 knockout mice. *Neuron* 35:121–33

Miki H, Yamaguchi H, Suetsugu S, Takenawa T. 2000. IRSp53 is an essential intermediate between Rac and WAVE in the regulation of membrane ruffling. *Nature* 408:732–35

Minamide LS, Striegl AM, Boyle JA, Meberg PJ, Bamburg JR. 2000. Neurodegenerative stimuli induce persistent ADF/cofilin-actin rods that disrupt distal neurite function. *Nat. Cell Biol.* 2:628–36

Miyamoto Y, Yamauchi J, Tanoue A, Wu C, Mobley WC. 2006. TrkB binds and tyrosine-phosphorylates Tiam1, leading to activation of Rac1 and induction of changes in cellular morphology. *Proc. Natl. Acad. Sci. USA* 103:10444–49

Mizui T, Takahashi H, Sekino Y, Shirao T. 2005. Overexpression of drebrin A in immature neurons induces the accumulation of F-actin and PSD-95 into dendritic filopodia, and the formation of large abnormal protrusions. *Mol. Cell Neurosci.* 30:149–57

Moresco EM, Koleske AJ. 2003. Regulation of neuronal morphogenesis and synaptic function by Abl family kinases. *Curr. Opin. Neurobiol.* 13:535–44

Murai KK, Nguyen LN, Irie F, Yamaguchi Y, Pasquale EB. 2003. Control of hippocampal dendritic spine morphology through ephrin-A3/EphA4 signaling. *Nat. Neurosci.* 6:153–60

Nadif Kasri N, Nakano-Kobayashi A, Malinow R, Li B, Van Aelst L. 2009. The Rho-linked mental retardation protein oligophrenin-1 controls synapse maturation and plasticity by stabilizing AMPA receptors. *Genes Dev* 23:1289–302

Nadif Kasri N, Van Aelst L. 2008. Rho-linked genes and neurological disorders. *Pflugers Arch.* 455:787–97

Naisbitt S, Kim E, Tu JC, Xiao B, Sala C, et al. 1999. Shank, a novel family of postsynaptic density proteins that binds to the NMDA receptor/PSD-95/GKAP complex and cortactin. *Neuron* 23:569–82

Nakayama AY, Harms MB, Luo L. 2000. Small GTPases Rac and Rho in the maintenance of dendritic spines and branches in hippocampal pyramidal neurons. *J. Neurosci.* 20:5329–38

Nelson WJ. 2008. Regulation of cell-cell adhesion by the cadherin-catenin complex. *Biochem. Soc. Trans.* 36:149–55

Nibuya M, Takahashi M, Russell DS, Duman RS. 1999. Repeated stress increases catalytic TrkB mRNA in rat hippocampus. *Neurosci. Lett.* 267:81–84

Niell CM, Meyer MP, Smith SJ. 2004. In vivo imaging of synapse formation on a growing dendritic arbor. *Nat. Neurosci.* 7:254–60

Ohira K, Homma KJ, Hirai H, Nakamura S, Hayashi M. 2006. TrkB-T1 regulates the RhoA signaling and actin cytoskeleton in glioma cells. *Biochem. Biophys. Res. Commun.* 342:867–74

Okamoto K, Nagai T, Miyawaki A, Hayashi Y. 2004. Rapid and persistent modulation of actin dynamics regulates postsynaptic reorganization underlying bidirectional plasticity. *Nat. Neurosci.* 7:1104–12

Okamoto K, Narayanan R, Lee SH, Murata K, Hayashi Y. 2007. The role of CaMKII as an F-actin-bundling protein crucial for maintenance of dendritic spine structure. *Proc. Natl. Acad. Sci. USA* 104:6418–23

Okuda T, Yu LM, Cingolani LA, Kemler R, Goda Y. 2007. Beta-catenin regulates excitatory postsynaptic strength at hippocampal synapses. *Proc. Natl. Acad. Sci. USA* 104:13479–84

Patil CG, Lad SP, Katznelson L, Laws ER Jr. 2007. Brain atrophy and cognitive deficits in Cushing's disease. *Neurosurg. Focus* 23:E11

Peacock JG, Miller AL, Bradley WD, Rodriguez OC, Webb DJ, Koleske AJ. 2007. The Abl-related gene tyrosine kinase acts through p190RhoGAP to inhibit actomyosin contractility and regulate focal adhesion dynamics upon adhesion to fibronectin. *Mol. Biol. Cell* 18:3860–72

Penzes P, Beeser A, Chernoff J, Schiller MR, Eipper BA, et al. 2003. Rapid induction of dendritic spine morphogenesis by trans-synaptic ephrinB-EphB receptor activation of the Rho-GEF kalirin. *Neuron* 37:263–74

Penzes P, Johnson RC, Alam MR, Kambampati V, Mains RE, Eipper BA. 2000. An isoform of kalirin, a brain-specific GDP/GTP exchange factor, is enriched in the postsynaptic density fraction. *J. Biol. Chem.* 275:6395–403

Penzes P, Johnson RC, Sattler R, Zhang X, Huganir RL, et al. 2001. The neuronal Rho-GEF kalirin-7 interacts with PDZ domain-containing proteins and regulates dendritic morphogenesis. *Neuron* 29:229–42

Petratos S, Li QX, George AJ, Hou X, Kerr ML, et al. 2008. The beta-amyloid protein of Alzheimer's disease increases neuronal CRMP-2 phosphorylation by a Rho-GTP mechanism. *Brain* 131:90–108

Pittenger C, Duman RS. 2008. Stress, depression, and neuroplasticity: a convergence of mechanisms. *Neuropsychopharmacology* 33:88–109

Proschel C, Blouin MJ, Gutowski NJ, Ludwig R, Noble M. 1995. Limk1 is predominantly expressed in neural tissues and phosphorylates serine, threonine and tyrosine residues in vitro. *Oncogene* 11:1271–81

Racz B, Weinberg RJ. 2006. Spatial organization of cofilin in dendritic spines. *Neuroscience* 138:447–56

Radley JJ, Rocher AB, Rodriguez A, Ehlenberger DB, Dammann M, et al. 2008. Repeated stress alters dendritic spine morphology in the rat medial prefrontal cortex. *J. Comp. Neurol.* 507:1141–50

Rajan I, Witte S, Cline HT. 1999. NMDA receptor activity stabilizes presynaptic retinotectal axons and postsynaptic optic tectal cell dendrites in vivo. *J. Neurobiol.* 38:357–68

Rajkowska G, Miguel-Hidalgo JJ, Wei J, Dilley G, Pittman SD, et al. 1999. Morphometric evidence for neuronal and glial prefrontal cell pathology in major depression. *Biol. Psychiatry* 45:1085–98

Rapraeger AC, Ott VL. 1998. Molecular interactions of the syndecan core proteins. *Curr. Opin. Cell Biol.* 10:620–28

Reichardt LF. 2006. Neurotrophin-regulated signaling pathways. *Philos. Trans. R. Soc. Lond. B Biol. Sci.* 361:1545–64

Rex CS, Lin CY, Kramar EA, Chen LY, Gall CM, Lynch G. 2007. Brain-derived neurotrophic factor promotes long-term potentiation-related cytoskeletal changes in adult hippocampus. *J. Neurosci.* 27:3017–29

Rodriguez MA, Pesold C, Liu WS, Kriho V, Guidotti A, et al. 2000. Colocalization of integrin receptors and reelin in dendritic spine postsynaptic densities of adult nonhuman primate cortex. *Proc. Natl. Acad. Sci. USA* 97:3550–55

Rosenmund C, Westbrook GL. 1993. Calcium-induced actin depolymerization reduces NMDA channel activity. *Neuron* 10:805–14

Roussignol G, Ango F, Romorini S, Tu JC, Sala C, et al. 2005. Shank expression is sufficient to induce functional dendritic spine synapses in aspiny neurons. *J. Neurosci.* 25:3560–70

Ruchhoeft ML, Ohnuma S, McNeill L, Holt CE, Harris WA. 1999. The neuronal architecture of Xenopus retinal ganglion cells is sculpted by rho-family GTPases in vivo. *J. Neurosci.* 19:8454–63

Ryu J, Liu L, Wong TP, Wu DC, Burette A, et al. 2006. A critical role for myosin IIb in dendritic spine morphology and synaptic function. *Neuron* 49:175–82

Sala C, Piech V, Wilson NR, Passafaro M, Liu G, Sheng M. 2001. Regulation of dendritic spine morphology and synaptic function by Shank and Homer. *Neuron* 31:115–30

Schmidt HD, Duman RS. 2007. The role of neurotrophic factors in adult hippocampal neurogenesis, antidepressant treatments and animal models of depressive-like behavior. *Behav. Pharmacol.* 18:391–418

Schober A, Wolf N, Huber K, Hertel R, Krieglstein K, et al. 1998. TrkB and neurotrophin-4 are important for development and maintenance of sympathetic preganglionic neurons innervating the adrenal medulla. *J. Neurosci.* 18:7272–84

Schubert V, Da Silva JS, Dotti CG. 2006. Localized recruitment and activation of RhoA underlies dendritic spine morphology in a glutamate receptor-dependent manner. *J. Cell Biol.* 172:453–67

Schubert V, Dotti CG. 2007. Transmitting on actin: synaptic control of dendritic architecture. *J. Cell Sci.* 120:205–12

Seil FJ. 1998. The extracellular matrix molecule, laminin, induces purkinje cell dendritic spine proliferation in granule cell depleted cerebellar cultures. *Brain Res.* 795:112–20

Sekino Y, Kojima N, Shirao T. 2007. Role of actin cytoskeleton in dendritic spine morphogenesis. *Neurochem. Int.* 51:92–104

Selemon LD, Goldman-Rakic PS. 1999. The reduced neuropil hypothesis: a circuit based model of schizophrenia. *Biol. Psychiatry* 45:17–25

Sfakianos MK, Eisman A, Gourley SL, Bradley WD, Scheetz AJ, et al. 2007. Inhibition of Rho via Arg and p190RhoGAP in the postnatal mouse hippocampus regulates dendritic spine maturation, synapse and dendrite stability, and behavior. *J. Neurosci.* 27:10982–92

Shankar GM, Bloodgood BL, Townsend M, Walsh DM, Selkoe DJ, Sabatini BL. 2007. Natural oligomers of the Alzheimer amyloid-beta protein induce reversible synapse loss by modulating an NMDA-type glutamate receptor-dependent signaling pathway. *J. Neurosci.* 27:2866–75

Shankar GM, Li S, Mehta TH, Garcia-Munoz A, Shepardson NE, et al. 2008. Amyloid-beta protein dimers isolated directly from Alzheimer's brains impair synaptic plasticity and memory. *Nat. Med.* 14:837–42

Shen K, Teruel MN, Subramanian K, Meyer T. 1998. CaMKIIbeta functions as an F-actin targeting module that localizes CaMKIIalpha/beta heterooligomers to dendritic spines. *Neuron* 21:593–606

Shi Y, Ethell IM. 2006. Integrins control dendritic spine plasticity in hippocampal neurons through NMDA receptor and Ca^{2+}/calmodulin-dependent protein kinase II-mediated actin reorganization. *J. Neurosci.* 26:1813–22

Shi Y, Pontrello CG, DeFea KA, Reichardt LF, Ethell IM. 2009. Focal adhesion kinase acts downstream of EphB receptors to maintain mature dendritic spines by regulating cofilin activity. *J. Neurosci.* 29:8129–42

Shirao T, Kojima N, Kato Y, Obata K. 1988. Molecular cloning of a cDNA for the developmentally regulated brain protein, drebrin. *Brain Res.* 464:71–74

Shirayama Y, Chen AC, Nakagawa S, Russell DS, Duman RS. 2002. Brain-derived neurotrophic factor produces antidepressant effects in behavioral models of depression. *J. Neurosci.* 22:3251–61

Silhol M, Bonnichon V, Rage F, Tapia-Arancibia L. 2005. Age-related changes in brain-derived neurotrophic factor and tyrosine kinase receptor isoforms in the hippocampus and hypothalamus in male rats. *Neuroscience* 132:613–24

Smith MA, Makino S, Kvetnansky R, Post RM. 1995. Stress and glucocorticoids affect the expression of brain-derived neurotrophic factor and neurotrophin-3 mRNAs in the hippocampus. *J. Neurosci.* 15:1768–77

Sorra KE, Harris KM. 2000. Overview on the structure, composition, function, development, and plasticity of hippocampal dendritic spines. *Hippocampus* 10:501–11

Star EN, Kwiatkowski DJ, Murthy VN. 2002. Rapid turnover of actin in dendritic spines and its regulation by activity. *Nat. Neurosci.* 5:239–46

Stockmeier CA, Mahajan GJ, Konick LC, Overholser JC, Jurjus GJ, et al. 2004. Cellular changes in the postmortem hippocampus in major depression. *Biol. Psychiatry* 56:640–50

Sudhof TC. 2008. Neuroligins and neurexins link synaptic function to cognitive disease. *Nature* 455:903–11

Takahashi H, Sekino Y, Tanaka S, Mizui T, Kishi S, Shirao T. 2003. Drebrin-dependent actin clustering in dendritic filopodia governs synaptic targeting of postsynaptic density-95 and dendritic spine morphogenesis. *J. Neurosci.* 23:6586–95

Tashiro A, Minden A, Yuste R. 2000. Regulation of dendritic spine morphology by the Rho family of small GTPases: antagonistic roles of Rac and Rho. *Cereb. Cortex* 10:927–38

Tashiro A, Yuste R. 2004. Regulation of dendritic spine motility and stability by Rac1 and Rho kinase: evidence for two forms of spine motility. *Mol. Cell Neurosci.* 26:429–40

Tata DA, Marciano VA, Anderson BJ. 2006. Synapse loss from chronically elevated glucocorticoids: relationship to neuropil volume and cell number in hippocampal area CA3. *J. Comp. Neurol.* 498:363–74

Terry RD, Masliah E, Salmon DP, Butters N, DeTeresa R, et al. 1991. Physical basis of cognitive alterations in Alzheimer's disease: synapse loss is the major correlate of cognitive impairment. *Ann. Neurol.* 30:572–80

Togashi H, Abe K, Mizoguchi A, Takaoka K, Chisaka O, Takeichi M. 2002. Cadherin regulates dendritic spine morphogenesis. *Neuron* 35:77–89

Tolias KF, Bikoff JB, Kane CG, Tolias CS, Hu L, Greenberg ME. 2007. The Rac1 guanine nucleotide exchange factor Tiam1 mediates EphB receptor-dependent dendritic spine development. *Proc. Natl. Acad. Sci. USA* 104:7265–70

Trachtenberg JT, Chen BE, Knott GW, Feng G, Sanes JR, et al. 2002. Long-term in vivo imaging of experience-dependent synaptic plasticity in adult cortex. *Nature* 420:788–94

Tu JC, Xiao B, Naisbitt S, Yuan JP, Petralia RS, et al. 1999. Coupling of mGluR/Homer and PSD-95 complexes by the Shank family of postsynaptic density proteins. *Neuron* 23:583–92

Uylings HBM, West MJ, Coleman PD, De Brabander JM, Flood DG. 2000. Neuronal and cellular changes in the aging brain. In *Neurodegenerative Dementias: Clinical Features and Pathological Mechanisms*, ed. CM Clark, JQ Trojanowski, pp. 61–76. New York: McGraw-Hill

Valverde F. 1967. Apical dendritic spines of the visual cortex and light deprivation in the mouse. *Exp. Brain Res.* 3:337–52

Van den Veyver IB, Zoghbi HY. 2000. Methyl-CpG-binding protein 2 mutations in Rett syndrome. *Curr. Opin. Genet. Dev.* 10:275–79

Varoqueaux F, Aramuni G, Rawson RL, Mohrmann R, Missler M, et al. 2006. Neuroligins determine synapse maturation and function. *Neuron* 51:741–54 von Bohlen und Halbach O, Minichiello L, Unsicker K. 2003. Haploinsufficiency in trkB and/or trkC neurotrophin receptors causes structural alterations in the aged hippocampus and amygdala. *Eur. J. Neurosci.* 18:2319–25

Wang JY, Wigston DJ, Rees HD, Levey AI, Falls DL. 2000. LIM kinase 1 accumulates in presynaptic terminals during synapse maturation. *J. Comp. Neurol.* 416:319–34

Watanabe Y, Gould E, McEwen BS. 1992. Stress induces atrophy of apical dendrites of hippocampal CA3 pyramidal neurons. *Brain Res.* 588:341–45

Webb DJ, Zhang H, Majumdar D, Horwitz AF. 2007. Alpha5 integrin signaling regulates the formation of spines and synapses in hippocampal neurons. *J. Biol. Chem.* 282:6929–35

Webster MJ, Herman MM, Kleinman JE, Shannon Weickert C. 2006. BDNF and trkB mRNA expression in the hippocampus and temporal cortex during the human lifespan. *Gene Expr. Patterns* 6:941–51

Wellman CL. 2001. Dendritic reorganization in pyramidal neurons in medial prefrontal cortex after chronic corticosterone administration. *J. Neurobiol.* 49:245–53

Wong WT, Faulkner-Jones BE, Sanes JR, Wong RO. 2000. Rapid dendritic remodeling in the developing retina: dependence on neurotransmission and reciprocal regulation by Rac and Rho. *J. Neurosci.* 20:5024–36

Woo TU, Whitehead RE, Melchitzky DS, Lewis DA. 1998. A subclass of prefrontal gamma-aminobutyric acid axon terminals are selectively altered in schizophrenia. *Proc. Natl. Acad. Sci. USA* 95:5341–46

Woolley CS, Gould E, McEwen BS. 1990. Exposure to excess glucocorticoids alters dendritic morphology of adult hippocampal pyramidal neurons. *Brain Res.* 531:225–31

Wu GY, Cline HT. 1998. Stabilization of dendritic arbor structure in vivo by CaMKII. *Science* 279:222–26

Wu GY, Zou DJ, Rajan I, Cline H. 1999. Dendritic dynamics in vivo change during neuronal maturation. *J. Neurosci.* 19:4472–83

Wyszynski M, Lin J, Rao A, Nigh E, Beggs AH, et al. 1997. Competitive binding of alpha-actinin and calmodulin to the NMDA receptor. *Nature* 385:439–42

Xiao B, Tu JC, Worley PF. 2000. Homer: a link between neural activity and glutamate receptor function. *Curr. Opin. Neurobiol.* 10:370–74

Xie Z, Photowala H, Cahill ME, Srivastava DP, Woolfrey KM, et al. 2008. Coordination of synaptic adhesion with dendritic spine remodeling by AF-6 and kalirin-7. *J. Neurosci.* 28:6079–91

Xie Z, Srivastava DP, Photowala H, Kai L, Cahill ME, et al. 2007. Kalirin-7 controls activity-dependent structural and functional plasticity of dendritic spines. *Neuron* 56:640–56

Xu B, Zang K, Ruff NL, Zhang YA, McConnell SK, et al. 2000. Cortical degeneration in the absence of neurotrophin signaling: dendritic retraction and neuronal loss after removal of the receptor TrkB. *Neuron* 26:233–45

Yamagata Y, Kobayashi S, Umeda T, Inoue A, Sakagami H, et al. 2009. Kinase-dead knock-in mouse reveals an essential role of kinase activity of Ca^{2+}/calmodulin-dependent protein kinase IIalpha in dendritic spine enlargement, long-term potentiation, and learning. *J. Neurosci.* 29:7607–18

Yang N, Higuchi O, Ohashi K, Nagata K, Wada A, et al. 1998. Cofilin phosphorylation by LIM-kinase 1 and its role in Rac-mediated actin reorganization. *Nature* 393:809–12

Youn H, Jeoung M, Koo Y, Ji H, Markesbery WR, et al. 2007. Kalirin is underexpressed in Alzheimer's disease hippocampus. *J. Alzheimers Dis.* 11:385–97

Zanni G, Saillour Y, Nagara M, Billuart P, Castelnau L, et al. 2005. Oligophrenin 1 mutations frequently cause X-linked mental retardation with cerebellar hypoplasia. *Neurology* 65:1364–9

Zha XM, Dailey ME, Green SH. 2009. Role of Ca^{2+}/calmodulin-dependent protein kinase II in dendritic spine remodeling during epileptiform activity in vitro. *J. Neurosci. Res.* 87:1969–79

Zhan SS, Kamphorst W, Van Nostrand WE, Eikelenboom P. 1995. Distribution of neuronal growth-promoting factors and cytoskeletal proteins in altered neurites in Alzheimer's disease and nondemented elderly. *Acta Neuropathol.* 89:356–62

Zhang S, Ehlers MD, Bernhardt JP, Su CT, Huganir RL. 1998. Calmodulin mediates calcium-dependent inactivation of N-methyl-D-aspartate receptors. *Neuron* 21:443–53

Zhang W, Benson DL. 2001. Stages of synapse development defined by dependence on F-actin. *J. Neurosci.* 21:5169–81

Zhao L, Ma QL, Calon F, Harris-White ME, Yang F, et al. 2006. Role of p21-activated kinase pathway defects in the cognitive deficits of Alzheimer disease. *Nat. Neurosci.* 9:234–42

Ziv NE, Smith SJ. 1996. Evidence for a role of dendritic filopodia in synaptogenesis and spine formation. *Neuron* 17:91–102

Zuo Y, Lin A, Chang P, Gan WB. 2005. Development of long-term dendritic spine stability in diverse regions of cerebral cortex. *Neuron* 46:181–89

Connecting Vascular and Nervous System Development: Angiogenesis and the Blood-Brain Barrier

Stephen J. Tam and Ryan J. Watts

Neurodegeneration Labs, Department of Neuroscience, Genentech, Inc., South San Francisco, California 94080; email: watts.ryan@gene.com

Annu. Rev. Neurosci. 2010. 33:379–408

First published online as a Review in Advance on April 1, 2010

The *Annual Review of Neuroscience* is online at neuro.annualreviews.org

This article's doi:
10.1146/annurev-neuro-060909-152829

Key Words

endothelial cells, axon guidance, neurovascular unit, astrocytes, tight junctions, Wnt signaling

Abstract

The vascular and nervous systems share a common necessity of circuit formation to coordinate nutrient and information transfer, respectively. Shared developmental principles have evolved to orchestrate the formation of both the vascular and the nervous systems. This evolution is highlighted by the identification of specific guidance cues that direct both systems to their target tissues. In addition to sharing cellular and molecular signaling events during development, the vascular and nervous systems also form an intricate interface within the central nervous system called the neurovascular unit. Understanding how the neurovascular unit develops and functions, and more specifically how the blood-brain barrier within this unit is established, is of utmost importance. We explore the history, recent discoveries, and unanswered questions surrounding the relationship between the vascular and nervous systems with a focus on developmental signaling cues that guide network formation and establish the interface between these two systems.

Contents

CNS: central nervous system

NVU: neurovascular unit

BBB: blood-brain barrier

INTRODUCTION

The anatomical similarities between the vascular and nervous systems have been recognized for centuries. Recent discoveries have continued to deepen this apparent relationship at both the cellular and the molecular levels. At a gross anatomical level, the pathways of blood vessels and nerves appear congruous. At the cellular level, extending vessels and neurons utilize similar specialized structures that drive target tissue guidance through sensing their local environment. Ultimately these similarities can be traced back to several families of shared molecular guidance cues that guide both endothelial cells and neurons (Carmeliet & Tessier-Lavigne 2005, Dickson 2002, Larrivée et al. 2009, O'Donnell et al. 2009, Tessier-Lavigne & Goodman 1996). Although still in its early years, the field of vascular guidance has taken many cues from past insights provided by the axonal guidance principles of attraction and repulsion.

Beyond the common anatomical, cellular, and molecular properties, the vascular and nervous systems have developed an intricate relationship within the central nervous system (CNS) itself. This connection is found at the interface between the vascular and nervous systems, generally known as the neurovascular unit (NVU). The NVU is composed of endothelial cells, pericytes, glia, and neurons, which are tightly coupled to control cerebrovascular function (Hawkins & Davis 2005, Iadecola 2004). A major component of this interface is the blood-brain barrier (BBB), which is formed by endothelial cells of the NVU. The specialized blood vessels of the BBB are characterized by the formation of tight junctions, expression of specialized transporters and pumps, and reduced basal endocytic activity as compared with vessels found in peripheral organs (Rubin & Staddon 1999). Although the barrier is generally termed the blood-brain barrier, there are in fact several distinct barriers between the periphery and the nervous system, including the blood-cerebral spinal fluid barrier, the blood-retinal barrier, the blood–spinal cord barrier, and other barriers. Furthermore, this "barrier" is, in fact, a dynamic site of molecular transport that constantly regulates the flow of essential components, such as amino acids, across endothelial walls. Although many fundamental questions regarding BBB development and maintenance remain unanswered, the crucial role of the BBB in the protection and proper functioning of the nervous system has begun to attract the attention of researchers.

Recent discoveries have uncovered several molecular cues that regulate barrier formation. For instance, the wingless-type protein (Wnt) signaling pathway was identified as a key activator of angiogenesis in the nervous system and subsequent BBB formation (Daneman et al. 2009, Liebner et al. 2008, Stenman et al. 2008). Notably, Wnt signaling has been extensively studied in nervous system development and function (Salinas & Zou 2008). The identification of Wnt signaling as a regulator of angiogenesis and BBB formation in the CNS provides molecular evidence to support the original observation that neuronal tissue provides instructive cues to advancing endothelial cells, forming specialized blood vessels in the nervous system (Stewart & Wiley 1981). These findings are also another example of the common signaling pathways that drive the development of both the vascular and the nervous systems.

Significant progress has been made in identifying how vascular and nervous systems develop and function; recent advances highlight the common cellular and molecular mechanisms shared by these two systems. We examine these similarities and highlight the close relationship between the vascular and nervous systems, with an emphasis on their shared common developmental principles and how the interface between these systems, the BBB, is formed.

ANATOMICAL PATTERN OF THE VASCULAR AND NERVOUS SYSTEMS

Evolution of the Nervous and Vascular Systems

The fundamental requirement for an organism to both sense its environment and broadly distribute nutrients throughout its body has driven the evolution of the nervous and vascular systems, respectively. Although the exact origin of the nervous system is debatable, investigators generally believe that modern nervous systems arose first in the form of diffuse nerve nets, in sea-dwelling organisms such as cnidarians (Miller 2009). These organisms lack a vascular system (defined as a system to carry both oxygen and nutrients) because resources were readily available through diffusion and ingestion. On the basis of these observations, the general consensus indicates that the modern nervous system evolved prior to the vascular system.

The major driving force behind vascular evolution is the need for oxygen to fuel aerobic metabolism in multicellular organisms, combined with the limited ability of oxygen to diffuse through tissue (Fisher & Burggren 2007). For arthropods, two systems seem to have evolved independently to distribute nutrients and oxygen. To deliver nutrients, hemolymph bathes all cells and is distributed by the coordinated movement of muscles and a primitive heart. Independent of nutrients, oxygen is delivered by the tracheal system. The tracheal system of insects is most akin to the modern vascular system as it relates to developmental principles, highlighted by the stereotyped formation of a tubular network that functions by directing air exchange of oxygen and carbon dioxide. In fact, many features of tube formation and sprouting are shared between the insect tracheal system and the modern vascular systems (Lubarsky & Krasnow 2003). As larger organisms evolved, however, a system to distribute oxygen and nutrients throughout dense tissue more efficiently became necessary.

Consistent with the chronology of neuronal and vascular evolution, several vascular-specific genes may have arisen later in evolution. Vascular endothelial growth factor (VEGF) was first discovered as a growth factor that selectively mediates endothelial proliferation and vessel permeability (Keck et al. 1989, Leung et al. 1989), and expression of this hypoxia-responsive gene is tightly regulated to control vasculogenesis and angiogenesis (Coultas et al. 2005, Fong 2009). Distant homologs to both mammalian VEGF and VEGFRs do exist

VEGF: vascular endothelial growth factor

in invertebrates (Cho et al. 2002). The functions of these molecules in *Drosophila* are limited to controlling blood cell migration, a function that is speculated to be a precursor to the many VEGF-driven functions in mammals. One of the coreceptors for VEGF, neuropilin-1 (Nrp1), arose after the nervous system in evolution. Nrp1 was first identified as a receptor for the axon guidance cue semaphorin (He & Tessier-Lavigne 1997, Kolodkin et al. 1997) and then subsequently shown also to be a receptor for VEGF (Soker et al. 1998). Nrp1 is found in vertebrates but is absent from invertebrates, whereas semaphorins and their other coreceptors, the plexins, are present in both invertebrates and vertebrates (Kolodkin 1996, Kolodkin et al. 1992, Tamagnone et al. 1999). Therefore, the evolution of the *Nrp1* gene may have been congruent with the origin of the modern vascular system, yet Nrp1 is an essential component in vertebrate neuronal development as well.

Nrp1 interaction with both semaphorin and VEGF is one example of how the vascular and nervous systems have co-opted the same molecular cascades for development. However, many of the guidance cues may have originated first in the nervous system, with family members likely evolving later to play a role in vascular development. These pathways are discussed in more detail in subsequent sections.

Shared Anatomical Patterning

Similar patterning between the vascular and nervous systems is apparent at the anatomical level (**Figure 1a**). These similarities were likely first observed at the macroscopic level (Vesalius 1543). With modern histological methodology, the coursing of artery and veins with nerve bundles can now be described at the microscopic level (Larrivée et al. 2009, Mukouyama et al. 2002). How can such highly stereotypic patterning in two diverse systems be achieved? Do developing vessels provide signals for growing axons, or vice versa? Looking at the developing mouse embryo, one can see that vascular outgrowth precedes axon outgrowth (**Figure 1b**). Consistent with this observation, many molecular cues are provided by the developing vasculature to guide growing axons. For example, endothelin-3 is secreted by smooth-muscle cells of the external carotid artery, providing an attractive signal for extending axons of the superior cervical ganglia (Makita et al. 2008). Furthermore, vascular smooth-muscle expression of Artemin, a glial cell–derived neurotrophic factor (GDNF) family member, is a signal attracting axons of sympathetic neurons (Honma et al. 2002). *Artemin*-deficient mice show axon outgrowth and patterning defects, an observation that is phenocopied in mice lacking the Artemin receptor *GFRalpha3*. These findings provide convincing evidence that vessels can direct neurons to form congruent patterns.

As part of normal vascular and neural development, exuberant primitive networks are initially formed followed by the selective remodeling of these networks to form functional circuits. In the case of vascular remodeling, a robust vascular plexus is generated in the embryo, followed by remodeling to arteries, veins, and fine capillary networks. Vascular remodeling takes place after peripheral nerve development and is directed by peripheral sensory neurons and Schwann cells (Mukouyama et al. 2002).

Figure 1

Anatomical resemblance between the vascular and nervous systems. (*a*) Depiction of the shared anatomical patterning between nervous and vascular networks in the human body. Recent evidence indicates that blood vessels and neurons direct their outgrowth using shared molecular mechanisms, ultimately resulting in a similarly patterned network. (*b*) Although developing mouse embryos exhibit shared nervous and vascular patterning as well, the vascular system develops at a faster pace than does the nervous system. In fact, vessels can direct neurons to form congruent patterns. *Left panel*: neurons visualized by neurofilament (Nf) staining of an E12.5 embryo. *Right panel*: blood vessels visualized by FLK1 in situ hybridization staining of an E10.5 embryo.

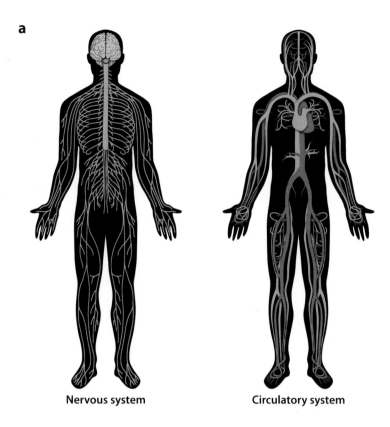

Nervous system   Circulatory system

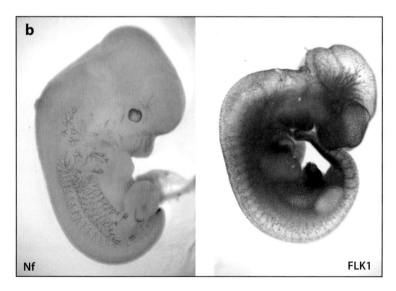

Nf   FLK1

These conclusions are derived from both necessity and sufficiency experiments. Mutant embryos that lack peripheral neurons do not undergo proper vascular remodeling, highlighted by defects in arterial differentiation. Moreover, mutants that misdirect peripheral axon outgrowth result in coalignment of blood vessels with these misguided neuronal patterns. Mukouyama et al. (2005) later showed that VEGF is one such cue involved in vascular remodeling and arteriogenesis that is provided by neurons and Schwann cells.

More than four and a half centuries since Vesalius first observed the overall anatomical similarities between vascular and neuronal networks, evidence now demonstrates that vessels and neurons codirect their patterning to achieve exquisite alignment at the anatomical level. For such guidance to take place, growing vessels and neurons must be able to sense the complex chemical cues that orchestrate the morphological changes necessary for proper patterning. These sensory mechanisms are found in the cellular structures that constitute the developing neuron and blood vessel.

SHARED CELLULAR MECHANISMS

The Cellular Organization of the Nervous and Vascular Systems

In 1906, Camillo Golgi and Santiago Ramón y Cajal presented differing theories about the cellular organization of the nervous system in a famous Nobel debate. Golgi postulated that the nervous system consisted of a syncytial system with multiple cells of the nervous system possibly connecting via protoplasmic fusions to form a reticular system (Golgi 1906). Cajal, on the other hand, was an ardent proponent of the "neuron doctrine," which argued that cells in the nervous system constitute individual components connected by "a granular cement, or special conducting substance" keeping neurons "intimately in contact," yet "in contiguity but not in continuity" (Ramón y Cajal 1906). History has largely proven both Cajal

and Waldeyer, the first to propose the neuron doctrine, to be correct. We understand now that the nervous system consists of a complex network of diverse neurons, with extending axons connecting to dendrites of target cells via specialized contacts known as synapses. Many neurons must send axonal processes long distances through a labyrinth of tissue to reach their final targets. The emergence of embryonic neurons from neuroblasts, followed by their migration and elaboration, results in single cellular units that possess the ability to receive and send information.

The nervous system also contains a battery of glial cells, which are required for a multitude of nervous system functions, including enhancement of nerve conductance, immune surveillance, and neurotransmitter processing and signaling (Barres 2008). Golgi himself first observed that glial cells make direct contacts with the vasculature, which were likely to be the astrocytic endfeet and their ensheathment of blood vessels (see The Neurovascular Unit, below). Oligodendrocytes and Schwann cells are responsible primarily for myelination of the central and peripheral nervous systems, respectively. They are functionally similar to pericytes and smooth-muscle cells of the vasculature because both are supporting cell types that enhance the function of the main constituent cell types, neurons and endothelial cells, respectively (Bergers & Song 2005, Betsholtz et al. 2005). However, it remains to be determined if similar signaling mechanisms are employed toward the formation of both endothelial-pericytes and neuron-glia interactions.

In contrast to network formation in the nervous system, which is the result of elaborated single neurons extending protoplasmic processes to form specific connections, the cellular components of the vascular system undergo blood vessel formation through two distinct processes. First, in vasculogenesis, endothelial cells differentiate from angioblasts to form a vascular plexus, which consists of a meshwork of connected strands of endothelial cells (Coultas et al. 2005). The vascular plexus serves as the starting substrate for further growth and refinement. Second, new vessels sprout from existing vessels, a process termed angiogenesis. VEGF is necessary for both vasculogenesis and angiogenesis; however, other molecules provide additional cues for further differentiation and refinement of the vascular system. Similar to the nervous system, where exuberant connectivity is followed by pruning to form a functional network, the vascular system also undergoes remodeling. This process is influenced by both blood flow and contact with surrounding tissues. Furthermore, vessels formed via angiogenesis must sense their environment to follow the correct trajectory toward their final target.

The Axon Growth Cone and Endothelial Tip Cell

Cajal's illustrious career was also highlighted by his discovery of the axon growth cone, a structure found at the tip of growing axons that guides outgrowth by sensing its local environment (de Castro et al. 2007). In Cajal's own words, a growth cone is "like a living battering-ram, soft and flexible, which advances mechanically, pushing aside the obstacles that it finds in its way until it reaches its peripheral destination" (Ramón y Cajal 1917). Modern imaging has since verified Cajal's assessment and has further defined the molecular components of the growth cone (Lowery & Van Vactor 2009). Specifically, an axonal growth cone is found at the end of an axon shaft and consists largely of cytoskeletal machinery (**Figure 2a**). The hand-like structure itself is composed of both actin-rich filopodia and lamellopodia, with actin bundles that are connected to the microtubule framework that continues down the axon shaft.

More than 100 years after Cajal's original description of the axon growth cone, Gerhardt

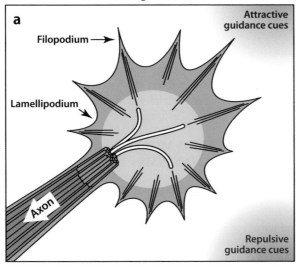
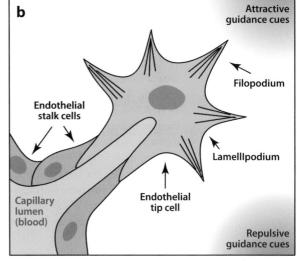

Figure 2

The axon growth cone and endothelial tip cell. Blood vessels and neurons must sense their environment to follow the correct trajectory toward their final target. Cellular projections of axons and endothelia that guide such outgrowth share morphological features. (*a*) The axon growth cone forms at the leading edge of migrating axons through an elaborate cytoskeletal actin network linked to axonal microtubules. (*b*) The endothelial tip cell caps the migrating end of stalk cells, a lumenized column of cells that form a budding or angiogenic blood vessel. Although the tip cell is its own cellular entity separate from the rest of the growing blood vessel, it is the functional analog of the axonal growth cone. Both filopodial and lamellopodial projections of the extending growth cone and tip cell are believed to form the motor that physically pulls cells toward attractive cues (*green*) and away from repulsive cues (*red*) as extending neurons and blood vessels sense their environment.

et al. (2003) described a similar structure in angiogenic sprouts of growing blood vessels in the mouse retina. This structure is composed of a specialized cell found at the growing end of extending vessels, termed the endothelial tip cell (**Figure 2b**). The shaft of the extending vessel is composed of a lumenized endothelial cell chain made of stalk cells. From an architectural standpoint, the tip cell and associated endothelial stalk cells are similar to the axon growth cone and its associated axonal shaft. Tip cells, akin to axon growth cones, are actin-rich structures that dynamically navigate their environment when sprouting from existing vasculature. This process was elegantly observed through live imaging of green fluorescent protein (GFP)-expressing vessels in transgenic zebrafish, where extending tip cells were described as functionally similar to axon growth cones (Lawson & Weinstein 2002).

Tip cells, at least in the retina, are exquisitely responsive to VEGF (Gerhardt et al. 2003). Using mice engineered to express distinct isoforms of VEGF that have different diffusion properties, investigators showed that endothelial tip cells migrate in response to VEGF gradients, whereas stalk cell proliferation is a property of total VEGF concentrations. These data are consistent with previous observations utilizing VEGF isoform mutants (Ruhrberg et al. 2002). Specifically, a shallow VEGF gradient with high total concentrations of VEGF exposure results in large-diameter vessels with reduced branching. VEGF receptor 2 (Flk1) is both expressed on the tip cell and necessary for filopodia formation (Gerhardt et al. 2003). Thus, in the retina, VEGF acts as the predominant mitogenic factor and is chemoattractive for blood vessels.

How is a tip cell selected to form an angiogenic sprout? The vascular system likely addressed this need by co-opting the Notch signaling cascade described first in *Drosophila*. In the case of fly tracheal development, a leading tracheal cell is selected by Notch-mediated lateral inhibition (Ghabrial & Krasnow 2006). In mammals, delta-like 4 (Dll4) appears to be the major Notch ligand regulating vascular development (Duarte et al. 2004, Gale et al. 2004, Krebs et al. 2004). Additional mechanistic studies have carefully dissected Dll4's role in vascular development and concluded that Dll4/Notch signaling is likely acting via lateral inhibition to establish the tip cell morphology, whereas adjacent cells are destined to become stalk cells (Hellstrom et al. 2007, Lobov et al. 2007, Suchting et al. 2007). These observations should not be surprising because Notch signaling is a major mechanism for cell fate determination in many tissues, including the nervous system (Gaiano & Fishell 2002).

The cellular parallels between an extending neuron's axon growth cone and an angiogenic blood vessel's tip cell suggest that these structures may respond in ways similar to chemical cues provided by their respective environments. In 1963, Roger Sperry summarized the chemospecificity hypothesis, which postulates that chemical cues in the target tissue provide specific directions for the developing and regenerating nervous system, resulting in stereotyped neural network formation (Sperry 1963). Do blood vessels also respond to chemical cues resulting in stereotyped development? Recent discoveries have addressed this hypothesis at the molecular level and provide strong evidence that the nervous and vascular systems share common principles of development.

SHARED MOLECULAR MECHANISMS

Axon Guidance Molecules and Angiogenesis

The possibility that the vascular and nervous systems respond to common molecular cues to guide their development is highly plausible owing to both their stereotyped circuitry and the shared common cellular structures used to sense their environment during development. We have described evidence that nerves may provide cues to growing vessels; likewise, the opposing relationship has also been established. A major theme of cellular guidance is the balance between attractive and repulsive forces

(**Figure 2**). In introducing the endothelial tip cell, we have discussed how VEGF is a major attractive molecule for extending blood vessels (Gerhardt et al. 2003). Do other attractive cues exist for blood vessels? Are there repulsive forces that also guide developing vessels? In searching for additional candidate molecules, we now turn our attention to the accumulating evidence indicating that the "classical" axon guidance cues also guide vascular tip cells.

Axon guidance molecules exist in four major families: (*a*) Slit/Robo, (*b*) semaphorin/plexin/neuropilin, (*c*) Netrin/Unc5/DCC, and (*d*) Ephrin/Eph (**Figure 3**). The role of these pathways in guidance of axons in the developing nervous system has been described extensively (Dickson 2002, O'Donnell et al. 2009, Tessier-Lavigne & Goodman 1996). Recent studies also indicate that various well-described morphogens, including wingless-type proteins (Wnts), bone morphogenic proteins (BMPs), and sonic hedgehog (Shh) also mediate axon guidance (Butler & Dodd 2003, Charron et al. 2003, Lyuksyutova et al. 2003, Yoshikawa et al. 2003). With the exception of the Wnts, which are discussed in detail below related to

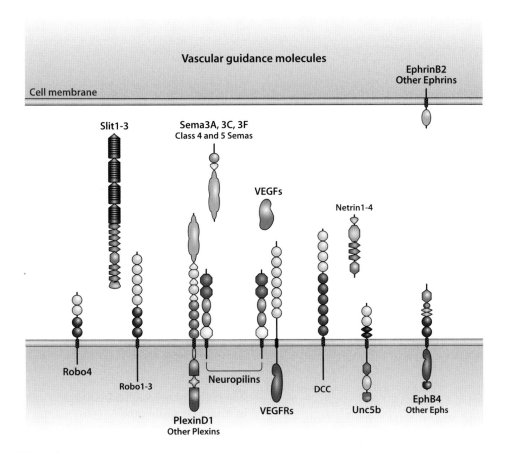

Figure 3

Axon guidance pathways involved in angiogenesis. Schematic representation of the four major families of axon guidance ligand-receptor pairs: Slit/Robo, semaphorin/plexin/neuropilin, Netrin/Unc5/DCC, and Ephrin/Eph. Axonal guidance cues mediate complex cellular navigational programs within axons, as both chemoattractants and repellents. Recent evidence has identified roles for several of these "classic" guidance molecules in directing angiogenic tip cells toward their final destination.

angiogenesis and BBB formation in the developing CNS, we discuss the role of the four major families of axon guidance molecules in angiogenesis.

The Slits and Robos in Vascular Guidance

The Slit/Robo family of axon guidance molecules was discovered as mediators of commissural axon crossing in *Drosophila* (Kidd et al. 1998, 1999; Seeger et al. 1993) and of axon guidance in *C. elegans* (Zallen et al. 1998). Wang et al. (1999) showed that Slit enhances axonal branching and elongation (Wang et al. 1999). There are three members of the Slit ligand family and four members of the Robo receptor family (**Figure 3**). Slits contain 4 leucine-rich repeats, 7–9 EGF repeats, and one laminin domain. The roundabout (Robo) receptors were named after their fly loss-of-function phenotype of aberrant midline crossing, including axons repeatedly crossing the midline, giving the appearance of a circular road junction, or roundabout, when analyzed by histology (Seeger et al. 1993). The Robo1-3 receptors consist of five Ig domains, three fibronectin type 3 domains, and one intracellular signaling domain. A fourth member of the Robo receptor family, Robo4 or magic roundabout, contains only two Ig domains and two fibronectin type 3 domains. Notably, Robo4 is almost exclusively expressed in the vasculature (Huminiecki et al. 2002).

During neural development, Slit/Robo interactions can result in either a repulsive or an attractive response (Dickson & Gilestro 2006). Most data indicate that Slit/Robo binding on neurites primarily regulates a repulsive signal. Thus it was intriguing when Wang et al. (2003) found Slit2 to be an attractive factor for extending angiogenic sprouts through binding to Robo1 on endothelial cells. In contrast to this data, others have argued that Slit2 is repulsive to endothelial cells via binding to Robo4 (Park et al. 2003). To complicate matters further, it is unclear whether Slit2 binds to Robo4 because in vitro binding studies showed a convincing interaction only between Slit2 and Robo1, but not to Robo4 (Suchting et al. 2005). Robo4 may also mediate a repulsive or attractive response on the basis of cellular context, including cell-type and/or ligand binding. For example, Sheldon et al. (2009) recently showed that Robo1 and Robo4 can form heterodimers, indicating that Slit2 binding to Robo1 may result in signaling via Robo4.

In addition to in vitro observations for Slit/Robo interactions in the context of the vascular system, studies in zebrafish have shown that a knockdown of Robo4 results in an intersomitic vessel guidance defect with a reduction in vessel sprouting and a misdirection of those sprouts that do form (Bedell et al. 2005). However, no example of a vascular developmental defect in mouse knockouts of the various Slit/Robo genes has been reported. A recent study has proposed that Robo4 mediates vascular integrity in adult animals by inhibiting angiogenesis and reducing vascular permeability (Jones et al. 2008). Robo4-deficient mice exhibited an increase in vascular sprouting and permeability in an oxygen-induced retinopathy model. Furthermore, injecting Slit2 reversed the increase in angiogenesis and vascular permeability in these models in a Robo4-dependent fashion. These data support a repulsive role for Robo4 in angiogenic sprouting. However, because Robo4 is expressed on endothelial stalk cells in the retina, it is intriguing to speculate that Robo4 may inhibit new tip cell formation from existing vessels. To date, the interaction between Robo4 and Slit2 remains to be fully verified, leaving open the possibility that other ligands mediate Robo4 function.

The Semaphorins/Plexins/Nrps in Vascular Guidance

Although semaphorins (Sema) were first discovered as mediators of axon growth cone collapse (Raper 2000), they are now more commonly recognized as proteins with robust effects on cellular morphology. There are eight classes of Sema proteins, including both membrane-bound and secreted forms. For the purposes of this review, we focus on one of the most versatile families, the class 3 semaphorins (Sema3).

The Sema3 family members are secreted proteins that contain a Sema domain, a plexin/semaphorin/integrin domain (PSI), and an Ig domain (**Figure 3**). The Plexin and Neuropilin receptor families collaborate to form multimeric receptor assemblies responsible for downstream Sema signaling. The extracellular region of plexins consists of one Sema domain, three PSI domains, four Ig-like/plexins/transcription factor domains (IPT), and two Ig domains. A transmembrane helix links this region to the cytoplasmic region containing one coiled-coil helix and one Rac-binding domain (RBD) that bisects a GTPase activating protein domain (GAP).

Mammals have two neuropilin (Nrp) proteins: Nrp1 and Nrp2. Nrps share two complement-binding domains (CUB or a1a2 domains), two coagulation factor V/VIII homology domains (b1b2), and a meprin/A5 protein/phosphatase-μ domain (MAM) in the extracellular space followed by a trans-membrane helix and a short cytoplasmic tail. Furthermore, Nrps can also be expressed as soluble isoforms lacking the MAM and transmembrane domains (Rossignol et al. 2000). In addition to being semaphorin receptors, Nrps bind multiple VEGF isoforms and mediate distinct VEGF functions (Pan et al. 2007; Soker et al. 1998, 2002). The Sema/Plexin/Nrp family is a prime example of a group of molecules with diverse functions in both neuronal and vascular development. Here, we focus on their roles in vascular biology.

Sema3s were originally classified as potent axonal chemorepellents that signal indirectly through Plexin receptors by binding to Nrps (Chen et al. 1997, Fujisawa 2004, He & Tessier-Lavigne 1997, Kolodkin et al. 1997). Of the seven Sema3s (A-G) reported in mammals, several function in more versatile ways, including a key role in guiding extending endothelial cells (Tran et al. 2007). Sema3A was first described in the vascular system as a mediator of vessel tip cell lamellipodial collapse and was thus an inhibitor of angiogenesis (Miao et al. 1999, Serini et al. 2003, Shoji et al. 2003). However, a role for Sema3A in endothelial biology and vascular development has been disputed. More recent studies have shown that Sema3A does not reduce endothelial cell migration (Pan et al. 2007), nor does it compete with VEGF for binding to Nrp1 (Appleton et al. 2007). Furthermore, *sema3a*$^{-/-}$ knockout mice do not have vascular development phenotypes (Vieira et al. 2007). Additional studies may be needed to determine if Sema3A plays a more specific or subtle role in vascular biology that has yet to be discovered. For example, although the overwhelming evidence suggests that semaphorins act as chemorepulsive cues in the nervous system, at least one example shows that the cellular context can switch Sema from a repulsive to an attractive signal (Song et al. 1998).

In contrast to Sema3A, Sema3E is emerging as a critical mediator of vascular biology through its direct interaction with the endothelial expressed receptor, PlexinD1 (Gitler et al. 2004, Gu et al. 2005, Torres-Vazquez et al. 2004). The vascular expression of *plexinD1* was first described in two separate studies, one examining all Plexins and focusing on their roles in the nervous system (Cheng et al. 2001) and the other looking specifically at *plexinD1* expression during embryonic development (van der Zwaag et al. 2002). *Sema3e* and *plexinD1* loss-of-function studies in both zebrafish and mouse result in a vessel guidance defect consistent with a repulsive role for Sema3E acting via PlexinD1. Somites express Sema3E, thereby preventing the PlexinD1-expressing intersomitic vessels from entering the somites. However, *plexinD1* knockout mice show a cardiac development defect not observed in *sema3E* knockouts, thus suggesting that another Sema, speculated to be Sema3C, may act through PlexinD1.

Nrp receptors bind both VEGF and class 3 semaphorins. It is therefore not surprising that knockouts of both *Nrp1* and *Nrp2*, which are expressed in developing vasculature and lymphatic systems respectively, lead to numerous vessel-patterning and lymphangiogenic defects (Gu et al. 2003, Kawasaki et al. 1999, Takashima et al. 2002). However, Nrps' interactions with both VEGFs and Sema3s complicate the interpretation of these results, making it difficult to pinpoint a particular phenotype as being related to Sema or VEGF. To address this issue Gu and colleagues (2002) first performed structure/function studies to identify residues responsible for Sema3 binding to Nrp1. Subsequently, knock-in mice expressing mutant Nrp1 lacking Sema binding and keeping VEGF binding intact were generated (Gu et al. 2003). These mice showed no vascular defects, suggesting that Nrp1's interaction with VEGF, but not Sema, is critical for vascular patterning. On the basis of Nrp1's multifaceted role in vascular biology, this molecule is now being therapeutically targeted to modulate angiogenesis in vivo.

The Netrin/Unc5/DCC Family in Vascular Guidance

Netrin was first identified as UNC-6 in *C. elegans* (Ishii et al. 1992), and three Netrin family members have been identified in mammals (Moore et al. 2007). The netrins are secreted proteins that contain one laminin domain, three EGF-like repeats, and a carboxyl-terminal domain (CRD) (**Figure 3**). Netrins act as both chemoattractants and chemorepellents using their interactions with respective receptors. During CNS development, Netrin1 is secreted from cells at the ventral midline to attract commissural axons (Kennedy et al. 1994, Serafini et al. 1994) via activation of the DCC (Deleted in Colorectal Cancer) receptor (Keino-Masu et al. 1996). In *netrin1* and *dcc* knockout mice, most commissural axons fail to reach and cross the midline (Fazeli et al. 1997, Serafini et al. 1996).

DCC and neogenin are members of the DCC family of receptors and contain four Ig domains, six Fn3 domains, and one transmembrane domain, followed by an intracellular signaling domain with putative protein interaction and phosphorylation sites (Moore et al. 2007). DCC can also interact with Robo

receptors to inactivate Netrin's attractant activity once midline crossing has been achieved (Stein & Tessier-Lavigne 2001). Conversely, Netrin-mediated axon repulsion involves the activation of Unc5 (uncoordinated 5) receptor family members, either as homodimers or Unc5-DCC heterodimers (Hong et al. 1999, Keleman & Dickson 2001, Leonardo et al. 1997). The Unc5 family contains two Ig domains, two thrombospondin type I domains, one transmembrane domain, one Zona occludens-5 domain, one DCC interacting domain, and one death domain.

Netrins and their cognate receptors have also been implicated in the regulation of several developmental processes in nonneuronal tissues. The first hints that Netrin/Unc5 signaling may regulate blood vessel morphogenesis were discovered as a consequence of knocking out the Unc5b receptor (Lu et al. 2004). Depletion of Unc5b in mice and zebrafish resulted in excessive tip cell filopodia formation and branching, suggesting that Unc5b activation drives endothelial repulsion. Investigators showed that Unc5b is present in the tip cell of extending blood vessels (Bouvrée et al. 2008, Larrivée et al. 2007). Knockdown of *netrin1a* in zebrafish results in a loss-of-function phenotype similar to the *unc5b* loss-of-function. Similar to Sema3A and PlexinD1, intersomitic paths express Netrin1a and thereby prevent the Unc5b-expressing tip cells from meandering off path into the somites. In mice, the ligand for Unc5b-mediated vessel tip cell repulsion remains to be identified because *netrin1* knockouts maintain normal vascular architecture (Salminen et al. 2000, Serafini et al. 1996).

The lack of Netrin1 involvement in mammalian blood vessel repulsion may be resolved by findings involving the Netrin4 family member. Netrin4 was upregulated in angiogenic endothelium and bound Unc5 receptors, possibly through direct interactions with the coreceptor Neogenin (Lejmi et al. 2008, Qin et al. 2007). Although in vivo mammalian loss-of-function data are lacking, recent in vitro endothelial migration results suggest that both Netrin1 and Netrin4 can inhibit VEGF-induced angiogenesis. Conversely, Netrin1- and Netrin4-mediated blood vessel chemoattraction is suggested on the basis of both in vitro vessel proliferation and migration studies and in vivo viral delivery studies. Nevertheless, the cognate in vivo receptors for Netrin1 and Netrin4 remain to be conclusively identified. In addition, these data remain controversial because knockout phenotypes for *Unc5b* reported by various other groups do not mimic those seen in these studies (Nguyen & Cai 2006, Park et al. 2004, Wilson et al. 2006). Moreover, the functional duality shared by many axon and vessel guidance cues suggests that Netrin/Unc5/DCC may mediate both repulsion and attraction in the vascular system (Yang et al. 2007). Thus, discovering the vascular in vivo receptor(s) for Netrin1 and Netrin4 and the ligand for Unc5b may help to clarify the existing data.

The Ephrin/Eph Family in Vascular Guidance

The Ephrin/Eph receptor tyrosine kinase family of short-range axon guidance molecules was first identified as chemorepellents (Cheng et al. 1995, Drescher et al. 1995). Their dual functional nature was subsequently elaborated, whereby Ephrin/Eph interactions serve as both attractive and repulsive cues in both the nervous and the vascular systems (Palmer & Klein 2003). Ephrin ligands contain one cupredoxin homolog domain (**Figure 3**), whereas EphrinA1-5 are anchored to plasma membranes via a glycosyl-phosphatidyl-inositol moiety and EphrinB1-3 have a transmembrane domain followed by a cytoplasmic tail containing a PDZ binding domain that can mediate reverse signaling (Holland et al. 1996).

The Eph receptors include 13 family members: EphA1-8, EphB1-4, and EphB6. Eph receptor tyrosine kinases contain one laminin domain, two tumor necrosis factor receptor (TNFR) CRD domains, two Fn3 domains, a transmembrane domain followed by a kinase domain, and a sterile alpha motif (SAM)

domain. In the nervous system, these molecules drive many wiring processes, including midline pathfinding, dendritic spine formation, and synaptic plasticity (Palmer & Klein 2003).

Ephrin/Eph signaling molecules also control blood vessel development. Ephrin/Eph participate in another key aspect of vascular development: the maintenance of vascular patterning through segregated expression in venous and arteriole vessels. EphB4 is expressed in developing veins, whereas EphrinB2 is found in arteries, providing a key mechanism for repulsive tissue boundary preservation within vascular networks (Adams et al. 1999, Gerety et al. 1999, Wang et al. 1998). These observations are best depicted in *ephrinB2* knockout mice, in which angiogenesis defects in both embryonic arteries and veins suggest that both forward and reverse Ephrin/Eph signaling takes place in developing vasculature. Conversely, Ephrin/Eph guidance factors function as positive cues in the context of tumor angiogenesis. For example, EphrinA1 stimulates angiogenesis when expressed on tumor cells, and treatment with the EphA2 receptor ectodomain inhibits neovascularization in these tumor models (Ogawa et al. 2000, Pandey et al. 1995). Therefore, like the other axon guidance family members described, the Ephrin/Eph family of vascular guidance cues also exhibits exquisite dynamic versatility, mediating complex cellular navigational programs within axons and blood vessels as both chemoattractants and repellents.

The past decade of discoveries has elucidated key roles for axon guidance molecules in vascular development and has shown that they utilize biological principles similar to those first observed in the nervous system. These pathways are now a focus of research in the disease setting, for modulating both blood vessel and axonal growth (see sidebar, Axon Guidance Molecules as Targets for Cancer). Having focused on how the vascular and nervous systems share common cues for development, we now turn our attention to the interaction between the vascular and the nervous systems, again with an emphasis on development.

AXON GUIDANCE MOLECULES AS TARGETS FOR CANCER

As solid tumors begin to grow and become hypoxic, tumor cells recruit blood vessels by secreting VEGF and other angiogenic factors. Blocking angiogenesis is now well established as a mechanism to treat cancer; this process slows tumor growth by reducing nutrient and oxygen transfer to tumor cells (Ferrara & Kerbel 2005). With the discovery of numerous axon guidance molecules playing an essential role in vessel growth during development, investigators have proposed that these mediators of angiogenesis represent a new set of potential cancer targets. Furthermore, several axon guidance molecules have also been studied in the context of tumor cell biology, for example in mediating tumor cell migration. Future studies in both animal models and cancer patients will determine the value of targeting axon guidance molecules as a means to slow tumor growth and disease progression.

THE INTERFACE BETWEEN THE VASCULAR AND NERVOUS SYSTEMS

The human brain features a large vascular network of ~400 miles of blood vessels that receives more than 20% of the body's cardiac output and utilizes most of its blood glucose (Begley & Brightman 2003). These outstanding characteristics underscore the evolved importance of efficient and thorough exchange of both nutrients and waste between blood and the CNS. This architecture also represents a daunting task for development of a complex

vascular network within an even more intricate neural network. Before describing what is known about vascular development in the CNS we first illustrate in detail the properties of the interface between the vascular and nervous systems, with a focus on the BBB, an essential component of the multicellular NVU.

Molecular Properties of the Blood-Brain Barrier

The now classic experiments demonstrating that a polar dye, trypan blue, injected into blood is specifically excluded from the CNS serve as the first known experimental paradigm in which a barrier between the CNS and peripheral tissue was described (Goldmann 1909, 1913). In the first set of experiments, investigators prematurely concluded that brain tissue had minimal dye-binding affinity compared with other non-CNS organs. The presence of a BBB was subsequently established with the converse experiment of dye injection into the cerebrospinal fluid, the result being staining of brain. Taken together, investigators concluded that various molecules, in this case typan blue (872.88 Da), are uniquely excluded from the CNS and thus that a biological barrier must be in place.

The BBB is an essential regulator of blood-brain exchange of nutrients and waste that operates on several distinct levels (**Figure 4**). First, endothelial intercellular assemblies such as tight and adherens junctions function as a physical barrier that restricts the paracellular movement of molecules across the BBB.

Figure 4

The cellular and molecular properties of the blood-brain barrier (BBB). Many proteins function at the blood-brain barrier (BBB) to regulate molecular exchange. (*a*) These components include small molecule transporters (P-gp/MRPs, amino-acid, and glucose transporters), large molecule transporters (insulin, LDL, transferrin, and leptin receptors), junctional assemblies (tight and adherens junctions), and metabolic enzymes (P450-related, MAO, and GGT). Tight junctions at the BBB limit most paracellular movement and are composed of several protein families, including occludin, claudin, e-cadherin, ZO, JAM, catenins, cingulin, and actin. (*b*) Small and/or lipophilic molecules can bypass the tight junctions but are rapidly degraded by intracellular enzymes and actively pumped back into the bloodstream by P-gp and other multidrug-resistant proteins (MRPs). Large molecule, amino-acid, and glucose transporters allow regulated passage of essential nutrients from blood to brain. (*c*) Electron micrograph of endothelial contacts within the blood-brain barrier (image courtesy of Reese & Karnovsky 1967). Arrows mark intercellular kissing-points indicative of tight junctions.

BBB: tight junctions, transporters, and transcytosis

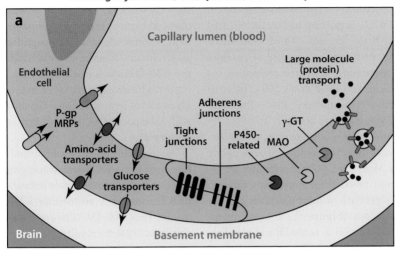

Barrier regulation: molecular exchange between blood and CNS

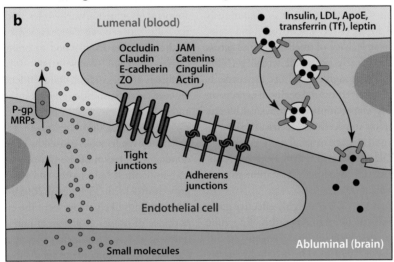

Ultrastructural image of endothelial tight junctions at the BBB

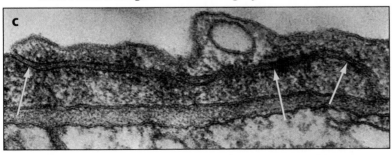

Second, several transport mechanisms resident to the BBB jettison unwanted molecules out of the CNS while importing molecules essential for proper brain function. Third, endothelial cells of the BBB express a variety of enzymes that can metabolize unwanted molecules, including toxins (de Boer et al. 2003, Iadecola 2004, Reese & Karnovsky 1967, Rubin & Staddon 1999, Zlokovic 2008).

The molecular exchange across the endothelium of the BBB is limited and tightly regulated. Most large, polar metabolites are excluded from blood-brain movement through a complex network of tight junctions and adherens junctions (**Figure 4**). Integral membrane proteins, such as occludin and claudins, physically associate with the actin cytoskeletal network through zonula occluden adaptor proteins. Claudin-5 is the most well characterized TJ protein within the BBB. It is ubiquitously expressed in all tissue vasculature, and deletion results in leakage of small molecules across the BBB and early postnatal lethality (Nitta et al. 2003). Although cell culture experiments demonstrate that truncated occludin expression functionally compromises the physical barrier because of low transcellular electrical resistance and increased leakage of small molecules across the BBB (Bamforth et al. 1999), occludin is not required for TJ structural formation in vivo (Saitou et al. 2000).

Although the BBB is a highly effective gatekeeper, substances that are either lipophilic or smaller than 400 Da may effectively bypass the physical barrier. The metabolic barrier eliminates such chemicals that would otherwise move from the blood into the CNS. As the primary route of drug metabolism, p450-related cytochrome enzyme oxidizes unwanted substances within the cytoplasm of endothelial cells (Alavijeh et al. 2005). Likewise, monoamine oxidase (MAO) also contributes to the metabolic barrier to protect the brain from circulating neurotoxins and biogenic amines. For instance, MAO metabolizes the neurotoxin MPTP and attenuates the parkinsonism associated with this drug (Kalaria & Harik 1987). In some cases, enzymatic activity within the BBB does not

remove unwanted molecules but instead facilitates the transport of essential substances from blood to brain. Through its transpeptidase activity, gamma-glutamyl transpeptidase (GGT) assists in the transfer of amino acids across the BBB (Hawkins et al. 2006). By modulating pyroglutamate levels, GGT indirectly stimulates a wide array of sodium-dependent amino acid transporters at the plasma membrane (**Figure 4**).

Amino acids (AA) represent one of many classes of nutrients essential for proper CNS function and cannot passively diffuse across the BBB because they are neither lipophilic nor smaller than 400 Da. Transport mechanisms tightly facilitate the regulated exchange of these substances across the BBB by utilizing a diverse array of membrane protein carriers and transcytotic receptors. Small metabolites including amino acids, glucose, nucleosides, and vitamins typically move down concentration gradients in the direction from blood to brain through carrier-mediated transporters. GLUT1, a member of the sodium-independent class of glucose carriers, shuttles glucose into the brain and is highly expressed in the BBB. The neurodevelopmental disorder known as GLUT1 deficiency syndrome is caused by an autosomal dominant mutation of the glucose transporter and is manifest by seizures and reduced head growth after birth (Wang et al. 2005). These genetic findings likely underscore the critical importance of proper exchange of nutrients, particularly glucose, across the BBB. The deficits observed in humans with this disorder can be modeled in mice that have lost one copy of *Glut1* (Wang et al. 2006). Conversely, a subclass of transporters called sodium-dependent AA transporters helps prevent the accumulation of potentially neurotoxic excitatory AAs such as glutamate and aspartate (Zlotnik et al. 2007).

Even trace amounts of metabolites that can passively diffuse across the plasma membrane can be highly neuroactive. In these cases, active efflux transport mechanisms are necessary to drive molecular exchange against concentration gradients from brain to blood (**Figure 4b**). The most well-characterized efflux conduit is the multidrug resistance protein P-glycoprotein, one of several multidrug resistance-associated proteins (MRPs). This membrane protein efficiently removes a diverse structural array of lipophilic molecules through an internal pore (Aller et al. 2009). Researchers propose that through the combined action of several MRPs, both potentially beneficial therapeutics and neurotoxic chemicals are rapidly cleared from the brain (Hermann & Bassetti 2007, Loscher & Potschka 2005).

Receptor-mediated transcytosis is responsible for the movement of large proteins across the BBB (Pardridge 1988). Leptin, insulin, transferrin, and low-density lipoproteins (LDL) are a few examples of proteins that gain entry into the CNS through binding to their cognate receptors within BBB endothelium and are internalized via endocytic vesicles (**Figure 4***b*). This biological transport is termed receptor-mediated transcytosis. How directionality is achieved when transporting these large proteins across the BBB is not well understood. We presume that the polarity of BBB endothelium and differences between CNS and blood environments drive the transport of these molecules in a direction that benefits the organism.

The Neurovascular Unit

For the past two decades, studies have thoroughly established that BBB development, regulation, and maintenance are controlled by various cells in the brain, including pericytes, astrocytes, and neurons (Ek et al. 2006, Janzer & Raff 1987, Stewart & Wiley 1981). The close juxtaposition and physical contact between these different cell types and endothelial cells form the neurovascular unit (NVU) (**Figure 5**). This functional unit enables efficient oxygen and nutrient flow between various cell types in the CNS, which promotes proper brain function (Banerjee & Bhat 2007, Iadecola 2004). Intercellular communication within the NVU allows the CNS to achieve tight temporal control of cerebral blood flow to

match metabolic needs of the surrounding neural tissue. During development, the neuroepithelium and pericytes actively regulate angiogenesis and vessel guidance. For example, neurons secrete proangiogenic factors such as VEGF (Haigh et al. 2003, Raab et al. 2004) and Wnt ligands (see below) to provide growth and differentiation signals to extending blood vessels. In the mature NVU, neurons further regulate vascular function via signaling through astrocytes, whereas smooth-muscle cells and pericytes control blood flow in response to this neuronal activity. We recognize that cerebral blood flow is regulated by neuronal activity, but the exact mechanisms of NVU function are still under investigation. The functioning of the NVU has been addressed in detail elsewhere (Banerjee & Bhat 2007, Iadecola 2004). Here we address the development of the NVU and its major component, the BBB.

In an elegant study utilizing the morphological differences between chick and quail endothelial cells, the unique features of BBB vasculature were shown to be extrinsically induced by the CNS microenvironment (Stewart & Wiley 1981). These studies demonstrated that avascular embryonic quail brain transplanted into chick gut is vascularized by chick peripheral vessels. These vessels take on features of the BBB including tight junctions and the ability to restrict molecular diffusion into the CNS tissue, thus strongly indicating that instructive cues are produced by neuroepithelium to drive barrier formation. It has also become increasingly clear that the barrier is established during the earliest stages of angiogenic development, while astrocytes further drive BBB maturation (Ek et al. 2006, Janzer & Raff 1987). Pericytes and neurons of the NVU are present at the onset of CNS angiogenesis and BBB formation, consistent with the idea that many features of the BBB are initiated during vascular sprouting into the CNS (Virgintino et al. 2007). Furthermore, hallmark BBB components such as Claudin-5 and GLUT1 are present at the earliest stages of CNS angiogenesis; newly formed vessels are established with a physical barrier that can exclude small molecules from entering the embryonic CNS (Daneman et al. 2009, Ek et al. 2006).

Whereas CNS angiogenesis and BBB formation occur during embryogenesis, formation of astrocytes and their subsequent ensheathment of vessels start postnatally in rodents, thus the NVU does not fully form until after birth. Detailed rat studies have observed complete

The neurovascular unit (NVU)

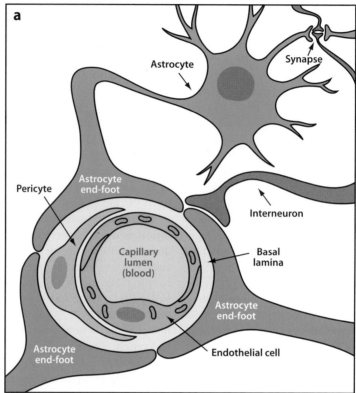

encirclement of vessels by astrocytes three weeks after birth (Caley & Maxwell 1970). Astrocytic regulation of BBB function has been characterized through in vitro coculture experiments and CNS disease studies (Argaw et al. 2009, Hurwitz et al. 1993, Janzer & Raff 1987). However, how astrocytic endfeet affect specific BBB properties during maturation remains to be determined. For example, astrocyte removal had no effect on the physical barrier of the BBB when using the tracer horseradish peroxidase (HRP) to investigate integrity (Krum & Rosenstein 1993).

Substantial progress has been made on understanding how the NVU functions in the adult animal (Banerjee & Bhat 2007, Iadecola 2004). Growing evidence also indicates that the dysfunction of the NVU may be a major contributor to neurological disease (Iadecola 2004, Neuwelt et al. 2008, Zlokovic 2008). In particular, altered cerebral blood flow and a compromised BBB have been described in numerous neurodegenerative conditions (see sidebar, The BBB and Neurodegenerative Disease). Nevertheless, to gain further insight into the function of the NVU and associated BBB, a better understanding of the development of these systems must be established.

CNS Angiogenesis and BBB Formation: A Role for Wnt Signaling

The canonical Wnt signaling pathway was recently identified as an essential regulator of CNS angiogenesis and BBB development (Daneman et al. 2009, Liebner et al. 2008, Stenman et al. 2008). This developmental

THE BBB AND NEURODEGENERATIVE DISEASE

In addition to the neuronal loss and inflammatory processes that are common hallmarks of neurodegenerative disease, accumulating evidence indicates that the BBB is disrupted in many neurodegenerative diseases, including Alzheimer's, Parkinson's, and ALS (Zlokovic 2008). It remains to be seen if the disruption of the BBB is a cause or consequence of neurodegenerative processes. Nevertheless, the fact that the BBB is disrupted is pushing researchers to understand BBB biology in the context of neurodegenerative disease, addressing the following questions, among others: Which molecular mechanisms drive barrier disruption? How does disruption of neurovascular coupling alter CNS function? Is drug delivery altered in neurodegenerative disease? And ultimately, will repairing the BBB slow neurodegenerative disease? This area of research is ripe for discovery and may lead to the development of therapeutics with a primary mechanism of barrier modulation to treat these diseases.

pathway involves Wnt ligand-driven activation of Frizzled and LRP cell surface coreceptors, which then stabilize cellular beta-catenin through activated Disheveled protein (**Figure 6**). The increase in beta-catenin levels then induces transcription of Wnt-responsive genes through Lef-1/TCF DNA binding proteins (van Amerongen & Nusse 2009). In situ hybridization experiments found that the Wnt ligands, *wnt7a* and *wnt7b*, are expressed in neural progenitor cells during CNS angiogenesis (**Figure 6a**). Functionally, Wnt7a/7b were required for developmental angiogenesis induced from the vascular plexus surrounding the spinal cord. Invading vessels utilize

Figure 5

The neurovascular unit (NVU). (*a*) The mature blood-brain barrier consists of close association and physical contact between endothelial cells, neurons, astrocytic endfeet, basal lamina, and pericytes. The NVU enables efficient cellular communication for proper brain function, such as regulating cerebral blood flow in response to neural activity. (*b*) Immunohistochemical staining of the cellular components within the mouse NVU. Antibody staining against BBB transporters P-gp and GLUT1 label endothelial cells, whereas NG2 detects the pericytes found along these endothelial cells, which tightly regulate blood flow. Note the tight ensheathment of endothelial cells by the basal lamina (Collagen-IV) and astrocytes (GFAP) visualized by costaining. Astrocyte endfeet (GFAP), in concert with connections from interneurons, regulate neurovascular calcium signaling, blood flow, ionic transport, and water transport.

a

Embryo neural tube cross-section

Angiogenesis and BBB formation

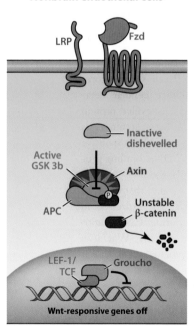
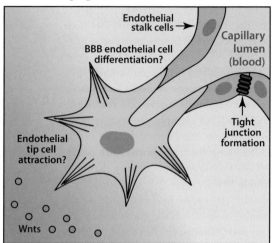

b

Nonbrain endothelial cells

Brain endothelial cells

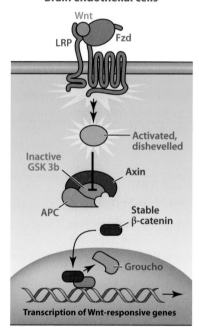

Figure 6

Wnts regulate angiogenesis and BBB formation in the nervous system. The canonical Wnt signaling pathway is essential for proper BBB development. (*a*) Wnt ligands 7a and 7b are specifically expressed in neuroepithelium to induce developmental angiogenesis from the perineural vascular plexus into the developing neural tube. These secreted proteins may form attractive gradients that directly shape endothelial tip cell morphology. Furthermore, Wnt signaling in the CNS drives BBB endothelial cell differentiation, thereby inducing glucose transporter-1 (GLUT1) and tight junction protein Claudin-3 expression. (*b*) The specificity of canonical Wnt signaling on CNS vasculature is highlighted by the observation that Wnt-responsive genes are enriched in BBB endothelium when compared with nonbrain endothelium. This developmental pathway involves Wnt ligand–driven activation of Frizzled and LRP cell surface coreceptors, which then activate Dishevelled. This protein inactivates GSK-3β activity in complex with APC and Axin, thereby stabilizing beta-catenin to drive Wnt target genes through Lef-1/TCF DNA binding proteins.

paracrine signaling derived from the neuroepithelium to coordinate proper BBB development because loss-of-function of either *wnt7a/7b* in neuroepithelium or of *beta-catenin* in endothelial cells leads to angiogenic defects and subsequent vascular hemorrhage. These mice exhibit angiogenic defects only in the CNS and not in the periphery, consistent with the observation that Wnt signaling is not activated in non-CNS endothelium (**Figure 6b**).

Although BBB formation parallels angiogenesis in the CNS, proper CNS vascular sprouting alone is not sufficient for complete barrier development. BBB formation requires the expression of a diverse array of gene products to establish barrier function. As previously discussed, GLUT1 is essential for proper brain development because autosomal dominant mutations lead to several neurological maladies and developmental delays (Wang et al. 2005). The tight junction protein Claudin-3, also found predominantly in the BBB vasculature, has been linked to BBB deficit under pathological conditions as well (Wolburg et al. 2003). Wnt signaling appears to induce not only CNS angiogenesis, but also expression of BBB components GLUT1 and Claudin-3 (Daneman et al. 2009, Liebner et al. 2008, Stenman et al. 2008). Either Wnt7a/7b or beta-catenin depletion results in a lack of GLUT1 and Claudin-3 expression at the BBB. Conversely, ectopic Wnt7a/7b expression induced GLUT1 expression outside the CNS, whereas addition of Wnt ligands to cultured brain endothelial cells induced both GLUT1 and Claudin-3 protein expression. These findings emphasize the dual functional nature of Wnt-mediated vascular development in the CNS, whereby both angiogenesis and BBB formation are tightly coupled. We do not yet know if Wnt also plays a role in the final maturation of the NVU, including driving astrocytic associations with CNS blood vessels.

Although the ligands required for Wnt-driven BBB development have been at least partially identified as Wnt7a/7b, the cognate receptor(s) have not. Most members of the Frizzled family (Fzd3, 4, 6, 8, 10) and both Lrp5 and Lrp6 are expressed in the CNS.

Daneman and colleagues (2009) speculate that Fzd8 may be important in CNS angiogenesis and BBB development on the basis of its specific enrichment in CNS vasculature. The molecular mechanisms of Wnt-driven BBB development in the CNS could be further elucidated by examining whether endothelial-specific deletion of particular Fzd or LRP isoforms can mimic defects observed in Wnt7a/7b knockout embryos.

The specificity of Wnt signaling on CNS vasculature is also highlighted by the observation that canonical Wnt signaling components, such as Lef-1, Tcf7, Axin2, Apcdd1, Ppard, Tbx3, Foxf2, and Stra6, are enriched in BBB endothelium when compared with peripheral tissue endothelium (Daneman et al. 2009). Because this Wnt expression profile was observed in adult mice, the tissue specificity of canonical Wnt signaling appears to be maintained in mature BBB. Could this result indicate that chronic Wnt/beta-catenin signaling is required to maintain proper BBB function within the NVU in the mature animal? The fact that neurons remain a consistent source of Wnt ligand from embryo to adulthood supports such a possibility. For instance, Wnt7a/7b are actively secreted during synaptic activity, which then mediates experience-related regulation of synapse abundance and network complexity (Gogolla et al. 2009).

A role for Wnt signaling in BBB maintenance is particularly attractive because it provides a new drug target for both selective opening of the barrier for entry of CNS therapeutics and selective tightening of the BBB in CNS diseases where barrier function may be compromised (Zlokovic 2008). Liebner and colleagues (2008) suggest that Wnt signaling activity is low by the time the BBB is fully mature on the basis of the observation that Wnt signaling activity per unit vessel length shows a precipitous drop after embryonic day 15.5. However, this result could be reflective of endothelial cells lengthening dramatically during development, such that the number of nuclei per unit length drops (Liebner et al. 2008). Therefore, BBB function will have to be examined upon suppression of

Wnt/beta-catenin signaling in the adult animal to determine the validity of this model.

In hindsight, the Wnt signaling pathway's involvement in BBB development within the CNS should not have come as a surprise. In the developing mouse, Wnt gradients direct axonal pathfinding, regulate synaptogenesis, and coordinate axon and dendrite development (Salinas & Zou 2008). Thus the theme of CNS guidance cues regulating vascular development is once again apparent. Several other morphogenic processes including the BMP and hedgehog signaling pathways act in concert with the Wnt pathway in the CNS (Charron & Tessier-Lavigne 2005). Specifically, overlapping BMP and Shh ligand gradients cooperate with Netrin1 to drive commissural axons from the roofplate down to the floorplate of the developing spinal cord (Augsburger et al. 1999, Charron et al. 2003). Overlapping Shh and Wnt4 gradients subsequently direct these axonal projections along the anterior/posterior axis to the final target tissue. Could these signaling pathways be required for proper CNS angiogenesis and BBB development as well? There is precedence for not only shared ligand-receptor paracrine signaling driving both neuronal and vascular patterning, but also Wnt morphogen involvement in CNS angiogenesis and BBB formation, as just described. Indeed, Araya et al. (2008) demonstrated that brain endothelial cells secrete BMP ligands, resulting in astrocyte-derived VEGF production and subsequent BBB breakdown.

In addition to the morphogenic signaling pathways, a wide array of axon guidance molecules are also potential candidates for CNS-specific angiogenic and BBB formation cues. Notably, Nrp1 is required for angiogenic sprouting in the CNS (Gerhardt et al. 2004). We need to identify which, if any, of the many axon guidance cues that have been implicated in vessel guidance, branching, development, or maintenance also contribute to CNS angiogenesis and BBB development.

The rhodopsin-family G protein–coupled receptor (GPCR), *Moody*, was recently shown to be required for formation and maintenance of the *Drosophila* hemolymph-brain barrier, the functional analog of the mammalian BBB (Bainton et al. 2005). Does GPCR signaling control BBB permeability in mammalian systems as well? Although many GPCRs are expressed in the mouse brain, there are no obvious mammalian orthologs to *Moody* that can be identified using sequence homology. Nevertheless, it is interesting to note that the adhesion GPCR, gpr125, is upregulated following brain injury and is specifically expressed in lung and cortical tissue (Pickering et al. 2008). In addition, the vasodilator bradykinin acutely modulates BBB permeability through its cognate GPCRs, bdkrb1 and b2 (Bartus et al. 1996, Su et al. 2009).

We are only just beginning to understand the signaling pathways that coordinate the formation of a fully functional BBB. Furthermore, specific pathways may drive BBB development within different regions of the CNS. Consistent with this possibility, some evidence indicates that Wnt/beta-catenin signaling is not necessary for pan-CNS angiogenesis and BBB formation. For instance, Wnt7a/7b ligands are not expressed in dorsal neural tubes, and as a consequence, Wnt7a/7b knockout mice can still form proper blood vessels in this region (Stenman et al. 2008). Similarly, beta-catenin knockouts exhibit normal BBB formation in the hindbrain (Daneman et al. 2009), and Fzd4 knockout mice show barrier deficits in the cerebellum, but not the cortex, suggesting that multiple Fzd receptors may coordinate BBB development in a region-specific fashion (Ye et al. 2009). Therefore, unique positions within the CNS seem to require activation of other signaling pathways.

Unbiased approaches toward uncovering novel molecular mechanisms of BBB development and maintenance are needed at this juncture. Signaling pathways coregulated with Wnt activation in CNS endothelium may suggest participation in BBB formation. Downstream of Wnt signaling, many of the BBB-related genes directly or indirectly regulated by Wnt/beta-catenin have yet to be identified. Investigation of BBB endothelial cell expression profiles along the developmental timeline may paint a more comprehensive picture of BBB development, maturation, and maintenance.

CONCLUSIONS AND FUTURE DIRECTIONS

Mirroring the sequential origin of the nervous and vascular systems, researchers have applied principles of nervous system development to address vascular biology questions. This approach stemmed from the original observation that the nervous and vascular systems share anatomical and cellular similarities. Moreover, both systems must be present throughout an organism to deliver and receive vital information and nutrients. It is apparent now that similar molecular mechanisms drive both vascular and neuronal network formation. These discoveries have been key to accelerating our understanding of how blood vessels develop. Also, the vascular system can directly impact nervous system development and vice versa; thus caution should be taken when characterizing phenotypes after disrupting signaling pathways in either system.

We have described examples of specific axon guidance molecules regulating vascular biology. Although not comprehensive, these examples illustrate the influence that discoveries in the field of axon guidance have had on understanding vascular development. However, much remains unknown. For instance, the role that VEGF plays in neural development and function is just beginning to be unraveled (Ruiz de Almodovar et al. 2009). In addition, we have not addressed in this review the close relationship between blood vessels and neural progenitor cells at the vascular niche (Barami 2008). Many important questions remain that would further our understanding of the relationship between vascular and nervous system development. Are there other guidance cues that mediate both vascular and neural development? How are developmentally sequential events, such as initial peripheral blood vessel innervation followed by neuronal process innervation, regulated? Are molecular mechanisms of vascular and neuronal remodeling also shared

or codependent? How did the vascular system co-opt nervous system cues? What role do these common cues play in pathological angiogenesis or neurological disease? After more than a decade of applying insight from neuronal development paradigms to the vascular system, it is clear that this path of investigation is yielding key findings.

Much less is understood about the development, maturation, and maintenance of the blood-brain barrier (BBB). Understanding how the BBB develops and functions is of enormous importance. This need is fueled by the fact that the barrier hinders therapeutic treatment of neurological diseases and because it may also be at the center of neurodegenerative disease biology. Subsequent to the original discovery that the CNS microenvironment directs blood vessel differentiation to form the BBB (Stewart & Wiley 1981), the recent identification of the Wnt/beta-catenin signaling pathway regulating CNS angiogenesis and BBB formation is likely just the first of many examples of regulatory cues derived from the CNS that drive brain endothelial cell development and differentiation. A unique Frizzled receptor ligand, Norrin, has recently been described to act via Fzd4, LRP5, and a tetraspanin molecule, TSPAN12, to regulate retinal vascular development specifically (Junge et al. 2009, Ye et al. 2009). In addition, *Norrin*, *Fzd4*, *Lrp5*, and *Tspan12* knockout mice show a vascular hemorrhage phenotype in the retina, consistent with the disruption of the retinal-blood barrier. These data suggest that unique Wnt signaling pathways have evolved in a regional-specific fashion to coordinate endothelial cell differentiation and barrier formation.

Beyond Wnt signaling in CNS angiogenesis and BBB formation, many questions remain about how the BBB forms, matures, and is maintained. The barrier also consists of other cell types, including pericytes, astrocytes, and neuronal components. How is this complex NVU formed? How is the barrier actively maintained in the adult animal? What are the roles of the various cellular components in the development, maturation, and maintenance of the BBB? Discovering the molecular components that address these broader questions will deepen our understanding of the BBB, thus enabling researchers to address unresolved issues of drug penetration into the CNS and barrier disruption in neurological disease paradigms. Using recent history as an example, clues to crack the mystery of BBB formation may be derived from applying principles of neuronal development to the vascular system.

DISCLOSURE STATEMENT

The authors are employees of Genentech, Inc.

ACKNOWLEDGMENTS

We apologize to our colleagues whose research we could not cite owing to space limitations. We thank Allison Bruce for the generation of illustrations and thank Kimberly Scearce-Levie, Weilan Ye, Joy Yu, Morgan Sheng, and Marc Tessier-Lavigne for their helpful comments on the manuscript.

LITERATURE CITED

Adams RH, Wilkinson GA, Weiss C, Diella F, Gale NW, et al. 1999. Roles of ephrinB ligands and EphB receptors in cardiovascular development: demarcation of arterial/venous domains, vascular morphogenesis, and sprouting angiogenesis. *Genes Dev.* 13:295–306

Alavijeh MS, Chishty M, Qaiser MZ, Palmer AM. 2005. Drug metabolism and pharmacokinetics, the blood-brain barrier, and central nervous system drug discovery. *NeuroRx* 2:554–71

Aller SG, Yu J, Ward A, Weng Y, Chittaboina S, et al. 2009. Structure of P-glycoprotein reveals a molecular basis for poly-specific drug binding. *Science* 323:1718–22

Appleton BA, Wu P, Maloney J, Yin J, Liang WC, et al. 2007. Structural studies of neuropilin/antibody complexes provide insights into semaphorin and VEGF binding. *EMBO J.* 26:4902–12

Araya R, Kudo M, Kawano M, Ishii K, Hashikawa T, et al. 2008. BMP signaling through BMPRIA in astrocytes is essential for proper cerebral angiogenesis and formation of the blood-brain-barrier. *Mol. Cell Neurosci.* 38:417–30

Argaw AT, Gurfein BT, Zhang Y, Zameer A, John GR. 2009. VEGF-mediated disruption of endothelial CLN-5 promotes blood-brain barrier breakdown. *Proc. Natl. Acad. Sci. USA* 106:1977–82

Augsburger A, Schuchardt A, Hoskins S, Dodd J, Butler S. 1999. BMPs as mediators of roof plate repulsion of commissural neurons. *Neuron* 24:127–41

Bainton RJ, Tsai LT, Schwabe T, DeSalvo M, Gaul U, Heberlein U. 2005. *moody* encodes two GPCRs that regulate cocaine behaviors and blood-brain barrier permeability in *Drosophila*. *Cell* 123:145–56

Bamforth SD, Kniesel U, Wolburg H, Engelhardt B, Risau W. 1999. A dominant mutant of occludin disrupts tight junction structure and function. *J. Cell Sci.* 112(Pt. 12):1879–88

Banerjee S, Bhat MA. 2007. Neuron-glial interactions in blood-brain barrier formation. *Annu. Rev. Neurosci.* 30:235–58

Barami K. 2008. Relationship of neural stem cells with their vascular niche: implications in the malignant progression of gliomas. *J. Clin. Neurosci.* 15:1193–97

Barres BA. 2008. The mystery and magic of glia: a perspective on their roles in health and disease. *Neuron* 60:430–40

Bartus RT, Elliott P, Hayward N, Dean R, McEwen EL, Fisher SK. 1996. Permeability of the blood brain barrier by the bradykinin agonist, RMP-7: evidence for a sensitive, auto-regulated, receptor-mediated system. *Immunopharmacology* 33:270–78

Bedell VM, Yeo SY, Park KW, Chung J, Seth P, et al. 2005. *roundabout4* is essential for angiogenesis in vivo. *Proc. Natl. Acad. Sci. USA* 102:6373–78

Begley DJ, Brightman MW. 2003. Structural and functional aspects of the blood-brain barrier. *Prog. Drug Res.* 61:39–78

Bergers G, Song S. 2005. The role of pericytes in blood-vessel formation and maintenance. *Neuro Oncol.* 7:452–64

Betsholtz C, Lindblom P, Gerhardt H. 2005. Role of pericytes in vascular morphogenesis. *Exs* 2005:115–25

Bouvrée K, Larrivée B, Lv X, Yuan L, DeLafarge B, et al. 2008. Netrin-1 inhibits sprouting angiogenesis in developing avian embryos. *Dev. Biol.* 318:172–83

Butler SJ, Dodd J. 2003. A role for BMP heterodimers in roof plate-mediated repulsion of commissural axons. *Neuron* 38:389–401

Caley DW, Maxwell DS. 1970. Development of the blood vessels and extracellular spaces during postnatal maturation of rat cerebral cortex. *J. Comp. Neurol.* 138:31–47

Carmeliet P, Tessier-Lavigne M. 2005. Common mechanisms of nerve and blood vessel wiring. *Nature* 436:193–200

Charron F, Stein E, Jeong J, McMahon AP, Tessier-Lavigne M. 2003. The morphogen sonic hedgehog is an axonal chemoattractant that collaborates with netrin-1 in midline axon guidance. *Cell* 113:11–23

Charron F, Tessier-Lavigne M. 2005. Novel brain wiring functions for classical morphogens: a role as graded positional cues in axon guidance. *Development* 132:2251–62

Chen H, Chédotal A, He Z, Goodman CS, Tessier-Lavigne M. 1997. Neuropilin-2, a novel member of the neuropilin family, is a high affinity receptor for the semaphorins Sema E and Sema IV but not Sema III. *Neuron* 19:547–59

Cheng HJ, Bagri A, Yaron A, Stein E, Pleasure SJ, Tessier-Lavigne M. 2001. Plexin-A3 mediates semaphorin signaling and regulates the development of hippocampal axonal projections. *Neuron* 32:249–63

Cheng HJ, Nakamoto M, Bergemann AD, Flanagan JG. 1995. Complementary gradients in expression and binding of ELF-1 and Mek4 in development of the topographic retinotectal projection map. *Cell* 82:371–81

Cho NK, Keyes L, Johnson E, Heller J, Ryner L, et al. 2002. Developmental control of blood cell migration by the *Drosophila* VEGF pathway. *Cell* 108:865–76

Coultas L, Chawengsaksophak K, Rossant J. 2005. Endothelial cells and VEGF in vascular development. *Nature* 438:937–45

Daneman R, Agalliu D, Zhou L, Kuhnert F, Kuo CJ, Barres BA. 2009. Wnt/beta-catenin signaling is required for CNS, but not non-CNS, angiogenesis. *Proc. Natl. Acad. Sci. USA* 106:641–46

de Boer AG, van der Sandt IC, Gaillard PJ. 2003. The role of drug transporters at the blood-brain barrier. *Annu. Rev. Pharmacol. Toxicol.* 43:629–56

de Castro F, López-Mascaraque L, De Carlos JA. 2007. Cajal: lessons on brain development. *Brain Res. Rev.* 55:481–89

Dickson BJ. 2002. Molecular mechanisms of axon guidance. *Science* 298:1959–64

Dickson BJ, Gilestro GF. 2006. Regulation of commissural axon pathfinding by slit and its Robo receptors. *Annu. Rev. Cell Dev. Biol.* 22:651–75

Drescher U, Kremoser C, Handwerker C, Loschinger J, Noda M, Bonhoeffer F. 1995. In vitro guidance of retinal ganglion cell axons by RAGS, a 25 kDa tectal protein related to ligands for Eph receptor tyrosine kinases. *Cell* 82:359–70

Duarte A, Hirashima M, Benedito R, Trindade A, Diniz P, et al. 2004. Dosage-sensitive requirement for mouse Dll4 in artery development. *Genes Dev.* 18:2474–78

Ek CJ, Dziegielewska KM, Stolp H, Saunders NR. 2006. Functional effectiveness of the blood-brain barrier to small water-soluble molecules in developing and adult opossum (*Monodelphis domestica*). *J. Comp. Neurol.* 496:13–26

Fazeli A, Dickinson SL, Hermiston ML, Tighe RV, Steen RG, et al. 1997. Phenotype of mice lacking functional Deleted in colorectal cancer (Dcc) gene. *Nature* 386:796–804

Ferrara N, Kerbel RS. 2005. Angiogenesis as a therapeutic target. *Nature* 438:967–74

Fisher SA, Burggren WW. 2007. Role of hypoxia in the evolution and development of the cardiovascular system. *Antioxid. Redox Signal.* 9:1339–52

Fong GH. 2009. Regulation of angiogenesis by oxygen sensing mechanisms. *J. Mol. Med.* 87:549–60

Fujisawa H. 2004. Discovery of semaphorin receptors, neuropilin and plexin, and their functions in neural development. *J. Neurobiol.* 59:24–33

Gaiano N, Fishell G. 2002. The role of notch in promoting glial and neural stem cell fates. *Annu. Rev. Neurosci.* 25:471–90

Gale NW, Dominguez MG, Noguera I, Pan L, Hughes V, et al. 2004. Haploinsufficiency of delta-like 4 ligand results in embryonic lethality due to major defects in arterial and vascular development. *Proc. Natl. Acad. Sci. USA* 101:15949–54

Gerety SS, Wang HU, Chen ZF, Anderson DJ. 1999. Symmetrical mutant phenotypes of the receptor EphB4 and its specific transmembrane ligand ephrin-B2 in cardiovascular development. *Mol. Cell* 4:403–14

Gerhardt H, Golding M, Fruttiger M, Ruhrberg C, Lundkvist A, et al. 2003. VEGF guides angiogenic sprouting utilizing endothelial tip cell filopodia. *J. Cell Biol.* 161:1163–77

Gerhardt H, Ruhrberg C, Abramsson A, Fujisawa H, Shima D, Betsholtz C. 2004. Neuropilin-1 is required for endothelial tip cell guidance in the developing central nervous system. *Dev. Dyn.* 231:503–9

Ghabrial AS, Krasnow MA. 2006. Social interactions among epithelial cells during tracheal branching morphogenesis. *Nature* 441:746–49

Gitler AD, Lu MM, Epstein JA. 2004. PlexinD1 and semaphorin signaling are required in endothelial cells for cardiovascular development. *Dev. Cell* 7:107–16

Gogolla N, Galimberti I, Deguchi Y, Caroni P. 2009. Wnt signaling mediates experience-related regulation of synapse numbers and mossy fiber connectivities in the adult hippocampus. *Neuron* 62:510–25

Goldmann EE. 1909. Die äussere und innere sekretion des gesunden und kranken organismus im lichte der 'vitalen Farbung.' *Beitr. Klin. Chir.* 64:192–265

Goldmann EE. 1913. Vitalfarbung am zentralnervensystem. *Abh. Preuss. Akad. Wiss. Phys.-Math* 1:1–60

Golgi C. 1906. *The neuron doctrine—theory and facts.* **http://nobelprize.org/nobel_prizes/medicine/laureates/1906/golgi-lecture.pdf**

Gu C, Limberg BJ, Whitaker GB, Perman B, Leahy DJ, et al. 2002. Characterization of neuropilin-1 structural features that confer binding to semaphorin 3A and vascular endothelial growth factor 165. *J. Biol. Chem.* 277:18069–76

Gu C, Rodriguez ER, Reimert DV, Shu T, Fritzsch B, et al. 2003. Neuropilin-1 conveys semaphorin and VEGF signaling during neural and cardiovascular development. *Dev. Cell* 5:45–57

Gu C, Yoshida Y, Livet J, Reimert DV, Mann F, et al. 2005. Semaphorin 3E and plexin-D1 control vascular pattern independently of neuropilins. *Science* 307:265–68

Haigh JJ, Morelli PI, Gerhardt H, Haigh K, Tsien J, et al. 2003. Cortical and retinal defects caused by dosage-dependent reductions in VEGF-A paracrine signaling. *Dev. Biol.* 262:225–41

Hawkins BT, Davis TP. 2005. The blood-brain barrier/neurovascular unit in health and disease. *Pharmacol. Rev.* 57:173–85

Hawkins RA, O'Kane RL, Simpson IA, Vina JR. 2006. Structure of the blood-brain barrier and its role in the transport of amino acids. *J. Nutr.* 136:S218–26

He Z, Tessier-Lavigne M. 1997. Neuropilin is a receptor for the axonal chemorepellent Semaphorin III. *Cell* 90:739–51

Hellstrom M, Phng LK, Hofmann JJ, Wallgard E, Coultas L, et al. 2007. Dll4 signalling through Notch1 regulates formation of tip cells during angiogenesis. *Nature* 445:776–80

Hermann DM, Bassetti CL. 2007. Implications of ATP-binding cassette transporters for brain pharmacotherapies. *Trends Pharmacol. Sci.* 28:128–34

Holland SJ, Gale NW, Mbamalu G, Yancopoulos GD, Henkemeyer M, Pawson T. 1996. Bidirectional signalling through the EPH-family receptor Nuk and its transmembrane ligands. *Nature* 383:722–25

Hong K, Hinck L, Nishiyama M, Poo MM, Tessier-Lavigne M, Stein E. 1999. A ligand-gated association between cytoplasmic domains of UNC5 and DCC family receptors converts netrin-induced growth cone attraction to repulsion. *Cell* 97:927–41

Honma Y, Araki T, Gianino S, Bruce A, Heuckeroth R, et al. 2002. Artemin is a vascular-derived neurotropic factor for developing sympathetic neurons. *Neuron* 35:267–82

Huminiecki L, Gorn M, Suchting S, Poulsom R, Bicknell R. 2002. Magic roundabout is a new member of the roundabout receptor family that is endothelial specific and expressed at sites of active angiogenesis. *Genomics* 79:547–52

Hurwitz AA, Berman JW, Rashbaum WK, Lyman WD. 1993. Human fetal astrocytes induce the expression of blood-brain barrier specific proteins by autologous endothelial cells. *Brain Res.* 625:238–43

Iadecola C. 2004. Neurovascular regulation in the normal brain and in Alzheimer's disease. *Nat. Rev. Neurosci.* 5:347–60

Ishii N, Wadsworth WG, Stern BD, Culotti JG, Hedgecock EM. 1992. UNC-6, a laminin-related protein, guides cell and pioneer axon migrations in *C. elegans*. *Neuron* 9:873–81

Janzer RC, Raff MC. 1987. Astrocytes induce blood-brain barrier properties in endothelial cells. *Nature* 325:253–57

Jones EA, Yuan L, Breant C, Watts RJ, Eichmann A. 2008. Separating genetic and hemodynamic defects in neuropilin 1 knockout embryos. *Development* 135:2479–88

Junge HJ, Yang S, Burton JB, Paes K, Shu X, et al. 2009. TSPAN12 regulates retinal vascular development by promoting Norrin- but not Wnt-induced FZD4/beta-catenin signaling. *Cell* 139:299–311

Kalaria RN, Harik SI. 1987. Blood-brain barrier monoamine oxidase: enzyme characterization in cerebral microvessels and other tissues from six mammalian species, including human. *J. Neurochem.* 49:856–64

Kawasaki T, Kitsukawa T, Bekku Y, Matsuda Y, Sanbo M, et al. 1999. A requirement for neuropilin-1 in embryonic vessel formation. *Development* 126:4895–902

Keck PJ, Hauser SD, Krivi G, Sanzo K, Warren T, et al. 1989. Vascular permeability factor, an endothelial cell mitogen related to PDGF. *Science* 246:1309–12

Keino-Masu K, Masu M, Hinck L, Leonardo ED, Chan SS, et al. 1996. Deleted in Colorectal Cancer (DCC) encodes a netrin receptor. *Cell* 87:175–85

Keleman K, Dickson BJ. 2001. Short- and long-range repulsion by the *Drosophila* Unc5 netrin receptor. *Neuron* 32:605–17

Kennedy TE, Serafini T, de la Torre JR, Tessier-Lavigne M. 1994. Netrins are diffusible chemotropic factors for commissural axons in the embryonic spinal cord. *Cell* 78:425–35

Kidd T, Bland KS, Goodman CS. 1999. Slit is the midline repellent for the robo receptor in *Drosophila*. *Cell* 96:785–94

Kidd T, Brose K, Mitchell KJ, Fetter RD, Tessier-Lavigne M, et al. 1998. Roundabout controls axon crossing of the CNS midline and defines a novel subfamily of evolutionarily conserved guidance receptors. *Cell* 92:205–15

Kolodkin AL. 1996. Growth cones and the cues that repel them. *Trends Neurosci.* 19:507–13

Kolodkin AL, Levengood DV, Rowe EG, Tai YT, Giger RJ, Ginty DD. 1997. Neuropilin is a semaphorin III receptor. *Cell* 90:753–62

Kolodkin AL, Matthes DJ, O'Connor TP, Patel NH, Admon A, et al. 1992. Fasciclin IV: sequence, expression, and function during growth cone guidance in the grasshopper embryo. *Neuron* 9:831–45

Krebs LT, Shutter JR, Tanigaki K, Honjo T, Stark KL, Gridley T. 2004. Haploinsufficient lethality and formation of arteriovenous malformations in Notch pathway mutants. *Genes Dev.* 18:2469–73

Krum JM, Rosenstein JM. 1993. Effect of astroglial degeneration on the blood-brain barrier to protein in neonatal rats. *Brain Res. Dev. Brain Res.* 74:41–50

Larrivée B, Freitas C, Suchting S, Brunet I, Eichmann A. 2009. Guidance of vascular development: lessons from the nervous system. *Circ. Res.* 104:428–41

Larrivée B, Freitas C, Trombe M, Lv X, Delafarge B, et al. 2007. Activation of the UNC5B receptor by Netrin-1 inhibits sprouting angiogenesis. *Genes Dev.* 21:2433–47

Lawson ND, Weinstein BM. 2002. In vivo imaging of embryonic vascular development using transgenic zebrafish. *Dev. Biol.* 248:307–18

Lejmi E, Leconte L, Pedron-Mazoyer S, Ropert S, Raoul W, et al. 2008. Netrin-4 inhibits angiogenesis via binding to neogenin and recruitment of Unc5B. *Proc. Natl. Acad. Sci. USA* 105:12491–96

Leonardo ED, Hinck L, Masu M, Keino-Masu K, Ackerman SL, Tessier-Lavigne M. 1997. Vertebrate homologues of *C. elegans* UNC-5 are candidate netrin receptors. *Nature* 386:833–38

Leung DW, Cachianes G, Kuang WJ, Goeddel DV, Ferrara N. 1989. Vascular endothelial growth factor is a secreted angiogenic mitogen. *Science* 246:1306–9

Liebner S, Corada M, Bangsow T, Babbage J, Taddei A, et al. 2008. Wnt/beta-catenin signaling controls development of the blood-brain barrier. *J. Cell Biol.* 183:409–17

Lobov IB, Renard RA, Papadopoulos N, Gale NW, Thurston G, et al. 2007. Delta-like ligand 4 (Dll4) is induced by VEGF as a negative regulator of angiogenic sprouting. *Proc. Natl. Acad. Sci. USA* 104:3219–24

Loscher W, Potschka H. 2005. Drug resistance in brain diseases and the role of drug efflux transporters. *Nat. Rev. Neurosci.* 6:591–602

Lowery LA, Van Vactor D. 2009. The trip of the tip: understanding the growth cone machinery. *Nat. Rev. Mol. Cell Biol.* 10:332–43

Lu X, Le Noble F, Yuan L, Jiang Q, De Lafarge B, et al. 2004. The netrin receptor UNC5B mediates guidance events controlling morphogenesis of the vascular system. *Nature* 432:179–86

Lubarsky B, Krasnow MA. 2003. Tube morphogenesis: making and shaping biological tubes. *Cell* 112:19–28

Lyuksyutova AI, Lu CC, Milanesio N, King LA, Guo N, et al. 2003. Anterior-posterior guidance of commissural axons by Wnt-frizzled signaling. *Science* 302:1984–88

Makita T, Sucov HM, Gariepy CE, Yanagisawa M, Ginty DD. 2008. Endothelins are vascular-derived axonal guidance cues for developing sympathetic neurons. *Nature* 452:759–63

Miao HQ, Soker S, Feiner L, Alonso JL, Raper JA, Klagsbrun M. 1999. Neuropilin-1 mediates collapsin-1/semaphorin III inhibition of endothelial cell motility: functional competition of collapsin-1 and vascular endothelial growth factor-165. *J. Cell Biol.* 146:233–42

Miller G. 2009. Origins. On the origin of the nervous system. *Science* 325:24–26

Moore SW, Tessier-Lavigne M, Kennedy TE. 2007. Netrins and their receptors. *Adv. Exp. Med. Biol.* 621:17–31

Mukouyama YS, Gerber HP, Ferrara N, Gu C, Anderson DJ. 2005. Peripheral nerve-derived VEGF promotes arterial differentiation via neuropilin 1-mediated positive feedback. *Development* 132:941–52

Mukouyama YS, Shin D, Britsch S, Taniguchi M, Anderson DJ. 2002. Sensory nerves determine the pattern of arterial differentiation and blood vessel branching in the skin. *Cell* 109:693–705

Neuwelt E, Abbott NJ, Abrey L, Banks WA, Blakley B, et al. 2008. Strategies to advance translational research into brain barriers. *Lancet Neurol.* 7:84–96

Nguyen A, Cai H. 2006. Netrin-1 induces angiogenesis via a DCC-dependent ERK1/2-eNOS feed-forward mechanism. *Proc. Natl. Acad. Sci. USA* 103:6530–35

Nitta T, Hata M, Gotoh S, Seo Y, Sasaki H, et al. 2003. Size-selective loosening of the blood-brain barrier in claudin-5-deficient mice. *J. Cell Biol.* 161:653–60

O'Donnell M, Chance RK, Bashaw GJ. 2009. Axon growth and guidance: receptor regulation and signal transduction. *Annu. Rev. Neurosci.* 32:383–412

Ogawa K, Pasqualini R, Lindberg RA, Kain R, Freeman AL, Pasquale EB. 2000. The ephrin-A1 ligand and its receptor, EphA2, are expressed during tumor neovascularization. *Oncogene* 19:6043–52

Palmer A, Klein R. 2003. Multiple roles of ephrins in morphogenesis, neuronal networking, and brain function. *Genes Dev.* 17:1429–50

Pan Q, Chanthery Y, Liang WC, Stawicki S, Mak J, et al. 2007. Blocking neuropilin-1 function has an additive effect with anti-VEGF to inhibit tumor growth. *Cancer Cell* 11:53–67

Pandey A, Shao H, Marks RM, Polverini PJ, Dixit VM. 1995. Role of B61, the ligand for the Eck receptor tyrosine kinase, in TNF-alpha-induced angiogenesis. *Science* 268:567–69

Pardridge WM. 1988. Recent advances in blood-brain barrier transport. *Annu. Rev. Pharmacol. Toxicol.* 28:25–39

Park KW, Crouse D, Lee M, Karnik SK, Sorensen LK, et al. 2004. The axonal attractant Netrin-1 is an angiogenic factor. *Proc. Natl. Acad. Sci. USA* 101:16210–15

Park KW, Morrison CM, Sorensen LK, Jones CA, Rao Y, et al. 2003. Robo4 is a vascular-specific receptor that inhibits endothelial migration. *Dev. Biol.* 261:251–67

Pickering C, Hagglund M, Szmydynger-Chodobska J, Marques F, Palha JA, et al. 2008. The Adhesion GPCR GPR125 is specifically expressed in the choroid plexus and is upregulated following brain injury. *BMC Neurosci.* 9:97

Qin S, Yu L, Gao Y, Zhou R, Zhang C. 2007. Characterization of the receptors for axon guidance factor netrin-4 and identification of the binding domains. *Mol. Cell Neurosci.* 34:243–50

Raab S, Beck H, Gaumann A, Yuce A, Gerber HP, et al. 2004. Impaired brain angiogenesis and neuronal apoptosis induced by conditional homozygous inactivation of vascular endothelial growth factor. *Thromb. Haemost.* 91:595–605

Ramón y Cajal S. 1906. *The structure and connexions of neurons.* **http://nobelprize.org/nobel_prizes/ medicine/laureates/1906/cajal-lecture.pdf**

Ramón y Cajal S. 1917. *Recuerdos de mi vida.* Madrid: Moya

Raper JA. 2000. Semaphorins and their receptors in vertebrates and invertebrates. *Curr. Opin. Neurobiol.* 10:88–94

Reese TS, Karnovsky MJ. 1967. Fine structural localization of a blood-brain barrier to exogenous peroxidase. *J. Cell Biol.* 34:207–17

Rossignol M, Gagnon ML, Klagsbrun M. 2000. Genomic organization of human neuropilin-1 and neuropilin-2 genes: identification and distribution of splice variants and soluble isoforms. *Genomics* 70:211–22

Rubin LL, Staddon JM. 1999. The cell biology of the blood-brain barrier. *Annu. Rev. Neurosci.* 22:11–28

Ruhrberg C, Gerhardt H, Golding M, Watson R, Ioannidou S, et al. 2002. Spatially restricted patterning cues provided by heparin-binding VEGF-A control blood vessel branching morphogenesis. *Genes Dev.* 16:2684–98

Ruiz de Almodovar C, Lambrechts D, Mazzone M, Carmeliet P. 2009. Role and therapeutic potential of VEGF in the nervous system. *Physiol. Rev.* 89:607–48

Saitou M, Furuse M, Sasaki H, Schulzke JD, Fromm M, et al. 2000. Complex phenotype of mice lacking occludin, a component of tight junction strands. *Mol. Biol. Cell* 11:4131–42

Salinas PC, Zou Y. 2008. Wnt signaling in neural circuit assembly. *Annu. Rev. Neurosci.* 31:339–58

Salminen M, Meyer BI, Bober E, Gruss P. 2000. Netrin 1 is required for semicircular canal formation in the mouse inner ear. *Development* 127:13–22

Seeger M, Tear G, Ferres-Marco D, Goodman CS. 1993. Mutations affecting growth cone guidance in *Drosophila*: genes necessary for guidance toward or away from the midline. *Neuron* 10:409–26

Serafini T, Colamarino SA, Leonardo ED, Wang H, Beddington R, et al. 1996. Netrin-1 is required for commissural axon guidance in the developing vertebrate nervous system. *Cell* 87:1001–14

Serafini T, Kennedy TE, Galko MJ, Mirzayan C, Jessell TM, Tessier-Lavigne M. 1994. The netrins define a family of axon outgrowth-promoting proteins homologous to *C. elegans* UNC-6. *Cell* 78:409–24

Serini G, Valdembri D, Zanivan S, Morterra G, Burkhardt C, et al. 2003. Class 3 semaphorins control vascular morphogenesis by inhibiting integrin function. *Nature* 424:391–97

Sheldon H, Andre M, Legg JA, Heal P, Herbert JM, et al. 2009. Active involvement of Robo1 and Robo4 in filopodia formation and endothelial cell motility mediated via WASP and other actin nucleation-promoting factors. *FASEB J.* 23:513–22

Shoji W, Isogai S, Sato-Maeda M, Obinata M, Kuwada JY. 2003. Semaphorin3a1 regulates angioblast migration and vascular development in zebrafish embryos. *Development* 130:3227–36

Soker S, Miao HQ, Nomi M, Takashima S, Klagsbrun M. 2002. VEGF165 mediates formation of complexes containing VEGFR-2 and neuropilin-1 that enhance VEGF165-receptor binding. *J. Cell Biochem.* 85:357–68

Soker S, Takashima S, Miao HQ, Neufeld G, Klagsbrun M. 1998. Neuropilin-1 is expressed by endothelial and tumor cells as an isoform-specific receptor for vascular endothelial growth factor. *Cell* 92:735–45

Song H, Ming G, He Z, Lehmann M, McKerracher L, et al. 1998. Conversion of neuronal growth cone responses from repulsion to attraction by cyclic nucleotides. *Science* 281:1515–18

Sperry RW. 1963. Chemoaffinity in the orderly growth of nerve fiber patterns and connections. *Proc. Natl. Acad. Sci. USA* 50:703–10

Stein E, Tessier-Lavigne M. 2001. Hierarchical organization of guidance receptors: silencing of netrin attraction by slit through a Robo/DCC receptor complex. *Science* 291:1928–38

Stenman JM, Rajagopal J, Carroll TJ, Ishibashi M, McMahon J, McMahon AP. 2008. Canonical Wnt signaling regulates organ-specific assembly and differentiation of CNS vasculature. *Science* 322:1247–50

Stewart PA, Wiley MJ. 1981. Developing nervous tissue induces formation of blood-brain barrier characteristics in invading endothelial cells: a study using quail–chick transplantation chimeras. *Dev. Biol.* 84:183–92

Su J, Cui M, Tang Y, Zhou H, Liu L, Dong Q. 2009. Blockade of bradykinin B2 receptor more effectively reduces postischemic blood-brain barrier disruption and cytokines release than B1 receptor inhibition. *Biochem. Biophys. Res. Commun.* 388:205–11

Suchting S, Freitas C, le Noble F, Benedito R, Breant C, et al. 2007. The Notch ligand Delta-like 4 negatively regulates endothelial tip cell formation and vessel branching. *Proc. Natl. Acad. Sci. USA* 104:3225–30

Suchting S, Heal P, Tahtis K, Stewart LM, Bicknell R. 2005. Soluble Robo4 receptor inhibits in vivo angiogenesis and endothelial cell migration. *FASEB J.* 19:121–23

Takashima S, Kitakaze M, Asakura M, Asanuma H, Sanada S, et al. 2002. Targeting of both mouse neuropilin-1 and neuropilin-2 genes severely impairs developmental yolk sac and embryonic angiogenesis. *Proc. Natl. Acad. Sci. USA* 99:3657–62

Tamagnone L, Artigiani S, Chen H, He Z, Ming GI, et al. 1999. Plexins are a large family of receptors for transmembrane, secreted, and GPI-anchored semaphorins in vertebrates. *Cell* 99:71–80

Tessier-Lavigne M, Goodman CS. 1996. The molecular biology of axon guidance. *Science* 274:1123–33

Torres-Vazquez J, Gitler AD, Fraser SD, Berk JD, Van NP, et al. 2004. Semaphorin-plexin signaling guides patterning of the developing vasculature. *Dev. Cell* 7:117–23

Tran TS, Kolodkin AL, Bharadwaj R. 2007. Semaphorin regulation of cellular morphology. *Annu. Rev. Cell Dev. Biol.* 23:263–92

van Amerongen R, Nusse R. 2009. Towards an integrated view of Wnt signaling in development. *Development* 136:3205–14

van der Zwaag B, Hellemons AJ, Leenders WP, Burbach JP, Brunner HG, et al. 2002. PLEXIN-D1, a novel plexin family member, is expressed in vascular endothelium and the central nervous system during mouse embryogenesis. *Dev. Dyn.* 225:336–43

Vesalius A. 1543. *De Humani Corporis Fabrica*. Basel: Oporinus

Vieira JM, Schwarz Q, Ruhrberg C. 2007. Selective requirements for NRP1 ligands during neurovascular patterning. *Development* 134:1833–43

Virgintino D, Girolamo F, Errede M, Capobianco C, Robertson D, et al. 2007. An intimate interplay between precocious, migrating pericytes and endothelial cells governs human fetal brain angiogenesis. *Angiogenesis* 10:35–45

Wang B, Xiao Y, Ding BB, Zhang N, Yuan X, et al. 2003. Induction of tumor angiogenesis by Slit-Robo signaling and inhibition of cancer growth by blocking Robo activity. *Cancer Cell* 4:19–29

Wang D, Pascual JM, Yang H, Engelstad K, Jhung S, et al. 2005. Glut-1 deficiency syndrome: clinical, genetic, and therapeutic aspects. *Ann. Neurol.* 57:111–18

Wang D, Pascual JM, Yang H, Engelstad K, Mao X, et al. 2006. A mouse model for Glut-1 haploinsufficiency. *Hum. Mol. Genet.* 15:1169–79

Wang HU, Chen ZF, Anderson DJ. 1998. Molecular distinction and angiogenic interaction between embryonic arteries and veins revealed by ephrin-B2 and its receptor Eph-B4. *Cell* 93:741–53

Wang KH, Brose K, Arnott D, Kidd T, Goodman CS, et al. 1999. Biochemical purification of a mammalian slit protein as a positive regulator of sensory axon elongation and branching. *Cell* 96:771–84

Wilson BD, Ii M, Park KW, Suli A, Sorensen LK, et al. 2006. Netrins promote developmental and therapeutic angiogenesis. *Science* 313:640–44

Wolburg H, Wolburg-Buchholz K, Kraus J, Rascher-Eggstein G, Liebner S, et al. 2003. Localization of claudin-3 in tight junctions of the blood-brain barrier is selectively lost during experimental autoimmune encephalomyelitis and human glioblastoma multiforme. *Acta Neuropathol.* 105:586–92

Yang Y, Zou L, Wang Y, Xu KS, Zhang JX, Zhang JH. 2007. Axon guidance cue Netrin-1 has dual function in angiogenesis. *Cancer Biol. Ther.* 6:743–48

Ye X, Wang Y, Cahill H, Yu M, Badea TC, et al. 2009. Norrin, frizzled-4, and Lrp5 signaling in endothelial cells controls a genetic program for retinal vascularization. *Cell* 139:285–98

Yoshikawa S, McKinnon RD, Kokel M, Thomas JB. 2003. Wnt-mediated axon guidance via the *Drosophila* Derailed receptor. *Nature* 422:583–88

Zallen JA, Yi BA, Bargmann CI. 1998. The conserved immunoglobulin superfamily member SAX-3/Robo directs multiple aspects of axon guidance in *C. elegans*. *Cell* 92:217–27

Zlokovic BV. 2008. The blood-brain barrier in health and chronic neurodegenerative disorders. *Neuron* 57:178–201

Zlotnik A, Gurevich B, Tkachov S, Maoz I, Shapira Y, Teichberg VI. 2007. Brain neuroprotection by scavenging blood glutamate. *Exp. Neurol.* 203:213–20

Motor Neuron Diversity in Development and Disease

Kevin C. Kanning, Artem Kaplan, and Christopher E. Henderson

Departments of Pathology, Neurology, and Neuroscience, Center for Motor Neuron Biology and Disease, Columbia University Medical Center, New York, NY 10032; email: ch2331@columbia.edu

Annu. Rev. Neurosci. 2010. 33:409–40

First published online as a Review in Advance on April 1, 2010

The *Annual Review of Neuroscience* is online at neuro.annualreviews.org

This article's doi: 10.1146/annurev.neuro.051508.135722

Key Words

amyotrophic lateral sclerosis (ALS), spinal muscular atrophy (SMA), motor pool, motor unit, neuromuscular junction, axonal degeneration

Abstract

Although often considered as a group, spinal motor neurons are highly diverse in terms of their morphology, connectivity, and functional properties and differ significantly in their response to disease. Recent studies of motor neuron diversity have clarified developmental mechanisms and provided novel insights into neurodegeneration in amyotrophic lateral sclerosis (ALS). Motor neurons of different classes and subtypes—fast/slow, alpha/gamma—are grouped together into motor pools, each of which innervates a single skeletal muscle. Distinct mechanisms regulate their development. For example, glial cell line–derived neurotrophic factor (GDNF) has effects that are pool-specific on motor neuron connectivity, column-specific on axonal growth, and subtype-specific on survival. In multiple degenerative contexts including ALS, spinal muscular atrophy (SMA), and aging, fast-fatigable (FF) motor units degenerate early, whereas motor neurons innervating slow muscles and those involved in eye movement and pelvic sphincter control are strikingly preserved. Extrinsic and intrinsic mechanisms that confer resistance represent promising therapeutic targets in these currently incurable diseases.

Contents

INTRODUCTION

Motor neurons are the sole means by which the brain can trigger contraction of skeletal muscle and thereby control the movements on which life depends, including swallowing and breathing. Because of the accessibility of the neuromuscular junction (NMJ), their synaptic connection with muscle, motor neurons have long served as model systems for understanding neuronal development and function. In addition, motor neurons are the principal target of incurable neurodegenerative diseases such as amyotrophic lateral sclerosis (ALS) and spinal muscular atrophy (SMA). Degeneration and death of motor neurons in these diseases leads to outcomes that are always severe and often fatal.

In spite of their shared function, motor neurons are not all created equal. Coordinated control of muscle contraction requires the involvement of multiple motor neuron classes and subtypes—fast and slow, alpha and gamma—grouped into muscle-specific motor pools. Motor neuron diversity is therefore a functional necessity. Whereas certain aspects of development and degeneration seem to be common to all motor neurons, others are restricted to

NMJ: neuromuscular junction

ALS: amyotrophic lateral sclerosis

SMA: spinal muscular atrophy

subsets. Our goal in this review is to explore how considering motor neuron diversity can illuminate the study of motor neuron development, survival, degeneration, and death. Recent approaches that focus on motor neuron subtypes have led to a clearer understanding, and in some cases a simplification, of the cellular and molecular changes involved. Such mechanistic data are important both for what they teach us about the healthy nervous system and because they can lead to the identification of novel therapeutic targets for slowing motor neuron degeneration in patients.

FUNCTIONAL AND MOLECULAR DIVERSITY AMONG SPINAL MOTOR NEURONS

The human body has more than 300 bilateral pairs of muscles containing more than 100 million muscle fibers, which are innervated by more than 120,000 motor neurons in the spinal cord alone. Motor neurons in the adult can be classified into functionally diverse classes and subtypes (**Figure 1**). One division—into alpha (α), beta (β), and gamma (γ) motor neurons—is made according to the type of muscle fiber that each class innervates (**Figure 1a**). Alpha motor neurons innervate extrafusal skeletal muscle and drive muscle contraction. Gamma motor or fusimotor neurons innervate intrafusal muscle fibers of the muscle spindle and play complex roles in motor control. A third less well-defined population called β-motor neurons innervates both intra- and extrafusal fibers. Alpha motor neurons are the most abundant of these classes, and they can in turn be classified into subtypes according to the contractile properties of the motor units that they form with target muscle fibers: fast-twitch fatigable (FF), fast-twitch fatigue-resistant (FR), and slow-twitch fatigue-resistant (S) (Burke et al. 1973). In this section, we review our current knowledge of the morphological, molecular, and functional characteristics that distinguish the different classes and subtypes because these are essential to understanding subtype-specific aspects of development and disease.

Functional and Molecular Distinctions between Alpha and Gamma Motor Neurons

Extrafusal muscle fibers innervated by alpha motor neurons (α-MNs) generate force to move the skeleton, whereas intrafusal fibers innervated by gamma motor neurons (γ-MNs) modulate the sensitivity of muscle spindles to stretch (Hunt & Kuffler 1951, Kuffler et al. 1951). Reflecting their different targets and functions, mature α- and γ-MNs show pronounced differences in size and form (**Figure 1b**, **Table 1**). Gamma motor neurons are smaller: Their average soma diameter is half that of the smallest α-MNs and axon conduction velocities are slower, reflecting their smaller axon caliber (Burke et al. 1977, Shneider et al. 2009, Westbury 1982). The dendritic trees of α- and γ-MNs are of similar length, but those of γ motor neurons are significantly less branched and are simpler overall (Westbury 1982). The differences also extend to connectivity within the spinal cord. All γ-MNs are devoid of monosynaptic Ia input from proprioceptive sensory neurons, whereas most (but not all) α-MNs receive direct Ia input (Eccles et al. 1960, Friese et al. 2009). Few γ-MNs have intraspinal axon collaterals, so they are unlikely to contribute to recurrent inhibition within the spinal cord (Westbury 1982). Thus α- and γ-MNs have largely distinct postsynaptic targets and presynaptic inputs.

These marked differences between the two populations imply the existence of separate programs for the determination of α- and γ-MN identity. Determining how and when the two populations diverge will require the identification of early markers because differences in cell diameter are not sufficient, especially during development (Gordon et al. 1991, Hoover & Durkovic 1991, Horcholle-Bossavit et al. 1990, Simon et al. 1996). The α₃ subunit of the Na⁺,K⁺-ATPase is selectively expressed in small myelinated axons of motor nerves and in spindle-innervating axons in muscle, but it is absent from nerves such as the facial nerve known not to contain fusimotor axons (Buss

α-MNs: alpha motor neurons

γ-MNs: gamma motor neurons

S: slow-twitch, fatigue-resistant

FR: fast-twitch, fatigue-resistant

FF: fast-twitch, fatigable

et al. 2006, Dobretsov et al. 2003). However, differences in expression are not apparent at the cell body level (Buss et al. 2006). The transcription factor Err3, an orphan nuclear hormone receptor, can distinguish γ- from α-MNs in the spinal cord because it is expressed in cell bodies; it becomes restricted to γ-MNs during the first two postnatal weeks (Friese et al. 2009). Gamma-MNs also express higher levels of the glial cell line–derived neurotrophic factor

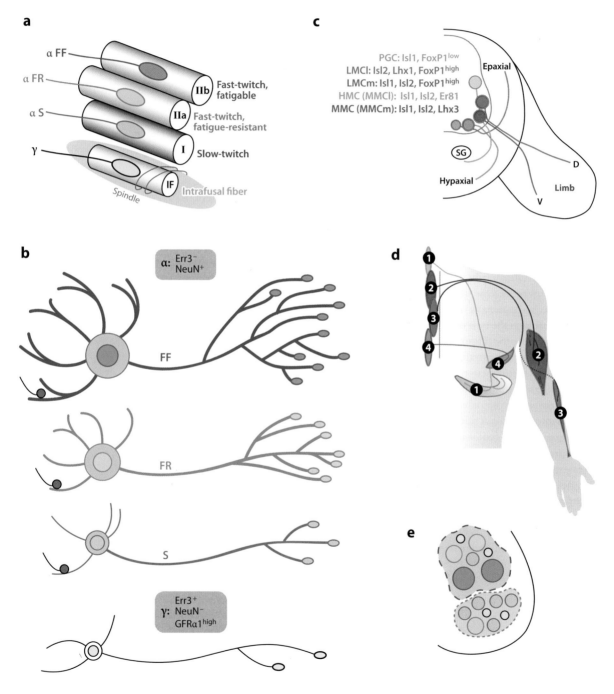

(GDNF) receptor subunit GFRα1 as detected using reporter mice (Shneider et al. 2009). Conversely, the NeuN antigen is strongly expressed by α-MNs but not by γ-MNs (Friese et al. 2009, Shneider et al. 2009).

Although these reports considerably advance the field, each of these markers begins to distinguish γ- from α-MNs only at postnatal stages. This likely reflects the need to screen candidate markers at postnatal stages when size differences are apparent. It is not clear whether they are late markers of populations that are genetically determined at earlier stages (Gould et al. 2008), or whether it is only at postnatal stages that differential interactions with the periphery finally establish a molecular distinction between γ- and α-MNs (Friese et al. 2009). Therefore, if γ-MNs exist as a distinct population at embryonic stages, we currently have no way of identifying them.

Functional and Molecular Differences between Slow and Fast Alpha Motor Neurons

Alpha motor units can be categorized into three subtypes—S (slow-twitch, fatigue-resistant), FR (fast-twitch, fatigue-resistant), and FF (fast-twitch, fatigable)—on the basis of their contractile properties. The corresponding subtypes of α-MNs possess morphological and functional characteristics that suit them to these different functions (**Figure 1b**, **Table 1**). In reality, muscle fiber types show a continuum of maximum force, isometric twitch speed, and endurance (Kernell 2003, Kernell et al. 1999). Correspondingly, α-MNs exhibit a continuous variation in size, excitability, firing patterns, and conduction speed. The distinctions between subtypes are therefore less sharp than between α- and γ-MNs, but the range of properties is considerable.

Size and morphology. On one end of the spectrum, type S motor neurons have smaller cell bodies and axons. At the other end, type FF motor neurons are large, with large-diameter, fast-conducting axons. The average membrane area for FF motor neurons is >20% larger than for S motor neurons, reflecting more axonal and dendritic branches and more presynaptic neuromuscular terminals per motor neuron (Cullheim et al. 1987). Properties of FR motor neurons are presumably intermediate, but some investigators have argued that the FF versus FR distinction relates only to the

---

Figure 1

Morphological characteristics and spatial organization of motor neuron classes and subtypes. (*a*) Distinct motor units within skeletal muscle. Alpha motor neurons innervate extrafusal muscle fibers to form three subtypes of motor unit: Slow-twitch (S) units control Type I fibers, fast-twitch fatigue-resistant (FR) units control Type IIa fibers, and fast-twitch fatigable (FF) units control Type IIb/x fibers. Intrafusal (IF) muscle fibers are innervated by γ-MNs and also by β-MNs (not shown). (*b*) The size and morphological complexity of alpha motor neurons diminish progressively from FF through FR to S motor units. Most α-MNs receive direct Ia innervation from VGLUT1⁺ (vesicular glutamate transporter 1) proprioceptive sensory neurons (*red terminals*). Gamma motor neurons are smaller still and do not receive Ia innervation. At postnatal stages, α- and γ-MN cell bodies can be distinguished by their size, connectivity, and the indicated molecular markers. (*c*) Motor columns in the embryonic spinal cord provide the basic framework for subsequent differentiation of motor classes, subtypes, and pools. They are distinguished by combinatorial patterns of expression of transcription factors and by the position of their peripheral targets. The medial motor column (MMC, previously referred to as MMCm) contains motor neurons that innervate epaxial muscles, the hypaxial motor column (HMC, previously referred to as MMCl) innervates hypaxial muscles, the medial and lateral subdivisions of the lateral motor column (LMCm and LMCl) innervate ventral and dorsal limb muscles, respectively, and the preganglionic column (PGC) innervates sympathetic ganglia (SG). (*d*) Motor pools at later stages are spatially situated in the spinal cord in a manner that overall reflects peripheral organization. Among limb-innervating pools (2: *biceps*; 3: *brachioradialis*), those innervating more proximal muscles are situated more rostrally. Intercostal muscles (4) are innervated by thoracic motor neurons (4), whereas the diaphragm (1) is innervated by phrenic motor neurons (1) in the cervical spinal cord, reflecting the more rostral position of diaphragm precursors during development. (*e*) Motor pools contain multiple motor neuron classes and subtypes. Two representative motor pools, each innervating a single muscle, are schematized using the color code from panel *b*. The upper pool innervates a fast muscle but also contains slow and gamma motor neurons. Similarly, the lower pool would be classified as slow.

Table 1 Summary of principal differences between motor neuron subtypes[a,b,c]

Motor unit characteristic	Motor neuron subtype and class			
	FAST	**SLOW**	**ALPHA**	**GAMMA**
Target fiber	Fast (IIa/b/x)	Slow (I)	Extrafusal	Intrafusal
Presence in different motor pools	Variable	Variable	All	Most
Cell body size	Larger	Smaller	Larger	Smaller
Dendrite length	Similar	Similar	Similar	Similar
Dendrite branching	More	Less	More	Less
Direct Ia proprioceptive input	Most	Most	Most	None
Synaptic density on proximal dendrites	Similar	Similar	Higher	Lower
Synaptic density on distal dendrites	Lower	Higher	Higher	Lower
C-type boutons on soma	More	Less	Present	Absent
EPSP magnitude from Ia afferents	Smaller	Larger and longer	Greater	None
Membrane input resistance	Smaller	Larger	Larger	Smaller
Order of recruitment	Late	Early	Size dependent	–
Firing behavior	Phasic	Tonic	Subtype dependent	Subtype dependent
Bistable behavior	No	Yes	Yes	Maybe
Axon conduction velocity	Faster	Slower	Faster	Slower
Post-spike after-hyperpolarization	Shorter	Longer	Subtype-dependent	Varied, debatable
Axonal recurrent collaterals	More	Less	Yes	No
Motor unit size (innervation ratio)	Larger	Smaller	Larger	Smaller
NMJ morphology	Larger, more complex	Smaller, less complex	Deep synaptic folds	Shallow synaptic folds
Synaptic vesicle density in terminals	Lower	Higher	–	–
Affected in ALS	Early	Late	Yes	Unclear
Affected in aging	Early	Late	Yes	No
Molecular correlates	–	SV2a (adult)	NeuN, Hb9::GFP[+]	Err3, Gfrα1, SDH[hi], TrkC[hi]

[a]The table summarizes the functional and molecular differences discussed in the text between fast and slow, and between alpha and gamma, motor neurons. The descriptors are for some comparisons relative trends rather than absolute rules and apply only within a given pairwise comparison, not across the whole table. This is because absolute values vary between species and at different ages.

[b]Criteria for which the results are not clearly established, e.g., direct contacts from corticospinal axons, are omitted. More specific details in each instance can be found in the references provided in **Supplemental Table 1**. Follow the **Supplemental Material link** from the Annual Reviews home page at **http://www.annualreviews.org**.

[c]Abbreviations: ALS, amyotrophic lateral sclerosis; EPSP, excitatory postsynaptic potential; NMJ, neuromuscular junction.

muscle fiber, especially in rodents in which motor neuron size differences are less broad (Bakels & Kernell 1993, Gardiner 1993).

Electrical properties. These simple differences in size have biophysical consequences for the recruitment order of motor units in response to a graded stimulation. Smaller S motor neurons have higher input resistance and therefore require less synaptic activation to initiate action potentials, meaning that during muscle contraction they reach threshold first and large motor neurons are activated last. This rule is known as the "size principle" (Mendell 2005). As a consequence, during normal motor behavior, the fast motor units—which are strongest—are mainly employed in short-lasting bouts of forceful contraction. For example, in the triceps surae the fast units fire maximally during running and jumping, whereas during postural tasks, such as standing, only S motor neurons are active (Burke 1980).

Firing rate is another characteristic that distinguishes FF and S motor neurons. Type S motor neurons show repetitive firing. Once initiated, this activity can persist even in the absence of presynaptic excitatory drive, amplifying and prolonging synaptic input signals and ceasing only when an inhibitory stimulus is applied (Lee & Heckman 1998a,b), in part owing to the longer-lasting persistent inward currents (PIC) on dendrites of S than of FF motor neurons (Heckman et al. 2008). Another physiologic parameter that differs between α-MN subtypes is the duration of the postspike after-hyperpolarization (AHP), which is shaped largely by a Ca^{2+}-dependent K^+ current (Eccles et al. 1957, Gardiner 1993). Fast motor neurons have a shorter AHP than do S motor neurons so that the firing frequency of each subtype is speed matched to the contractile frequency of the target muscle fiber (Bakels & Kernell 1993, Gardiner 1993).

Molecular markers. In contrast to the defined characteristics of slow and fast muscle fibers, little is known about the molecular differences between the corresponding motor neurons. Levels of succinate dehydrogenase (SDH) and calcitonin gene-related peptide (CGRP) have been proposed to differ between fast and slow motor neurons, but this likely reflects size differences rather than a genetic distinction (Kernell et al. 1999, Piehl et al. 1993). Chakkalakal et al. (2008) reported that synaptic vesicle protein SV2A is restricted to slow neuromuscular synapses, although this distinction emerges only gradually over the postnatal period. So, as with γ-MNs, the lack of early markers has hindered the study of the emergence of the slow-fast distinction during development. Studies using chick embryos demonstrate that slow motor axons selectively fasciculate with each other, as do fast axons, and innervate appropriate muscles from the earliest stages (Rafuse et al. 1996). However, in these experiments, the fast and slow axons belong to different pools, and this, rather than their contractile properties, may be the relevant difference between them.

Motor Neurons of All Classes and Subtypes Are Organized into Motor Columns and Pools

The motor neuron subtypes discussed so far—alpha and gamma, slow and fast—are found at all levels of the spinal cord. However, for their functions to be exerted in a coordinated manner, they are spatially grouped in a way that reflects both their developmental history and their adult function.

During development, newly postmitotic motor neurons are grouped into motor columns, which stretch along the rostrocaudal extent of the neural tube (**Figure 1c**). In mammals, these are composed of the medial motor column (MMC), which projects to epaxial muscles of the dorsal body region, the hypaxial motor column (HMC), which projects to hypaxial muscles of the ventral body wall, and the lateral motor columns (LMC), which project to the limbs. At thoracic levels, the preganglionic column (PGC) innervates sympathetic ganglia. Postural muscles are heavily concentrated in midline areas innervated by MMC motor neurons, respiratory muscles are innervated predominantly by HMC, and limb muscles exclusively receive LMC innervation. Motor columns are generated through the combined actions of diffusible factors expressed in a polarized manner in and around the neural tube and transcriptional interactions within it (reviewed in Dasen & Jessell 2009). Each expresses a characteristic profile of transcription factors (**Figure 1c**) and provides the framework within which muscle-specific innervation can emerge.

Within a given column, the collective of motor neurons that innervate a single skeletal muscle is defined as a motor pool (**Figure 1d,e**). Pool organization within the spinal cord has an overall topological logic: More proximal muscles in a limb are innervated by more rostral pools, whereas dorsal and ventral limb muscles are innervated by lateral and medial motor neurons, respectively (Landmesser 1978, Romanes 1951, Vanderhorst & Holstege 1997). Hence a transformed three-dimensional coordinate system recreates the body's muscle plan within the ventral horn of the spinal cord (**Figure 1d**). A typical motor pool contains all types of motor neuron—fast and slow, alpha and gamma (**Figure 1e**). Gamma-MNs typically represent roughly one-third of the motor neurons (Burke et al. 1977, Friese et al. 2009), but the ratio of fast to slow α-MNs can vary considerably. The nomenclature for a given pool and muscle reflects the dominant population: The mouse soleus pool, for example, contains as many as 10% FF motor units but is referred to as "slow" because 50% of its motor units are of the S type (Hegedus et al. 2007).

Motor neurons in a given pool, whatever their subtype, share a number of molecular and morphological properties. The former include characteristic profiles of transcription factor expression (Dasen & Jessell 2009), downstream of which are pool-specific combinations of cell-surface molecules such as axonal guidance receptors and adhesion molecules (Dalla Torre di Sanguinetto et al. 2008) and neurotransmitter receptors (Rekling et al. 2000). During development, the molecular characteristics of each pool are established through a complex interplay between peripheral inductive factors and the coordinated expression and mutual inhibition of endogenous transcription factors such as the *Hox* genes (reviewed in Dasen & Jessell 2009). These genes in turn contribute to determining the characteristic morphology and connectivity of motor neurons in the pool. The links between transcriptional profile, morphology, and connectivity have begun to be unraveled in the case of the cutaneus maximus motor pool (see below), but many general questions remain about how pool-specific molecular characteristics shape target innervation and presynaptic input from sensory afferents, spinal interneurons, and supraspinal centers.

Motor Neuron Diversity Reflects Multiple Parameters

The term motor neuron is therefore a generalization that covers many different overlapping populations that perform variations on a common function. A given motor neuron likely has a unique set of molecular, functional, and connectional properties that are only grossly characterized by the above classifications. All these attributes need to be considered when analyzing specific responses during development or in disease.

DEVELOPMENTAL REGULATION OF GROWTH AND SURVIVAL OF MOTOR NEURON SUBSETS

The final number and subtype distribution of motor neurons in a given motor pool are essential for its functional specialization. However, although our knowledge of the genetic mechanisms of pool specification continues to increase (Dasen & Jessell 2009), many questions about the control of motor neuron numbers and characteristics remain to be addressed. Motor neurons of larger pools are initially generated in higher numbers during development (Lin et al. 1998, Oppenheim et al. 1989). Numbers are subsequently modulated by the competing influences of cell death pathways and survival factors (reviewed in Oppenheim 1991; Pettmann & Henderson 1998). Recent progress in our understanding of how this equilibrium is regulated has relied strongly on concepts related to motor neuron diversity.

The survival requirements of motor neurons during development have been intensively studied, yet exactly how these map onto motor neuron diversity remains unclear. Certain facts are well established by multiple studies (reviewed in Gould & Enomoto 2009). First, survival of all LMC motor neurons depends on limb-derived trophic factors (Oppenheim 1991). Second, more than 15 polypeptide factors expressed in the environment of the motor neuron can keep motor neurons alive in vivo and in vitro (Henderson 1996, Oppenheim 1996). This number continues to grow with the recent demonstration of neurotrophic activity for factors such as the cardiotrophin-like cytokine factor-1 (CLCF1) complex (Elson et al. 2000, Forger et al. 2003), vascular endothelial growth factor (VEGF) (Van Den Bosch et al.

2004), pleiotrophin (Mi et al. 2007), neuropoietin (Derouet et al. 2004), Mullerian inhibiting substance (MSH) (Wang et al. 2005b), growth differentiation factor-15 (GDF-15) (Strelau et al. 2009), progranulin (Van Damme et al. 2008), and angiogenin (Kieran et al. 2008). Third, receptors for neurotrophic factors are expressed in subsets of motor neurons throughout the spinal cord, and individual motor pools express multiple receptors (Garces et al. 2000, Gould & Oppenheim 2004). Last, knockout mice for several neurotrophic factors show reductions in total motor neuron numbers at the end of embryogenesis, which suggests that the factors in question do regulate developmental cell death. In the postnatal period, motor neurons continue to rely on surrounding tissues for trophic support, although the specific molecular dependencies shift (Holtmann et al. 2005, Strelau et al. 2009). In none of the ligand/receptor knockout strains, however, is there a complete loss of motor neurons, which suggests that individual neurotrophic factors act on subsets of motor neurons (reviewed in Gould & Enomoto 2009). In spite of this wealth of data, we do not have a clear idea of how most neurotrophic factors act, which subsets of motor neurons are dependent on them, and over what timeframe. Recent data concerning the role of GDNF provide a first response to these questions and a potential model for the action of other factors, and so are reviewed here in more detail.

Diverse Roles for GDNF during Motor Neuron Development

GDNF first attracted interest in the motor neuron context because of its potent survival activity for motor neurons in vitro and in vivo (Henderson et al. 1994). It was therefore postulated to be a target-derived neurotrophic factor for α-MNs. Many subsequent studies have revealed the diversity and complexity of the roles it plays during development. GDNF can act in manners that are pool-, column-, or subtype-specific, depending on the developmental process under consideration (**Figure 2**).

Cellular sources of GDNF. The expression pattern of GDNF in the environment of the motor neuron is highly dynamic (**Figure 2a**). When motor neurons first send axons out of the spinal cord at e10 in the mouse, the only site of GDNF expression in their peripheral environment is a small group of mesenchymal cells at the base of each limb (Haase et al. 2002, Kramer et al. 2006, Wright & Snider 1996). Over the following days (e11.5-e12.5), brachial GDNF-expressing cells appear to migrate to colonize two specific flank muscles: cutaneus maximus (CM) and latissimus dorsi (LD). At the same stage, two muscles of the developing hindlimb—*gluteus* and *iliopsoas*—also express GDNF (Gould et al. 2008). In parallel, by e12.5, Schwann cells express high levels in the developing peripheral nerve (Gould et al. 2008, Henderson et al. 1994). At later stages, around e15, GDNF expression appears in all muscles but is limited to muscle spindles and the fibers that surround them (Gould et al. 2008, Shneider et al. 2009, Whitehead et al. 2005). The roles thus far determined for GDNF correspond well to these sites of expression.

Pool-specific effects of GDNF on motor neuron morphology and connectivity. Most of our knowledge of the role of GDNF in motor development has come from the study of null mutants (**Figure 2b**). In *gdnf* knockout embryos at e12.5, motor neuron survival is unaffected, but the motor neurons of the CM and LD pools are misplaced within the spinal cord and their axons fail to invade the target muscles normally (Haase et al. 2002). An indistinguishable phenotype is seen in knockout mice for the transcription factor PEA3, which is specifically expressed by the CM and LD motor pools at forelimb levels (Lin et al. 1998, Livet et al. 2002). These and other data led to the following model for normal development. At early stages, GDNF at the base of the limb acts through the Ret receptor on CM and LD axons to activate transcription of PEA3. Expansion of the PEA3 pool then occurs by a noncell-autonomous process of recruitment triggered by hepatocyte growth factor (HGF) in the periphery (Helmbacher et al.

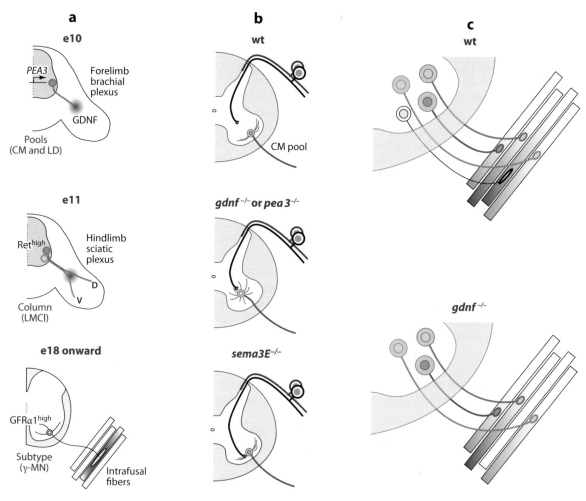

Figure 2

Roles of glial cell line–derived neurotrophic factor (GDNF) in motor neuron development. (*a*) Developmental sites of expression of GDNF (*purple*) related to its pool-specific, column-specific, and subtype-specific actions. At e10, GDNF at the base of the forelimb induces PEA3 expression and innervation of cutaneus maximus (CM) and latissimus dorsi muscles. At e11, GDNF at the base of the hindlimb leads LMCl axons to project dorsally. From e18 onward, GDNF expressed in spindles is available to γ-MNs. (*b*) Effects of GDNF and its downstream effectors on motor neuron morphology, position and connectivity in the CM motor pool. Sensory neurons are shown in black/green (proprioceptive Ia afferents) and brown/orange (nociceptive). The characteristic dendritic morphology and position of CM motor neurons in wt mice is altered in the absence of GDNF or PEA3, but not in null mutants for *sema3E*. In all three mutants, but not wt, CM motor neurons receive monosynaptic Ia input. In *sema3E* mutants this is from homonymous CM afferents, whereas in *gdnf* and *pea3* nulls, the afferents originate in the triceps muscle. (*c*) Selective loss of gamma motor neurons (*yellow/black*; shown innervating GDNF-expressing intrafusal muscle fibers, *purple*) in null mutants for GDNF.

2003). Although it is expressed only after axon outgrowth has begun, PEA3 subsequently governs several major features of the CM/LD motor pools and corresponding lumbar motor neurons: cell body position and muscle target innervation, dendrite morphology, and afferent synapse formation (Haase et al. 2002, Livet et al. 2002, Vrieseling & Arber 2006).

The molecular mechanisms through which this occurs are beginning to be understood. An unusual property of CM motor neurons is that they do not receive direct monosynaptic

Ia afferents from proprioceptive sensory neurons (Pecho-Vrieseling et al. 2009). One PEA3-dependent gene expressed specifically in the CM pool is semaphorin 3E, which is known to repel axons that express its receptor PlexinD1; these include Ia afferents. In *sema3E* null mutants, many CM motor neurons receive direct functional Ia input, whereas in transgenic mice overexpressing Sema3E in all motor neurons, sensory afferents are massively excluded from the ventral horn (Pecho-Vrieseling et al. 2009). Thus, acting through PEA3 and its downstream effectors, GDNF not only influences cell body position and morphology but can also dictate the connectivity of motor neurons that respond to it (**Figure 2b**).

Column-specific effects of GDNF on motor axon projections. Within the lateral motor column at hindlimb levels, motor neurons of the lateral (LMCl) division, which projects to the dorsal limb, express higher levels of the GDNF receptor Ret than do medial LMCm motor neurons (Kramer et al. 2006). The choice of LMCl axons to project dorsally at the base of the hindlimb is known to be driven in part by repulsive signals from ephrins in the ventral limb acting through the EphA4 receptor (Helmbacher et al. 2000, Kania & Jessell 2003). However, in mutants for *gdnf* or *ret*, LMCl axons misproject ventrally, which suggests that GDNF expressed in the dorsal limb cooperates with ventrally expressed ephrins to ensure correct column-specific projection to muscle targets. A further level of control of GDNF function in this system comes about through physical interactions of the Ret receptor with the leucine-rich repeat and immunoglobulin (LIG) family member ISLR2/LINX, which modulates the Ret response to ligand. In the absence of ISLR2, as with GDNF and Eph mutants, the Ret-dependent growth of the peroneal nerve is defective (Mandai et al. 2009).

Subtype-specific effects of GDNF on motor neuron survival. Experimentally, GDNF can keep all motor neurons alive for at least one week in vitro and in vivo (Henderson et al.

1994). At longer times, however, GDNF maintains only ~25% of motor neurons (Pennica et al. 1996, Vejsada et al. 1998). In an apparent parallel, knockout mice for GDNF show a reduction of only ~25% in motor neuron numbers in many motor pools (Moore et al. 1996, Oppenheim et al. 2000). These and other data suggested that, although exogenous GDNF can acutely save large numbers of motor neurons, physiological concentrations and availability may limit its actions to subsets of them.

Closer analysis of postnatal mice mutant for *gdnf* and *ret* reveals a selective loss of smaller motor neuron cell bodies, a reduction in spindle innervation, and a marked reduction of small myelinated axons in ventral roots (Gould et al. 2008, Whitehead et al. 2005). Therefore, most of the motor neurons lost at this stage in GDNF mutants appear to be fusimotor γ-motor neurons (**Figure 2c**). This observation may be correlated with a pronounced increase in motor neuron cell death between e12.5 and e14.5 in mutant embryos (Gould et al. 2008). However, in the absence of markers, we cannot determine whether the neurons lost are presumptive γ- or α-MNs because the survival effect of GDNF on γ-MNs may be indirect (Gould et al. 2008). Indeed, although γ-MNs express higher levels of GFRα1 than do α-MNs at postnatal stages, making them a likely target for GDNF action, this distinction is not yet apparent at e14 (Shneider et al. 2009). A gain-of-function argument for a role of GDNF in γ-MN survival comes from transgenic mice overexpressing GDNF in muscle under the control of the MyoD promoter (*MyoGDNF*). In the presence of increased GDNF, all motor neurons are induced to sprout exuberantly at the NMJ (Keller-Peck et al. 2001, Nguyen et al. 1998), but counting of axons in the ventral root, proximal to the site of sprouting, shows that only the small, presumptive fusimotor, axons are increased in number (Whitehead et al. 2005).

Thus a single factor, GDNF, initially characterized as a survival factor, can regulate pool-specific cell migration, axonal growth, and synaptic connectivity by inducing PEA3-dependent transcriptional programs. It also controls column-specific axonal outgrowth and subtype-specific survival in a manner that appears to be PEA3-independent. If other neurotrophic factors were to play similarly selective roles during development of other pools and subtypes, this might help to explain the considerable number of distinct neurotrophic factors known to be active on motor neurons (Henderson 1996).

Cellular and Molecular Mechanisms Underlying Developmental Motor Neuron Death

The example of GDNF shows that neurotrophic factors can play roles that are not restricted to neuronal survival. Nevertheless, their role in keeping motor neurons alive remains critical. The programmed cell death process they combat may also vary according to motor neuron subtype.

Early studies clearly defined the timing and morphological aspects of motor neuron death and suggested that all motor neurons, even those that die, initially contact their target muscle (reviewed in Pettmann & Henderson 1998). Which motor neurons subsequently die? One view considers cell death to be stochastic and potentially to affect all motor neuron subtypes in the pool to the same extent. This proposal is a tenet of the neurotrophic hypothesis, in which competition for survival factors with limited availability is considered to determine the final outcome (Davies 1996). An alternative view is that half the motor neurons are predisposed to die—although not irreversibly so—and that they have molecular or functional characteristics that distinguish them from those that normally survive. If cell death were blocked, the former hypothesis predicts that the composition of the motor pool would be essentially unchanged, whereas the second implies that the rescued neurons would, in some way, be distinct from those that normally survive. This hypothesis can, in principle, be tested by studying genetic models in which motor neuron death is reduced. A recent study of supernumerary neurons in adult *Bax* knockout and *MyoGDNF*

transgenic mice sheds interesting light on the question (Buss et al. 2006). In support of the first hypothesis, the contractile properties of slow and fast muscles are unaffected in the mutants, which suggests that the distribution of α-MN subtypes is not radically changed by cell death. In contrast, the properties of many supernumerary motor neurons are similar to those of γ-MNs: small diameter of myelinated axons and cell body, muscle spindle innervation, and expression of α_3 Na$^+$,K$^+$-ATPase (Buss et al. 2006). Conversely to the *Bax* mutants, mice lacking the prosurvival factor *Bcl-2* show a selective deficiency of fusimotor neurons (Hui et al. 2008). There are at least two potential explanations for these findings. The first, which would support the second hypothesis, posits that presumptive γ-MNs constitute most of the neurons that die during development. However, a second explanation would be that the neurons saved by *Bax* inactivation are initially identical to the others but adopt a γ-MN phenotype to accommodate the increased number of muscle spindles in the mutants (Buss et al. 2006). The question will be resolved only once unambiguous early markers for motor neuron subtypes are identified.

How do different motor neurons die during development? Inactivation of the proapoptotic gene *Bax* provides remarkably complete protection, arguing strongly for a role of one of the many programmed cell death mechanisms with a mitochondrial decision point (Deckwerth et al. 1996, Galluzzi et al. 2009, Sun et al. 2003). It is surprising, however, that several mutations that inactivate the classical apoptosome—the complex formed between Apaf-1, caspase-9, and cytochrome *c*—do not modify the extent of developmental motor neuron death, although its morphological features are changed (Kanungo et al. 2008; Oppenheim et al. 2001, 2008). The failure of caspase inactivation to prevent cell death may reflect redundancy among family members, but this explanation is unlikely for cytochrome *c* or Apaf-1. Therefore, the apoptosome may normally be involved during development, but in its absence, motor neurons may upregulate other genes and die by other pathways. Alternatively, Bax-dependent mechanisms other than the apoptosome may underlie naturally occurring motor neuron death. For instance, cultured embryonic motor neurons express the Fas death receptor, and when Fas is activated, 50% of them are triggered to die through a motor neuron–specific mechanism involving p38 and caspase-8 (Raoul et al. 2002). However, which motor neuron subset is Fas sensitive and whether motor neuron numbers are changed in null mutants for Fas and FasL in vivo remain to be determined (Haase et al. 2008).

MOTOR NEURON DEGENERATION AND DEATH IN DISEASE

The developmental processes outlined above contribute to establishing the functional diversity of motor neuron subtypes that is necessary for motor control. In most individuals, the numbers of motor neurons subsequently remain approximately constant for many years, and the overall profile of motor neuron subtypes remains unchanged. Unfortunately, in the childhood genetic disease spinal muscular atrophy (SMA) and the adult-onset neurodegenerative disease amyotrophic lateral sclerosis (ALS), motor function is lost as motor neurons degenerate and die (Mitsumoto et al. 2006). In both diseases, specific pools and subtypes of motor neurons are differentially affected, providing new clues about the pathogenic mechanisms involved. As an illustration of the selectivity of degenerative processes, we focus on the mutant Cu,Zn-superoxide dismutase (SOD1) mouse model of familial ALS. In addition, we will note striking parallels with what is known for sporadic ALS in human patients, SMA, and normal aging.

Mutant SOD1 Mice as a Model of ALS

ALS is characterized by axonal degeneration and cell death of both α- and γ-MNs in the spinal cord (Saito et al. 1978, Swash et al. 1986, Swash & Fox 1974) and motor cortex,

leading progressively to fatal paralysis (Cleveland & Rothstein 2001, Radunovic et al. 2007). Most cases (~90%) of ALS are sporadic—having no known genetic linkage—and have therefore been difficult to model in the laboratory (Valdmanis & Rouleau 2008). However, ~10% of cases are familial (fALS), and identification of multiple genes bearing mutations in these families has provided new insights into the disease mechanisms (Pasinelli & Brown 2006, Valdmanis & Rouleau 2008). Of the familial forms, the most studied and most common involve toxic gain-of-function mutations in SOD1. Many (>140) individual point mutations scattered throughout the primary structure of SOD1 trigger ALS in patients, which suggests that protein misfolding must be a common initial trigger (reviewed in Boillee et al. 2006). Transgenic mice expressing mutant forms of human SOD1 (mSOD1) in all tissues show selective motor neuron degeneration and death, which in many respects resemble the pathology of human sporadic and familial ALS (Wong et al. 2002). Recent work has shown that the full disease phenotype depends on expression of mSOD1 not only in motor neurons, but also in other cell types such as astrocytes and microglia (reviewed in Papadimitriou et al. 2009). The SOD1 mutation therefore acts in both cell-autonomous and non-cell-autonomous manners. These mSOD1 mice, and emerging genetic models of other forms of fALS, therefore provide investigators a unique opportunity to observe the presymptomatic sequence of molecular and cellular events that lead successively to weakness, paralysis, and death and to understand the cellular and molecular mechanisms involved in pathogenesis.

Natural History of Disease Progression in the Mutant SOD1 Mouse Model of ALS

A natural history of ALS provides the clearest way in which to compare the selective susceptibility and resistance of diverse motor neuron subtypes to degeneration and death. It is also a necessary foundation on which to base

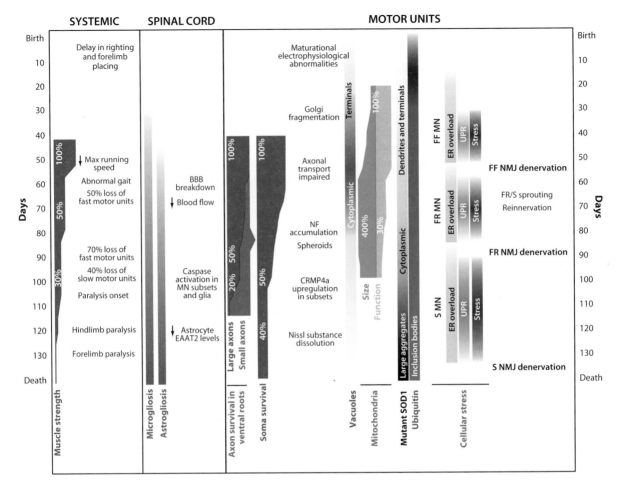

Figure 3

Time course of neurodegeneration in the SOD1^{G93A} mouse model of amyotrophic lateral sclerosis (ALS). The diagram provides an overview of the complex ballet of cellular and molecular mechanisms that lead over six months to the death of this severe model of ALS. It is based on detailed data in **Supplemental Table 2** (follow the **Supplemental Material link** from the Annual Reviews home page at **http://www.annualreviews.org**). Many changes occur before muscle strength is reduced by half, including initial alterations in electrophysiology and behavior followed by ubiquitination and ER stress in susceptible FF motor neurons leading to axonal dieback and microgliosis and astrogliosis in the spinal cord. These are accompanied by subcellular changes such as Golgi fragmentation and mitochondrial swelling. During the following months, these changes become exacerbated and generalized to other motor units, leading to extensive motor neuron loss and muscle paralysis. Indicated stages (scale in days) represent those in the G93A high-expressor line. Some parameters have not been studied at earlier stages, so the indicated dates represent the latest possible onset. The overall layout progresses from systemic and behavioral changes on the left toward molecular and cellular changes in motor units on the right.

pathogenic and therapeutic hypotheses. We have therefore compiled a full sequential timeline of the disease process from multiple publications over the past ten years [**Figure 3, Supplemental Table 2** (follow the **Supplemental Material link** from the Annual Reviews home page at **http://www.annualreviews.org**)]. We focus on data from the widely used SOD1^{G93A} high-expressor model, but results from other SOD1 mutant models are, in general, consistent when corrected for the overall life span of each line. The timeline cannot be perfectly accurate since the phenotype may vary between laboratories even for a given strain and some reports do not define a date of onset for the phenomenon under study. However, parallel presentation of the known systemic and cellular phenotypes provides a holistic impression of the degenerative process (**Figure 3**).

Sequence of behavioral and systemic changes in mutant SOD1 mice. The first behavioral changes reported are delays in the righting reflex and errors in forelimb placement that can occur as early as P10 (van Zundert et al. 2008). It is only after P50 that changes in gait and running speed can be detected (Veldink et al. 2003, Wooley et al. 2005) corresponding to a rapid decrease of overall muscle strength and dropout in the number of intact motor units in muscles with high percentages of FF fibers, descending to 50% of control levels. Around P90, another dramatic decrease of intact motor units is noted in all muscles but again particularly in fast muscles (Hegedus et al. 2007). Overt signs of muscle paralysis are first detected clinically around P100. Complete hindlimb paralysis is observed around P125, and the final rapidly evolving process of degeneration leads to forelimb paralysis around P135 and euthanasia of the animal soon after (Chiu et al. 1995). Because these behavioral changes provide potential early readouts for the testing of therapeutic strategies, there is a clear need for further development of more quantitative noninvasive parameters to measure changes over the first two months.

Sequence of cellular changes in mutant SOD1 mice. In the spinal cord, the earliest detectable pathological alterations occur before muscle denervation (**Figure 3**). Onset of microgliosis can be detected around P30 (Saxena et al. 2009) and is followed by astrocytosis around P50 (Fischer et al. 2004). Both increase progressively in intensity until death. It is not known at which stage secretion of toxic factors by microglia and astrocytes begins, but it may well precede these overt signs of activation. Partial breakdown of the blood-brain barrier (BBB) and a reduction in blood

flow occur around P60 (Zhong et al. 2008), although reductions in capillary length have been detected as early as P7 (Yoshikawa et al. 2009). Following the onset of symptoms, long-term activation of caspases (Martin et al. 2007, Pasinelli et al. 2000, Vukosavic et al. 2000) and loss of astrocyte-specific glutamate transporters (Warita et al. 2002) are observed in the ventral horn of the spinal cord. Excitotoxicity due to the consequently increased stimulation of glutamate receptors on motor neurons and subsequent rise in calcium ion influx may contribute to motor neuron degeneration in ALS patients and mutant SOD1 mice (reviewed in Van Den Bosch et al. 2006).

Much effort has been devoted to characterizing changes in motor neurons and in their contacts with skeletal muscle (**Figure 3**). The earliest reported phenotypes concern cultured embryonic motor neurons. Cultured mSOD1 motor neurons show remarkably increased sensitivity to external stressors such as Fas ligand, which triggers their death by activating caspase-8 and p38 (Raoul et al. 2002). They are also hyperexcitable owing to abnormally strong persistent inward currents (Kuo et al. 2004, 2005). Mitochondrial function appears to be impaired in these cells (De Vos et al. 2007). Axonal transport defects (both anterograde and retrograde) are also found in cultured motor neurons (De Vos et al. 2007, Kieran et al. 2005). However, Perlson et al. (2009) report that such defects are also found in neurons with dynein mutations, which show milder pathologies in vivo. In contrast, retrograde transport of specific stress signal–related cargos—such as caspase-8 and fragments of p75^{NTR}—is increased in mSOD1 but not in dynein mutant neurons, and inhibitors of these cargos protect against the toxic effects of mutant SOD1 (Perlson et al. 2009). Thus, changes in specific death pathways and the transport of their effectors are an early effect of expression of mutant SOD1 at levels that cause disease.

In vivo, the earliest change detected to date is an increase in the electrical excitability of hypoglossal motor neurons at P4, reflecting an abnormally strong persistent inward sodium current. In parallel, early pruning of dendrites may reflect premature functional maturation (van Zundert et al. 2008). By P14, there are abnormal microvacuoles in the cytoplasm and swollen mitochondria in cytoplasm and axoplasm (Bendotti et al. 2001). From P30 onward, fragmentation of the Golgi apparatus (Mourelatos et al. 1996), vacuolization of the mitochondria in axon terminals (Gould et al. 2006), and accumulation of ubiquitin increase with age (Saxena et al. 2009, Vlug et al. 2005). By early adulthood (P50–90), both mitochondrial function (Martin et al. 2007) and retrograde (Ligon et al. 2005) and anterograde (Pun et al. 2006, Zhang et al. 1997) axonal transport are impaired, leading to accumulation of neurofilaments in the cytoplasm and proximal axons (Tu et al. 1996). These changes are accompanied by swelling of the endoplasmic reticulum (ER), mitochondria, and cytoplasm (Kong & Xu 1998, Martin et al. 2007). From P50 onward, there is progressive denervation and reinnervation of motor endplates through a process of axonal dieback and compensatory sprouting (Schaefer et al. 2005) that is discussed below. After P100, with the onset of paralysis, motor neurons show impaired function of the ubiquitin-proteasome system (Cheroni et al. 2009), and by P110 accumulate aggregates containing ubiquitin and mutant SOD1 first in processes then in the cytoplasm (Kato 2008, Sumi et al. 2006). At the same time, significant motor neuron loss is observed, with surviving motor neurons showing dissolution of the Nissl substance in the cytoplasm (Martin et al. 2007). Most of the above changes are not seen in all motor neurons at a given time point, but rather are detected in small subsets, reflecting differential susceptibility of motor neuron subpopulations to neurodegenerative processes. For example, vacuolation and swelling of mitochondria can be detected in particularly vulnerable motor neurons of the FF type as early as P14 (see below and Bendotti et al. 2001, Yoshikawa et al. 2009).

Subtype-Specific Differences in Disease Susceptibility in mSOD1 Motor Neurons

Not all motor neuron subtypes are equally affected by the disease (**Figures 3** and **4**). This is first apparent at the morphological level. FF motor units undergo atrophy earliest in mutant SOD mice, with near total loss of FF terminals from type IIb muscle fibers in the triceps surae of SOD1[G93A] mice by P50 (Frey et al. 2000). This remarkably synchronous dieback is followed by a delayed but also rapid and synchronous fallout in FR units, whereas S motor units are well preserved late into the disease process (Pun et al. 2006). In agreement with this, ventral roots show preferential loss of large caliber axons, whereas in the spinal cord, 90% of large motor neurons expressing CGRP disappear as compared with the average 50% of total motor neurons (Fischer et al. 2004, Kong & Xu 1998). Electrophysiological studies and fiber type analysis corroborate the general sequence of degeneration from FF to FR to S deduced from the morphological data (Hegedus et al. 2007) but suggest that an initial switch in motor unit phenotype from FF to FR may precede the loss of FF motor axons (Gordon et al. 2009, Hegedus et al. 2008, Kieran et al. 2005). The relative resistance of FR and S motor units may reflect their high sprouting capacity (Duchen & Tonge 1973, Frey et al. 2000), which may allow them to reinnervate motor endplates vacated by the degenerating FF axons (**Figure 4**).

Data from ALS patients suggest that the early axonal dieback and selective susceptibility of FF axons found in mice are also seen in human disease. Muscle pathology consistent with denervation can be observed prior to motor neuron loss (Fischer et al. 2004). Moreover, electromyogram (EMG) patterns are consistent with cycles of denervation/reinnervation (de Carvalho et al. 2008), and the twitch force of fast motor units is affected earliest in patients with sporadic ALS (Dengler et al. 1990). Thus, marked differential vulnerability of

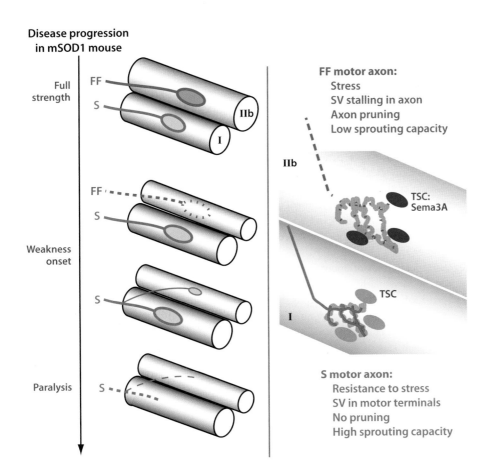

Disease progression in mSOD1 mouse

Full strength

FF

S

IIb

I

Weakness onset

FF

S

S

Paralysis

S

FF motor axon:
Stress
SV stalling in axon
Axon pruning
Low sprouting capacity

IIb

TSC:
Sema3A

TSC

I

S motor axon:
Resistance to stress
SV in motor terminals
No pruning
High sprouting capacity

Figure 4

Neuromuscular junction phenotypes in fast-fatigable (FF) and slow (S) motor units in mutant SOD1 mice. *Left panel:* In healthy adult mice, although there is a limited degree of synaptic remodeling, slow Type I and FF Type IIb/x fibers remain stably innervated by S and FF motor neurons, respectively. During the presymptomatic period in SOD1^{G93A} mice, there is selective dieback of FF motor axons. However, denervated endplates can, for a period, be reinnervated by axonal regrowth or sprouting from either FR or S motor neurons (only S shown for simplicity). At late stages, even S axons die back. *Right panel:* Differences between FF and S neuromuscular junctions and terminal axons at late presymptomatic stages. In addition to the indicated axonal characteristics, terminal Schwann cells at FF NMJs express higher levels of Sema3A. SV, synaptic vesicles.

different motor neuron subtypes is likely a feature common to all forms of ALS.

Intrinsic Molecular Mechanisms Underlying Selective Dieback of FF Axons

How is it that a ubiquitously expressed toxic protein such as mSOD1 can have such selective effects? Because ALS can be triggered by

so many different point mutants of SOD1, research has focused on the ways that misfolded SOD1 impacts the cell. The unfolded protein response (UPR), also known as ER stress signaling, is the name given to a complex set of transcriptional and posttranscriptional events, which are triggered by excessive accumulation of misfolded proteins within the ER. The initial function of the UPR is to restore homeostasis to allow the cell to function normally again; however, when the process is not successful, the UPR can also trigger cell death through programs of apoptosis or autophagy. It has been recognized for several years that many molecular changes characteristic of the UPR—activation of transcription factors, relocation of signaling molecules, cleavage of procaspases—can be detected in both mutant SOD1 rodents and human patients with sporadic ALS (Atkin et al. 2008, Kieran et al. 2007, Kikuchi et al. 2006, Wootz et al. 2004). However, there has been disagreement about the degree of UPR activation and whether it can be detected presymptomatically. Moreover, deletion in SOD1^{G93A} mice of PUMA (p53-upregulated modulator of apoptosis), a BH3-only protein particularly associated with ER stress, provided only transitory protection to motor neurons and did not prolong life span of the mice (Kieran et al. 2007).

Three recent sets of data suggest that the UPR may nevertheless be an important early player in the disease process. First, proteomic analysis of the spinal cord of presymptomatic SOD1^{G93A} rats at P60 showed that protein-disulfide isomerase (PDI), an ER chaperone upregulated during the UPR, was among the five most strongly upregulated proteins (Atkin et al. 2006). Second, deletion of ASK-1 (apoptosis signal regulated kinase 1), which is recruited to the ER as part of the stress response and is also an intermediate in the Fas/NO motor neuron–specific cell death pathway (Raoul et al. 2002), confers a 3.5-week increase in survival in SOD1^{G93A} mice (Nishitoh et al. 2008). Last, Saxena et al. (2009) have provided evidence for an early and selective upregulation of UPR-related genes in the set of motor neurons that are most susceptible to the disease. Using laser-capture microdissection of retrogradely labeled FF, S, or FR motor neurons from the lateral gastrocnemius and soleus pools of SOD1^{G93A} mice and controls at different presymptomatic stages, they show that at early stages (P20) all mutant motor neurons transiently upregulate ubiquitin levels. However, only in the vulnerable FF motor neurons is this process followed at P30 by the activation of a full UPR and expression of general markers of cellular stress such as activating transcription factor 3 (ATF3). This directly precedes synaptic vesicle stalling in axons, vesicle depletion from terminals, and denervation in the FF motor units. A similar upregulation of UPR genes occurs in FR motor units, but later, and also appears to directly precede denervation (Saxena et al. 2009). That the ER stress response should be detected earlier than in other studies is likely the result of focusing on a small group of vulnerable neurons (discussed in Kikuchi et al. 2006). Nevertheless, why only FF motor neurons should trigger a UPR following the initial wave of ubiquitination signaling remains unclear.

Additional candidate intrinsic effectors of motor axon dieback may not be directly involved in the UPR but are downstream of stress-induced cellular changes such as the UPR, reactive oxygen species, or reduced mitochondrial function. Collapsin response mediator protein 4a (CRMP4a) is specifically upregulated in cultured mSOD1 motor neurons exposed to nitric oxide and can be detected in vivo in a fraction of SOD1^{G93A} motor neurons by P60 (Duplan et al. 2010). Silencing of CRMP4a protects against NO toxicity in vitro, whereas viral overexpression either in vitro or in vivo triggers axonal dieback and motor neuron cell death. It will be interesting to determine whether the fraction of motor neurons selectively upregulating CRMP4a reflects a particular subtype and whether CRMP4a plays an essential role in the disease process in vivo.

Candidate Extrinsic Factors Involved in Subtype-Selective Axonal Dieback

Which factors trigger the intrinsic degenerative process in FF axons? Like the effects of mutant SOD1 on motor neuron survival, they are likely both cell autonomous and non-cell-autonomous. It seems reasonable to suppose that exogenous factors acting locally on axonal retraction may be expressed either in the target muscle or in Schwann cells. Here we review four potential candidates—Nogo-A, Sema3A, galectin-1, and N-APP—each of which acts on motor axons in vivo or is expressed in their vicinity, but only two of which have yet been studied in the context of ALS models.

The most direct evidence exists for Nogo-A (reticulon-4A), an axon growth inhibitor, but there is disagreement about its role. Nogo-A is upregulated in skeletal muscle of sporadic ALS patients and mutant SOD1 mice (Jokic et al. 2005). This may be a secondary effect of muscle denervation. Nevertheless, overexpression of Nogo-A in the slow soleus muscle in adult wildtype mice led to axonal dieback at the NMJ (Jokic et al. 2006). Moreover, using the G86R strain of SOD1 mice, the same authors found that genetic ablation of Nogo-A protected against denervation of the soleus, as judged by reduced upregulation of nicotinic acetylcholine receptor and MuSK (muscle-specific receptor tyrosine kinase). In contrast, Yang et al. (2009) found that Nogo-A ablation led to a reduction of motor axon numbers and life span in the G93A strain. They argue that the protective effect in the Jokic et al. study may be attributable to the increased levels of Nogo-B in their strain of knockout mouse (Yang et al. 2009). Further analysis of the difference between these studies will be needed to determine the role of Nogo-A in ALS.

Another repulsive factor for axons, semaphorin 3A (Sema3A), is expressed in a fashion that suggestively matches the selective degeneration of FF axons, although its functional involvement in axonal degeneration has not been tested. Within skeletal muscle, Sema3A is expressed only by terminal Schwann cells (De Winter et al. 2006). The numbers of Sema3A-expressing Schwann cells are markedly increased in mutant SOD1 mice, especially at presymptomatic stages. This upregulation is limited to subsets of the IIx/b fibers innervated by FF motor neurons (De Winter et al. 2006). An intriguing possibility is that localized expression of Sema3A prevents regenerative sprouting in fast axons and may even actively trigger FF-specific axonal degeneration.

Two other proteins have recently been shown to act directly on motor axonal de-generation, although a role in pathological axon dieback remains to be demonstrated. The first is the lactose-binding protein galectin-1 (Lgals1). Galectin-1 is found in neurofilamentous spheroids in both sporadic and familial forms of ALS (Kato et al. 2001). Moreover, in its reduced form, galectin-1 induces degeneration of cultured neurons, acting as a necessary downstream effector of p75^{NTR} (Plachta et al. 2007). At the NMJ, *Lgals1* null mutants show a delay in the degeneration of motor nerve terminals following axotomy (Plachta et al. 2007). Because both p75^{NTR} and galectin-1 are expressed in terminal Schwann cells and can contribute to motor axon dieback, their potential role in ALS will be interesting to determine.

Axon pruning in cultured motor neurons can be triggered by activating the death receptor DR6, a member of the TNF receptor superfamily that is expressed by motor neurons throughout life (Allen Spinal Cord Atlas; Nikolaev et al. 2009). The N-terminal fragment of amyloid precursor protein (N-APP) was recently identified as a ligand of this orphan receptor. In cultured motor neurons deprived of trophic support, N-APP is released from the cell surface, binds to DR6, and triggers axonal degeneration in a Bax- and caspase-6-dependent manner, which is followed by caspase-3-dependent apoptosis (Nikolaev et al. 2009). Moreover, in DR6 knockout mice, as well as in double knockouts of APP and its homolog APLP2, early postnatal motor axon terminals exhibit aberrant protrusions past endplates, indicating either reduced retraction or excessive sprouting (Nikolaev et al. 2009, Wang et al. 2005a). The role of the APP/DR6/caspase-6 pathway in mutant SOD1 mice has not been tested. It seems worthy of study because APP is strongly expressed at healthy human NMJs (Askanas et al. 1992) and because its levels increase in muscles of presymptomatic SOD1^{G93A} mice and sporadic ALS patients (Koistinen et al. 2006).

As suggested by the sprouting and axonal regeneration observed during the period of axonal dieback (Schaefer et al. 2005), other factors must act on the motor axons to compete against the degenerative process. MicroRNA miR-206, a small RNA expressed in muscle, is one of these, probably acting through histone deacetylase 4 and fibroblast growth factor signaling pathways (Williams et al. 2009). Like several neurotrophic factors, it is upregulated in denervated muscle of mSOD1 mice and may slow disease progression by promoting sprouting and regeneration of neuromuscular synapses.

Pool-Specific Resistance to Neurodegeneration in ALS

In addition to the subtype-selective vulnerability to axonal dieback that we have discussed so far, there are also highly significant differences among different motor pools in their response to the disease. In late-stage ALS patients, nearly all voluntary movement is lost, reflecting massive degeneration and death of motor neurons at multiple levels of the spinal cord. However, clinical studies show that in most patients, ocular movement and voluntary control of eliminative functions remain unimpaired until terminal stages (Mitsumoto et al. 2006). These functions are controlled by motor neurons of the oculomotor, trochlear, and abducens nuclei in the midbrain/hindbrain, and by Onuf's nucleus in the lumbosacral spinal cord, respectively. Correspondingly, in ALS autopsy material, motor neurons in oculomotor and Onuf's nuclei are almost completely preserved at stages at which most spinal motor neurons have undergone cell death (Gizzi et al. 1992, Kaminski et al. 2002, Mannen 2000, Schroder & Reske-Nielsen 1984).

Mouse mSOD1 models also show almost complete resistance of oculomotor, trochlear, and abducens (Ferrucci et al. 2009) and Onuf's (A. Kaplan and C.E. Henderson, unpublished results) motor neurons to degeneration. This observation provides an opportunity to identify molecular and cellular differences between these pools and to test their roles in disease resistance. Resistant motor neurons expressed higher levels of calcium-buffering proteins such as calbindin-D28K and parvalbumin, which may confer resistance to excitotoxic stimuli (Alexianu et al. 1994, Obal et al. 2006). Overexpression of parvalbumin in the SOD1 mouse model of ALS did show modest benefit (Beers et al. 2001). However, the expression data have been challenged (Laslo et al. 2000), and in more recent studies, expression of calcium-binding proteins has even been correlated with greater vulnerability (Sasaki et al. 2006). Further studies are therefore required to establish the molecular basis of this remarkable degree of natural neuroprotection.

SMA and Motor Unit Aging Show the Same Subtype- and Pool-Specificity as in ALS

Are the stereotyped patterns of susceptibility and resistance seen in ALS patients and mouse models a disease-specific phenomenon, or do they reflect intrinsic differences between motor neuron subtypes that are relevant in general? Selective vulnerability of FF is also seen in acute injury paradigms. In mutant SOD1 mice, ischemia/reperfusion injury of the muscle accelerated end plate denervation in FF but not S motor units (David et al. 2007). Moreover, even in wildtype mice, FF motor neurons were selectively vulnerable to stress caused by axonal injury (Saxena et al. 2009).

Spinal muscular atrophy (SMA) also involves neuromuscular dysfunction and denervation followed by motor neuron death and likely also has a strong non-cell-autonomous component (Mentis et al. 2008, Monani 2005). In spite of clear differences with ALS, striking parallels can be seen in terms of pool and subtype specificity. In muscle biopsies from severe SMA patients, there is widespread atrophy of type II (fast) fibers, whereas type I (slow) fibers show compensatory hypertrophy (Dubowitz 1978). Thus, as in ALS, slow motor units are more resistant. Clinically, children with SMA show a sparing of many facial muscles including those involved in eye movement (Kubota et al. 2000). Moreover, SMA patients do not show symptoms of urethral or anal sphincter disturbances, and correspondingly, the neurons of Onuf's nucleus are preserved in Type 1 patients (Iwata & Hirano 1978, Sung & Mastri 1980).

As healthy mammals age, beginning in midlife a reduction of strength is correlated with a gradual (~5% per annum) loss of motor neurons and muscle fibers (Larsson & Ansved 1995, Vandervoort 2002). Later in life (in humans, at ~60 years of age), there is a precipitous drop in the number of functioning motor units (McComas 1991). This later loss of 30% of motor neurons selectively affects the largest motor neurons, which are presumably of the FF type (Hashizume et al. 1988, Hirofuji et al. 2000). This occurrence is mirrored in the pattern of muscle denervation: In the medial gastrocnemius there is a 34% decrease in the number of FF motor units and a 14% decrease in their size (Kadhiresan et al. 1996, Kanda & Hashizume 1989). In contrast, S motor units are not lost, but their unit size increases threefold, accompanied by fiber type grouping and changes in innervation ratio consistent with compensatory sprouting by S motor neurons (Kanda & Hashizume 1989). Thus, just as in ALS and SMA, the larger FF motor neurons are more susceptible to degeneration as animals age. Another parallel is found in the observation that motor neuron numbers in oculomotor nucleus are stable in aged mice, whereas numbers in the facial nucleus decline significantly (Sturrock 1988, 1991).

The parallels between ALS and other degenerative contexts suggest that some motor units (slow) and some motor pools (oculomotor and other pools innervating extraocular muscle, Onuf's) have intrinsic properties that confer upon them resistance to multiple challenges. Understanding the molecular and cellular basis for such resistance may provide important clues about therapeutic targets.

THERAPEUTIC STRATEGIES FOR PREVENTING MOTOR NEURON DEGENERATION

One major reason for studying mechanisms of motor neuron degeneration is the hope that this work will lead to rational strategies for preventing disease progression. In ALS and SMA, multiple translational initiatives using gene therapy, antisense oligonucleotides, chemical compounds, and cell therapy are currently underway, targeting multiple levels of the disease process (for reviews see Boillee & Cleveland 2004, Nayak et al. 2006, Traynor et al. 2006). We limit our discussion to strategies based on mechanisms discussed elsewhere in the review: (*a*) application of exogenous neurotrophic factors, and (*b*) inhibition of intrinsic mechanisms of cell death and degeneration. We also explore the need to target such strategies to specific motor neuron subtypes.

Neurotrophic Factors as Protective Agents in Motor Neuron Disease

The concept of using natural survival factors as tools to prevent motor neuron death is not a new one: It drove much of the early research on neurotrophic factors (Henderson 1995). More recent data showing that neurotrophic factors can enhance axonal growth and neuronal maturation have further strengthened the rationale. Nevertheless, the difficulty of administering these proteins as drugs and the lack of success with the first clinical trials has diminished enthusiasm for this approach. Recently, however, Turner & Talbot (2008) compiled 226 different experiments testing therapeutic agents in mutant SOD1 mice. Other than SOD1 modifiers, only 6 treatments prolonged survival by more than 30%: four of these were neurotrophic factors and one was an inhibitor of programmed cell death. The same authors analyzed data from a total of 113 crosses with other mouse strains to detect genetic modifiers (Turner & Talbot 2008). Beyond lines that directly modify mSOD1 expression and therefore remove the disease trigger, 11 genetically modified lines showed an increase in survival of 15% or more. Of the latter, three involved overexpression of neurotrophic factors, two targeted programmed cell death, and one was the NogoA knockout discussed above. These data prompt reinvestigation of the potential of neurotrophic and neuroprotective approaches in motor neuron disease.

The greatest effects were obtained using intramuscular viral delivery of factors such as IGF-1 (insulin-like growth factor-1), VEGF, cardiotrophin-1, and GDNF (Acsadi et al. 2002, Azzouz et al. 2004, Bordet et al. 2001, Kaspar et al. 2003). These treatments clearly provide far from complete protection: Even VEGF and IGF did not produce more than a 30% increase in survival. Although this result may reflect selective actions of each factor on a given subset of motor neurons, exogenous neurotrophic factors are administered at levels that should target a large fraction of motor neurons. Another reason for prudence is that a recent trial of IGF-1 in human ALS patients once again failed to show benefit (Howe et al. 2009). However, since the mouse studies were performed, there has been considerable progress in developing viral vectors for delivery of recombinant proteins, although scale-up from mouse to man still represents a challenge.

Inhibition of Cell Death and Degeneration Pathways as a Therapeutic Strategy in ALS

Initial excitement about the use of cell death inhibitors has been tempered by considerations such as the risk of oncogenesis, the difficulty of designing small molecules with sufficient specificity, and the need to protect against other aspects of the disease process such as axonal degeneration (Guegan & Przedborski 2003). Nevertheless, the rationale for pursuing such approaches seems strong. Crossing mSOD1 lines to Bax knockout mice did not provide long-term benefit, leading the authors to cast doubt on the utility of antiapoptotic therapies for ALS (Gould et al. 2006). However, this strategy, as with overexpression of the antiapoptotic Bcl-2 (Kostic et al. 1997), is among the most successful genetic manipulations in mSOD1 mice, other than those which inactivate the disease trigger (Turner & Talbot 2008). Both Bax inactivation and

Bcl-2 overexpression led to a 15% increase in overall survival. Correspondingly, end-stage $Bax^{-/-}$;SOD1^{G93A} mice had double the number of motor neurons of uncrossed mSOD1 controls. Even more strikingly for an antiapoptotic approach, after an early loss of 40% of innervated NMJs, the remaining axonal dieback was significantly slowed in double mutant mice (Gould et al. 2006). Thus it is tempting to speculate that coupling cell death inhibition with agents that can promote axonal stability and increase neuronal size may provide significant functional benefit.

Classical drugs and oligonucleotides can be used to target specific cell death pathways within motor neuron subsets. The Fas-activated pathway involving p38 and caspase-8 is highly activated in 50% of mSOD1 motor neurons in vitro and in vivo (Raoul et al. 2002, 2006). A single icv administration of siRNA to Fas to presymptomatic mSOD1 mice inhibited activation of p38 and caspase-8 and led to a 14% increase in life span (Locatelli et al. 2007). To inhibit subtype-specific activation of ER stress, Saxena et al. (2009) treated a small cohort of mSOD1 mice with salubrinal, a chemical compound that protects against ER stress by blocking dephosphorylation of eukaryotic initiation factor 2a (Boyce et al. 2005). Salubrinal prevented upregulation of ER stress markers in FF motor neurons and extended the life span of SOD G93A mice by 22%. Although salubrinal has deleterious side effects including pancreatic toxicity (Cnop et al. 2007) and therefore may not be an optimal drug candidate, these results suggest that the ER stress pathway in FF motor neurons is a rational therapeutic target in ALS.

Overall, therefore, although strategies in mSOD1 mice focused on death and survival mechanisms by no means constitute a cure, they represent one of the most successful options currently available. A better understanding of the motor neuron subtypes affected by disease and those targeted by each strategy may provide a potential explanation for the partial effects observed and constitute the basis for a rational approach to combination therapy.

SUMMARY POINTS

1. Although they share core functions, motor neurons are not created equal. Multiple subtypes—fast/slow, alpha/gamma—are required for the graded control of muscle contraction. Motor neurons of these classes have significantly different molecular, morphological, and functional properties (**Figure 1; Table 1**).

2. A motor pool is the group of motor neurons that innervate a single muscle. A given motor pool contains a characteristic ratio of fast, slow, alpha, and gamma motor neurons adapted to its function. Although they retain their subtype-specific characteristics, motor neurons within a pool also share certain molecular, morphological, and connectional properties (**Figure 1**).

3. Although many questions about the control of motor neuron numbers during development remain to be addressed, a focus on motor neuron subtypes makes the biology of neurotrophic factors simpler to decipher. GDNF (glial cell line-derived neurotrophic factor) was originally thought to have partial effects on motor neuron survival. We now know that it acts in a pool-specific manner to regulate motor neuron connectivity in a column-specific manner to direct axonal outgrowth and in a subtype-specific manner to support motor neuron survival (**Figure 2**).

4. The mouse mSOD1 model may not be a perfect model of all forms of ALS, but it allows for close analysis of the sequence and nature of cellular interactions over the six-month period of degeneration. There is strong evidence for non-cell-autonomous mechanisms in this model, but motor neuron degeneration remains a core element of the phenotype. Disease-related functional changes in axonal transport and cell death can already be detected in cultured embryonic neurons, whereas the first electrophysiological and behavioral changes in vivo are detected soon after birth. There is increasing evidence for a role of ER stress (endoplasmic reticulum stress) in motor neuron degeneration, and this is linked to axonal dieback from the neuromuscular junction (NMJ).

5. Motor neuron subtypes differ strikingly in the degree to which they are affected by ALS. Fast-fatigable (FF) motor units are the most sensitive and are affected earliest in the disease. In contrast, several subsets are still intact at the disease endpoint. These include slow (S) motor units and motor neurons involved in eye movement and sphincter contraction. A very similar pattern of resistance is seen in spinal muscular atrophy (SMA) and aging and in response to other stressors.

6. There are exciting developments in therapeutic strategies that target the genes with causal mutations in ALS (*SOD1*) and SMA (*SMN*). In parallel, it is important to develop strategies focused on downstream targets. Although no such treatment has been found to rescue mutant SOD1 mice, the most significant benefits reported in mice, to date, involve administration of neurotrophic factors or inhibition of cell death pathways. However, these remain problematic for clinical use.

FUTURE ISSUES

1. In contrast with motor column and pool biology, we know little about the developmental mechanisms underlying the emergence of motor neuron classes and subtypes. This lack of information reflects a need for more molecular markers of motor neuron diversity—fast versus slow, gamma versus alpha—at embryonic and early postnatal stages.

2. Much of the developmental biology of motor neurons stops at birth, before many of the maturational events that remodel the motor system and adapt it to adult function. Because ALS and SMA affect later stages, there is a clear need for markers and mechanisms involved in motor neuron maturation.

3. In spite of intensive study of the process of naturally occurring cell death of motor neurons, much remains to be learned about the nature of the 50% of motor neurons that die, and the molecular pathways involved, and the ways in which cell death is regulated. The multiple, yet selective, actions of GDNF may provide a model for the role of other neurotrophic factors.

4. The cell types involved in extrinsic cellular influences on motor neuron degeneration in mSOD1 mice are well characterized, but how they act is unclear. Definition of the molecular nature of toxic factors that trigger axonal dieback and cell death would be a major step forward.

5. Whatever the success of therapeutic approaches based on the *SOD1* gene, further definition of the molecular and cellular events downstream of the initial triggers is needed if robust therapeutic targets are to be defined for sporadic ALS. New in vitro models based on induced pluripotent stem (iPS) cells constitute a promising approach to this problem. Alternatively, if the molecular basis for the intrinsic resistance of slow, oculomotor and Onuf's motor neurons could be determined, it would likely point to potent new therapeutic targets.

DISCLOSURE STATEMENT

The authors are not aware of any affiliations, memberships, funding, or financial holdings that might be perceived as affecting the objectivity of this review.

ACKNOWLEDGMENTS

We thank T.M. Jessell, R.W. Oppenheim, P. Caroni, C. Milligan, N. Shneider, and S. Przedborski for comments and helpful suggestions on this manuscript. Members of the Motor Neuron Center also influenced our thinking through many enjoyable discussions. The work in our laboratory is supported by the SMA Foundation, the Claire and Leonard Tow Charitable Trust, Project A.L.S., Wings Over Wall Street/MDA, New York State Foundation for Science, Technology, and Innovation (NYSTAR), New York State Department of Health (NYSDOH), and the National Institute of Neurological Disorders and Stroke (NINDS).

LITERATURE CITED

Acsadi G, Anguelov RA, Yang H, Toth G, Thomas R, et al. 2002. Increased survival and function of SOD1 mice after glial cell-derived neurotrophic factor gene therapy. *Hum. Gene Ther.* 13:1047–59

Alexianu ME, Ho BK, Mohamed AH, La Bella V, Smith RG, Appel SH. 1994. The role of calcium-binding proteins in selective motoneuron vulnerability in amyotrophic lateral sclerosis. *Ann. Neurol.* 36:846–58

Askanas V, Engel WK, Alvarez RB. 1992. Strong immunoreactivity of beta-amyloid precursor protein, including the beta-amyloid protein sequence, at human neuromuscular junctions. *Neurosci. Lett.* 143:96–100

Atkin JD, Farg MA, Turner BJ, Tomas D, Lysaght JA, et al. 2006. Induction of the unfolded protein response in familial amyotrophic lateral sclerosis and association of protein-disulfide isomerase with superoxide dismutase 1. *J. Biol. Chem.* 281:30152–65

Atkin JD, Farg MA, Walker AK, McLean C, Tomas D, Horne MK. 2008. Endoplasmic reticulum stress and induction of the unfolded protein response in human sporadic amyotrophic lateral sclerosis. *Neurobiol. Dis.* 30:400–7

Azzouz M, Ralph GS, Storkebaum E, Walmsley LE, Mitrophanous KA, et al. 2004. VEGF delivery with retrogradely transported lentivector prolongs survival in a mouse ALS model. *Nature* 429:413–17

Bakels R, Kernell D. 1993. Matching between motoneurone and muscle unit properties in rat medial gastrocnemius. *J. Physiol.* 463:307–24

Beers DR, Ho BK, Siklos L, Alexianu ME, Mosier DR, et al. 2001. Parvalbumin overexpression alters immune-mediated increases in intracellular calcium, and delays disease onset in a transgenic model of familial amyotrophic lateral sclerosis. *J. Neurochem.* 79:499–509

Bendotti C, Calvaresi N, Chiveri L, Prelle A, Moggio M, et al. 2001. Early vacuolization and mitochondrial damage in motor neurons of FALS mice are not associated with apoptosis or with changes in cytochrome oxidase histochemical reactivity. *J. Neurol. Sci.* 191:25–33

Boillee S, Cleveland DW. 2004. Gene therapy for ALS delivers. *Trends Neurosci.* 27:235–38

Boillee S, Vande Velde C, Cleveland DW. 2006. ALS: a disease of motor neurons and their nonneuronal neighbors. *Neuron* 52:39–59

Bordet T, Lesbordes JC, Rouhani S, Castelnau-Ptakhine L, Schmalbruch H, et al. 2001. Protective effects of cardiotrophin-1 adenoviral gene transfer on neuromuscular degeneration in transgenic ALS mice. *Hum. Mol. Genet.* 10:1925–33

Boyce M, Bryant KF, Jousse C, Long K, Harding HP, et al. 2005. A selective inhibitor of eIF2alpha dephosphorylation protects cells from ER stress. *Science* 307:935–39

Burke RE. 1980. Motor unit types: functional specializations in motor control. *Trends Neurosci.* 3:255–58

Burke RE, Levine DN, Tsairis P, Zajac FE 3rd. 1973. Physiological types and histochemical profiles in motor units of the cat gastrocnemius. *J. Physiol.* 234:723–48

Burke RE, Strick PL, Kanda K, Kim CC, Walmsley B. 1977. Anatomy of medial gastrocnemius and soleus motor nuclei in cat spinal cord. *J. Neurophysiol.* 40:667–80

Buss RR, Gould TW, Ma J, Vinsant S, Prevette D, et al. 2006. Neuromuscular development in the absence of programmed cell death: phenotypic alteration of motoneurons and muscle. *J. Neurosci.* 26:13413–27

Chakkalakal J, Nishimune H, Sanes JR. 2008. The synaptic vesicle protein SV2A selectively marks slow neuromuscular junctions and motor units. Progr. No. 859.1. *2008 Neuroscience Meeting Planner*. Washington, DC: Soc. Neurosci. **http://www.sfn.org/am2008/**

Cheroni C, Marino M, Tortarolo M, Veglianese P, De Biasi S, et al. 2009. Functional alterations of the ubiquitin-proteasome system in motor neurons of a mouse model of familial amyotrophic lateral sclerosis. *Hum. Mol. Genet.* 18:82–96

Chiu AY, Zhai P, Dal Canto MC, Peters TM, Kwon YW, et al. 1995. Age-dependent penetrance of disease in a transgenic mouse model of familial amyotrophic lateral sclerosis. *Mol. Cell Neurosci.* 6:349–62

Cleveland DW, Rothstein JD. 2001. From Charcot to Lou Gehrig: deciphering selective motor neuron death in ALS. *Nat. Rev. Neurosci.* 2:806–19

Cnop M, Ladriere L, Hekerman P, Ortis F, Cardozo AK, et al. 2007. Selective inhibition of eukaryotic translation initiation factor 2 alpha dephosphorylation potentiates fatty acid-induced endoplasmic reticulum stress and causes pancreatic beta-cell dysfunction and apoptosis. *J. Biol. Chem.* 282:3989–97

Cullheim S, Fleshman JW, Glenn LL, Burke RE. 1987. Membrane area and dendritic structure in type-identified triceps surae alpha motoneurons. *J. Comp. Neurol.* 255:68–81

Dalla Torre di Sanguinetto SA, Dasen JS, Arber S. 2008. Transcriptional mechanisms controlling motor neuron diversity and connectivity. *Curr. Opin. Neurobiol.* 18:36–43

Dasen JS, Jessell TM. 2009. Hox networks and the origins of motor neuron diversity. *Curr. Top. Dev. Biol.* 88:169–200

David G, Nguyen K, Barrett EF. 2007. Early vulnerability to ischemia/reperfusion injury in motor terminals innervating fast muscles of SOD1-G93A mice. *Exp. Neurol.* 204:411–20

Davies AM. 1996. The neurotrophic hypothesis: Where does it stand? *Philos. Trans. R. Soc. Lond. B Biol. Sci.* 351:389–94

de Carvalho MA, Pinto S, Swash M. 2008. Paraspinal and limb motor neuron involvement within homologous spinal segments in ALS. *Clin. Neurophysiol.* 119:1607–13

Deckwerth TL, Elliott JL, Knudson CM, Johnson EM Jr, Snider WD, Korsmeyer SJ. 1996. BAX is required for neuronal death after trophic factor deprivation and during development. *Neuron* 17:401–11

Dengler R, Konstanzer A, Kuther G, Hesse S, Wolf W, Struppler A. 1990. Amyotrophic lateral sclerosis: macro-EMG and twitch forces of single motor units. *Muscle Nerve* 13:545–50

Derouet D, Rousseau F, Alfonsi F, Froger J, Hermann J, et al. 2004. Neuropoietin, a new IL-6-related cytokine signaling through the ciliary neurotrophic factor receptor. *Proc. Natl. Acad. Sci. USA* 101:4827–32

De Vos KJ, Chapman AL, Tennant ME, Manser C, Tudor EL, et al. 2007. Familial amyotrophic lateral sclerosis-linked SOD1 mutants perturb fast axonal transport to reduce axonal mitochondria content. *Hum. Mol. Genet.* 16:2720–28

De Winter F, Vo T, Stam FJ, Wisman LA, Bar PR, et al. 2006. The expression of the chemorepellent Semaphorin 3A is selectively induced in terminal Schwann cells of a subset of neuromuscular synapses that display limited anatomical plasticity and enhanced vulnerability in motor neuron disease. *Mol. Cell Neurosci.* 32:102–17

Dobretsov M, Hastings SL, Sims TJ, Stimers JR, Romanovsky D. 2003. Stretch receptor-associated expression of alpha 3 isoform of the Na$^+$, K$^+$-ATPase in rat peripheral nervous system. *Neuroscience* 116:1069–80

Dubowitz V. 1978. Muscle disorders in childhood. *Major Probl. Clin. Pediatr.* 16:iii–xiii, 1–282

Duchen LW, Tonge DA. 1973. The effects of tetanus toxin on neuromuscular transmission and on the morphology of motor end-plates in slow and fast skeletal muscle of the mouse. *J. Physiol.* 228:157–72

Duplan L, Bernard N, Casseron WD, Dudley K, Thouvenot E, et al. 2010. Collapsin response mediator protein 4a (CRMP4a) is upregulated in motoneurons of mutant SOD1 mice and can trigger motoneuron death. *J. Neurosci.* 30:785–96

Eccles JC, Eccles RM, Iggo A, Lundberg A. 1960. Electrophysiological studies on gamma motoneurones. *Acta Physiol. Scand.* 50:32–40

Eccles JC, Eccles RM, Lundberg A. 1957. Durations of after-hyperpolarization of motoneurones supplying fast and slow muscles. *Nature* 179:866–68

Elson GC, Lelievre E, Guillet C, Chevalier S, Plun-Favreau H, et al. 2000. CLF associates with CLC to form a functional heteromeric ligand for the CNTF receptor complex. *Nat. Neurosci.* 3:867–72

Ferrucci M, Spalloni A, Bartalucci A, Cantafora E, Fulceri F, et al. 2010. A systematic study of brainstem motor nuclei in a mouse model of ALS, the effects of lithium. *Neurobiol. Dis.* 37:370–83

Fischer LR, Culver DG, Tennant P, Davis AA, Wang M, et al. 2004. Amyotrophic lateral sclerosis is a distal axonopathy: evidence in mice and man. *Exp. Neurol.* 185:232–40

Forger NG, Prevette D, deLapeyriere O, de Bovis B, Wang S, et al. 2003. Cardiotrophin-like cytokine/cytokine-like factor 1 is an essential trophic factor for lumbar and facial motoneurons in vivo. *J. Neurosci.* 23:8854–58

Frey D, Schneider C, Xu L, Borg J, Spooren W, Caroni P. 2000. Early and selective loss of neuromuscular synapse subtypes with low sprouting competence in motoneuron diseases. *J. Neurosci.* 20:2534–42

Friese A, Kaltschmidt JA, Ladle DR, Sigrist M, Jessell TM, Arber S. 2009. Gamma and alpha motor neurons distinguished by expression of transcription factor Err3. *Proc. Natl. Acad. Sci. USA* 106:13588–93

Galluzzi L, Blomgren K, Kroemer G. 2009. Mitochondrial membrane permeabilization in neuronal injury. *Nat. Rev. Neurosci.* 10:481–94

Garces A, Haase G, Airaksinen MS, Livet J, Filippi P, deLapeyriere O. 2000. GFRalpha 1 is required for development of distinct subpopulations of motoneuron. *J. Neurosci.* 20:4992–5000

Gardiner PF. 1993. Physiological properties of motoneurons innervating different muscle unit types in rat gastrocnemius. *J. Neurophysiol.* 69:1160–70

Gizzi M, DiRocco A, Sivak M, Cohen B. 1992. Ocular motor function in motor neuron disease. *Neurology* 42:1037–46

Gordon DC, Loeb GE, Richmond FJ. 1991. Distribution of motoneurons supplying cat sartorius and tensor fasciae latae, demonstrated by retrograde multiple-labeling methods. *J. Comp. Neurol.* 304:357–72

Gordon T, Tyreman N, Li S, Putman CT, Hegedus J. 2010. Functional overload saves motor units in the SOD1-G93A transgenic mouse model of amyotrophic lateral sclerosis. *Neurobiol. Dis.* 37:412–22

Gould TW, Buss RR, Vinsant S, Prevette D, Sun W, et al. 2006. Complete dissociation of motor neuron death from motor dysfunction by Bax deletion in a mouse model of ALS. *J. Neurosci.* 26:8774–86

Gould TW, Enomoto H. 2009. Neurotrophic modulation of motor neuron development. *Neuroscientist* 15:105–16

Gould TW, Oppenheim RW. 2004. The function of neurotrophic factor receptors expressed by the developing adductor motor pool in vivo. *J. Neurosci.* 24:4668–82

Gould TW, Yonemura S, Oppenheim RW, Ohmori S, Enomoto H. 2008. The neurotrophic effects of glial cell line-derived neurotrophic factor on spinal motoneurons are restricted to fusimotor subtypes. *J. Neurosci.* 28:2131–46

Guegan C, Przedborski S. 2003. Programmed cell death in amyotrophic lateral sclerosis. *J. Clin. Invest.* 111:153–61

Haase G, Dessaud E, Garces A, de Bovis B, Birling M, et al. 2002. GDNF acts through PEA3 to regulate cell body positioning and muscle innervation of specific motor neuron pools. *Neuron* 35:893–905

Haase G, Pettmann B, Raoul C, Henderson CE. 2008. Signaling by death receptors in the nervous system. *Curr. Opin. Neurobiol.* 18:284–91

Hashizume K, Kanda K, Burke RE. 1988. Medial gastrocnemius motor nucleus in the rat: age-related changes in the number and size of motoneurons. *J. Comp. Neurol.* 269:425–30

Heckman CJ, Johnson M, Mottram C, Schuster J. 2008. Persistent inward currents in spinal motoneurons and their influence on human motoneuron firing patterns. *Neuroscientist* 14:264–75

Hegedus J, Putman CT, Gordon T. 2007. Time course of preferential motor unit loss in the SOD1 G93A mouse model of amyotrophic lateral sclerosis. *Neurobiol. Dis.* 28:154–64

Hegedus J, Putman CT, Tyreman N, Gordon T. 2008. Preferential motor unit loss in the SOD1 G93A transgenic mouse model of amyotrophic lateral sclerosis. *J. Physiol.* 586:3337–51

Helmbacher F, Dessaud E, Arber S, deLapeyriere O, Henderson CE, et al. 2003. Met signaling is required for recruitment of motor neurons to PEA3-positive motor pools. *Neuron* 39:767–77

Helmbacher F, Schneider-Maunoury S, Topilko P, Tiret L, Charnay P. 2000. Targeting of the EphA4 tyrosine kinase receptor affects dorsal/ventral pathfinding of limb motor axons. *Development* 127:3313–24

Henderson CE. 1995. Neurotrophic factors as therapeutic agents in amyotrophic lateral sclerosis. Potential and pitfalls. *Adv. Neurol.* 68:235–40

Henderson CE. 1996. Role of neurotrophic factors in neuronal development. *Curr. Opin. Neurobiol.* 6:64–70

Henderson CE, Phillips HS, Pollock RA, Davies AM, Lemeulle C, et al. 1994. GDNF: a potent survival factor for motoneurons present in peripheral nerve and muscle. *Science* 266:1062–64

Hirofuji C, Ishihara A, Roy RR, Itoh K, Itoh M, et al. 2000. SDH activity and cell size of tibialis anterior motoneurons and muscle fibers in SAMP6. *Neuroreport* 11:823–28

Holtmann B, Wiese S, Samsam M, Grohmann K, Pennica D, et al. 2005. Triple knock-out of CNTF, LIF, and CT-1 defines cooperative and distinct roles of these neurotrophic factors for motoneuron maintenance and function. *J. Neurosci.* 25:1778–87

Hoover JE, Durkovic RG. 1991. Morphological relationships among extensor digitorum longus, tibialis anterior, and semitendinosus motor nuclei of the cat: an investigation employing the retrograde transport of multiple fluorescent tracers. *J. Comp. Neurol.* 303:255–66

Horcholle-Bossavit G, Jami L, Thiesson D, Zytnicki D. 1990. Postnatal development of peroneal motoneurons in the kitten. *Brain Res. Dev. Brain Res.* 54:205–15

Howe CL, Bergstrom RA, Horazdovsky BF. 2009. Subcutaneous IGF-1 is not beneficial in 2-year ALS trial. *Neurology* 73:1247; author reply p. 1248

Hui K, Kucera J, Henderson JT. 2008. Differential sensitivity of skeletal and fusimotor neurons to Bcl-2-mediated apoptosis during neuromuscular development. *Cell Death Differ.* 15:691–99

Hunt CC, Kuffler SW. 1951. Further study of efferent small-nerve fibers to mammalian muscle spindles; multiple spindle innervation and activity during contraction. *J. Physiol.* 113:283–97

Iwata M, Hirano A. 1978. Sparing of the Onufrowicz nucleus in sacral anterior horn lesions. *Ann. Neurol.* 4:245–49

Jokic N, Gonzalez de Aguilar JL, Dimou L, Lin S, Fergani A, et al. 2006. The neurite outgrowth inhibitor Nogo-A promotes denervation in an amyotrophic lateral sclerosis model. *EMBO Rep.* 7:1162–67

Jokic N, Gonzalez de Aguilar JL, Pradat PF, Dupuis L, Echaniz-Laguna A, et al. 2005. Nogo expression in muscle correlates with amyotrophic lateral sclerosis severity. *Ann. Neurol.* 57:553–56

Kadhiresan VA, Hassett CA, Faulkner JA. 1996. Properties of single motor units in medial gastrocnemius muscles of adult and old rats. *J. Physiol.* 493(Pt. 2):543–52

Kaminski HJ, Richmonds CR, Kusner LL, Mitsumoto H. 2002. Differential susceptibility of the ocular motor system to disease. *Ann. N. Y. Acad. Sci.* 956:42–54

Kanda K, Hashizume K. 1989. Changes in properties of the medial gastrocnemius motor units in aging rats. *J. Neurophysiol.* 61:737–46

Kania A, Jessell TM. 2003. Topographic motor projections in the limb imposed by LIM homeodomain protein regulation of ephrin-A:EphA interactions. *Neuron* 38:581–96

Kanungo AK, Hao Z, Elia AJ, Mak TW, Henderson JT. 2008. Inhibition of apoptosome activation protects injured motor neurons from cell death. *J. Biol. Chem.* 283:22105–12

Kaspar BK, Llado J, Sherkat N, Rothstein JD, Gage FH. 2003. Retrograde viral delivery of IGF-1 prolongs survival in a mouse ALS model. *Science* 301:839–42

Kato S. 2008. Amyotrophic lateral sclerosis models and human neuropathology: similarities and differences. *Acta Neuropathol.* 115:97–114

Kato T, Kurita K, Seino T, Kadoya T, Horie H, et al. 2001. Galectin-1 is a component of neurofilamentous lesions in sporadic and familial amyotrophic lateral sclerosis. *Biochem. Biophys. Res. Commun.* 282:166–72

Keller-Peck CR, Feng G, Sanes JR, Yan Q, Lichtman JW, Snider WD. 2001. Glial cell line-derived neurotrophic factor administration in postnatal life results in motor unit enlargement and continuous synaptic remodeling at the neuromuscular junction. *J. Neurosci.* 21:6136–46

Kernell D. 2003. Principles of force gradation in skeletal muscles. *Neural. Plast.* 10:69–76

Kernell D, Bakels R, Copray JC. 1999. Discharge properties of motoneurones: How are they matched to the properties and use of their muscle units? *J. Physiol. Paris* 93:87–96

Kieran D, Hafezparast M, Bohnert S, Dick JR, Martin J, et al. 2005. A mutation in dynein rescues axonal transport defects and extends the life span of ALS mice. *J. Cell Biol.* 169:561–67

Kieran D, Sebastia J, Greenway MJ, King MA, Connaughton D, et al. 2008. Control of motoneuron survival by angiogenin. *J. Neurosci.* 28:14056–61

Kieran D, Woods I, Villunger A, Strasser A, Prehn JH. 2007. Deletion of the BH3-only protein puma protects motoneurons from ER stress-induced apoptosis and delays motoneuron loss in ALS mice. *Proc. Natl. Acad. Sci. USA* 104:20606–11

Kikuchi H, Almer G, Yamashita S, Guegan C, Nagai M, et al. 2006. Spinal cord endoplasmic reticulum stress associated with a microsomal accumulation of mutant superoxide dismutase-1 in an ALS model. *Proc. Natl. Acad. Sci. USA* 103:6025–30

Koistinen H, Prinjha R, Soden P, Harper A, Banner SJ, et al. 2006. Elevated levels of amyloid precursor protein in muscle of patients with amyotrophic lateral sclerosis and a mouse model of the disease. *Muscle Nerve* 34:444–50

Kong J, Xu Z. 1998. Massive mitochondrial degeneration in motor neurons triggers the onset of amyotrophic lateral sclerosis in mice expressing a mutant SOD1. *J. Neurosci.* 18:3241–50

Kostic V, Jackson-Lewis V, de Bilbao F, Dubois-Dauphin M, Przedborski S. 1997. Bcl-2: prolonging life in a transgenic mouse model of familial amyotrophic lateral sclerosis. *Science* 277:559–62

Kramer ER, Knott L, Su F, Dessaud E, Krull CE, et al. 2006. Cooperation between GDNF/Ret and ephrinA/EphA4 signals for motor-axon pathway selection in the limb. *Neuron* 50:35–47

Kubota M, Sakakihara Y, Uchiyama Y, Nara A, Nagata T, et al. 2000. New ocular movement detector system as a communication tool in ventilator-assisted Werdnig-Hoffmann disease. *Dev. Med. Child Neurol.* 42:61–64

Kuffler SW, Hunt CC, Quilliam JP. 1951. Function of medullated small-nerve fibers in mammalian ventral roots; efferent muscle spindle innervation. *J. Neurophysiol.* 14:29–54

Kuo JJ, Schonewille M, Siddique T, Schults AN, Fu R, et al. 2004. Hyperexcitability of cultured spinal motoneurons from presymptomatic ALS mice. *J. Neurophysiol.* 91:571–75

Kuo JJ, Siddique T, Fu R, Heckman CJ. 2005. Increased persistent Na(+) current and its effect on excitability in motoneurones cultured from mutant SOD1 mice. *J. Physiol.* 563:843–54

Landmesser L. 1978. The distribution of motoneurones supplying chick hind limb muscles. *J. Physiol.* 284:371–89

Larsson L, Ansved T. 1995. Effects of ageing on the motor unit. *Prog. Neurobiol.* 45:397–458

Laslo P, Lipski J, Nicholson LF, Miles GB, Funk GD. 2000. Calcium binding proteins in motoneurons at low and high risk for degeneration in ALS. *Neuroreport* 11:3305–8

Lee RH, Heckman CJ. 1998a. Bistability in spinal motoneurons in vivo: systematic variations in persistent inward currents. *J. Neurophysiol.* 80:583–93

Lee RH, Heckman CJ. 1998b. Bistability in spinal motoneurons in vivo: systematic variations in rhythmic firing patterns. *J. Neurophysiol.* 80:572–82

Ligon LA, LaMonte BH, Wallace KE, Weber N, Kalb RG, Holzbaur EL. 2005. Mutant superoxide dismutase disrupts cytoplasmic dynein in motor neurons. *Neuroreport* 16:533–36

Lin JH, Saito T, Anderson DJ, Lance-Jones C, Jessell TM, Arber S. 1998. Functionally related motor neuron pool and muscle sensory afferent subtypes defined by coordinate ETS gene expression. *Cell* 95:393–407

Livet J, Sigrist M, Stroebel S, De Paola V, Price SR, et al. 2002. ETS gene Pea3 controls the central position and terminal arborization of specific motor neuron pools. *Neuron* 35:877–92

Locatelli F, Corti S, Papadimitriou D, Fortunato F, Del Bo R, et al. 2007. Fas small interfering RNA reduces motoneuron death in amyotrophic lateral sclerosis mice. *Ann. Neurol.* 62:81–92

Mandai K, Guo T, St Hillaire C, Meabon JS, Kanning KC, et al. 2009. LIG family receptor tyrosine kinase-associated proteins modulate growth factor signals during neural development. *Neuron* 63:614–27

Mannen T. 2000. Neuropathological findings of Onuf's nucleus and its significance. *Neuropathology* 20(Suppl.):S30–33

Martin LJ, Liu Z, Chen K, Price AC, Pan Y, et al. 2007. Motor neuron degeneration in amyotrophic lateral sclerosis mutant superoxide dismutase-1 transgenic mice: mechanisms of mitochondriopathy and cell death. *J. Comp. Neurol.* 500:20–46

McComas AJ. 1991. Invited review: motor unit estimation: methods, results, and present status. *Muscle Nerve* 14:585–97

Mendell LM. 2005. The size principle: a rule describing the recruitment of motoneurons. *J. Neurophysiol.* 93:3024–26

Mentis GZ, Sumner CJ, O'Donovan MJ. 2008. Altered synaptic input and excitability of motor neurons in SMA mice. Progr. No. 643.22. *Neuroscience Meeting Planner.* Washington, DC: Soc. Neurosci. **http://www.sfn.org/am2008/**

Mi R, Chen W, Hoke A. 2007. Pleiotrophin is a neurotrophic factor for spinal motor neurons. *Proc. Natl. Acad. Sci. USA* 104:4664–69

Mitsumoto H, Przedborski S, Gordon PH. 2006. *Amyotrophic Lateral Sclerosis.* Boca Raton, FL: Taylor & Francis. xxvi, 830 pp.

Monani UR. 2005. Spinal muscular atrophy: a deficiency in a ubiquitous protein; a motor neuron-specific disease. *Neuron* 48:885–96

Moore MW, Klein RD, Farinas I, Sauer H, Armanini M, et al. 1996. Renal and neuronal abnormalities in mice lacking GDNF. *Nature* 382:76–79

Mourelatos Z, Gonatas NK, Stieber A, Gurney ME, Dal Canto MC. 1996. The Golgi apparatus of spinal cord motor neurons in transgenic mice expressing mutant Cu,Zn superoxide dismutase becomes fragmented in early, preclinical stages of the disease. *Proc. Natl. Acad. Sci. USA* 93:5472–77

Nayak MS, Kim YS, Goldman M, Keirstead HS, Kerr DA. 2006. Cellular therapies in motor neuron diseases. *Biochim. Biophys. Acta* 1762:1128–38

Nguyen QT, Parsadanian AS, Snider WD, Lichtman JW. 1998. Hyperinnervation of neuromuscular junctions caused by GDNF overexpression in muscle. *Science* 279:1725–29

Nikolaev A, McLaughlin T, O'Leary DD, Tessier-Lavigne M. 2009. APP binds DR6 to trigger axon pruning and neuron death via distinct caspases. *Nature* 457:981–89

Nishitoh H, Kadowaki H, Nagai A, Maruyama T, Yokota T, et al. 2008. ALS-linked mutant SOD1 induces ER stress- and ASK1-dependent motor neuron death by targeting Derlin-1. *Genes Dev.* 22:1451–64

Obal I, Engelhardt JI, Siklos L. 2006. Axotomy induces contrasting changes in calcium and calcium-binding proteins in oculomotor and hypoglossal nuclei of Balb/c mice. *J. Comp. Neurol.* 499:17–32

Oppenheim RW. 1991. Cell death during development of the nervous system. *Annu. Rev. Neurosci.* 14:453–501

Oppenheim RW. 1996. Neurotrophic survival molecules for motoneurons: an embarrassment of riches. *Neuron* 17:195–97

Oppenheim RW, Blomgren K, Ethell DW, Koike M, Komatsu M, et al. 2008. Developing postmitotic mammalian neurons in vivo lacking Apaf-1 undergo programmed cell death by a caspase-independent, non-apoptotic pathway involving autophagy. *J. Neurosci.* 28:1490–97

Oppenheim RW, Cole T, Prevette D. 1989. Early regional variations in motoneuron numbers arise by differential proliferation in the chick embryo spinal cord. *Dev. Biol.* 133:468–74

Oppenheim RW, Flavell RA, Vinsant S, Prevette D, Kuan CY, Rakic P. 2001. Programmed cell death of developing mammalian neurons after genetic deletion of caspases. *J. Neurosci.* 21:4752–60

Oppenheim RW, Houenou LJ, Parsadanian AS, Prevette D, Snider WD, Shen L. 2000. Glial cell line-derived neurotrophic factor and developing mammalian motoneurons: regulation of programmed cell death among motoneuron subtypes. *J. Neurosci.* 20:5001–11

Papadimitriou D, Le Verche V, Jacquier A, Ikiz B, Przedborski S, Re DB. 2010. Inflammation in ALS and SMA: sorting out the good from the evil. *Neurobiol. Dis.* 37:493–502

Pasinelli P, Brown RH. 2006. Molecular biology of amyotrophic lateral sclerosis: insights from genetics. *Nat. Rev. Neurosci.* 7:710–23

Pasinelli P, Houseweart MK, Brown RH Jr, Cleveland DW. 2000. Caspase-1 and -3 are sequentially activated in motor neuron death in Cu,Zn superoxide dismutase-mediated familial amyotrophic lateral sclerosis. *Proc. Natl. Acad. Sci. USA* 97:13901–6

Pecho-Vrieseling E, Sigrist M, Yoshida Y, Jessell TM, Arber S. 2009. Specificity of sensory-motor connections encoded by Sema3e-Plxnd1 recognition. *Nature* 459:842–46

Pennica D, Arce V, Swanson TA, Vejsada R, Pollock RA, et al. 1996. Cardiotrophin-1, a cytokine present in embryonic muscle, supports long-term survival of spinal motoneurons. *Neuron* 17:63–74

Perlson E, Jeong GB, Ross JL, Dixit R, Wallace KE, et al. 2009. A switch in retrograde signaling from survival to stress in rapid-onset neurodegeneration. *J. Neurosci.* 29:9903–17

Pettmann B, Henderson CE. 1998. Neuronal cell death. *Neuron* 20:633–47

Piehl F, Arvidsson U, Hokfelt T, Cullheim S. 1993. Calcitonin gene-related peptide-like immunoreactivity in motoneuron pools innervating different hind limb muscles in the rat. *Exp. Brain Res.* 96:291–303

Plachta N, Annaheim C, Bissiere S, Lin S, Ruegg M, et al. 2007. Identification of a lectin causing the degeneration of neuronal processes using engineered embryonic stem cells. *Nat. Neurosci.* 10:712–19

Pun S, Santos AF, Saxena S, Xu L, Caroni P. 2006. Selective vulnerability and pruning of phasic motoneuron axons in motoneuron disease alleviated by CNTF. *Nat. Neurosci.* 9:408–19

Radunovic A, Mitsumoto H, Leigh PN. 2007. Clinical care of patients with amyotrophic lateral sclerosis. *Lancet Neurol.* 6:913–25

Rafuse VF, Milner LD, Landmesser LT. 1996. Selective innervation of fast and slow muscle regions during early chick neuromuscular development. *J. Neurosci.* 16:6864–77

Raoul C, Buhler E, Sadeghi C, Jacquier A, Aebischer P, et al. 2006. Chronic activation in presymptomatic amyotrophic lateral sclerosis (ALS) mice of a feedback loop involving Fas, Daxx, and FasL. *Proc. Natl. Acad. Sci. USA* 103:6007–12

Raoul C, Estevez AG, Nishimune H, Cleveland DW, deLapeyriere O, et al. 2002. Motoneuron death triggered by a specific pathway downstream of Fas. potentiation by ALS-linked SOD1 mutations. *Neuron* 35:1067–83

Rekling JC, Funk GD, Bayliss DA, Dong XW, Feldman JL. 2000. Synaptic control of motoneuronal excitability. *Physiol. Rev.* 80:767–852

Romanes GJ. 1951. The motor cell columns of the lumbo-sacral spinal cord of the cat. *J. Comp. Neurol.* 94:313–63

Saito M, Tomonaga M, Narabayashi H. 1978. Histochemical study of the muscle spindles in parkinsonism, motor neuron disease and myasthenia. An examination of the pathological fusimotor endings by the acetylcholinesterase technic. *J. Neurol.* 219:261–71

Sasaki S, Warita H, Komori T, Murakami T, Abe K, Iwata M. 2006. Parvalbumin and calbindin D-28k immunoreactivity in transgenic mice with a G93A mutant SOD1 gene. *Brain Res.* 1083:196–203

Saxena S, Cabuy E, Caroni P. 2009. A role for motoneuron subtype-selective ER stress in disease manifestations of FALS mice. *Nat. Neurosci.* 12:627–36

Schaefer AM, Sanes JR, Lichtman JW. 2005. A compensatory subpopulation of motor neurons in a mouse model of amyotrophic lateral sclerosis. *J. Comp. Neurol.* 490:209–19

Schroder HD, Reske-Nielsen E. 1984. Preservation of the nucleus X-pelvic floor motosystem in amyotrophic lateral sclerosis. *Clin. Neuropathol.* 3:210–16

Shneider NA, Brown MN, Smith CA, Pickel J, Alvarez FJ. 2009. Gamma motor neurons express distinct genetic markers at birth and require muscle spindle-derived GDNF for postnatal survival. *Neural Dev.* 4:42

Simon M, Destombes J, Horcholle-Bossavit G, Thiesson D. 1996. Postnatal development of alpha- and gamma-peroneal motoneurons in kittens: an ultrastructural study. *Neurosci Res.* 25:77–89

Strelau J, Strzelczyk A, Rusu P, Bendner G, Wiese S, et al. 2009. Progressive postnatal motoneuron loss in mice lacking GDF-15. *J. Neurosci.* 29:13640–48

Sturrock RR. 1988. Loss of neurons from the motor nucleus of the facial nerve in the ageing mouse brain. *J. Anat.* 160:189–94

Sturrock RR. 1991. Stability of motor neuron number in the oculomotor and trochlear nuclei of the ageing mouse brain. *J. Anat.* 174:125–29

Sumi H, Nagano S, Fujimura H, Kato S, Sakoda S. 2006. Inverse correlation between the formation of mitochondria-derived vacuoles and Lewy-body-like hyaline inclusions in G93A superoxide-dismutase-transgenic mice. *Acta Neuropathol.* 112:52–63

Sun W, Gould TW, Vinsant S, Prevette D, Oppenheim RW. 2003. Neuromuscular development after the prevention of naturally occurring neuronal death by Bax deletion. *J. Neurosci.* 23:7298–310

Sung JH, Mastri AR. 1980. Spinal autonomic neurons in Werdnig-Hoffmann disease, mannosidosis, and Hurler's syndrome: distribution of autonomic neurons in the sacral spinal cord. *J. Neuropathol. Exp. Neurol.* 39:441–51

Swash M, Fox KP. 1974. The pathology of the human muscle spindle: effect of denervation. *J. Neurol. Sci.* 22:1–24

Swash M, Leader M, Brown A, Swettenham KW. 1986. Focal loss of anterior horn cells in the cervical cord in motor neuron disease. *Brain* 109(Pt. 5):939–52

Traynor BJ, Bruijn L, Conwit R, Beal F, O'Neill G, et al. 2006. Neuroprotective agents for clinical trials in ALS: a systematic assessment. *Neurology* 67:20–27

Tu PH, Raju P, Robinson KA, Gurney ME, Trojanowski JQ, Lee VM. 1996. Transgenic mice carrying a human mutant superoxide dismutase transgene develop neuronal cytoskeletal pathology resembling human amyotrophic lateral sclerosis lesions. *Proc. Natl. Acad. Sci. USA* 93:3155–60

Turner BJ, Talbot K. 2008. Transgenics, toxicity and therapeutics in rodent models of mutant SOD1-mediated familial ALS. *Prog. Neurobiol.* 85:94–134

Valdmanis PN, Rouleau GA. 2008. Genetics of familial amyotrophic lateral sclerosis. *Neurology* 70:144–52

Van Damme P, Van Hoecke A, Lambrechts D, Vanacker P, Bogaert E, et al. 2008. Progranulin functions as a neurotrophic factor to regulate neurite outgrowth and enhance neuronal survival. *J. Cell Biol.* 181:37–41

Van Den Bosch L, Storkebaum E, Vleminckx V, Moons L, Vanopdenbosch L, et al. 2004. Effects of vascular endothelial growth factor (VEGF) on motor neuron degeneration. *Neurobiol. Dis.* 17:21–28

Van Den Bosch L, Van Damme P, Bogaert E, Robberecht W. 2006. The role of excitotoxicity in the pathogenesis of amyotrophic lateral sclerosis. *Biochim. Biophys. Acta* 1762:1068–82

Vanderhorst VG, Holstege G. 1997. Organization of lumbosacral motoneuronal cell groups innervating hindlimb, pelvic floor, and axial muscles in the cat. *J. Comp. Neurol.* 382:46–76

Vandervoort AA. 2002. Aging of the human neuromuscular system. *Muscle Nerve* 25:17–25

van Zundert B, Peuscher MH, Hynynen M, Chen A, Neve RL, et al. 2008. Neonatal neuronal circuitry shows hyperexcitable disturbance in a mouse model of the adult-onset neurodegenerative disease amyotrophic lateral sclerosis. *J. Neurosci.* 28:10864–74

Vejsada R, Tseng JL, Lindsay RM, Acheson A, Aebischer P, Kato AC. 1998. Synergistic but transient rescue effects of BDNF and GDNF on axotomized neonatal motoneurons. *Neuroscience* 84:129–39

Veldink JH, Bar PR, Joosten EA, Otten M, Wokke JH, van den Berg LH. 2003. Sexual differences in onset of disease and response to exercise in a transgenic model of ALS. *Neuromuscul. Disord.* 13:737–43

Vlug AS, Teuling E, Haasdijk ED, French P, Hoogenraad CC, Jaarsma D. 2005. ATF3 expression precedes death of spinal motoneurons in amyotrophic lateral sclerosis-SOD1 transgenic mice and correlates with c-Jun phosphorylation, CHOP expression, somato-dendritic ubiquitination and Golgi fragmentation. *Eur. J. Neurosci.* 22:1881–94

Vrieseling E, Arber S. 2006. Target-induced transcriptional control of dendritic patterning and connectivity in motor neurons by the ETS gene Pea3. *Cell* 127:1439–52

Vukosavic S, Stefanis L, Jackson-Lewis V, Guegan C, Romero N, et al. 2000. Delaying caspase activation by Bcl-2: a clue to disease retardation in a transgenic mouse model of amyotrophic lateral sclerosis. *J. Neurosci.* 20:9119–25

Wang P, Yang G, Mosier DR, Chang P, Zaidi T, et al. 2005a. Defective neuromuscular synapses in mice lacking amyloid precursor protein (APP) and APP-Like protein 2. *J. Neurosci.* 25:1219–25

Wang PY, Koishi K, McGeachie AB, Kimber M, Maclaughlin DT, et al. 2005b. Mullerian inhibiting substance acts as a motor neuron survival factor in vitro. *Proc. Natl. Acad. Sci. USA* 102:16421–25

Warita H, Manabe Y, Murakami T, Shiote M, Shiro Y, et al. 2002. Tardive decrease of astrocytic glutamate transporter protein in transgenic mice with ALS-linked mutant SOD1. *Neurol. Res.* 24:577–81

Westbury DR. 1982. A comparison of the structures of alpha and gamma-spinal motoneurones of the cat. *J. Physiol.* 325:79–91

Whitehead J, Keller-Peck C, Kucera J, Tourtellotte WG. 2005. Glial cell-line derived neurotrophic factor-dependent fusimotor neuron survival during development. *Mech. Dev.* 122:27–41

Williams AH, Valdez G, Moresi V, Qi X, McAnally J, et al. 2009. MicroRNA-206 delays ALS progression and promotes regeneration of neuromuscular synapses in mice. *Science* 326:1549–54

Wong PC, Cai H, Borchelt DR, Price DL. 2002. Genetically engineered mouse models of neurodegenerative diseases. *Nat. Neurosci.* 5:633–39

Wooley CM, Sher RB, Kale A, Frankel WN, Cox GA, Seburn KL. 2005. Gait analysis detects early changes in transgenic SOD1(G93A) mice. *Muscle Nerve* 32:43–50

Wootz H, Hansson I, Korhonen L, Napankangas U, Lindholm D. 2004. Caspase-12 cleavage and increased oxidative stress during motoneuron degeneration in transgenic mouse model of ALS. *Biochem. Biophys. Res. Commun.* 322:281–86

Wright DE, Snider WD. 1996. Focal expression of glial cell line-derived neurotrophic factor in developing mouse limb bud. *Cell Tissue Res.* 286:209–17

Yang YS, Harel NY, Strittmatter SM. 2009. Reticulon-4A (Nogo-A) redistributes protein disulfide isomerase to protect mice from SOD1-dependent amyotrophic lateral sclerosis. *J. Neurosci.* 29:13850–59

Yoshikawa M, Vinsant S, Mansfield CM, Moreno RJ, Gifondorwa DJ, et al. 2009. Identification of changes in muscle, neuromuscular junctions and spinal cord at early presymptomatic stages in the mutant SOD1 mouse model of ALS may provide novel insight for diagnosis and treatment development. Progr. No. 632.5. *2009 Neuroscience Meeting Planner.* Chicago, IL: Soc. Neurosci. **http://www.sfn.org/am2009/**

Zhang B, Tu P, Abtahian F, Trojanowski JQ, Lee VM. 1997. Neurofilaments and orthograde transport are reduced in ventral root axons of transgenic mice that express human SOD1 with a G93A mutation. *J. Cell Biol.* 139:1307–15

Zhong Z, Deane R, Ali Z, Parisi M, Shapovalov Y, et al. 2008. ALS-causing SOD1 mutants generate vascular changes prior to motor neuron degeneration. *Nat. Neurosci.* 11:420–22

The Genomic, Biochemical, and Cellular Responses of the Retina in Inherited Photoreceptor Degenerations and Prospects for the Treatment of These Disorders

Alexa N. Bramall,[1,2] Alan F. Wright,[4]
Samuel G. Jacobson,[5] and Roderick R. McInnes[1,2,3]

[1] Programs in Genetics and Developmental Biology, The Research Institute, The Hospital for Sick Children, Toronto, Canada M5G 1L7; email: alexa.bramall@utoronto.ca

[2] Department of Molecular Genetics, University of Toronto, Toronto, Canada M5S 1A1

[3] Lady Davis Institute, Jewish General Hospital, McGill University, Montreal, Canada H3T 1E2; email: rod.mcinnes@mcgill.ca

[4] MRC Human Genetics Unit, Institute of Genetics and Molecular Medicine, Edinburgh, United Kingdom EH4 2XU; email: alan.wright@hgu.mrc.ac.uk

[5] Department of Ophthalmology, Scheie Eye Institute, University of Pennsylvania, Philadelphia 19104; email: jacobsos@mail.med.upenn.edu

Annu. Rev. Neurosci. 2010. 33:441–72

The *Annual Review of Neuroscience* is online at neuro.annualreviews.org

This article's doi:
10.1146/annurev-neuro-060909-153227

Key Words

retinal degeneration, pathogenesis of photoreceptor death, genomic, biochemical, and cellular responses to mutation, human histology, gene therapy

Abstract

The association of more than 140 genes with human photoreceptor degenerations, together with studies of animal models of these monogenic diseases, has provided great insight into their pathogenesis. Here we review the responses of the retina to photoreceptor mutations, including mechanisms of photoreceptor death. We discuss the roles of oxidative metabolism, mitochondrial reactive oxygen species, metabolic stress, protein misfolding, and defects in ciliary proteins, as well as the responses of Müller glia, microglia, and the retinal vasculature. Finally, we report on potential pharmacologic and biologic therapies, the critical role of histopathology as a prerequisite to treatment, and the exciting promise of gene therapy in animal models and in phase 1 trials in humans.

Contents

INTRODUCTION

Over the past two decades, 148 genes and 186 loci have been associated with inherited photoreceptor degenerations (IPDs). The proteins encoded by the IPD genes are entirely diverse in their functions, including not only many required for phototransduction, but also structural and cytoskeletal proteins,

RNA splicing factors, intracellular trafficking molecules, and proteins involved in phagocytosis and the regulation of intracellular pH. Up-to-date documentation of IPD genes and loci, with references, continues at RetNet (**http://www.sph.uth.tmc.edu/RetNet/**).

The genotypic and phenotypic heterogeneity of IPDs was examined in detail in the last *Annual Review* article on retinal degeneration (Pacione et al. 2003) and in more recent articles, as well (Daiger et al. 2007, Hartong et al. 2006). In addition, as the prevalence of age-related macular degeneration (AMD) has increased over the past decade, now affecting almost 9% of Americans 40 years of age or older; knowledge of the genetic contribution to this disease has also grown. Excellent reviews on the genetics and pathophysiology of AMD are available (Jager et al. 2008, Rattner & Nathans 2006, Swaroop et al. 2007); we do not replicate that information here.

THE STRUCTURE OF THE MAMMALIAN RETINA: PHOTORECEPTORS, GLIA, AND VESSELS

The neural retina is composed of three morphologically discrete layers of cell bodies, bridged by synaptic terminals that guide the transmission of light-initiated neuronal and electrical impulses (**Figure 1**). We briefly discuss not only the photoreceptors (PRs), but also retinal glia and the retinal vasculature, both of which have been increasingly implicated in the pathogenesis of IPDs.

The Photoreceptors

The PR layer is organized as a dense array of light-sensing cells in the retina, with cones (5% of total PRs) responsible for high intensity and contrast vision under photopic or bright light conditions, and rods (95% of total PRs) functioning under scotopic or dark-adapted conditions (Delyfer et al. 2004, Pacione et al. 2003). The conversion of visual cues to biochemical signals via phototransduction occurs within a

IPD: inherited photoreceptor degenerations

PR: photoreceptor

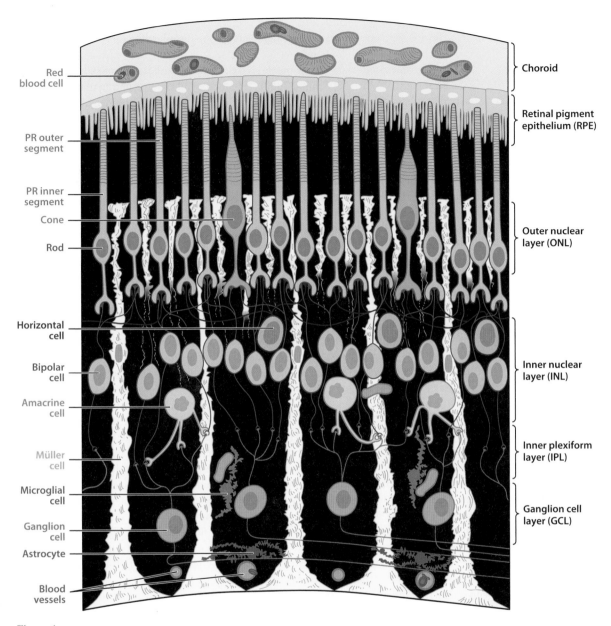

Red blood cell

PR outer segment

PR inner segment

Cone

Rod

Horizontal cell

Bipolar cell

Amacrine cell

Müller cell

Microglial cell

Ganglion cell

Astrocyte

Blood vessels

Choroid

Retinal pigment epithelium (RPE)

Outer nuclear layer (ONL)

Inner nuclear layer (INL)

Inner plexiform layer (IPL)

Ganglion cell layer (GCL)

Figure 1

A schematic representation of the human retina showing photoreceptors (PRs), other retinal neurons, Müller glia, microglia, astrocytes, and vessels. The outer nuclear layer (ONL) is composed of the cell bodies of rod and cone PRs; the inner nuclear layer (INL) contains the cell bodies of several types of neurons (horizontal cells, bipolar cells, and amacrine cells) as well as the bodies of Müller glia. The ganglion cell layer (GCL) contains ganglion cells. Adapted from (Krsti**).

stack of flattened membranous discs in the outer segments of PRs, linked to the PR cell body by a connecting cilium from which the outer segments evaginate. These discs are continu-

ously replenished at the base of the outer segment, necessitating the phagocytic function of the retinal pigment epithelium (RPE) at the distal tip to ensure a constant turnover rate and

RPE: retinal pigment epithelium

maintain the viability of the PR cell (Young 1971). The PR segment is a modified cilium that is connected to the cell body by a narrow <0.2 μm diameter connecting cilium overlying the basal body (Wolfrum & Schmitt 2000). About 2000 rhodopsin molecules and 0.1 μm² of membrane must pass through the connecting cilium every minute because the outer segments account for most of the cell's surface area. The connecting cilium constantly regulates the influx and efflux of soluble and membranous proteins and lipids. It is crucial for all outer segment functions because it maintains the polarized distribution of all soluble and membrane-bound components.

Interestingly, the nuclei of rod PRs differ between nocturnal and diurnal animals: Nocturnal, unlike diurnal animals, exhibit an inverted pattern of chromatin organization compared with other cells in the body, with euchromatin at the nuclear periphery and heterochromatin at the center (Solovei et al. 2009). Solovei et al. postulated that this inverted pattern optimizes the channeling of light through rod PR bodies toward the rod outer segments.

Ciliary gene mutations and IPDs. The one area of genetics that has provided substantial new insight into the biology of the normal and mutant PR in recent years has been the frequent association of mutations in ciliary genes with IPD. Given the significance of the cilium to the PR, it is therefore not entirely surprising that over the past decade at least 38 genes associated with IPDs either have a ciliary or basal body location or are thought to influence ciliary function (Adams et al. 2007, Lancaster & Gleeson 2009; **http://www.sph.uth.tmc.edu/RetNet/**) (**Table 1**). Mutations in these genes give rise to a wide range of IPD phenotypes, some of which also cause kidney, inner ear, brain, and developmental defects. There is genetic overlap between these disorders as well as evidence for epistatic interactions between loci (Beales et al. 2003) owing to the presence of multiprotein complexes within cilia, such as the BBSome and the Usher or RPGRIP1 interactomes. Different

mutations in a single gene often cause more than one disorder, the most extreme example being *NPHP6/CEP290* mutations, which can be associated with Leber congenital amaurosis, Joubert syndrome, nephronophthisis, Meckel syndrome (type 4), or Bardet-Biedl syndrome (type 14) (**Table 1**). Cilia are not just passive transport vehicles; they are also involved in at least five signaling cascades (Shh, canonical and noncanonical Wnt, calcium, PDGFRα), which may explain the variety of extraocular defects in the associated syndromes (**Table 1**).

Retinal Glia

In addition to the three major classes of neurons, the retina is also populated by three types of glia: Müller cells, microglia, and astrocytes. The importance of glia in conserving the structural and functional integrity of the retina is often underestimated or considered secondary to neuronal function. Rather than acting as bystanders, however, glia have recently been implicated as critical mediators in cell death and disease. Müller cells constitute 90% of all retinal glia (de Melo Reis et al. 2008) and extend radial processes from the base of the ganglion cell layer to the PR inner segments, ensheathing neurons along their path. In addition to serving a fundamental physiological function as optical fibers transmitting light while minimizing light scatter (Franze et al. 2007), Müller cells can be neuroprotective after retinal injury by releasing neurotrophic factors such as basic fibroblast growth factor (bFGF) and ciliary neurotrophic factor (CNTF) (Wilson et al. 2007) or antioxidants such as glutathione (Schutte & Werner 1998). However, Müller cells may also contribute to neuronal cell death by dedifferentiating in response to prolonged or intense retinal injury. This dedifferentiation may impair neurotransmitter removal, as well as ion and water homeostasis (Bringmann et al. 2006). Astrocytes provide many of the same functions as Müller glia, but are restricted almost entirely to the nerve fiber layer of the retina. Unlike astrocytes and Müller cells, the third glial subtype, microglia, may arise from myeloid tissue.

Table 1 Ciliopathy genes associated with human retinal degeneration (modified from Lancaster & Gleeson 2009). Ciliary trafficking includes either intraflagellar transport (IFT) or the docking and selection of transport substrates (see also Jin & Nachury 2009)[a]

Clinical phenotype	Gene/protein	Retinal localizations	Proposed function
Retinitis pigmentosa	RPGR	CC, BB, (OS)	Ciliary trafficking
	RP1	CC	Disc morphogenesis
Cone-rod dystrophy	RPGR	CC, BB, (OS)	Ciliary trafficking
Leber congenital amaurosis	TULP1	CC, IS, SYN	Ciliary, trafficking
	LCA5/Lebercilin	CC, BB	Ciliary, trafficking
	NPHP6/CEP290	CC, BB	Ciliary, trafficking
	RPGRIP1	CC, BB	Ciliary, trafficking
Nephronophthisis	NPHP1/Nephrocystin	BB, PC	Ciliary structure
	NPHP4/Nephroretinin	BB, PC	Ciliary trafficking
	NPHP5/IQCB1	PC	Ciliary trafficking
	NPHP6/CEP290	BB	Ciliary trafficking
	NPHP8/RPGRIP1L	BB	Shh signaling
Joubert syndrome	NPHP8/RPGRIP1L	BB	Shh signaling
	AHI1/Jouberin	BB, PC	Wnt signaling
	ARL13B	BB, PC	Cilia structure
	NPHP1/Nephrocystin	BB, PC	Cilia structure
	NPHP6/CEP290	BB	Ciliary trafficking
Bardet-Biedl syndrome	BBS1	BB, PC	BBSome, Wnt/PCP, ciliary trafficking
	BBS2	BB, PC	BBSome, ciliary trafficking
	BBS3/ARL6	BB	Vesicle trafficking
	BBS4	BB, PC	BBSome, Wnt/PCP, ciliary trafficking
	BBS5	BB, PC	BBSome
	BBS6/MKKS	BB	Chaperonin
	BBS7	BB, PC	BBSome, ciliary trafficking
	BBS8/TTC8	BB, PC	BBSome, ciliary trafficking
	BBS9/PTHB1	BB, PC	BBSome
	BBS10	BB	Chaperonin, ciliogenesis, Wnt
	BBS11/TRIM32	BB	E3 ubiquitin ligase
	BBS12	BB	Chaperonin, ciliogenesis, Wnt
	BBS13/MKS1	BB	Ciliary trafficking
	BBS14/CEP290	BB	Ciliary trafficking
Alstrom syndrome	ALMS1	BB	Cilia maintenance
Usher syndome	USH1A	CC	Ciliary trafficking
	USH1B/Myosin VIIa	CC	Ciliary trafficking
	USH1C/Harmonin	SYN, IS, OS	Uncertain
	USH1D/Cadherin 23	CC	Ciliary trafficking
	USH1E/Protocadherin 15	CC	Ciliary trafficking
	USH1F/SANS	CC	Ciliary trafficking
	USH2A/Usherin	CC	Ciliary trafficking
	USH2C/GPR98 (VLGR1)	CC	Ciliary trafficking
	USH2D/Whirlin	CC	Ciliary trafficking
	USH3A/Clarin	CC, IS, SYN	Ciliary trafficking

[a]BB, basal body; CC, connecting cilium; IS, photoreceptor inner segment; OS, photoreceptor outer segment (parentheses indicate species-specific localizations); PC, primary cilium; SYN, synapse.

PUFA:
polyunsaturated fatty
acid

Microglia play a central role in the elimination of neuronal debris and in the initiation of immunological signaling through the release of chemokines, cytokines, or neurotoxic factors (Langmann 2007).

Retinal Vessels

Two independent vascular networks supply oxygen, glucose, and other nutrients to the retina. The central retinal artery proceeds from the optic nerve head, dividing into three capillary networks to deliver nutrients to the innermost layers. The choroidal blood vessels, separated from the RPE by Bruch's membrane, are vital to maintain the outer retina and PRs, have little capacity to autoregulate, and have an extremely high rate of blood flow. The oxygen content of the choroidal vascular bed is also unusually high: The arterio-venous difference in oxygen saturation is ~35% in retinal vessels, whereas it is only 2%–4% in the choroid (Bill et al. 1983).

MECHANISMS OF PHOTORECEPTOR DEATH: HOW CELLS DIE IN IPDs

Programmed cell death (PCD) is an evolutionarily conserved feature of development, tissue turnover, and homeostasis found in all metazoans. PCD can be accomplished in four main ways—caspase-dependent and -independent apoptosis, autophagy, and the granzyme B pathway—although mixed forms also occur (Kroemer et al. 2009, Tait & Green 2008). Autophagy is currently classed as a PCD pathway, but in a physiological context it probably acts as a protective response to stress and death occurs only when protection fails (Boya & Kroemer 2008).

Apoptosis is the major form of cell death in the nervous system. PR death during development is less prominent than in other retinal layers, but it involves caspase activation and is abolished or delayed in Bax/Bak double knockout or Bim single knockout mice (Donovan et al. 2006, Hahn et al. 2003). As developmental cell death

declines, PRs become progressively less sensitive to proapoptotic stimuli, in parallel with the downregulation of caspases and proapoptotic Bcl-2-related proteins, making cells more prone to caspase-independent cell death or autophagy in the face of genetic insults (Donovan et al. 2006).

The earlier literature emphasized the importance of classic apoptotic cell death in IPDs (Chang et al. 1993, Doonan & Hunt 1996, Portera-Cailliau et al. 1994, Xu et al. 1996). These studies relied heavily on the TUNEL assay, which was later shown to be nonspecific, because it labels both caspase-dependent and -independent apoptotic cells, as well as autolytic and necrotic cells (Colicos & Dash 1996, Grasl-Kraupp et al. 1995). More recent work has showed that at least some rodent models of retinal degeneration, including *Prph2^{rds/rds}* and *Pde6b^{rd1/rd1}* mice, involve caspase-independent apoptotic cell death (Doonan et al. 2005, Lohr et al. 2006, Paquet-Durand et al. 2007, Rohrer et al. 2004, Sancho-Pelluz et al. 2008). Features of macroautophagy can also be present (Ahuja et al. 2008, Doonan et al. 2005).

THE RESPONSES OF THE RETINA TO PHOTORECEPTOR MUTATIONS

In this section, we focus principally on IPD genes within the broader context of the molecular and cellular pathogenesis of PR death. In many cases, the biological role of the corresponding wild-type protein is often unclear, and the link between the altered function of the mutant protein and the ultimate death of the PR is not evident. Nevertheless, significant insights are emerging.

Given the large number of genes associated with IPDs, it is not surprising that the biochemical, cellular, and clinical phenotypes observed in these disorders, and in animal models of them, are extremely heterogeneous. The singular sensitivity of PRs to environmental and genetic insults appears to result from many factors: exposure to light, high polyunsaturated fatty acid (PUFA) content, high oxidative metabolism, limited oxygen concentrations when functional demand rises and excessive oxygen exposure when it falls, the structurally narrow connecting cilium linking inner and outer segments, and the high protein and lipid turnover associated with outer segment renewal, among other factors (Travis 1998). The only common pathogenetic factor behind the diversity of the mutant genes associated with IPDs may be the shared ability to stress one of the most genetically vulnerable cells in the body.

High-throughput microarray gene expression analysis of the retinas in animal (mainly mouse) models of IPDs has been widely used to identify the secondary retinal genomic responses to the presence of a PR mutation. One of the most interesting outcomes of these studies has been the considerable degree of overlap in the resulting data sets; many of the same genes are differentially expressed in diverse IPD models. This finding suggests that IPDs share pathogenetic mechanisms, irrespective of the identity of the mutant gene or the clinical phenotype.

Mechanisms of Photoreceptor Death—Why Cells Die in IPDs

PR cell death appears to be a stochastic phenomenon in which the probability of cell death is increased by a broad range of cellular stresses, which may vary over time as the *milieu interieur* of the retina changes owing to both adaptive and maladaptive responses to PR loss. These stresses appear to converge on a final common death pathway, which is primarily apoptotic (whether caspase-dependent or -independent), but it remains unclear how this convergence takes place. Some of the proximal stresses and distal apoptotic triggers are now discussed.

The role of mitochondria and oxidative metabolism in photoreceptor death. PR cells have one of the highest rates of oxygen consumption in the body (Graymore 1960), and oxygen consumption is particularly high in the parafoveal region of primates, where the

rod density is highest (Ahmed et al. 1993, Yu & Cringle 2005). PR cell loss or reductions in energy-demanding activities, such as phototransduction, can lead to elevated tissue oxygen concentrations because choroidal blood vessels are not autoregulated by local oxygen levels. Increases in outer retinal oxygen concentrations have been confirmed in $Mertk^{RCS/RCS}$ rats (Yu et al. 2000), Tg.$Rho^{Pro23His}$ rats (Yu et al. 2004), and $CEP290$ mutant cats (Padnick-Silver et al. 2006). The attenuation of inner retinal arterioles (which do autoregulate) has been attributed to the high oxygen levels present in degenerating retinas (Noell 1953). The oxygen toxicity hypothesis suggests that hyperoxia is a significant contributor to degeneration only in the later stages of disease progression (Stone et al. 1999), but this proposal may underestimate the impact of small changes in oxygen, especially in the presence of light and potentially destructive retinoid derivatives.

Mitochondrial reactive oxygen species and PR death. Oxygen metabolism within mitochondria may also play an important and generic role in PR cell death. If each gene product and its associated pathway is evolutionarily well conserved—as visual pathways generally are—then the degeneration rates should be comparable across mammalian species, other factors being equal. The latter is clearly not the case because mammals differ markedly in size, metabolic rate, longevity, and other factors. Wright et al. (2004) explored this issue by contrasting rates of retinal degeneration caused by equivalent mutations in five genes (RHO, RDS, $RPGR$, $TULP1$, and $ATXN7$) across five mammalian species. This process was repeated for five inherited neurodegenerations affecting the brain (APP, $SNCA$, $SOD1$, HD, and $PRNP$). The results showed an inverse relationship (exponent 1.1 ± 0.2) between the degeneration rate and the maximum life span potential of the species. Two observations were particularly noteworthy: First, the relationship across genes and species is remarkably consistent; second, the degeneration rates differed by 10- to 100-fold. If such large differences in degeneration

ROS: reactive oxygen species

RP: retinitis pigmentosa

rate could be reproduced by genetic manipulation, the result would provide a truly worthwhile generic treatment intervention. The proposed explanation lay in the strong correlation between in vitro mitochondrial reactive oxygen species (ROS) formation and life span (Wright et al. 2004). Other factors, such as basal metabolic rate, body mass index, phylogenetic group, and chance, were shown to be unlikely, a conclusion later reinforced by the work of Lambert et al. (2007).

About 90% of cellular oxygen consumption occurs in mitochondria, where ROS are formed primarily as a by-product of oxidative phosphorylation (Murphy 2009). The in vivo rates of constitutive mitochondrial ROS formation are not accurately known because of their brief existence and low concentrations (Murphy 2009). Nevertheless, mammals such as the mouse clearly have high rates of constitutive mitochondrial ROS formation and high rates of retinal degeneration, whereas humans have correspondingly low rates, which probably reflects the differing selection pressures to extend longevity (Partridge 1997).

This situation begs the question as to why mitochondrial ROS have such a powerful influence on cell death. One proposal is that gradual, age-related, and subliminal increases in constitutive (tonic) mitochondrial ROS formation influence retrograde (mitochondria-to-nucleus) and anterograde (nucleus-to-mitochondria) redox-sensitive signaling pathways and essentially set the probability of apoptosis accordingly for each cell (Wright et al. 2009). Such pathways have been well characterized in a wide variety of organisms and commonly involve ROS-induced posttranslational modification of critical thiols in enzymes such as phosphatases and a variety of transcription factors, which in turn affects both survival and proapoptotic signaling pathways (Janssen-Heininger et al. 2008, Woodson & Chory 2008). Mitochondria integrate cellular stress responses and control apoptotic cell death (Green & Kroemer 2004), so the tonic regulation of cell death by interorganellar signaling could provide an additional level of long-term cellular quality control.

Two signaling pathways that influence PR cell death rates and contain known or potential redox-sensitive signaling components are p53 (Ali et al. 1998), and HDAC4 (Chen & Cepko 2009), but many other pro- and anti-apoptotic pathways are likely involved, depending on the type of PR stress. These pathways are proposed to set the balance of apoptotic signaling within cells, each of which has a unique redox history. The probability of apoptosis will vary among cells around the mean value for each genotype, species, and cell type. This proposal readily accounts for the observed kinetics of PR cell death and for its progressive nature (Clarke et al. 2000, Clarke & Lumsden 2005, Wright et al. 2009).

Intermediary metabolism, RNA processing, and photoreceptor death. A number of IPDs result directly or indirectly from metabolic stress. Metabolic stress—a disturbance of metabolic homeostasis—is generally one of the most common triggers for apoptotic cell death (Jin et al. 2007). One example is mutation in the nicotinamide adenine dinucleotide (NAD) specific mitochondrial enzyme isocitrate dehydrogenase (IDH3) (Hartong et al. 2008). In most tissues, this reaction is catalyzed by the product of another gene, NADP-specific IDH2, suggesting that PRs have a particular requirement for IDH3, perhaps because its kinetic characteristics are more suited to the high substrate and cofactor concentrations found in PRs. Mutations in mitochondrial oxidative phosphorylation components can also give rise to syndromic forms of retinitis pigmentosa (RP) (e.g., *MTATP6*) (Childs et al. 2007). The retinal degeneration resulting from mutations in another metabolic enzyme, IMPDH1, which catalyzes a key step in guanine nucleotide biosynthesis, may be unrelated to this function however; instead it may involve loss of IMPDH's ability to regulate the translation of specific retinal mRNAs at polyribosomes, including the rhodopsin transcript (Mortimer et al. 2008). This finding provides a strong connection to retinal degeneration because mutations influencing rhodopsin expression are a known cause of IPD (Daiger et al. 2007, Kosmaoglou et al. 2008). This connection also raises the question of whether mutations in the ubiquitous precursor mRNA-processing factors PRPF31, PRPF8, PRPF3, and PAP1, which also cause IPDs, impose a similar type of cellular stress because of the unusual processing requirements of rhodopsin, which is produced at the rate of 10^6–10^7 molecules/day/cell (Wolfrum & Schmitt 2000), or because of other abundant PR-specific transcripts (Deery et al. 2002, Mordes et al. 2007).

Protein misfolding in mutant photoreceptors. Oxidative and nitrosative stress associated with protein misfolding is probably the single most important cause of neurodegenerative brain disorders (Nakamura & Lipton 2007). Some classes of mutation in the IPD genes *CA4*, *BBS6*, *BEST1*, *EFMP1*, *RPE65*, *AIPL1*, *SCA7*, *ELOVL4*, *ABCA4*, *IMPDH1*, *RPGR*, *RS1*, and *RHO* can be associated with protein misfolding, but many others will likely be identified in the future (Kosmaoglou et al. 2008).

Rhodopsin accounts for 85%–90% of rod outer segment protein mass and may be the most extreme example of polarized protein distribution. The most common class of rhodopsin mutation (type II) involves the misfolding and retention of rhodopsin in the endoplasmic reticulum (ER) (Saliba et al. 2002). This causes ER stress, which is detected by sensor proteins that can initiate an unfolded protein response (UPR) by activating adaptive signaling pathways (IRE1, ATF6, and PERK) (Lin et al. 2007, Ron & Walter 2007). These pathways reduce the translation, enhance the folding capacity, or facilitate the clearance of misfolded proteins. Transgenic rats expressing the type II *Rho^Pro23His* mutation show UPR activation and upregulation of protective chaperones that later decrease in abundance, in parallel with increased CCAAT/enhancer-binding protein homologous protein (CHOP) transcription factor expression, which induces transcriptional activation of proapoptotic genes (Lin et al. 2007). A plausible sequence of events is the generation of a ROS signal by endoplasmic reticulum oxidase 1 (ERO1) as a result of futile attempts to correct high levels of misfolded protein, which in turn activates UPR signaling and apoptotic cell death pathways (Malhotra et al. 2008).

Chaperones are essential for maintaining the continuous high throughput protein folding required by the activity of PR ribbon synapses. The cysteine string protein α (CSPα) cochaperone activates the ATPase activity of the HSC70 chaperone and is located in the presynaptic compartment of ribbon synapses (Tobaben et al. 2001). CSPα knockout mice develop a rapidly progressive PR degeneration preceded by impaired presynaptic membrane traffic and synaptic terminal shrinkage (Schmitz et al. 2006). The CSPα chaperone complex prevents the unfolding and subsequent toxicity of abundant synaptic proteins, possibly including certain SNAREs. The TULP1 protein has also been localized to ribbon synapses, as well as to PR connecting cilia, and compelling evidence from *Tulp1^−/−* mice now indicates that both ciliary trafficking and functional ribbon synapse defects contribute to the severity of the PR death in this model (Grossman et al. 2009).

The role of light in IPDs. Human ocular media transmit visible light to the retina in the range of 410–700 nm because the lens effectively absorbs most ultraviolet light (Wyszecki & Stiles 1982). Visible and ultraviolet light are insufficiently energetic to ionize most biomolecules, but oxygen enhances the ionizing effects of light, especially in the presence of photosensitizers such as retinoids (Halliwell & Gutteridge 2007, Quintiliani 1986). Light exposure can damage the PR and/or RPE in wild-type animals, and the damage is influenced by many factors (Organisciak et al. 2000). In many IPDs, however, accumulating evidence indicates that there is increased retinal susceptibility to light damage, even at ambient light levels (Cideciyan et al. 2005, Paskowitz et al. 2006, Wu et al. 2009). Acute light damage may be due to the accumulation of all-*trans* retinal (Maeda et al. 2009, Sun & Nathans 2001), which readily generates singlet oxygen (Delmelle 1977,

Lion et al. 1976). In addition, a small fraction of all-*trans* retinal molecules released by light spontaneously condenses to form either all-*trans*-retinal dimers or a variety of potentially toxic bis-retinoids including A2E (N-retinylidene-N-retinylethanolamine), iso-A2E, and A2-DHP-PE, which accumulate in RPE cells following disc phagocytosis (Sparrow et al. 1999, Wu et al. 2009). These compounds absorb light maximally at the lower end of the visible spectrum (428–510 nm), and many are capable of generating ROS (Wu et al. 2009). Moreover, they are not easily degraded by lysosomal enzymes and are major constituents of RPE lipofuscin.

Stargardt disease caused by *ABCA4* mutations is the IPD most clearly associated with increased formation of toxic bis-retinoids, owing to defective transport of all-*trans* retinal out of rod and cone outer segment discs by the ABCA4 flippase (Allikmets et al. 1997, Travis et al. 2007). The predominantly macular location of the degeneration may relate to the photoreceptor/RPE ratio in this region of the retina (Snodderly et al. 2002).

The relevance of all the above mechanisms to other genetically diseased retinas is still debated. Some retinal degenerations (e.g., $Rho^{Thr4Arg}$) are dramatically accelerated by light (Cideciyan et al. 2005). Three different rhodopsin mutations are light sensitive, whereas another three are not, but at least seven other models of human retinal degeneration are exacerbated by light, protected by dark rearing, or both (Paskowitz et al. 2006). Investigators have also reported clinically or histologically detected increases in RPE lipofuscin in a growing number of retinopathies, including mutations in *GUCA1A*, *GUCY2D*, *RPGR*, *RIMS1*, *KCNV2*, *MERTK*, *RS1*, and others of unknown genotype (Birnbach et al. 1994, Bunt-Milam et al. 1983, Robson et al. 2008, Szamier & Berson 1977). It therefore seems likely that a variety of genetic defects that either increase the formation of toxic retinoids or slow their removal may be exacerbated by light exposure. These defects include mutations that deform or destabilize discs, slow disc turnover, or inhibit their phagocytosis.

Lipid mediators of PR death. Outer segments are lipid rich because of their tightly packed discs; the lipid content of outer segments is 15% of wet weight, compared with 1% in most cells (Whikehart 2003). The amounts of very long chain PUFAs in outer segments are particularly high. Mutations in the *ELOVL4* gene impair the elongation of C28 and C30 saturated and polyunsaturated fatty acids (Agbaga et al. 2008) and are associated with an autosomal dominant form of Stargardt disease (Bernstein et al. 2001). C28–C36 fatty acids are found together with another abundant PUFA, docosahexaenoic acid, predominantly in rod and cone outer segments where they are closely associated with opsin (McMahon & Kedzierski 2009). The disease mechanism in PRs with *ELOVL4* mutations is related to selective deficiency of C28–C36 fatty acids, altered phototransduction, and excessive accumulation of the toxic retinoid A2E, for reasons that remain unclear (McMahon & Kedzierski 2009). This observation raises the possibility that dietary supplementation of long-chain PUFAs to type 3 Stargardt patients may be beneficial.

PR-specific responses in IPDs: Endothelin-2. It is absolutely remarkable that although PRs commonly carry the primary mutation in IPDs, the only PR gene identified to date that is exclusively differentially expressed in mutant PRs, and not in other retinal cell types, is *Endothelin-2 (Edn2)*. The EDN2 peptide is one of three endothelins, vasoactive peptides implicated in diverse pathological processes from pulmonary hypertension to oncogenesis. The PR *Edn2* mRNA increases in response to a spectrum of retinal insults, with significantly higher expression levels in models of IPD (Chen et al. 2004, Cottet et al. 2006, Rattner & Nathans 2005), retinal detachment (Rattner & Nathans 2005, Zacks et al. 2006), increased intraocular pressure (Ahmed et al. 2004), hyperoxia (Natoli et al. 2008), and ischemia (Kamphuis et al. 2007). Because the Endothelin B receptor (EDNRB) is concurrently upregulated in light damage, Rattner & Nathans (2005) postulated that EDN2 signals through EDNRB, which is localized to Müller cells. EDN2 signaling through Müller cell EDNRB may be protective because intravitreal administration of a selective EDNRB receptor agonist in a mouse model of IPD reduced PR death, whereas an EDNRB antagonist increased PR death (Joly et al. 2008). In apparently contradictory studies, however, the dual EDNRA-EDNRB receptor antagonist Tezosentan reduced both caspase-3 cleavage in the outer nuclear layer (ONL) and glial fibrillary acidic protein (GFAP) expression in retinal extracts after light damage, suggesting a pathogenetic role for EDN2 (Torbidoni et al. 2006).

Altogether, these studies indicate that the production of *Edn2* mRNA by PRs is a general and perhaps universal response to PR stress. It is currently unclear whether the increased levels of the *Edn2* message contribute to PR death or reduce it. If increased EDN2 is detrimental to PRs, blocking EDN2 signaling may be a novel therapeutic avenue for the treatment of IPDs because a myriad of pharmaceutical endothelin pathway antagonists exist and are being further developed.

The Response of Other Neuroretinal Cell Types to Photoreceptor Mutations

Although the PRs carry the mutation, many if not most of the changes in retinal gene expression in IPDs occur in glia. These secondary responses may influence the rate of PR death.

The Müller glia response. The apparently universal glial response to retinal insult is the increased expression of the proteins GFAP and VIMENTIN, which are regarded as the hallmarks of reactive gliosis. Gene expression studies have confirmed *Gfap* mRNA upregulation in IPDs (Hackam et al. 2004, Rattner & Nathans 2005), light damage (Chen et al. 2004, Rattner & Nathans 2005), and other models of stress-

induced retinal injury (Kamphuis et al. 2007, Natoli et al. 2008). Although Müller cells are a source of growth and survival factors in the retina, recent work indicates that Müller cell gliosis may contribute to PR death. Müller glia of *Gfap*$^{-/-}$ *Vim*$^{-/-}$ mice in retinal detachment–induced PR death did not become hypertrophic and exhibited attenuated activation of downstream pathways, with a significant reduction in the number of infiltrating monocytes and 70% fewer TUNEL-positive PRs (Nakazawa et al. 2007, Verardo et al. 2008).

A cardinal feature of the retinal response to a PR mutation or insult is increased expression of the transcription factor gene, *Stat3*, in Müller glia. STAT3 lies at the nexus of the IL-6 family cytokine signaling cascade (Hirano et al. 2000). Several lines of evidence suggest that increased IL-6 family signaling in the retina, leading to increased STAT3 expression in Müller cells, is a protective response in IPDs and other retinal stresses. First, like *Gfap*, *Stat3* mRNA is upregulated in both IPD and environmentally induced models of PR death (Hackam et al. 2004, Rattner & Nathans 2005). STAT3 expression in the INL is highest in the nuclei of Müller glia following light injury (Kassen et al. 2007). Second, exogenous delivery of the IL-6 family cytokine CNTF or its analog Axokine increases STAT3 expression in Müller glia (Kassen et al. 2009, Peterson et al. 2000) and reduces PR death (Adamus et al. 2003, Huang et al. 2004, Tao et al. 2002). However, CNTF can impair visual function when injected at high doses (McGill et al. 2007). Third, the expression of another IL-6 family cytokine, *Lif*, is also consistently upregulated in mouse models of IPD, with increased expression in a subset of Müller cells (Joly et al. 2008). LIF may bind its receptor, LIFR, on the membrane of Müller cells (Sarup et al. 2004) in an autocrine fashion to activate STAT3. Intravitreal injection of LIF also rescues PRs (Ueki et al. 2008), whereas genetic ablation of the *Lif* gene increased PR death in a mouse model of IPD (Joly et al. 2008) and in light damage (Burgi et al. 2009). In summary, although STAT3 signaling can induce GFAP expression in Müller glia (Wang et al. 2002), implying an exacerbation of reactive gliosis, the net effect of STAT3 signaling in IPDs appears to be protective of mutant PRs. However, it remains to be established that the IL-6-family induced increase in Müller cell STAT3 and the protective effect are causally linked. For example, in a light damage model in zebrafish, CNTF activated the mitogen-activated protein kinase (MAPK) pathway for neuroprotection and a STAT3 pathway for Müller cell proliferation (Kassen et al. 2009).

Increasing evidence suggests that the complement system, which is required to clear foreign particles and antigens and can trigger the release of cytokines and promote inflammation, is a significant participant in the retinal response to a PR mutation. First, transcripts for components of the complement cascade, including *C4*, *C3*, and *C1q*, are upregulated in the retina in mouse models of IPD (Demos et al. 2008, Rattner & Nathans 2005, Rohrer et al. 2004) and glaucoma (Ahmed et al. 2004, Steele et al. 2006). Moreover, polymorphisms in the genes for *CFH*, *CFB*, *C3*, and *C2* are associated with AMD (Gold et al. 2006, Haines et al. 2005, Klein et al. 2005, Rattner & Nathans 2006, Yates et al. 2007). Second, the deletion of complement factor D, part of the alternative complement pathway, slowed PR death in a mouse light damage model (Rohrer et al. 2007b), which is further evidence that the complement system has an influential role in the retinal response to injury. In contrast, loss of C1qa, a member of the classical complement activation pathway, or loss of B- and T-cell function did not alter the rate of PR death in a *Pde6b^{rd1/rd1}* model (Rohrer et al. 2007a). In a mouse model of glaucoma, C1q expression was localized mainly to Müller glia and to the area of the retinal inner limiting membrane, which is partially composed of Müller cell end feet (Stasi et al. 2006). Further study of the role of the complement system in mouse models will clarify the importance of this system in IPDs.

The microglia response to photoreceptor mutations. The microglia response in mouse models of IPDs can be either pathogenetic to or protective of PRs, apparently depending on the context. The canonical function of retinal microglia is to phagocytose dying PRs (Roque & Caldwell 1993). A pathologic role in IPDs is suggested by the migration of activated retinal microglia to the degenerating PR layer and the release of neurotoxic factors, including TNF-alpha (Zeng et al. 2005) and perhaps IL-1beta (Ni et al. 2008). Moreover, minocycline treatment, which suppresses microglial activation, was protective of mutant PRs (Yang et al. 2007). In contrast, a protective role for microglia in IPDs is suggested by the secretion of the neurotrophic factors GDNF and CNTF, thereby increasing bFGF and BDNF expression in Müller cells and enhancing PR survival (Harada et al. 2002). Bone marrow-derived cells can be recruited into degenerating retina, differentiate into microglia, and migrate toward both regressing vessels and dying PRs. A reduction in the number of bone-marrow derived microglia increased PR death, and the activation of microglial progenitors by systemic administration of granulocyte colony stimulating factor (G-CSF) or erythropoietin resulted in a dramatic rescue of cone PRs (Sasahara et al. 2008). In conclusion, the evidence demonstrates that microglia are critically involved in the retinal response to PR mutations. Manipulation of the retinal microglia response in IPDs is a potential avenue for therapeutic intervention.

Cell-autonomous and non-cell-autonomous cell death in IPDs. Most PR mutations that cause IPDs also result in the non-cell-autonomous death of other cells. For example, RPE cells die in response to primary PR defects in Stargardt disease (Travis et al. 2007), and cones die as a late consequence of primary rod PR loss. Humans spend most of their lives in ambient light levels that are well within the dynamic range of cones, but levels at which rods are saturated, so secondary cone loss accounts for much of the disability in IPDs. Various mechanisms have been proposed to explain secondary cone loss: (*a*) release of toxic compounds by dying rods (Ripps 2002); (*b*) loss of a rod-derived survival factor (Leveillard et al. 2004); (*c*) microglia activation, which kills adjacent cones (Gupta et al. 2003); (*d*) oxidative stress due to retinal hyperoxia (Stone et al. 1999); and (*e*) starvation of cones resulting from perturbed cone-RPE interactions due to rod loss (Punzo et al. 2009).

The retinal vasculature and IPDs. The temporal relationship between vascular and PR degeneration in the mouse was first examined by Blanks & Johnson (1986), who concluded that PR death appears to precede vascular regression. Current hypotheses on the role of the vasculature in PR death therefore rest on the assumption that the vascular changes are secondary to decreased metabolic demand as the mutant PRs die and that increasing outer retinal oxygen concentrations, concomitant with the loss of PR outer segments and PR cell loss, result in the gradual ablation of retinal vessels (Yu & Cringle 2005).

Recently, the association of vascular regression and PR death was examined by injecting hematopoietic stem cells into two mouse models of IPD. Otani et al. (2002) injected Lin⁻ bone marrow–derived cells into neonatal mouse eyes and observed colocalization of the injected cells with developing vessels and astrocytes (Otani et al. 2002). When Lin⁻ bone marrow–derived cells were then intravitreally injected in two separate mouse models of IPD before the onset of PR death, the severe vascular regression normally associated with PR death was reduced. Concomitantly, a substantial rescue of cone PRs and mild restoration of retinal electroretinogram (ERG) responses were observed (Otani et al. 2004). The investigators remarked on the elongated appearance of the injected cells incorporated into the vascular network and their association with astrocytes, and they interpreted these features to be characteristic of endothelial cells. In contrast, Sasahara et al. (2008) demonstrated that transplanted bone marrow cells were predominantly microglial, in the same models of IPD. Regardless of whether vascular stability is primary or secondary to neuronal rescue, these results suggest that there is an important relationship between the fitness of the retinal vasculature and PR death in IPDs. The role of the vasculature in degeneration, although often overlooked, is therefore an avenue of research that deserves further exploration.

INHERITED PHOTORECEPTOR DEGENERATIONS: STEPS TOWARD TREATMENT

Pharmacologic and Biologic Interventions as Therapies for IPDs

A number of agents, including trophic factors, insulin, antioxidants, and vitamin A, slow PR death to a limited extent in IPDs (Stone et al. 1999). The following synopsis focuses on novel factors and recent advances. Some of these approaches illustrate the non-cell-autonomous nature of IPDs. The heterogeneous nature of these therapeutic strategies suggests that multiple factors reduce the viability of the PRs in IPDs and that multifaceted pharmacological approaches may be required to facilitate PR rescue.

Antioxidants. The possibility that dying rods impose oxidative stress on adjacent cones was investigated by daily injection of an antioxidant mixture to the *Pde6b^rd1/rd1* mouse (Komeima et al. 2006). In this RP mouse model, rod death is almost complete by postnatal day 21 (P21), after which cones die slowly over the subsequent two to three months. Treated animals showed a reduction in oxidative damage and a twofold increase in cone density compared with untreated controls. This effect was reproduced in two other mouse models of human RP (Komeima et al. 2007). Rod survival was also prolonged in the slower degenerating *Pde6b^rd10/rd10* model. Upregulation of antioxidant defenses by joint overexpression of SOD2 and catalase in mitochondria also protected cones (Usui et al. 2009). Increased superoxide concentrations were found during cone cell degeneration, and nitric oxide synthase inhibitors both reduced peroxynitrite-induced damage and increased cone survival in *Pde6b^{rd1/rd1}* mice (Komeima et al. 2008).

The *Pde6b^{rd1/rd1}* mouse model of IPD is associated with severe oxidative stress due to calcium dysregulation (Farber 1995). Individual antioxidants have little effect on disease progression, but feeding a mixture of antioxidants to mutant and control mice resulted in reduced apoptotic rod cell death (Sanz et al. 2007). The oxidizing environment of the outer retina appears therefore to be a significant factor in a range of retinal degenerations.

The deleterious effect of ROS is illustrated in one article that describes the use of cerium oxide nanoparticles (nanoceria) to scavenge reactive oxygen species in a rat light damage model (Chen et al. 2006a). Intravitreal pretreatment with nanoceria preserved retinal ERG A- and B-wave amplitudes to 79% and 87% of controls, compared with 22% and 26% for vehicle.

Neurotrophins and survival signaling. The idea that cones die from loss of trophic support by rods has a long history (Faktorovich et al. 1990, Steinberg 1994, Streichert et al. 1999). Four neurotrophic factors (bFGF, NGF, BDNF, CNTF) slow retinal degeneration in at least some animal models of RP. Retinal delivery of CNTF shows efficacy in 13 different animal models (MacDonald et al. 2007). The study employs encapsulated CNTF-expressing cells that are surgically implanted into the vitreous (Sieving et al. 2006). The use of nanoparticles as a noninflammatory and nonimmunogenic genetic delivery technology, in addition to their use as a source of survival factors, is an interesting and potentially promising alternative to more conventional methods (Cai et al. 2009). A recent publication showed that vascular endothelial growth factor (VEGF) may promote PR survival in a paracrine fashion, with Müller-derived VEGF signaling through the VEGFR2 receptor on PRs. VEGF being protective has important implications for anti-VEGF-directed therapies for IPDs (Saint-Geniez et al. 2008).

Punzo et al. (2009) analyzed nonautonomous cone degeneration in four different mouse models of human RP and found that activation of the insulin/mTOR pathway coincided with the activation of chaperone-mediated autophagy. When mice were treated with daily insulin injections for 4 weeks, cone survival increased by 20%. The authors proposed a model in which autophagy is induced mainly in response to nutrient starvation and only to a lesser extent by oxidative stress.

Chen & Cepko (2009) recently identified another promising means of rescuing rods and cones using analysis of the nuclear corepressor histone deacetylase 4 (HDAC4) in the *Pde6b^{rd1/rd1}* mouse. Downregulation of HDAC4 expression increased the developmental apoptosis of PRs, whereas upregulation led to substantial neuroprotection from both developmental and degenerative PR death. The protective effect was due to retention of HDAC4 in the cytoplasm rather than the nucleus and was at least partly mediated by stabilizing HIF1α.

Rod-derived cone viability factor (RdCVF) also improves cone survival and function as shown by ERG in the rat Tg.*Rho^{P23H}* model of RP (Leveillard et al. 2004, Yang et al. 2009). RdCVF is secreted by rods and is encoded by the nucleoredoxin-like 1(*Nxnl1*) gene, which is a member of the thioredoxin family. However, it lacks thiol oxidoreductase activity because it lacks the thioredoxin active site motif. Paradoxically, an alternatively spliced transcript that does contain a thioredoxin active site lacks trophic activity for cones (Fridlich et al. 2009). Serial subretinal injections of human RdCVF protein into transgenic *Rho^{Pro23His}* rats increased cone density by 19%–20% compared with controls (Fridlich et al. 2009).

The promise of cell replacement therapy. Two recent studies have provided impressive evidence for the potential of cell replacement therapy for human IPDs. MacLaren et al. (2006) found that rod precursor cells transplanted into the subretinal space of rhodopsin null mice could integrate normally into the retina and differentiate into rod PRs, forming synaptic connections and leading to improved visual function. Successfully integrated PRs were unexpectedly derived only from postmitotic rod precursors harvested during a narrow window of development (postnatal days 1–7) and not from proliferating retinal progenitor cells. Lamba et al. (2009) subsequently demonstrated that human embryonic stem cells (hESCs) that had been directed to a retina cell fate in vitro (using a combination of a BMP inhibitor, a Wnt inhibitor, and IGF-1) migrated into wild-type mouse retinas after intraocular injection, localized to the appropriate layers, and expressed markers for differentiated cells, including both rod and cone photoreceptor cells (Lamba et al. 2009) (**Figure 2**). When transplanted into the subretinal space of an IPD mouse model lacking the PR transcription factor CRX, the hESC-derived retinal cells differentiated into functional PRs, integrated into the

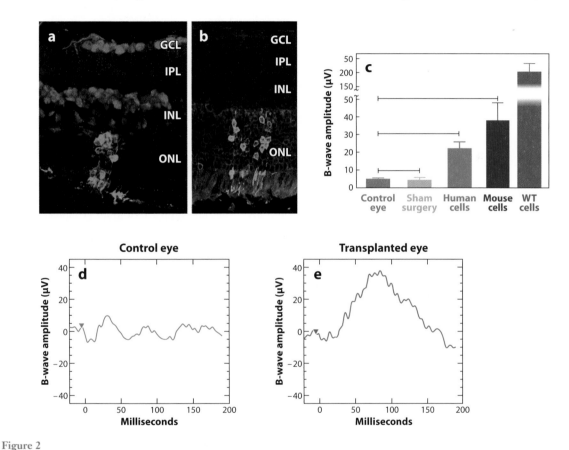

Figure 2

Subretinal transplantation of hESCs in adult wild-type mice and functional rescue in $Crx^{-/-}$ retinas. (*a*) GFP-expressing human photoreceptors (PRs) that have migrated into the outer nuclear layer (ONL) from the subretinal space. Cells show highly differentiated PR morphology with outer segments. The section is co-stained for Pax6 in red (1 mm confocal slice). (*b*) Similar image of transplanted mouse GFP+ retinal cells into wild-type mouse. Panel *c* shows the B-wave amplitudes following light flash in control uninjected eyes subjected to sham surgery with no subretinal transplant, subretinal transplanted human cells, subretinal transplanted wild-type mouse cells, and finally, traces from wild-type mice. Eyes that received subretinal human and mouse retinal cells were light responsive (***$p <$ 0.0001, error bars represent standard error of mean). (*d*) A representative electroretinogram (ERG) B-wave from a control nontransplanted eye with no response to light stimulus. (*e*) Averaged traces from the eye of the same animal after it had received the transplanted human embryonic stem cell–derived retinal cells. Reprinted from Lamba DA, Gust J, Reh TA. 2009. Cell stem cell, transplantation of human embryonic stem cell-derived photoreceptors restores some visual function in Crx-deficient mice. Reprinted from *Cell Stem. Cell* 4:73–79, with permission from Elsevier.

retina, and restored light response (**Figure 2**). Together with the discovery that pluripotent stem cells (PSCs) can be induced from human embryonic fibroblasts (Woltjen et al. 2009), the findings of Lamba et al. indicate that hESCs or induced PSCs may be a source of PRs for cell replacement therapy in humans with IPDs.

Understanding the Histopathology of Human IPDs: A Prerequisite to Treatment

For more than a century, clinicians described a group of progressive blinding retinal diseases termed RP (Bird 1981). *En face* views of the ocular fundus in RP had abnormal pigmentation, changes in blood vessel caliber, and discoloration of the optic nerve head. Later, these diseases were recognized to follow classic genetic patterns.

En face fundus appearances of RP did not easily sort with genetic pattern and could not define disease mechanisms. A basic question remained: Which retinal cells are involved primarily in RP? Early histopathology of postmortem donor eyes from patients with RP pointed to PRs and the RPE as primary sites of degeneration (Cideciyan et al. 1998). Noninvasive studies, such as ERG, confirmed that photoreceptor dysfunction was a major part of RP (Hartong et al. 2006). Psychophysical measurement of rod- and cone-mediated vision has been a surrogate for retinal histopathology in RP for decades (Cideciyan et al. 1998, Jacobson et al. 1986; Lyness et al. 1985; Massof & Finkelstein 1979, 1981; Wald & Zeavin 1956).

Molecular genetic understanding of IPDs led to different directions for histopathological study (Cideciyan et al. 1998). For example, we learned about elucidation of the morphological basis of adRP phenotypes secondary to *rhodopsin* mutations (Cideciyan et al. 1998, Li et al. 1994); proof that there are indeed excess short-wavelength cones relative to other cone subtypes and rods in the enhanced S-cone syndrome caused by *NR2E3* mutations (Milam et al. 2002); demonstration of thick sub-RPE deposits in late-onset retinal degeneration caused by *C1QTNF5/CTRP5* mutation (Hayward et al. 2003, Kuntz et al. 1996), supporting the hypothesis that secondary vitamin A deprivation causes slowed rhodopsin regeneration in this disease (Jacobson et al. 1995, 2001); and the finding of photoreceptor cilial abnormalities in Usher syndrome retinas (Barrong et al. 1992, Berson & Adamian 1992, Hunter et al. 1986).

As scientific attention shifted toward identifying gene causation, molecular mechanisms were sought using in vitro methods and in vivo studies of genetically engineered or naturally occurring animals with gene mutations comparable to those in human diseases (Baehr & Frederick 2009, Pacione et al. 2003, Rattner et al. 1999). Investigators could then advance to preclinical proof-of-concept studies and then human therapy. Identification of molecular causes in man and investigations of animal models, however, still left a gap in understanding whether human and animal diseases were sufficiently similar to warrant extrapolation from one to the other.

The gap was filled by the advent of optical methods to assess the living human retina, specifically optical coherence tomography (OCT). The laminar architecture of the retina in human IPDs could be visualized at all stages of disease. Imaging modalities that measure integrity of the RPE have provided further information (Cideciyan et al. 2007b, Gibbs et al. 2009). A more complete understanding of human disease has now become possible using in vivo histopathology of the human retina taken together with clinical data, psychophysics, and ERGs and results from studies of animal models with comparable genotypes and phenotypes.

We illustrate cross-sectional images using OCT and colocalized rod and cone visual psychophysical results in different human retinal degenerations in **Figure 3**. Normal human retinal laminar architecture (**Figure 3a**) has hypo- and hyper-reflective layers with predictable relationships to histology (Aleman et al. 2008, Huang et al. 1998, Jacobson et al. 2003). The photoreceptor or ONL (**Figure 3**) is thickest at

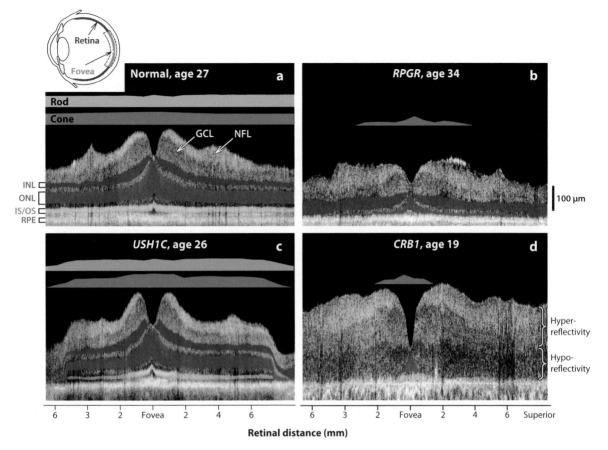

Figure 3

Retinal laminar architecture by in vivo microscopy in (*a*) a normal subject compared with (*b–d*) three patients representing different molecularly defined retinal degenerative diseases with: Images are obtained using optical coherence tomography (OCT) across the vertical meridian through the fovea. OCTs are shown in grayscale with lowest reflectivity as black and highest reflectivity as white. Brackets at left edge of normal scan define inner nuclear layer (INL, *purple*), outer nuclear layer (ONL, *blue*), inner and outer segment junction (IS/OS, *yellow*), and retinal pigment epithelium (RPE). Arrows point to ganglion cell layer (GCL) and nerve fiber layer (NFL). Inset (*a, upper left*) shows the location of the vertical retinal section on a schematic of the eye. Psychophysically determined rod (*gray*) and cone (*red*) sensitivities are shown above the scans. (*d*) Yellow brackets at right edge of the coarsely laminated retina in a patient with *CRB1*-LCA shows regions of hyporeflectivity (deeper) and hyperreflectivity (more superficial) with uncertain layer identity.

the fovea, the site of highest cone density. The junction between inner and outer segments of the PRs (IS/OS) is deep to the ONL. The inner nuclear layer (INL), ganglion cell layer (GCL), and nerve fiber layer (NFL) are visible.

Dramatic abnormalities are evident in X-linked (XL) RP owing to mutation in the *RPGR* (retinitis pigmentosa GTPase regulator) gene (Aleman et al. 2007, Shu et al. 2007, Wright & Shu 2007) (**Figure 3b**). The ONL is thinned at the fovea and barely detectable past 4–6 mm

eccentricity. Thinned ONL is accompanied by INL thickening; GCL and NFL are retained. Psychophysics indicates that only reduced cone function remains. These noninvasive results are consistent with the histopathology of RP and XLRP (Cideciyan et al. 1998, Szamier et al. 1979). Thickening of the INL is a marker for retinal remodeling in retinal degenerative diseases and has been documented in histopathology of animal models (Aleman et al. 2008, Cideciyan et al. 2007a, Jacobson et al. 2006b).

USH: Usher syndromes

LCA: Leber congenital amaurosis

If this XLRP-*RPGR* patient was a candidate for a clinical treatment trial of, for example, a gene-based and cone-specific therapy, then focal delivery to the central 10 mm of the retina may be a strategy (Jacobson et al. 2006a).

The Usher syndromes (USH) are syndromic forms of RP with hearing loss and possible vestibular abnormalities. Investigators have associated at least 12 loci with USH. The disease-causing genes encode proteins of different classes, thought to be integrated into a network (Maerker et al. 2008, Williams 2008). Testing hypotheses about USH pathogenetic mechanisms in the mammalian retina has been difficult because most mouse models of human USH genotypes have no retinal degeneration phenotype (Williams 2008). The onus is thus on human studies for information about USH retinal degeneration. We recently presented evidence for a common PR disease phenotype among human USH genotypes (Jacobson et al. 2008a).

Our in vivo histopathology in patients with USH1B (caused by mutations in *MYO7A*, Myosin VIIa) or USH1C showed the surprising finding that there can be disease stages in which patients have central regions of structurally and functionally normal retina. Adjacent regions show severe laminopathy and visual loss (Jacobson et al. 2009b, Williams et al. 2009). Unlike XLRP-*RPGR*, the USH1C retina illustrated (**Figure 3c**) is normal in structure and function from the fovea to ~4–6 mm eccentricity. At greater eccentricities, ONL thickness decreases, lamination becomes abnormal, and there is visual dysfunction. Degeneration of the central region will likely occur in later life, but we need to ponder whether we should be treating this normal retina with experimental therapies in early-phase safety trials (Hashimoto et al. 2007, Williams 2008).

Markedly abnormal retinal lamination is exemplified in a form of Leber congenital amaurosis (LCA). LCA is a group of childhood-onset retinal degenerations associated with many different genes (den Hollander et al. 2008, Stone 2007). *CRB1* (crumbs homolog-1) mutations account for ~10% of LCA and also cause a rare

form of RP. The Crumbs molecules are transmembrane proteins involved in apico-basal cell polarity and in the morphogenesis and maintenance of the retina (Gosens et al. 2008). A *CRB1*-mutant retina is thickened, and there are coarse outer and inner laminar zones; a small cone-mediated island of vision corresponds with a limited island of foveal ONL (**Figure 3d**). The abnormal retina in *CRB1*-LCA resembles that of immature normal retina, and the disease may have interrupted naturally occurring apoptosis (Jacobson et al. 2003). Are these retinas treatable with current therapeutic strategies? A retinal prosthesis may be helpful (Yanai et al. 2007), pending evidence that the visual pathways from ganglion cells to cortex are sufficiently intact to warrant such an approach.

In summary, investigators can now study the many human retinal degenerations and the many disease stages of each with in vivo histopathology. Quantitative retinal structural data coupled with focal psychophysics can be used to elucidate mechanisms, perform natural history studies, compare human disease expression with those of animal models, and develop inclusion and exclusion criteria for clinical trials and outcomes of therapy. Advances will continue to occur in this technology with higher-resolution instrumentation, wider-angle views, and eventual single-cell visualization; but this clinically feasible methodology should be used now to advance our understanding of IPDs, even if footnotes with revisions of interpretation will be necessary in the future.

From Proof-of-Concept Gene-Therapy Experiments to Human Clinical Trials

For decades, clinician researchers specializing in orphan IPDs played the disappointing role of disproving claims of treatment success from various sources and trying to discourage their patients from endangering their poorly functioning eyes and possibly their general health by rushing to accept unproven therapies.

The current era is more positive, given many successful proof-of-concept experiments using gene therapy in animal models of human IPDs. Early successes in the lab did not move to the clinic (Ali et al. 2000, Bennett et al. 1996). The autosomal recessive disease resulting in RPE65 (retinal pigment epithelium-specific-65-kDa) deficiency has mainly made this transition. This disorder interrupts function of the visual-retinoid cycle (Travis et al. 2007), visual pigment is not available to PRs through this key pathway (**Figure 4a**), and vision is severely compromised. In a canine model of RPE65 deficiency, subretinal delivery of a gene-viral vector agent led to a remarkable activation of retinal and postretinal function (Acland et al. 2001). Mice with RPE65 deficiency showed the same dramatic treatment effect (Bemelmans et al. 2006, Chen et al. 2006b, Dejneka et al. 2004, Jacobson et al. 2005, Lai et al. 2004, Pang et al. 2006). Following further proof-of-concept studies, dose-response data, and toxicity testing (Acland et al. 2005; Jacobson et al. 2006a,b; Jacobson et al. 2005), human gene therapy in RPE65–LCA seemed worthwhile to consider.

A pretrial concern was whether human and animal versions of RPE65 deficiency were identical. Human RPE65-LCA differed because there was considerable retinal degeneration from early life (Jacobson et al. 2005, 2007, 2008b, 2009a). Did the remaining PRs in human RPE65–LCA show evidence of visual cycle dysfunction? Animals with Rpe65 deficiency have a disproportionate functional abnormality (i.e., severely reduced visual responses) when compared with the degree of retinal structural abnormality. Also, the animal diseases were not complicated by significant degeneration at the ages when treatment proved efficacious. A human trial would not be justifiable if the residual PRs in patients were behaving as if they were quantum-catch limited, the relationship of structure and function that is traditionally found in most other animal and human IPDs (Jacobson et al. 2005). Colocalized measurements of the ONL by OCT and photoreceptor-mediated function by psychophysics showed that RPE65–LCA patients did have disproportionate functional loss for the amount of PRs

retained (Jacobson et al. 2005). These observations opened the door to consider a clinical trial more seriously, and investigators devised strategies to optimize delivery of the gene-vector to retinal areas with PRs (Jacobson et al. 2005, 2008b, 2009a).

Another pretrial concern involved cone PRs. The visual cycles of rod and cone PRs have differences, and the exact pathways are still debated (Travis et al. 2007). Mouse models with Rpe65 deficiency show early and severe cone PR loss. Was this also occurring in RPE65–LCA (Jacobson et al. 2007), and would the outcome of gene therapy be only rod-based efficacy? The canine model, unlike the mouse, did show improved cone function posttreatment (Acland et al. 2001, 2005). A specific analysis of the cone foveas of RPE65–LCA patients over a wide age range indicated that there was definitely some loss of cone PRs from the earliest ages. Yet, some cones were remaining and functioning, albeit abnormal in function (Jacobson et al. 2007), which implied that residual cones in the fovea of humans with RPE65–LCA may be using an alternative visual cycle and that treating the fovea may not lead to fovea-specific increases in visual acuity. This rather important issue has still not been resolved despite some claims of increased visual acuity posttreatment; such results are equally consistent with extrafoveal rod or cone recovery.

Three contemporaneous clinical trials began at this time, and in 2008, all three studies reported preliminary results of safety and modest efficacy after uniocular subretinal injections of AAV2-RPE65 (Bainbridge et al. 2008, Cideciyan et al. 2008, Hauswirth et al. 2008, Maguire et al. 2008; **Figure 4b**). Cone and rod function increased (Cideciyan et al. 2008; **Figure 4c,d**), but no foveal cone increase in function has been reported to date. Systemic and ocular safety was demonstrated, but the surgical technique is still one of the weaker elements in this multidisciplinary application of science to medicine. Foveal thinning and a macular hole have occurred in subfoveal injections (Hauswirth et al. 2008, Maguire et al. 2008). A formal analysis of the relationship between

a

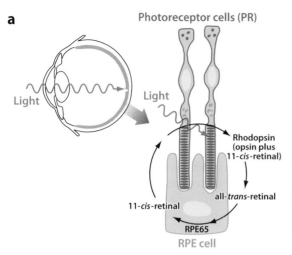

Photoreceptor cells (PR)

Light

Light

Rhodopsin (opsin plus 11-*cis*-retinal)

all-*trans*-retinal

11-*cis*-retinal

RPE65

RPE cell

b

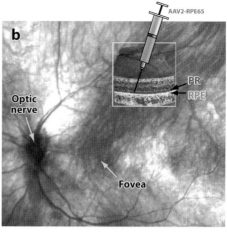

AAV2-RPE65

Optic nerve

Fovea

PR
RPE

c

RODS

CONES

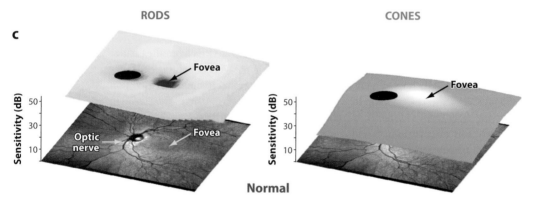

Sensitivity (dB)

50
30
10

Fovea

Optic nerve

Fovea

Sensitivity (dB)

50
30
10

Fovea

Normal

d

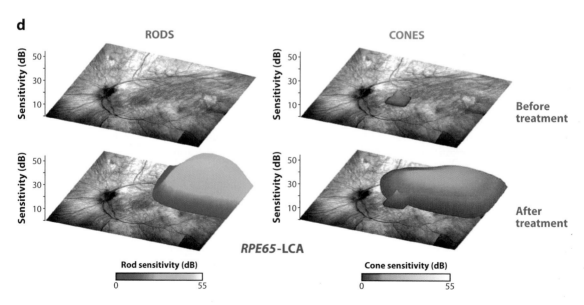

RODS

CONES

Sensitivity (dB)

50
30
10

Sensitivity (dB)

50
30
10

Before treatment

Sensitivity (dB)

50
30
10

Sensitivity (dB)

50
30
10

After treatment

RPE65-LCA

Rod sensitivity (dB)

0   55

Cone sensitivity (dB)

0   55

visual psychophysical improvement after treatment and its PR layer potential showed that the complex enzymatic and degenerative condition in the patients had been simplified by correction of the visual cycle abnormality (Cideciyan et al. 2008). Unfortunately, what remained was the retinal degeneration, and its natural history is uncertain.

An unexpected outcome was discovered and replicated in our treated patients (Cideciyan et al. 2008, 2009b). Patients self-reported that their vision was definitely brighter upon awakening compared with later in the day, after exposure to ambient light. This observation suggested that rod visual recovery could be better if the treated eye had more time to dark adapt. A slowed dark adaptation, implying a slowed visual cycle, was documented, and full adaptation took ~8 h in treated retina compared with <1 h in normal retina. This effect has been persistent and unchanged for a year (Cideciyan et al. 2009b). The basis of the abnormality remains unknown. The practical consequence of the result is that therapeutic outcome of the trial should not be measured after bright light photography or ophthalmoscopy. Such clinical assessment tools should be reserved for times after rod-based outcomes are measured. Without thoughtful consideration of protocol, underestimates of efficacy or apparent variability will result. Also, for the patients to take full advantage of any recovered vision, they should be instructed not to expose their eyes to too much light; light exposure would bleach their newly acquired night vision and they would have to wait for hours to regain full sensitivity. After conversion of this complex human

disease to a simpler retinal degeneration by gene therapy, we should also consider whether the newly acquired vision will subsequently be susceptible to the deleterious effects of light damage (Cideciyan et al. 2005, Paskowitz et al. 2006).

Questions about the longevity of safety and efficacy in gene therapy for *RPE65–LCA* have started to be addressed (Cideciyan et al. 2009a,b). At one year postprocedure, our first cohort of patients continued to show no vector-related serious adverse events. Immunological assays to identify reaction to AAV2 capsid were unchanged from baseline, and clinical eye examinations remained as they were at three months postprocedure (Hauswirth et al. 2008). The dramatic improvements in visual sensitivity at 3 months were unchanged at 12 months. Specifically, the retinal extent and magnitude of rod and cone components of the newly enhanced vision remained the same. A fascinating self-report by one of the patients at one year postprocedure was also studied to understand its basis. The young adult noted for the first time that a numerical display on the dashboard of the family vehicle was readable. A visual acuity change was not present, so this was not simply slowly improving spatial resolution; visual sensitivity in the treated retinal region had also not varied. What had changed, however, was the fixation in this eye under different conditions: There was a slow and progressive movement of fixation over many months from the anatomical fovea to the treated retinal region. The region of therapy had become a preferred locus for use in this eye under certain conditions (Cideciyan et al. 2009b).

Figure 4

Spatial distribution of rod and cone vision displayed in three dimensional representation across a normal eye and a gene-therapy-treated eye of a patient with *RPE65*-LCA. (*a*) Schematic of the eye and retina illustrating the visual-retinoid cycle. Light enters the normal eye and is absorbed by visual pigments (rhodopsin) in photoreceptor cells. The visual-retinoid cycle with its isomerase RPE65 (located in the RPE cells) is responsible for converting all-*trans*- to 11-*cis*-retinal to regenerate the visual pigment. (*b*) Infrared ocular fundus view in a patient with *RPE65*–LCA. Overlaid 3D optical coherence tomography (OCT) image depicts the superior retinal site of the subretinal injection of vector-gene. (*c*) Normal rod (*left panel*) and cone (*right panel*) visual function displayed in 3D and overlaid onto an ocular fundus view of a normal subject. (*d,e*) Rod and cone visual function before and after treatment in the same *RPE65*-LCA patient depicted in panel *b*. Noteworthy is the lack of measurable rod function before treatment; also, only a foveal island of reduced cone function exists before treatment. After treatment, rod and cone visual sensitivity is increased in the region of the subretinal injection of vector-gene.

The new visual experiences originating from the treated retinal area in our patient could be driving cortical adaptations (Butz et al. 2009). From a clinical perspective, this slow experience-dependent process in the treated eye may be accelerated by visual training (Taub et al. 2002).

The teams of gene therapists who have initiated this new direction of treatment for IPDs have major obligations. Owing to the rarity of *RPE65*–LCA and other IPDs, there will continue to be limited numbers of patients to enroll in these trials. Therefore, we must understand as much as possible about each treated patient: not only what we have done but also what we could not accomplish with this novel therapy. Ignoring or minimizing failure and complexities and publicizing only success is a trap we should avoid, although so many decades without treatment success makes everyone keen to announce recent progress. To date, the human studies in *RPE65*–LCA have been performed safely and with some efficacy. The daunting next major step for gene therapy trials is to devise treatment for primary PR degenerations. Unlike *RPE65*—LCA, these disorders include patients with expanses of useful and sometimes only moderately abnormal vision. High expectations for remarkable visual recoveries may have to be replaced by the less exciting but still useful outcome of slowed progression of these ultimately blinding disorders.

CONCLUDING REMARKS

The photoreceptor may have the distinction of being the most genetically vulnerable cell in mammals: Mutations in more than 140 genes have been associated with a single phenotype, cell death. The sensitivity of the PR to mutation provides a large number of perspectives from which the critical pathophysiological pathways culminating in PR cell death may be identified. Given the parsimony that generally underlies biology, however, we suggest that research be directed to identifying the pathophysiological processes common to most or all mutant PRs. These shared responses are most likely to lead

to the deepest understanding of the molecular mechanisms that culminate in the "one hit" that ultimately kills the cell (Clarke et al. 2000, Pacione et al. 2003).

We have highlighted features of PR biology that likely contribute to its genetic susceptibility, including, for example, the inordinate demand for (*a*) PUFAs, (*b*) the visual pigment rhodopsin and other abundant proteins of the outer segment, and (*c*) high metabolic activity to sustain the rapid turnover rate of PR outer segments. Whether any one feature of PR biology is uniquely responsible for the susceptibility of PRs to mutation-induced cell death is unknown. Studies of mutant PRs have suggested that these cells are singularly vulnerable to a variety of stresses, including (*a*) the disruption of metabolic homeostasis, (*b*) oxidative stress associated with protein misfolding or elevated oxygen concentrations, and (*c*) constant exposure to light. Novel insights into the stresses imposed by mutation have been acquired from unbiased gene expression array studies, including a group of responses to a PR mutation that occurs commonly, perhaps even invariably, such as (*a*) the upregulation of *Edn2* mRNA by mutant PRs, (*b*) the expression of intermediate filament proteins implied in the exacerbation of reactive gliosis, and (*c*) the activation of signaling cascades and upregulation of immune markers. Out of this complexity of altered gene expression, we must dissect the critical responses that promote cell death or survival. Is there a single biochemical event that, if prevented, would arrest PR death, or is the cell altered in so many ways that such simplicity is not possible? Determination of the significance of any change in a mutant PR will be greatly facilitated by genetic analyses (e.g. does a gene's loss of function improve PR survival?), and by examining the proteome of the mutant PR, which is now becoming technically feasible.

One can see several major outcomes of current IPD research. First, therapeutic opportunities will increase as we continue to reveal the pathophysiology of the mutant PR. For example, if the knockdown or ablation of a gene whose expression is increased in mutant retina prolongs mutant PR survival, then the small molecule–mediated inhibition of the activity of the corresponding gene product holds promise as a drug therapy. Second, the identification of responses that occur in all mutant PRs, irrespective of the mutant gene, may lead to general therapies. Even if the promise of gene replacement therapy is fully realized, the basic research and clinical trials that precede successful gene therapy take years, and a more general treatment for IPDs would fill an enormous therapeutic gap. Third, AAV-based or other forms of gene replacement are likely to continue to be valuable strategies to slow or arrest vision loss and in some diseases improve vision in patients with mutations in specific genes. Fourth, advances in the generation of induced pluripotent stem cells suggest that lost PRs may be replaced by genetically corrected PRs derived from, say, a patient's skin fibroblasts, thereby avoiding problems of immune rejection. Altogether, the future for treating this group of devastating inherited diseases has never looked brighter.

DISCLOSURE STATEMENT

The authors are not aware of any affiliations, memberships, funding, or financial holdings that might be perceived as affecting the objectivity of this review.

ACKNOWLEDGMENTS

This work was supported by an award to A.B. from the Foundation Fighting Blindness of Canada; and by grants to R.R.M. from the Canadian Institutes of Health Research and The Macula Vision Research Foundation (MVRF), to A.W. from the UK Medical Research Council, Fight for Sight (UK), British Retinitis Pigmentosa Society and the MVRF, and to S.G.J. from the MVRF and Hope for Vision. We thank Dr. Alexander Sumaroka, Dr. Artur Cideciyan, Dr. Tomas Aleman, Dr. Dianna Martin, and Ms. Lori Atkinson for helpful advice.

LITERATURE CITED

Acland GM, Aguirre GD, Bennett J, Aleman TS, Cideciyan AV, et al. 2005. Long-term restoration of rod and cone vision by single dose rAAV-mediated gene transfer to the retina in a canine model of childhood blindness. *Mol. Ther.* 12:1072–82

Acland GM, Aguirre GD, Ray J, Zhang Q, Aleman TS, et al. 2001. Gene therapy restores vision in a canine model of childhood blindness. *Nat. Genet.* 28:92–95

Adams NA, Awadein A, Toma HS. 2007. The retinal ciliopathies. *Ophthalmic Genet.* 28:113–25

Adamus G, Sugden B, Shiraga S, Timmers AM, Hauswirth WW. 2003. Anti-apoptotic effects of CNTF gene transfer on photoreceptor degeneration in experimental antibody-induced retinopathy. *J. Autoimmun.* 21:121–29

Agbaga MP, Brush RS, Mandal MN, Henry K, Elliott MH, Anderson RE. 2008. Role of Stargardt-3 macular dystrophy protein (ELOVL4) in the biosynthesis of very long chain fatty acids. *Proc. Natl. Acad. Sci. USA* 105:12843–48

Ahmed F, Brown KM, Stephan DA, Morrison JC, Johnson EC, Tomarev SI. 2004. Microarray analysis of changes in mRNA levels in the rat retina after experimental elevation of intraocular pressure. *Invest. Ophthalmol. Vis. Sci.* 45:1247–58

Ahmed J, Braun RD, Dunn R Jr, Linsenmeier RA. 1993. Oxygen distribution in the macaque retina. *Invest. Ophthalmol. Vis. Sci.* 34:516–21

Ahuja S, Ahuja-Jensen P, Johnson LE, Caffe AR, Abrahamson M, et al. 2008. rd1 Mouse retina shows an imbalance in the activity of cysteine protease cathepsins and their endogenous inhibitor cystatin C. *Invest. Ophthalmol. Vis. Sci.* 49:1089–96

Aleman TS, Cideciyan AV, Sumaroka A, Schwartz SB, Roman AJ, et al. 2007. Inner retinal abnormalities in X-linked retinitis pigmentosa with RPGR mutations. *Invest. Ophthalmol. Vis. Sci.* 48:4759–65

Aleman TS, Cideciyan AV, Sumaroka A, Windsor EA, Herrera W, et al. 2008. Retinal laminar architecture in human retinitis pigmentosa caused by Rhodopsin gene mutations. *Invest. Ophthalmol. Vis. Sci.* 49:1580–90

Ali RR, Reichel MB, Kanuga N, Munro PM, Alexander RA, et al. 1998. Absence of p53 delays apoptotic photoreceptor cell death in the rds mouse. *Curr. Eye Res.* 17:917–23

Ali RR, Sarra GM, Stephens C, Alwis MD, Bainbridge JW, et al. 2000. Restoration of photoreceptor ultra-structure and function in retinal degeneration slow mice by gene therapy. *Nat. Genet.* 25:306–10

Allikmets R, Singh N, Sun H, Shroyer NF, Hutchinson A, et al. 1997. A photoreceptor cell-specific ATP-binding transporter gene (ABCR) is mutated in recessive Stargardt macular dystrophy. *Nat. Genet.* 15:236–46

Baehr W, Frederick JM. 2009. Naturally occurring animal models with outer retina phenotypes. *Vis. Res.* 49:2636–52

Bainbridge JW, Smith AJ, Barker SS, Robbie S, Henderson R, et al. 2008. Effect of gene therapy on visual function in Leber's congenital amaurosis. *N. Engl. J. Med.* 358:2231–39

Barrong SD, Chaitin MH, Fliesler SJ, Possin DE, Jacobson SG, Milam AH. 1992. Ultrastructure of connecting cilia in different forms of retinitis pigmentosa. *Arch. Ophthalmol.* 110:706–10

Beales PL, Badano JL, Ross AJ, Ansley SJ, Hoskins BE, et al. 2003. Genetic interaction of BBS1 mutations with alleles at other BBS loci can result in non-Mendelian Bardet-Biedl syndrome. *Am. J. Hum. Genet.* 72:1187–99

Bemelmans AP, Kostic C, Crippa SV, Hauswirth WW, Lem J, et al. 2006. Lentiviral gene transfer of RPE65 rescues survival and function of cones in a mouse model of Leber congenital amaurosis. *PLoS Med.* 3:e347

Bennett J, Tanabe T, Sun D, Zeng Y, Kjeldbye H, et al. 1996. Photoreceptor cell rescue in retinal degeneration (rd) mice by in vivo gene therapy. *Nat. Med.* 2:649–54

Bernstein PS, Tammur J, Singh N, Hutchinson A, Dixon M, et al. 2001. Diverse macular dystrophy phenotype caused by a novel complex mutation in the ELOVL4 gene. *Invest. Ophthalmol. Vis. Sci.* 42:3331–36

Berson EL, Adamian M. 1992. Ultrastructural findings in an autopsy eye from a patient with Usher's syndrome type II. *Am. J. Ophthalmol.* 114:748–57

Bill A, Sperber G, Ujiie K. 1983. Physiology of the choroidal vascular bed. *Int. Ophthalmol.* 6:101–7

Bird AC. 1981. The Duke-Elder Lecture, 1981. Retinal receptor dystrophies. *Trans. Ophthalmol. Soc. U. K.* 101:39–47

Birnbach CD, Jarvelainen M, Possin DE, Milam AH. 1994. Histopathology and immunocytochemistry of the neurosensory retina in fundus flavimaculatus. *Ophthalmology* 101:1211–19

Blanks JC, Johnson LV. 1986. Vascular atrophy in the retinal degenerative rd mouse. *J. Comp. Neurol.* 254:543–53

Boya P, Kroemer G. 2008. Lysosomal membrane permeabilization in cell death. *Oncogene* 27:6434–51

Bringmann A, Pannicke T, Grosche J, Francke M, Wiedemann P, et al. 2006. Muller cells in the healthy and diseased retina. *Prog. Retin. Eye Res.* 25:397–424

Bunt-Milam AH, Kalina RE, Pagon RA. 1983. Clinical-ultrastructural study of a retinal dystrophy. *Invest. Ophthalmol. Vis. Sci.* 24:458–69

Burgi S, Samardzija M, Grimm C. 2009. Endogenous leukemia inhibitory factor protects photoreceptor cells against light-induced degeneration. *Mol. Vis.* 15:1631–37

Butz M, Worgotter F, van Ooyen A. 2009. Activity-dependent structural plasticity. *Brain Res. Rev.* 60:287–305

Cai X, Nash Z, Conley SM, Fliesler SJ, Cooper MJ, Naash MI. 2009. A partial structural and functional rescue of a retinitis pigmentosa model with compacted DNA nanoparticles. *PLoS One* 4:e5290

Chang GQ, Hao Y, Wong F. 1993. Apoptosis: final common pathway of photoreceptor death in rd, rds, and rhodopsin mutant mice. *Neuron* 11:595–605

Chen B, Cepko CL. 2009. HDAC4 regulates neuronal survival in normal and diseased retinas. *Science* 323:256–59

Chen J, Patil S, Seal S, McGinnis JF. 2006a. Rare earth nanoparticles prevent retinal degeneration induced by intracellular peroxides. *Nat. Nanotechnol.* 1:142–50

Chen L, Wu W, Dentchev T, Zeng Y, Wang J, et al. 2004. Light damage induced changes in mouse retinal gene expression. *Exp. Eye Res.* 79:239–47

Chen Y, Moiseyev G, Takahashi Y, Ma JX. 2006b. RPE65 gene delivery restores isomerohydrolase activity and prevents early cone loss in Rpe65-/- mice. *Invest. Ophthalmol. Vis. Sci.* 47:1177–84

Childs AM, Hutchin T, Pysden K, Highet L, Bamford J, et al. 2007. Variable phenotype including Leigh syndrome with a 9185T > C mutation in the MTATP6 gene. *Neuropediatrics* 38:313–16

Cideciyan AV, Aleman TS, Boye SL, Schwartz SB, Kaushal S, et al. 2008. Human gene therapy for RPE65 isomerase deficiency activates the retinoid cycle of vision but with slow rod kinetics. *Proc. Natl. Acad. Sci. USA* 105:15112–17

Cideciyan AV, Aleman TS, Jacobson SG, Khanna H, Sumaroka A, et al. 2007a. Centrosomal-ciliary gene CEP290/NPHP6 mutations result in blindness with unexpected sparing of photoreceptors and visual brain: implications for therapy of Leber congenital amaurosis. *Hum. Mutat.* 28:1074–83

Cideciyan AV, Hauswirth WW, Aleman TS, Kaushal S, Schwartz SB, et al. 2009a. Human RPE65 gene therapy for Leber congenital amaurosis: persistence of early visual improvements and safety at 1 year. *Hum. Gene Ther.* 20:999–1004

Cideciyan AV, Hauswirth WW, Aleman TS, Kaushal S, Schwartz SB, et al. 2009b. Vision 1 year after gene therapy for Leber's congenital amaurosis. *N. Engl. J. Med.* 361:725–27

Cideciyan AV, Hood DC, Huang Y, Banin E, Li ZY, et al. 1998. Disease sequence from mutant rhodopsin allele to rod and cone photoreceptor degeneration in man. *Proc. Natl. Acad. Sci. USA* 95:7103–8

Cideciyan AV, Jacobson SG, Aleman TS, Gu D, Pearce-Kelling SE, et al. 2005. In vivo dynamics of retinal injury and repair in the rhodopsin mutant dog model of human retinitis pigmentosa. *Proc. Natl. Acad. Sci. USA* 102:5233–38

Cideciyan AV, Swider M, Aleman TS, Roman MI, Sumaroka A, et al. 2007b. Reduced-illuminance autofluorescence imaging in ABCA4-associated retinal degenerations. *J. Opt. Soc. Am. A* 24:1457–67

Clarke G, Collins RA, Leavitt BR, Andrews DF, Hayden MR, et al. 2000. A one-hit model of cell death in inherited neuronal degenerations. *Nature* 406:195–99

Clarke G, Lumsden CJ. 2005. Scale-free neurodegeneration: cellular heterogeneity and the stretched exponential kinetics of cell death. *J. Theor. Biol.* 233:515–25

Colicos MA, Dash PK. 1996. Apoptotic morphology of dentate gyrus granule cells following experimental cortical impact injury in rats: possible role in spatial memory deficits. *Brain Res.* 739:120–31

Cottet S, Michaut L, Boisset G, Schlecht U, Gehring W, Schorderet DF. 2006. Biological characterization of gene response in Rpe65-/- mouse model of Leber's congenital amaurosis during progression of the disease. *FASEB J.* 20:2036–49

Daiger SP, Bowne SJ, Sullivan LS. 2007. Perspective on genes and mutations causing retinitis pigmentosa. *Arch. Ophthalmol.* 125:151–58

Deery EC, Vithana EN, Newbold RJ, Gallon VA, Bhattacharya SS, et al. 2002. Disease mechanism for retinitis pigmentosa (RP11) caused by mutations in the splicing factor gene PRPF31. *Hum. Mol. Genet.* 11:3209–19

Dejneka NS, Surace EM, Aleman TS, Cideciyan AV, Lyubarsky A, et al. 2004. In utero gene therapy rescues vision in a murine model of congenital blindness. *Mol. Ther.* 9:182–88

Delmelle M. 1977. Retinal damage by light: possible implication of singlet oxygen. *Biophys. Struct. Mech.* 3:195–98

Delyfer MN, Leveillard T, Mohand-Said S, Hicks D, Picaud S, Sahel JA. 2004. Inherited retinal degenerations: therapeutic prospects. *Biol. Cell* 96:261–69

de Melo Reis RA, Ventura AL, Schitine CS, de Mello MC, de Mello FG. 2008. Müller glia as an active compartment modulating nervous activity in the vertebrate retina: neurotransmitters and trophic factors. *Neurochem. Res.* 33:1466–74

Demos C, Bandyopadhyay M, Rohrer B. 2008. Identification of candidate genes for human retinal degeneration loci using differentially expressed genes from mouse photoreceptor dystrophy models. *Mol. Vis.* 14:1639–49

den Hollander AI, Roepman R, Koenekoop RK, Cremers FP. 2008. Leber congenital amaurosis: genes, proteins and disease mechanisms. *Prog. Retin. Eye Res.* 27:391–419

Donovan M, Doonan F, Cotter TG. 2006. Decreased expression of proapoptotic Bcl-2 family members during retinal development and differential sensitivity to cell death. *Dev. Biol.* 291:154–69

Doonan F, Donovan M, Cotter TG. 2005. Activation of multiple pathways during photoreceptor apoptosis in the rd mouse. *Invest. Ophthalmol. Vis. Sci.* 46:3530–38

Doonan J, Hunt T. 1996. Cell cycle. Why don't plants get cancer? *Nature* 380:481–82

Faktorovich EG, Steinberg RH, Yasumura D, Matthes MT, LaVail MM. 1990. Photoreceptor degeneration in inherited retinal dystrophy delayed by basic fibroblast growth factor. *Nature* 347:83–86

Farber DB. 1995. From mice to men: the cyclic GMP phosphodiesterase gene in vision and disease. The Proctor Lecture. *Invest. Ophthalmol. Vis. Sci.* 36:263–75

Franze K, Grosche J, Skatchkov SN, Schinkinger S, Foja C, et al. 2007. Muller cells are living optical fibers in the vertebrate retina. *Proc. Natl. Acad. Sci. USA* 104:8287–92

Fridlich R, Delalande F, Jaillard C, Lu J, Poidevin L, et al. 2009. The thioredoxin-like protein rod-derived cone viability factor (RdCVFL) interacts with TAU and inhibits its phosphorylation in the retina. *Mol. Cell Proteomics* 8:1206–18

Gibbs D, Cideciyan AV, Jacobson SG, Williams DS. 2009. Retinal pigment epithelium defects in humans and mice with mutations in MYO7A: imaging melanosome-specific autofluorescence. *Invest. Ophthalmol. Vis. Sci.* 50:4386–93

Gold B, Merriam JE, Zernant J, Hancox LS, Taiber AJ, et al. 2006. Variation in factor B (BF) and complement component 2 (C2) genes is associated with age-related macular degeneration. *Nat. Genet.* 38:458–62

Gosens I, den Hollander AI, Cremers FP, Roepman R. 2008. Composition and function of the Crumbs protein complex in the mammalian retina. *Exp. Eye Res.* 86:713–26

Grasl-Kraupp B, Ruttkay-Nedecky B, Koudelka H, Bukowska K, Bursch W, Schulte-Hermann R. 1995. In situ detection of fragmented DNA (TUNEL assay) fails to discriminate among apoptosis, necrosis, and autolytic cell death: a cautionary note. *Hepatology* 21:1465–68

Graymore C. 1960. Metabolism of the developing retina. III. Respiration in the developing normal rat retina and the effect of an inherited degeneration of the retinal neuroepithelium. *Br. J. Ophthalmol.* 44:363–69

Green DR, Kroemer G. 2004. The pathophysiology of mitochondrial cell death. *Science* 305:626–29

Grossman GH, Pauer GJ, Narendra U, Peachey NS, Hagstrom SA. 2009. Early synaptic defects in tulp1-/- mice. *Invest. Ophthalmol. Vis. Sci.* 50:3074–83

Gupta N, Brown KE, Milam AH. 2003. Activated microglia in human retinitis pigmentosa, late-onset retinal degeneration, and age-related macular degeneration. *Exp. Eye Res.* 76:463–71

Hackam AS, Strom R, Liu D, Qian J, Wang C, et al. 2004. Identification of gene expression changes associated with the progression of retinal degeneration in the rd1 mouse. *Invest. Ophthalmol. Vis. Sci.* 45:2929–42

Hahn P, Lindsten T, Ying GS, Bennett J, Milam AH, et al. 2003. Proapoptotic bcl-2 family members, Bax and Bak, are essential for developmental photoreceptor apoptosis. *Invest. Ophthalmol. Vis. Sci.* 44:3598–605

Haines JL, Hauser MA, Schmidt S, Scott WK, Olson LM, et al. 2005. Complement factor H variant increases the risk of age-related macular degeneration. *Science* 308:419–21

Halliwell B, Gutteridge JMC, eds. 2007. *Free Radicals in Biology and Medicine.* London: Oxford Univ. Press. 4th ed.

Harada T, Harada C, Kohsaka S, Wada E, Yoshida K, et al. 2002. Microglia-Muller glia cell interactions control neurotrophic factor production during light-induced retinal degeneration. *J. Neurosci.* 22:9228–36

Hartong DT, Berson EL, Dryja TP. 2006. Retinitis pigmentosa. *Lancet* 368:1795–809

Hartong DT, Dange M, McGee TL, Berson EL, Dryja TP, Colman RF. 2008. Insights from retinitis pigmentosa into the roles of isocitrate dehydrogenases in the Krebs cycle. *Nat. Genet.* 40:1230–34

Hashimoto T, Gibbs D, Lillo C, Azarian SM, Legacki E, et al. 2007. Lentiviral gene replacement therapy of retinas in a mouse model for Usher syndrome type 1B. *Gene Ther.* 14:584–94

Hauswirth WW, Aleman TS, Kaushal S, Cideciyan AV, Schwartz SB, et al. 2008. Treatment of Leber congenital amaurosis due to RPE65 mutations by ocular subretinal injection of adeno-associated virus gene vector: short-term results of a phase I trial. *Hum. Gene Ther.* 19:979–90

Hayward C, Shu X, Cideciyan AV, Lennon A, Barran P, et al. 2003. Mutation in a short-chain collagen gene, CTRP5, results in extracellular deposit formation in late-onset retinal degeneration: a genetic model for age-related macular degeneration. *Hum. Mol. Genet.* 12:2657–67

Hirano T, Ishihara K, Hibi M. 2000. Roles of STAT3 in mediating the cell growth, differentiation and survival signals relayed through the IL-6 family of cytokine receptors. *Oncogene* 19:2548–56

Huang SP, Lin PK, Liu JH, Khor CN, Lee YJ. 2004. Intraocular gene transfer of ciliary neurotrophic factor rescues photoreceptor degeneration in RCS rats. *J. Biomed. Sci.* 11:37–48

Huang Y, Cideciyan AV, Papastergiou GI, Banin E, Semple-Rowland SL, et al. 1998. Relation of optical coherence tomography to microanatomy in normal and rd chickens. *Invest. Ophthalmol. Vis. Sci.* 39:2405–16

Hunter DG, Fishman GA, Mehta RS, Kretzer FL. 1986. Abnormal sperm and photoreceptor axonemes in Usher's syndrome. *Arch. Ophthalmol.* 104:385–89

Jacobson SG, Acland GM, Aguirre GD, Aleman TS, Schwartz SB, et al. 2006a. Safety of recombinant adeno-associated virus type 2-RPE65 vector delivered by ocular subretinal injection. *Mol. Ther.* 13:1074–84

Jacobson SG, Aleman TS, Cideciyan AV, Heon E, Golczak M, et al. 2007. Human cone photoreceptor dependence on RPE65 isomerase. *Proc. Natl. Acad. Sci. USA* 104:15123–28

Jacobson SG, Aleman TS, Cideciyan AV, Roman AJ, Sumaroka A, et al. 2009a. Defining the residual vision in leber congenital amaurosis caused by RPE65 mutations. *Invest. Ophthalmol. Vis. Sci.* 50:2368–75

Jacobson SG, Aleman TS, Cideciyan AV, Sumaroka A, Schwartz SB, et al. 2005. Identifying photoreceptors in blind eyes caused by RPE65 mutations: prerequisite for human gene therapy success. *Proc. Natl. Acad. Sci. USA* 102:6177–82

Jacobson SG, Aleman TS, Sumaroka A, Cideciyan AV, Roman AJ, et al. 2009b. Disease boundaries in the retina of patients with Usher syndrome caused by MYO7A gene mutations. *Invest. Ophthalmol. Vis. Sci.* 50:1886–94

Jacobson SG, Cideciyan AV, Aleman TS, Pianta MJ, Sumaroka A, et al. 2003. Crumbs homolog 1 (CRB1) mutations result in a thick human retina with abnormal lamination. *Hum. Mol. Genet.* 12:1073–78

Jacobson SG, Cideciyan AV, Aleman TS, Sumaroka A, Roman AJ, et al. 2008a. Usher syndromes due to MYO7A, PCDH15, USH2A or GPR98 mutations share retinal disease mechanism. *Hum. Mol. Genet.* 17:2405–15

Jacobson SG, Cideciyan AV, Aleman TS, Sumaroka A, Windsor EA, et al. 2008b. Photoreceptor layer topography in children with Leber congenital amaurosis caused by RPE65 mutations. *Invest. Ophthalmol. Vis. Sci.* 49:4573–77

Jacobson SG, Cideciyan AV, Regunath G, Rodriguez FJ, Vandenburgh K, et al. 1995. Night blindness in Sorsby's fundus dystrophy reversed by vitamin A. *Nat. Genet.* 11:27–32

Jacobson SG, Cideciyan AV, Sumaroka A, Aleman TS, Schwartz SB, et al. 2006b. Remodeling of the human retina in choroideremia: rab escort protein 1 (REP-1) mutations. *Invest. Ophthalmol. Vis. Sci.* 47:4113–20

Jacobson SG, Cideciyan AV, Wright E, Wright AF. 2001. Phenotypic marker for early disease detection in dominant late-onset retinal degeneration. *Invest. Ophthalmol. Vis. Sci.* 42:1882–90

Jacobson SG, Voigt WJ, Parel JM, Apathy PP, Nghiem-Phu L, et al. 1986. Automated light- and dark-adapted perimetry for evaluating retinitis pigmentosa. *Ophthalmology* 93:1604–11

Jager RD, Mieler WF, Miller JW. 2008. Age-related macular degeneration. *N. Engl. J. Med.* 358:2606–17

Janssen-Heininger YM, Mossman BT, Heintz NH, Forman HJ, Kalyanaraman B, et al. 2008. Redox-based regulation of signal transduction: principles, pitfalls, and promises. *Free Radic. Biol. Med.* 45:1–17

Jin H, Nachury MV. 2009. The BBSome. *Curr. Biol.* 19:R472–73

Jin S, DiPaola RS, Mathew R, White E. 2007. Metabolic catastrophe as a means to cancer cell death. *J. Cell Sci.* 120:379–83

Joly S, Lange C, Thiersch M, Samardzija M, Grimm C. 2008. Leukemia inhibitory factor extends the lifespan of injured photoreceptors in vivo. *J. Neurosci.* 28:13765–74

Kamphuis W, Dijk F, van Soest S, Bergen AA. 2007. Global gene expression profiling of ischemic preconditioning in the rat retina. *Mol. Vis.* 13:1020–30

Kassen SC, Ramanan V, Montgomery JE, Burket T, Liu CG, et al. 2007. Time course analysis of gene expression during light-induced photoreceptor cell death and regeneration in albino zebrafish. *Dev. Neurobiol.* 67:1009–31

Kassen SC, Thummel R, Campochiaro LA, Harding MJ, Bennett NA, Hyde DR. 2009. CNTF induces photoreceptor neuroprotection and Muller glial cell proliferation through two different signaling pathways in the adult zebrafish retina. *Exp. Eye Res.* 88:1051–64

Klein RJ, Zeiss C, Chew EY, Tsai JY, Sackler RS, et al. 2005. Complement factor H polymorphism in age-related macular degeneration. *Science* 308:385–89

Komeima K, Rogers BS, Campochiaro PA. 2007. Antioxidants slow photoreceptor cell death in mouse models of retinitis pigmentosa. *J. Cell. Physiol.* 213:809–15

Komeima K, Rogers BS, Lu L, Campochiaro PA. 2006. Antioxidants reduce cone cell death in a model of retinitis pigmentosa. *Proc. Natl. Acad. Sci. USA* 103:11300–5

Komeima K, Usui S, Shen J, Rogers BS, Campochiaro PA. 2008. Blockade of neuronal nitric oxide synthase reduces cone cell death in a model of retinitis pigmentosa. *Free Radic. Biol. Med.* 45:905–12

Kosmaoglou M, Schwarz N, Bett JS, Cheetham ME. 2008. Molecular chaperones and photoreceptor function. *Prog. Retin. Eye Res.* 27:434–49

Kroemer G, Galluzzi L, Vandenabeele P, Abrams J, Alnemri ES, et al. 2009. Classification of cell death: recommendations of the Nomenclature Committee on Cell Death 2009. *Cell Death Differ.* 16:3–11

Krsti** RV. 1991. *Human Microscopic Anatomy: An Atlas for Students of Medicine and Biology.* Berlin/New York: Springer-Verlag. xvi, 616 pp.

Kuntz CA, Jacobson SG, Cideciyan AV, Li ZY, Stone EM, et al. 1996. Sub-retinal pigment epithelial deposits in a dominant late-onset retinal degeneration. *Invest. Ophthalmol. Vis. Sci.* 37:1772–82

Lai CM, Yu MJ, Brankov M, Barnett NL, Zhou X, et al. 2004. Recombinant adeno-associated virus type 2-mediated gene delivery into the Rpe65-/- knockout mouse eye results in limited rescue. *Genet. Vaccines Ther.* 2:3

Lamba DA, Gust J, Reh TA. 2009. Transplantation of human embryonic stem cell-derived photoreceptors restores some visual function in Crx-deficient mice. *Cell Stem. Cell* 4:73–79

Lambert AJ, Boysen HM, Buckingham JA, Yang T, Podlutsky A, et al. 2007. Low rates of hydrogen peroxide production by isolated heart mitochondria associate with long maximum lifespan in vertebrate homeotherms. *Aging Cell* 6:607–18

Lancaster MA, Gleeson JG. 2009. The primary cilium as a cellular signaling center: lessons from disease. *Curr. Opin. Genet. Dev.* 19:220–29

Langmann T. 2007. Microglia activation in retinal degeneration. *J. Leukoc. Biol.* 81:1345–51

Leveillard T, Mohand-Said S, Lorentz O, Hicks D, Fintz AC, et al. 2004. Identification and characterization of rod-derived cone viability factor. *Nat. Genet.* 36:755–59

Li ZY, Jacobson SG, Milam AH. 1994. Autosomal dominant retinitis pigmentosa caused by the threonine-17-methionine rhodopsin mutation: retinal histopathology and immunocytochemistry. *Exp. Eye Res.* 58:397–408

Lin JH, Li H, Yasumura D, Cohen HR, Zhang C, et al. 2007. IRE1 signaling affects cell fate during the unfolded protein response. *Science* 318:944–49

Lion Y, Delmelle M, van de Vorst A. 1976. New method of detecting singlet oxygen production. *Nature* 263:442–43

Lohr HR, Kuntchithapautham K, Sharma AK, Rohrer B. 2006. Multiple, parallel cellular suicide mechanisms participate in photoreceptor cell death. *Exp. Eye Res.* 83:380–89

Lyness AL, Ernst W, Quinlan MP, Clover GM, Arden GB, et al. 1985. A clinical, psychophysical, and electroretinographic survey of patients with autosomal dominant retinitis pigmentosa. *Br. J. Ophthalmol.* 69:326–39

MacDonald IM, Sauve Y, Sieving PA. 2007. Preventing blindness in retinal disease: ciliary neurotrophic factor intraocular implants. *Can. J. Ophthalmol.* 42:399–402

MacLaren RE, Pearson RA, MacNeil A, Douglas RH, Salt TE, et al. 2006. Retinal repair by transplantation of photoreceptor precursors. *Nature* 444:203–7

Maeda A, Maeda T, Golczak M, Chou S, Desai A, et al. 2009. Involvement of all-trans-retinal in acute light-induced retinopathy of mice. *J. Biol. Chem.* 284:15173–83

Maerker T, van Wijk E, Overlack N, Kersten FF, McGee J, et al. 2008. A novel Usher protein network at the periciliary reloading point between molecular transport machineries in vertebrate photoreceptor cells. *Hum. Mol. Genet.* 17:71–86

Maguire AM, Simonelli F, Pierce EA, Pugh EN Jr, Mingozzi F, et al. 2008. Safety and efficacy of gene transfer for Leber's congenital amaurosis. *N. Engl. J. Med.* 358:2240–48

Malhotra JD, Miao H, Zhang K, Wolfson A, Pennathur S, et al. 2008. Antioxidants reduce endoplasmic reticulum stress and improve protein secretion. *Proc. Natl. Acad. Sci. USA* 105:18525–30

Massof RW, Finkelstein D. 1979. Rod sensitivity relative to cone sensitivity in retinitis pigmentosa. *Invest. Ophthalmol. Vis. Sci.* 18:263–72

Massof RW, Finkelstein D. 1981. Two forms of autosomal dominant primary retinitis pigmentosa. *Doc. Ophthalmol.* 51:289–346

McGill TJ, Lund RD, Douglas RM, Wang S, Lu B, et al. 2007. Syngeneic Schwann cell transplantation preserves vision in RCS rat without immunosuppression. *Invest. Ophthalmol. Vis. Sci.* 48:1906–12

McMahon A, Kedzierski W. 2009. Polyunsaturated extremely long chain C28-C36 fatty acids and retinal physiology. *Br. J. Ophthalmol.* In press

Milam AH, Rose L, Cideciyan AV, Barakat MR, Tang WX, et al. 2002. The nuclear receptor NR2E3 plays a role in human retinal photoreceptor differentiation and degeneration. *Proc. Natl. Acad. Sci. USA* 99:473–78

Mordes D, Yuan L, Xu L, Kawada M, Molday RS, Wu JY. 2007. Identification of photoreceptor genes affected by PRPF31 mutations associated with autosomal dominant retinitis pigmentosa. *Neurobiol. Dis.* 26:291–300

Morgan BP, Gasque P. 1996. Expression of complement in the brain: role in health and disease. *Immunol. Today* 17:461–66

Mortimer SE, Xu D, McGrew D, Hamaguchi N, Lim HC, et al. 2008. IMP dehydrogenase type 1 associates with polyribosomes translating rhodopsin mRNA. *J. Biol. Chem.* 283:36354–60

Murphy MP. 2009. How mitochondria produce reactive oxygen species. *Biochem. J.* 417:1–13

Nakamura T, Lipton SA. 2007. Molecular mechanisms of nitrosative stress-mediated protein misfolding in neurodegenerative diseases. *Cell Mol. Life Sci.* 64:1609–20

Nakazawa T, Takeda M, Lewis GP, Cho KS, Jiao J, et al. 2007. Attenuated glial reactions and photoreceptor degeneration after retinal detachment in mice deficient in glial fibrillary acidic protein and vimentin. *Invest. Ophthalmol. Vis. Sci.* 48:2760–68

Natoli R, Provis J, Valter K, Stone J. 2008. Expression and role of the early-response gene Oxr1 in the hyperoxia-challenged mouse retina. *Invest. Ophthalmol. Vis. Sci.* 49:4561–67

Ni YQ, Xu GZ, Hu WZ, Shi L, Qin YW, Da CD. 2008. Neuroprotective effects of naloxone against light-induced photoreceptor degeneration through inhibiting retinal microglial activation. *Invest. Ophthalmol. Vis. Sci.* 49:2589–98

Noell WK. 1953. Experimentally induced toxic effects on structure and function of visual cells and pigment epithelium. *Am. J. Ophthalmol.* 36:103–16

Organisciak DT, Darrow RM, Barsalou L, Kutty RK, Wiggert B. 2000. Circadian-dependent retinal light damage in rats. *Invest. Ophthalmol. Vis. Sci.* 41:3694–701

Otani A, Dorrell MI, Kinder K, Moreno SK, Nusinowitz S, et al. 2004. Rescue of retinal degeneration by intravitreally injected adult bone marrow-derived lineage-negative hematopoietic stem cells. *J. Clin. Invest.* 114:765–74

Otani A, Kinder K, Ewalt K, Otero FJ, Schimmel P, Friedlander M. 2002. Bone marrow-derived stem cells target retinal astrocytes and can promote or inhibit retinal angiogenesis. *Nat. Med.* 8:1004–10

Pacione LR, Szego MJ, Ikeda S, Nishina PM, McInnes RR. 2003. Progress toward understanding the genetic and biochemical mechanisms of inherited photoreceptor degenerations. *Annu. Rev. Neurosci.* 26:657–700

Padnick-Silver L, Derwent JJ, Giuliano E, Narfstrom K, Linsenmeier RA. 2006. Retinal oxygenation and oxygen metabolism in Abyssinian cats with a hereditary retinal degeneration. *Invest. Ophthalmol. Vis. Sci.* 47:3683–89

Pang JJ, Chang B, Kumar A, Nusinowitz S, Noorwez SM, et al. 2006. Gene therapy restores vision-dependent behavior as well as retinal structure and function in a mouse model of RPE65 Leber congenital amaurosis. *Mol. Ther.* 13:565–72

Paquet-Durand F, Johnson L, Ekstrom P. 2007. Calpain activity in retinal degeneration. *J. Neurosci. Res.* 85:693–702

Partridge L. 1997. Evolutionary biology and age-related mortality. In *Between Zeus and the Salmon*, ed. KW Wachter, CE Finch, pp. 78–95. Washington, DC: Natl. Acad. Press

Paskowitz DM, LaVail MM, Duncan JL. 2006. Light and inherited retinal degeneration. *Br. J. Ophthalmol.* 90:1060–66

Peterson WM, Wang Q, Tzekova R, Wiegand SJ. 2000. Ciliary neurotrophic factor and stress stimuli activate the Jak-STAT pathway in retinal neurons and glia. *J. Neurosci.* 20:4081–90

Portera-Cailliau C, Sung CH, Nathans J, Adler R. 1994. Apoptotic photoreceptor cell death in mouse models of retinitis pigmentosa. *Proc. Natl. Acad. Sci. USA* 91:974–78

Punzo C, Kornacker K, Cepko CL. 2009. Stimulation of the insulin/mTOR pathway delays cone death in a mouse model of retinitis pigmentosa. *Nat. Neurosci.* 12:44–52

Quintiliani M. 1986. The oxygen effect in radiation inactivation of DNA and enzymes. *Int. J. Radiat. Biol. Relat. Stud. Phys. Chem. Med.* 50:573–94

Rattner A, Nathans J. 2005. The genomic response to retinal disease and injury: evidence for endothelin signaling from photoreceptors to glia. *J. Neurosci.* 25:4540–49

Rattner A, Nathans J. 2006. Macular degeneration: recent advances and therapeutic opportunities. *Nat. Rev. Neurosci.* 7:860–72

Rattner A, Sun H, Nathans J. 1999. Molecular genetics of human retinal disease. *Annu. Rev. Genet.* 33:89–131

Ripps H. 2002. Cell death in retinitis pigmentosa: gap junctions and the 'bystander' effect. *Exp. Eye Res.* 74:327–36

Robson AG, Michaelides M, Saihan Z, Bird AC, Webster AR, et al. 2008. Functional characteristics of patients with retinal dystrophy that manifest abnormal parafoveal annuli of high density fundus autofluorescence; a review and update. *Doc. Ophthalmol.* 116:79–89

Rohrer B, Demos C, Frigg R, Grimm C. 2007a. Classical complement activation and acquired immune response pathways are not essential for retinal degeneration in the rd1 mouse. *Exp. Eye Res.* 84:82–91

Rohrer B, Guo Y, Kunchithapautham K, Gilkeson GS. 2007b. Eliminating complement factor D reduces photoreceptor susceptibility to light-induced damage. *Invest. Ophthalmol. Vis. Sci.* 48:5282–89

Rohrer B, Pinto FR, Hulse KE, Lohr HR, Zhang L, Almeida JS. 2004. Multidestructive pathways triggered in photoreceptor cell death of the rd mouse as determined through gene expression profiling. *J. Biol. Chem.* 279:41903–10

Ron D, Walter P. 2007. Signal integration in the endoplasmic reticulum unfolded protein response. *Nat. Rev. Mol. Cell Biol.* 8:519–29

Roque RS, Caldwell RB. 1993. Isolation and culture of retinal microglia. *Curr. Eye Res.* 12:285–90

Saint-Geniez M, Maharaj AS, Walshe TE, Tucker BA, Sekiyama E, et al. 2008. Endogenous VEGF is required for visual function: evidence for a survival role on Müller cells and photoreceptors. *PLoS One* 3:e3554

Saliba RS, Munro PM, Luthert PJ, Cheetham ME. 2002. The cellular fate of mutant rhodopsin: quality control, degradation and aggresome formation. *J. Cell Sci.* 115:2907–18

Sancho-Pelluz J, Arango-Gonzalez B, Kustermann S, Romero FJ, van Veen T, et al. 2008. Photoreceptor cell death mechanisms in inherited retinal degeneration. *Mol. Neurobiol.* 38:253–69

Sanz MM, Johnson LE, Ahuja S, Ekström PA, Romero J, van Veen T. 2007. Significant photoreceptor rescue by treatment with a combination of antioxidants in an animal model for retinal degeneration. *Neuroscience* 145:1120–29

Sarup V, Patil K, Sharma SC. 2004. Ciliary neurotrophic factor and its receptors are differentially expressed in the optic nerve transected adult rat retina. *Brain Res.* 1013:152–58

Sasahara M, Otani A, Oishi A, Kojima H, Yodoi Y, et al. 2008. Activation of bone marrow-derived microglia promotes photoreceptor survival in inherited retinal degeneration. *Am. J. Pathol.* 172:1693–703

Schmitz F, Tabares L, Khimich D, Strenzke N, de la Villa-Polo P, et al. 2006. CSPalpha-deficiency causes massive and rapid photoreceptor degeneration. *Proc. Natl. Acad. Sci. USA* 103:2926–31

Schutte M, Werner P. 1998. Redistribution of glutathione in the ischemic rat retina. *Neurosci. Lett.* 246:53–56

Shu X, Black GC, Rice JM, Hart-Holden N, Jones A, et al. 2007. RPGR mutation analysis and disease: an update. *Hum. Mutat.* 28:322–28

Sieving PA, Caruso RC, Tao W, Coleman HR, Thompson DJ, et al. 2006. Ciliary neurotrophic factor (CNTF) for human retinal degeneration: phase I trial of CNTF delivered by encapsulated cell intraocular implants. *Proc. Natl. Acad. Sci. USA* 103:3896–901

Snodderly DM, Sandstrom MM, Leung IY, Zucker CL, Neuringer M. 2002. Retinal pigment epithelial cell distribution in central retina of rhesus monkeys. *Invest. Ophthalmol. Vis. Sci.* 43:2815–18

Solovei I, Kreysing M, Lanctot C, Kosem S, Peichl L, et al. 2009. Nuclear architecture of rod photoreceptor cells adapts to vision in mammalian evolution. *Cell* 137:356–68

Sparrow JR, Parish CA, Hashimoto M, Nakanishi K. 1999. A2E, a lipofuscin fluorophore, in human retinal pigmented epithelial cells in culture. *Invest. Ophthalmol. Vis. Sci.* 40:2988–95

Stasi K, Nagel D, Yang X, Wang RF, Ren L, et al. 2006. Complement component 1Q (C1Q) upregulation in retina of murine, primate, and human glaucomatous eyes. *Invest. Ophthalmol. Vis. Sci.* 47:1024–29

Steele MR, Inman DM, Calkins DJ, Horner PJ, Vetter ML. 2006. Microarray analysis of retinal gene expression in the DBA/2J model of glaucoma. *Invest. Ophthalmol. Vis. Sci.* 47:977–85

Steinberg RH. 1994. Survival factors in retinal degenerations. *Curr. Opin. Neurobiol.* 4:515–24

Stone EM. 2007. Leber congenital amaurosis—a model for efficient genetic testing of heterogeneous disorders: LXIV Edward Jackson Memorial Lecture. *Am. J. Ophthalmol.* 144:791–811

Stone J, Maslim J, Valter-Kocsi K, Mervin K, Bowers F, et al. 1999. Mechanisms of photoreceptor death and survival in mammalian retina. *Prog. Retin. Eye Res.* 18:689–735

Streichert LC, Birnbach CD, Reh TA. 1999. A diffusible factor from normal retinal cells promotes rod photoreceptor survival in an in vitro model of retinitis pigmentosa. *J. Neurobiol.* 39:475–90

Sun H, Nathans J. 2001. ABCR, the ATP-binding cassette transporter responsible for Stargardt macular dystrophy, is an efficient target of all-trans-retinal-mediated photooxidative damage in vitro. Implications for retinal disease. *J. Biol. Chem.* 276:11766–74

Swaroop A, Branham KE, Chen W, Abecasis G. 2007. Genetic susceptibility to age-related macular degeneration: a paradigm for dissecting complex disease traits. *Hum. Mol. Genet.* 16(Spec. No. 2):R174–82

Szamier RB, Berson EL. 1977. Retinal ultrastructure in advanced retinitis pigmentosa. *Invest. Ophthalmol. Vis. Sci.* 16:947–62

Szamier RB, Berson EL, Klein R, Meyers S. 1979. Sex-linked retinitis pigmentosa: ultrastructure of photoreceptors and pigment epithelium. *Invest. Ophthalmol. Vis. Sci.* 18:145–60

Tait SW, Green DR. 2008. Caspase-independent cell death: leaving the set without the final cut. *Oncogene* 27:6452–61

Tao W, Wen R, Goddard MB, Sherman SD, O'Rourke PJ, et al. 2002. Encapsulated cell-based delivery of CNTF reduces photoreceptor degeneration in animal models of retinitis pigmentosa. *Invest. Ophthalmol. Vis. Sci.* 43:3292–98

Taub E, Uswatte G, Elbert T. 2002. New treatments in neurorehabilitation founded on basic research. *Nat. Rev. Neurosci.* 3:228–36

Tobaben S, Thakur P, Fernández-Chacón R, Südhof TC, Rettig J, Stahl B. 2001. A trimeric protein complex functions as a synaptic chaperone machine. *Neuron* 31:987–99

Torbidoni V, Iribarne M, Suburo AM. 2006. Endothelin receptors in light-induced retinal degeneration. *Exp. Biol. Med.* 231:1095–100

Travis GH. 1998. Mechanisms of cell death in the inherited retinal degenerations. *Am. J. Hum. Genet.* 62:503–8

Travis GH, Golczak M, Moise AR, Palczewski K. 2007. Diseases caused by defects in the visual cycle: retinoids as potential therapeutic agents. *Annu. Rev. Pharmacol. Toxicol.* 47:469–512

Ueki Y, Wang J, Chollangi S, Ash JD. 2008. STAT3 activation in photoreceptors by leukemia inhibitory factor is associated with protection from light damage. *J. Neurochem.* 105:784–96

Usui S, Komeima K, Lee SY, Jo YJ, Ueno S, et al. 2009. Increased expression of catalase and superoxide dismutase 2 reduces cone cell death in retinitis pigmentosa. *Mol. Ther.* 17:778–86

Verardo MR, Lewis GP, Takeda M, Linberg KA, Byun J, et al. 2008. Abnormal reactivity of muller cells after retinal detachment in mice deficient in GFAP and vimentin. *Invest. Ophthalmol. Vis. Sci.* 49:3659–65

Wald G, Zeavin BH. 1956. Rod and cone vision in retinitis pigmentosa. *Am. J. Ophthalmol.* 42:253–69

Wang Y, Smith SB, Ogilvie JM, McCool DJ, Sarthy V. 2002. Ciliary neurotrophic factor induces glial fibrillary acidic protein in retinal Muller cells through the JAK/STAT signal transduction pathway. *Curr. Eye Res.* 24:305–12

Whikehart DR. 2003. *Biochemistry of the Eye*. Philadelphia: Butterworth-Heinemann. 2nd ed.

Williams DS. 2008. Usher syndrome: animal models, retinal function of Usher proteins, and prospects for gene therapy. *Vis. Res.* 48:433–41

Williams DS, Aleman TS, Lillo C, Lopes VS, Hughes LC, et al. 2009. Harmonin in the murine retina and the retinal phenotypes of Ush1c-mutant mice and human USH1C. *Invest. Ophthalmol. Vis. Sci.* 50:3881–89

Wilson RB, Kunchithapautham K, Rohrer B. 2007. Paradoxical role of BDNF: BDNF+/- retinas are protected against light damage-mediated stress. *Invest. Ophthalmol. Vis. Sci.* 48:2877–86

Wolfrum U, Schmitt A. 2000. Rhodopsin transport in the membrane of the connecting cilium of mammalian photoreceptor cells. *Cell Motil. Cytoskelet.* 46:95–107

Woltjen K, Michael IP, Mohseni P, Desai R, Mileikovsky M, et al. 2009. piggyBac transposition reprograms fibroblasts to induced pluripotent stem cells. *Nature* 458:766–70

Woodson JD, Chory J. 2008. Coordination of gene expression between organellar and nuclear genomes. *Nat. Rev. Genet.* 9:383–95

Wright AF, Jacobson SG, Cideciyan AV, Roman AJ, Shu X, et al. 2004. Lifespan and mitochondrial control of neurodegeneration. *Nat. Genet.* 36:1153–58

Wright AF, Murphy MP, Turnbull DM. 2009. Do organellar genomes function as long-term redox damage sensors? *Trends Genet.* 25:253–61

Wright AF, Shu X. 2007. Focus on molecules: RPGR. *Exp. Eye Res.* 85:1–2

Wu Y, Fishkin NE, Pande A, Pande J, Sparrow JR. 2009. Novel lipofuscin bisretinoids prominent in human retina and in a model of recessive Stargardt disease. *J. Biol. Chem.* 284:20,155–66

Wyszecki G, Stiles WS. 1982. *Color Science, Concepts and Methods, Quantitative Data and Formulas*, pp. 108–11. New York: Wiley. 2nd ed.

Xu GZ, Li WW, Tso MO. 1996. Apoptosis in human retinal degenerations. *Trans. Am. Ophthalmol. Soc.* 94:411–30; discussion 30–31

Yanai D, Weiland JD, Mahadevappa M, Greenberg RJ, Fine I, Humayun MS. 2007. Visual performance using a retinal prosthesis in three subjects with retinitis pigmentosa. *Am. J. Ophthalmol.* 143:820–27

Yang LP, Li Y, Zhu XA, Tso MO. 2007. Minocycline delayed photoreceptor death in rds mice through iNOS-dependent mechanism. *Mol. Vis.* 13:1073–82

Yang Y, Mohand-Said S, Danan A, Simonutti M, Fontaine V, et al. 2009. Functional cone rescue by RdCVF protein in a dominant model of retinitis pigmentosa. *Mol. Ther.* 17:787–95

Yates JR, Sepp T, Matharu BK, Khan JC, Thurlby DA, et al. 2007. Complement C3 variant and the risk of age-related macular degeneration. *N. Engl. J. Med.* 357:553–61

Young RW. 1971. Shedding of discs from rod outer segments in the rhesus monkey. *J. Ultrastruct. Res.* 34:190–203

Yu DY, Cringle S, Valter K, Walsh N, Lee D, Stone J. 2004. Photoreceptor death, trophic factor expression, retinal oxygen status, and photoreceptor function in the P23H rat. *Invest. Ophthalmol. Vis. Sci.* 45:2013–19

Yu DY, Cringle SJ. 2005. Retinal degeneration and local oxygen metabolism. *Exp. Eye Res.* 80:745–51

Yu DY, Cringle SJ, Su EN, Yu PK. 2000. Intraretinal oxygen levels before and after photoreceptor loss in the RCS rat. *Invest. Ophthalmol. Vis. Sci.* 41:3999–4006

Zacks DN, Han Y, Zeng Y, Swaroop A. 2006. Activation of signaling pathways and stress-response genes in an experimental model of retinal detachment. *Invest. Ophthalmol. Vis. Sci.* 47:1691–95

Zeng HY, Zhu XA, Zhang C, Yang LP, Wu LM, Tso MO. 2005. Identification of sequential events and factors associated with microglial activation, migration, and cytotoxicity in retinal degeneration in rd mice. *Invest. Ophthalmol. Vis. Sci.* 46:2992–99

Genetics and Cell Biology of Building Specific Synaptic Connectivity

Kang Shen[1] and Peter Scheiffele[2]

[1]Howard Hughes Medical Institute, Department of Biology and Pathology, Stanford University, Stanford, California 94305; email: kangshen@stanford.edu

[2]Department of Cell Biology, Biozentrum of the University of Basel, Basel 4056, Switzerland; email: peter.scheiffele@unibas.ch

Annu. Rev. Neurosci. 2010. 33:473–507

First published online as a Review in Advance on April 1, 2010

The *Annual Review of Neuroscience* is online at neuro.annualreviews.org

This article's doi: 10.1146/annurev.neuro.051508.135302

Key Words

recognition, trans-synaptic signaling, guidepost cell, synapse elimination

Abstract

The assembly of specific synaptic connections during development of the nervous system represents a remarkable example of cellular recognition and differentiation. Neurons employ several different cellular signaling strategies to solve this puzzle, which successively limit unwanted interactions and reduce the number of direct recognition events that are required to result in a specific connectivity pattern. Specificity mechanisms include the action of contact-mediated and long-range signals that support or inhibit synapse formation, which can take place directly between synaptic partners or with transient partners and transient cell populations. The molecular signals that drive the synaptic differentiation process at individual synapses in the central nervous system are similarly diverse and act through multiple, parallel differentiation pathways. This molecular complexity balances the need for central circuits to be assembled with high accuracy during development while retaining plasticity for local and dynamic regulation.

Contents

INTRODUCTION

The central nervous system is a gigantic network of neurons connected by an astronomical number of synapses. The connection specificity is critically important for the function of neuronal circuits. The structural organization of neuronal circuits emerges from a complex set of developmental events that include cell fate determination, cell migration, axon guidance, axonal and dendritic branch layer formation, synapse formation, and the activity-dependent maturation of synaptic circuits. Each of these successive steps contributes to wiring specificity

by gradually restricting the availability of potential synaptic partners.

In this review, we focus on the cellular and molecular mechanisms that generate synapse specificity during the process of synapse formation. We define synapse specificity as the cellular process that enables pre- and postsynaptic cells to select each other as synaptic partners among the surrounding cells and that allows synapses to form at particular subcellular locations. Neuronal activity plays profound roles in modifying synaptic circuits after their initial establishment. This topic has been reviewed extensively in several previous articles (Cohen &

Greenberg 2008, Katz & Shatz 1996); therefore, we do not delve deeply into this issue. Instead, we focus on the recent advances in our understanding of the molecular mechanisms that specify connectivity.

Many anatomical studies suggest that neurons have the exquisite ability to choose their appropriate synaptic partners from a dense array of potential targets. Quantitative analyses of a neuron's ability to select synaptic targets come from serial electron microscopy reconstruction studies. A reconstruction of an arbor from a retinal ganglion axon in the lateral geniculate nucleus revealed that it only selects 4 cells as synaptic partners from a total of 43 contacting cells (Hamos et al. 1987). A systematic survey of the *Caenorhabditis elegans* nervous system showed that one out of six contacting neurons form synapses onto each other (White 1986). Both of these studies highlight the fact that neurons actively choose synaptic partners from surrounding cells.

Neurons are highly differentiated cells with complex morphologies and distinct functional compartments. Emerging anatomical and physiological evidence suggests that synapses are often formed onto specific subcellular compartments between neurons. For example, inhibitory synapses formed onto the perisomatic domain of a postsynaptic neuron have profound impact on the action potential that fires in the postsynaptic cell, whereas inhibitory synapses that are formed onto distal dendrites primarily affect dendritic calcium spikes (Miles et al. 1996, Pouille & Scanziani 2004). Developmentally, this subcellular specificity phenomenon begs the question of how such a precise innervation pattern is generated during development.

Finally, the stoichiometry of synaptic connectivity is precisely regulated at two levels: first, with respect to the number of different presynaptic partners that innervate a single postsynaptic cell; and second, with respect to the number of synapses that are formed between a single afferent and its target cell. Some afferents sparsely innervate select targets but achieve reliable activation of the postsynaptic partners through efficacious synapses, whereas

other inputs require activation in concert with other inputs to be effective. The observed reproducibility of synaptic connection stoichiometry implies that this parameter is tightly controlled in developing circuits. In this article, we first describe different cell recognition events to shed light on how specificity is achieved during development. We detail examples of direct pre- and postsynaptic matching decisions and the functions of glia and guidepost cells in circuit formation. Then, we discuss molecular mechanisms that drive assembly and remodeling of synaptic connectivity and provide examples of the regulation of wiring molecules at the transcriptional level. Throughout the article, we combine information obtained in vertebrate and invertebrate model systems, using genetic and cell biological approaches, to identify general principles underlying the generation of synaptic specificity in the nervous system.

HOW IS SPECIFICITY GENERATED?

Conceptually, developing axons and dendrites can use a variety of cellular and molecular mechanisms to choose their synaptic targets and appropriate subcellular compartments. First, positive selection cues on the membranes of pre- and postsynaptic neurons can locally induce the assembly of pre- and postsynaptic differentiation. Under this mode of selection, synaptic partners express distinct sets of adhesion molecules or secreted anterograde and retrograde signals that drive the synaptic differentiation process (**Figure 1a**). Indeed, several synaptic signaling molecules are sufficient to induce the assembly of the pre- and postsynaptic apparatus. In the case of the vertebrate neuromuscular junction, interactions between axon, muscle, and components of the basal lamina can achieve a similar outcome of patterned synaptic connections.

Second, inhibitory cues presented on neurites or released from the local environment can prevent synapses from forming between particular neurons or at certain subcellular locations. Similar to the concept of repellents in

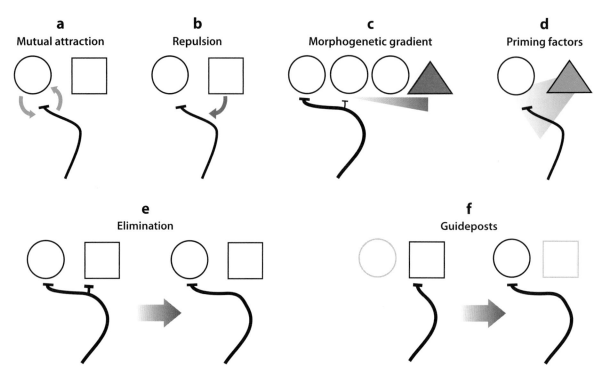

a Mutual attraction

b Repulsion

c Morphogenetic gradient

d Priming factors

e Elimination

f Guideposts

Figure 1

Model mechanisms for synaptic specificity during development. Interactions with appropriate synaptic partners can be accomplished by mutual attraction through positive regulators (*green*), or selective repulsion through negative regulators (*red*) derived from an inappropriate target cell or released by other cell types in the target territory in the form of a morphogenetic gradient. The synapse formation competence of afferents can be locally controlled by priming factors that are released in the target territory. During development, synaptic specificity can emerge through the elimination of contacts with inappropriate targets. The formation of transient synapses with guidepost cells (*squares*) provide a means of prepatterning synaptic structures before final target cells have arrived or matured in the target area. In some systems, guidepost cells are transient populations that are eliminated by cell death once the final wiring pattern has been accomplished.

axon guidance, diffusible and membrane tethered cues may exert negative constraints against synapse formation (**Figure 1b**). This is a relatively new area with less experimental evidence. However, a few studies show that negative regulators of synapse formation might also be used to encode connection specificity (discussed below). Interestingly, negative cues can be presented by a potential interaction partner or act in the form of a morphogenetic gradient that is generated by adjacent cells in the target area (**Figure 1c**).

Third, synapse elimination is a powerful way of achieving synaptic specificity (**Figure 1e**). This has been well documented in classical studies of neuromuscular junction maturation

as well as for many connections in the central nervous system. This type of regulation allows an initial, promiscuous phase of synapse formation, which leads to excessive synapse formation onto many targets, followed by a second phase of synapse elimination during which specific groups of synapses are eliminated. In many cases, this elimination process is driven by neuronal activity and serves as a way for neural activity to carve connectivity during the maturation of the synaptic circuit.

Last, besides the spatial regulation mentioned above, the temporal regulation of competence can serve as a way to limit synaptic partner choices (**Figure 1**). For example, axons establish many cellular contacts during axonal migration. However, neurons might not be competent to assemble synapses during this early phase of development, therefore avoiding ectopic connections. Once axons reach their target field, molecular programs for synapse formation are then upregulated. Indeed, several glia- and target-derived priming factors increase the number of synapses and induce the maturation of synapses and changes in gene expression in afferents (Christopherson et al. 2005, Diaz et al. 2002, Kalinovsky & Scheiffele 2004, Umemori et al. 2004). Importantly, the above-mentioned mechanisms are not mutually exclusive. Multiple strategies are likely used by the same neuron at different stages of development to establish, maintain, or remodel its synaptic connectivity.

Direct Matching of Synaptic Partners

The complex cellular environment during synapse development in vivo makes it difficult to discern whether synaptogenesis is initiated by contact between pre- and postsynaptic neurons and to exclude the involvement of surrounding cells. However, the fact that dissociated pure neuronal cultures can form functional synapses argues strongly that synapse formation is a cellular process that can be executed by synaptic partner cells alone (Bartlett & Banker 1984). Hence, one intuitive model for synapse specificity is that the recognition between synaptic partners induces synapse formation. This model suggests that designated synaptic partners possess molecular tags that direct synaptic connectivity. In this scenario, the recognition tags might either initiate synapse formation themselves or they promote selective cell-cell contacts and thereby selectively engage a common synaptogenic core machinery. The possibility that nonpartner cells in the target area might inhibit synapse formation, therefore avoiding the formation of inappropriate synapses, is less appreciated. Examples for each of these mechanisms, recognition and inhibitory signals, are discussed below.

Mutual attraction of synaptic partners by synaptic adhesion molecules. A simple model for the selective attraction between synaptic partner cells is the adhesion molecule-mediated match making accomplished by the coordinated expression of homophilic adhesion molecules in pre- and postsynaptic partners (Fannon & Colman 1996, Shapiro et al. 2007). An array of homophilic adhesion molecules has been identified, which includes cadherins and immunoglobulin-superfamily (IgSF) proteins. The substantial molecular diversification of isoforms within such gene families provides a molecular basis for an adhesive code with the potential for encoding a multitude of selective cell-cell interactions. Therefore, appropriate axons and target dendrites might be sorted into a common synaptic domain in the same way that cells with different cadherin expression can be sorted into different clusters of cells during early embryonic development (reviewed in Steinberg 2007).

This homophilic adhesion model for wiring has been tested extensively in the retina in which synapses between amacrine and bipolar cell interneurons that are formed with retinal ganglion cell dendrites are organized in a highly ordered laminar arrangement in the inner plexiform layer (IPL). Subpopulations of interneurons and retinal ganglion cells express different variants of a closely related group of IgSF-proteins: Sidekick-1, Sidekick-2, Dscam, and DscamL (Yamagata & Sanes 2008, Yamagata et al. 2002). Each of the IgSF proteins engages in strictly homophilic interactions. Interneuron processes and ganglion cell dendrites that express the same IgSF isoform elaborate arbors and synapses in the same IPL sublaminae. Importantly, the IgSF protein repertoire of synaptic partners is instructive for their connectivity. When the IgSF content of cells is perturbed by either RNA interference or the misexpression of other isoforms, the processes misproject into inappropriate sublaminae (Yamagata & Sanes 2008). Therefore, this homophilic adhesion system directs the lamina-specific targeting of neuronal processes. Whether Sidekick and Dscam proteins trigger the initiation of synapse assembly in these laminae remains to be explored. Notably, Sidekick and Dscam proteins are not restricted to the retina but are broadly expressed in neuronal populations throughout the CNS, in which they may have analogous roles in the formation of lamina-specific synapses.

Mutual attraction through synaptic adhesion molecules has also been implicated in sculpting the subcellular specificity of neuronal connections. The vast majority of excitatory synapses are formed on dendritic spines, whereas specific inhibitory presynaptic inputs are restricted to perisomatic domains, dendritic domains, or the axon initial segment. In principal neurons of the primary visual cortex, this innervation specificity persists even in the absence of functional sensory and thalamic inputs, which suggests that subcellular specificity is genetically encoded (Di Cristo et al. 2004). Subcellular domain-specific inputs are also observed in Purkinje cells in the cerebellum where two classes of GABAergic inputs from stellate and basket cells are segregated into the dendritic domain and the axon initial segment, respectively. The restriction of basket cell inputs to the initial segment depends on the IgSF protein NrCAM and the cytoplasmic ankyrin scaffold because the position of basket cell terminals becomes more diffuse in knockout mice that lack the protein (Ango et al. 2004). Notably, only the initial segment restriction, but not the formation of basket cell-Purkinje cell synapses, is abolished in the absence of NrCAM, which suggests that subcellular specificity and synapse formation are controlled by different signaling systems. Notably, the dendritic domain innervation of Purkinje cells by stellate cells does not appear to be mediated by a similar mutual attraction between synaptic partners but by glial-derived signals (discussed below).

Repulsion from nonsynaptic partners in the innervation field. Although mutual attractive mechanisms between synaptic partners can be used to specify connectivity as discussed above, repulsive signals derived from inappropriate target cells also prevent abnormal innervation.

The best understood examples of such mechanisms come from studies of the development of neuromuscular connectivity of the fruit fly. Each *Drosophila* abdominal hemisegment contains 30 muscles that are innervated by approximately 40 axons. Each of these axons establishes muscle-specific neuromuscular junctions. Many membrane-tethered and secreted molecules contribute to the specificity of these synaptic connections. Some cues, such as FasII and SemaII, are expressed in all muscles and act as general pro- and antisynaptogenic forces, respectively. Interestingly, NetrinB is only expressed by a subset of muscles, in which it attracts certain axons while repelling others (Winberg et al. 1998). The interplay of multiple positive and negative cues led the authors to propose that synaptic specificity in this system does not depend on unique, synapse-specific signals that act similar to a key-lock mechanism. Instead, growth cones assess the relative balance of attractive and repulsive forces and establish synapses with the best available partner.

In a more recent study using the same system, Nose and colleagues focused on two similar adjacent muscles, M12 and M13, that only differ in their innervation patterns by different motor neurons. Through single-cell microarray analysis, they found that several genes are differentially expressed between these two cells. For example, Wnt4 is enriched in M13 but not in M12. In the absence of Wnt4 or its receptor Drizzled 2, neurons that normally only innervate M12 also synapse onto M13. The ectopic expression of Wnt4 in M12 inhibits synapse formation by MN12. These data strongly suggest that Wnt4 determines synaptic specificity by inhibiting synapse formation with inappropriate targets (Inaki et al. 2007).

Guideposts in Specificity

Although direct interactions between synaptic partners can transform into synapses, pre- and postsynaptic neurons are likely to contact many other cells during the time of synapse formation. Is it possible that these other cell-cell interactions between nonpartner cells also impact synapse formation? Guidepost cells, or intermediate targets for axon guidance events, were discovered previously. These cells are located at critical positions along the axonal trajectory where turns are often made (Bate 1976). From studies on synapse formation, there is now accumulating evidence that similar transient target interactions with cells other than pre- and postsynaptic partners might also play critical roles in synaptic partner choices, the regulation of synapse numbers, and synapse elimination.

Transient neuronal populations as synaptic placeholders. During the development of several systems, presynaptic neurons extend their axonal processes to the appropriate target field before postsynaptic partners are fully differentiated. In such cases, transient neuronal populations can act as placeholders to provide cues for target field selection by presynaptic axons.

Examples of placeholder cells are Cajal-Retzius cells and GABAergic interneurons in the hippocampus (reviewed in Sanes & Yamagata 1999). In mature hippocampus, afferent axons from the entorhinal cortex form synapses on the distal dendrites of pyramidal neurons in the stratum lacunosum-moleculare (SLM) layer, while commissural/associational fibers form synapses on the more proximal part of these same pyramidal dendrites in the stratum radiatum (SR). Developmentally, entorhinal axons enter the SLM layer, before the majority of pyramidal dendrites have arrived in this area. Two classes of early-born neurons, calretinin-positive Cajal-Retzius (CR) cells and calbindin-positive GABAergic interneurons, are present in the SLM and SR, respectively, when entorhinal and commissural axons enter the target field (Soriano et al. 1994, Super et al. 1998). Cajal-Retzius and GABAergic interneurons form transient synaptic contacts with entorhinal axons and commissural axons, respectively, and both cell populations undergo cell death later during development (Super et al. 1998). The ablation of CR cells prevents layer-specific innervation by entorhinal axons, which suggests that CR cells are required for the layer-specific innervation of these axons

(Del Rio et al. 1997). Therefore, CR cells serve as transient placeholders that allow for layer-specific innervation in the hippocampus. Interestingly, Cajal-Retzius cells also form transient synaptic contacts with pyramidal cells in the cortex (Frotscher 1998), which suggests that CR cells may also act as transient placeholders for cortical maturation.

Another example of such placeholder cells is the subplate cell in the developing visual cortex. In the mature mammalian visual system, axons that originated from the lateral geniculate nucleus (LGN) innervate layer 4 neurons in the visual cortex. However, during development, LGN axons arrive in the cortical target field long before layer 4 neurons have migrated into the region. While LGN axons wait for layer 4 neurons to migrate and mature, they form transient synaptic connections with subplate cells. Ablation studies have demonstrated that subplate neurons are necessary for the proper axon guidance of LGN neurons into layer 4 cortex (Ghosh et al. 1990), as well as for the formation of ocular dominance columns (Ghosh & Shatz 1992, Kanold et al. 2003). Thus, the transient synaptic interaction between subplate cells and thalamic axons is critical for the establishment of mature visual cortical circuits.

Glia and glia-like cells in local synaptic connectivity. In the mammalian nervous system, glial cells outnumber neurons by tenfold. It is not surprising that glia play important roles in the function and development of the nervous system (Barres 2008). Glial cells can secrete critical axon guidance molecules and serve as intermediate targets to specify axon trajectory (Learte & Hidalgo 2007). The role of glia or glia-like cells in regulating local connectivity and synapse formation is starting to be elucidated by recent studies.

How might glia cells coordinate the recognition between pre- and postsynaptic partners? One elegant example was reported by Josh Huang and colleagues during their studies in the development of cerebellar circuits. They showed that axons from stellate interneurons are organized and guided towards Purkinje cell dendrites by an intermediate scaffold of Bergmann glial (BG) cells. In the absence of an L1 type molecule that is required for the association between the stellate axon and the BG, synapse formation between mistargeted stellate axons and Purkinje dendrites is reduced and synapses cannot be maintained, which leads to a progressive atrophy of axon terminals. Hence, the BG provides a substrate for the meeting between the stellate axon and the postsynaptic dendrite (Ango et al. 2008).

A remarkable parallel to these mechanisms can be found in the nematode *C. elegans*, despite its much simpler nervous system. A recent study found that two glia-like sheath cells coordinate the innervation between the interneurons AIY and RIA in the thermotaxis circuit of *C. elegans*. The processes of sheath cells converge at a location where AIY forms synapses onto RIA. Sheath cells secrete the axon guidance molecule, UNC-6/Netrin, which elicits distinct UNC-40/DCC-dependent responses in the RIA and AIY neurons, regulating axon guidance of RIA and synapse formation of AIY, respectively (Colon-Ramos et al. 2007).

A third example of glia-like cells that act as scaffolds for synapse formation comes from studies of egg-laying HSN motor neurons in *C. elegans*. A group of guidepost epithelial cells determines the subcellular localization of HSN synapses through the heterologous interaction between two transmembrane immunoglobulin superfamily proteins, SYG-1 and SYG-2, which bind to each other and function as receptor and ligand to mediate the recognition between guidepost cells and the HSNL axon. SYG-2 is expressed transiently by guidepost cells during the early stages of HSNL synaptogenesis. SYG-1 functions in the presynaptic HSNL neuron and localizes to synapses early during synapse formation. In loss-of-function *syg-1* and *syg-2* mutants, the HSNL axon fails to form synaptic connections with its normal targets and instead forms synapses to adjacent cells that do not normally receive synaptic input from the HSNL. Therefore, guidepost cells function as a placeholder of the presynaptic specialization of HSN through the action of a pair of Ig super-

family proteins (Shen & Bargmann 2003, Shen et al. 2004).

Besides serving as guidepost cells to specify location and partner selection, glial cells can also determine the number and functionality of synapses. For example, astrocyte-conditioned medium greatly enhances the number of synapses and the synaptic transmission of purified retinal ganglion cells (Ullian et al. 2001). Astrocytes secrete multiple factors to control various aspects of synapse formation. Glia produce cholesterol that facilitates the maturation of dendrites and the presynaptic terminal (Mauch et al. 2001). Astrocytes also secrete thrombospondin (TSP), which is sufficient to induce morphologically defined synapses in vitro (Christopherson et al. 2005). TSP1/2 double-knockout animals showed significantly reduced synapse numbers, which suggests that these astrocyte-derived molecules are required for synapse formation in vivo. Recent studies indicate that two synaptic proteins, neuroligin-1 (Xu et al. 2009) and the accessory calcium channel subunit alpha2delta-1, serve as TSP receptors (Eroglu et al. 2009). The glial cell line-derived neurotrophic factor (GDNF) and its GPI-anchored receptor GFRa1 induce the differentiation of presynaptic specializations (Ledda et al. 2007). Collectively, these data argue that glial cells play important roles in determining the partnership, subcellular localization, and density of synapses during development.

Morphogenetic gradients regulate local connectivity through the inhibition of synapse formation. Organization centers and graded diffusible signals play critical roles in cell fate determination (Lee & Jessell 1999) and axon guidance (Tessier-Lavigne & Goodman 1996). Recent studies in *C. elegans* suggest that similar gradients may shape synaptic connections directly. Worm neurons form en passant synapses on specific segments of their axons. For example, a particular motor neuron DA9 synapses onto the postsynaptic target muscle in a restricted domain, which only spans a small area of the entire axon. Although the posterior axonal segment contacts target cells, it is devoid of presynaptic terminals. A putative Wnt gradient formed by two Wnts, LIN-44 and EGL-20, is responsible for inhibiting synapse formation in this local area. Because both Wnts are only expressed in a small group of cells in the tail and wild-type DA9, synapses are not localized at the most posterior segment of the axon where the Wnt concentration is high. Furthermore, the loss of Wnts leads to a shift of synaptic distribution into the posterior region. When LIN-44 is ectopically expressed, it inhibits synapse formation in adjacent axon segments. Therefore, a local Wnt gradient shapes the DA9 synaptic domain through its antisynaptogenic activity (Klassen & Shen 2007).

Interestingly, another morphogenetic gradient, formed by the classic axon guidance molecule UNC-6/netrin, appears to exclude presynapse from the ventral axon of DA9. UNC-6 is expressed by ventral tissues and forms a ventral high-dorsal low gradient (Wadsworth et al. 1996). In the absence of *unc-6* or its repulsive receptor *unc-5*, a significant amount of presynaptic vesicle precursors and active zone proteins are localized to the ventral dendrite and axon (Poon et al. 2008). Hence, the Wnts and netrin, with different expression patterns, inhibit presynapse formation along different parts of the DA9 axon.

Morphogenetic gradients are well known for their functions in determining cell fate and guiding axons. How could their activity in shaping synaptic connections be distinguished from their early roles in the developing nervous system? A divergence of their activities might be at the level of receptor or at the level of downstream players. At least in the case of *C. elegans* Wnts, the diverse activities of Wnts appear to be mediated by the same receptor LIN-17/Frizzled. LIN-44/EGL-20 gradients exhibit distinct developmental roles in several classes of neurons. EGL-20 controls the migration of HSN and the Q neuroblast along the anterior-posterior (A-P) axis (Maloof et al. 1999), whereas several mechanosensory neurons utilize Wnt signaling to orient their anteroposterior polarity (Hilliard & Bargmann 2006, Pan et al. 2006, Prasad & Clark 2006). In many cases, the same receptor LIN-17 mediates these diverse cellular behaviors (Hilliard & Bargmann 2006, Klassen & Shen 2007, Pan et al. 2006, Prasad & Clark 2006). These results argue that diverse downstream signaling pathways dictate the various cellular responses of Wnts.

MOLECULAR MECHANISMS OF SPECIFIC SYNAPSE ASSEMBLY

After the initial recognition events, pre- and postsynaptic cells undergo a series of changes, which transform the incipient membrane contacts into highly specialized pre- and postsynaptic domains. In the process of this transformation, pre- and postsynaptic-derived factors drive structural and functional changes in developing synapses (**Figure 2**). Many of these factors are sufficient to drive synaptic differentiation to a remarkable extent and, accordingly, are considered synaptogenic molecules. Below, we review our knowledge of such synaptogenic molecules in a few established experimental systems.

BIDIRECTIONAL ORGANIZATION OF SYNAPTIC STRUCTURES BY ADHESION COMPLEXES

The close apposition of synaptic membranes and the mechanical stability of synaptic structures have long highlighted the substantial trans-synaptic adhesive interactions that are partly mediated through trans-synaptic adhesion molecules. Cell biological studies have demonstrated that adhesion molecules not only glue synaptic partners together but that they organize functional aspects of the synaptic structure (Biederer et al. 2002, Linhoff et al. 2009, Scheiffele et al. 2000, Woo et al. 2009). This ability of synaptic adhesion molecules to drive synaptic differentiation was tested in fibroblast-neuron coculture assays in which individual neuronal adhesion molecules expressed in non-neuronal cells were tested for their ability to recruit pre- or postsynaptic components in

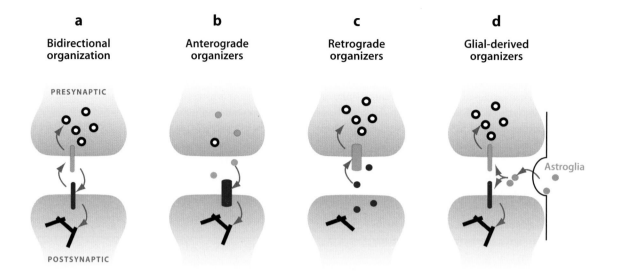

a	b	c	d
Bidirectional organization	**Anterograde organizers**	**Retrograde organizers**	**Glial-derived organizers**

PRESYNAPTIC

POSTSYNAPTIC

Astroglia

Figure 2

Synapse organizing signals. Differentiation of pre- and postsynaptic domains, recruitment of synaptic vesicles, and pre- and postsynaptic receptors and scaffolds can be driven by multiple trans-synaptic signals, derived from either a synaptic partner or astroglia that flank synaptic sites.

cultured neurons (Biederer & Scheiffele 2007, Scheiffele et al. 2000). Although most adhesion molecules only promote the formation of contacts between transfected fibroblasts and neuronal cells, a specific subset of proteins exhibits synaptogenic activities in such assays. The neurexin-neuroligin complex provides one such bidirectional synapse differentiation signal. Postsynaptic neuroligins expressed in fibroblasts are sufficient to trigger the accumulation of active zone components and synaptic vesicles in axons downstream of the presynaptic neurexin receptor. These presynaptic specializations are functional, contain a pool of recycling synaptic vesicles, and release neurotransmitters upon depolarization (Dean et al. 2003, Fu et al. 2003, Scheiffele et al. 2000). Transgenic mice that overexpress NL1, a neuroligin isoform primarily associated with glutamatergic synapses, exhibit a twofold increase in the size of glutamatergic presynaptic specializations and an increase in the number of synaptic vesicles per terminal (Dahlhaus et al. 2009, Hines et al. 2008). Importantly, this synaptogenic activity of the neurexin-neuroligin complex is not solely

a retrograde signal that organizes the presynaptic compartment but acts bidirectionally. The expression of NL1 in hippocampal neurons in vitro and in vivo promotes the formation of dendritic spines, drives the recruitment of postsynaptic scaffolding molecules and NMDA-receptors to synaptic sites, and thereby, promotes the assembly of silent synapses (Chih et al. 2005, Dahlhaus et al. 2009, Gerrow et al. 2006, Sara et al. 2005). In vitro, this postsynaptic differentiation process can be elicited solely by the contact of dendrites with the presynaptic NL-receptor neurexin expressed in fibroblasts(Chih et al. 2006, Graf et al. 2004, Nam & Chen 2005).

Several other synaptogenic adhesion complexes have been identified using assay systems that are similar to those used for the characterization of the neurexin-neuroligin complex. An unbiased expression library screen using the fibroblast coculture assay isolated LRRTM1, a member of the leucine-rich repeat family of synaptic adhesion proteins with synaptogenic activity (Linhoff et al. 2009). Further candidate gene approaches characterized synapse-organizing activities for SynCAMs (IgSF proteins from the Nectin protein family) (Biederer et al. 2002) and for additional leucine-rich repeat proteins, which include netrin-G-ligands (NGL1, -2, and -3) and LRRTM2 (Kim et al. 2006, Linhoff et al. 2009, Woo et al. 2009). Interestingly, LRRTM2 binds with nanomolar affinity to neurexin-1, and knock-down of neurexin-1 in cultured hippocampal neurons abrogates LRRTM2-induced presynaptic differentiation (de Wit et al. 2009, Ko et al. 2009). These recent findings highlight an unexpected cross-talk between the neuroligin-neurexin complex and leucine-rich repeat proteins.

How can a single trans-synaptic adhesion complex drive a near-complete program for pre- and postsynaptic differentiation? The synaptic signaling mechanisms downstream of any synaptogenic adhesion molecule discussed above are poorly understood. One model for the assembly of pre- and postsynaptic specializations is the nucleation of cytoplasmic subsynaptic scaffolds, which facilitates the recruitment and retention of ion channels at incipient synaptic sites. Similar to NGLs and SynCAM, the neuroligin and neurexin proteins carry PDZ-binding motifs on their C-termini that couple the adhesion molecules to scaffolding proteins (Ichtchenko et al. 1995, Ushkaryov et al. 1992). In the case of presynaptic neurexins, interactions with adapter proteins Lin2/CASK and Lin10/Mint may provide a link to voltage-gated calcium channels at glutamatergic synapses (Maximov & Bezprozvanny 2002). In the postsynaptic compartment, NL1 may bind through cytoplasmic PDZ-domain interactions to PSD95 and thereby recruit NMDA-Rs to glutamatergic synapses (Irie et al. 1997). Although the relevance of PSD95 or other scaffolding molecules for the NL1-mediated synaptic recruitment of NMDA-Rs remains to be tested, a conceptually similar model has been examined for NL2, a member of the neuroligin protein family that is concentrated at glycinergic and GABAergic synapses (Varoqueaux et al. 2004). Via its cytoplasmic tail, NL2 binds and activates col-libystin, which in turn tethers gephyrin at the postsynaptic membrane and recruits glycine and GABA-A-receptors (Poulopoulos et al. 2009). In the hippocampus of neuroligin-2 knockout mice, the perisomatic accumulation of gephyrin and GABA-A receptors is reduced, which highlights an essential role for NL2 in nucleating a cytoplasmic protein platform of GABAergic and glycinergic synapses.

Although loss-of-function studies in mice and invertebrate systems have uncovered severe structural phenotypes at neuromuscular synapses (Sanes & Lichtman 1999; see below), alterations in central synapse numbers for most of the bi-directional synaptic organizers described so far have been modest. The loss of the synaptogenic leucine-rich repeat protein LR-RTM1 does not appear to result in a significant change in synapse density in the hippocampus (Linhoff et al. 2009). Similarly, the ablation of netrinG1 and netrinG2, respective ligands of NGL1 and -2 receptors, perturbs NGL localization but leaves synaptic connectivity intact (Nishimura-Akiyoshi et al. 2007). Neuroligin-1, -2, -3 triple-knockout animals exhibit only a 15% decrease in synapse density in the brain stem, a much smaller alteration than observed with acute perturbation experiments in vitro (Chih et al. 2005, Levinson et al. 2005, Nam & Chen 2005, Varoqueaux et al. 2006). In addition, KO mice exhibit more severe reductions in synaptic transmission owing to defects in the recruitment of postsynaptic neurotransmitter receptors (Chubykin et al. 2007, Varoqueaux et al. 2006). Alpha-neurexin knockout mice exhibit synaptic transmission defects owing to the reduced function of presynaptic voltage-gated calcium channels (Missler et al. 2003). On the structural level, these mice still form central synapses, although symmetric synapse density in the brain is reduced by 50%, and the elaboration of dendrites and the number of dendritic spines in glutamatergic neurons appear to be significantly reduced (Dudanova et al. 2007, Missler et al. 2003). This means that although synaptogenic adhesion complexes can drive a broad program of pre- and postsynaptic differentiation, only certain aspects of these functions are essential for synapse formation in vivo. Notably, neurexin loss-of-function phenotypes in *Drosophila* are more dramatic. Glutamatergic neuromuscular junctions in *Drosophila* mutants that lack neurexin have severely reduced numbers of synaptic butons and detaching pre- and postsynaptic membranes (Li et al. 2007). The comparably milder phenotypes with respect to the structural assembly of individual synapses in vertebrates highlight the increased molecular complexity of the vertebrate system characterized by the presence of multiple, partially redundant, parallel trans-synaptic pathways. Most likely, these pathways endow the synapse assembly process in the mammalian brain with the necessary robustness to ensure functional connectivity while still allowing for the necessary plasticity observed in CNS networks. Several such parallel pathways that act as anterograde or retrograde trans-synaptic signals are discussed below.

ANTEROGRADE ORGANIZERS

Anterograde Organizers at Mammalian Neuromuscular Junctions

We refer to presynaptic-derived factors that drive postsynaptic maturation as anterograde organizers (**Figure 2b**). Arguably, the best understood factor of this group is agrin, an extracellular matrix protein found at the synaptic cleft of neuromuscular junctions (NMJs) in vertebrate animals (Nitkin et al. 1987). The motor axon derived agrin is indispensable for the clustering of postsynaptic acetyl choline receptor (AchR) at the postsynaptic membrane. Experimental evidence that supports agrin's role as an anterograde inducer has been reviewed extensively by Kummer and colleagues (Kummer et al. 2006). Although the agrin-MuSK-rapsyn pathway has been well established on the basis of cell biology and genetic experiments, MuSK may not constitute the receptor for agrin because it does not bind to agrin. Recently, two groups reported the discovery that a member of the LDL receptor family, Lrp4, associates with MuSK and serves

as a receptor for agrin in the process of NMJ formation (Kim et al. 2008, Zhang et al. 2008). Agrin binds directly to the extracellular domain of LRP4 and promotes the formation of an LRP4-MuSK complex. Agrin binding to LRP4 also triggers the phosphorylation of LRP4 and MuSK intracellular domains, which suggests that it might initiate the signaling of these receptor complexes. The activation of MuSK stabilizes AchR clusters at the postsynaptic membrane through a rapsyn-dependent mechanism; however, the exact molecular pathway downstream of MuSK has not been elucidated (Gautam et al. 1995).

Interestingly, motor neurons also release ACh, which acts as a negative signal to disperse noninnervated clusters of receptors and refine synaptic receptor clusters. In choline acetyltransferase (ChAT) mutant mice in which ACh is absent, AchR clusters are abnormally large. In ChAT/agrin double mutants, many AChR clusters are maintained, whereas they are largely absent in agrin single mutants (Lin et al. 2005, Misgeld et al. 2005). These genetic data strongly suggest that ACh and agrin antagonize each other and fine tune the size and location of AChR clusters. Downstream of ACh, the cyclin-dependent kinase, Cdk5, may be an effector in dispersing AChRs. The activity of Cdk5 is regulated by p35 and its proteolytic product p25, which is an even stronger activator of Cdk5 (Lin et al. 2005). One recent study showed that the ACh-induced activation of muscle cells stimulated the activity of a protease, Calpain, which cleaves p35 and produces p25. This calpain-dependent activity links the release of ACh and the activation of Cdk5 in the process of dispersing AChR clusters (Chen et al. 2007). Why would motor neurons release two antagonistic factors for postsynaptic development? The answer may be that the ranges of action for these two factors are different. Whereas ACh depolarizes the entire postsynaptic membrane and disperses AChR clusters across the entire postsynaptic muscle, the activity of agrin may be restricted to synapses. Therefore, through a global dispersal activity and a local clustering activity, the motor axons mold the postsynaptic AChRs juxtaposed to the presynaptic terminal.

Anterograde organizers and direct interactions with neurotransmitter receptors. The identification of trans-synaptic signaling pathways that orchestrate the synaptic recruitment of AChR at neuromuscular synapses prompted the search for proteins with comparable activities in the CNS. Several anterograde inducers released from the presynaptic partner have been identified that organize postsynaptic components. Similar to the agrin-MuSK-rapsyn cascade, researchers anticipated that such factors would act through the recruitment of cytoplasmic scaffolds, which in turn tether neurotransmitter receptors at postsynaptic sites. However, more recently, for glutamatergic synapses, a different mechanism of action emerged for several of these anterograde inducers, namely a direct extracellular interaction of the differentiation signals with postsynaptic glutamate receptors.

Ionotropic glutamate receptors are tetramers, generally composed of dimers of two different subunits. In addition to ligand-binding and pore-forming domains, glutamate receptor subunits contain a large extracellular N-terminal domain (NTD) that shares sequence homology with the bacterial periplasmic amino acid-binding protein (also called the LIVBP-domain). Remarkably, the direct attachment of trans-synaptic differentiation signals to these NTDs provides an anterograde signal for the synaptic recruitment of postsynaptic glutamate receptors. The first such factors to be identified are neuronal pentraxins, which consist of two secreted proteins, the neuronal activity regulated pentraxin (NARP) and neuronal pentraxin 1 (NP1), and the trans-membrane neuronal pentraxin receptor (NPR). NARP and NP1 associate directly with AMPA-receptor (GluA) subunits (O'Brien et al. 1999). NP1 released from glutamatergic axons is sufficient to drive the synaptic aggregation of GluA4-containing receptors through interaction with the NTD (Sia et al. 2007). In vivo, this NARP/NP1 synaptic differentiation machinery localizes

and acts selectively at glutamatergic shaft synapses formed on GABAergic interneurons. Specifically, the number of GluA4-containing synaptic structures is reduced (although not abolished) within the hippocampal dentate gyrus in pentraxin triple knockout mice (Sia et al. 2007). Similar mechanisms of NTD-dependent synapse formation might also apply for glutamatergic spine synapses. In particular, in vitro experiments provided evidence that the NTD of the AMPA-receptor subunit GluA2 promotes the differentiation of spine synapses in hippocampal neurons (Passafaro et al. 2003). This function may be mediated through direct extracellular coupling between GluA2 and the adhesion molecule N-cadherin (Nuriya & Huganir 2006, Saglietti et al. 2007). The exact role for these interactions in regulating synaptic structure awaits further study because, surprisingly, synaptic morphology is largely unperturbed in cells that lack GluA2 or all GluA receptor isoforms in vivo (Lu et al. 2009).

Importantly, NTD-mediated synaptic differentiation signals are not unique to AMPA-type glutamate receptors but are emerging as a general principle for postsynaptic neurotransmitter recruitment in the CNS. LIVBP-like N-terminal domains are also contained in the GluN1 and GluN2 subunits of NMDA-type glutamate receptors. In vitro studies identified the GluN1 NTD as a direct interaction site with EphB-receptor tyrosine kinase receptors (Dalva et al. 2000). Whereas EphB-receptors and their ephrinB ligands have been primarily characterized as signaling molecules in axonal guidance, ligands and receptors are expressed during synapse formation and in the adult CNS (Klein 2001). Presynaptic ephrinB-ligands may cluster postsynaptic EphB receptors, which in turn nucleate a tripartite complex of ephrinB, EphB-receptor, and the NMDA-receptor complex. The functional relevance of the GluN1 NTD in the synaptic recruitment of receptors has not been tested in vivo. However, the ablation of full-length EphB2 expression in mice results in reduced NMDA-receptor synaptic localization and function (Grunwald et al. 2001,

Henderson et al. 2001), which is consistent with a critical role for EphB-receptors in NMDA-receptor incorporation in glutamatergic synapses. Moreover, synapse density and dendritic spine development are decreased in knockout mice that lack multiple EphB receptors, which further emphasizes an important role for EphB tyrosine kinase receptors in synapse development (Henkemeyer et al. 2003, Kayser et al. 2008).

Finally, the GluD2 receptor is probably the most remarkable example of the importance of NTD-interactions with neurotransmitter receptors in synaptic differentiation. GluD2 shares the same domain organization with AMPA- and NMDA-type receptors but, owing to amino acid alterations in the pore, it does not mediate currents (Yuzaki 2003). GluD2 is selectively expressed in cerebellar Purkinje cells. Mice that lack GluD2 suffer from severe ataxia and exhibit two major phenotypes on the synaptic level: First, parallel fiber synapses form between cerebellar granule cells and Purkinje cells detach from Purkinje cell dendritic spines, which indicates a severe weakening of trans-synaptic interactions; second, the remaining parallel fiber synapses exhibit diminished long-term depression. Both phenotypes can be rescued by the reintroduction of full-length GluD2, but GluD2 that lacks the NTD only restores LTD (Hirano et al. 1994; Kakegawa et al. 2008, 2009). In cultured cerebellar granule cells, the GluD2 NTD is sufficient to trigger the assembly of presynaptic terminals, which suggests that it is part of a trans-synaptic signaling complex (Kakegawa et al. 2009, Uemura & Mishina 2008). A hint at the identity of the presynaptic ligand for the GluD2 NTD comes from an analysis of mutant mice that lack cerebellin-1 (Cbln1), a small, secreted protein that is part of the C1q/TNFalpha superfamily (Yuzaki 2008). Cbln1 mutant mice closely phenocopy the GluD2 knockout phenotype (Hirai et al. 2005, Kurihara et al. 1997). Whereas the GluD2 receptor is accumulated in Purkinje cell dendritic spines, Cbln1 is secreted from cerebellar granule cells and concentrates at parallel fiber synapses with Purkinje cells. Therefore,

Cbln1 may bind directly to the GluD2 NTD and thereby exert its synaptogenic activity in the cerebellum (Kakegawa et al. 2009).

Recently, a similar mechanism has also been identified for the synaptic concentration of AChRs in *C. elegans* in which a complement control-like protein interacts directly with the extracellular domain of a neurotransmitter receptor subunit (Gendrel et al. 2009). These examples illustrate an alternative mechanism to the receptor recruitment via cytoplasmic scaffolding molecules. Neurotransmitter NTD interactions may drive the equilibrium of extrasynaptic, dispersed receptors towards the clustering of receptors at synaptic sites. The focal presentation of axon-derived differentiation signals at sites of cell contact and the direct action onto receptor extracellular domains represent a particularly effective and elegant way to achieve a rapid synaptic accumulation. In this model, cytoplasmic scaffolding molecules act downstream of neurotransmitter receptors by enlarging and stabilizing postsynaptic assemblies, and neurotransmitter receptors serve a structural role in synapse assembly that is independent from their ionotropic action.

Wingless as an Anterograde Signal in *Drosophila* Neuromuscular Junction Formation

Another anterograde regulator was discovered through studies of glutamatergic NMJs in *Drosophila*. Wnt-1 wingless (wg) was found at presynaptic terminals, and its receptor Frizzled (DFz) is concentrated at the postsynaptic membrane as well as on the presynaptic side. Wg secreted from the presynaptic terminals is critical for the development of fly NMJs. The loss of wg causes reduced bouton size and the incomplete development of boutons, whereas the overexpression of Wg leads to enhanced bouton proliferation (Packard et al. 2002). Interestingly, the cytoplasmic region of DFz is cleaved by proteases that lead to the generation of an 8kd polypeptide fragment. Wg activation of DFz appears to promote the translocation of this fragment into the nucleus, which then regulates the differentiation of postsynaptic development (Mathew et al. 2005).

Wg loss-of-function mutants also display presynaptic defects. Mutant boutons have enhanced unbundled microtubules, which may explain bouton proliferation defects because MT dynamics affect the proliferation of boutons (Roos et al. 2000, Ruiz-Canada & Budnik 2006). This hints that Wg can act not only as an anterograde inducer to regulate postsynaptic development but also might act to modify presynaptic development through an autocrine loop because the DFz receptor is also found on the presynaptic membrane. Presynaptic effects may be caused by a retrograde inducer whose activity is regulated by Wg. Further experiments are required to distinguish these possibilities.

The growth of the *Drosophila* NMJ is stimulated by neuronal activity, and Wg might mediate this effect. To support this notion, Wg secretion was enhanced by activity and coincided with rapid activity-dependent NMJ growth. Furthermore, heterozygous wingless mutants suppressed activity-dependent synaptic growth; the overexpression of Wg reduces the strength of electric stimuli that are required to reach a certain level of growth (Ataman et al. 2008).

Wnts have also been implicated as modifiers of agrin-induced AChR clusters in vertebrate neuromuscular junctions. Stimulatory and inhibitory actions of wnts have been reported in the literature. In some contexts, the readout of the trans-synaptic Wnt signal might depend on the activation of canonical or noncanonical wnt signaling pathways (Davis et al. 2008). A detailed discussion on this topic can be found in a recent review (Korkut & Budnik 2009). Moreover, in the vertebrate CNS, wnts act as retrograde signals, which are discussed below.

RETROGRADE ORGANIZERS

Retrograde organizers are factors that are secreted from postsynaptic cells that influence the differentiation and maturation of presynaptic terminals (**Figure 2c**). The precise apposition of synaptic membranes argues that synaptogenesis involves bidirectional communication between synaptic partners. Moreover, as animals grow in size, the dimension and number of synapses grow accordingly so that the presynaptic neuron can efficiently excite postsynaptic cells (Davis 2006). This homeostasis process strongly suggests that the activation state of the postsynaptic cell must be sensed by presynaptic terminals, which is then converted to a growth signal to regulate the size and function of presynaptic terminals. Therefore, understanding the action of retrograde signals will not only provide insight into how synapses form but also how synapses grow. Indeed, genetic analyses of different synapses identified several retrograde signals.

The Transforming Growth Factor Beta Family Member gbb Activates Neuromuscular Junction Growth in *Drosophila*

Forward genetic screens in the *Drosophila* NMJ mutant led to the isolation of the Wishful thinking (Wit) mutant, which exhibits reduced numbers of synaptic boutons and abnormalities in ultrastructure that include the detachment of pre- and postsynaptic membranes and floating T-bars. Electrophysiologically, these mutant synapses showed drastically reduced quantal content and frequency of spontaneous release, consistent with a predominantly presynaptic defect (Aberle et al. 2002, Marques et al. 2002). Wit encodes the *Drosophila* ortholog of the human BMP type II receptor, is expressed in the axon, and is required for the activation of the BMP downstream signaling pathway (Marques et al. 2002, 2003). These observations argue that Wit receptors on axon terminals receive a ligand that stimulates the growth and maturation of the NMJ.

The search for the ligand for Wit resulted in the discovery of BMP-7 glass boat bottom (Gbb), which is expressed in muscles. Gbb binds to Wit and activates its signaling pathway in the axon (McCabe et al. 2003). The loss of gbb results in a reduced number of synapses and defective synaptic transmission, similar to phenotypes found in Wit mutants. The loss of other components of the BMP pathway, which include type I receptors (Tkv and Sax), the R-smad (Mad), and the co-Smad (Med), causes similar synaptic-growth phenotypes (McCabe et al. 2004, Rawson et al. 2003). Moreover, these signal transduction genes appear to be required in the motor neuron for their function in synaptic growth, which further argues that the neuron is the receiving end of the gbb pathway. Therefore, as animals grow in size, muscles secrete the TGF family protein gbb to promote the growth of synapses and to scale with the increase of muscle mass through the activation of the Wit receptor on nerve terminals.

Growth Factors as Synaptic Differentiation Signals in the Cerebellum

In many parts of the nervous system, synapses are organized into specialized synaptic complexes. For example, the cerebellar glomerular rosette is a multisynaptic structure that is formed between a glutamatergic mossy fiber axon, dendrites from numerous cerebellar granular cells, and Golgi cell terminals that form inhibitory synapses onto granule cell dendrites (Altman & Bayer 1997). The large size of this specialized synapse makes it an excellent model synapse to examine the mechanisms of synaptic differentiation in the vertebrate CNS. During development, upon contacting its postsynaptic targets, the typical developing mossy fiber axon spreads out into a fan-like structure that is eventually converted into the synaptic rosette with multiple pre- and postsynaptic elements (Mason et al. 1997). Signaling through several growth factors has been implicated in the development of this synaptic complex. The first to be identified was Wnt7a, which is released from postsynaptic granule cells at the time of synapse formation. The addition of recombinant Wnt7a is sufficient to induce changes in growth-cone morphology in mossy fiber explant culture, whereas blocking endogenous Wnts inhibits the maturation

process (Hall et al. 2000). In *Wnt7a* knock-out mice, rosette formation is delayed, which suggests that Wnt7a is an important player in this process and that there might be additional partially redundant inducers besides Wnt7a. Several Wnt downstream molecules have been implicated in mediating the action of Wnt7a in this context. For example, *Dvl* single mutants exhibit defects in mossy fiber synapses that are similar to those in *Wnt7a* mutants. Dvl and Wnt7 double mutants exhibit more severe phenotypes compared with single mutants (Ahmad-Annuar et al. 2006). Moreover, inhibitors of GSK3β mimic the effects of wnt in vitro (Hall et al. 2000, 2002). Taken together, this indicates that target-derived Wnt7a induces changes in the presynaptic terminal through a Dvl- and GSK3β-mediated signaling pathway.

A second granule cell-derived retrograde signal is the fibroblast growth factor FGF22. Similar to Wnt7a, FGF22 and its close homologs induce morphological changes and the accumulation of synaptic vesicles in mossy fiber axons in vitro (Umemori et al. 2004). Blocking FGF22 signaling with recombinant proteins or genetic ablation of its receptor FGFR2c results in a reduction in synaptic vesicle accumulation in mossy fiber rosettes, whereas the synaptic structures themselves appear to persist. This indicates that trans-synaptic FGF-signaling controls selectively presynaptic vesicle accumulation at mossy synapses in the cerebellum. The similarity of Wnt7a and FGF22 activities on mossy fiber terminals raises the question of whether these two signals have redundant function or whether they each control specific aspects of mossy fiber differentiation. Wnt7a signaling has been proposed to alter microtubule stability. In contrast, FGF signaling has been primarily linked to the distribution of synaptic vesicles in axons. However, the delineation of cytoplasmic mediators of FGF signaling at synapses is required to understand whether FGF and Wnt signaling regulate presynaptic assembly through parallel pathways or whether they converge in common effectors. Notably, the integration of FGF and Wnt signaling is observed during cell fate decisions in earlier development (Fuentealba et al. 2007), and it remains to be investigated whether an analogous integration occurs during trans-synaptic signaling in the central nervous system.

Within the glomerular structure, Wnt7a and FGF22 trans-synaptic signaling may act specifically on glutamatergic mossy fiber terminals. What are the signals that control the formation of inhibitory Golgi cell terminals within the same glomerular structure? Mutant mice that lack the neurotrophin receptor TrkB in the cerebellum and cerebellar afferents exhibit a strong reduction in the number of GABAergic Golgi cell synapses in cerebellar glomeruli (Rico et al. 2002). Notably, the density of asymmetric synapses within the glomerulus is not altered, which indicates that TrkB signaling selectively regulates GABAergic synapse assembly in this structure. Although the source and role for the TrkB ligand BDNF in this system remains to be determined, BDNF signaling has been linked to the maturation of GABAergic synapses in other CNS circuits (Vicario-Abejon et al. 2002). In summary, these studies highlight the importance of multiple retrograde growth factors in the differentiation of specific synaptic structures in the cerebellar glomerulus.

Laminin β2 Organizes Active Zones in the Vertebrate Neuromuscular Junction

Although the essential roles of agrin and ACh in patterning postsynaptic AChRs argue strongly that anterograde axonal signals play the lead role in NMJ formation, there is also experimental evidence that supports instructive roles for muscle-derived retrograde signals in synapse formation. For example, Burden and colleagues showed that the overexpression of MuSK leads to ectopic AChR clusters outside of the normal synaptic region on the muscle surface. This manipulation also leads to abnormal axon branch-

ing and the formation of ectopic synapses (Kim & Burden 2008). However, the nature of the retrograde inducer is not clear yet.

During the maturation phase of NMJ formation, the axon terminal and muscle continue to interact with each other. One retrograde factor involved is an extracellular matrix protein, laminin β2. Laminins are secreted glycoproteins and major components of the basal lamina. Laminin β2 is made by the muscle and localizes to small stretches of basal lamina that extend through the synaptic cleft at vertebrate neuromuscular junctions (Patton et al. 1997). The loss of laminin β2 causes multiple presynaptic defects that include a reduced number of active zones and an abnormal distribution of synaptic vesicles, which suggests that laminin β2 is essential for the maturation of presynaptic terminals (Noakes et al. 1995). One interesting question is how muscle-secreted laminin β2 clusters synaptic vesicles and builds active zones near a postsynaptic specialization. A critical issue in our understanding of laminin's function at synapses is to define its receptor. Laminin β2 binds directly to the voltage-gated calcium channels, a component of the active zone (Nishimune et al. 2004). More importantly, the perturbation of this interaction in vivo results in the disassembly of neurotransmitter release sites. Artificial beads coated with laminin β2 are sufficient to cluster calcium channels at a point of contact on the axon, as well as to accumulate other active zone proteins and synaptic vesicles. These data suggest that laminin β2 is secreted by muscle cells and deposited at the synaptic basal lamina where it induces presynaptic differentiation through direct binding to the voltage-gated calcium channels. One outstanding question is how laminin β2 achieves its specific localization at the synaptic cleft.

SYNAPSE ELIMINATION

A major mode of cellular behavior that contributes critically to connection specificity is synapse elimination. Anatomical experiments on the developing nervous system in vertebrate animals demonstrate many examples of synapse elimination and axon pruning (Lichtman & Colman 2000). Axon pruning can lead to the loss of major axon projections or local branches (Luo & O'Leary 2005), both of which might be intimately related to synapse elimination. However, how synapse disassembly leads to the retraction of axons is unclear.

In principle, there are two classes of synapse elimination processes: one in which the stoichiometry of innervation is reduced and one in which the cell type or subcellular specificity of synaptic connections is refined. The best understood example of synapse elimination that results in a change in innervation stoichiometry is found in the maturation of the vertebrate neuromuscular junction. In immature NMJs, multiple axons innervate each myotube. As development proceeds, all but one of the axons are eliminated, which results in the mature monoinnervation pattern between axons and muscles (Lichtman & Colman 2000). This elimination process appears to depend on neuronal activity in which active axons are more likely to win in the competition among innervating axons. Other examples of synapse elimination with axon loss include climbing fiber-Purkinje neuron connections (Crepel et al. 1976), thalamocortical axon-layer 4 neuron synapses (Hubel et al. 1977), and infrapyramidal mossy fiber axon-CA3 synapses in the hippocampus (Liu et al. 2005).

Although the phenomenon of activity-dependent synapse elimination is well established in the vertebrate, the molecular mechanism of synapse disassembly is poorly understood. Invertebrate nervous systems also show stereotyped synapse elimination. The most prominent case is the pruning of axons during metamorphosis (Luo & O'Leary 2005). There are also examples of synapse elimination without the loss of axons in *C. elegans* in which most of the synaptic connections are made en passant. Below we discuss recent work that is starting to elucidate the molecular mechanisms that underlie the stereotyped synapse elimination programs.

Mono-Innervation of Vertebrate Neuromuscular Junctions

Synapse elimination is a critical step in building the mono-innervated, mature neuromuscular junction. This is a protracted process with many inputs that gradually decrease in synaptic strength as a single input increases its strength (Colman et al. 1997). Genetic manipulations that generate situations in which active axons compete with inactive axons showed that active axons always win the competition, which suggests that axons must be effective in depolarizing the muscle in order to be maintained (Buffelli et al. 2003). Furthermore, the blockade of synaptic activity throughout an entire junction prevents synapse elimination, whereas increasing activity accelerates the process (Sanes & Lichtman 1999). Forced synchronous activity in all the fibers that innervate a single NMJ prevents synapse elimination, which indicates that the unequal ability of axons to depolarize muscle is the driving force for synapse elimination in the vertebrate NMJ (Busetto et al. 2000).

Synapse Elimination Through the Complement Cascade in the Central Nervous System

As discussed above, synapse elimination at the NMJ is critical to generate the mature one neuron—one fiber innervation pattern. Similarly, synapse elimination is important during the maturation of the visual circuits. The visualization of developing circuits in the visual systems of *Xenopus* and zebrafish provides a picture of how mature synaptic connections are established. Live imaging studies revealed the highly dynamic neuronal arbors that undergo frequent phases of synapse formation and synapse elimination (Hua & Smith 2004). In immature circuits of the mouse, each dorsal lateral geniculate nucleus (dLGN) neuron receives synaptic input from multiple retinal ganglion cell (RGC) axons. As the circuit matures, each dLGN neuron receives stable input from only one or two RGC axons, owing to synapse elimination (Hooks & Chen 2006). Unexpectedly, the glia-derived complement pathway proteins C1q and C3 are required for this synapse elimination event. C1q and C3 are found locally at developing synapses but are absent from mature circuits. In C1q mutants, the eye-specific segregation of RGC axons in LGN is defective, and LGN neurons remain multiply innervated at a mature age. Similar defects were also observed in C3-deficient mice (Stevens et al. 2007). These results strongly suggest that synapse elimination is a critical process in the peripheral nervous system as well as the central nervous system.

Synaptic Pruning Mediated by Semaphorins

The infrapyramidal bundle (IPB) of the mossy fiber pathway in the hippocampus is present in newborn animals and significantly shortened by P30. This pruning is mediated by repulsive axon guidance receptors, plexin A3 and neuropilin-2, because IPB persists in plexinA3- and neuropilin-2-deficient mice. Semaphorin 3F, a ligand for the plexinA3 and neuropilin-2 receptor, appears to be expressed in the right place at the right time to trigger the pruning of IPB (Yaron et al. 2005). This genetic evidence strongly suggests that sema3F and its receptors prevent the maintenance of the IPB. However, the evidence does not offer cell biological insight into what cellular process initiates and underlies the pruning event. Further studies using electron microscopy showed that IPB pruning is intimately associated with synapse elimination. Cheng and colleagues found that IPB axon collaterals form transient synaptic complexes with basal dendrites of CA3 pyramidal cells in the early postnatal mouse hippocampus. At later postnatal ages, these synaptic complexes stop maturing and are removed before stereotyped pruning. In knockout mice that lack plexin-A3 signaling, the synaptic complexes continue to mature, and, as a result, the collaterals are not pruned. Thus, intact plexin-A3 signaling triggers axon pruning by promoting synaptic elimination (Liu et al. 2005). The same ligand and receptors also mediate the pruning of corti-

cospinal axons from the visual cortex (Low et al. 2008).

Recent studies further elucidated functions of semaphorins in specifying synaptic connectivity in vertebrate neurons. Overexpression and knockdown studies in hippocampal neurons demonstrated that Sema5B reduces synapse numbers, presumably by destabilizing presynaptic terminals that are associated with a Sema5B-expressing postsynaptic cell (O'Connor et al. 2009). In mouse cortex, Tran and colleagues found that Sema3F negatively regulated the number and size of dendritic spines in dentate gyrus (DG), granule cell (GC), and cortical layer V pyramidal neurons, possibly by promoting the loss of spines and synapses. In contrast, a distinct Sema3A pathway controls basal dendritic arborization in layer V cortical neurons. These disparate effects of secreted semaphorins are reflected in the restricted dendritic localization of Npn-2 (a Sema3F receptor) to apical dendrites and of Npn-1 (a Sema3A receptor) to all dendrites of cortical pyramidal neurons (Tran et al. 2009).

Molecular Insights into Synapse Elimination from *C. elegans*

The extracellular ligand-receptor interaction provides the logic for the specificity and timing of synaptic elimination. Only certain specific synapses are eliminated at particular developmental stages because the expression of ligands and receptors is spatially and temporally controlled. The next questions are to understand how the synapse elimination program is coordinated with the development of the organism and how receptor signaling leads to the disassembly of the synaptic structure. Two studies in *C. elegans*, discussed below, are starting to provide some insights into these mechanisms.

In *C. elegans*, DD-type GABAergic motor neurons (DDs) remodel their synaptic circuits during larvae development. During the first larvae stage, DDs eliminate their embryonic presynaptic terminals and form new en passant synapses without changing their cell shape. Although the molecular mechanism of this remodeling is not yet clear, its coordination with organismal development is mediated by the heterochronic gene, LIN-14, which was previously known to specify the timing of cell division patterns in the development of non-neuronal tissues (Ambros & Moss 1994). In *lin-14* mutants, synapse elimination occurs precociously—a phenotype that can be rescued by LIN-14 expression in DD neurons, which suggests that LIN-14 functions cell autonomously to prevent synapse elimination (Hallam & Jin 1998). Consistent with this notion, the expression level of LIN-14 is drastically reduced right before the synapse elimination starts. Hence, the controlled expression of a heterochronic gene drives the elimination process.

Transient presynaptic terminals are also found in egg-laying motor neuron HSNs. Synapse formation of this neuron takes place in the L4 stage, whereas HSN only becomes functionally active several hours later in the adult stage (Desai et al. 1988). In early L4 animals, numerous en passant synapses form in a segment of the axon near the vulva organ. As animals mature, stereotyped synapse elimination converts this immature pattern of synapses to a more restricted, mature distribution pattern. Immature synapses at the anterior location are always eliminated, whereas synapses at the center of the vulval area persist and grow. Thus, the decision of synapse elimination is executed by the SYG-1 and SYG-2 family of immunoglobulin superfamily proteins (Shen & Bargmann 2003, Shen et al. 2004). First, in *syg-1* or *syg-2* mutants, the synapse elimination process is defective, which leads to persistent anterior synapses. Second, SYG-1 functions cell autonomously in HSN to execute synapse elimination and is specifically localized to the center of the vulval region. Synapses that colocalize with SYG-1 are spared from elimination, whereas synapses that fall out of the SYG-1 zone are doomed to disappear. Third, the overexpression of SYG-1 leads to the precocious elimination of anterior synapses. Therefore, the decision of where to eliminate synapses is made

by the SYG-1 protein in HSN (Ding et al. 2007).

Then, how can SYG-1 locally control the fate of synapses? The search for SYG-1 interaction proteins led to the discovery of SKR-1, an ortholog of SKP1, which is a central component of the SKP-Cullin-F-box (SCF) ubiquitin ligase complex. The SCF complexes are E3 ubiquitin ligases that ubiquinate protein substrates destined for degradation (Cardozo & Pagano 2004). The loss of function of SKR-1, Cullin, or F-box protein sel-10 leads to delayed and incomplete synapse elimination in HSN, which suggests that the SCF complex is required for this process. Interestingly, SYG-1 binding to SKR-1 disrupts the assembly of the SCF complex, which indicates that SYG-1 protects synapses by inhibiting the synapse elimination mediated by the SCF complex (Ding et al. 2007). Taken together, these studies suggest that synapse elimination activity is distributed across the entire axon, whereas synapse protection activity is localized by specific recognition events between synaptic partners or guidepost cells. The balance between synapse elimination and assembly at different subcellular locations leads to the mature synapse pattern.

TRANSCRIPTIONAL CONTROL OF SYNAPTIC CONNECTIVITY

The emergence of molecular signaling systems that either promote or inhibit the assembly and/or stability of synaptic junctions raises the question of how the expression of such components is controlled during neuronal development. Afferent neurons undergo substantial transcriptional changes upon entry in the target area; moreover, dendritic growth is regulated by transcriptional programs (Diaz et al. 2002, Polleux et al. 2007). In recent studies, transcriptional regulators have been identified that control synapse formation programs at two levels: First, there are cell-specific transcriptional programs that are linked to the fate determination of the cell that also control key wiring decisions at later developmental stages. Thus, many aspects of axonal trajectories and synaptic interactions are prespecified by the expression of guidance and specificity factors in a cell-specific manner. Second, the transcription of signaling factors can be regulated dynamically depending on neuronal activity and signals from the environment. This second mechanism has been primarily implicated in the regulation of the number of synaptic connections that are formed by a cell, whereas the first mechanism is primarily involved in tying cells into specific neuronal circuits.

Transcriptional Control of Wiring Specificity

The remarkable reproducibility of neuronal connectivity between individuals and the substantial preservation of connection specificity in the absence of neuronal activity suggest that many aspects of neuronal connectivity are genetically encoded. Recent studies have provided several examples that single transcription factors can directly control the expression of individual surface receptors that endow cells with specific responsiveness to molecular cues for migration, axon guidance, and synaptic partner selection. The best-established examples for this type of regulation are found in axonal guidance, specifically, the response of axonal projections to midline-derived guidance cues. The zinc-finger transcription factor Zic2 binds directly to promotor elements of the axon guidance receptor EphB1 and thereby directs the ipsilateral growth of retinal ganglion cell axons at the optic chiasm (Herrera et al. 2003). Similarly, in the mouse spinal cord, the LIM homeodomain transcription factor Lhx2 is required for the expression of the Robo-receptor RIG1 and the immunoglobulin superfamily member TAG-1 in a population of commissural interneurons (Wilson et al. 2008). The loss of Lhx2 results in a loss of RIG1 and thereby the ipsilateral misprojection of axons in mutant mice. Importantly, RIG1 expression in other populations of commissural interneurons does not depend on Lhx2. That means, there is not simply one transcriptional

program that defines the RIG1-dependent commissural trajectory; instead, the interpretation of transcriptional programs is complex and occurs in a cell-type-specific manner.

Also, other binary choices in axonal trajectories are directly regulated by transcription factors. The LIM homeodomain transcription factor Lhx3 regulates FGF-receptor 1 expression in medial-class spinal motor neurons. In the absence of FGF receptor 1, axons fail to respond to target-derived FGF signals from the dermomyotome and misproject to limb muscles (Shirasaki et al. 2006). Another LIM-homeodomain transcription factor, Lhx1, instructs the outgrowth of motor axons towards dorsal limb muscles through regulating the expression of Eph-receptors (Kania et al. 2000, Kania & Jessell 2003).

These examples of transcriptional regulation of axon guidance decisions exemplify how single transcription factors directly regulate the expression of individual cell surface receptors, which in turn mediate growth decisions at choice points of the neuronal trajectory. There is emerging evidence that the same principle of transcriptionally controlled wiring programs also underlies key aspects of synaptic target selection. For example, the synaptic layer specificity of two closely related groups of *Drosophila* photoreceptor neurons, R7 and R8, is controlled by the interplay of the transcriptional regulators senseless, prospero, and NF-Y. Here, the R8-specific transcription factor senseless directly regulates the cell surface receptor Capricious, a leucine-rich repeat protein that is essential for R8 receptor targeting (Shinza-Kameda et al. 2006). The R8-specific targeting program is repressed in R7 neurons by NF-Y, and the loss of NF-Y results in senseless-dependent mistargeting of R7 neurons to the R8 target lamina. Notably, NF-Y mutant R7 neurons continue to express many R7 markers, and targeting defects only occur at a late stage of R7 development, which indicates that the loss of NF-Y does not result in a complete fate switch of the neurons but only a selective defect in their targeting specificity (Morey et al. 2008).

In vertebrate systems, the best evidence for a transcriptionally-controlled wiring mechanism is derived from studies on sensory-motor reflex circuits, in which the analysis of target specificity is greatly facilitated by transcriptional markers for specific motor neuron pools and highly selective labeling approaches through dye injections into muscle targets. Combinations of Hox gene transcription factors expressed in motor neurons specify motor neuron pools and provide a transcriptional framework for connection specificity (Dasen et al. 2003, 2005). The Hox code restricts the competence of cell pools to activate downstream wiring programs. Interestingly, surface molecules that control synaptic specificity are often not directly under Hox transcription factor control and, therefore, not immediately expressed upon the specification of neuronal populations. Instead, these wiring molecules are controlled by transcriptional regulators that depend on target-derived signals, similar to ETS transcription factors Pea3 and Er81 (Lin et al. 1998). Target-derived GDNF is required for the upregulation of Pea3 (Haase et al. 2002), whereas a yet unknown target-area-derived signal leads to the upregulation of Er81 in motor neurons. Motor neurons in Pea3-deficient mice exhibit alterations in several critical aspects of terminal differentiation: cell body positioning, dendrite morphology, and the pattern of sensory-motor connectivity (Vrieseling & Arber 2006). Specifically, Pea3 is selectively expressed in a pool of motor neurons that innervate the cutaneous maximus muscle (CM). These CM motor neurons are devoid of monosynaptic input from proprioceptive sensory neurons. In contrast, other motor neurons that are Pea3-negative receive direct monosynaptic proprioceptive inputs. In *Pea3* knockout mice, proprioceptive sensory neurons form aberrant monosynaptic connections with CM neurons, which indicates the loss of a signaling system that controls synaptic specificity in this reflex arc (Vrieseling & Arber 2006). The repellent signal Sema3e and its receptor PlexinD1 are selectively expressed in the CM reflex arc, with Sema3e present in the Pea3-positive CM

motor neurons and PlexinD1 in the proprioceptive sensory neurons that innervate CM. Notably, Sema3e expression is lost in Pea3 mutants. The genetic ablation of either Sema3e or PlexinD1 phenocopies the connectivity defect seen in Pea3 mutant mice, with CM motor neurons that receive direct, monosynaptic proprioceptive input (Pecho-Vrieseling et al. 2009). These findings establish a repellent Sema3e-PlexinD1 signaling system and its control through the transcription factor Pea3 in the establishment of synaptic specificity in this CNS circuit. Notably, the ablation of Sema3e does not result in a complete loss of specificity because proprioceptive inputs remain restricted to CM motor neurons and do not innervate inappropriate motor neuron pools. This highlights the fact that, in this system, synaptic specificity is encoded by multiple signals that each control specific aspects of the connectivity program.

In summary, analyses of transcriptional programs for neuronal wiring specificity support a model of layered specificity in which a hierarchy of transcription factors regulate different aspects of connectivity. Transcriptional programs are established in a step-wise fashion with some intermediate transcription factors that are activated downstream of an initial transcription factor code and others that are regulated by target-derived signals. Initial programs direct the growth trajectory of neuronal processes, whereas later aspects of neuronal connectivity can be modulated by neuronal activity or target-derived signals.

The Control of Synapse Numbers by Transcriptional Programs

The transcriptional regulation of synaptic connectivity controls not only the selectivity of axonal trajectories and synaptic partners but also the numbers of synapses formed by a single cell. The function of neuronal circuits critically depends on precisely balanced numbers of synapses formed on a target cell. In several systems, homeostatic mechanisms have been described that result in an adaptation of synapse numbers as well as synaptic function in response to acute alterations in neuronal activity (Davis 2006). These adaptive processes may rely on transcription factors that are directly controlled by cellular depolarization. In contrast to the cell type-specific transcriptional programs described above, these adaptive programs may be common to many neuronal cell populations and activate common programs that drive the formation or elimination of synapses. Over the past years, several activity-regulated transcription factors have been identified (Polleux et al. 2007); two recent examples are discussed below.

Work by the Greenberg laboratory identified myocyte enhancer factor 2 (MEF2A-D) as a family of transcription factors that are regulated by calcium signaling as negative regulators of synapse number (Flavell 2006). Calcium-dependent dephosphorylation of MEF2 leads to its activation and transcription of target genes. Gain- and loss-of-function studies on MEF2 in cultured hippocampal neurons revealed that the suppression of MEF2 results in an increase in glutamatergic synapse numbers, whereas MEF2 activation or overexpression results in decreased synapse numbers. Cowan and colleagues demonstrated that the suppression and overactivation of MEF2 in the nucleus accumbens in vivo result in an increase and decrease, respectively, of dendritic spine density and thereby presumably synapse numbers (Pulipparacharuvil et al. 2008). A genome-wide search for MEF2 targets revealed several genes that encode proteins that are implicated in the destabilization of glutamatergic synapses, such as arc, homer1a, and delta-protocadherins (Flavell & Greenberg 2008). Targets identified in the nucleus accumbens appear to differ significantly from those identified in hippocampal cells (Pulipparacharuvil et al. 2008). This might imply that although MEF2-mediated transcription in both systems results in a reduction in synapse numbers, it might not mediate those effects through the same transcriptional targets. Interestingly, the activity of MEF2 might be further regulated by sumoylation, a small ubiquitin-like covalent modification. Whereas nonsumoylated MEF2A acts as a tran-

scriptional activator, the conjugation of a sumo-residue to an arginine residue converts MEF2A to a transcriptional repressor. This function is critical for the formation of dendritic claws in cerebellar granule cells, a specific aspect of post-synaptic differentiation in these cells (Shalizi et al. 2006).

Another example of the regulation of synapse numbers at the transcriptional level is the basic-helix-loop-helix transcription factor Npas4 (Lin et al. 2008). Although MEF2 is primarily regulated by posttranslational modifications, Npas4 expression is strongly increased in response to neuronal activity. Increased Npas4 expression stimulates the formation of GABAergic synapses on Npas4-expressing cells (Lin et al. 2008). A genome-wide analysis of Npas4 target genes identified 270 candidate targets that include brain-derived neurotrophic factor (BDNF), which has been implicated in promoting the development of GABAergic synapses. The suppression of BDNF expression in Npas4-expressing cells attenuates the increase in GABAergic innervation downstream of Npas4. This suggests that the upregulation of BDNF represents one of the relevant target genes that mediate a Npas4-induced increase in GABAergic synapse formation.

In summary, recent studies have provided a first glimpse at transcriptional programs that dynamically regulate glutamatergic and GABAergic synapse numbers. Although these transcription factors do not appear to be essential for the formation of synapses during development, loss-of-function phenotypes reveal important functions specifically in altering synapse numbers in response to acute stimulation or long-lasting changes in network activity. Given that several trans-synaptic signaling systems have been identified that can act as positive and negative regulators of synapse formation, it will be interesting to see which of these regulators are controlled through factors such as MEF2 and Npas4. Another unresolved question is whether there are specific, rate-limiting synaptic proteins that are common to all neuronal populations or whether different cell types and synapses formed with select synaptic partners have evolved unique molecular mechanisms that differ between neuronal and synapse populations.

MOLECULAR DIVERSITY BEYOND ANATOMICALLY DEFINED CELL TYPES

Recent efforts to unravel the molecular mechanism of synaptic specificity have also brought up renewed questions regarding the diversity of neurons in the nervous system (Masland 2004, Nelson et al. 2006). Classically, neuronal cell types have been defined by morphological criteria. More recently, additional criteria for the characterization of neuronal subpopulations have been added, which include: specific electrophysiological properties, responsiveness to specific stimuli (such as direction selectivity of cells in the retina), specific connectivity, and molecular markers that have become a key approach for classifying neuronal populations.

Although some studies on individual genes or gene families have highlighted molecular heterogeneity within cell populations, primarily, genome-scale studies have revealed new aspects of cellular diversification. Two key technologies are the combination of single-cell transcript analysis with microarray technology as well as genome-wide in situ hybridization studies with cellular resolution (Thompson et al. 2008, Tietjen et al. 2003). A surprising finding from these studies is that apparently morphologically identical neurons can express highly divergent molecular markers. These findings are notable for several reasons: First, they reveal a previously unappreciated diversity of neuronal populations; second, the identification of regulatory elements of genes with highly selective expression patterns will facilitate the selective genetic manipulation of these populations; and third, some of the products encoded by differentially expressed genes may contribute to selective wiring patterns or cell-specific functional properties. We discuss several examples below.

The Molecular Diversity of *Drosophila* Dscam1 Encodes Self-Recognition

Arguably, the most impressive example of molecular diversity is provided by the immunoglobulin superfamily protein Dscam1 in *Drosophila*. The combination of alternative exons at three positions in dscam mRNA generates Dscam1 variants with 19,000 different extracellular domains. Above, we have discussed the functions of vertebrate Dscam and Dscam-like-1 (DscamL), which control laminar specificity in the chick retina through homophilic adhesive interactions (Yamagata & Sanes 2008). In systematic biochemical studies of *Drosphilia* Dscam1, Zipursky and colleagues uncovered exclusively homophilic binding between Dscam1 variants—this means alternative splicing enables a single gene to encode 19,000 unique recognition events (Sawaya et al. 2008; Wojtowicz et al. 2004, 2007). A detailed analysis of Dscam1 variant expression in different cell populations and a comparison of the Dscam1 content of single photoreceptor cells leads to two important conclusions: First, single cells express multiple (14–50) distinct Dscam1 isoforms; and second, splice isoform choice is apparently stochastic, with single photoreceptor cells of the same type that express differt Dscam1 isoforms (Neves et al. 2004). Given the large number of potential isoforms, the stochastic expression of variants will leave every cell with a unique signature of Dscam1 variants. In mutant flies that lack five of the twelve possible alternative exons at one of the alternative splicing sites, the branching pattern of mechanosensory neurons was severely disrupted, which highlights that even a modest reduction of Dscam1 diversity results in anatomical phenotypes (Chen et al. 2006). Whether this function of Dscam1 involves trans-synaptic interactions between molecules located on axons and their target cells or primarily axo-axonal interactions between Dscam1 molecules on the neuronal processes of a single cell remains unknown.

A second process that critically requires Dscam1 function is in the dendrite development of *Drosophila* da sensory neurons. Here, homophilic Dscam1 interactions mediate self-recognition in individual sensory neurons that cover the body walls of *Drosophila* larvae. Self-recognition between two branches of a given dendrite is converted into repulsion between the two dendritic processes and thereby ensures the spreading of dendrites to evenly cover a field of body-wall surface (Hughes et al. 2007, Matthews et al. 2007, Soba et al. 2007). This self-avoidance function is conserved in mouse retina where Dscam is expressed in a subset of amacrine cells that use the transmitter dopamine. In mice with a spontaneous loss-of-function mutation in Dscam, dendrites of these dopaminergic amacrine cells exhibit an unusual degree of self-crossings, which subsequently develop into large fascicles with amacrine cell bodies that aggregate into clumps (Fuerst et al. 2008). Importantly, these defects in self-avoidance are specific to this Dscam-positive population of cells because other amacrine cell populations develop their normal mosaic spacing within the retina. This indicates that, similar to *Drosophila* Dscam1, vertebrate Dscam is required for iso-neuronal self-avoidance.

The Molecular Diversity of Synaptic Proteins in Vertebrates

One protein family that has provided an early example of expression in neuronal subpopulations is the group of clustered protocadherins (Morishita & Yagi 2007). These proteins are encoded in the mouse genome within three closely related protocadherin gene clusters (α, β, and γ) that generate 58 protocadherin variants that differ in their extracellular domains. Each beta-protocadherin protein is encoded by a single exon, whereas the molecular complexity of alpha- and gamma-protocadherins is achieved by the combination of variable exons, which encode the extracellular domain of proteins, with constant exons that encode the cytoplasmic tail (Wu & Maniatis 1999).

Although alpha protocadherin isoforms (originally termed cadherin neuronal receptors) are expressed in grossly similar patterns in the nervous system, individual isoforms are not detected in all neurons of one cell type but in divergent subpopulations (Kohmura et al. 1998). A single-cell analysis of individual Purkinje cells in the mouse cerebellum revealed that single cells within this morphologically homogeneous population express multiple but divergent variants of the protocadherin family of surface receptors (Esumi et al. 2005). Similarly, neighboring periglomerular cells in the olfactory bulb express different combinations of protocadherin exons (Kohmura et al. 1998). The alpha- and gamma-protocadherin proteins are localized to synapses, although they are not restricted to synaptic sites (Kohmura et al. 1998, Phillips et al. 2003, Wang et al. 2002). Moreover, similar to classical cadherins, protocadherin variants were proposed to mediate isoform-specific homophilic interactions (Fernandez-Monreal et al. 2008). The ablation of the entire protocadherin-gamma cluster, which encodes 22 of the described protocadherin isoforms, did not result in dramatic changes in neuronal migration and process outgrowth but uncovered a requirement for these proteins in neuronal survival (Wang et al. 2002). When apoptotic cell death in protocadherin-gamma neurons was prevented in double mutants that lacked the proapoptotic gene Bax, synapse numbers were still reduced (Weiner et al. 2005). This suggests that gamma-protocadherins are indeed required for some aspect of synaptic differentiation. Interestingly, this synaptic function appears to involve multiple modes of cellular interactions because gamma-protocadherin proteins are not only concentrated at neuron-neuron junctions but also contribute to synapse formation at neuron-glia junctions, adjacent to synaptic sites (Garrett & Weiner 2009). Further studies on neuronal connectivity in the retinae of conditional gamma-protocadherin knockout mice did not reveal alterations in the specificity of laminar connectivity but primarily suggest a function for gamma-protocadherins in coordinating the numbers of specific cell populations in the retina (Lefebvre et al. 2008). In contrast, mutant mice in which the constant region of the alpha-protocadherin gene cluster was ablated appear to show wiring-specificity defects. In these mice, olfactory sensory neurons form small, ectopic glomerular structures rather than coalescing into one larger glomerular structure per hemisphere as seen in wild-type mice (Hasegawa et al. 2008). Although currently available data support important functions for clustered protocadherins at synapses as well as in cell survival, the relevance of their molecular diversity has remained obscure and requires further study.

Other examples of molecular diverse synaptic proteins are neurexins, synaptic cell surface receptors that have potent synapse-organizing functions (discussed above), and takusans (a Japanese word that means many), which are cytoplasmic scaffolding proteins at glutamatergic synapses. For neurexins, more than 3900 variants are predicted to be generated from three genes (NRXN1, -2, and -3), through the expression from an alternative promoter and alternatively splicing at five sites (Missler & Sudhof 1998, Ushkaryov et al. 1992). Intriguingly, similar to Dscams and protocadherin family members, the molecular diversity of neurexins is restricted to the extracellular domain of the protein, which provides another example of a cell surface receptor that might connect diverse extracellular interactions into a common intracellular signaling pathway or structure. Some biochemical isoform-specific interaction partners and functions have been identified (Boucard et al. 2005, Chih et al. 2006, de Wit et al. 2009, Graf et al. 2006, Ko et al. 2009), but the relevance of neurexin diversity for synapse formation and function remains largely unexplored.

Takusans further add to the perplexing complexity of synaptic components. This gene family consists of more than 400 variants of a cytoplasmic protein, which were identified on the basis of their upregulation in mouse brains with decreased NMDA-receptor function (Tu et al. 2007). Some Takusan isoforms inter-act directly with the glutamatergic scaffolding protein PSD95 and regulate the surface transport of GluA1-containing AMPA-type glutamate receptors. Single, cortical pyramidal cells exhibit significantly divergent contents of Takusan splice variants. Therefore, individual cells might fine tune synaptic AMPA-receptor function through the expression of different Takusan isoforms.

Notably, alternative splicing is only one of many gene expression mechanisms to achieve genetically preprogrammed or stochastic selection of molecular repertoires (see review by Muotri & Gage 2006). Many neuronal gene products underlie complex transcriptional regulation from multiple alternative promoters, but gene regulation may also occur through epigenetic and posttranscriptional mechanisms.

Although the functional relevance of molecular diversity in neuronal proteins is only beginning to emerge, current examples allow us to advance hypotheses on how molecular diversity in cellular subpopulations underlies specific connectivity or functional properties. First, molecularly diverse proteins might contribute to wiring and tie specific cellular subpopulations into functional circuits. Such a model would be most conceivable for homophilic cell adhesion molecules in which subpopulations of cells that express a given isoform are preferentially connected. Second, in other cases, the molecular diversification might provide a mechanism for self-recognition, through homophilic signaling similar to Dscam1, or through coupling to heterophilic ligands for the cell-autonomous functional regulation of single cells. Third, the molecular heterogeneity of splice variants of cell surface or cytoplasmic proteins in neuronal cell populations might represent the fine tuning of functional properties, which sets excitability and plasticity properties of cells within a network. Notably, electrophysiological studies reveal considerable heterogeneity in mRNA expression for ion channels, even in apparently identical neurons (Schulz et al. 2006, 2007). In the latter case, the molecular diversification would enable cells to achieve certain electrophysiological properties through a well-balanced but variable expression of different components.

CONCLUSIONS

The extensive literature cited in this review reveals two overarching features of the molecular programs for synaptogenesis. First, diverse molecular and cellular mechanisms encode the specificity of synaptic connections, which not only testifies to the complexity of the brain but also affords the flexibility for any particular neural circuit to achieve precise synaptic connections. None of the mechanisms are mutually exclusive, which makes it likely that the sequential recruitment of positive and negative selections may be used to achieve specificity in a step-wise manner. Second, multiple redundant pathways exist to ensure the completion of the synapse formation process. Despite an extensive search for the agrin-like molecule for central synapses, no genetic manipulation of a single molecule or several molecules eliminates synapses. Interestingly, convincing evidence shows that several synaptic adhesion molecules are sufficient to drive synapse formation in vitro. Together, these findings highlight the possibility that multiple, redundant trans-synaptic pathways cooperate in vivo, although each of them has the capacity to support an extensive program for synapse formation.

Our current knowledge on synapse specificity and synapse formation raises many new questions and begs for more definitive answers. The diversity of molecular strategies employed at different populations of synapses emphasizes the need to focus loss-of-function analyses on specific synapses between reproducibly identifiable partners. A dissection of redundant synaptogenetic programs in vivo should be feasible by the combination of genetic knockouts and viral silencing methods that target multiple trans-synaptic signaling systems and will provide us with an opportunity to test whether synaptogenesis can be completely blocked. Using simple model organisms such as flies and worms, we might be able study particular model synapses to gain a full understanding of the interplay of extracellular signals that mediate synaptic target choices and determine subcellular synaptic localization, as well as the intracellular events that eventually assemble the pre- and post-synaptic apparatus.

DISCLOSURE STATEMENT

The authors are not aware of any affiliations, memberships, funding, or financial holdings that might be perceived as affecting the objectivity of this review.

ACKNOWLEDGMENTS

The authors thank members of their research groups for discussions and comments on the manuscript. Work in the authors' laboratories was supported by grants DA20844 (to P.S.) and grants from the Human Frontier Science Foundation and Howard Hughes Medical Institute (to K.S.).

LITERATURE CITED

Aberle H, Haghighi AP, Fetter RD, McCabe BD, Magalhaes TR, Goodman CS. 2002. wishful thinking encodes a BMP type II receptor that regulates synaptic growth in *Drosophila*. *Neuron* 33:545–58

Ahmad-Annuar A, Ciani L, Simeonidis I, Herreros J, Fredj NB, et al. 2006. Signaling across the synapse: a role for Wnt and Dishevelled in presynaptic assembly and neurotransmitter release. *J. Cell Biol.* 174:127–39

Altman J, Bayer SA. 1997. Development of the cerebellar system. In *Relation to Its Evolution, Structure, and Functions*. Boca Raton: CRC Press

Ambros V, Moss EG. 1994. Heterochronic genes and the temporal control of *C. elegans* development. *Trends Genet.* 10:123–27

Ango F, di Cristo G, Higashiyama H, Bennett V, Wu P, Huang ZJ. 2004. Ankyrin-based subcellular gradient of neurofascin, an immunoglobulin family protein, directs GABAergic innervation at Purkinje axon initial segment. *Cell* 119:257–72

Ango F, Wu C, Van der Want JJ, Wu P, Schachner M, Huang ZJ. 2008. Bergmann glia and the recognition molecule CHL1 organize GABAergic axons and direct innervation of Purkinje cell dendrites. *PLoS Biol.* 6:e103

Ataman B, Ashley J, Gorczyca M, Ramachandran P, Fouquet W, et al. 2008. Rapid activity-dependent modifications in synaptic structure and function require bidirectional Wnt signaling. *Neuron* 57:705–18

Barres BA. 2008. The mystery and magic of glia: a perspective on their roles in health and disease. *Neuron* 60:430–40

Bartlett WP, Banker GA. 1984. An electron microscopic study of the development of axons and dendrites by hippocampal neurons in culture. II. Synaptic relationships. *J. Neurosci.* 4:1954–65

Bate CM. 1976. Pioneer neurones in an insect embryo. *Nature* 260:54–56

Biederer T, Sara Y, Mozhayeva M, Atasoy D, Liu X, et al. 2002. SynCAM, a synaptic adhesion molecule that drives synapse assembly. *Science* 297:1525–31

Biederer T, Scheiffele P. 2007. Mixed-culture assays for analyzing neuronal synapse formation. *Nat. Protoc.* 2:670–76

Boucard AA, Chubykin AA, Comoletti D, Taylor P, Sudhof TC. 2005. A splice code for trans-synaptic cell adhesion mediated by binding of neuroligin 1 to alpha- and beta-neurexins. *Neuron* 48:229–36

Buffelli M, Burgess RW, Feng G, Lobe CG, Lichtman JW, Sanes J. 2003. Genetic evidence that relative synaptic efficacy biases the outcome of synaptic competition. *Nature* 424:430–34

Busetto G, Buffelli M, Tognana E, Bellico F, Cangiano A. 2000. Hebbian mechanisms revealed by electrical stimulation at developing rat neuromuscular junctions. *J. Neurosci.* 20:685–95

Cardozo T, Pagano M. 2004. The SCF ubiquitin ligase: insights into a molecular machine. *Nat. Rev. Mol. Cell Biol.* 5:739–51

Chen BE, Kondo M, Garnier A, Watson FL, Puettmann-Holgado R, et al. 2006. The molecular diversity of Dscam is functionally required for neuronal wiring specificity in *Drosophila*. *Cell* 125:607–20

Chen F, Qian L, Yang ZH, Huang Y, Ngo ST, et al. 2007. Rapsyn interaction with calpain stabilizes AChR clusters at the neuromuscular junction. *Neuron* 55:247–60

Chih B, Engelman H, Scheiffele P. 2005. Control of excitatory and inhibitory synapse formation by neuroligins. *Science* 307:1324–28

Chih B, Gollan L, Scheiffele P. 2006. Alternative splicing controls selective trans-synaptic interactions of the neuroligin-neurexin complex. *Neuron* 51:171–78

Christopherson KS, Ullian EM, Stokes CC, Mullowney CE, Hell JW, et al. 2005. Thrombospondins are astrocyte-secreted proteins that promote CNS synaptogenesis. *Cell* 120:421–33

Chubykin AA, Atasoy D, Etherton MR, Brose N, Kavalali ET, et al. 2007. Activity-dependent validation of excitatory versus inhibitory synapses by neuroligin-1 versus neuroligin-2. *Neuron* 54:919–31

Cohen S, Greenberg ME. 2008. Communication between the synapse and the nucleus in neuronal development, plasticity, and disease. *Annu. Rev. Cell Dev. Biol.* 24:183–209

Colman H, Nabekura J, Lichtman JW. 1997. Alterations in synaptic strength preceding axon withdrawal. *Science* 275:356–61

Colon-Ramos DA, Margeta MA, Shen K. 2007. Glia promote local synaptogenesis through UNC-6 (netrin) signaling in *C. elegans*. *Science* 318:103–6

Crepel F, Mariani J, Delhaye-Bouchaud N. 1976. Evidence for a multiple innervation of Purkinje cells by climbing fibers in the immature rat cerebellum. *J. Neurobiol.* 7:567–78

Dahlhaus R, Hines RM, Eadie BD, Kannangara TS, Hines DJ, et al. 2009. Overexpression of the cell adhesion protein neuroligin-1 induces learning deficits and impairs synaptic plasticity by altering the ratio of excitation to inhibition in the hippocampus. *Hippocampus* 20(2):305–22

Dalva MB, Takasu MA, Lin MZ, Shamah SM, Hu L, et al. 2000. EphB receptors interact with NMDA receptors and regulate excitatory synapse formation. *Cell* 103:945–56

Dasen JS, Liu JP, Jessell TM. 2003. Motor neuron columnar fate imposed by sequential phases of Hox-c activity. *Nature* 425:926–33

Dasen JS, Tice BC, Brenner-Morton S, Jessell TM. 2005. A Hox regulatory network establishes motor neuron pool identity and target-muscle connectivity. *Cell* 123:477–91

Davis EK, Zou Y, Ghosh A. 2008. Wnts acting through canonical and noncanonical signaling pathways exert opposite effects on hippocampal synapse formation. *Neural Dev.* 3:32

Davis GW. 2006. Homeostatic control of neural activity: from phenomenology to molecular design. *Annu. Rev. Neurosci.* 29:307–23

de Wit J, Sylwestrak E, O'Sullivan ML, Otto S, Tiglio K, et al. 2009. LRRTM2 interacts with Neurexin1 and regulates excitatory synapse formation. *Neuron* 64:799–806

Dean C, Scholl FG, Choih J, DeMaria S, Berger J, et al. 2003. Neurexin mediates the assembly of presynaptic terminals. *Nat. Neurosci.* 6:708–16

Del Rio JA, Heimrich B, Borrell V, Forster E, Drakew A, et al. 1997. A role for Cajal-Retzius cells and reelin in the development of hippocampal connections. *Nature* 385:70–74

Desai C, Garriga G, McIntire SL, Horvitz HR. 1988. A genetic pathway for the development of the *Caenorhabditis elegans* HSN motor neurons. *Nature* 336:638–46

Di Cristo G, Wu C, Chattopadhyaya B, Ango F, Knott G, et al. 2004. Subcellular domain-restricted GABAergic innervation in primary visual cortex in the absence of sensory and thalamic inputs. *Nat. Neurosci.* 7:1184–86

Diaz E, Ge Y, Yang YH, Loh KC, Serafini TA, et al. 2002. Molecular analysis of gene expression in the developing pontocerebellar projection system. *Neuron* 36:417–34

Ding M, Chao D, Wang G, Shen K. 2007. Spatial regulation of an E3 ubiquitin ligase directs selective synapse elimination. *Science* 317:947–51

Dudanova I, Tabuchi K, Rohlmann A, Sudhof TC, Missler M. 2007. Deletion of alpha-neurexins does not cause a major impairment of axonal pathfinding or synapse formation. *J. Comp. Neurol.* 502:261–74

Eroglu C, Allen NJ, Susman MW, O'Rourke NA, Park CY, et al. 2009. Gabapentin receptor alpha2delta-1 is a neuronal thrombospondin receptor responsible for excitatory CNS synaptogenesis. *Cell* 139:380–92

Esumi S, Kakazu N, Taguchi Y, Hirayama T, Sasaki A, et al. 2005. Monoallelic yet combinatorial expression of variable exons of the protocadherin-alpha gene cluster in single neurons. *Nat. Genet.* 37:171–76

Fannon AM, Colman DR. 1996. A model for central synaptic junctional complex formation based on the differential adhesive specificities of the cadherins. *Neuron* 17:423–34

Fernandez-Monreal M, Kang S, Phillips GR. 2008. Gamma-protocadherin homophilic interaction and intracellular trafficking is controlled by the cytoplasmic domain in neurons. *Mol. Cell Neurosci.* 40(3):344–53

Flavell EA. 2006. Activity-dependent regulation of MEF2 transcription factors suppresses excitatory synapse number. *Science* 311(5763):1008–12

Flavell SW, Greenberg ME. 2008. Signaling mechanisms linking neuronal activity to gene expression and plasticity of the nervous system. *Annu. Rev. Neurosci.* 31:563–90

Frotscher M. 1998. Cajal-Retzius cells, reelin, and the formation of layers. *Curr. Opin. Neurobiol.* 8:570–75

Fu Z, Washbourne P, Ortinski P, Vicini S. 2003. Functional excitatory synapses in HEK293 cells expressing neuroligin and glutamate receptors. *J. Neurophysiol.* 90:3950–57

Fuentealba LC, Eivers E, Ikeda A, Hurtado C, Kuroda H, et al. 2007. Integrating patterning signals: Wnt/GSK3 regulates the duration of the BMP/Smad1 signal. *Cell* 131:980–93

Fuerst PG, Koizumi A, Masland RH, Burgess RW. 2008. Neurite arborization and mosaic spacing in the mouse retina require DSCAM. *Nature* 451:470–74

Garrett AM, Weiner JA. 2009. Control of CNS synapse development by {gamma}-protocadherin-mediated astrocyte-neuron contact. *J. Neurosci.* 29:11723–31

Gautam M, Noakes PG, Mudd J, Nichol M, Chu GC, et al. 1995. Failure of postsynaptic specialization to develop at neuromuscular junctions of rapsyn-deficient mice. *Nature* 377:232–36

Gendrel M, Rapti G, Richmond JE, Bessereau JL. 2009. A secreted complement-control-related protein ensures acetylcholine receptor clustering. *Nature* 461:992–96

Gerrow K, Romorini S, Nabi SM, Colicos MA, Sala C, El-Husseini A. 2006. A preformed complex of postsynaptic proteins is involved in excitatory synapse development. *Neuron* 49:547–62

Ghosh A, Antonini A, McConnell SK, Shatz CJ. 1990. Requirement for subplate neurons in the formation of thalamocortical connections. *Nature* 347:179–81

Ghosh A, Shatz CJ. 1992. Involvement of subplate neurons in the formation of ocular dominance columns. *Science* 255:1441–43

Graf ER, Kang Y, Hauner AM, Craig AM. 2006. Structure function and splice site analysis of the synaptogenic activity of the neurexin-1 beta LNS domain. *J. Neurosci.* 26:4256–65

Graf ER, Zhang X, Jin SX, Linhoff MW, Craig AM. 2004. Neurexins induce differentiation of GABA and glutamate postsynaptic specializations via neuroligins. *Cell* 119:1013–26

Grunwald IC, Korte M, Wolfer D, Wilkinson GA, Unsicker K, et al. 2001. Kinase-independent requirement of EphB2 receptors in hippocampal synaptic plasticity. *Neuron* 32:1027–40

Haase G, Dessaud E, Garces A, de Bovis B, Birling M, et al. 2002. GDNF acts through PEA3 to regulate cell body positioning and muscle innervation of specific motor neuron pools. *Neuron* 35:893–905

Hall AC, Brennan A, Goold RG, Cleverley K, Lucas FR, et al. 2002. Valproate regulates GSK-3-mediated axonal remodeling and synapsin I clustering in developing neurons. *Mol. Cell Neurosci.* 20:257–70

Hall AC, Lucas FR, Salinas PC. 2000. Axonal remodeling and synaptic differentiation in the cerebellum is regulated by WNT-7a signaling. *Cell* 100:525–35

Hallam SJ, Jin Y. 1998. Lin-14 regulates the timing of synaptic remodelling in *Caenorhabditis elegans*. *Nature* 395:78–82

Hamos JE, Van Horn SC, Raczkowski D, Sherman SM. 1987. Synaptic circuits involving an individual retinogeniculate axon in the cat. *J. Comp. Neurol.* 259:165–92

Hasegawa S, Hamada S, Kumode Y, Esumi S, Katori S, et al. 2008. The protocadherin-alpha family is involved in axonal coalescence of olfactory sensory neurons into glomeruli of the olfactory bulb in mouse. *Mol. Cell Neurosci.* 38:66–79

Henderson JT, Georgiou J, Jia Z, Robertson J, Elowe S, et al. 2001. The receptor tyrosine kinase EphB2 regulates NMDA-dependent synaptic function. *Neuron* 32:1041–56

Henkemeyer M, Itkis OS, Ngo M, Hickmott PW, Ethell IM. 2003. Multiple EphB receptor tyrosine kinases shape dendritic spines in the hippocampus. *J. Cell Biol.* 163:1313–26

Herrera E, Brown L, Aruga J, Rachel RA, Dolen G, et al. 2003. Zic2 patterns binocular vision by specifying the uncrossed retinal projection. *Cell* 114:545–57

Hilliard MA, Bargmann CI. 2006. Wnt signals and frizzled activity orient anterior-posterior axon outgrowth in *C. elegans*. *Dev. Cell* 10:379–90

Hines RM, Wu L, Hines DJ, Steenland H, Mansour S, et al. 2008. Synaptic imbalance, stereotypies, and impaired social interactions in mice with altered neuroligin 2 expression. *J. Neurosci.* 28:6055–67

Hirai H, Pang Z, Bao D, Miyazaki T, Li L, et al. 2005. Cbln1 is essential for synaptic integrity and plasticity in the cerebellum. *Nat. Neurosci.* 8:1534–41

Hirano T, Kasono K, Araki K, Shinozuka K, Mishina M. 1994. Involvement of the glutamate receptor delta 2 subunit in the long-term depression of glutamate responsiveness in cultured rat Purkinje cells. *Neurosci. Lett.* 182:172–76

Hooks BM, Chen C. 2006. Distinct roles for spontaneous and visual activity in remodeling of the retinogeniculate synapse. *Neuron* 52:281–91

Hua JY, Smith SJ. 2004. Neural activity and the dynamics of central nervous system development. *Nat. Neurosci.* 7:327–32

Hubel DH, Wiesel TN, LeVay S. 1977. Plasticity of ocular dominance columns in monkey striate cortex. *Philos. Trans. R. Soc. Lond. B Biol. Sci.* 278:377–409

Hughes ME, Bortnick R, Tsubouchi A, Baumer P, Kondo M, et al. 2007. Homophilic Dscam interactions control complex dendrite morphogenesis. *Neuron* 54:417–27

Ichtchenko K, Hata Y, Nguyen T, Ullrich B, Missler M, et al. 1995. Neuroligin 1: a splice site-specific ligand for beta-neurexins. *Cell* 81:435–43

Inaki M, Yoshikawa S, Thomas JB, Aburatani H, Nose A. 2007. Wnt4 is a local repulsive cue that determines synaptic target specificity. *Curr. Biol.* 17:1574–79

Irie M, Hata Y, Takeuchi M, Ichtchenko K, Toyoda A, et al. 1997. Binding of neuroligins to PSD-95. *Science* 277:1511–15

Kakegawa W, Miyazaki T, Emi K, Matsuda K, Kohda K, et al. 2008. Differential regulation of synaptic plasticity and cerebellar motor learning by the C-terminal PDZ-binding motif of GluRdelta2. *J. Neurosci.* 28:1460–68

Kakegawa W, Miyazaki T, Kohda K, Matsuda K, Emi K, et al. 2009. The N-terminal domain of GluD2 (GluRdelta2) recruits presynaptic terminals and regulates synaptogenesis in the cerebellum in vivo. *J. Neurosci.* 29:5738–48

Kalinovsky A, Scheiffele P. 2004. Transcriptional control of synaptic differentiation by retrograde signals. *Curr. Opin. Neurobiol.* 14:272–79

Kania A, Jessell TM. 2003. Topographic motor projections in the limb imposed by LIM homeodomain protein regulation of ephrin-A:EphA interactions. *Neuron* 38:581–96

Kania A, Johnson RL, Jessell TM. 2000. Coordinate roles for LIM homeobox genes in directing the dorsoventral trajectory of motor axons in the vertebrate limb. *Cell* 102:161–73

Kanold PO, Kara P, Reid RC, Shatz CJ. 2003. Role of subplate neurons in functional maturation of visual cortical columns. *Science* 301:521–25

Katz LC, Shatz CJ. 1996. Synaptic activity and the construction of cortical circuits. *Science* 274:1133–38

Kayser MS, Nolt MJ, Dalva MB. 2008. EphB receptors couple dendritic filopodia motility to synapse formation. *Neuron* 59:56–69

Kim N, Burden SJ. 2008. MuSK controls where motor axons grow and form synapses. *Nat. Neurosci.* 11:19–27

Kim N, Stiegler AL, Cameron TO, Hallock PT, Gomez AM, et al. 2008. Lrp4 is a receptor for agrin and forms a complex with MuSK. *Cell* 135:334–42

Kim S, Burette A, Chung HS, Kwon SK, Woo J, et al. 2006. NGL family PSD-95-interacting adhesion molecules regulate excitatory synapse formation. *Nat. Neurosci.* 9:1294–301

Klassen MP, Shen K. 2007. Wnt signaling positions neuromuscular connectivity by inhibiting synapse formation in *C. elegans. Cell* 130:704–16

Klein R. 2001. Excitatory Eph receptors and adhesive ephrin ligands. *Curr. Opin. Cell Biol.* 13:196–203

Ko J, Fuccillo MV, Malenka RC, Sudhof TC. 2009. LRRTM2 functions as a neurexin ligand in promoting excitatory synapse formation. *Neuron* 64:791–98

Kohmura N, Senzaki K, Hamada S, Kai N, Yasuda R, et al. 1998. Diversity revealed by a novel family of cadherins expressed in neurons at a synaptic complex. *Neuron* 20:1137–51

Korkut C, Budnik V. 2009. WNTs tune up the neuromuscular junction. *Nat. Rev. Neurosci.* 10:627–34

Kummer TT, Misgeld T, Sanes JR. 2006. Assembly of the postsynaptic membrane at the neuromuscular junction: paradigm lost. *Curr. Opin. Neurobiol.* 16:74–82

Kurihara H, Hashimoto K, Kano M, Takayama C, Sakimura K, et al. 1997. Impaired parallel fiber–>Purkinje cell synapse stabilization during cerebellar development of mutant mice lacking the glutamate receptor delta2 subunit. *J. Neurosci.* 17:9613–23

Learte AR, Hidalgo A. 2007. The role of glial cells in axon guidance, fasciculation and targeting. *Adv. Exp. Med. Biol.* 621:156–66

Ledda F, Paratcha G, Sandoval-Guzman T, Ibanez CF. 2007. GDNF and GFRalpha1 promote formation of neuronal synapses by ligand-induced cell adhesion. *Nat. Neurosci.* 10:293–300

Lee KJ, Jessell TM. 1999. The specification of dorsal cell fates in the vertebrate central nervous system. *Annu. Rev. Neurosci.* 22:261–94

Lefebvre JL, Zhang Y, Meister M, Wang X, Sanes JR. 2008. {gamma}-Protocadherins regulate neuronal survival but are dispensable for circuit formation in retina. *Development* 135:4141–51

Levinson JN, Chery N, Huang K, Wong TP, Gerrow K, et al. 2005. Neuroligins mediate excitatory and inhibitory synapse formation: involvement of PSD-95 and neurexin-1beta in neuroligin induced synaptic specificity. *J. Biol. Chem.* 280:17312–19

Li J, Ashley J, Budnik V, Bhat MA. 2007. Crucial role of *Drosophila* neurexin in proper active zone apposition to postsynaptic densities, synaptic growth, and synaptic transmission. *Neuron* 55:741–55

Lichtman JW, Colman H. 2000. Synapse elimination and indelible memory. *Neuron* 25:269–78

Lin JH, Saito T, Anderson DJ, Lance-Jones C, Jessell TM, Arber S. 1998. Functionally related motor neuron pool and muscle sensory afferent subtypes defined by coordinate ETS gene expression. *Cell* 95:393–407

Lin W, Dominguez B, Yang J, Aryal P, Brandon EP, et al. 2005. Neurotransmitter acetylcholine negatively regulates neuromuscular synapse formation by a Cdk5-dependent mechanism. *Neuron* 46:569–79

Lin Y, Bloodgood BL, Hauser JL, Lapan AD, Koon AC, et al. 2008. Activity-dependent regulation of inhibitory synapse development by Npas4. *Nature* 455:1198–204

Linhoff MW, Lauren J, Cassidy RM, Dobie FA, Takahashi H, et al. 2009. An unbiased expression screen for synaptogenic proteins identifies the LRRTM protein family as synaptic organizers. *Neuron* 61:734–49

Liu XB, Low LK, Jones EG, Cheng HJ. 2005. Stereotyped axon pruning via plexin signaling is associated with synaptic complex elimination in the hippocampus. *J. Neurosci.* 25:9124–34

Low LK, Liu XB, Faulkner RL, Coble J, Cheng HJ. 2008. Plexin signaling selectively regulates the stereotyped pruning of corticospinal axons from visual cortex. *Proc. Natl. Acad. Sci. USA* 105:8136–41

Lu W, Shi Y, Jackson AC, Bjorgan K, During MJ, et al. 2009. Subunit composition of synaptic AMPA receptors revealed by a single-cell genetic approach. *Neuron* 62:254–68

Luo L, O'Leary DD. 2005. Axon retraction and degeneration in development and disease. *Annu. Rev. Neurosci.* 28:127–56

Maloof JN, Whangbo J, Harris JM, Jongeward GD, Kenyon C. 1999. A Wnt signaling pathway controls hox gene expression and neuroblast migration in *C. elegans*. *Development* 126:37–49

Marques G, Bao H, Haerry TE, Shimell MJ, Duchek P, et al. 2002. The *Drosophila* BMP type II receptor wishful thinking regulates neuromuscular synapse morphology and function. *Neuron* 33:529–43

Marques G, Haerry TE, Crotty ML, Xue M, Zhang B, O'Connor MB. 2003. Retrograde Gbb signaling through the Bmp type 2 receptor wishful thinking regulates systemic FMRFa expression in *Drosophila*. *Development* 130:5457–70

Masland RH. 2004. Neuronal cell types. *Curr. Biol.* 14:R497–500

Mason CA, Morrison ME, Ward MS, Zhang Q, Baird DH. 1997. Axon-target interactions in the developing cerebellum. *Perspect. Dev. Neurobiol.* 5:69–82

Mathew D, Ataman B, Chen J, Zhang Y, Cumberledge S, Budnik V. 2005. Wingless signaling at synapses is through cleavage and nuclear import of receptor DFrizzled2. *Science* 310:1344–47

Matthews BJ, Kim ME, Flanagan JJ, Hattori D, Clemens JC, et al. 2007. Dendrite self-avoidance is controlled by Dscam. *Cell* 129:593–604

Mauch DH, Nagler K, Schumacher S, Goritz C, Muller EC, et al. 2001. CNS synaptogenesis promoted by glia-derived cholesterol. *Science* 294:1354–57

Maximov A, Bezprozvanny I. 2002. Synaptic targeting of N-type calcium channels in hippocampal neurons. *J. Neurosci.* 22:6939–52

McCabe BD, Hom S, Aberle H, Fetter RD, Marques G, et al. 2004. Highwire regulates presynaptic BMP signaling essential for synaptic growth. *Neuron* 41:891–905

McCabe BD, Marques G, Haghighi AP, Fetter RD, Crotty ML, et al. 2003. The BMP homolog Gbb provides a retrograde signal that regulates synaptic growth at the *Drosophila* neuromuscular junction. *Neuron* 39:241–54

Miles R, Toth K, Gulyas AI, Hajos N, Freund TF. 1996. Differences between somatic and dendritic inhibition in the hippocampus. *Neuron* 16:815–23

Misgeld T, Kummer TT, Lichtman JW, Sanes JR. 2005. Agrin promotes synaptic differentiation by counter-acting an inhibitory effect of neurotransmitter. *Proc. Natl. Acad. Sci. USA* 102:11088–93

Missler M, Sudhof TC. 1998. Neurexins: three genes and 1001 products. *Trends Genet.* 14:20–26

Missler M, Zhang W, Rohlmann A, Kattenstroth G, Hammer RE, et al. 2003. Alpha-neurexins couple Ca^{2+} channels to synaptic vesicle exocytosis. *Nature* 423:939–48

Morey M, Yee SK, Herman T, Nern A, Blanco E, Zipursky SL. 2008. Coordinate control of synaptic-layer specificity and rhodopsins in photoreceptor neurons. *Nature* 456:795–99

Morishita H, Yagi T. 2007. Protocadherin family: diversity, structure, and function. *Curr. Opin. Cell Biol.* 19:584–92

Muotri AR, Gage FH. 2006. Generation of neuronal variability and complexity. *Nature* 441:1087–93

Nam CI, Chen L. 2005. Postsynaptic assembly induced by neurexin-neuroligin interaction and neurotrans-mitter. *Proc. Natl. Acad. Sci. USA* 102:6137–42

Nelson SB, Sugino K, Hempel CM. 2006. The problem of neuronal cell types: a physiological genomics approach. *Trends Neurosci.* 29:339–45

Neves G, Zucker J, Daly M, Chess A. 2004. Stochastic yet biased expression of multiple Dscam splice variants by individual cells. *Nat. Genet.* 36:240–46

Nishimune H, Sanes JR, Carlson SS. 2004. A synaptic laminin-calcium channel interaction organizes active zones in motor nerve terminals. *Nature* 432:580–87

Nishimura-Akiyoshi S, Niimi K, Nakashiba T, Itohara S. 2007. Axonal netrin-Gs transneuronally determine lamina-specific subdendritic segments. *Proc. Natl. Acad. Sci. USA* 104:14801–6

Nitkin RM, Smith MA, Magill C, Fallon JR, Yao YM, et al. 1987. Identification of agrin, a synaptic organizing protein from Torpedo electric organ. *J. Cell Biol.* 105:2471–78

Noakes PG, Gautam M, Mudd J, Sanes JR, Merlie JP. 1995. Aberrant differentiation of neuromuscular junctions in mice lacking s-laminin/laminin beta 2. *Nature* 374:258–62

Nuriya M, Huganir RL. 2006. Regulation of AMPA receptor trafficking by N-cadherin. *J. Neurochem.* 97:652–61

O'Brien RJ, Xu D, Petralia RS, Steward O, Huganir RL, Worley P. 1999. Synaptic clustering of AMPA receptors by the extracellular immediate-early gene product Narp. *Neuron* 23:309–23

O'Connor TP, Cockburn K, Wang W, Tapia L, Currie E, Bamji SX. 2009. Semaphorin 5B mediates synapse elimination in hippocampal neurons. *Neural. Dev.* 4:18

Packard M, Koo ES, Gorczyca M, Sharpe J, Cumberledge S, Budnik V. 2002. The *Drosophila* Wnt, wingless, provides an essential signal for pre- and postsynaptic differentiation. *Cell* 111:319–30

Pan CL, Howell JE, Clark SG, Hilliard M, Cordes S, et al. 2006. Multiple Wnts and frizzled receptors regulate anteriorly directed cell and growth cone migrations in *Caenorhabditis elegans*. *Dev. Cell* 10:367–77

Passafaro M, Nakagawa T, Sala C, Sheng M. 2003. Induction of dendritic spines by an extracellular domain of AMPA receptor subunit GluR2. *Nature* 424:677–81

Patton BL, Miner JH, Chiu AY, Sanes JR. 1997. Distribution and function of laminins in the neuromuscular system of developing, adult, and mutant mice. *J. Cell Biol.* 139:1507–21

Pecho-Vrieseling E, Sigrist M, Yoshida Y, Jessell TM, Arber S. 2009. Specificity of sensory-motor connections encoded by Sema3e-Plxnd1 recognition. *Nature* 459:842–46

Phillips GR, Tanaka H, Frank M, Elste A, Fidler L, et al. 2003. Gamma-protocadherins are targeted to subsets of synapses and intracellular organelles in neurons. *J. Neurosci.* 23:5096–104

Polleux F, Ince-Dunn G, Ghosh A. 2007. Transcriptional regulation of vertebrate axon guidance and synapse formation. *Nat. Rev. Neurosci.* 8:331–40

Poon VY, Klassen MP, Shen K. 2008. UNC-6/netrin and its receptor UNC-5 locally exclude presynaptic components from dendrites. *Nature* 455:669–73

Pouille F, Scanziani M. 2004. Routing of spike series by dynamic circuits in the hippocampus. *Nature* 429:717–23

Poulopoulos A, Aramuni G, Meyer G, Soykan T, Hoon M, et al. 2009. Neuroligin 2 drives postsynaptic assembly at perisomatic inhibitory synapses through gephyrin and collybistin. *Neuron* 63:628–42

Prasad BC, Clark SG. 2006. Wnt signaling establishes anteroposterior neuronal polarity and requires retromer in *C. elegans*. *Development* 133:1757–66

Pulipparacharuvil S, Renthal W, Hale CF, Taniguchi M, Xiao G, et al. 2008. Cocaine regulates MEF2 to control synaptic and behavioral plasticity. *Neuron* 59:621–33

Rawson JM, Lee M, Kennedy EL, Selleck SB. 2003. *Drosophila* neuromuscular synapse assembly and function require the TGF-beta type I receptor saxophone and the transcription factor Mad. *J. Neurobiol.* 55:134–50

Rico B, Xu B, Reichardt LF. 2002. TrkB receptor signaling is required for establishment of GABAergic synapses in the cerebellum. *Nat. Neurosci.* 5:225–33

Roos J, Hummel T, Ng N, Klambt C, Davis GW. 2000. *Drosophila* Futsch regulates synaptic microtubule organization and is necessary for synaptic growth. *Neuron* 26:371–82

Ruiz-Canada C, Budnik V. 2006. Synaptic cytoskeleton at the neuromuscular junction. *Int. Rev. Neurobiol.* 75:217–36

Saglietti L, Dequidt C, Kamieniarz K, Rousset MC, Valnegri P, et al. 2007. Extracellular interactions between GluR2 and N-cadherin in spine regulation. *Neuron* 54:461–77

Sanes JR, Lichtman JW. 1999. Development of the vertebrate neuromuscular junction. *Annu. Rev. Neurosci.* 22:389–442

Sanes JR, Yamagata M. 1999. Formation of lamina-specific synaptic connections. *Curr. Opin. Neurobiol.* 9:79–87

Sara Y, Biederer T, Atasoy D, Chubykin A, Mozhayeva MG, et al. 2005. Selective capability of SynCAM and neuroligin for functional synapse assembly. *J. Neurosci.* 25:260–70

Sawaya MR, Wojtowicz WM, Andre I, Qian B, Wu W, et al. 2008. A double S shape provides the structural basis for the extraordinary binding specificity of Dscam isoforms. *Cell* 134:1007–18

Scheiffele P, Fan J, Choih J, Fetter R, Serafini T. 2000. Neuroligin expressed in nonneuronal cells triggers presynaptic development in contacting axons. *Cell* 101:657–69

Schulz DJ, Goaillard JM, Marder E. 2006. Variable channel expression in identified single and electrically coupled neurons in different animals. *Nat. Neurosci.* 9:356–62

Schulz DJ, Goaillard JM, Marder EE. 2007. Quantitative expression profiling of identified neurons reveals cell-specific constraints on highly variable levels of gene expression. *Proc. Natl. Acad. Sci. USA* 104:13187–91

Shalizi A, Gaudilliere B, Yuan Z, Stegmuller J, Shirogane T, et al. 2006. A calcium-regulated MEF2 sumoylation switch controls postsynaptic differentiation. *Science* 311:1012–17

Shapiro L, Love J, Colman DR. 2007. Adhesion molecules in the nervous system: structural insights into function and diversity. *Annu. Rev. Neurosci.* 30:451–74

Shen K, Bargmann CI. 2003. The immunoglobulin superfamily protein SYG-1 determines the location of specific synapses in *C. elegans*. *Cell* 112:619–30

Shen K, Fetter RD, Bargmann CI. 2004. Synaptic specificity is generated by the synaptic guidepost protein SYG-2 and its receptor, SYG-1. *Cell* 116:869–81

Shinza-Kameda M, Takasu E, Sakurai K, Hayashi S, Nose A. 2006. Regulation of layer-specific targeting by reciprocal expression of a cell adhesion molecule, capricious. *Neuron* 49:205–13

Shirasaki R, Lewcock JW, Lettieri K, Pfaff SL. 2006. FGF as a target-derived chemoattractant for developing motor axons genetically programmed by the LIM code. *Neuron* 50:841–53

Sia GM, Beique JC, Rumbaugh G, Cho R, Worley PF, Huganir RL. 2007. Interaction of the N-terminal domain of the AMPA receptor GluR4 subunit with the neuronal pentraxin NP1 mediates GluR4 synaptic recruitment. *Neuron* 55:87–102

Soba P, Zhu S, Emoto K, Younger S, Yang SJ, et al. 2007. *Drosophila* sensory neurons require Dscam for dendritic self-avoidance and proper dendritic field organization. *Neuron* 54:403–16

Soriano E, Del Rio JA, Martinez A, Super H. 1994. Organization of the embryonic and early postnatal murine hippocampus. I. Immunocytochemical characterization of neuronal populations in the subplate and marginal zone. *J. Comp. Neurol.* 342:571–95

Steinberg MS. 2007. Differential adhesion in morphogenesis: a modern view. *Curr. Opin. Genet. Dev.* 17:281–86

Stevens B, Allen NJ, Vazquez LE, Howell GR, Christopherson KS, et al. 2007. The classical complement cascade mediates CNS synapse elimination. *Cell* 131:1164–78

Super H, Martinez A, Del Rio JA, Soriano E. 1998. Involvement of distinct pioneer neurons in the formation of layer-specific connections in the hippocampus. *J. Neurosci.* 18:4616–26

Tessier-Lavigne M, Goodman CS. 1996. The molecular biology of axon guidance. *Science* 274:1123–33

Thompson CL, Pathak SD, Jeromin A, Ng LL, MacPherson CR, et al. 2008. Genomic anatomy of the hippocampus. *Neuron* 60:1010–21

Tietjen I, Rihel JM, Cao Y, Koentges G, Zakhary L, Dulac C. 2003. Single-cell transcriptional analysis of neuronal progenitors. *Neuron* 38:161–75

Tran TS, Rubio ME, Clem RL, Johnson D, Case L, et al. 2009. Secreted semaphorins control spine distribution and morphogenesis in the postnatal CNS. *Nature* 462:1065–69

Tu S, Shin Y, Zago WM, States BA, Eroshkin A, et al. 2007. Takusan: a large gene family that regulates synaptic activity. *Neuron* 55:69–85

Uemura T, Mishina M. 2008. The amino-terminal domain of glutamate receptor delta2 triggers presynaptic differentiation. *Biochem. Biophys. Res. Commun.* 377:1315–19

Ullian EM, Sapperstein SK, Christopherson KS, Barres BA. 2001. Control of synapse number by glia. *Science* 291:657–61

Umemori H, Linhoff MW, Ornitz DM, Sanes JR. 2004. FGF22 and its close relatives are presynaptic organizing molecules in the mammalian brain. *Cell* 118:257–70

Ushkaryov YA, Petrenko AG, Geppert M, Sudhof TC. 1992. Neurexins: synaptic cell surface proteins related to the alpha- latrotoxin receptor and laminin. *Science* 257:50–56

Varoqueaux F, Aramuni G, Rawson RL, Mohrmann R, Missler M, et al. 2006. Neuroligins determine synapse maturation and function. *Neuron* 51:741–54

Varoqueaux F, Jamain S, Brose N. 2004. Neuroligin 2 is exclusively localized to inhibitory synapses. *Eur. J. Cell Biol.* 83:449–56

Vicario-Abejon C, Owens D, McKay R, Segal M. 2002. Role of neurotrophins in central synapse formation and stabilization. *Nat. Rev. Neurosci.* 3:965–74

Vrieseling E, Arber S. 2006. Target-induced transcriptional control of dendritic patterning and connectivity in motor neurons by the ETS gene Pea3. *Cell* 127:1439–52

Wadsworth WG, Bhatt H, Hedgecock EM. 1996. Neuroglia and pioneer neurons express UNC-6 to provide global and local netrin cues for guiding migrations in *C. elegans*. *Neuron* 16:35–46

Wang X, Weiner JA, Levi S, Craig AM, Bradley A, Sanes JR. 2002. Gamma protocadherins are required for survival of spinal interneurons. *Neuron* 36:843–54

Weiner JA, Wang X, Tapia JC, Sanes JR. 2005. Gamma protocadherins are required for synaptic development in the spinal cord. *Proc. Natl. Acad. Sci. USA* 102:8–14

White JG, Southgate E, Thomson JN, Brenner S. 1986. The structure of the nervous system of the nematode *Caenorhabditis elegans*. *Philos. Trans. R. Soc. Lond. B Biol. Sci.* 314:1–340

Wilson SI, Shafer B, Lee KJ, Dodd J. 2008. A molecular program for contralateral trajectory: Rig-1 control by LIM homeodomain transcription factors. *Neuron* 59:413–24

Winberg ML, Mitchell KJ, Goodman CS. 1998. Genetic analysis of the mechanisms controlling target selection: complementary and combinatorial functions of netrins, semaphorins, and IgCAMs. *Cell* 93:581–91

Wojtowicz WM, Flanagan JJ, Millard SS, Zipursky SL, Clemens JC. 2004. Alternative splicing of *Drosophila* Dscam generates axon guidance receptors that exhibit isoform-specific homophilic binding. *Cell* 118:619–33

Wojtowicz WM, Wu W, Andre I, Qian B, Baker D, Zipursky SL. 2007. A vast repertoire of Dscam binding specificities arises from modular interactions of variable Ig domains. *Cell* 130:1134–45

Woo J, Kwon SK, Choi S, Kim S, Lee JR, et al. 2009. Trans-synaptic adhesion between NGL-3 and LAR regulates the formation of excitatory synapses. *Nat. Neurosci.* 12:428–37

Wu Q, Maniatis T. 1999. A striking organization of a large family of human neural cadherin-like cell adhesion genes. *Cell* 97:779–90

Xu J, Xiao N, Xia J. 2009. Thrombospondin 1 accelerates synaptogenesis in hippocampal neurons through neuroligin 1. *Nat. Neurosci.* 13(1):22-24

Yamagata M, Sanes JR. 2008. Dscam and Sidekick proteins direct lamina-specific synaptic connections in vertebrate retina. *Nature* 451:465–69

Yamagata M, Weiner J, Sanes J. 2002. Sidekicks: synaptic adhesion molecules that promote lamina-specific connectivity in the retina. *Cell* 110:649–60

Yaron A, Huang PH, Cheng HJ, Tessier-Lavigne M. 2005. Differential requirement for Plexin-A3 and -A4 in mediating responses of sensory and sympathetic neurons to distinct class 3 semaphorins. *Neuron* 45:513–23

Yuzaki M. 2003. The delta2 glutamate receptor: 10 years later. *Neurosci. Res.* 46:11–22

Yuzaki M. 2008. Cbln and C1q family proteins: new transneuronal cytokines. *Cell Mol. Life Sci.* 65:1698–705

Zhang B, Luo S, Wang Q, Suzuki T, Xiong WC, Mei L. 2008. LRP4 serves as a coreceptor of agrin. *Neuron* 60:285–97

Cumulative Indexes

Contributing Authors, Volumes 24–33

Chapter Titles, Volumes 24–33